Fungi in Extreme Environments: Ecological Role and Biotechnological Significance

Sonia M. Tiquia-Arashiro • Martin Grube

Editors

Fungi in Extreme Environments: Ecological Role and Biotechnological Significance

 Springer

Editors
Sonia M. Tiquia-Arashiro (iD)
Department of Natural Sciences
University of Michigan-Dearborn
Dearborn, MI, USA

Martin Grube (iD)
Institute of Biology
University of Graz
Graz, Austria

ISBN 978-3-030-19029-3   ISBN 978-3-030-19030-9   (eBook)
https://doi.org/10.1007/978-3-030-19030-9

This Springer imprint is published by the registered company Springer Nature Switzerland AG
The registered company address is: Gewerbestrasse 11, 6330 Cham, Switzerland

Preface

Over the last decades, scientists have been intrigued by the fascinating microorganisms that inhabit extreme environments including extreme temperature, pressure, salinity, pH, etc. They grow optimally under one or several of these diverse conditions for their growth and are termed as extremophiles, which include acidophiles, alkaliphiles, halophiles, psychrophiles, thermophiles, hyperthermophiles, radioresistant microbes, barophiles, and endoliths. They thrive in habitats that are intolerably hostile or even lethal to other life-forms. Based on technological advances, the study of extremophiles has provided ground-breaking discoveries that challenge the paradigms of modern biology. In view of the nearly unlimited reservoir of extremophilic organisms existing in nature and the exciting achievements of modern biotechnology, there remains an enormous potential waiting for further progress in synthetic biology, drug discovery, bioenergy, and bioremediation. It is envisaged that biotransformations employing extremophiles will be increasingly exploited as a useful and often a unique tool for biotechnological and industrial applications.

Extremophiles have been identified to belong mainly in the domain archaea. However, extremophiles have also been identified in eubacterial and eukaryotic organisms such as fungi. Most fungi that are able to grow or survive under extreme conditions are in fact extremotolerant species, meaning they can also thrive under mesic conditions. It is still unclear how many fungi might clearly grow better under extreme conditions such as *Wallemia ichthyophaga*, which cannot grow without salt. In any case, the ecological versatility of these extremophilic/extremotolerant fungi may have potential impact in biotechnology. Fungi in general have played a very role in addressing major global challenges, being instrumental for improved resource efficiency, making renewable substitutes for products from fossil resources, upgrading waste streams to valuable food and feed ingredients, counteracting lifestyle diseases and antibiotic resistance through strengthening the gut biota (Lange, 2014: IMA Fungus 5, 463–471), making crop plants more robust to survive climate change conditions, and functioning as host organisms for the production of new biological drugs. This range of new uses of fungi all stand on the shoulders of the efforts of mycologists over generations. The next in demand are regimes of enzymes active under low-temperature conditions and enzymes active and stable at high temperatures for decomposition of both plant- and animal-derived biomasses. The expansion of commercialized enzymes from highly extreme ecological niches therefore calls for new breakthroughs. Here, mycologists specialized in fungi adapted to the extremes can contribute significantly. We follow the developments in this area and see how contemporary technologies help scientists to achieve a better understanding of biology at the extremes of life. This book builds on a platform of knowledge generated through the combined efforts of scientists and academics in the field of mycological research, where extremophilic fungi are seen as having the potential to contribute significantly toward a more sustainable world. This book puts together a rapidly growing and often scattered information on fungal life in the whole range of extreme environments and explores their habitats, biodiversity, ecology, evolution, genetics, biochemistry, and biotechnological applications in a collection of exciting reviews and original articles. It is a comprehensive and reliable source of information on the recent advances and challenges of extremophilic fungal research and a dependable reference text for readers interested in this field of research.

The book will be organized in five parts. Part I is designed to underpin the biodiversity, ecology, genetics, and physiology of extremophilic fungi. It is aimed to provide a sufficient overview of the fungal world found in extreme environments and to enable readers to fully understand and appreciate the diversity of these organisms and their metabolic capabilities. It reviews the literature on the diversity of fungi growing at extreme conditions. The readers should soon come to recognize the versatility of fungi, their ability to grow on a wide range of extreme environments.

Parts II and III explore the biotechnological potential of these fascinating organisms. It introduces the variety of extremophilic fungi as well as their osmolytes and enzymes. These parts also discuss the problems of experimental design associated with extremophiles/extremotolerants in biotechnological implications and the challenges and possibilities of developing extremolytes and novel biomolecules for commercial purposes. Several research programs have also focused on natural product isolation from microbes dwelling in marine environments, which, in this book, is defined as being in extreme environments with high salinity, extreme temperatures, minimal light, as well as variable acidities and pressures. With improved sampling, culturing, and molecular-based techniques, the number of bioactive metabolites reported from marine fungi has increased significantly over the last 30 years. For instance, cephalosporins and plinabulin are bioactive compounds inspired by marine fungal natural products and either have been clinically approved or are currently in phase III clinical trials. An important area of research is the synthesis of nanoparticles by halophilic/halotolerant fungi. These nanoparticles have found uses in biomedical and environmental fields. It is expected that with the advancement of the understanding of bionanoparticle synthesis pathways in fungi, the application of bionanoparticles will expand to many more fields than biomedical and environmental and will be potentially applied in diverse nanotechnological industries.

Part IV is devoted to the applications of extremophilic fungi in bioenergy and biofuel synthesis. Thermophilic fungi produce many thermostable enzymes that are of biotechnological importance, particularly for degradation of lignocellulosic biomass to make value-added biomaterials. This section reviews the function and utility of thermophilic enzymes and highlights the potential application of ligno-cellulose-degrading thermophilic fungi and their thermozymes in the extraction of natural rubber and biofuels.

Part V examines the application of extensive degradative capabilities of extremo-philic/extremotolerant fungi, particularly the harnessing of these properties in waste treatment, bioremediation, and pollution control.

Dearborn, MI, USA   Sonia M. Tiquia-Arashiro
Graz, Austria   Martin Grube

Contents

Contributors

Imran Afzal Faculty of Biology, Department of Biology, Lahore Garrison University, Lahore, Pakistan

Angeles Aguilera Centro de Astrobiología (CSIC-INTA), Madrid, Spain

Ali Akbar Department of Microbiology, University of Balochistan, Quetta, Pakistan

Jennifer Alcaíno Facultad de Ciencias, Departamento de Ciencias Ecológicas, Universidad de Chile, Santiago, Chile

Imran Ali School of Life Sciences and Engineering, Southwest University of Science and Technology, Mianyang, Sichuan, China

Plant Biomass Utilization Research Unit, Botany Department, Chulalongkorn University, Bangkok, Thailand

Institute of Biochemistry, University of Balochistan, Quetta, Pakistan

Naeem Ali Faculty of Biological Sciences, Department of Microbiology, Quaid-i-Azam University, Islamabad, Pakistan

Claudio Gennaro Ametrano Department of Life Science, University of Trieste, Trieste, Italy

Muhammad Zain Ul Arifeen State Key Laboratory of Pharmaceutical Biotechnology, School of Life Sciences Nanjing University, Nanjing, People's Republic of China

Marcelo Baeza Facultad de Ciencias, Departamento de Ciencias Ecológicas, Universidad de Chile, Santiago, Chile

Wai Kit Chan Faculty of Science and Technology, Department of Natural Sciences, Middlesex University, London, UK

Víctor Cifuentes Facultad de Ciencias, Departamento de Ciencias Ecológicas, Universidad de Chile, Santiago, Chile

Walter Patricio Mac Cormack Instituto Antártico Argentino (IAA), General San Martín, Buenos Aires, Argentina

Universidad de Buenos Aires, Ciudad Autónoma de Buenos Aires, Argentina

Katrina Cornish Food, Agricultural and Biological Engineering, Ohio Agricultural Research and Development Center, The Ohio State University, Wooster, OH, USA

Horticulture and Crop Science, Ohio Agricultural Research and Development Center, The Ohio State University, Wooster, OH, USA

Ekaterina Dadachova College of Pharmacy and Nutrition, University of Saskatchewan, Saskatoon, Canada

Lucía Inés Figueroa de Castellanos Planta Piloto de Procesos Industriales Microbiológicos (PROIMI-CONICET), San Miguel de Tucumán, Tucumán, Argentina

Universidad Nacional de Tucumán (UNT), San Miguel de Tucumán, Tucumán, Argentina

Tássio Brito de Oliveira Department of Biology, University of São Paulo (USP), Ribeirão Preto, SP, Brazil

Ángela De Sisto Área de Energía y Ambiente, Fundación Instituto de Estudios Avanzados (IDEA), Carretera Nacional Baruta-Hoyo de la Puerta, Valle de Sartenejas, CP, Caracas, Venezuela

Syeda Warisul Fatima Department of Chemistry, Indian Institute of Technology Delhi, New Delhi, India

Oriana Flores Facultad de Ciencias, Departamento de Ciencias Ecológicas, Universidad de Chile, Santiago, Chile

Hemda Garelick Faculty of Science and Technology, Department of Natural Sciences, Middlesex University, London, UK

Lesley-Ann Giddings Middlebury College, Middlebury, VT, USA

Meralys González Área de Energía y Ambiente, Fundación Instituto de Estudios Avanzados (IDEA), Carretera Nacional Baruta-Hoyo de la Puerta, Valle de Sartenejas, CP, Caracas, Venezuela

Elena González-Toril Centro de Astrobiología (CSIC-INTA), Madrid, Spain

Isabella Grishkan Institute of Evolution, University of Haifa, Haifa, Israel

Martin Grube Institute of Biology, University of Graz, Graz, Austria

Milan Gryndler Faculty of Science, Department of Biology, Jan Evangelista Purkyně University in Ústí nad Labem, Ústí nad Labem, Czech Republic

Gurram Shyam Prasad Department of Microbiology, Vaagdevi Degree and Post Graduate College, Warangal, Telangana State, India

Fariha Hasan Faculty of Biological Sciences, Department of Microbiology, Quaid-i-Azam University, Islamabad, Pakistan

Noor Hassan Faculty of Biological Sciences, Department of Microbiology, Quaid-i-Azam University, Islamabad, Pakistan

José Herrera Mercy College, Dobbs Ferry, NY, USA

Tamotsu Hoshino Faculty of Engineering, Department of Life and Environmental Science, The Hachinohe Institute of Technology, Hachinohe, Aomori, Japan

Martina Hujslová Laboratory of Fungal Biology, Institute of Microbiology ASCR, Prague 4, Czech Republic

Miriam I. Hutchinson Department of Biology, University of New Mexico, Albuquerque, NM, USA

Ysvic Inojosa Área de Energía y Ambiente, Fundación Instituto de Estudios Avanzados (IDEA), Carretera Nacional Baruta-Hoyo de la Puerta, Valle de Sartenejas, CP, Caracas, Venezuela

Kimiyasu Isobe Department of Biological Chemistry and Food Science, Iwate University, Morioka, Japan

Samira Khaliq Institute of Biochemistry, University of Balochistan, Quetta, Pakistan

Sunil K. Khare Department of Chemistry, Indian Institute of Technology Delhi, New Delhi, India

Sakae Kudoh National Institute of Polar Research (NIPR), Tokyo, Japan

Department of Polar Science, SOKENDAI (The Graduate University for Advanced Studies), Tokyo, Japan

Ramesh Chander Kuhad Department of Microbiology, Central University of Haryana, Mahendergarh, India

Vladimir León Área de Energía y Ambiente, Fundación Instituto de Estudios Avanzados (IDEA), Carretera Nacional Baruta-Hoyo de la Puerta, Valle de Sartenejas, CP, Caracas, Venezuela

Duo-Chuan Li Department of Mycology, Shandong Agricultural University, Taian, Shandong, China

Nelson Lima CEB—Centre of Biological Engineering, Campus de Gualtar, University of Minho, Braga, Portugal

Chang-Hong Liu State Key Laboratory of Pharmaceutical Biotechnology, School of Life Sciences Nanjing University, Nanjing, People's Republic of China

Irena Maček Biotechnical Faculty, University of Ljubljana, Ljubljana, Slovenia

Faculty of Mathematics, Natural Sciences and Information Technologies (FAMNIT), University of Primorska, Koper, Slovenia

Kapil Mahabare National Centre for Microbial Resource (NCMR), National Centre for Cell Science, S.P. Pune University, Pune, Maharashtra, India

Mackenzie E. Malo College of Pharmacy and Nutrition, University of Saskatchewan, Saskatoon, Canada

María Martha Martorell Instituto Antártico Argentino (IAA), General San Martín, Buenos Aires, Argentina

Universidad de Buenos Aires, Ciudad Autónoma de Buenos Aires, Argentina

Gorji Marzban Department of Biotechnology, University of Natural Resources and Life Sciences, Vienna, Austria

Lucia Muggia Department of Life Science, University of Trieste, Trieste, Italy

Lata Nain Division of Microbiology, ICAR-Indian Agricultural Research Institute, New Delhi, India

Leopoldo Naranjo-Briceño Área de Energía y Ambiente, Fundación Instituto de Estudios Avanzados (IDEA), Carretera Nacional Baruta-Hoyo de la Puerta, Valle de Sartenejas, CP, Caracas, Venezuela

Grupo de Microbiología Aplicada, Universidad Regional Amazónica Ikiam, CP, Tena, Ecuador

Donald O. Natvig Department of Biology, University of New Mexico, Albuquerque, NM, USA

David J. Newman Newman Consulting LLC, Wayne, PA, USA

Anastassios C. Papageorgiou Turku Centre for Biotechnology, University of Turku and Åbo Akademi University, Turku, Finland

Robert Russell M. Paterson Department of Plant Protection, Universiti Putra Malaysia, Serdang, Selangor, Malaysia

CEB—Centre of Biological Engineering, Campus de Gualtar, University of Minho, Braga, Portugal

Trigal Perdomo Área de Energía y Ambiente, Fundación Instituto de Estudios Avanzados (IDEA), Carretera Nacional Baruta-Hoyo de la Puerta, Valle de Sartenejas, CP, Caracas, Venezuela

Beatriz Pernía Área de Energía y Ambiente, Fundación Instituto de Estudios Avanzados (IDEA), Carretera Nacional Baruta-Hoyo de la Puerta, Valle de Sartenejas, CP, Caracas, Venezuela

Facultad de Ciencias Naturales, Universidad de Guayaquil, CP, Guayaquil, Ecuador

Amy J. Powell Sandia National Laboratories, Albuquerque, NM, USA

Om Prakash National Centre for Microbial Resource (NCMR), National Centre for Cell Science, S.P. Pune University, Pune, Maharashtra, India

Diane Purchase Faculty of Science and Technology, Department of Natural Sciences, Middlesex University, London, UK

Muhammad Rafiq Faculty of Biological Sciences, Department of Microbiology, Quaid-i-Azam University, Islamabad, Pakistan

Maliha Rehman Faculty of Life Sciences and Informatics, Department of Microbiology, Balochistan University of Information Technology, Engineering and Management Sciences (BUITEMS), Quetta, Pakistan

Andre Rodrigues Department of Biochemistry and Microbiology, São Paulo State University (UNESP), Rio Claro, SP, Brazil

Lucas Adolfo Mauro Ruberto Instituto Antártico Argentino (IAA), General San Martín, Buenos Aires, Argentina

Universidad de Buenos Aires, Ciudad Autónoma de Buenos Aires, Argentina

Instituto de Nanobiotecnología (NANOBIOTEC-UBA-CONICET), Ciudad Autónoma de Buenos Aires, Argentina

Ayesha Sadaf Department of Chemistry, Indian Institute of Technology Delhi, New Delhi, India

Sumbal Sajid Institute of Biochemistry, University of Balochistan, Quetta, Pakistan

Abha Sharma Division of Microbiology, ICAR-Indian Agricultural Research Institute, New Delhi, India

Anamika Sharma Division of Microbiology, ICAR-Indian Agricultural Research Institute, New Delhi, India

Rohit Sharma National Centre for Microbial Resource (NCMR), National Centre for Cell Science, S.P. Pune University, Pune, Maharashtra, India

Shomaila Sikandar Faculty of Biology, Department of Biology, Lahore Garrison University, Lahore, Pakistan

Faculty of Biological Sciences, Department of Microbiology, Quaid-i-Azam University, Islamabad, Pakistan

Surender Singh Division of Microbiology, ICAR-Indian Agricultural Research Institute, New Delhi, India

Department of Microbiology, Central University of Haryana, Mahendergarh, India

Katja Sterflinger Department of Biotechnology, University of Natural Resources and Life Sciences, Vienna, Austria

Yukiko Tanabe National Institute of Polar Research (NIPR),Tokyo, Japan

Department of Polar Science, SOKENDAI (The Graduate University for Advanced Studies), Tokyo, Japan

Donatella Tesei Department of Biotechnology, University of Natural Resources and Life Sciences, Vienna, Austria

Sonia M. Tiquia-Arashiro Department of Natural Sciences, University of Michigan-Dearborn, Dearborn, MI, USA

Masaharu Tsuji National Institute of Polar Research (NIPR), Tokyo, Japan

Héctor Urbina Área de Energía y Ambiente, Fundación Instituto de Estudios Avanzados (IDEA), Carretera Nacional Baruta-Hoyo de la Puerta, Valle de Sartenejas, CP, Caracas, Venezuela

Division of Plant Industry, Florida Department of Agriculture, Gainesville, FL, USA

Yi Wei College of Plant Sciences, Jilin University, Changchun, People's Republic of China

Dirk Wildeboer Faculty of Science and Technology, Department of Natural Sciences, Middlesex University, London, UK

Ya-Rong Xue State Key Laboratory of Pharmaceutical Biotechnology, School of Life Sciences Nanjing University, Nanjing, People's Republic of China

Krishna Kumar Yadav National Centre for Microbial Resource (NCMR), National Centre for Cell Science, S.P. Pune University, Pune, Maharashtra, India

Miwa Yamada Department of Biological Chemistry and Food Science, Iwate University, Morioka, Japan

Shi-Hong Zhang College of Plant Sciences, Jilin University, Changchun, People's Republic of China

Part I
Biodiversity, Ecology, Genetics and Physiology of Extremophilic Fungi

Chapter 1
Diversity and Ecology of Fungi in Mofettes

Irena Maček (iD)

1.1 Introduction

Mofette fields are areas with gas vents of ambient temperature geological CO_2 and consequent permanent soil hypoxia. These specific and extreme ecosystems have been used to investigate soil microbial communities with the majority of the existent studies focusing on soil archaea and bacteria while much less work has been published on soil fungal diversity (Maček et al. 2016b). In research of soil fungi, particular emphasis has been on the ubiquitous and ancient symbiotic interaction between plants and arbuscular mycorrhizal (AM) fungi subjected to soil hypoxia (low O_2 concentration), while other filamentous fungal groups from mofette sites have not yet been described, and a single report on the diversity of yeasts in wet and dry mofettes in Slovenia has been published (Šibanc et al. 2018). Nevertheless, all the existent studies confirm that mofette fields can be a rich source of information on how organisms, their populations and communities cope with long-term environmental pressures in situ in their natural habitats. There is substantial evidence demonstrating that organisms in mofette areas are subject to intense selection pressures from the abiotic environment, such as soil hypoxia (Maček et al. 2011, 2016b; Šibanc et al. 2018). The objective of this chapter is to summarise the current knowledge on the subject of soil fungal diversity and ecology in mofette sites, with the emphasis on the two groups of fungi that have been the focus of the published reports, yeasts and AM fungi. With this work the first review that is fully focused on mofette fungal biology, including the recently published findings on the yeast diversity in Slovenian mofettes (Šibanc et al. 2018), is presented. Finally, future

I. Maček (✉)
Biotechnical Faculty, University of Ljubljana, Ljubljana, Slovenia

Faculty of Mathematics, Natural Sciences and Information Technologies (FAMNIT), University of Primorska, Koper, Slovenia
e-mail: irena.macek@bf.uni-lj.si

© Springer Nature Switzerland AG 2019
S. M. Tiquia-Arashiro, M. Grube (eds.), *Fungi in Extreme Environments: Ecological Role and Biotechnological Significance*,
https://doi.org/10.1007/978-3-030-19030-9_1

directions in microbial ecology research in these specific extreme ecosystems and advances in the use of locally extreme environments for long-term ecological studies are addressed and discussed, with an acknowledgement of the possibilities for widening the scope of research, which could include environmental impact assessments in case of CO_2 leakage from carbon capture and storage (CCS) systems, bioprospecting for industrially important microbes, research into potential hypoxia-tolerant fungal pathogens as well as research of specific mofette food webs and mofette ecological networks (Maček et al. 2016b).

1.2 Mofettes or Natural CO_2 Springs

Mofettes are extreme ecosystems present in tectonically or volcanically active areas worldwide (Pfanz et al. 2004; Maček et al. 2016b). In mofettes, ambient temperature geological CO_2 of deep mantle origin reaches the surface, resulting in a severe and relatively constant change in concentrations of soil gases, in particular CO_2 and O_2 (Fig. 1.1). Traces of other gases can also be present, including methane (CH_4), nitrogen (N_2), hydrogen sulphide (H_2S) or noble gases, but concentrations of those

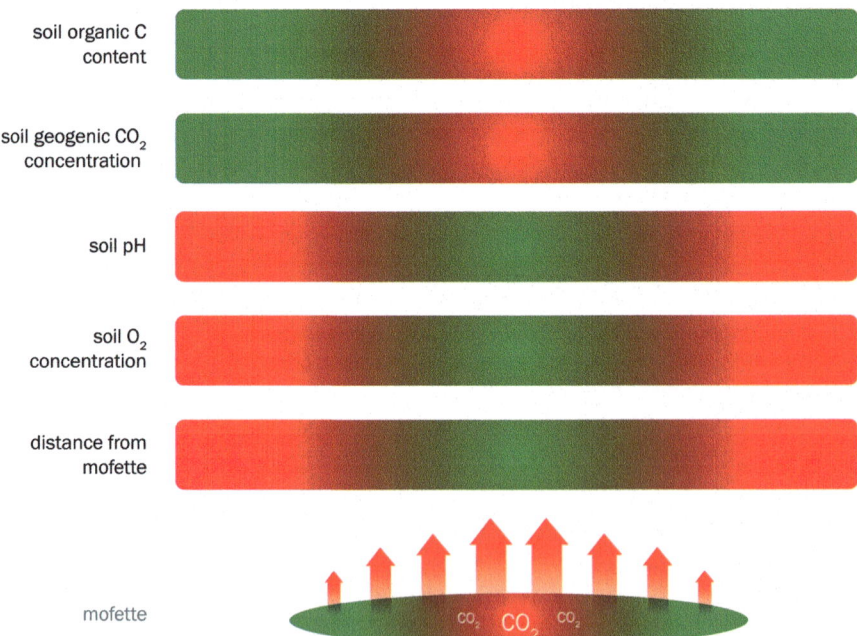

Fig. 1.1 Schematic of the effect of CO_2 venting at mofettes on abiotic factors. The gradients in different factors are indicated by different colours. Red is indicative of a higher value of the specific parameter closer to the mofette centre (e.g. CO_2 concentration); green indicates the opposite (e.g. O_2 concentration)

differ in different mofette sites with some locations emitting very pure CO_2 (e.g. in Slovenian, German and Czech mofettes with 99.9 vol. % pure CO_2) (Pfanz et al. 2004; Vodnik et al. 2006; Beulig et al. 2015).

The first reports of possibilities of using mofettes in environmental and biological studies were published at the beginning of the 1990s (e.g. Miglietta et al. 1993; Raschi et al. 1997). Following the initial use (primarily for research of the vegetation and plant above-ground responses to elevated, atmospheric CO_2 concentrations—e.g. Raschi et al. 1997), a second feature, the importance of changed gas regime in mofette soils and its impact on soil biota, was observed (Maček et al. 2005). The major characteristics of practically all mofette sites are relatively stable and high CO_2 concentrations and hypoxia (low O_2 concentration) in soil horizons, while aboveground CO_2 concentrations are lower and more variable. At 1–2 m, the surface atmosphere is most frequently only slightly enriched with CO_2 (van Gardingen et al. 1995; Kies et al. 2015). However, as it is heavier than air, mofette CO_2 may accumulate within depressions in the landscape, which if large enough may even form gas lakes with concentrations ranging from 5% to nearly 100% CO_2, forming deadly animal CO_2 traps (Fig. 1.2, see Maček et al. (2016b) for detailed descriptions on the mofette fauna and flora). In locations with an open topology and common windy conditions (e.g. Stavešinci mofettes in Slovenia), such lakes rarely form and are transient as CO_2 quickly dilutes due to high concentration gradients in the Earth's atmosphere (0.04% CO_2). As a consequence of CO_2 solubility in soil water and its conversion to carbonic acid, the pH also decreases in high geological CO_2-impacted soils. A linear correlation between CO_2 concentrations and pH values in soil was reported (Maček et al. 2012, 2016b), with high CO_2 soils and groundwater typically reported pH between 4 and 5 (Maček et al. 2016b; Šibanc et al. 2018).

Apart from mofettes, soil hypoxia also affects a range of other more common ecosystems, e.g. submerged, flooded or compacted soils (Perata et al. 2011; Maček 2017a). Furthermore, as projected by climate change models, the frequency and severity of flooding events will dramatically increase in the future (Hirabayashi et al. 2013). Therefore, understanding the response of different organisms to soil

Fig. 1.2 CO_2 traps serving as sampling sites for the local mofette faunal community, Stavešinci mofette, Slovenia (see Maček et al. 2016b for more details on mofette fauna and flora)

hypoxia, including crop plants, and their interaction with symbiotic and ubiquitous AM fungi is becoming increasingly important in order to enhance plant yield and to promote sustainable agriculture in the future (Maček 2017a). In this respect, mofettes can serve as model ecosystems to study the impacts of long-term levels of soil hypoxia and other soil abiotic factors on soil biota (Maček et al. 2016b).

1.3 Mofette Soil Biodiversity and Microbial Communities

With the new biotechnological breakthroughs in the recent decades, soil life has become the focus of biological studies in mofettes with an increasing number of papers reporting on soil biota responses to long-term elevated CO_2 and soil hypoxia, which include mesofauna (Collembola, Nematoda (Frerichs et al. 2013; Hohberg et al. 2015)) and microorganisms: fungi (Maček et al. 2011; Šibanc et al. 2018), microalgae (Collins and Bell 2006; Beulig et al. 2016), archaea and bacteria (e.g. Krüger et al. 2011; Šibanc et al. 2014; Beulig et al. 2016). The soil gas regime at mofettes, and in particular soil hypoxia, has a substantial impact on the communities of predominantly obligatory aerobic eukaryotic organisms, such as plants (e.g. Maček et al. 2016b), soil fauna (e.g. Hohberg et al. 2015) and fungi (Maček et al. 2011, 2016b; Šibanc et al. 2018). Soil O_2 concentration was also the strongest abiotic predictor of the composition of soil archaeal and bacterial communities in the Slovenian Stavešinci mofettes, with the secondary effects of other soil factors, such as CO_2 concentrations, soil pH and nutrient availability (Šibanc et al. 2014). Bacterial and archaeal communities from all mofettes exhibit increased abundance of methanogenic and anaerobic taxa, and in some cases also acidophiles (e.g. Krüger et al. 2009, 2011; Oppermann et al. 2010; Frerichs et al. 2013; Šibanc et al. 2014). Even though most of the existing mofette studies examining the community composition of different organisms represent single snapshots or, at best, a few time points, this suggests a general and relatively stable pattern in the development of mofette soil microbial communities. Furthermore, this also appears to extend to other groups of organisms, e.g. fungi (Maček et al. 2011, 2016b) with recently described novel species of soil yeast *Occultifur mephitis* sp. nov. (Šibanc et al. 2018) (Fig. 1.3) and invertebrates (collembolans) (Schulz and Potapov 2010).

In contrast to more available published data on soil prokaryotes, the diversity of fungi at mofette sites remains mostly unexplored (Maček et al. 2016b; Maček 2017b), and it has even been reported that the majority of soil fungi are potentially excluded from mofette soil food webs due to sensibility to soil hypoxia (Beulig et al. 2016). Indeed, most fungi are considered to be aerobes, although their habitats can often be hypoxic or even anoxic (e.g. due to soil infiltration with water or metabolic activity during infection of other organisms, in the deep biosphere) (Simon and Keith 2008; Drake et al. 2017). However, the existent studies show that diverse fungal communities inhabit also hypoxic mofette sites (Maček et al. 2011, 2016b; Šibanc et al. 2018) with the fungal gene number copies only slightly decreasing at high geological CO_2

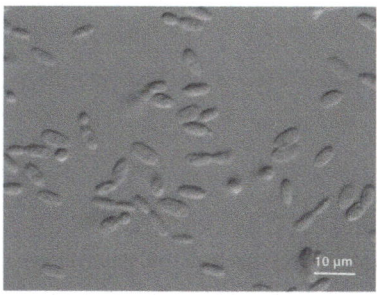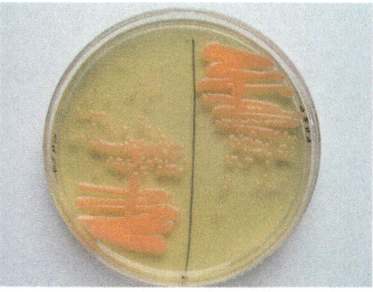

Fig. 1.3 A newly described yeast species *Occultifur mephitis* f.a., sp. nov., isolated from soil at Stavešinci mofette (NE Slovenia) and named after Mephitis, a Roman goddess of gasses emitting from ground (*me.phi¢tis.* L. fem. Gen. n. *mephitis*). Yeast morphology (left). Colonies in 9 cm diameter petri dish (right). All from EXF-6436 (holotype); see Šibanc et al. (2018) for details. Photo: N. Šibanc

fluxes and with severe decrease of fungal gene copies only when exposed to the most extreme sites, as reported for Spanish mofettes (Fernández-Montiel et al. 2016). There are only a few published studies of fungi in mofettes (e.g. Maček et al. 2011, 2016b; Šibanc et al. 2018), and further investigation is urgently needed to understand the complex biological processes and ecological interactions of soil biota in this extreme environment that also involves the fungal part.

1.4 Fungi in Mofette Habitats

Fungi are eukaryotes with complex cell structures; some have the ability to make tissues and organs typical for higher organisms. They are a large, diverse and widespread group, consisting of different forms that spread from the filamentous fungi, such as moulds, mushrooms and mycorrhizal fungi, to unicellular yeasts. Like other multicellular organisms, fungi are considered to be aerobes; however, their habitats can often be hypoxic or even anoxic, e.g. due to soil infiltration with water, metabolic activity in their environments or during infection of other organisms (Maček et al. 2011, 2016b; Grahl et al. 2012). The majority of the known fungal taxa is associated with terrestrial environments where they play crucial roles in C-cycling and mineralisation. Soilborne organisms, including fungi, have evolved to tolerate low or rapidly changing oxygen levels (e.g. Maček et al. 2011) and are exposed to occasional hypoxia, where the hypoxic response also starts in fungi (similar as in multicellular eukaryotes) at an oxygen level of about 6% (Simon and Keith 2008). Yeasts are an exception in the fungal kingdom, as many are known to have fermentation ability (e.g. Šibanc et al. 2018). Fungal diversity and community composition have only been reported for the mofettes in Slovenia, including the existent research on the fungal ecology of plant symbiotic AM fungi (Fig. 1.4). Yeasts are a second diverse group of fungi that are known to be competitive in hypoxic

Fig. 1.4 Spores of arbuscular mycorrhizal fungi isolated from Stavešinci mofette soil. *Glomus* sp. (left, middle), *Acaulospora* sp. (right). Photo: F. Oehl

environments and have been inventoried on the mofette sites in Slovenia (Table 1.1) (Šibanc et al. 2018). Apart from those, there are no published data of any other specific fungal group from mofettes.

1.4.1   Yeasts

Yeasts are a widespread group of microfungi that colonise terrestrial and aquatic systems. Soil yeasts are essential in ecosystem functioning as they are involved in the mineralisation of organic material and the assimilation of plant-derived carbohydrates as well as nutrient cycling (Botha 2006, 2011). Recently, a report on the yeast diversity from mofette sites in northeast Slovenia has been published, including an inventory of cultivable yeasts from the mofette soil (dry mofettes) and water (wet mofettes) (Table 1.1, Šibanc et al. 2018). In total, 142 yeast strains were isolated and identified from high geological CO_2-exposed soil and groundwater in a meadow, a forest pond and stream water. They were assigned to six basidiomycetous yeast genera (six species) and 11 ascomycetous genera (18 species) (Table 1.1, Šibanc et al. 2018). Using high dilution plating of a mofette soil sample, four strains of an unknown basidiomycetous species were isolated and were newly described as *Occultifur mephitis* f.a., sp. nov. (Fig. 1.3), based on molecular phylogenetic and phenotypic criteria (Šibanc et al. 2018). *O. mephitis* did not show any fermentative abilities and was not able to grow in 100% CO_2 atmosphere in vitro conditions (Šibanc et al. 2018). Nevertheless, it was growing in a hypoxic atmosphere generated by 100% N_2 in an anaerobic jar, where respiration resulted in an accumulation of up to 32% CO_2, while the remaining O_2 level was up to 5% (Table 1.1, Šibanc et al. 2018). This suggests that the hypoxic conditions in mofette soils probably include microenvironments with a minimum O_2 supply required for this and probably also other species to survive.

In the Slovenian mofettes, the highest yeast species richness (15 species), all ascomycetes, was found in water from forest mofette (pond and stream) that included yeast species only encountered in forest water: *Candida boleticola, Debaryomyces hansenii, Kazachstania exigua, Kluyveromyces dobzhanskii*, a representative of the *Metschnikowia pulcherrima* species complex (Lachance 2016),

Table 1.1 A list of yeast species isolated from the Stavešinci mofette site (according to Šibanc et al. 2018)

Substrate	DNA barcode-based identification retrieved from isolated strains	Growth in CO_2 atmosphere	Growth in N_2 atmosphere
High CO_2-exposed soil meadow mofette	*Meyerozyma guilliermondii*; A	+	+
	Occultifur mephitis; B	−	Weak
	Phaeotremella species; B	+	Weak
	Rhodotorula glutinis sensu stricto; B	−	
	Saitozyma podzolica; B	−	
	Wickerhamomyces anomalus; A	+	+
Control soil meadow	**Candida sake**; A	+	+
	Candida vartiovaarae; A	+	+
	Cutaneotrichosporon moniliiforme; B	−	
	Cyberlindnera saturnus; A	+	+
	Rhodotorula glutinis sensu stricto; B	−	
	Schwanniomyces capriottii; A	+	+
	Ustilago-Sporisorium-Macalpinomyces species complex; B	−	
Meadow mofette (water)	*Candida pseudolambica*; A	+	+
	Candida sophiae-reginae; A	+	+
	Pichia fermentans; A	+	+
Forest mofette (water)	*Candida boleticola*; A	+	+
	Candida pseudolambica; A	+	+
	Candida sake; A	+	+
	Candida sophiae-reginae; A	+	+
	Candida vartiovaarae; A	+	+
	Debaryomyces hansenii; A	−	
	Kazachstania exigua; A	+	+
	Kluyveromyces dobzhanskii; A	+	+
	Metschnikowia pulcherrima species complex; A	+	+
	Metschnikowia pulcherrima; A	+	+
	Pichia fermentans; A	+	+
	Pichia kudriavzevii; A	+	+
	Suhomyces species; A	+	
	Torulaspora delbrueckii; A	+	+
	Wickerhamomyces anomalus; A	+	+

Ascomycetous (second column, A) and basidiomycetous yeast species (B) including results from in vitro tests (growth in CO_2 and N_2 atmospheres). Most abundantly isolated yeast species are in bold. See Šibanc et al. (2018) for more details on isolation methods, in vitro tests, ex-stain accession numbers and GenBank accession numbers

Metschnikowia pulcherrima, *Pichia kudriavzevii*, *Suhomyces species* and *Torulaspora delbrueckii* (Šibanc et al. 2018). Among those, isolates identified as *Metschnikowia pulcherrima* species complex differ in 11/462 nucleotide positions of the D1/D2 (98% identity) from the sequence of the type strain of *Metschnikowia fructicola* and might represent another not-yet-described species. From the isolated taxa, all ascomycetous yeasts, with the exception of *Debaryomyces hansenii*, were able to grow under elevated CO_2 and fermented glucose. *Candida sophiae-reginae*, *Pichia fermentans* and *Candida vartiovaarae* were the dominating species in meadow and forest high CO_2-exposed water. *Meyerozyma guilliermondii* and *Wickerhamomyces anomalus* predominated in high CO_2-exposed soils (Šibanc et al. 2018). The frequent occurrence of *M. guilliermondii* and *W. anomalus* and their in vitro ability to grow in high CO_2 and N_2 atmospheres and fermentative ability suggest that they might be well-adapted to ecological niches characterised by increased CO_2 and consequently decreased O_2. The same may apply to the majority of other ascomycetous yeast species isolated from mofettes (Table 1.1). The most abundant ascomycetous yeast species found in meadow mofette soils are also described as fermentative taxa in the literature (Kurtzman et al. 2011). Among isolated yeast species, all ascomycetous taxa except *Debaryomyces hansenii* that were able to ferment glucose were also able to grow under elevated CO_2 concentrations (incubation under initial 100% CO_2) (Table 1.1). Tested strains representing these species were also able to grow under 100% N_2 (Table 1.1, Šibanc et al. 2018). The low pH encountered in high CO_2 mofette environments may also favour yeast species that are known to survive in environments with an at least moderately reduced pH value. This proves that mofette habitats enable new insights into microbial responses and adaptations to long-term changes in the soil abiotic environment and are a valuable source for the discovery of novel taxa.

1.4.2 Arbuscular Mycorrhizal (AM) Fungi

AM fungi (Fig. 1.4) were the first group of fungi whose diversity and community have been characterised from any mofette site (the Slovenian Stavešinci mofette) (Maček et al. 2011). Research in mofette fungal ecology has primarily focused on these functionally important and ubiquitous soil fungi with diverse communities in the hypoxic mofette environment (Maček et al. 2011, 2016b). The widely distributed AM fungi, obligately biotrophic plant root endosymbionts, are a ubiquitous functional group in soils. They are present in all terrestrial ecosystems and are estimated to colonise roots of around two-thirds of plant species (Fitter and Moyersoen 1996). They are also the most common fungal group in agroecosystems and may be important in promoting sustainable agriculture (Smith and Read 2008). The AM fungi have the potential to exert a profound influence on ecosystem function. For their host plants, they are the principal conduit for phosphorus uptake, and they can influence them in a variety of other ways, including defence against pathogens, improved water relations and uptake of micronutrients and nitrogen

(Smith and Read 2008). Roots of plants of several species growing in the most extreme locations in mofette area Stavešinci (NE Slovenia) have consistently been shown to be highly colonised by AM fungi, despite the prolonged stress and the carbon cost of colonisation (Maček et al. 2011, 2012; Maček 2013). Fungal hyphae are finer than roots by at least an order of magnitude and the costs to a plant of acquiring nutrients symbiotically will be lower than those of doing so by new root growth, although the functional aspects of arbuscular mycorrhiza in mofette sites are not understood. It is not yet clear how these fungi cope but the extraradical mycelium is likely to be severely restricted in highly hypoxic soils, and it is unknown whether the plants benefit from the symbiosis in this environment (Maček et al. 2011; Maček 2017a). However, the mofette fungal types are probably not being subsidised by mycelium in surrounding soil, which might explain how these aerobic fungi survive. Therefore they must be adapted to, or at least competitive in, hypoxic conditions, presumably either tolerating low O_2 or acquiring sufficient O_2 from the roots; either explanation has profound implications for their biology (Maček et al. 2011). Acquiring sufficient O_2 from roots (Maček et al. 2011) is a novel and unexplored concept of facilitation in AM fungal ecology (Maček et al. 2016b; Maček 2017a), not only relevant for mofette fungi (Maček et al. 2011), but also for microbes colonising submerged plants in aquatic environments. Therefore, arising from physically extreme environments, plant-AM fungal interactions could, in addition to nutrient (trophic) interaction, be expanded to the additional benefit of a positive effect of one species on another by reducing physical or biotic stress in existing habitats and by creating new habitats for AM fungi (Maček et al. 2016b; Maček 2017a). In the context of an extreme environment, this means that some species modify conditions sufficiently to make life more hospitable for others that otherwise would not be able to survive in this environment. The concept is in plant literature well known under the term 'facilitation' and is used to beneficial (non-trophic) interactions that occur between physiologically independent plants and that are mediated through changes in the abiotic environment (Brooker and Callaway 2009). The cross-trophic and cross-system level interactions in this concept of facilitation are still a matter of debate and a challenge to a current 'working' definition of facilitation as being limited to the plant-plant interactions, mostly in terrestrial environments (Brooker and Callaway 2009; Maček et al. 2016b).

Regarding AM fungal community ecology, different studies suggest that when an extreme environmental stress occurs in soils, there are a small number of AM fungal lineages that are better able to tolerate those conditions, which results in unique, adapted populations (Helgason and Fitter 2009; Dumbrell et al. 2010; Maček et al. 2011). AM fungi form an extensive mycelial network in soil and, therefore, will be subject to strong selection pressures from the abiotic soil environment (e.g. Dumbrell et al. 2010; Maček et al. 2011). However, reports on molecular community analyses and diversity studies of AM fungi in extreme ecosystems remain scarce (e.g. Appoloni et al. 2008; Maček et al. 2011, 2016a; Maček 2017b). Maček et al. (2011) report on significant levels of AM fungal community turnover (beta diversity) between soil types and the numerical dominance of specific AM fungal taxa in hypoxic mofette soils. This work shows that direct environmental selection acting on AM fungi is a significant factor in regulating AM fungal communities and their phylogeographic patterns. Consequently, some AM fungi are more strongly associated with local variations in the soil environment than with their host plant's distribution (Dumbrell et al. 2010; Maček et al. 2011). The higher temporal predictability (stability) is also evident from the preliminary results on AM fungal communities, which suggest that under permanent (long-term) selective pressure, community composition is more constant relative to control sites (Maček et al. 2011, 2016b). In the latter, stochastic processes and other environmental factors (e.g. vegetation) play a much more significant role in structuring communities over time. The major shifts in AM fungal community composition within and between consecutive years happen each spring, when the winter community supported by low photosynthetic carbon flux into roots is shifted to the summer community (with high photosynthetic carbon flux into roots), and the pattern of new community assembly during each year is largely stochastic (Dumbrell et al. 2011). However, this pattern may be less prominent in extreme environments, such as mofettes. The case of AM fungal community composition shows the potential of the mofettes to serve as model ecosystems to study some of the significant unresolved questions in microbial community ecology (Maček et al. 2011, 2016b). Importantly, the observed stability may be more ubiquitous than currently acknowledged in many other environments with long-term disturbances or specific selective pressures (Maček et al. 2016b; Maček 2017b). Thus, the strong environmental gradients of mofette sites, which include extreme and lethal conditions, make them ideal models for further study of community assembly rules, temporal dynamics (e.g. interannual variability), facilitation (plant O_2 supply) and symbiosis in AM fungi and provide an insight into different pathways of plant mineral assimilation and the role of the fungal partner in this process in hypoxic environments (Maček et al. 2016b).

1.4.3   Other Fungi

Apart from AM fungi and yeasts from the Slovenian mofette site at Stavešinci, there is only a single published study reporting on fungal abundance from mofette sites (Fernández-Montiel et al. 2016). In this study, microbial communities have been studied in a range of CO_2 fluxes from a natural volcanic vent in Campo de Calatrava, Spain, using a range of different techniques (quantitative PCR, DGGE and Biolog EcoPlates™) to assess changes in the abundance, diversity and functionality of the main groups of soil microbiota (bacteria, archaea and fungi). A general decrease for all studied variables (gene copies and band richness of bacteria, archaea and fungi, and Biolog activities) was observed from control to high CO_2 fluxes. In contrast, at extreme CO_2 fluxes, the communities of archaea and bacteria increased their abundance and activity but remained less diverse. The fungal community, however, showed a decrease in the number of fungal gene copies as CO_2 fluxes increased. In the case of fungi, extreme CO_2 fluxes diminished fungal gene copy numbers, with a significant decrease of three orders of magnitude in gene copy numbers in the high

CO_2 flux exposed sampling point. There are no detailed data available on the overall fungal community composition and taxa identity for this site.

1.5 Frontiers in Mofette Microbial Ecology Research

Mofettes show much potential for further investigation in a range of different fields, from soil ecology and biodiversity research to bioprospecting for new taxa with potential for biotechnological applications, research into hypoxia-tolerant human pathogens and others. Below is a description of recent and future advances in mofette research and potential applications, which is by no means exclusive, as further developments in this field may occur in the future.

1.5.1 Impact Assessment of Carbon Capture and Storage (CCS) Systems

Mofettes have been used as models that enable the evaluation of the potential risks to native ecosystems of geological carbon capture and storage (CCS) (CO_2 leakage) (Holloway et al. 2007), which has significant potential as a mitigation technique for climate change, both within the EU and internationally (see Maček et al. 2016b). Mofette studies can be of use to researchers and policy-makers in assessing the value of this technology and have the potential to inform the design of future experiments and models.

1.5.2 Mofettes and Bioprospecting for New Taxa

To date, apart from initial studies on AM fungi and yeasts (Maček et al. 2011, 2016b; Šibanc et al. 2018), there have been no reports on diversity, ecology or function of fungi from mofette areas. In general, reports on any aspect of AM fungal biology from extreme habitats or hypoxic environment are relatively scarce (Maček 2017a, b; Drake et al. 2017) while other fungal groups (e.g. yeasts) have been more widely studied in extremes (see Cantrell et al. 2011; Rangel et al. 2018). Extreme environments can serve as novel study systems to examine how long-term abiotic selection pressures drive natural communities and their evolution and possibly result in new specialist taxa (e.g. *Occultifur mephitis*, Šibanc et al. 2018). Isolation of hypoxia (stress)-tolerant microbes and microbial communities can have great potential in biotechnology (e.g. new drug discovery). As many biotechnological applications, such as industrial fermentation, require the capacity to grow in high-CO_2/low-O_2 environments, mofettes are likely ideal locations for bioprospecting for industrially important microbes. For now, the biotechnological and medical potential of mofette sites and their biota remains mostly unknown and untapped.

1.5.3 Hypoxia and Fungal Human Pathogen Research

Hypoxia/anoxia-tolerant microbes are among the most resistant human fungal pathogens. Several moulds, typically found in soil and decaying organic matter like *Aspergillus fumigatus* and *Fusarium oxysporum*, can cause human disease in immunodeficient individuals, when O_2 levels in their host tissues can be very low (Grahl et al. 2012). As most eukaryotic human fungal pathogens are generally considered obligate aerobes, O_2 availability during fungal pathogenesis may play a critical role in the outcome of infection from the perspective of both the human host and the fungus (Grahl et al. 2012). In the context of microbial pathogenesis, it is generally accepted that hypoxia occurs at sites of infection, thus generating significant environmental stress on most host and microbial pathogen cells (e.g. Cramer et al. 2003; Peyssonaux and Johnson 2004; Nizet and Johnson 2009). In healthy tissues in the human body, O_2 levels of 2.5–9% are considered normal, while oxygen levels of 1%, which have been described in tumours and wounds, are considered hypoxic (Arnold et al. 1987; Simmen et al. 1994; Nizet and Johnson 2009). Furthermore, the human gastrointestinal tract, where one of the most frequently occurring human fungal pathogens *Candida albicans* is normally located, contains significant regions of hypoxia (He et al. 1999; Karhausen et al. 2004). Oxygen concentrations in the human brain are also significantly lower than in the atmosphere, indicating that *Cryptococcus neoformans*, causing cryptococcal meningitis, is also exposed to reduced O_2 levels during infection (Erińska and Silver 2001; Sharp and Bernaudin 2004). Thus, pathogenic fungi like *C. albicans*, *C. neoformans* and *A. fumigatus* can be exposed to oxygen-limited or even hypoxic microenvironments during fungal pathogenesis. Soil environments where levels of O_2 are low (e.g. submerged, flooded and compacted soil and mofettes) or can rapidly change with the microbial metabolic activity (e.g. in compost piles) could be a source of human pathogenic fungi, adapted to hypoxia (Grahl et al. 2012). Although most moulds are traditionally considered obligate aerobes, *A. fumigatus* has been observed to tolerate O_2 levels as low as 0.1%, and some studies even suggest that *A. fumigatus* can survive and grow anaerobically (Tabak and Cooke 1968; Park et al. 1992; Hall and Denning 1994). In addition, *Fusarium* species seem particularly adept at tolerating hypoxic and even anoxic conditions, which is consistent with their resident ecological niche in soil (Gunner and Alexander 1964; Hollis 1948). Thus, these studies strongly suggest that many fungal taxa, which cause human disease, may not be typical obligate aerobes but rather are likely to be facultative anaerobes. Therefore, with the research into mofette and other hypoxic environments, potential risks for hypoxic habitats in nature serving as reservoirs of pathogens can be identified and further studied.

1.5.4 Mofettes as Long-Term Experiments in Ecology

As many natural phenomena and ecological processes take place exceptionally slowly, long-term observations and experiments to investigate them are required (Franklin 1989), but are still mostly missing for many microbial groups. Mofettes can serve as long-term natural experiments in ecology and evolution and thus have a potential to enable better predictions of the impacts on ecosystems due to induced long-term environmental change (Maček et al. 2016b). Soil is one of the most bio-diverse habitats on Earth, and soil microbes are an essential driver of many vital biogeochemical cycles and processes (Fitter 2005). Different taxa of soil microbes have different benefits and/or negative interactions with plants and human and are functionally differentiated. Changes in population densities of soil microbes in response to long-term environmental factors have the potential to impact the productivity of plant communities and affect human health. Importantly, questions about long-term-related changes in soil microbial communities are not only relevant for research into hypoxia as a stress factor, but also many other long-term anthropogenic drivers, including nutrient input, soil pollution, land-use change and more (e.g. Maček et al. 2016b; Maček 2017b). Studying mofettes can be one of the ways to increase the knowledge of microbial and fungal ecology under long-term environmental changes.

1.5.5 Mofettes and Ecological Networks

Aided by advances in sequencing technologies in recent years (e.g. Dumbrell et al. 2016), more and more studies also focus belowground, and this has been particularly true for the microbial taxa. Not only soil community composition but also the various interactions (networks) that occur among taxa are helping in monitoring the response of taxa interactions to human alterations in the environment, which is crucial in preserving ecosystems (Vacher et al. 2016). Ecological networks are now becoming a standard method for representing and simultaneously analysing interactions among taxa (e.g. Coyte et al. 2015; Vacher et al. 2016; Tylianakis and Morris 2017). In particular, the question on how networks change across environmental gradients, motivated by the need to understand how communities respond to the environment, is becoming increasingly important in the face of the global change as it may allow us to predict how networks might respond to future environmental change (Tylianakis and Morris 2017). However, ecological networks in soils are still mostly unknown. The reduction of the soil community to the microbial component makes the mofette a valuable model environment for studying diversity effects on specific soil functions. This all highlights the need for studies examining mofette systems within a broader food-web context. Namely, mofettes are characterised by a permanent exclusion of higher trophic levels and their associated physical and ecological traits from the local food webs (Maček et al. 2016b). As a significant portion of the specific ecology of these systems is microbial, mofettes provide an ideal opportunity to explore network-based approaches for incorporating next-generation sequencing-based data into food-web ecology (e.g. Vacher et al. 2016; Maček et al. 2016b).

1.6   Conclusions and Future Perspectives

Mofettes have previously been identified as a locally extreme environment that can be used as a natural analogue for the research of long-term ecological and evolutionary processes in soil biota, including fungi (Maček et al. 2016b). There remains an extensive research space in which a blend of state-of-the-art technology (e.g. next-generation sequencing), ecological network analysis and the power of the long-term natural model ecosystems (like mofettes) could open new horizons in the study of the long-term effects of press disturbance on soil microbial communities, their stability, ecological networks, as well as bioprospecting for new taxa and industrially important microbes, and the studying of ecology of human pathogens. All these research topics are of great importance for the future directions of soil ecological explorations and the maintenance of well-being and diversity on our planet. Mofette research is one possible step towards the realisation of that aim.

Acknowledgements This work was supported by the Slovenian Research Agency (ARRS) projects J4-5526, J4-7052 and programme P4-0085. We gratefully acknowledge all of the support given.

References

Appoloni S, Lekberg Y, Tercek MT, Zabinski CA, Redecker D (2008) Molecular community analysis of arbuscular mycorrhizal fungi in roots of geothermal soils in Yellowstone National Park (USA). Microb Ecol 56:649–659

Arnold F, West D, Kumar S (1987) Wound healing: the effect of macrophage and tumour derived angiogenesis factors on skin graft vascularization. Br J Exp Pathol 68:569–574

Beulig F, Heuer VB, Akob DM, Viehweger B, Elvert M, Herrmann M, Hinrichs K-U, Küsel K (2015) Carbon flow from volcanic CO_2 into soil microbial communities of a wetland mofette. ISME J 9:746–759

Beulig F, Urich T, Nowak M, Trumbore SE, Gleixner G, Gilfillan GD, Fjelland KE, Küsel K (2016) Altered carbon turnover processes and microbiomes in soils under long-term extremely high CO_2 exposure. Nat Microbiol 1:1–9

Botha A (2006) Yeast in soil. In: Rosa CA, Péter G (eds) The yeast handbook biodiversity and ecophysiology of yeasts. Springer-Verlag, Berlin, pp 221–240

Botha A (2011) The importance and ecology of yeasts in soil. Soil Biol Biochem 43:1–8

Brooker RW, Callaway RM (2009) Facilitation in the conceptual melting pot. J Ecol 97:1117–1120

Cantrell SA, Dianese JC, Fell J, Gunde-Cimerman N, Zalar P (2011) Unusual fungal niches. Mycologia 103:1161–1174

Collins S, Bell G (2006) Evolution of natural algal populations at elevated CO_2. Ecol Lett 9:129–135

Coyte KZ, Schluter J, Foster KR (2015) The ecology of the microbiome: networks, competition, and stability. Science 350:663–666

Cramer T, Yamanishi Y, Clausen BE, Förster I, Pawlinski R, Mackman N, Haase VH, Jaenisch R, Corr M, Nizet V, Firestein GS, Gerber HP, Ferrara N, Johnson RS (2003) HIF-1alpha is essential for myeloid cell-mediated inflammation. Cell 112:645–657

Drake H, Ivarsson M, Bengtson S, Heim C, Siljeström S, Whitehouse MJ, Broman C, Belivanova V, Åström ME (2017) Anaerobic consortia of fungi and sulfate reducing bacteria in deep granite fractures. Nat Commun 8:55

Dumbrell AJ, Nelson M, Helgason T, Dytham C, Fitter AH (2010) Relative roles of niche and neutral processes in structuring a soil microbial community. ISME J 4:337–345

Dumbrell AJ, Ashton PD, Aziz N, Feng G, Nelson M, Dytham C, Fitter AH, Helgason T (2011) Distinct seasonal assemblages of arbuscular mycorrhizal fungi revealed by massively parallel pyrosequencing. New Phytol 190:794–804

Dumbrell AJ, Ferguson RMW, Clark DR (2016) Microbial community analysis by single-amplicon high-throughput next generation sequencing: data analysis – from raw output to ecology. In: McGenity TJ, Timmis KN, Nogales B (eds) Hydrocarbon and lipid microbiology protocols, Springer protocols handbooks. Springer, Heidelberg

Erińska M, Silver IA (2001) Tissue oxygen tension and brain sensitivity to hypoxia. Respir Physiol 128:263–276

Fernández-Montiel I, Pedescoll A, Bécares E (2016) Microbial communities in a range of carbon dioxide fluxes from a natural volcanic vent in Campo de Calatrava, Spain. Int J Greenhouse Gas Control 50:70–79

Fitter AH (2005) Darkness visible: reflections on underground ecology. J Ecol 93:231–243

Fitter AH, Moyersoen B (1996) Evolutionary trends in root-microbe symbioses. Philosop Transac Roy Soc B: Biol Sci 351:1367–1375

Franklin JF (1989) Importance and justification of long-term studies in ecology. In: Likens GE (ed) Long-term studies in ecology: approaches and alternatives. Springer, New York, pp 3–19

Frerichs J, Oppermann BI, Gwosdz S, Möller I, Herrmann M, Krüger M (2013) Microbial community changes at a terrestrial volcanic CO_2 vent induced by soil acidification and anaerobic microhabitats within the soil column. FEMS Microbiol Ecol 84:60–74

Grahl N, Shepardson KM, Chung D, Cramer RA (2012) Hypoxia and fungal pathogenesis: to air or not to air? Eukaryot Cell 11:560–570

Gunner HB, Alexander M (1964) Anaerobic growth of Fusarium oxysporum. J Bacteriol 87:1309–1316

Hall LA, Denning DW (1994) Oxygen requirements of Aspergillus species. J Med Microbiol 41:311–315

He G, Shankar RA, Chzhan M, Samouilov A, Kuppusamy P, Zweier JL (1999) Noninvasive measurement of anatomic structure and intraluminal oxygenation in the gastrointestinal tract of living mice with spatial and spectral EPR imaging. Proc Natl Acad Sci U S A 96:4586–4591

Helgason T, Fitter AH (2009) Natural selection and the evolutionary ecology of the arbuscular mycorrhizal fungi (Phylum Glomeromycota). J Exp Bot 60:2465–2480

Hirabayashi Y, Mahendran R, Koirala S et al (2013) Global flood risk under climate change. Nat Clim Chang 3:816–821

Hohberg K, Schulz H-J, Balkenhol B, Pilz M, Thomalla A, Russell DJ, Pfanz H (2015) Soil faunal communities from mofette fields: effects of high geogenic carbon dioxide concentration. Soil Biol Biochem 88:420–429

Hollis JP (1948) Oxygen and carbon dioxide relations of Fusarium oxysporum Schlecht and Fusarium eumartii Carp. Phytopathology 38:761–775

Holloway S, Pearce JM, Hards VL, Ohsumi T, Gale J (2007) Natural emissions of CO_2 from the geosphere and their bearing on the geological storage of carbon dioxide. Energy 32:1194–1201

Karhausen J, Furuta GT, Tomaszewski JE, Johnson RS, Colgan SP, Haase VH (2004) Epithelial hypoxia-inducible factor-1 is protective in murine experimental colitis. J Clin Invest 114:1098–1106

Kies A, Hengesch O, Tosheva Z, Raschi A, Pfanz H (2015) Diurnal CO_2-cycles and temperature regimes in a natural CO_2 gas lake. Int J Greenhouse Gas Control 37:142–145

Krüger M, West J, Frerichs J, Oppermann B, Dictor M-C, Jouliand C, Jones D, Coombs P, Green K, Pearce J, May F, Möller I (2009) Ecosystem effects of elevated CO_2 concentrations on microbial populations at a terrestrial CO_2 vent at Laacher See, Germany. Energy Procedia 1:1933–1939

Krüger M, Jones D, Frerichs J, Oppermann BI, West J, Coombs P, Green K, Barlow T, Lister R, Shaw R, Strutt M, Möller I (2011) Effects of elevated CO_2 concentrations on the vegetation and microbial populations at a terrestrial CO_2 vent at Laacher See, Germany. Int J Greenhouse Gas Control 5:1093–1098

Kurtzman CP, Fell JW, Boekhoutm T (2011) The yeasts a taxonomic study, 5th edn. Elsevier, Amsterdam

Lachance MA (2016) Metschnikowia: half tetrads, a regicide and the fountain of youth. Yeast 33:563–574

Maček I (2013) A decade of research in mofette areas has given us new insights into adaptation of soil microorganisms to abiotic stress. Acta Agricult Slovenica 101:209–217

Maček I (2017a) Arbuscular mycorrhizal fungi in hypoxic environments. In: Varma A, Prasad R, Tuteja N (eds) Mycorrhiza – function, diversity, state of the art. Springer, New York, pp 329–348

Maček I (2017b) Arbuscular mycorrhizal fungal communities pushed over the edge – lessons from extreme ecosystems. In: Lukac M, Grenni P, Gamboni M (eds) Soil biological communities and ecosystem resilience, Book Series: Sustainability in Plant and Crop Protection. Springer, New York, pp 157–172

Maček I, Pfanz H, Francetič V, Batič F, Vodnik D (2005) Root respiration response to high CO_2 concentrations in plants from natural CO_2 springs. Environ Exp Bot 54:90–99

Maček I, Dumbrell AJ, Nelson M, Fitter AH, Vodnik D, Helgason T (2011) Local adaptation to soil hypoxia determines the structure of an arbuscular mycorrhizal fungal community in roots from natural CO_2 springs. AEM 77:4770–4777

Maček I, Kastelec D, Vodnik D (2012) Root colonization with arbuscular mycorrhizal fungi and glomalin-related soil protein (GRSP) concentration in hypoxic soils from natural CO_2 springs. Agric Food Sci 21:62–71

Maček I, Šibanc N, Kavšček M, Lestan D (2016a) Diversity of arbuscular mycorrhizal fungi in metal polluted and EDTA washed garden soils before and after soil revitalization with commercial and indigenous fungal inoculum. Ecol Eng 95:330–339

Maček I, Vodnik D, Pfanz H, Low-Décarie E, Dumbrell AJ (2016b) Locally extreme environments as natural long-term experiments in ecology. In: Dumbrell AJ, Kordas R, Woodward G, Large scale ecology: model systems to global perspectives. Adv Ecol Res 55:283–323

Miglietta F, Raschi A, Bettarini I, Resti R, Selvi F (1993) Natural CO_2 springs in Italy: a resource for examining long-term response to rising atmospheric CO_2 concentrations. Plant Cell Environ 16:873–878

Nizet V, Johnson RS (2009) Interdependence of hypoxic and innate immune responses. Nat Rev Immunol 9:609–617

Oppermann BI, Michaelis W, Blumenberg M, Frerichs J, Schulz HM, Schippers A, Beaubien SE, Krüger M (2010) Soil microbial community changes as a result of long-term exposure to a natural CO_2 vent. Geochim Cosmochim Acta 74:2697–2716

Park MK, Myers RA, Marzella L (1992) Oxygen tensions and infections: modulation of microbial growth, activity of antimicrobial agents, and immunologic responses. Clin Infect Dis 14:720–740

Perata P, Armstrong W, Voesenek LACJ (2011) Plants and flooding stress. New Phytol 190:269–273

Peyssonaux C, Johnson RS (2004) An unexpected role for hypoxic response: oxygenation and inflammation. Cell Cycle 3:168–171

Pfanz H, Vodnik D, Wittmann C, Aschan G, Raschi A (2004) Plants and geothermal CO_2 exhalations—survival in and adaption to high CO_2 environment. Prog Bot 65:499–538

Rangel DEN, Finlay RD, Hallsworth JE, Dadachova E, Gadd GM (2018) Fungal strategies for dealing with environment- and agriculture-induced stresses. Fungal Biol 122:602–612

Raschi A, Miglietta F, Tognetti R, van Gardingen P (1997) Plant responses to elevated CO_2: evidence from natural springs. Cambridge University Press, Cambridge, p 272

Schulz H-J, Potapov MB (2010) A new species of *Folsomia* from mofette fields of the Northwest Czechia (Collembola, Isotomidae). Zootaxa 2553:60–64

Sharp FR, Bernaudin M (2004) HIF1 and oxygen sensing in the brain. Nat Rev Neurosci 5:437–448

Šibanc N, Dumbrell AJ, Mandić-Mulec I, Maček I (2014) Impacts of naturally elevated soil CO_2 concentrations on communities of soil archaea and bacteria. Soil Biol Biochem 68:348–356

Šibanc N, Zalar P, Schroers H, Zajc J, Pontes A, Sampaio JP, Maček I (2018) *Occultifur mephitis* f.a., sp. nov. and other yeast species from hypoxic and elevated CO_2 mofette environments. Int J Syst Evol Microbiol 68:2285–2298

Simmen HP, Battaglia H, Giovanoli P, Blaser J (1994) Analysis of pH, pO_2 and pCO_2 in drainage fluid allows for rapid detection of infectious complications during the follow-up period after abdominal surgery. Infection 22:386–389

Simon MC, Keith B (2008) The role of oxygen availability in embryonic development and stem cell function. Nat Rev Mol Cell Biol 4:285–296

Smith SE, Read DJ (2008) Mycorrhizal symbiosis, 3rd edn. Academic Press, London, pp 11–145

Tabak HH, Cooke WB (1968) Growth and metabolism of fungi in an atmosphere of nitrogen. Mycologia 60:115–140

Tylianakis JM, Morris RJ (2017) Ecological networks across environmental gradients. Futuyma DJ (Ed.). Annu Rev Ecol Evol Syst 48:25–48

Vacher C, Alireza Tamaddoni-Nezhad A, Kamenova S et al (2016) Learning ecological networks from next-generation sequencing data. In: Woodward G, Bohan DA (Eds.) Ecosystem services: from biodiversity to society, PT2. Adv Ecol Res 54:1–39

van Gardingen PR, Grace J, Harkness DD, Miglietta F, Raschi A (1995) Carbon dioxide emissions at an Italian mineral spring: measurements of average CO_2 concentration and air temperature. Agric For Meteorol 73:17–27

Vodnik D, Kastelec D, Pfanz H, Maček I, Turk B (2006) Small-scale spatial variation in soil CO_2 concentration in a natural carbon dioxide spring and some related plant responses. Geoderma 133:309–319

Chapter 2
Eukaryotic Life in Extreme Environments: Acidophilic Fungi

Angeles Aguilera ⓘ and Elena González-Toril

2.1 Introduction

Exploration of the biosphere has led to continued discoveries of life in environments that were previously considered uninhabitable. Thus, life can survive and sometimes thrive under what seem to be harsh environmental conditions. Extreme environments (defined from our anthropocentric view) usually possess various factors incompatible with most life forms. Thus, certain environmental conditions such as low water availability in hyperarid deserts or high temperatures seem to be close to the limit of biological activity. On the contrary, other parameters such as radiation or pressure remain well within the limits of life on Earth. An example is pressure in the deep oceans which does not seem to affect life's abundance and diversity. This does not mean, however, that all localities are habitats, places where life is actively metabolizing and reproducing. In many localities life simply survives in a dormant state, waiting for environmental conditions to change and become more suitable (Schulze-Makuch et al. 2017). However, in spite of the apparent hostility of these extreme habitats, they contain a higher level of biodiversity than expected.

On the other hand, extreme habitats are large sources of biodiversity and new adaptation mechanisms. In these habitats, evolution works with a special intensity. These are extreme, greatly selective, and confined habitats, which constitute a favorable environment for the creation of a unique type of biodiversity and specific adaptation mechanisms. Extreme ecosystems are real sources of biological uniqueness. This makes these ecosystems especially interesting for the study of biodiversity and for the protection of the biological patrimony. Uniqueness and specificity also make these extreme habitats especially fragile. Two factors can end with this type

A. Aguilera (✉) · E. González-Toril
Centro de Astrobiología (CSIC-INTA), Madrid, Spain
e-mail: aguileraba@cab.inta-csic.es

© Springer Nature Switzerland AG 2019
S. M. Tiquia-Arashiro, M. Grube (eds.), *Fungi in Extreme Environments: Ecological Role and Biotechnological Significance*,
https://doi.org/10.1007/978-3-030-19030-9_2

of biodiversity specifically linked to the environment: breaking the balance in ecosystems which are already under extreme conditions and reestablishing the normal or non-extreme conditions in the environment.

The number of different organisms known to reside and thrive in these environmentally extreme conditions has grown rapidly in recent years. For example, we find robust microbial communities at high temperature ranges, i.e., the hot springs acidophilic algae (*Cyanidiaceae*) grow at 45–56 °C (Skorupa et al. 2013), while the hyperthermophilic archaea tolerate a temperature range above the boiling point (>100 °C) (Antranikian et al. 2017). Other organisms live in cold polar zones, such as the psychrophiles of the Antarctica, which are able to live in briny waters, with several times the salinity of seawater and temperatures below −10 °C, beneath 20 m thick ice (Lopatina et al. 2013; Bakermans et al. 2014). Recently, multicellular eukaryotes such as the lichen *Umbilicaria* and the yeast *Rhodotorula glutinis* have been shown to still grow at −17 and −18 °C, respectively (De Maayer et al. 2014). Similarly, there are microbes living in very alkaline environments (as high as pH 12) (Kambura et al. 2016). On the other end of the pH scale there are the acidophilic archaea (i.e., *Thermoplasma acidophilum*) (González-Toril et al. 2003), or the unicellular alga *Cyanidium caldarium* thriving in very acidic habitats (pH ranges from 0–4) (Seckbach 1994). Some microbial communities have been isolated from hypersaline areas, lakes, or mines containing saturated salt solutions such as those present in the Dead Sea, or are able to grow under potent ionizing radiation fields in nuclear reactors. We can also find some microbes that grow under extremely dry conditions and we find others that grow in the deepest parts of the oceans and require 500–1000 bars of hydrostatic pressure. To survive, organisms can assume forms that enable them to withstand freezing, complete desiccation, starvation, high levels of radiation exposure, and other physical or chemical challenges. Furthermore, they can survive exposure to such conditions for weeks, months, years, or even centuries (Rothschild and Mancinelli 2001).

In addition, interest in discovering extreme environments and the organisms that inhabit them has grown over the past years due to both basic, the idea that extreme environments are believed to reflect early Earth conditions in which prokaryotes originally evolved and adapted, and applied aspects, i.e., extremophiles as sources of enzymes and other cell products. Although originally considered to be nothing more than "scientific curiosities," the biotechnological potential of extremophiles and their cellular products is now a major impetus driving research. The fields of biotechnology that could benefit from mining the extremophiles are numerous and include the search for new bioactive compounds for industrial, agricultural, environmental, and pharmaceutical uses (Tiquia and Mormile 2010, Tiquia-Arashiro 2014; Krüger et al. 2018).

In this chapter, we will review the general trends concerning the diversity and ecophysiology of extremophilic fungi, paying a special attention to acidophilic ones, because unlike many other extremophiles that can adapt to diverse geophysical constraints, acidophiles actually thrive in the extreme conditions their chemolithotrophic metabolisms generate. In addition, the inorganic products of this metabolism may play an important part in the formation of specific minerals which are, in turn, extremely important biosignatures that very well may lead to the detection of similar microorganisms in remote locations.

2.2 Extremophilic Fungi

When we think of extremophiles, prokaryotes come to mind first. Thomas Brock's pioneering studies of extremophiles carried out in Yellowstone's hydrothermal environments set the focus of life in extreme environments on prokaryotes and their metabolisms (Brock 1978). While archaea and bacteria are mostly the record holder for adapting to a particular kind of extreme environmental stress or a combination of stresses, there are also eukaryotic organisms, even relatively complex animals and plants that can withstand or even metabolize and reproduce in harsh environments (Schulze-Makuch et al. 2017). Thus, eukaryotic microbial life may be found actively growing in almost any extreme condition where there is a source of energy to sustain it, with the only exception of high temperature (>70 °C) (Roberts 1999) and the deep subsurface biosphere. The development of molecular technologies and their application to microbial ecology has increased our knowledge of eukaryotic diversity in many different environments (Caron et al. 2004). This is particularly relevant in extreme environments, generally more difficult to replicate in the laboratory.

Recent studies based on molecular ecology have demonstrated that eukaryotic organisms are exceedingly adaptable and not notably less so than the prokaryotes, although most habitats have not been sufficiently well explored for sound generalizations to be made. In fact, molecular analysis has also revealed novel protist genetic diversity in different extreme environments (Caron et al. 2004; Ragon et al. 2012). Extremophiles are promising models to further our understanding of the functional evolution of stress adaptation. Their biology widens our views on the diversity of terrestrial life and it has come as a surprise that not only prokaryotes but also eukaryotes have a great capacity to adapt to extreme conditions. Particularly successful examples can be found in the fungal kingdom. Thus, specialized fungi have been discovered in extreme cold, dry, salty, acidic, and deep-sea habitats (Gostincar et al. 2010) (Fig. 2.1).

Cold Environments: Several fungal species have been isolated in considerable numbers from subglacial ice of polythermal glaciers (Gunde-Cimerman et al. 2003). Among these, there are cosmopolitan species belonging to common mold genera that do not, at first glance, look much different from isolates from elsewhere. Several *Penicillium* species are found in cold environments. Thus, *Penicillium crustosum* populations isolated from glaciers (Svalbard, Norway) demonstrated that the majority of these Arctic isolates cluster into two main groups that are distinct from strains isolated in other parts of the world (Sonjak et al. 2009). One of these groups cannot use creatine as the sole carbon source and produces a secondary metabolite, andrastin A, two properties never found previously in this species (Sonjak et al. 2007a). Another *Penicillium* species reported in cold environments is closely related to

Fig. 2.1 Images of extremophilic fungi species. (**a**) *Thelebolus microspores*, psychrophilic fungi isolated from Antarctic soils samples, (**b**) *Penicillium crustosum* micrograph, psychrophilic fungi isolated from glaciers in Svalbard, Norway, (**c**) colony of the acidophilic fungi *Purpureocillium lilacinum* isolated from Rio Tinto (SW Spain) waters, (**d**) *Penicillium crustosum* 2 weeks colonies, (**e**) micrograph of *Thelebolus microspores*

other temperate *Penicillium* species, but differs in the production of secondary metabolites and in the morphology of its conidia and penicilli; it has been named *Penicillium svalbardense* (Sonjak et al. 2007b). Species of the genus *Thelebolus* tend to be psychrophilic (Wicklow and Malloch 1971). *Thelebolus microsporus* occurs globally in boreal climate zones, while in the extreme climate of the Antarctic, it has evolved into two endemic genotypes; these have a strongly reduced morphology and cannot undergo sexual interactions. These two genotypes were described as novel species: *Thelebolus ellipsoideus* and *Thelebolus globosus* (De Hoog et al. 2005).

Saline Habitats: Until recently, it was believed that microbial communities at high salinities are dominated exclusively by Archaea and Bacteria and one eukaryotic species, the alga *Dunaliella salina* (Oren 2002). However, it has become evident that there is a higher diversity of eukaryotic microorganisms in hypersaline waters of solar salterns than previously presumed (Casamayor et al. 2002).

In this regard, studies of fungal populations in natural hypersaline environments on several continents have revealed the abundant and consistent occurrence of several specialized fungal species (Gunde-Cimerman et al. 2000). These species are characterized by extensive and complex molecular adaptations to low water activities and high concentrations of toxic ions (for reviews, see, e.g., Gunde-Cimerman et al. 2005a, b; Plemenitas et al. 2008).

Melanized fungi are a new group of eukaryotic halophiles, represented by black, yeast-like hyphomycetes: *Hortaea werneckii*, *Phaeotheca triangularis*, *Trimmatostroma salinum*, and *Aureobasidium pullulans*, together with phylogenetically closely related *Cladosporium* species, all belonging to the order Dothideales (De Hoog et al. 1999, Sterflinger et al. 1999, Gunde-Cimerman et al. 2000). Melanized fungi have been isolated from hypersaline waters on three different continents, indicating their global presence in hypersaline waters of man-made salterns. Besides, these fungi display some distinctive features that help them to adapt both to high and low salt concentrations. They are able to survive periods of extreme environmental stress in a viable, resting state. When conditions change, they respond immediately with increased metabolic activity, growth, and propagation. Their pleomorphism and adaptive halophilic behavior enables a continual colonization of salterns (Butinar et al. 2005).

The species *Hortaea werneckii* is one of the most halotolerant fungi, with a broad growth optimum from 1.0 to 3.0 M NaCl (Gunde-Cimerman et al. 2000), and it can grow in nearly saturated salt solutions, as well as without sodium chloride. Hypersaline waters appear to be its primary ecological niche in nature, such as those found in salterns (Gunde-Cimerman et al. 2000; Butinar et al. 2005). Despite its ability to grow without salt, it has been isolated only occasionally in NaCl concentrations <1.0 M, while at 3.0–4.5 M NaCl, this species can represent as much as 85–90% of all of the fungal isolates from salterns. *Hortaea werneckii* is well adapted to environments with low water activities through several of its traits: plasma membrane composition, enzymes involved in fatty acid modifications (Turk et al. 2004, 2007; Gostincar et al. 2010), osmolyte composition and accumulation of ions (Petrovic et al. 2002; Kogej et al. 2005, 2006), melanization of the cell wall (Kogej et al. 2004), differences in the high osmolarity glycerol signalling pathway (Turk and Plemenitas 2002), and differential gene expression (Petrovic et al. 2002; Vaupotic and Plemenitas 2007).

Extreme dry ecosystems: Water is an essential component of all active cells as it is the matrix in which cellular reactions occur. Availability of water can be limited by a low relative humidity, e.g., in hot, dry deserts, or when water is bound up in ice or by a high concentration of solutes (e.g., in salterns and at high sugar concentrations) (Williams and Hallsworth, 2009; Gostincar et al. 2010). Microorganisms have developed various strategies to grow in each of these conditions, and fungi are among those best adapted to growth when little water is available (Leong et al. 2015). *Xeromyces bisporus* is an ascomycete filamentous fungus that has the unique trait of being, arguably, the most xerophilic organism discovered to date (Grant 2004; Williams and Hallsworth 2009; Leong et al. 2011). *X. bisporus* actively grows in conditions of decreased water availability. Indeed, it has an absolute requirement for lowered water availability in order to grow and has an optimal water activity for growth around 0.85 (where water activity, aw, is the vapor pressure of water above a sample divided by that of pure water in the sample, pure water having $aw = 1$) (Grant 2004). Only a small number of microbial systems can retain activity at <0.710 water activity (Stevenson et al. 2015a).

In this way, *X. bisporus* shares its preference for decreased water availability with other extremophiles, such as the halophilic microbiota of salterns (bacteria, e.g., *Salinibacter ruber*; archaea, e.g., *Haloquadratum walsbyi*; the alga *Dunaliella salina*; yeasts, e.g., *Hortaea werneckii*) (Ma et al. 2010). But unlike the halophiles, *X. bisporus* prefers sugars or glycerol as a solute in the growth medium and given such conditions can even grow at 0.61 aw (Leong et al. 2011), lower than any other organism reported to date. The majority of *X. bisporus* strains have been isolated from high-sugar foods, including dried fruits (Pitt and Hocking 2009), and thus, wizened berries and fruits are likely to be the natural habitat for this fungus. *Xeromyces bisporus* increased glycerol production during hypo- and hyper-osmotic stress, and much of its wet weight comprised water and rinsable solutes; leaked solutes may form a protective slime. *X. bisporus* and other food-borne molds increased membrane fatty acid saturation as water activity decreased. Such modifications did not appear to be transcriptionally regulated in *X. bisporus*; however, genes modulating sterols, phospholipids, and the cell wall were differentially expressed.

Additionally, a number of recent studies indicate that *Aspergillus penicillioides* is active close to the water-activity limit of Earth's biosphere (Stevenson et al. 2015a, b). *A. penicillioides* is at the same time xerophilic, osmophilic, and halophilic (in relation to low water activity, high sugar- and NaCl concentrations, respectively), grows close to 0 °C (and almost certainly at sub-zero temperatures, as well), and can function anaerobically (Chin et al. 2010; Zhang et al. 2013; Nazareth and Gonsalves 2014). When growing in saline conditions, xerophilic aspergilli synthesize and accumulate molar concentrations of glycerol at low water activity (Nazareth and Gonsalves 2014; de Lima Alves et al. 2015).

2.3 Acidic Extreme Environments and Acidophilic Fungi

As mentioned before, eukaryotic organisms are exceedingly adaptable, and they are present in all the extreme environments reported until now. In this regard, acidophilic environments are not an exception. Although it is usually assumed that high metal concentrations in acidic habitats limit eukaryotic growth and diversity due to their toxicity, most of these extreme environments showed an unexpectedly high degree of eukaryotic diversity. Extreme acidic ecosystems usually include as well different abiotic extremes than low pH (Rothschild and Mancinelli 2001; Tiquia-Arashiro and Rodrigues 2016). Thus, eukaryotes thriving at these habitats are often also exposed to low nutrient levels (Brake and Hasiotis 2010), high concentrations of toxic metals (Aguilera et al. 2007a), and/or extreme temperatures (González-Toril et al. 2015). Additionally, several studies have revealed representatives from multiple evolutionary eukaryotic lineages, suggesting that the ability to adapt to pH extremes may be widespread (Amaral-Zettler et al. 2002). This raises the question of whether there are cosmopolitan eukaryotic taxa that have adapted to a wide range of pH extremes. Which environmental parameters are most

Fig. 2.2 Extreme acidic environments. (**a**) Seltun acidic geothermal area, SW Iceland, (**b**) elemental sulfur forming from gases venting the Soufriere Hills volcano on Montserrat, West Indies, (**c**) terraces formed by iron precipitates in Río Tinto, SW Spain, (**d**) Río Tinto, SW Spain, at Salinas site

influential in shaping eukaryotic microbial diversity patterns at pH extremes also remains underexplored.

Most aquatic acidic habitats have two major origins, one associated to volcanic activities and the other to metal and coal mining (Johnson 1998) (Fig. 2.2). In the first case, acidity is mainly generated by the biological oxidation of elemental sulfur produced as a result of the condensation reactions among sulfur containing volcanic gases. These can result from the generation of sulfuric acid by the microbial oxidation of reduced forms of sulfur (reviewed by Johnson and Aguilera 2016). Both hydrogen sulfide and sulfur dioxide are common in many volcanic gasses. Condensation of these two gasses produces elemental sulfur $(2H_2S + SO_2 \rightarrow 3S^0 + 2H_2O)$ which can occasionally be seen as prismatic sulfur particles forming around the peripheries of volcanic vents (Fig. 2.2a, b). Dissimilatory oxidation of sulfide, sulfite (from SO_2), elemental S, and other forms of sulfur with oxidation states of <+6, by chemolithotrophic archaea and bacteria, generates sulfuric acid (e.g., $S^0 + 3 H_2O + 1.5 O_2 \rightarrow 2 H_3O^+ + SO_4^{2-}$). In the second case, metal and coal mining expose sulfidic minerals to the combined action of water and oxygen, which facilitate microbial attack (Aguilera et al. 2010; Vera et al. 2013). The most abundant sulfidic mineral, pyrite, is of particular interest in this context. In low pH environments, the main oxidant of this mineral is ferric iron, which attacks pyrite liberating ferrous iron and oxidizing the reduced sulfur moiety (oxidation state -1) to thiosulfate ($FeS_2 + 6 Fe^{3+} \rightarrow 7 Fe^{2+} + S_2O_3^{2-} + 6 H^+$). This is an acid-generating reaction that does not involve oxygen. For the reaction to continue, the ferrous iron generated needs to be re-oxidized to ferric, which does require oxygen and is also acid-consuming ($Fe^{2+} + 0.25 O_2 + H_3O^+ \rightarrow Fe^{3+} + 1.5 H_2O$) (Fig. 2.2c, d). Both habitats vary greatly in their physicochemical characteristics and, as a consequence, in their microbial ecology (Table 2.1). Acidic environments associated to mining operations are very recent on the geological and evolutionary scale, although some metal mining activities have a relatively long history.

On the other side, in the acidic habitats related to mining activity, the extreme conditions found in the environment are the product of the metabolic activity of chemolithotrophic microorganisms, mostly iron- and sulfur-oxidizing bacteria that can be found in high numbers in their waters (Johnson 2009). Their iron-oxidizing microorganisms are responsible for the solubilization of sulfidic minerals, mainly pyrite, and the correspondent high concentration of ferric iron, sulfate, and protons in the water column. Most of these chemolithotrophic prokaryotes are autotrophic. Thus, in addition to promoting the extreme conditions of the habitat they are primary producers (González-Toril et al. 2003). Acidophilic chemolithotrophs, especially *Acidithiobacillus ferrooxidans* and *Leptospirillum* spp., accelerate the rate of

Table 2.1 Physicochemical parameters measured at different acidic environments

Acidic origin	Location	pH	Fe (mM)	Cu (mM)	As (µM)	Cd (µM)	Zn (mM)	Cr (µM)
Acidic mine drainage	Río Tinto (SW Spain)	2.3	52.4	100.3	48.6	43.9	120.6	13.1
Acidic mine drainage	La Zarza (SW, Spain)	1.9	47.7	0.70	30.6	62.6	1.85	8.5
Acidic rock drainages	Pachacoto River (Perú)	2.8	6.8	12.5	0.7	1.8	12.8	2.9
Acidic rock drainages	Bjørndalen (Svalbard)	2.8	15.7	0.03	3.6	0.6	1.1	0.6
Acidic volcanic	Río Agrio (NE, Argentina)	2.4	5.6	1.5	0.3	0.04	0.75	0.5
Acidic geothermal	Seltun (SW, Iceland)	2.9	21.4	117.8	0.1	0.2	78.2	27.8
Acidic geothermal	Hveradalir (SW, Iceland)	3.4	28.5	5.7	0.02	0.07	30.6	2.9

The table shows the average values for pH and concentration of different metals measured in the water at each location

pyrite oxidation. At the same time, low pH facilitates metal solubility. Therefore acidic water tends to have high concentrations of heavy metals.

Despite these extreme environmental conditions, most acidic environments showed an unexpectedly high degree of eukaryotic diversity. Thus, chlorophyta such as *Chlamydomonas, Chlorella,* and *Dunaliella* are frequently found in acidic environments, as well as the photosynthetic protist *Euglena mutabilis* (Aguilera et al. 2006, 2007a; González-Toril et al. 2015). Within the decomposers, fungi are very abundant and exhibit great diversity, including yeast and filamentous forms. In fact, until now, only four eukaryotic organisms were known to grow near pH 0: one algae, *Cyanidium caldarium,* and three fungi, *Acontium velatum, Cephalosporium* spp., and *Trichosporon cerebriae* (Schleper et al. 1995).

2.4 Acidophilic Fungi Diversity

Among acidophilic eukaryotic organisms, algae and protozoans have received more attention than fungi and yeasts, although fungi have long been recognized as active participants in acidification of sulfide-rich environments (Armstrong 1921; Gross and Robbins 2000). Additionally, most studies related to these organisms have been focused on biodiversity, and little is known about their physiology or their interactions with other species. Most fungi living in acidic habitats should be regarded as acid tolerant rather than strictly acidophilic because they are also able to grow under neutral or even alkaline pH (Gross and Robbins 2000). Although field studies on acidophilic fungi are usually carried out only in soil samples, up to 81 fungal species have been described (Gross and Robbins 2000). Molecular analysis have shown that acidophilic basidiomycetes phylotypes had more than 97% sequence identity to known taxa, whereas the phylotypes of the acidophilic Zygomycota/Chytridiomycota had less than 93% sequence identity to sequences available in the GenBank database (Gadanho and Sampaio 2006). Some of those phylotypes were allocated at the base of the fungal clade, being their closest relatives situated at the base of the fungal radiation.

Generally, fungi occur over a wide pH range (pH 1.0–11.0) and have been detected in acid habitats like volcanic springs, acid mine drainage, or acid industrial wastewaters (Gross and Robbins 2000). Many of them are primarily acid tolerant, but truly acidophilic species have also been detected. From a stream carrying acid mine drainage 189 species of fungi including yeast as well as filamentous fungi were isolated (Cooke 1976). Two classes of fungi, Dothideomycetes and Eurotiomycetes, have been isolated from the highly acidic (pH 0.8) and metal-rich acid mine drainage from Richmond Iron Mountain, California (Baker et al. 2004, 2009). The presence of *Geotrichum* sp. and *Aspergillus* sp. in Sar Cheshmeh Copper Mine is also reported (Orandi et al. 2007). Additionally, near hundred strains of yeast have been isolated from the Río Tinto Basin, one of the largest acidic ecosystems reported until now (Oggerin et al. 2014). Fifty-two percent are able to grow in media amended with river water (pH 2.3). They belong to five genera of basidiomycetes (*Rhodotorula, Cryptococcus, Tremella, Holtermannia,* and *Mrakia*) and two ascomycetes (*Candida* and *Williopsis*). In addition, over thousand strains of hyphomycetes have been isolated from the Tinto ecosystem and most of them characterized phenotypically. Around 50% of the isolated filamentous fungi are able to grow in the extreme acidic conditions of the river. Of this, 19% belong to the genus *Penicillium* and the rest have been identified as members of the *Scytalidium, Bahusakala, Phoma,* and *Heteroconium* genera or showed dark sterile mycelia (probably dematiaceous hyphomycetes). In addition, some isolates have been identified as strains of the ascomycete genera *Lecythophora* and *Acremonium* and the zygomycete genus *Mortierella*. Besides, fungi were the dominant group in Carnoulès sediments (France) and in the Richmond Mine (Iron Mountain, Los Angeles, CA, United States) biofilms (Baker et al. 2004; Volant et al. 2016).

Species related to Opisthokonta fungi, mainly affiliated within the Ascomycota *Helotiales* and *Dothideomycetes*, have been identified in Los Rueldos (NW Spain), an abandoned mercury underground mine, as well as the genus *Paramicrosporidium* and the Zygomycota genus *Mucoromycotina* (Mesa et al. 2017). Fungi detected in water samples in this area mainly belonged to the groups *Pezizomycotina*, LKM11 clade, and *Nucletmycea*. The environmental fungal clade LKM11 belongs to a group of fungi located near the phylogenetic root of the fungal kingdom, Rozellomycota. The clade is primarily comprised of *Rozellida*, a group of parasites with algal and fungal hosts (Masquelier et al. 2010; Auld et al. 2016). Phylogeny reconstruction indicated that the LKM11 group shared high similarity with *Paramicrosporidium* fungi, which are endonuclear parasites of free-living amoebae (Corsaro et al. 2014). Many of the fungi detected using molecular techniques have sequences that probably correspond to novel genera (Amaral-Zettler et al. 2002; Gross and Robbins 2000). Concerning metal tolerance Eurotiomycetes and Sordariomycetes isolates showed the highest resistance to toxic heavy metals, much higher than the concentrations detected in the river, while members of the Dothideomycetes showed a level of resistance to concentrations similar to those existing in the water column, and those from Basidiomycetes were in general less metal tolerant (Oggerin et al. 2014).

As mentioned before, three of the most acidophilic microorganisms described until now are fungi. The acidophilic bacteria *Thiobacillus thiooxidans* had been considered unique in its tolerance to acid until Starkey and Waksman (1943) demonstrated that two fungi, *Acontium velatum* and a *Dematiaceae* related species, were able to grow in a medium initially as acid as 2.5 N sulfuric acid solution saturated with copper sulfate, growing at pH values considerably below pH 1. Additionally, a fungus provisionally identified as *Trichosporon cerebriforme* was found to grow in 2.5 N sulfuric acid and 280 g CuSO4 in solution (Sletten and Skinner 1948). Growth was evident in about 10 days. All these species were isolated from acid solutions employed in industrial plants.

Studies on yeast in acidic environments are even scarcer and also related to biodiversity description. Usually, those that formed pink colonies (*Rhodotorula* sp.), arthroconidia (*Trichosporon* sp.), and *Candida* are the most prominent. Among the ascomycetes, species belonging to the Hemiascomycetes and Euascomycetes have been described in the Iberian Pyrite Belt (*Glomerella* sp. and *Lecythophora* sp.), as well as basidiomycetous yeasts distributed in the classes Hymenomycetes and Urediniomycetes (Gadanho et al. 2006). These species have also been described in natural acidic geothermal areas (Russo et al. 2008). Three novel asexual basidiomycetous yeast species have been recently described in the Iberian Pyrite Belt (*Cryptococcus aciditolerans* sp. nov., *Cryptococcus ibericus* sp. nov., and *Cryptococcus metallitolerans* sp. nov.) belonging to the order Filobasidiales and form a well-separated clade (Gadanho and Sampaio 2009). These *Cryptococcus* species are apparently specialists of acidic aquatic environments since they require low pH for growth, a property that has not been observed before in yeasts.

2.5 Acidophilic Fungi and Metal Immobilization

2.5.1 Ecological Role of Fungi in Extreme Acidic Environments

Fungi seem to play an important role in these environments because, together with other microorganisms, they form biofilms on the surface of rocks. These biofilms are the site of metal and mineral precipitation and provide a substrate for other microbial populations. Fungi can display metal resistance and can sequester specific metals, allowing less tolerant species to exist. Most of the eukaryotic microbial communities found in extreme acidic environments are distributed in extensive biofilms, mainly formed by microalgal species and fungi (Aguilera et al. 2007b). As in other habitats, monospecies biofilms are relatively rare and thus most biofilms are composed of mixtures of microorganisms. These biofilms are organized multicellular systems with a structural and functional architecture which influences metabolic processes, response to nutrients, predation, and other factors of the ecosystem. Moreover, it is important to study how biofilms in highly acidic conditions affect geochemical processes as metal immobilization and influence the ecophysiological rates of the microorganisms when compared to microorganisms growing in a planktonic form (Aguilera et al. 2008). This is even more important in extreme environments where forming a structured biofilm might protect the organisms from external stress conditions and allow them to resist more extreme conditions. Furthermore, analysis regarding the microcolonization sequence in Río Tinto showed an initial accumulation of amorphous particles composed of bacteria and inorganic grains of minerals. By the end of the second month, the organic matrix was also populated by fungi, bacteria, and a few eukaryotic heterotrophs such as amoebae and small flagellates. Diatoms only showed significant colonization in regions where mycelial matrices were first established (Aguilera et al. 2007b). This fact supports the idea that fungi provide a suitable substrate for less tolerant species to low pH and metals, allowing them to grow in such adverse environmental conditions. The diversity of eukaryotic microorganisms inhabiting extreme acidic systems includes microscopic algae, which are primary producers; protozoans (ciliates, flagellates, rotifers, amoebae), contributing to primary or secondary production; and fungi, which act as decomposers and contribute to carbon recycling (Méndez-García et al. 2015). Fungi and protists confer structure to the biofilms and impact the community composition by grazing on resident bacteria and archaea (Baker et al. 2004).

The ecological role of fungi in acid environments or affected streams and lakes is not adequately studied (reviewed by Das et al. 2009). Extreme acidic waters are often very low in easily degradable organic carbon. The degradation of terrestrial organic carbon sources, such as leaf litter, is impaired by the absence of invertebrates that actively shred the leaves at pH values below 3.5 (Siefert and Mutz 2001). Under these conditions, fungi may become important as primary degraders of complex organic matter. At the same time, the fungi will contribute to oxygen consumption, thereby limiting oxidative stress for the SRB. Moreover, fungi can be directly involved in the reduction of ferric iron or sulphate (Ottow and von Klopotek 1969) which is an important electron transport process in acid mine drainage and contributes to biological alkalinity generation if appropriate solid products are formed.

2.5.2   Metal Resistance Mechanisms

Acidophilic fungi can display metal resistance and can sequester specific metals allowing less tolerant species to exist (reviewed by Das et al. 2009). Fungi can absorb metals in their cell wall or adsorb in extracellular polysaccharide slime. This capacity enables them to grow in the presence of high amounts of heavy metals. Fungi can also neutralize by excreting basic substances (Shiomi et al. 2004). Thus, fungi isolated from Río Tinto (pH between 2 and 2.5) are capable of growing in the presence of metal concentrations as high as 0.4 M (Duran et al. 1999a). In general, these isolates are more resistant to heavy metals than the reference systems obtained from type collections, exhibiting a characteristic polyresistance profile. Some of the fungi can sequester heavy metals with rather high efficiency, which is normally associated with metal resistance, showing also specific heavy metal sequestering. Preliminary data using *Penicillium* isolates suggested that the mechanism of specific copper sequestering (33% at 100 mM of Cu^{2+}) depends on active cell growth, involving metal transport and formation of cellular inclusions.

The high concentration of heavy metals in solution results toxic to numerous aquatic organisms (Bowman et al. 2018). Therefore acidophilic fungi require suitable mechanisms to develop in these extreme conditions. At low pH, the cellular components must be adapted to the high concentration of extracellular acidity and be able to maintain the cytoplasm near neutrality using different homeostatic mechanisms (Baker-Austin and Dopson 2007; Magan 2007). In addition, heavy metals may induce denaturation of proteins and disruption of cellular membranes, act as antimetabolites of essential cellular functions, and generate very reactive free radicals (Gadd 1993; Tiquia-Arashiro 2018). Thus, fungal survival and growth in these environments are possible through the development of several strategies including the synthesis of metallothioneins, extracellular precipitation, biosorption to cell walls, impermeability, and intracellular compartmentation among others (Gad and Griffiths 1978; Brown and Hall 1990; Mehra and Winge 1991; Gadd 2008). The best understood mechanism that fungi have developed to survive in the presence of toxic heavy metals are their sequestering on the cellular envelops or active transport to eliminate metals from the intracellular media. Moreover, it has been detected the capacity of different acidophilic fungi to specifically sequester toxic metals intracellularly, resulting in the base of a methodology to eliminate and recover valuable metals from industrial contaminated wastewaters (Duran et al. 1999a, b). Some of the best examples of microbial metal tolerance are also found in the genus *Penicillium*, underlining the fact that metal responses may be strain specific. As noted above, *P. ochro-chloron* can grow in saturated $CuSO_4$ and it is frequently isolated from industrial effluents (Stokes and Lindsay 1979), whereas *P. lilacinum* comprised 23% of all fungi isolated from soil polluted by mine drainage (Tatsuyama et al. 1975).

Moreover, it has been described that ferrous iron oxidation carried out by *Thiobacillus ferrooxidans* was stimulated by the basidiomycetous *Rhodotorula mucilaginosa*. Recent studies have shown that biogenic formation of jarosite, in natural acidic environments, usually related to bacterial activity, is also induced by acidophilic fungi. *Purpureocillium lilacinum*, a fungal strain isolated from Río Tinto, specifically precipitates hydronium-jarosite. The mineral starts to nucleate on the fungal wall of both living and dead cells as well as on the extrapolymeric substances. This may act as heterogeneous crystallization nuclei for a metabolism independent process of jarosite precipitation, possibly promoting local ion oversaturation for mineral crystallization when suitable physicochemical conditions are present and helping to shape and control the geochemical properties of the environment (Oggerin et al. 2014).

Understanding the evolution and adaptive mechanisms of microorganisms to thrive in extreme environments will increase our basic knowledge of evolutionary processes and allow a better evaluation of the potential ecological consequences of environmental changes.

2.6 Conclusions and Future Perspectives

Knowledge of the phylogenetic and physiological diversities of acidophilic microorganisms has expanded greatly in the past 25 years. Data from biomolecular studies of extremely acidic sites, however, suggest that a large number of extremophilic fungi in general and particularly acidophilic ones still await isolation and characterization. There is a great deal of interest in acidophiles, not only from the standpoint of understanding how these microorganisms can thrive in conditions that are hostile to most life forms, but also due to their importance in environmental pollution and in biotechnology. To date little is known regarding the role of acidophilic fungi in shaping the varied ecosystems that occur in acidic environments and less about whether these microorganisms can support microenvironmental conditions that increase the survival of other members of the microbial community, or the interaction between different organisms enhances colonization of others. Acidophilic fungi have to play an even more important role in the development of microbial communities in extreme environments by providing a suitable architectural structure, mechanical stability, and protection against external conditions, and be able to selectively accumulate metals from the surrounding water.

Acknowledgements Funding was provided by the Spanish Ministry of Economy and Competitivity (MINECO) under Grant No. CGL2015-69758.

References

Aguilera A, Manrubia SC, Gómez F, Rodriguez N, Amils R (2006) Eukaryotic community distribution and their relationship to water physicochemical parameters in an extreme acidic environment, Río Tinto (SW, Spain). Appl Environ Microbiol 72:5325–5330

Aguilera A, Zettler E, Gomez F, Amaral-Zettler L, Rodrıguez N, Amilsa R (2007a) Distribution and seasonal variability in the benthic eukaryotic community of Río Tinto (SW, Spain), an acidic, high metal extreme environment. Syst Appl Microbiol 30:531–546

Aguilera A, Souza-Egipsy V, Gomez F, Amils R (2007b) Development and structure of eukaryotic biofilms in an extreme acidic environment, Río Tinto (SW, Spain). Microb Ecol 53:294–305

Aguilera A, Souza-Egipsy V, San Martín-Úriz P, Amils R (2008) Extracellular matrix assembly in extreme acidic eukaryotic biofilms and their possible implications in heavy metal adsorption. Aquat Toxicol 88:257–266

Aguilera A, Souza-Egipsy V, González-Toril E, Rendueles O, Amils R (2010) Eukaryotic microbial diversity of phototrophic microbial mats in two icelandic geothermal hot Springs. Int Microbiol 13:29–40

Amaral-Zettler L, Gomez F, Zettler E, Keenan B, Amils R, Sogin M (2002) Eukaryotic diversity in Spain's river of fire. Nature 417:137

Antranikian G, Suleiman M, Schäfers C, Adams MW, Bartolucci S (2017) Diversity of bacteria and archaea from two shallow marine hydrothermal vents from Vulcano Island. Extremophiles 21:733–742

Armstrong GM (1921) Studies in the physiology of the fungi-sulfur nutrition, the use of thiosulphate as influenced by hydrogen ion concentration. Ann Missouri Bot Garden 8:237–248

Auld R, Mykytczuk N, Leduc L, Merritt T (2016) Seasonal variation in an acid mine drainage microbial community. Can J Microbiol 63:137–152

Baker BJ, Lutz MA, Dawson SC, Bond PL, Banfield JF (2004) Metabolically active eukaryotic communities in extremely acidic mine drainage. Appl Environ Microbiol 70:6264–6271

Baker BJ, Tyson GW, Goosherst L, Banfield JF (2009) Insights into the diversity of eukaryotes in acid mine drainage biofilm communities. Appl Environ Microbiol 75:2192–2199

Baker-Austin C, Dopson M (2007) Life in acid: pH homeostasis in acidophiles. Trends Microbiol 15:165–171

Bakermans C, Skidmore ML, Douglas S, McKay CP (2014) Molecular characterization of bacteria from permafrost of the Taylor Valley, Antarctica. FEMS Microbiol Ecol 89:331–346

Bowman N, Patel P, Sanchez S, Xu W, Alsaffar A, Tiquia-Arashiro SM (2018) Lead-resistant bacteria from Saint Clair River sediments and Pb removal in aqueous solutions. Appl Microbiol Biotechnol 102:2391–2398

Brake SS, Hasiotis ST (2010) Eukaryote-dominated biofilms and their significance in acidic environments. Geomicrobiol J 27:534–558

Brock T (1978) Thermophilic microorganisms and life at high temperatures. Springer-Verlag, New York, NY, p 432

Brown MT, Hall IR (1990) Metal tolerance in fungi. In: Shaw J (ed) Heavy metal tolerance in plants: evolutionary aspects. CRC Press, Boca Raton, FL, pp 95–104

Butinar L, Sonjak S, Zalar P, Plemenitas A, Cimerman NG (2005) Melanized halophilic fungi are eukaryotic members of microbial communities in hypersaline waters of solar salterns. Bot Mar 48:73–79

Caron DA, Countway PD, Brown MV (2004) The growing contributions of molecular biology and immunology to protistan ecology: molecular signatures as ecological tools. J Eukaryot Microbiol 51:38–48

Casamayor EO, Massana R, Benlloch S, Ovreas L, Díez B, Goddard VJ, Gasol JM, Joint I, Rodríguez F, Pedrós-Alió C (2002) Changes in archaeal, bacterial and eukaryal assemblages along a salinity gradient by comparison of genetic fingerprinting methods in a multipond solar saltern. Environ Microbiol 4:338–348

Chin JP, Megaw J, Magill CL, Nowotarski K, Williams JP, Bhaganna P (2010) Solutes determine the temperature windows for microbial survival and growth. Proc Natl Acad Sci U S A 107:7835–7840

Cooke WB (1976) Fungi in and near streams carrying acid mine-drainage. Ohio J Sci 76:231–240

Corsaro D, Walochnik J, Venditti D, Steinmann J, Müller KD, Michel R (2014) Microsporidia-like parasites of amoebae belong to the early fungal lineage Rozellomycota. Parasitol Res 113:1909–1918

Das BK, Roy A, Koschorreck M, Mandal SM, Wendt-Potthoff K, Bhattacharya J (2009) Occurrence and role of algae and fungi in acid mine drainage environment with special reference to metals and sulfate Immobilization. Water Res 43:883–894

De Hoog GS, Zalar P, Urzı C, de Leo F, Yurlova NA, Sterflinger K (1999) Relationships of dothideaceous black yeasts and meristematic fungi based on 5.8S and ITS2 rDNA sequence comparison. Stud Mycol 43:31–37

De Hoog GS, Göttlich E, Platas G, Genilloud O, Leotta G, van Brummelen J (2005) Evolution, taxonomy and ecology of the genus Thelebolus in Antarctica. Stud Mycol 51:33–76

De Lima Alves F, Stevenson A, Baxter E, Gillion JL, Hejazi F, Hayes S (2015) Concomitant osmotic and chaotropicity-induced stresses in *Aspergillus wentii*: compatible solutes determine the biotic window. Curr Genet 61:457–477

De Maayer P, Anderson D, Cary C, Cowan DA (2014) Some like it cold: understanding the survival strategies of psychrophiles. EMBO Rep 15:508–517

Duran C, Marín I, Amils R (1999a) Specific metal sequestering acidophilic fungi. Process Metallurgy 9:521–530

Duran C, Marín I, Amils R (1999b) Specific metal sequestering acidophilic fungi. In: Amils R, Ballester A (eds) Biohydrometallurgy and the environment toward the mining of the 21st century, vol B. Elsevier, Amsterdam, pp 521–530

Gad GM, Griffiths AJ (1978) Microorganisms and heavy metal toxicity. Microb Ecol 4:303–317

Gadanho M, Sampaio JP (2006) Microeukaryotic diversity in the extreme environments of the Iberian Pyrite Belt: a comparison between universal and fungi-specific primer sets, temperature gradient gel electrophoresis and cloning. FEMS Microbiol Ecol 57:139–148

Gadanho M, Sampaio JP (2009) *Cryptococcus ibericus* sp. nov., *Cryptococcus aciditolerans* sp. nov. and *Cryptococcus metallitolerans* sp. nov., a new ecoclade of anamorphic basidiomycetous yeast species from an extreme environment associated with acid rock drainage in Sao Domingos pyrite mine, Portugal. Int J Syst Evol Microbiol 59:2375–2379

Gadanho M, Libkind D, Sampaio JP (2006) Yeast diversity in the extreme acidic environments of the Iberian Pyrite Belt. Microb Ecol 2:552–563

Gadd GM (1993) Interactions of fungi with toxic metals. New Phytol 124:25–60

Gadd GM (2008) Bacterial and fungal geomicrobiology: a problem with communities? Geobiology 6:278–284

González-Toril E, Llobet-Brossa E, Casamayor EO, Amann R, Amils R (2003) Microbial ecology of an extreme acidic environment, the Tinto River. Appl Environ Microbiol 69:4853–4865

González-Toril E, Santofimia E, Blanco Y, López-Pamo E, Gómez MJ, Bobadilla M, Cruz R, Palomino EJ, Aguilera A (2015) Pyrosequencing-based assessment of the microbial community structure of Pastoruri Glacier area (Huascarán Park, Perú), a natural extreme acidic environment. Microb Ecol 70:936–947

Gostincar C, Grube M, De Hoog S, Zalar P, Gunde-Cimerman N (2010) Extremotolerance in fungi: evolution on the edge. FEMS Microbiol Ecol 71:2–11

Grant WD (2004) Life at low water activity. Philos Trans R Soc Lond B 359:1249–1267

Gross S, Robbins EI (2000) Acidophilic and acid-tolerant fungi and yeasts. Hydrobiologia 433:91–109

Gunde-Cimerman N, Zalar P, De Hoog GS, Plemenitas A (2000) Hypersaline waters in salterns – natural ecological niches for halophilic black yeasts. FEMS Microbiol Ecol 32:235–240

Gunde-Cimerman N, Sonjak S, Zalar P, Frisvad JC, Diderichsen B, Plemenitas A (2003) Extremophilic fungi in Arctic ice: a relationship between adaptation to low temperature and water activity. Phys Chem Earth 28:1273–1278

Gunde-Cimerman N, Frisvad JC, Zalar P, Plemenitas A (2005a) Halotolerant and halophilic fungi. In: Deshmukh SK, Rai MK (eds) Biodiversity of fungi: their role in human life. Oxford and IBH Publishing Co. Pvt. Ltd, New Delhi, pp 69–127

Gunde-Cimerman N, Oren A, Plemenitas A (2005b) Adaptation to Life in High Salt Concentrations in Archaea, Bacteria, and Eukarya. Springer, Dordrecht, p 233

Johnson DB (1998) Biodiversity and ecology of acidophilic microorganisms. FEMS Microb Ecol 27:307–317

Johnson DB (2009) Extremophiles: acid environments. In: Schaechter M (ed) Encyclopaedia of microbiology. Elsevier, Oxford, pp 107–126

Johnson BD, Aguilera A (2016) Environmental microbiology in acidophilic environments. In: Manual of environmental microbiology, 4th edn. (MEM4). ASM Press, Washington, DC, pp 34–47

Kambura AK, Mwirichia RK, Kasili RW, Karanja EN, Makonde HM, Boga HI (2016) Bacteria and Archaea diversity within the hot springs of Lake Magadi and Little Magadi in Kenya. BMC Microbiol 16:136

Kogej T, Wheeler MH, Lanisnik Rizner T, Gunde-Cimerman N (2004) Evidence for 1,8-dihydroxynaphthalene melanin in three halophilic black yeasts grown under saline and nonsaline conditions. FEMS Microbiol Lett 232:203–209

Kogej T, Ramos J, Plemenitas A, Gunde-Cimerman N (2005) The halophilic fungus *Hortaea werneckii* and the halotolerant fungus *Aureobasidium pullulans* maintain low intracellular cation concentrations in hypersaline environments. Appl Environ Microbiol 71:6600–6605

Kogej T, Gostincar C, Volkmann M, Gorbushina AA, Gunde-Cimerman N (2006) Mycosporines in extremophilic fungi – novel complementary osmolytes? Environ Chem 3:105–110

Krüger A, Schäfers C, Schröder C, Antranikian G (2018) Towards a sustainable biobased industry: highlighting the impact of extremophiles. New Biotechnol 40:144–153

Leong SL, Pettersson OV, Rice T, Hocking AD, Schnürer J (2011) The extreme xerophilic mould *Xeromyces bisporus* – growth and competition at various water activities. Int J Food Microbiol 145:57–63

Leong SL, Lantz H, Pettersson OV, Frisvad JC, Thrane U, Heipieper HJ, Dijksterhuis J, Grabherr M, Pettersson M, Tellgren-Roth C, Schnürer J (2015) Genome and physiology of the ascomycete filamentous fungus *Xeromyces bisporus*, the most xerophilic organism isolated to date. Environ Microbiol 17:496–513

Lopatina A, Krylenkov V, Severinov K (2013) Activity and bacterial diversity of snow around Russian Antarctic stations. Res Microbiol 164:949–958

Ma Y, Galinski EA, Grant WD, Oren A, Ventosa A (2010) Halophiles 2010: life in saline environments. Appl Environ Microbiol 76:6971–6981

Magan N (2007) Fungi in extreme environments. In: Magan N (ed) Mycota: environmental and microbial relationships, vol 4. Elsevier, Amsterdam, pp 4–85

Masquelier S, Lepe C, Domaizon I, Curie M, Lepère C, Masquelier S (2010) Vertical structure of small eukaryotes in three lakes that differ by their trophic status: a quantitative approach. ISME J 4:1509–1519

Mehra RK, Winge DR (1991) Metal ion resistance in fungi: molecular mechanisms and their regulated expression. J Cell Biochem 45:30–40

Méndez-García C, Peláez AI, Mesa V, Sánchez J, GolyshinaOV FM (2015) Microbial diversity and metabolic networks in acid mine drainage habitats. Front Microbiol 6:475

Mesa V, Gallego JL, González-Gil R, Lauga B, Sánchez J, Méndez-García C, Peláez AI (2017) Bacterial, archaeal, and eukaryotic diversity across distinct microhabitats in an acid mine drainage. Front Microbiol 8:1756

Nazareth S, Gonsalves V (2014) *Aspergillus penicillioides* – a true halophile existing in hypersaline and polyhaline econiches. Front Microbiol 5:412

Oggerin M, Rodríguez M, del Moral C, Amils R (2014) Fungal jarosite biomineralization in Río Tinto. Res Microbiol 165:719–725

Orandi S, Yaghubpur A, Sahraei H (2007) Influence of AMD on aquatic life at Sar Cheshmeh copper mine. Abstract Goldschmidt Conference, Cologne, August 2007

Oren A (2002) Halophilic microorganisms and their environments. Kluwer Academic Publishers, Dordrecht, p 575

Ottow JGG, von Klopotek A (1969) Enzymatic reduction of iron oxide by fungi. Appl Microbiol 18:41–43

Petrovic U, Gunde-Cimerman N, Plemenitas A (2002) Cellular responses to environmental salinity in the halophilic black yeast *Hortaea werneckii*. Mol Microbiol 45:665–672

Pitt JI, Hocking AD (2009) Fungi and food spoilage, 3rd edn. Springer, Dordrecht, p 366

Plemenitas A, Vaupotic T, Lenassi M, Kogej T, Gunde-Cimerman N (2008) Adaptation of extremely halotolerant black yeast *Hortaea werneckii* to increased osmolarity: a molecular perspective at a glance. Stud Mycol 61:67–75

Ragon M, Fontaine MC, Moreira D, López-García P (2012) Different biogeographic patterns of prokaryotes and microbial eukaryotes in epilithic biofilms. Mol Ecol 21:3852–3868

Roberts DML (1999) Eukaryotic cells under extreme conditions. In: Seckbach J (ed) Enigmatic microorganisms and life in extreme environments. Kluwer Academic Publishers, London, pp 165–173

Rothschild LJ, Mancinelli RL (2001) Life in extreme environments. Nature 409:1092–1101

Russo G, Libkind D, Sampaio JP, VanBrock MR (2008) Yeast diversity in the acidic Río Agrio-Lake Caviahue volcanic environment (Patagonia, Argentina). FEMS Microbiol Ecol 65:415–424

Schleper C, Pühler G, Kühlmorgen B, Zillig W (1995) Life at extremely low pH. Nature 375:741–742

Schulze-Makuch D, Airo A, Schirmack J (2017) The adaptability of life on Earth and the diversity of planetary habitats. Front Microbiol 8:2011

Seckbach J (1994) Evolutionary pathways and enigmatic algae: *Cyanidium caldarium* (*Rhodophyta*) and related cells. In: Developments in hydrobiology, vol 91. Kluwer Academic Publishers, Dordrecht, p 349

Shiomi N, Yasuda T, Inoue Y, Kusumoto N, Iwasaki S, Katsuda T, Katoh S (2004) Characteristics of neutralization of acids by newly isolated fungal cells. Journal of Bioscience and Bioengineering 97(1):54–58

Siefert J, Mutz M (2001) Processing of leaf litter in acid waters of the post-mining landscape in Lusatia, Germany. Ecol Eng 17:297–306

Skorupa DJ, Reeb V, Castenholz RW, Bhattacharya D, McDermott TR (2013) Cyanidiales diversity in Yellowstone National Park. Lett Appl Microbiol 57:459–466

Sletten O, Skinner CE (1948) Fungi capable of growing in strongly acid media and in concentrated copper sulfate solutions. J Bacteriol 56:679–681

Sonjak S, Frisvad JC, Gunde-Cimerman N (2007a) Genetic variation among *Penicillium crustosum* isolates from the arctic and other ecological niches. Microb Ecol 54:298–305

Sonjak S, Ursic V, Frisvad JC, Gunde-Cimerman N (2007b) *Penicillium svalbardense*, a new species from Arctic glacial ice. Antonie Van Leeuwenhoek 92:43–51

Sonjak S, Frisvad JC, Gunde-Cimerman N (2009) Fingerprinting using extrolite profiles and physiological data shows sub-specific groupings of *Penicillium crustosum* strains. Mycol Res 113:836–841

Sterflinger K, De Hoog GS, Haase G (1999) Phylogeny and ecology of meristematic ascomycetes. Stud Mycol 43:5–22

Stevenson A, Cray JA, Williams JP, Santos R, Sahay R, Neuenkirchen N (2015a) Is there a common water-activity limit for the three domains of life? ISME J 9:1333–1351

Stevenson A, Burkhardt J, Cockell CS, Cray JA, Dijksterhuis J, Fox-Powell M (2015b) Multiplication of microbes below 0.690 water activity: implications for terrestrial and extraterrestrial life. Environ Microbiol 2:257–277

Stokes PM, Lindsay JE (1979) Copper tolerance and accumulation in *Penicillium ochro-chloron* isolated from copper-plating solution. Mycologia 71:796–806

Starkey RL, Waksman SA (1943) Arthur Trautmein Henrici. J Bacteriol 46:i2-490

Tatsuyama K, Egawa H, Senmaru H, Yamamoto H, Ishioka S, Tamatsukuri T, Saito K (1975) *Penicillium lilacinum:* its tolerance to cadmium. Experientia 31:1037–1038

Tiquia SM, Mormile M (2010) Extremophiles–A source of innovation for industrial and environmental applications. Environ Technol 31(8–9):823

Tiquia-Arashiro SM (2014) Thermophilic carboxydotrophs and their biotechnological applications. In: Springerbriefs in microbiology: extremophilic microorganisms, vol 131. Springer, New York

Tiquia-Arashiro SM (2018) Lead absorption mechanisms in bacteria as strategies for lead bioremediation. Appl Microbiol Biotechnol 102:5437–5444

Tiquia-Arashiro SM, Rodrigues D (2016) Alkaliphiles and acidophiles in nanotechnology. In: Extremophiles: applications in nanotechnology. Springer, New York, pp 129–162

Turk M, Plemenitas A (2002) The HOG pathway in the halophilic black yeast *Hortaea werneckii*: isolation of the HOG1 homologue gene and activation of HwHog1p. FEMS Microbiol Lett 216:193–199

Turk M, Mejanelle L, Sentjurc M, Grimalt JO, Gunde-Cimerman N, Plemenitas A (2004) Salt-induced changes in lipid composition and membrane fluidity of halophilic yeast-like melanized fungi. Extremophiles 8:53–61

Turk M, Abramovic Z, Plemenitas A, Gunde-Cimerman N (2007) Salt stress and plasma-membrane fluidity in selected extremophilic yeasts and yeast-like fungi. FEMS Yeast Res 7:550–557

Vaupotic T, Plemenitas A (2007) Differential gene expression and Hog1 interaction with osmoresponsive genes in the extremely halotolerant black yeast *Hortaea werneckii*. BMC Genomics 8:280–295

Vera M, Schippers A, Sand W (2013) Progress in bioleaching: fundamentals and mechanisms of bacterial metal sulfide oxidation – Part A. Appl Microbiol Biotechnol 97:7529–7541

Volant A, Héry M, Desoeuvre A, Casiot C, Morin G, Bertin PN (2016) Spatial distribution of eukaryotic communities using high-throughput sequencing along a pollution gradient in the arsenic-rich creek sediments of carnoulès mine, France. Microb Ecol 72:608–620

Wicklow D, Malloch D (1971) Studies in the genus *Thelebolus* temperature optima for growth and ascocarp development. Mycologia 63:118–131

Williams JP, Hallsworth JE (2009) Limits of life in hostile environments: no barriers to biosphere function? Environ Microbiol 11:3292–3308

Zhang XY, Zhang Y, Xu XY, Qi SH (2013) Diverse deep-sea fungi from the South China Sea and their antimicrobial activity. Curr Microbiol 67:525–530

Chapter 3
Ecology of Thermophilic Fungi

Tássio Brito de Oliveira and Andre Rodrigues ⓘ

3.1 Introduction

One fascinating property of microorganisms is their ability to adapt to extreme environments, in which factors such as pH, temperature, pressure, and salt concentration exceed the values that most living beings can survive. Among these factors, temperature alone can influence the function of the majority of biomolecules and the maintenance of biological structures. In fact, most of the currently known organisms can only sustain growth within a narrow temperature range. However, the existence of thermally stable environments has allowed the selection or the persistence of microorganisms that not only resist but also require high temperatures to survive: the thermophilic organisms.

Among the thermophilic microbes, fungi that sustain growth at high temperatures have attracted interest not only of biologists and ecologists but also for their wide applications in biotechnology and industry. While biologists investigate the adaptations of thermophilic fungi to their heated environment (Cooney and Emerson 1964; Crisan 1973), applied microbiologists explore these adaptations for economical purposes (Johri et al. 1999; Gomes et al. 2016; Singh et al. 2016). Regardless of the approach, the taxonomy and systematics of these fungi as well as their interactions with the environment are usually not considered in these fields. Here, our goal was to provide a taxonomic background of thermophilic fungi for users that explore these fungi but want to keep up-to-date with name changes of this group of microorganisms. Thus, we review the 46 currently known thermophilic fungal species

T. B. de Oliveira
Department of Biology, University of São Paulo (USP), Ribeirão Preto, SP, Brazil

A. Rodrigues (✉)
Department of Biochemistry and Microbiology, São Paulo State University (UNESP),
Rio Claro, SP, Brazil
e-mail: andrer@rc.unesp.br

© Springer Nature Switzerland AG 2019
S. M. Tiquia-Arashiro, M. Grube (eds.), *Fungi in Extreme Environments:
Ecological Role and Biotechnological Significance*,
https://doi.org/10.1007/978-3-030-19030-9_3

belonging to 23 genera. A second goal of this work is to discuss the various concepts of thermophilic fungi and how this depicts their ecology and lifestyles in the natural substrate. We go further and also explore biogeography and the mechanisms for adaptations to thermophily based on the current knowledge. Although several works reviewed the applications of thermophilic fungi (Cooney and Emerson 1964, Johri et al. 1999, Maheshwari et al. 2000, Gomes et al. 2016, Singh et al. 2016), here, we bring an updated background on the taxonomy and ecology.

3.2 Thermophilic Fungi

In general, thermophilic organisms can be classified as either moderate thermophilic or hyperthermophilic. The former exhibits growth temperatures ranging from a minimum of 20 °C up to a maximum of 60 °C and with optimal growth above 40 °C. These moderate thermophiles include species from the domains Bacteria and Archaea and representatives from Eukarya (mostly filamentous fungi), whose maximum temperature limit has been recorded to be 62 °C (Tansey and Brock 1972). On the other hand, the hyperthermophiles are organisms able to grow at temperatures between 65 and 110 °C. They contain several representatives from the domains Bacteria and Archaea, but do not include organisms from the domain Eukarya (Vieille and Zeikus 2001).

Although the first report of a thermophilic fungus, named *Mucor pusillus* (currently named as *Rhizomucor pusillus*), dates back to more than a century ago, the ability of fungi to grow at high temperatures was unknown at that point (Lindt 1886). Only a few years later that a second fungus (*Thermomyces lanuginosus*), isolated from potato disks, was described and its ability to grow at temperatures above 50 °C was demonstrated for the first time (Tsiklinsky 1899). In fact, the early systematic studies on thermophilic fungi are attributed to Miehe (1907) who described two important species, *Malbranchea pulchella* var. *sulfurea* (currently *Malbranchea cinnamomea*) and *Thermoascus aurantiacus*. However, it was only decades later when the first definition of thermophilic fungi was proposed (Apinis 1953).

Apinis was the first to use the term "thermophilus" to define all fungal species with good growth in the range of 35–40 °C (Apinis 1953). Thereafter, Crisan (1959) stated that all fungal species with regular growth at 40 °C or above are considered thermophiles. Then, Apinis (1963) proposed a classification using the cardinal points, classifying thermophilic fungi as having optimum temperatures for growth between 40 and 50 °C and a maximum up to 60 °C, but unable to grow at 20 °C. In the following year, Craveri et al. (1964) suggested a broad interpretation of thermophily in which a thermophilic fungus would be regarded as having the minimum growth temperature higher than 25 °C.

Considering that the end points (maximum and minimum) are easily demonstrated in comparison to the optimum growth, Cooney and Emerson (1964) proposed a less elaborate definition, separating these fungi into two subgroups:

thermophilic and thermotolerant. These authors considered thermophilic fungi as those with a maximum growth at 50 °C or above and a minimum growth temperature at 20 °C or above, separating them from the thermotolerant fungi, which could grow up to 50 °C and below 20 °C. Although this classification was generally used in the literature for many years, authors found several exceptions that do not fit to these criteria. As an example, *Aspergillus fumigatus* is a thermotolerant fungus able to grow at temperatures above 50 °C and below 20 °C (Mouchacca 2000a). Mostly because of these exceptions and the several different systems for distinguishing between thermophilic and thermotolerant fungi, many thermotolerant fungi are continuously classified as thermophilic (Mouchacca 2000a, 2007; Oliveira et al. 2015).

Evans (1971), noticing that the Cooney and Emerson definition was becoming artificial and obscure, delimited the heat-tolerant fungi in several groups. Briefly, group 1 (strong thermophiles) constitutes the obligate thermophiles, while species in groups 2 (weak thermophiles) and 3 (strong thermotolerants) form a transitional stage between true thermophily and general thermotolerance, the latter term embracing those species included in group 4 (thermotolerant in general). Then, Evans (1971) suggested that strains of certain fungi are transitional between the two groups: thermophilic and thermotolerant.

Alternatively, a few new definitions of thermophily in fungi came out in the last decades. Maheshwari et al. (2000) proposed a simpler classification, mostly used as a working model in applied research, where thermophilic fungi are defined as those species with an optimum growth temperature of 45 °C or above. Later, Morgenstern et al. (2012) used the criterion that a thermophilic fungus is that the one which grows faster at 45 °C than at 34 °C. The most recent definition was suggested by Oliveira et al. (2015), where thermophilic fungi are those with optimum growth ranging from 40 to 50 °C, separating them from thermotolerant species by the inability of growing below 20 °C. The inability to grow at low temperatures implies that thermophilic fungi are the only species that require higher minimum growth temperature ($\geq$20 °C) to exclude the gap presented for thermotolerant species (e.g., *A. fumigatus*).

3.3 Taxonomy

Thermophilic fungi comprise a paraphyletic group and, although they are distributed among taxonomically distinct lineages, they constitute an ecologically well-defined group. The first species reported as thermophilic was named *Thermomyces lanuginosus* (Tsiklinsky 1899). However, this species was relocated into different genera over the years, such as *Humicola*, *Monotospora*, and *Sepedonium*, before it has finally been relocated to *Thermomyces*. Likewise, many of the thermophilic fungi described until the last decades have been successively reclassified and renamed. The outcome of such name changes was the chaotic nomenclatural state of several members of this group as pointed by Mouchacca (2000b) and Oliveira et al. (2015).

Recently, an important step towards the stabilization of scientific names in fungal taxonomy was taken. A nomenclature code for fungi was proposed by mycologists resulting in the new *International Code of Nomenclature for Algae, Fungi and Plants* (ICN, McNeill et al. 2012). The development of this code was based on the "One fungus = One name" movement, seeking to stabilize fungal nomenclature by designating one name for pleomorphic fungal species (Taylor 2011; Wingfield et al. 2012). After that, Oliveira et al. (2015) published a full list of thermophilic fungal species and their nomenclatural status to date.

Eukaryotes can grow only up to 62 °C (Tansey and Brock 1972), and growth at this temperature is represented by a small number of fungal species. A large contribution on the taxonomy, biology, and economic importance of thermophilic fungi was given by Cooney and Emerson (1964) in their monograph. Eleven thermophilic species were documented with few being new to science (*Rhizomucor pusillus, R. miehei*, three varieties of *Chaetomium thermophilum* and *Chaetomium virginicum, Thermoascus aurantiacus, Melanocarpus albomyces, Malbranchea cinnamomea, Mycothermus thermophilus, Thermomyces lanuginosus*). Currently, the total number of fungal species described is approximately 120,000 (Hawksworth and Lücking 2017). From that, according to the last review on the taxonomy of thermophilic fungi, only 44 species are thermophilic, belonging to 20 genera (Oliveira et al. 2015). Thus, according to the authors, there are representatives of Mucoromycota (*Rhizomucor* and *Thermomucor*), Ascomycota (*Acremonium, Arthrinium, Canariomyces, Chaetomidium, Chaetomium, Humicola, Malbranchea, Melanocarpus, Myceliophthora, Myriococcum, Rasamsonia, Remersonia, Scytalidium, Sordaria, Thermoascus, Thermomyces,* and *Thielavia*), and one Basidiomycota (*Thermophymatospora*, Oliveira et al. 2015).

However, since the publication of the review by Oliveira et al. (2015) a few new species were described, renamed, or reclassified. A new *Rasamsonia* species from compost in China was described, *R. composticola* (Su and Cai 2013). The taxonomy of the genus *Myceliophthora* was re-evaluated through multilocus phylogenetic analysis (Marin-Felix et al. 2015). Four species, *Myceliophthora guttulata, M. hinnulea, M. heterothallica,* and *M. thermophila*, were reclassified to the new genus *Thermothelomyces* and *M. fergusii* to the new genus *Crassicarpon* as *C. thermophilum*. As described by Natvig et al. (2015), the species previously known as *Scytalidium thermophilum* is distantly related to the type species of the genus *Scytalidium, S. lignicola*, and could not be assigned to any existing genus. Then a new genus and combination was proposed, *Mycothermus thermophilus*. The thermophilic species in the phylum Basidiomycota, *Thermophymatospora fibuligera*, was renamed to *Ganoderma colossus*, following the new rules for fungal taxonomy (for more details, see Oliveira et al. 2015). Considering the recent changes, there are a total of 46 thermophilic species belonging to 23 genera (Table 3.1).

Thermophily is not a monophyletic character, because it is found in taxa in different phylogenetic lineages in the fungal tree of life. A recent phylogenetic analysis showed the paraphyletic nature of heat tolerance in fungi (Morgenstern et al. 2012). It is clear that this ability had multiple origins in the kingdom Fungi. However, in Chaetomiaceae (Sordariales), thermophily probably had a single origin and then

Table 3.1 Classification of accepted thermophilic fungal species (sensu Oliveira et al. 2015) according to the current literature

Phylum	Order	Genus	Species
Mucoromycota	Mucorales	*Rhizomucor*	*R. miehei*
			R. pusillus
		Thermomucor	*T. indicae-seudaticae*
Ascomycota	Eurotiales	*Rasamsonia*	*R. composticola*
			R. emersonii
			R. byssochlamydoides
		Thermoascus	*T. egyptiacus*
			T. aurantiacus
			T. crustaceus
			T. taitungiacus
			T. thermophilus
		Thermomyces	*T. dupontii*
			T. ibadanensis
			T. lanuginosus
			T. stellatus
			T. thermophilus
			T. verrucosus
	Hypocreales	*Acremonium*	*A. thermophilum*
	Incertae sedis	*Arthrinium*	*A. pterospermum*
		Malbranchea	*M. cinnamomea*
		Myriococcum	*M. thermophilum*
		Scytalidium	*S. indonesiacum*
	Microascales	*Canariomyces*	*C. thermophilus*
	Sordariales	*Chaetomidium*	*C. pingtungium*
		Chaetomium	*C. britannicum*
			C. mesopotamicum
			C. senegalense
			C. thermophilum
			C. virginicum
		Crassicarpon	*C. thermophilum*
		Humicola	*H. hyalothermophila*
		Melanocarpus	*M. albomyces*
			M. thermophilus
		Myceliophthora	*M. fusca*
			M. sulphurea
		Mycothermus	*M. thermophilus*
		Remersonia	*R. thermophila*
		Sordaria	*S. thermophila*
		Thermothelomyces	*T. guttulatus*
			T. hinnuleus
			T. heterothallicus

(continued)

Table 3.1 (continued)

Phylum	Order	Genus	Species
			T. thermophila
		Thielavia	*T. australiensis*
			T. terrestris
			T. terricola
Basidiomycota	Polyporales	*Ganoderma*	*G. colossus*

multiple losses subsequently occurred within the family, whereas multiple independent gains seem to be more likely in Trichocomaceae (Eurotiales) (Morgenstern et al. 2012; van Noort et al. 2013).

Overall, most representatives belong to the phylum Ascomycota, particularly in Sordariales and Eurotiales, orders that comprise the largest number of thermophilic species. There are also representatives in the orders Hypocreales and Microascales. In Mucoromycota, thermophily is restricted to the Mucorales order and in Basidiomycota to the order Polyporales (Table 3.1). The phylogenetic position of a few species is still unclear, such as species from the genera *Arthrinium*, *Malbranchea*, *Myriococcum*, and *Scytalidium*.

3.4   Ecology, Evolution, and Biogeography

Thermophilic fungi are cosmopolitan and they may occur either as propagules or as active mycelia in both natural and human-made environments. Their growth and activity are mainly regulated by the temperature and availability of nutrients. The evolution of thermophily in the kingdom Fungi is not clarified; however, a few speculations had come to light.

Some authors suggest that the heat tolerance in fungi evolved from mesophilic ancestors associated with nests of birds able to thermoregulate their nests, as those found in Australia (Megapodiidae) (Cooney and Emerson 1964; Rajasekaran and Maheshwari 1993). These birds use decaying plant material to stimulate the growth of microorganisms and warm up their nest. Fungi play an important role in this system; their exothermic metabolism raises the nest temperature to approximately 45 °C, similar to the heating process in natural composting (Seymour and Bradford 1992; Tiquia et al. 1996, 2002; Tiquia 2005). In contraposition, some authors suggested that thermophily arose as an adaptation to seasonal changes and high daytime temperatures rather than as an adaptation for the occupation of new high-temperature niches (Powell et al. 2012).

It is generally believed that their wide distribution is due to the propagules, which are easily transported by air, such as dispersal spores (sexually or asexually produced), resting spores, chlamydospores, and sclerotia (Thakur 1977; Rajasekaran and Maheshwari 1993; Le Goff et al. 2010). Thermophilic fungi have been recovered even from Antarctic soils (Ellis 1980a; Satyanarayana et al. 1992). Their widespread occurrence could well be explained by their wide dissemination machinery. These fungi occur mainly from man-made environments, such as composting systems, due to the production of aerosols carrying mycelia or reproductive propagules when revolving the piles (Le Goff et al. 2010). Most species do not show any geographical restrictions. According to Salar and Aneja (2006), soils in tropical countries do not appear to have a higher population of thermophilic fungi than soils in temperate countries as believed earlier. It is likely, however, that inoculum density in tropical soils and piles of plant material is higher than in temperate soils.

Thermophilic fungi are common in habitats wherever decomposition of organic matter takes place. According to several reports, they have been found in a variety of environments (Table 3.2). The habitats for the recovery of such fungi are not exotic like those of prokaryotes. The temperature, humidity, and atmosphere of these environments are favorable substrates for fungal development (Salar and Aneja 2006). There are no records of thermophilic fungi in many countries; however, this lack of information is not likely due to the absence of these fungi but most likely for lack of investigations in these sites.

Thermomyces, *Chaetomium*, and *Mycothermus* are the most recurrent genera in different environments (Table 3.2). Among them, *Thermomyces lanuginosus* and *Mycothermus thermophilus* (syn. *S. thermophilum*) are not only present but also often found as the most abundant (Pan et al. 2010; Powell et al. 2012; De Gannes et al. 2013; Langarica-Fuentes et al. 2014a, b, 2015; Oliveira et al. 2016). On the other hand, some species have not been recorded since they were identified for the first time, such as the Basidiomycota *G. colossus*.

Table 3.2 Thermophilic fungi currently known from natural and man-made environments

Environment	Genus	References
Soil	*Chaetomium, Thermothelomyces, Mycothermus, Rasamsonia, Rhizomucor, Thermoascus, Thermomyces*	Ellis (1980a), Salar and Aneja (2006), Pan et al. (2010), Powell et al. (2012)
Sediment	*Chaetomium, Thermomyces*	Tubaki et al. (1974); Ellis (1980b)
Pile of plant material	*Chaetomium, Rasamsonia, Rhizomucor, Thermoascus, Thermomyces, Thermothelomyces*	Tansey (1971), Pereira et al. (2015)
Nest of birds	*Chaetomium, Rasamsonia, Rhizomucor, Thermoascus, Thermomyces*	Korniłowicz-Kowalska and Kitowski (2013)
Composting	*Chaetomium, Crassicarpon, Mycothermus, Myriococcum, Rasamsonia, Rhizomucor, Thermoascus, Thermomyces, Thermomucor, Thermothelomyces, Thielavia*	Kane and Mullins (1973); Klamer and Søchting (1998); Straatsma et al. (1994); Hultman et al. (2010); De Gannes et al. (2013); Langarica-Fuentes et al. (2014a); Oliveira et al. (2016)
Compost	*Mycothermus, Rasamsonia, Thermomyces*	Langarica-Fuentes et al. (2014b)

Thermophilic fungi are usually found in self-heating environment. The compost-ing system, where the temperature rises due to the exothermic metabolism of micro-organisms, is by far the most suitable environment for their growth and dispersal. In a pressmud composting system, Oliveira et al. (2016) observed that the relative load of thermophilic species increases, from the thermophilic stage, over the reduction of mesophilic ones.

According to Paterson and Lima (2017), because of the increase in the average global temperature caused by climate change, it is expected that more fungi that tolerate or prefer higher temperatures can be found in areas where crops are being grown. They could turn into pathogens and infect crops, as they are adapted to simi-lar substrates, and would have little competition for this new niche against other fungi at such higher temperatures.

3.5   Ecological Roles

Lignocellulosic biomass consists of cellulose (32–50%), hemicellulose (19–25%), and lignin (23–32%) polymers, as well as a small part of organic acids, salts, and minerals (Pandey et al. 2000; Hamelinck et al. 2005). In nature, biomass-degrading microbes play a crucial role in plant biomass decomposition and in nutrient cycling (Plecha et al. 2013). The complexity of lignocellulosic biomass influences multiple microorganisms to produce equally a vast complex of enzymes, which act synergi-cally. The breakdown of lignocellulosic biomass involves the formation of long-chain polysaccharides, mainly cellulose and hemicellulose, and the subsequent hydrolysis of these polysaccharides into readily soluble saccharides, their 5- and 6-carbon monomers (Zhou and Ingram 2000; Sandgren et al. 2005).

Many microorganisms can degrade cellulose and other plant cell wall fibers, and fungi are known to be natural plant decomposers. There is a close relationship between the niche occupied by a microorganism and the characteristics of its intra and extracellular enzymes. It is expected that thermophilic microorganisms produce extracellular enzymes capable of tolerating temperatures corresponding to at least the optimum temperature for their growth, i.e., above 45 °C. Thus, these fungi are suitable to secrete a wide variety of lignocellulolytic enzymes, which act under this special environmental condition, significantly contributing to biomass decay in nature.

Many reports showed the ability of thermophilic fungi to produce a wide range of enzymes involved in biomass decomposition and nutrient recycling (Table 3.3). These fungi have a high capacity to secrete enzymes with a variety of mechanisms of action and substrate specificity. Most strains produce various enzymes in large amounts which are released in the environment and act in a synergistic manner (Dashtban et al. 2009). A plethora of studies screened various thermophilic fungi to find new and promising strains able to produce enzymes of biotechnological inter-est. Most of these studies employed agricultural residues to stimulate the production

Table 3.3 Cell wall-degrading and nutrient-recycling enzymes from thermophilic fungi

Species	Enzymes	References
Acremonium thermophilum	Cellulase	Voutilainen et al. (2008)
Canariomyces thermophilus	Protease	Srilakshmi et al. (2014)
Chaetomium mesopotamicum	Protease	Srilakshmi et al. (2014)
Chaetomium senegalense	Cellulase	Kolet (2010)
Chaetomium thermophilum	Cellulase, laccase, xylanase	Chefetz et al. (1998), Maheshwari et al. (2000), Venturi et al. (2002), Voutilainen et al. (2008)
Chaetomium virginicum	Cellulase	Kolet (2010)
Malbranchea cinnamomea	Amylase, protease, xylanase	Ong and Gaucher (1976), Gupta and Gautam (1993), Katapodis et al. (2003)
Melanocarpus albomyces	Amylase, cellulase, lipase, protease, xylanase	Prabhu and Maheshwari (1999), Narang et al. (2001), Hirvonen and Papageorgiou (2003), Srilakshmi et al. (2014)
Mycothermus thermophilus	Amylase, cellulase, pectinase, phosphatase, protease	Johri et al. (1999), Arifoğlu and Ögel (2000), Roy et al. (2000), Aquino et al. (2001), Guimarães et al. (2001), Ifrij and Ogel (2002)
Myriococcum thermophilum	Amylase, cellulase, lipase, protease	Srilakshmi et al. (2014)
Rasamsonia byssochlamydoides	Xylanase	Hayashida et al. (1988)
Rasamsonia emersonii	Cellulase	Murray et al. (2004)
Remersonia thermophila	Xylanase	McPhillips et al. (2014)
Rhizomucor miehei	Lipase, protease	Maheshwari et al. (2000), Rao and Divakar (2002), da Silva et al. (2016)
Rhizomucor pusillus	Pectinase, phytase, protease	Arima et al. (1968), Johri et al. (1999), Chadha et al. (2004)
Thermoascus aurantiacus	Cellulase, phytase, polygalacturonase, xylanase	Gomes et al. (2000), Nampoothiri et al. (2004), dos Santos et al. (2003), Kalogeris et al. (1998), Leite et al. (2007), Martins et al. (2007)
Thermoascus crustaceus	Cellulase, lipase, phytase, protease	Srilakshmi et al. (2014)
Thermoascus thermophilus	Phytase	Pasamontes et al. (1997)
Thermomucor indicae-seudaticae	Amylase, cellulase, pectinase	Kumar and Satyanarayana (2003), Martin et al. (2010), Pereira et al. (2014)
Thermomyces dupontii	Pectinase, protease	Hashimoto et al. (1972), Johri et al. (1999)

(continued)

Table 3.3 (continued)

Species	Enzymes	References
Thermomyces ibadanensis	Protease, lipase	Srilakshmi et al. (2014)
Thermomyces lanuginosus	Amylase, lipase, pectinase, phytase, protease, xylanase	Arima et al. (1968), Mishra and Maheshwari (1996), Chadha et al. (1997), Berka et al. (1998), Johri et al. (1999), Lin et al. (1999), Singh et al. (2000), Nguyena et al. (2002)
Thermomyces stellatus	Pectinase	Johri et al. (1999), Jensen et al. (2002)
Thermomyces thermophilus	Cellulases, phytase, protease, xylanase	Pasamontes et al. (1997), Srilakshmi et al. (2014)
Thermothelomyces heterothallicus	Cellulase, xylanase	van den Brink et al. (2013)
Thermothelomyces thermophilus	Cellulase, pectinase, phytase, protease, xylanase	Bhat and Maheshwari (1987), Mitchell et al. (1997), Kaur et al. (2004), Singh and Satyanarayana (2006a, b, 2008a, b, c), Moretti et al. (2012), Srilakshmi et al. (2014), Pereira et al. (2015)
Thielavia australiensis	Amylase, cellulase, protease	Srilakshmi et al. (2014)

of enzymes to target biotechnological process. From this perspective, we can perceive their ecological roles on biomass decomposition in nature.

Some thermophilic fungi such as *Thermomyces lanuginosus*, *T. dupontii*, *Rhizomucor pusillus*, and *R. miehei* cannot use cellulose as a carbon source (Table 3.3). However, the inability to hydrolyze a particular polymer, such as cellulose, does not mean that the fungus does not have an enzymatic system for the hydrolysis of another polymer. For instance, *T. lanuginosus* does not degrade cellulose, but it is able to use xylan as a carbon source and grow faster on this polymer than on medium with simpler sugars (Mchunu et al. 2013; Oliveira et al. 2015). Likewise, thermophilic pectinolytic fungi are not always high producers of hemicellulolytic enzymes. On the other hand, organisms that do not depolymerize the organic matter can grow as a commensal, using sugars released by other organisms.

Pereira et al. (2015) evaluated the production of cellulases and xylanases from heat-tolerant fungi by solid-state fermentation using lignocellulosic materials as substrate. *Thermothelomyces thermophila* (syn. *M. thermophila*) was the best producer of endoglucanase (357.51 U g^{-1}), β-glucosidase (45.42 U g^{-1}), xylanase (931.11 U g^{-1}), and avicelase (3.58 U g^{-1}). Martin et al. (2010) evaluated the production of polygalacturonase from *Thermomucor indicae-seudaticae* using a mixture of carbon sources; the highest production (108 U g^{-1}) was obtained in a mixture consisting of 40% orange bagasse, 40% wheat bran, and 20% sugarcane bagasse (Martin et al. 2010). *Mycothermus thermophilus* was able to produce high amounts of cellulases in a medium containing a mixture of rice straw and wheat bran (1:3) such as endoglucanase (64.7 U g^{-1}), avicelase (23.5 U g^{-1}), β-glucosidase

(160 U g^{-1}), and FPase (3.48 U g^{-1}). This fungus was also able to produce xylanase (199.2 U g^{-1}, Jatinder et al. 2006).

The majority of fungi growing on plant residues in nature usually produce both cellulolytic and xylanolytic enzymes due to the close association of cellulose and xylan in plant cell walls. However, so far *T. lanuginosus* was found to be a non-cellulolytic hyperproducer of xylanases. Alam et al. (1994) obtained high amount of xylanases (1889.6 U g^{-1}) and pectinase (673.2 U g^{-1}) in solid-state fermentation using wheat bran and no cellulose activity was detected. In the same study, the species *T. aurantiacus* was demonstrated to be able to produce cellulose (705.7 U g^{-1}), xylanase (306.5 U g^{-1}), and cellulase (215.8 U g^{-1}).

Considering the increased number of available genomes, new rational approaches, as genome mining, have been applied as an alternative to find target and new enzymes from thermophilic fungi. For instance, *T. lanuginosus* had its genome sequenced to obtain additional information for a better assessment in the industry (Mchunu et al. 2013). Zhou et al. (2014) reported the existence of a large number of genes coding for proteolytic, amylolytic and lipolytic enzymes including also xylanase and β-glucanase in the genome of *R. miehei*.

A genomic study of *T. thermophila* and *Thielavia terrestris* suggested that they were capable of hydrolyzing all the major polysaccharides present in the plant biomass (Berka et al. 2011). *Thielavia thermophila* has the highest number of hemicellulolytic enzymes and accessory enzymes observed to date; it contains 8 genes encoding endoglucanases, 7 cellobiohydrolases, 9 β-glucosidases, 25 lytic polysaccharide monooxygenases (LPMOs), and other enzymes of the group including xylanase, arabinases, mannanase, pectinases, and esterases (Karnaouri et al. 2014). Oliveira et al. (2018) evaluated 13 thermophilic species and listed the presence of peptidase-coding genes in their genome. All species have putative peptidase-coding genes, ranging from 241 to 347.

3.6 Fungal Adaptations to Thermophily

Sustaining growth at high temperatures involves adaptation of the cytoplasmic membrane, proteins, and DNA to temperatures above mesophilic range. These adaptations provoked great interest from both biological and evolutionary perspectives. However, it is in biotechnology that this interest is significantly explored, considering that the thermoresistance mechanisms of biomolecules of these microorganisms may be interesting models for bioengineering or to directly use in bioprocesses.

In general, all the features observed in thermophilic fungi are similar to those of mesophiles. Thermophilic fungi do not appear to have any specific organelles, structural modifications, or developmental patterns, which are not seen in their mesophilic counterparts (Singh et al. 2016). However, the adaptation of a particular microorganism allowing it to survive and grow at elevated temperatures involves crucial aspects as modifications of existing structures.

Crisan (1973) discusses four hypotheses that help explaining the ability of thermophiles to grow at high temperatures: (i) lipid solubilization, (ii) rapid resynthesis of essential metabolites, (iii) macromolecular thermostability, and (iv) ultrastructural thermostability. Only the latter two still appear to be of major importance. The importance of macromolecular thermostability is questionable since the existence of a single, thermostable, essential macromolecule, common to all thermophiles, has not been demonstrated. On the other hand, the author concluded that the hypothesis of ultrastructural thermostability appears to be the most promising to explain the existence of thermophilism in a variety of different organisms.

To date, no fungi have been identified with a growth above 62 °C. This may be related to the greater thermolability of their membrane systems than to the thermostability of enzymes or other cellular structures. The adaptation of thermophilic microbial membranes corresponds to the process named homeoviscous adaptation, which consists of the replacement of unsaturated fatty acids with saturated fatty acids. With these shifts, membrane acquires the balance between density and fluidity necessary for the maintenance of physical and functional integrity at elevated temperatures. The saturated fatty acids generate a stronger hydrophobic environment than the unsaturated ones, aiding in the stability of the membrane. This adaptation occurs in the Bacteria and Eukarya domains, while the latter is so far found only in the kingdom Fungi (Adams 1993).

The thermophilic fungi have experienced genome size reduction compared to the closest mesophilic species (van Noort et al. 2013). This process involves loss of protein-coding genes, transposable elements, and reductions in the size of introns and intergenic regions. Oliveira et al. (2018) found that the number of peptidase-coding genes in the genome of thermophilic fungi is reduced, when compared to peptidases from phylogenetically related mesophilic species, in contrast with the observations for cellulolytic enzymes, which reflects an increased wood degradation capacity (van Noort et al. 2013).

Although most of the studies exploring enzymes from thermophilic organisms are focused on their production and characterization, only a few focused in comparing the differences between enzymes from thermophilic and mesophilic species (Niehaus et al. 1999; van Noort et al. 2013; Oliveira et al. 2018). Some differences in the sequence, structure, function, dynamics, and thermodynamic properties were observed when making such kind of comparison (Niehaus et al. 1999; Oliveira et al. 2018).

Oliveira et al. (2018) evaluating the peptidases from thermophilic fungi found that they contain a larger proportion of Ala, Glu, Gly, Pro, Arg, and Val residues and a lower number of Cys, His, Ile, Lys, Met, Asn, Gln, Ser, Thr, and Trp residues when compared to the mesophilic ones. The authors also found that peptidases from thermophilic fungi had a reduction in the number of internal cavities. They suggested two possible evolutionary scenarios for these changes: in the course of genome reduction, (i) they have lost peptidases with large number of cavities and kept only those that are compactly folded or (ii) the enzymes were optimized to contain fewer.

In contrast to the reductionist genome tendency, the duplication of intriguing genes may provide insights into the evolution of thermophily, such as identifying the genes responsible for hyphal melanization, which are involved in resistance to high temperatures, desiccation, and UV radiation (van Noort et al. 2013).

3.7 Conclusions and Future Perspectives

Thermophily in fungi evolved several times in nature; however, they are an ecological well-defined group. Distributed worldwide, these fungi occur in several habitats, but they develop when high temperatures allow their growth and survival. Over the years, many concepts were employed to characterize these organisms, but here we showed some concepts that may be ecologically relevant. Moreover, the taxonomy of these fungi changed in recent years prompting to the several name changes reported in this review. As the search for fungi continues in unusual niches (Cantrell et al. 2011) including several self-heated environments, it is possible that new thermophilic fungi may be discovered. These taxonomic novelties may reveal new mechanisms on how eukaryotic life may thrive in high temperatures.

Thermophilic fungi are efficient in biomass degradation, what makes them target species to prospect for enzymes of biotechnological interest. They have evolved adaptations to high temperatures, and the knowledge of such characters might help to better understand their biology and make a better use of them to our welfare. Also, the increasing number of available genomes of thermophilic fungi brings new perspectives to their exploitation, such as genome mining. Nowadays, there are 46 known species and the use of molecular techniques (i.e., metagenomics) could increase our knowledge regarding their diversity and distribution. Moreover, the find of new species is linked to the discovery of new molecules with intrinsic properties.

References

Adams MW (1993) Enzymes and proteins from organisms that grow near and above 100 °C. Annu Rev Microbiol 47:627–658

Alam M, Gomes I, Mohiuddin G, Hoq MM (1994) Production and characterization of thermostable xylanases by *Thermomyces lanuginosus* and *Thermoascus aurantiacus* grown on lignocelluloses. Enzym Microb Technol 16:298–302

Apinis AE (1953) Distribution, classification and biology of certain soil inhabiting fungi. Ph.D. Thesis. Nottingham University, United Kingdom

Apinis AE (1963) Occurrence of thermophilous microfungi in certain alluvial soils near Nottingham. Nova Hedwigia 5:57–78

Aquino AC, Jorge JA, Terenzi HF, Polizeli ML (2001) Thermostable glucose-tolerant glucoamylase produced by the thermophilic fungus *Scytalidium thermophilum*. Folia Microbiol 46:11–16

Arifoğlu N, Ögel ZB (2000) Avicel-adsorbable endoglucanase production by the thermophilic fungus *Scytalidium thermophilum* type culture *Torula thermophila*. Enzym Microb Technol 27:560–569

Arima K, Iwasaki S, Tamura G (1968) Milk clotting enzymes from microorganisms. V. Purification and crystallization of *Mucor* rennin from *Mucor pusillus* Lindt. Appl Microbiol 16:1727–1733

Berka RM, Rey MW, Brown KM, Byun T, Klotz AV (1998) Molecular characterization and expression of a phytase gene from the thermophilic fungus *Thermomyces lanuginosus*. Appl Environ Microbiol 64:4423–4427

Berka RM, Grigoriev IV, Otillar R, Salamov A, Grimwood J, Reid I, Ishmael N, John T, Darmond C, Moisan MC, Henrissat B, Coutinho PM, Lombard V, Natvig DO, Lindquist E, Schmutz J, Lucas S, Harris P, Powlowski J, Bellemare A, Taylor D, Butler G, de Vries RP, Allijn IE, van den Brink J, Ushinsky S, Storms R, Powell AJ, Paulsen IT, Elbourne LD, Baker SE, Magnuson J, Laboissiere S, Clutterbuck AJ, Martinez D, Wogulis M, de Leon AL, Rey MW, Tsang A (2011) Comparative genomic analysis of the thermophilic biomass-degrading fungi *Myceliophthora thermophila* and *Thielavia terrestris*. Nat Biotechnol 29:922–929

Bhat KM, Maheshwari R (1987) *Sporotrichum thermophile*: growth, cellulose degradation, and cellulase activity. Appl Environ Microbiol 53:2175–2182

Cantrell SA, Dianese JC, Fell J, Gunde-Cimerman N, Zalar P (2011) Unusual fungal niches. Mycologia 103:1161–1174

Chadha BS, Singh S, Vohra G, Saini HS (1997) Shake culture studies for the production of amylases by *Thermomyces lanuginosus*. Acta Microbiol Immunol Hung 44:181–185

Chadha BS, Gulati H, Minhas M, Saini HS, Singh N (2004) Phytase production by the thermophilic fungus *Rhizomucor pusillus*. World J Microbiol Biotechnol 20:105–109

Chefetz B, Chen Y, Hadar Y (1998) Purification and characterization of laccase from *Chaetomium thermophilium* and its role in humification. Appl Environ Microbiol 64:3175–3179

Cooney DG, Emerson R (1964) Thermophilic fungi: an account of their biology, activities and classification. W.H. Freeman, San Francisco, CA

Craveri R, Manachini PL, Craveri A (1964) Eumiceti termofili presenti nel suolo. Ann Microbiol Enzimol 14:13–26

Crisan EV (1959) The isolation and identification of thermophilic fungi. M.Sc. thesis, Purdue University, Lafayette, Indiana

Crisan EV (1973) Current concepts of thermophilism and the thermophilic fungi. Mycologia 65:1171–1198

da Silva RR, Souto TB, de Oliveira TB, de Oliveira LC, Karcher D, Juliano MA, Juliano L, de Oliveira AH, Rodrigues A, Rosa JC, Cabral H (2016) Evaluation of the catalytic specificity, biochemical properties, and milk clotting abilities of an aspartic peptidase from *Rhizomucor miehei*. J Ind Microbiol Biotechnol 43:1059–1069

dos Santos E, Piovan T, Roberto IC, Milagres AM (2003) Kinetics of the solid state fermentation of sugarcane bagasse by *Thermoascus aurantiacus* for the production of xylanase. Biotechnol Lett 25:13–16

Dashtban M, Schraft H, Qin W (2009) Fungal bioconversion of lignocellulosic residues; opportunities and perspectives. Int J Biol Sci 5:578–595

De Gannes V, Eudoxie G, Hickey WJ (2013) Insights into fungal communities in composts revealed by 454-pyrosequencing: implications for human health and safety. Front Microbiol 4:1–9

Ellis DH (1980a) Thermophilic fungi isolated from some Antarctic and sub-Antarctic soils. Mycologia 72:1033–1036

Ellis DH (1980b) Thermophilic fungi isolated from a heated aquatic habitat. Mycologia 72:1030–1033

Evans HC (1971) Thermophilous fungi of coal spoil tips. II Occurrence, distribution and temperature relationships. Trans Br Mycol Soc 57:255–266

Gomes I, Gomes J, Gomes DJ, Steiner W (2000) Simultaneous production of high activities of thermostable endoglucanase and beta-glucosidase by the wild thermophilic fungus *Thermoascus aurantiacus*. Appl Microbiol Biotechnol 53:461–468

Gomes E, Souza AR, Orjuela GL, Silva R, Oliverira TB, Rodrigues A (2016) Applications and benefits of thermophilic microorganisms and their enzymes for industrial biotechnology. In: Schmoll M, Dattenböck C (eds) Gene expression systems in fungi: advancements and applications. Springer, Cham, p 459–462

Guimarães LHS, Terenzi HF, Jorge JA, Polizeli ML (2001) Thermostable conidial and mycelial alkaline phosphatases from the thermophilic fungus *Scytalidium thermophilum*. J Ind Microbiol Biotechnol 27:265–270

Gupta AK, Gautam SP (1993) Purification and properties of an extracellular α-glucosidase from thermophilic fungus *Malbranchea sulfurea*. J Gen Microbiol 139:963–967

Hamelinck CN, van Hooijdonk G, Faaij APC (2005) Ethanol from lignocellulosic biomass: techno-economic performance in short-, middle- and long-term. Biomass Bioenergy 28:384–410

Hashimoto H, Iwaasa T, Yokotsuka T (1972) Thermostable acid protease produced by *Penicillium duponti* K1014, a true thermophilic fungus newly isolated from compost. Appl Microbiol 24:986–992

Hawksworth D, Lücking R (2017) Fungal diversity revisited: 2.2 to 3.8 million species. Microbiol Spectr 5:1–17

Hayashida S, Ohta K, Mo K (1988) Xylanase of *Talaromyces byssochlamydoides*. Methods Enzymol 160:675–678

Hirvonen M, Papageorgiou AC (2003) Crystal structure of a family 45 endoglucanase from *Melanocarpus albomyces* mechanistic implications based on the free and cellobiose-bound forms. J Mol Biol 329:403–410

Hultman J, Vasara T, Partanen P, Kurola J, Kontro MH, Paulins L, Auvinen P, Romantschuk M (2010) Determination of fungal succession during municipal solid waste composting using a cloning-based analysis. J Appl Microbiol 108:472–487

Ifrij H, Ogel ZB (2002) Production of neutral and alkaline extracellular proteases by the thermophilic fungus, *Scytalidium thermophilum*, grown on microcrystalline cellulose. Biotechnol Lett 24:1107–1110

Jatinder K, Chadha BS, Saini HS (2006) Optimization of culture conditions for production of cellulases and xylanases by *Scytalidium thermophilum* using response surface methodology. World J Microbiol Biotechnol 22:169–176

Jensen B, Nebelong P, Olsen J, Reeslev M (2002) Enzyme production in continuous cultivation by the thermophilic fungus, *Thermomyces lanuginosus*. Biotechnol Lett 24:41–45

Johri BN, Satyanarayana T, Olsen J (1999) Thermophilic moulds in biotechnology. Springer, Netherland

Kalogeris E, Christakopoulos P, Kekos D, Macris BJ (1998) Studies on the solid state production of thermostable endoxylanases from *Thermoascus aurantiacus*, characterization of two isozymes. J Biotechnol 60:155–163

Kane BE, Mullins JT (1973) Thermophilic fungi in a municipal waste compost system. Mycologia 65:1087–1100

Karnaouri A, Topakas E, Antonopoulou I, Christakopoulos P (2014) Genomic insights into the fungal lignocellulolytic system of *Myceliophthora thermophila*. Front Microbiol 5:281

Katapodis P, Vrsanská M, Kekos D, Nerinckx W, Biely P, Claeyssens M, Macris BJ, Christakopoulos P (2003) Biochemical and catalytic properties of an endoxylanase purified from the culture filtrate of *Sporotrichum thermophile*. Carbohydr Res 338:1881–1890

Kaur G, Kumar S, Satyanarayana T (2004) Production, characterization and application of a thermostable polygalacturonase of a thermophilic mold *Sporotrichum thermophile* Apinis. Bioresour Technol 94:239–243

Klamer M, Søchting U (1998) Fungi in a controlled compost system-with special emphasis on the thermophilic fungi. Acta Hortic 469:405–413

Kolet M (2010) Quantitative analysis of cellulolytic activity of the genus *Chaetomium*. Bionano Front 3:304–306

Korniłłowicz-Kowalska T, Kitowski I (2013) *Aspergillus fumigatus* and other thermophilic fungi in nests of wetland birds. Mycopathologia 175:43–56

Kumar S, Satyanarayana T (2003) Purification and kinetics of a raw starch-hydrolyzing, thermostable, and neutral glucoamylase of the thermophilic mold *Thermomucor indicae-seudaticae*. Biotechnol Prog 19:936–944

Langarica-Fuentes A, Zafar U, Heyworth A, Brown T, Fox G, Robson GD (2014a) Fungal succession in an in-vessel composting system characterized using 454 pyrosequencing. FEMS Microbiol Ecol 88:296–308

Langarica-Fuentes A, Handley PS, Houlden A, Fox G, Robson GD (2014b) An investigation of the biodiversity of thermophilic and thermotolerant fungal species in composts using culture-based and molecular techniques. Fungal Ecol 11:132–144

Langarica-Fuentes A, Fox G, Robson GD (2015) Metabarcoding analysis of home composts reveals distinctive fungal communities with a high number of unassigned sequences. Microbiology 161:1921–1932

Le Goff O, Bru-Adan V, Bacheley H, Godon JJ, Wéry N (2010) The microbial signature of aerosols produced during the thermophilic phase of composting. J Appl Microbiol 108:325–340

Leite RSR, Gomes E, da Silva R (2007) Characterization and comparison of thermostability of purified β-glucosidases from a mesophilic *Aureobasidium pullulans* and a thermophilic *Thermoascus aurantiacus*. Process Biochem 42:1101–1106

Lin J, Ndlovu LM, Singh S, Pillay B (1999) Purification and biochemical characteristics of β-D-xylanase from a thermophilic fungus, *Thermomyces lanuginosus*-SSBP. Biotechnol Appl Biochem 30:73–79

Lindt W (1886) Mitteilungen über einige neue pathogene Shimmelpilze. Arch Exp Pathol Pharmakol 21:269–298

Maheshwari R, Bharadwaj G, Bhat MK (2000) Thermophilic fungi: their physiology and enzymes. Microbiol Mol Biol Rev 64:461–488

Marin-Felix Y, Stchigel AM, Miller AN, Guarro J, Cano-Lira JF (2015) A re-evaluation of the genus *Myceliophthora* (Sordariales, Ascomycota): its segregation into four genera and description of *Corynascus fumimontanus* sp. nov. Mycologia 107:619–632

Martin N, Guez MAU, Sette LD, da Silva R, Gomes E (2010) Pectinase production by a brazilian thermophilic fungus *Thermomucor indicae-seudaticae* N31 in solid-state and submerged fermentation. Microbiology 79:306–313

Martins E, Silva D, Leite RSR, Gomes E (2007) Purification and characterization of polygalacturonase produced by thermophilic *Thermoascus aurantiacus* CBMAI-756 in submerged fermentation. Antonie Van Leeuwenhoek 91:291–299

Mchunu NP, Permaul K, Rahman AYA, Saiot JA, Singh S, Alam M (2013) Xylanase superproducer: genome sequence of a compost-loving thermophilic fungus, *Thermomyces lanuginosus* strain SSBP. Genome Announc 1:e00388–e00313

McNeill J, Barrie FR, Buck WR, Demoulin V, Greuter W, Hawksworth DL, Herendeen PS, Knapp S, Marhold K., Prado J, Prud'homme van Reine WF, Smith GE, Wiersema JH, Turland NJ (eds) (2012) International Code of Nomenclature for algae, fungi, and plants (Melbourne Code) adopted by the Eighteenth International Botanical Congress Melbourne, Australia, July 2011 [Regnum Vegetabile no. 154.], A.R.G. Ganter Verlag, Ruggell

McPhillips K, Waters DM, Parlet C, Walsh DJ, Arendt EK, Murray PG (2014) Purification and characterisation of a β-1,4-xylanase from *Remersonia thermophila* CBS 540.69 and its application in bread making. Appl Biochem Biotechnol 172:1747–1762

Miehe H (1907) Die selbsterhitzung des Heus. Ene Biologische Studie. Gustav Fischer, Jena

Mishra R, Maheshwari R (1996) Amylases of the thermophilic fungus *Thermomyces lanuginosus*, their purification, properties, action on starch and response to heat. J Biosci 21:653–672

Mitchell DB, Vogel K, Weimann BJ, Pasamontes L, van Loon AP (1997) The phytase subfamily of histidine acid phosphatase; isolation of genes for two novel phytases from the *Aspergillus terreus* and *Myceliophthora thermophila*. Microbiology 143:245–252

Moretti MMS, Martins DAB, Silva R, Rodrigues A, Sette LD, Gomes E (2012) Selection of thermophilic and thermotolerant fungi for the production of cellulases and xylanases under solid-state fermentation. Braz J Microbiol 43:1062–1071

Morgenstern I, Powlowski J, Ishmael N, Darmond C, Marqueteau S, Moisan M, Quenneville G, Tsang A (2012) A molecular phylogeny of thermophilic fungi. Fungal Biol 116:489–502

Mouchacca J (2000a) Thermotolerant fungi erroneously reported in applied research work as possessing thermophilic attributes. World J Microbiol Biotechnol 16:869–880

Mouchacca J (2000b) Thermophilic fungi and applied research: a synopsis of name changes and synonymies. World J Microbiol Biotechnol 16:881–888

Mouchacca J (2007) Heat tolerant fungi and applied research: addition to the previously treated group of strictly thermotolerant species. World J Microbiol Biotechnol 23:1755–1770

Murray P, Aro N, Collins C, Grassick A, Penttilä M, Saloheimo M, Tuohy M (2004) Expression in *Trichoderma reesei* and characterization of a thermostable family 3 beta-glucosidase from the moderately thermophilic fungus *Talaromyces emersonii*. Protein Expr Purif 38:248–257

Nampoothiri KM, Tomes GJ, Roopesh K, Szakacs G, Nagy V, Soccol CR, Pandey A (2004) Thermostable phytase production by *Thermoascus aurantiacus* in submerged fermentation. Appl Biochem Biotechnol 118:205–214

Narang S, Sahai V, Bisaria VS (2001) Optimization of xylanase production by *Melanocarpus albomyces* IIS68 in solid state fermentation using response surface methodology. J Biosci Bioeng 91:425–427

Natvig DO, Taylor JW, Tsang A, Hutchinson MI, Powell AJ (2015) *Mycothermus thermophilus* gen. et comb. nov., a new home for the itinerant thermophile *Scytalidium thermophilum* (*Torula thermophila*). Mycologia 107:319–327

Nguyena QD, Judit M, Claeyssens RM, Stals I, Ágoston H (2002) Purification and characterization of amylolytic enzymes from thermophilic fungus *Thermomyces lanuginosus* strain ATCC 34626. Enzym Microb Technol 31:345–352

Niehaus F, Bertoldo C, Kahler M, Antranikian G (1999) Extremophiles as a source of novel enzymes for industrial applications. Appl Microbiol Biotechnol 51:711–729

Oliveira TB, Gomes E, Rodrigues A (2015) Thermophilic fungi in the new age of fungal taxonomy. Extremophiles 19:31–37

Oliveira TB, Lopes VCP, Barbosa FN, Ferro M, Meirelles LA, Sette LD, Gomes E, Rodrigues A (2016) Fungal communities in pressmud composting harbour beneficial and detrimental fungi for human welfare. Microbiology 162:1147–1156

Oliveira TB, Gostinčar C, Gunde-Cimerman N, Rodrigues A (2018) Genome mining for peptidases in heat-tolerant and mesophilic fungi and putative adaptations for thermostability. BMC Genomics 19:152

Ong PS, Gaucher GM (1976) Production, purification and characterization of thermomycolase, the extracellular serine protease of the thermophilic fungus *Malbranchea pulchella* var. *sulfurea*. Can J Microbiol 22:165–176

Pan WZ, Huang XW, Wei KB, Zhang CM, Yang DM, Ding JM, Zhang KG (2010) Diversity of thermophilic fungi in Tengchong Rehai National Park revealed by ITS nucleotide sequence analyses. J Microbiol 48:146–152

Pandey A, Soccol CR, Mitchell D (2000) New developments in solid state fermentation: I-bioprocesses and products. Process Biochem 35:1153–1169

Pasamontes L, Haiker M, Henriquez-Huecas M, Mitchell DB, van Loon AP (1997) Cloning of the phytases from *Emericella nidulans* and the thermophilic fungus *Talaromyces thermophilus*. Biochim Biophys Acta 1353:217–223

Paterson RRM, Lima N (2017) Thermophilic fungi to dominate aflatoxigenic/mycotoxigenic fungi on food under global warming. Int J Environ Res Public Health 14:199

Pereira JC, Leite RSR, Alves-Prado HF, Bocchini-Martins DA, Gomes E, Silva R (2014) Production and characterization of β-glucosidase obtained by the solid-state cultivation of the thermophilic fungus *Thermomucor indicae-seudaticae* N31. Appl Biochem Biotechnol 174:1–8

Pereira JC, Marques NP, Rodrigues A, Oliveira TB, Boscolo M, Silva R, Gomes E, Martins DAB (2015) Thermophilic fungi as new sources for production of cellulases and xylanases with potential use in sugarcane bagasse saccharification. J Appl Microbiol 118:928–939

Plecha S, Hall D, Tiquia-Arashiro SM (2013) Screening and characterization of soil microbes capable of degrading cellulose from switchgrass (*Panicum virgatum* L.). Environ Technol 34:1895–1904

Powell AJ, Parchert KJ, Bustamante JM, Ricken JB, Hutchinson MI, Natvig DO (2012) Thermophilic fungi in an aridland ecosystem. Mycologia 104:813–825

Prabhu K, Maheshwari R (1999) Biochemical properties of xylanases from a thermophilic fungus, *Melanocarpus albomyces*, and their action on plant cell walls. J Biosci 24:461–470

Rajasekaran AK, Maheshwari R (1993) Thermophilic fungi: an assessment of their potential for growth in soil. J Biosci 18:345–354

Rao P, Divakar S (2002) Response surface methodological approach for *Rhizomucor miehei* lipase-mediated esterification of α-terpineol with propionic acid and acetic anhydride. World J Microbiol Biotechnol 18:345–349

Roy I, Sastry MSR, Johri BN, Gupta MN (2000) Purification of alpha-amylase isoenzymes from *Scytalidium thermophilum* on a fluidized bed of alginate beads followed by concanavalin A-agarose column chromatography. Protein Expr Purif 20:162–168

Salar RK, Aneja KR (2006) Thermophilous fungi from temperate soils of northern India. J Agric Technol 2:49–58

Sandgren M, Ståhlberg J, Mitchinson C (2005) Structural and biochemical studies of GH family 12 cellulases: improved thermal stability, and ligand complexes. Prog Biophys Mol Biol 89:246–291

Satyanarayana T, Johri BN, Klein J (1992) Biotechnological potential of thermophilic fungi. In: Arora DK, Elander RP, Mukherji KG (eds) Handbook of applied mycology, vol 4. Marcel Dekker, New York, p 729–761

Seymour RS, Bradford DF (1992) Temperature regulation in the incubation mounds of the Australian brush-turkey. Condor 94:134–150

Singh B, Satyanarayana T (2006a) Phytase production by thermophilic mold *Sporotrichum thermophile* in solid-state fermentation and its application in dephytinization of sesame oil cake. Appl Biochem Biotechnol 133:239–250

Singh B, Satyanarayana T (2006b) A marked enhancement in phytase production by a thermophilic mold *Sporotrichum thermophile* using statistical designs in a cost-effective cane molasses medium. J Appl Microbiol 101:344–352

Singh B, Satyanarayana T (2008a) Improved phytase production by a thermophilic mold *Sporotrichum thermophile* in submerged fermentation due to statistical optimization. Bioresour Technol 99:824–830

Singh B, Satyanarayana T (2008b) Phytase production by a thermophilic mold *Sporotrichum thermophile* in solid state fermentation and its potential applications. Bioresour Technol 99:2824–2830

Singh B, Satyanarayana T (2008c) Phytase production by *Sporotrichum thermophile* in a cost-effective cane molasses medium in submerged fermentation and its application in bread. J Appl Microbiol 105:1858–1865

Singh S, Pillay B, Prior BA (2000) Thermal stability of beta-xylanases produced by different *Thermomyces lanuginosus* strains. Enzym Microb Technol 26:502–508

Singh B, Pocas-Fonseca MJ, Johri BN, Satyanarayana T (2016) Thermophilic molds: biology and applications. Crit Rev Microbiol 42:1–22

Srilakshmi J, Madhavi J, Lavanya S, Ammani K (2014) Commercial potential of fungal protease: past, present and future prospects. J Pharm Chem Biol Sci 2:218–234

Straatsma G, Samson RA, Olijnsma TW, Camp HJM, Gerrits JPG, Griensven LJLD (1994) Ecology of thermophilic fungi in mushroom compost, with emphasis on *Scytalidium thermophilum* and growth stimulation of *Agaricus bisporus* mycelium. Appl Environ Microbiol 60:454–458

Su YY, Cai L (2013) *Rasamsonia composticola*, a new thermophilic species isolated from compost in Yunnan, China. Mycol Prog 12:213–221

Tansey MR (1971) Isolation of thermophilic fungi from self-heated, industrial wood chip piles. Mycologia 63:537–547

Tansey MR, Brock TD (1972) The upper temperature limit for eukaryotic organisms. Proc Natl Acad Sci U S A 69:2426–2428

Taylor JW (2011) One fungus = one name: DNA and fungal nomenclature 20 years after PCR. IMA Fungus 2:113–120

Thakur SB (1977) Occurrence of spores of thermophilic fungi in the air at Bombay. Mycologia 69:197–199

Tiquia SM (2005) Microbial community dynamics in manure composts based on 16S and 18S rDNA T-RFLP profiles. Environ Technol 26:1104–1114

Tiquia SM, Tam NFY, Hodgkiss IJ (1996) Effects of composting on phytotoxicity of spent pig-manure sawdust litter. Environ Pollut 93:249–256

Tiquia SM, Wan JHC, Tam NFY (2002) Microbial population dynamics and enzyme activities during composting. Compost Sci Util 10:150–161

Tsiklinsky P (1899) Sur les mucédinées thermophiles. Ann Inst Pasteur 13:500–505

Tubaki K, Ito T, Natsuda Y (1974) Aquatic sediment as a habitat of thermophilic fungi. Ann Microbiol 24:199–207

van den Brink J, van Muiswinkel GCJ, Theelen B, Hinzb SWA, de Vries RP (2013) Efficient plant biomass degradation by thermophilic fungus *Myceliophthora heterothallica*. Appl Environ Microbiol 79:1316–1324

van Noort V, Bradatsch B, Arumugam M, Amlacher S, Bange G, Creevey C, Falk S, Mende DR, Sinning I, Hurt E, Bork P (2013) Consistent mutational paths predict eukaryotic thermostability. BMC Evol Biol 13:7

Venturi LL, Polizeli LM, Terenzi HF, Furriel Rdos P, Jorge JA (2002) Extracellular beta-D-glucosidase from *Chaetomium thermophilum* var. *coprophilum*: production, purification and some biochemical properties. J Basic Microbiol 42:55–66

Vieille C, Zeikus GJ (2001) Hyperthermophilic enzymes: sources, uses, and molecular mechanisms for thermostability. Microbiol Mol Biol Rev 65:1–43

Voutilainen S, Puranen T, Siika-Aho M, Lappalainen A, Alapuranen M, Kallio J, Hooman S, Viikari L, Vehmaanperä J, Koivula A (2008) Cloning, expression, and characterization of novel thermostable family 7 cellobiohydrolases. Biotechnol Bioeng 101:515–528

Wingfield MJ, De Beer ZW, Slippers B, Wingfield BD, Groenewald JZ, Lombard L, Crous PW (2012) One fungus, one name promotes progressive plant pathology. Mol Plant Pathol 13:604–613

Zhou S, Ingram LO (2000) Synergistic hydrolysis of carboxymethyl cellulose and acid-swollen cellulose by two endoglucanases (CelZ and CelY) from *Erwinia chrysanthemi*. J Bacteriol 182:5676–5682

Zhou P, Zhang G, Chen S, Jiang Z, Tang Y, Henrissat B, Yan Q, Yang S, Chen CF, Zhang B, Du Z (2014) Genome sequence and transcriptome analyses of the thermophilic zygomycete fungus *Rhizomucor miehei*. BMC Genomics 15:294

Chapter 4
New Perspectives on the Distribution and Roles of Thermophilic Fungi

Miriam I. Hutchinson, Amy J. Powell, José Herrera, and Donald O. Natvig (iD)

4.1 Introduction

The goal of this chapter is twofold. First, we briefly review the history, basic biology, evolution, and industrial relevance of thermophilic fungi. Second, we address ongoing questions concerning the ecology of these organisms. In the past two decades, several excellent reviews have considered one or more of these topics (Oliveira and Rodrigues, this volume; Maheshwari et al. 2000; Mouchacca 2000a, b; Salar and Aneja 2007; Salar 2018). Here, we give particular attention to topics for which there has been some difference of opinion. These include a discussion of the definition of thermophily as it pertains to fungi and an evaluation of the types of microhabitats that are most relevant to the growth and distribution of these organisms. We argue that the microenvironments capable of supporting the growth of thermophilic fungi are widespread and often transient. In the latter context, we present the results of a recent previously unpublished survey of thermophilic fungi in diverse ecosystems of the western United States, Mexico, and Canada.

Definition. While thermophilic fungi do not grow at the extreme temperatures that are optimal for many thermophilic bacteria and archaea, they are the only eukaryotes demonstrated to grow at temperatures up to 60 °C (Tansey and Brock 1972). In practice, the term thermophilic, when applied to fungi, has sometimes been used quite loosely, and there is no universally accepted definition.

M. I. Hutchinson · D. O. Natvig (✉)
Department of Biology, University of New Mexico, Albuquerque, NM, USA

A. J. Powell
Sandia National Laboratories, Albuquerque, NM, USA

J. Herrera
Mercy College, Dobbs Ferry, NY, USA

© Springer Nature Switzerland AG 2019
S. M. Tiquia-Arashiro, M. Grube (eds.), *Fungi in Extreme Environments: Ecological Role and Biotechnological Significance*,
https://doi.org/10.1007/978-3-030-19030-9_4

Cooney and Emerson (1964), who wrote the first monograph for thermophilic fungi, considered such fungi to be those that have "a maximum temperature for growth at or above 50 °C and a minimum temperature for growth at or above 20 °C." We have adopted a simpler working definition (Powell et al. 2012; Hutchinson et al. 2016). Namely, we consider a thermophilic fungus to be one that grows better at 45 °C than at 25 °C. One practical advantage of this latter definition is that it permits easy evaluation of fungal isolates.

Less consistent in the literature is the distinction between thermotolerance and thermophily. Cooney and Emerson considered thermotolerant fungi to be those with a maximum growth temperature near 50 °C while having a minimum growth temperature "well below" 20 °C. This definition is quite restrictive on the high end. Although it permits inclusion of the ubiquitous *Aspergillus fumigatus*, it excludes many fungi, for example, the model organism *Neurospora crassa*, that can grow at temperatures near or above 45 °C while having temperature optima below 50 °C. From a practical point of view, 45 °C is a temperature that is lethal or stress-inducing for most organisms, and we consider fungi that can grow at 45 °C to be thermotolerant.

History. The first reported thermophilic fungus, *Rhizomucor pusillus*, was isolated from bread by Lindt in the 1880s (Lindt 1886). Later, Tsiklinsky (1899) identified another thermophile, *Thermomyces lanuginosus*, growing on potatoes. In the early 1900s, Hugo Miehe (1907a, b; 1930a, b) published a series of papers derived from his investigations regarding the role of living organisms in the self-heating of stored hay. One result was the description of two new thermophiles, *Thermoidium sulfureum* (*Malbranchea cinnamomea*) and *Thermoascus aurantiacus*.

The study of these organisms languished for several decades before they were discovered to be part of the composting process associated with the production of rubber from the desert shrub Guayule (*Parthenium argentatum*). During World War II, the United States and allies lost access to rubber-plant plantations in the Pacific, which hindered the manufacture of rubber badly needed for the war effort. The US Department of Agriculture had a large-scale program aimed at developing Guayule latex as an alternative source of rubber. One of the experimental approaches involved chopping the shrub into pieces and composting it in piles. These "rets" were strongly thermogenic as a result of microbial activity, and the characterization of the organisms involved led to the identification of new and previously recognized thermophilic fungi (Cooney and Emerson 1964). The single publication by Allen and Emerson (1949) that resulted from the study of the effects of microbial activity on rubber quality did not detail the organisms involved in the process. The importance of the Guayule project in the "rediscovery" of thermophilic fungi as the basis for the studies that led to the Cooney and Emerson (1964) monograph of thermophilic fungi was recounted in the latter.

Industry. In recent decades, much of the attention given to thermophilic fungi has been in industry. This interest stems in large part from the ability of these fungi to yield thermostable enzymes, especially those that are cellulose-active. These enzymes function at temperatures high enough to exclude contaminants, and they accelerate reactions that convert cellulose into fermentable sugars for bioethanol

(Beckner et al. 2011; Rubin 2008; van den Brink et al. 2013). To understand the genetic mechanisms of thermophily and thermostability, the genomes of several fungal thermophiles have been sequenced (Berka et al. 2011).

4.2 Evolution

Of the more than 100 thousand described species of fungi, only approximately 50 species are thermophilic, representing a small fraction of the 2.2–3.8 million estimated fungal species (Salar and Aneja 2007; Hawksworth and Lücking 2017). Thermophilic fungi are known from two phyla, the Ascomycota and the Mucoromycota. In the Ascomycota, thermophiles are restricted to the orders Sordariales, Eurotiales, and Onygenales. Thermophiles in the Mucoromycota occur in the Mucorales (Salar 2018) and a recently created order, the Calcarisporiellales (Hirose et al. 2012; Morgenstern et al. 2012; Tedersoo et al. 2018). The order Mucorales contains two families with thermophiles, the Rhizopodaceae and the Lichtheimiaceae (Hoffmann et al. 2013). The Calcarisporiellales contains the thermophilic species *Calcarisporiella thermophile*. In the Sordariales, all known thermophilic species belong to the family Chaetomiaceae, which contains the greatest diversity of thermophilic fungi (Morgenstern et al. 2012). Among the Eurotiales, two families are considered to possess thermophilic members, the Trichocomaceae and the Thermoascaceae (Houbraken et al. 2014, 2016). A sole species of thermophilic fungus, *Malbranchea cinnamomea*, is found in the Onygenales (Morgenstern et al. 2012). Thermophilic Basidiomycota have been described by Straatsma et al. (1994) and Fergus (1971) but these species have either not been confirmed to be thermophilic or, as in the case of *Myriococcum thermophilum*, have been found to belong in the Ascomycota instead (Morgenstern et al. 2012; Koukol 2016).

Taxonomy for thermophilic fungi is in a state of considerable flux (Mouchacca 2000b; Oliveira et al. 2015; Natvig et al. 2015). This results in part from the fact that under the "One Fungus = One Name" convention recently adopted by *the International Code of Nomenclature for Algae, Fungi, and Plants*, the names for many thermophiles in the fungal kingdom need to be revised (Oliveira et al. 2015). This convention requires that the asexual and sexual nomenclature be unified and that a single name be assigned to a single species. In addition to name changes that have been required by changes in nomenclatural codes, in many cases, thermophilic fungi have simply been misclassified because of the failure to identify correct taxonomic affinities. The genus *Myceliophthora* provides examples of name changes required by new nomenclatural rules and by molecular phylogenetic studies that reveal true relationships (van den Brink et al. 2012). For example, the species recently recognized as *Myceliophthora heterothallica* was previously known under the teleomorphic names *Theilavia heterothallica* and *Corynascus heterothallicus*. To add to the confusion, as *T. heterothallica*, this species was once thought to be the teleomorph of *Chrysosporium thermophilum*, now recognized as *M. thermophila* (von Klopotek 1976; Hutchinson et al. 2016; van den Brink et al. 2012). A similar case exists for *Rasamsonia*, a genus erected to accommodate teleomorphs of *Geosmithia* and *Talaromyces* species, which were improperly identified (Houbraken et al. 2012). As a final example, the genus *Mycothermus* was recently erected to accommodate fungal strains previously known as *Scytalidium thermophilum*, placed in a genus (*Scytalidium*) that is appropriate for organisms in a different fungal class (Natvig et al. 2015).

4.3   Ecology

Despite advances in industry and genetics, comparatively little is known about the natural role and distribution of thermophilic fungi. Although commonly isolated from compost, these fungi are known to exhibit a variety of lifestyles, including as animal and plant associates and as saprotrophs (Salar 2018). For example, the thermophilic species *Myceliophthora thermophila* was identified as an endophyte of foliar tissue from a desert tree, *Parkinsonia microphylla* (Massimo et al. 2015). Another thermophile, *Rhizomucor pusillus*, has been reported to cause human infections, especially in immune-compromised individuals (St-Germain et al. 1993; Andrey et al. 2017). Cooney and Emerson (1964) noted that thermophilic fungi often remain unrecognized in culture when moderate incubation temperatures are used. As such, it may be that many thermophilic fungi remain undescribed.

A debate exists regarding how broadly distributed are the habitats in which thermophilic fungi can thrive. One hypothesis suggests that most thermophilic fungi are specialists of insulated compost-like substrates and that the presence of these fungi in soil and other non-compost substrates represents dispersal of aerial propagules (Maheshwari et al. 2000). Support for this idea has been presented for *Thermomyces lanuginosus*, which though common in soil was not competitive with mesophilic and thermotolerant fungi in soil microcosm experiments performed under fluctuating temperature regimes, unless temperatures were maintained above 40 °C. In addition, spores of *T. lanuginosus* failed to germinate in soil under conditions favorable for growth (Rajasekaran and Maheshwari 1993).

On the other hand, it is possible to wonder if understanding the role of thermophilic fungi in soil requires consideration of specific microhabitats and substrates suitable for growth. The proportion of physiologically active microorganisms in soil can be small compared to the total microbial biomass, and the level of activity for a microorganism or microbial group is dependent on substrate availability (Blagodatskaya and Kuzyakov 2013). Moreover, microcosm experiments performed with only mesophilic "soil" fungi demonstrate that the performance of one species relative to another is substrate dependent (e.g., Deacon et al. 2006). Therefore, while previous studies have reported thermophiles from diverse compost or pseudo-compost materials such as animal nests, manure compost, mushroom compost, and self-heating hay bales (Fergus and Sinden 1969; Tansey 1971, 1973, 1975, 1977; Tiquia 2005), it is likely that even a small 5-cm mass of leaf litter can be sufficiently insulated, moist and solar-heated to encourage growth of thermophilic fungi

(Subrahmanyam 1999). Indeed, recent studies of arid ecosystems (where sizeable composts are rare, if not absent), including the Sevilleta Long-Term Ecological Research (LTER) site in New Mexico, have demonstrated that thermophilic fungi are common in certain microhabitats (Powell et al. 2012). We recovered isolates from a variety of substrates including soil, biological soil crusts, leaf litter, and herbivore droppings. While these and other previous studies have shed light on microhabitats and distributions, the extent to which thermophilic fungi exhibit habitat specificity is unclear, as is the prevalence of thermophilic fungi on a regional scale.

Microhabitats Suitable for the Growth of Thermophilic Fungi Are Common in Diverse Ecosystems. Although the early studies of thermophilic fungi examined substrates that were self-heating as a result of microbial activity (Miehe 1907a, b; Cooney and Emerson 1964), soil and other substrates can achieve temperature and moisture conditions suitable for thermophiles as a result of solar gain (Tansey and Jack 1976; Powell et al. 2012). In reality, soil, litter, and herbivore droppings in temperate ecosystems often reach temperatures at or above those suitable for thermophilic fungi. In an experiment designed to follow the succession of thermophiles in a natural setting, we monitored temperatures in the droppings of three herbivores (elk, oryx, and rabbit) over a period of approximately 1 year (Fig. 4.1) at the Sevilleta National Wildlife Refuge. Even during winter months, daytime temperatures were often near or above 40 °C. In warmer months, daytime temperatures often reached 60–75 °C, temperatures at which fungal growth has ceased. In a single 24-h period, temperatures could swing from 15 °C to above 60 °C (Fig. 4.1). Droppings in this environment therefore represent an extreme microhabitat with dramatic and rapid changes in temperature and moisture. Thermophilic fungi are common in this microenvironment, and they participate in decomposition along with a complex community of bacteria, non-thermophilic fungi, and microfauna.

4.4 A Survey of Thermophilic Fungi from Across the Western United States

In a previously unpublished study, we surveyed thermophilic fungi in soils, plant litter, and herbivore droppings from a wide range of latitudes, elevations, and distinct climatic regions across sites from central Mexico to southern Canada. One goal was to evaluate the extent to which these fungi are common in locations where the opportunities for natural compost are rare. A second goal was to evaluate whether there exist geographical, latitudinal, or substrate differences in the distributions of major thermophile groups. Our sampling focused on soil, litter, and herbivore droppings. In addition, deep-frozen (−80 °C) rhizosphere soil samples collected from under blue grama grass (*Bouteloua gracilis*) were tested for the presence of thermophilic fungi.

Experimental Approach. Samples were collected in two phases. From May through June of 2008, 10 samples of rhizosphere soil were collected from each of five stands of *Bouteloua gracilis* in western North America as part of a separate

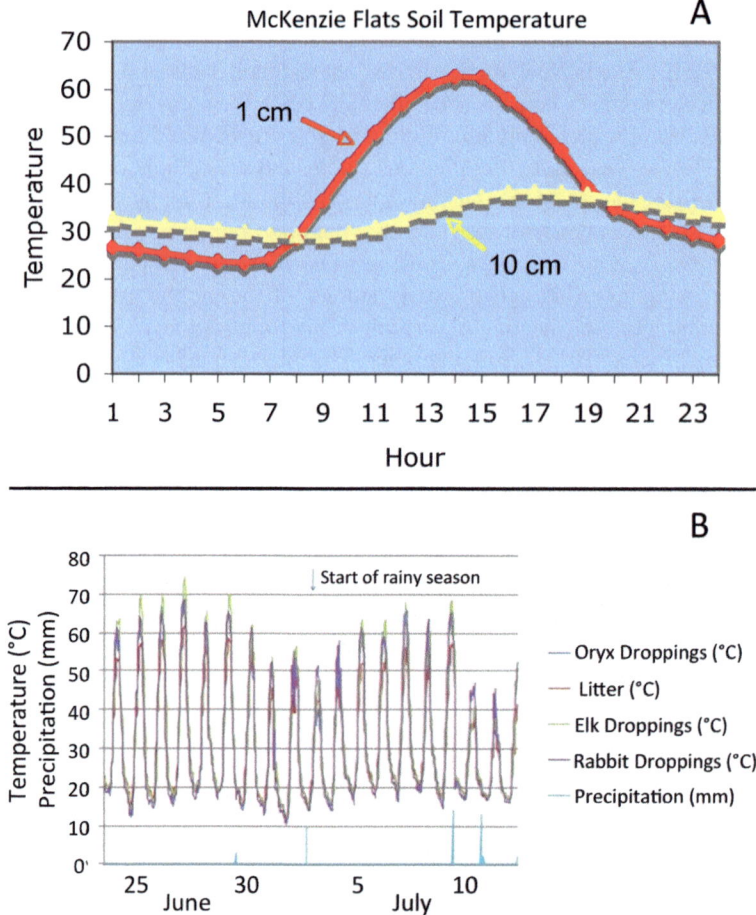

Fig. 4.1 Extreme microenvironments are common in temperate ecosystems. (**a**) Variation in soil temperature for a typical 23-h period (1:00 AM to midnight) in July at the Sevilleta National Wildlife Refuge in central New Mexico (adapted from Fig. 1 in Powell et al. 2012, copyright © Mycological Society of America, https://msafungi.org/, reprinted with permission from Taylor & Francis Ltd., http://www.tandfonline.com on behalf of the Mycological Society of America). (**b**) Dramatic swings in internal temperatures for herbivore droppings and litter in the foothills of the Los Pinos mountains in central New Mexico over 19 days surrounding the transition to the monsoon season in 2013. The temperature swings were frequently from 12 °C to 15 °C in early morning to over 70 °C at midday. The high temperatures were driven by solar gain. Air temperatures did not exceed 35 °C. Temperatures were measured with a small thermocouple and recorded on a Campbell Scientific CR1000 datalogger

study of root-associated fungi (Herrera et al. 2010). Soils were transported from the field on ice within 48 h and ultimately stored at −80 °C. These samples were plated in January of 2013. In a second effort, soil, herbivore droppings, and leaf litter samples were collected from each of 10 locations in the western United States

Fig. 4.2 Locations of soil, litter, and herbivore dropping samples employed for the thermophile survey presented here. Details of the samples are given in Table 4.1

between March 2012 and May 2013 (Fig. 4.2, Table 4.1). These samples were stored at 4–5 °C for no more than 4 days before plating.

All samples were plated onto malt extract agar (MEA) with 50 μg/mL ampicillin (to exclude bacteria) and incubated up to 10 days at 50 °C (see Bustamante 2006). Approximately 0.5–1.0 g of substrate was used for each plate. Rhizosphere soils from the Herrera et al. (2010) study were plated in replicates of 3. Resulting colonies from all cultures were then sub-cultured to obtain axenic isolates.

A cetyl trimethylammonium bromide (CTAB) DNA extraction procedure modified from Winnepenninckx et al. (1993) was used to isolate DNA from cultures, using methods previously described (Hutchinson et al. 2016). DNA was amplified by PCR of the ribosomal internal transcribed spacer (ITS) region using the fungal-specific primers ITS4 and ITS1F (White et al. 1990; Gardes and Bruns 1993). Each reaction consisted of 6.5 μL ExTaq polymerase (Takara, Mountain View, CA), 1 μL of each (5 μM) primers, 2 μL of 2% bovine serum albumin (Sigma-Aldrich, St. Louis, MO), 2 μL milliQ purified water, and 1 μL of template DNA, for a total of 13.5 μL. The following thermocycler settings were used: 95 °C for 5 min, 30 cycles at 94 °C for 30 s, annealing at 50 °C for 30 s, and extension at 72 °C for 45 s, followed by a final extension of 72 °C for 7 min. After PCR, reactions were purified by an enzyme procedure using the ExoSAP-IT kit (Affymetrix, Santa Clara, CA) and manufacturer's specifications.

Table 4.1 Isolate identifications and collection sites

OTU (putative species)	Isolate(s) represented	Substrate type	Collection date	GPS coordinates	Location	Elevation (m)
OTU 1 (*Thermomyces lanuginosus*)	6II	Litter	May 2013	44°56.593′N 110°38.397′W	Undine Falls, WY	2033
	10B_I	Droppings	May 2013	44°30.693′N 110°16.338′W	Lake Butte, WY	2614
	12I	Soil	May 2013	44°30.693′N 110°16.338′W	Lake Butte, WY	2614
	13F	Litter	May 2013	44°24.653′N 108°59.557′W	Cody, WY	2579
	24B	Litter	May 2013	38°51.423′N 105°03.795′W	Pike's Peak, CO	3951
	28C	Droppings	May 2013	38°54.032′N 105°04.058′W	Pike's Peak, CO	3033
	32D_I	Droppings	May 2013	36°33.535′N 104°34.692′W	Maxwell Wildlife Refuge, NM	1835
	OBg1_1	Rhizosphere soil	May 2008	21°46.860′N 101°36.721′W	Ojuelos, JAL, Mexico	2230
	OBg2_3	Rhizosphere soil	May 2008	21°46.860′N 101°36.721′W	Ojuelos, JAL, Mexico	2230
	OBg3_1	Rhizosphere soil	May 2008	21°46.860′N 101°36.721′W	Ojuelos, JAL, Mexico	2230
	SBg8_2	Rhizosphere soil	May 2008	34°24.094′N 106°40.662′W	Sevilleta, NM	1544
	Th002	Litter	March 2012	34°15.267′N 109°24.267′W	Apache County, AZ	1958
	Th047	Soil	March 2012	34°49.183′N 118°56.683′W	Los Padres National Forest, CA	1413
	ThUS015	Litter	September 2012	44°30.753′N 110°15.897′W	Lake Butte, WY	2679
	ThUS028	Litter	September 2012	43°31.117′N 108°10.917′W	Wind River Canyon, WY	1451
	ThUS057	Litter	September 2012	45°00.183′N 109°24.867′W	Beartooth Highway, WY	315

OTU	ID	Substrate	Date	Coordinates	Location	Elevation
OTU 2 (*Chaetomium thermophilum* var. *dissitum*)	3I	Soil	May 2013	45°45.950'N 109°47.583'W	Grey Cliff Prairie Dog State Park, MT	1208
	10B_II	Droppings	May 2013	44°30.693'N 110°16.338'W	Lake Butte, WY	2614
	12II	Soil	May 2013	44°30.693'N 110°16.338'W	Lake Butte, WY	2614
	16_I	Litter	May 2013	43°44.752'N 108°23.502'W	Thermopolis, WY	2580
	32A_II	Droppings	May 2013	36°33.535'N 104°34.692'W	Maxwell Wildlike Refuge, NM	1835
	WBg1_MH1	Rhizosphere soil	May 2008	43°34.236'N 103°23.210'W	Wind Cave, SD	1121
OTU 3 (*Thielavia arenaria*)	15C_L1	Droppings	May 2013	44°24.653'N 108°59.557'W	Cody, WY	2579
	18E_L1	Droppings	May 2013	43°44.752'N 108°23.502'W	Thermopolis, WY	2580
	GBg10_1	Soil	June 2008	49°10.705'N 107°33.634'W	Grasslands, SK, Canada	785
	Th044-2	Soil	March 2012	34°05.484'N 110°10.632'W	Gila National Forest, AZ	1793
OTU 4 (*Myceliophthora heterothallica*)	Th021	Soil	March 2012	34°27.833'N 118°41.017'W	Val Verde, CA	554
	Th022	Litter	March 2012	34°05.484'N 110°10.632'W	Gila National Forest, AZ	1793
	Th041	Litter	March 2012	37°00.117'N 120°50.367'W	Central Valley, CA	40
	Th044	Soil	March 2012	34°05.484'N 110°10.632'W	Gila National Forest, AZ	1793

(continued)

Table 4.1 (continued)

OTU (putative species)	Isolate(s) represented	Substrate type	Collection date	GPS coordinates	Location	Elevation (m)
OTU 5 (*Talaromyces thermophilus*)	5B	Droppings	May 2013	45°45.950′N 109°47.583′W	Grey Cliff Prairie Dog State Park, MT	1208
	10A	Droppings	May 2013	44°30.693′N 110°16.338′W	Lake Butte, WY	2614
	13B	Droppings	May 2013	44°30.693′N 110°16.338′W	Lake Butte, WY	2614
	19I	Litter	May 2013	42°34.898′N 106°41.133′W	Grey Reef, WY	1648
	ThUS017	Litter	September 2012	45°00.183′N 109°24.867′W	Beartooth Highway, WY	315
OTU 6 (*Aspergillus fumigatus*)	1E	Dropping	May 2013	45°45.950′N 109°47.583′W	Grey Cliff Prairie Dog State Park, MT	1208
	9	Soil	May 2013	44°50.328′N 110°26.528′W	Mount Washburn, WY	2529
	13D	Droppings	May 2013	44°30.693′N 110°16.338′W	Lake Butte, WY	2614
	17	Soil	May 2013	43°44.752′N 108°23.502′W	Thermopolis, WY	2580
	GBg6_1	Rhizosphere soil	June 2008	49°10.705′N 107°33.634′W	Grasslands, SK, Canada	785
	GBg9_1	Rhizosphere soil	June 2008	49°10.705′N 107°33.634′W	Grasslands, SK, Canada	785
	JBg11_2	Rhizosphere soil	May 2008	30°53.878′N 108°26.057′W	Janos, CHH, Mexico	1391
	OBg6_1	Rhizosphere soil	May 2008	21°46.860′N 101°36.721′W	Ojuelos, JAL, Mexico	2230

OTU	Code	Substrate		Date	Coordinates	Location	
OTU 7 (*Rasamsonia emersonii*)	2 Pike's Peak	Soil		August 2013	38°51.292'N 105°05.253'W	Pike's Peak, CO	3041
	6I	Litter		May 2013	44°56.593'N 110°38.397'W	Undine Falls, WY	2033
	Th008	Droppings		March 2012	37°44.300'N 121°36.7'W	Altamont Pass, CA	160
OTU 8 (*Rhizopus microsporus*)	JBg17_2	Rhizosphere soil		May 2008	30°53.878'N 108°26.057'W	Janos, CHH, Mexico	1391
	SBg6_3	Rhizosphere soil		May 2008	34°24.094'N 106°40.662'W	Sevilleta NWR, NM	1544
	SBg10_2	Rhizosphere soil		May 2008	34°24.094'N 106°40.662'W	Sevilleta NWR, NM	1544
OTU 9 (*Aspergillus nidulans*)	SBg3_3	Rhizosphere soil		May 2008	34°24.094'N 106°40.662'W	Sevilleta NWR, NM	1544
OTU 10 (*Thielavia gigaspora*)	WBg9_2	Rhizosphere soil		May 2008	43°34.236'N 103°23.210'W	Wind Cave, SD	1121
OTU 11 (*Thermoascus aurantiacus var. levisporus*)	16II	Litter		May 2013	43°44.752'N 108°23.502'W	Thermopolis, WY	2580
OTU 12 (*Mycothermus thermophilus*)	13C	Droppings		May 2013	44°30.693'N 110°16.338'W	Lake Butte, WY	2614
OTU 13 (*Chaetomium jodhpurense*)	WBg10_2	Rhizosphere soil		May 2008	43°34.236'N 103°23.210'W	Wind Cave, SD	1121
OTU 14 (*Rhizopus microsporus*)	Th040	Soil		March 2012	33°15.6'N 111°17.317'W	Near Phoenix, AZ	1740

Amplicons were Sanger sequenced with a BigDye Terminator v3.1 Cycle Sequencing kit (Applied Biosystems) in 10 μL reactions containing 0.5 μL BigDye Terminator v3.1, 2 μL of 5X Sequencing Buffer (Life Technologies/Applied Biosystems, Carlsbad, CA) 1 μL of 3 μM primer, and 5.5 μL of milliQ water. A BigDye STeP protocol was used with the following parameters: 96 °C for 60 s followed by 15 cycles of 96 °C for 10 s, 50 °C for 5 s, and 60 °C for 1 min 15 s; then 5 cycles of 96 °C for 10 s, 50 °C for 5 s, and 60 °C for 1 min30 s; and a final 5 cycles of 96 °C for 10 s, 50 °C for 5 s, 60 °C for 2 min/s (Platt et al. 2007).

Chromatogram files for the forward and reverse reads were edited and assembled into contigs using Sequencher v5.1 (Gene Codes, Ann Arbor MI). To determine the overall species richness among the isolates, ITS sequences were assembled into Operational Taxonomic Units (OTUs) using UPARSE 9.0 (Edgar 2013). OTU cutoffs were set to 97% identity. To obtain taxonomic information, the resulting OTUs were then queried at National Center for Biotechnology Information (NCBI) GenBank with Basic Local Alignment Search Tool Nucleotide (BLASTN) searches using the option to exclude uncultured and environmental samples.

Phylogenetic Analyses. ITS sequences were aligned in MUSCLE implemented through the European Bioinformatics Institute web interface (Edgar 2004; Li et al. 2015). Alignments were then visualized and trimmed in AliView v1.2.1 (Larsson 2014). Reference sequences from GenBank were included as a comparison to the newly acquired sequences, and type strains were selected as references when possible (Tables 4.2 and 4.3). Trees were constructed with the Randomized Axelerated Maximum Likelihood (RaxML) program v7.3.2 using 1000 bootstrap replicates (Stamatakis 2006). Because ITS sequences align poorly across distant phylogenetic groups, we built separate trees for each of the three orders to which the sequences were classified. Trees were visualized and edited with Mesquite v2.75 (Maddison and Maddison 2010).

Results. Thermophilic and thermotolerant fungi were recovered from every substrate type and nearly every location. Notably, propagules of thermophilic fungi from the rhizosphere soil were also able to survive storage at −80 °C for nearly 5 years. Sixty-two total isolates were recovered. After excluding duplicates from the same sample, 55 isolates were characterized at the sequence level, resulting in 14 putative OTUs, 10 genera, and 13 known species. The identity of each of the OTUs is summarized in Table 4.2, and relationships among the isolates are shown in Fig. 4.3. Most isolates fell into the fungal orders Eurotiales (34 isolates) and Sordariales (17 isolates). Only 4 isolates belonged to the Mucorales, and no isolates from the Onygenales were identified. The lack of isolates from the Onygenales may owe to the types of substrates and media used, as this group of fungi is known to be keratinophilic (Sharpton et al. 2009). The most common species was *Thermomyces lanuginosis*, represented by 16 isolates, followed by *Aspergillus fumigatus*, represented by 8 isolates, and *Chaetomium thermophilum var. dissitum,* represented by 6 isolates.

Several of the isolates were from species viewed as thermotolerant rather than thermophilic. Mouchacca (2000a) suggests that *A. fumigatus*, *A. nidulans*, and *C. jodhpurense* have been erroneously reported as thermophiles when they actually

Table 4.2 Isolate abundance and best BLAST hits

OTU	Abundance	Best blast hit (species)	Order	Family	Accession number
OTU 1	16	*Thermomyces lanuginosus* isolate TCSB341	Eurotiales	Trichocomaceae	KT365217.1
OTU 2	6	*Chaetomium thermophilum var. dissitum* strain: NBRC 31807	Sordariales	Chaetomiaceae	AB746179.1
OTU 3	4	*Thielavia arenaria* strain CBS 507.74	Sordariales	Chaetomiaceae	JN709489.1
OTU 4	4	*Myceliophthora heterothallica* CBS 202.75	Sordariales	Chaetomiaceae	JN659478.1
OTU 5	5	*Talaromyces thermophilus* strain NRRL 2155	Eurotiales	Trichocomaceae	JF412001.1
OTU 6	8	*Aspergillus fumigatus* strain IHEM 13935 isolate ISHAM-ITS_ID MITS168	Eurotiales	Aspergillaceae	KP131565.1
OTU 7	3	*Rasamsonia emersonii* strain CBS 396.64	Eurotiales	Trichocomaceae	JF417479.1
OTU 8	3	*Rhizopus microsporus* strain: TISTR 3518	Mucorales	Rhizopodaceae	AB381937.1
OTU 9	1	*Aspergillus nidulans* isolate KZR-132	Eurotiales	Aspergillaceae	KX878986.1
OTU 10	1	*Thielavia gigaspora* strain CBS 112062	Sordariales	Chaetomiaceae	MH862888.1
OTU 11	1	*Thermoascus aurantiacus var. levisporus* strain T81	Eurotiales	Thermoascaceae	FJ548834.1
OTU 12	1	*Mycothermus thermophilus* isolate A74	Sordariales	Chaetomiaceae	KX611046.1
OTU 13	1	*Chaetomium jodhpurense* strain CBS 602.69	Sordariales	Chaetomiaceae	MH859386.1
OTU 14	1	*Rhizopus microsporus* isolate VPCI 128/P/10	Mucorales	Rhizopodaceae	KJ417570.1

possess lower temperature optima than true thermophiles. Additionally, *Thielavia gigaspora* is a thermotolerant species previously isolated in Egypt (Moustafa and Abdel-Azeem 2008). Mouchacca (2000a) also reported *Rhizopus microsporus* as a misattributed thermophile, but (Peixoto-Nogueira et al. 2008) demonstrated that isolates grow optimally at 45 °C. Overall, thermotolerant species represented 29% of all of our isolates. Excluding the thermotolerant species, there were 25 isolates from the Eurotiales and 14 from the Sordariales.

Independent-samples Welch's t-tests were employed to compare elevation and latitude specificity for thermophilic isolates in the Eurotiales and Sordariales. Because the Mucorales were comparatively rare, they were not included in statistical analyses. For elevation, there was no significant difference between the distributions of Eurotiales and Sordariales ($M_{EUROTIALES}$ = 2038.28 m, SD = 900.51; $M_{SORDARIALES}$ = 1765 m, SD = 823.66; t(29) = 0.96, p = 0.05). For latitude, again,

Table 4.3 Reference strains used for phylogenetic analyses

Order	Strain	Species	Thermophile?	Accession number
Eurotiales	CBS 525.65	Aspergillus fischeri	No	MH858698.1
	CBS 139343	Aspergillus fumigatus	No	KU296268.1
	CBS 467.88	Aspergillus nidulans	No	KU866630.1
	CBS DTO_283-D3	Aspergillus udagawae	No	KY808744.1
	CBS 393.64	Rasamsonia emersonii[a]	Yes	JF417478.1
	CBS 398.64	Thermoascus aurantiacus	Yes	MH858464.1
	CBS 181.67	Thermoascus crustaceus[a]	Yes	FJ389925.1
	CBS 236.58	Thermomyces dupontii	Yes	MH857768.1
	CBS 632.91	Thermomyces lanuginosus	Yes	MH862287.1
Onygenales	CBS 120936	Coccidioides immitis[a]	No	NR_157446.1
Mucorales	ATCC 36186	Pilobolus crystallinus	No	FJ160949.1
	CBS 130158	Rhizopus microsporus	No	MH865595.1
	CBS 182.67	Rhizomucor miehei[a]	Yes	JF412011.1
Sordariales	CBS 160.62	Chaetomium globosum[a]	No	MH858130.1
	CBS 602.69	Chaetomium jodhpurense	No	MH859386.1
	LC4128	Chaetomium thermophilum var. dissitum	Yes	KP336781.1
	NBRC 31807	Chaetomium thermophilum var. dissitum	Yes	AB746179.1
	CBS 202.75	Myceliophthora heterothallica[a]	Yes	JN659478.1
	CBS 629.91	Mycothermus thermophilus	Yes	MH862286.1
	CBS 709.71	Neurospora crassa	No	MH860307.1
	CBS 507.74	Thielavia arenaria[a]	Yes	JN709489.1
	CBS 112062	Thielavia gigaspora[a]	No	MH862888.1
	CBS 125981	Thielavia subthermophila	No	MH863860.1

[a]Type strain

there was no significant difference between the distributions of Eurotiales and Sordariales ($M_{EUROTIALES} = 38.79°$, SD = 16.35; $M_{SORDARIALES} = 41.44°$, SD = 28.92; $t(18) = -0.32$, $p = 0.05$).

In terms of substrate preference, thermophilic samples in Eurotiales were most frequently isolated from litter (44%), while for samples in the Sordariales, the top sources were droppings (35.71%) and top soil (35.71%). Overall, the most thermophilic isolates originated from litter substrates (35.9%), followed by droppings (30.7%), soil (20%), and finally rhizosphere, which represented 12.8% of the samples.

For the soils collected in 2008 and stored at −80 °C, there appeared to be a latitudinal gradient in terms of the success of platings. Just over half (62.5%) of soils collected in Saskatchewan, Canada, were positive for thermophiles, compared to 80% of soils from Custer, South Dakota; 86.7% from Socorro, New Mexico; 93.9% from Janos (Chihuahua), Mexico; and 89.7% from Ojuelos (Jalisco), Mexico. With the exception of the soils from Janos (which showed a higher percentage than

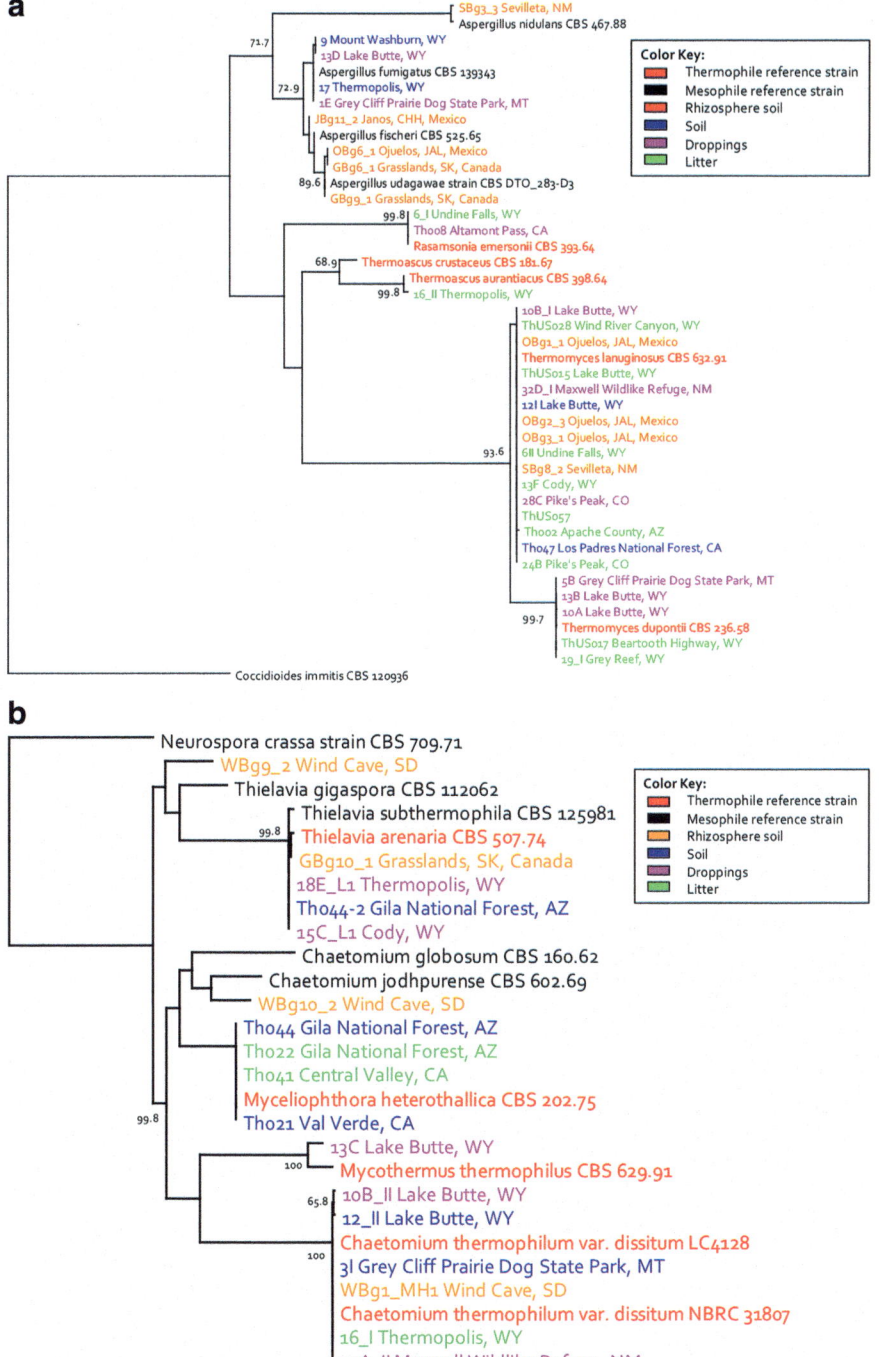

a

SBg3_3 Sevilleta, NM
Aspergillus nidulans CBS 467.88
71.7
9 Mount Washburn, WY
13D Lake Butte, WY
Aspergillus fumigatus CBS 139343
72.9 17 Thermopolis, WY
1E Grey Cliff Prairie Dog State Park, MT
JBg11_2 Janos, CHH, Mexico
Aspergillus fischeri CBS 525.65
OBg6_1 Ojuelos, JAL, Mexico
GBg6_1 Grasslands, SK, Canada
89.6 Aspergillus udagawae strain CBS DTO_283-D3
GBg9_1 Grasslands, SK, Canada
99.8 6_I Undine Falls, WY
Th008 Altamont Pass, CA
Rasamsonia emersonii CBS 393.64
68.9 Thermoascus crustaceus CBS 181.67
Thermoascus aurantiacus CBS 398.64
99.8 16_II Thermopolis, WY
10B_I Lake Butte, WY
ThUS028 Wind River Canyon, WY
OBg1_1 Ojuelos, JAL, Mexico
Thermomyces lanuginosus CBS 632.91
ThUS015 Lake Butte, WY
32D_I Maxwell Wildlike Refuge, NM
12I Lake Butte, WY
OBg2_3 Ojuelos, JAL, Mexico
OBg3_1 Ojuelos, JAL, Mexico
6II Undine Falls, WY
SBg8_2 Sevilleta, NM
13F Cody, WY
28C Pike's Peak, CO
ThUS057
Th002 Apache County, AZ
Th047 Los Padres National Forest, CA
24B Pike's Peak, CO
5B Grey Cliff Prairie Dog State Park, MT
13B Lake Butte, WY
10A Lake Butte, WY
99.7 Thermomyces dupontii CBS 236.58
ThUS017 Beartooth Highway, WY
19_I Grey Reef, WY
93.6

Coccidioides immitis CBS 120936

Color Key:
■ Thermophile reference strain
■ Mesophile reference strain
■ Rhizosphere soil
■ Soil
■ Droppings
■ Litter

b

Neurospora crassa strain CBS 709.71
WBg9_2 Wind Cave, SD
Thielavia gigaspora CBS 112062
Thielavia subthermophila CBS 125981
99.8 Thielavia arenaria CBS 507.74
GBg10_1 Grasslands, SK, Canada
18E_L1 Thermopolis, WY
Th044-2 Gila National Forest, AZ
15C_L1 Cody, WY
Chaetomium globosum CBS 160.62
Chaetomium jodhpurense CBS 602.69
WBg10_2 Wind Cave, SD
Th044 Gila National Forest, AZ
Th022 Gila National Forest, AZ
Th041 Central Valley, CA
Myceliophthora heterothallica CBS 202.75
99.8 Th021 Val Verde, CA
13C Lake Butte, WY
100 Mycothermus thermophilus CBS 629.91
65.8 10B_II Lake Butte, WY
12_II Lake Butte, WY
Chaetomium thermophilum var. dissitum LC4128
100 3I Grey Cliff Prairie Dog State Park, MT
WBg1_MH1 Wind Cave, SD
Chaetomium thermophilum var. dissitum NBRC 31807
16_I Thermopolis, WY
32A_II Maxwell Wildlike Refuge, NM

Color Key:
■ Thermophile reference strain
■ Mesophile reference strain
■ Rhizosphere soil
■ Soil
■ Droppings
■ Litter

Fig. 4.3 Ribosomal RNA ITS gene trees for three orders of thermophilic fungi recovered from a recent survey (collection sites are presented in Fig. 4.2 and Table 4.1): Eurotiales (**a**), Sordariales (**b**), Mucorales (**c**). Trees were rooted with *Coccidioides immitis*, *Neurospora crassa*, and *Pilobolus crystallinus*, respectively. New isolates are color coded by substrate type, while reference strains are colored by temperature optimum. Bootstrap values (1000 replicates) are displayed for all nodes receiving 65% or greater support. All new isolates form well-supported clades with previously identified species, and represent diverse substrate types and locations

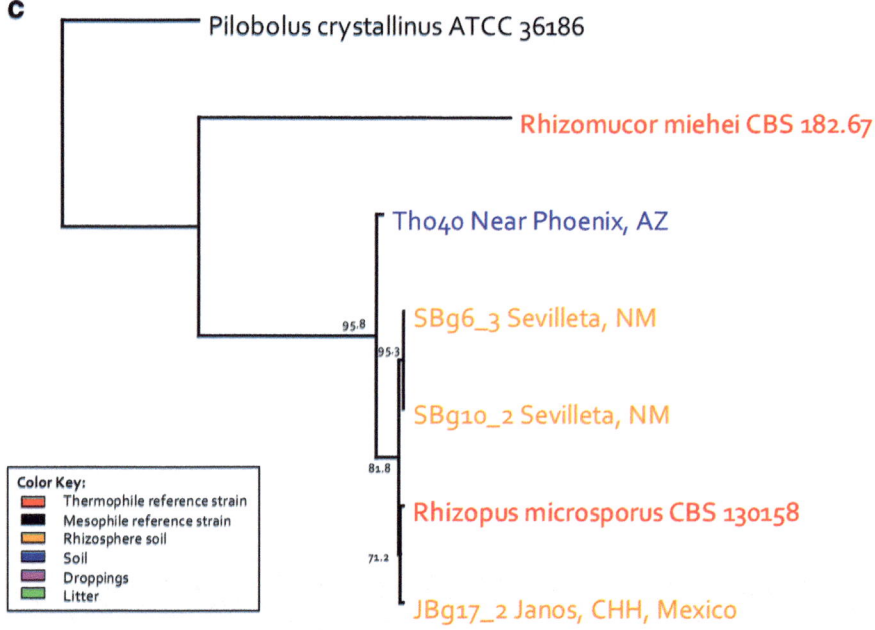

Fig. 4.3 (continued)

Ojuelos to the south), plating success declined with increasing latitude. In pairwise comparisons, plating success for Saskatchewan was an outlier, and significantly different from all other locations except South Dakota according to a Pearson's N-1 chi-square test [$\chi^2_{\text{SOUTHDAKOTA}}(1,N=54)=2.00, p=0.16; \chi^2_{\text{NEWMEXICO}}(1,N=54)=4.20, p=0.04; \chi^2_{\text{JALISCO}}(1,N=63)=6.60, p=0.003; \chi^2_{\text{CHIHUAHUA}}(1,N=57)=8.63, p=0.01$]. No other pairwise comparisons were significantly different.

Discussion. Our results indicate that thermophilic fungi are readily isolated from various substrates, from elevations as low as 40 m above sea level to as high as 3951 m and from a great range of latitudes between Mexico and Canada. We observed no correlation between phylogeny and environment. Specifically, isolates from the Eurotiales and Sordariales did not differ significantly for substrate preference, elevation, or latitude. Even within a single OTU cluster, constituent sequences were derived from diverse locations and substrates. For example, OTU1 (*Thermomyces lanuginosus*) represents isolates from as far south as Ojuelos, Jalisco, to as far north as the Beartooth Highway in Wyoming. This cluster also consisted of multiple isolates from every substrate type and of elevations from 315 m to above timberline at 3951 m. Indeed, at the resolution of OTUs at the 97% level, there appears to be no specificity of thermophilic fungi to a particular habitat. It is possible, however, that the 97% cutoff is too generous and blurs the finer distinctions among the isolates. To develop a better sense of the phylogenetic relationships between the isolates, one might also collect data for functional DNA regions that are less variable and more reliable at predicting deeper levels of taxonomy.

Studies show that members of the Chaetomiaceae (Sordariales) are proficient in decomposing cellulosic biomass, so they are thought to associate with plant-based substrates in nature (Ames 1963; Mehrotra and Aneja 1990). They have been previously isolated from herbivore droppings, leaf litter, and even from live plants (Kerekes et al. 2013; Richardson 2001; Abou Alhamed and Shebany 2012). Chaetomiaceae are also prevalent in composts (Cooney and Emerson 1964; Kane and Mullins 1973; Straatsma et al. 1994). For example, using an ITS barcoding approach, Neher et al. (2013) showed *Chaetomium* species to be dominant members of the fungal OTUs across all of the compost recipes they tested, especially in the earlier stages of composting. As discussed previously, composts have been proposed as the primary habitats for thermophilic fungi, with the suggestion that specimens found on other substrates are likely inactive propagules dispersed from compost (Rajasekaran and Maheshwari 1993). However, soil is also sufficiently rich in cellulose as it is one of the top sources of complex carbon polymers (Kögel-Knabner 2002; López-Mondéjar et al. 2016). Thus, it is perhaps unsurprising that many thermophilic species in the Chaetomiaceae have been identified from soil (Tansey and Jack 1976; Pan et al. 2010; Powell et al. 2012). Mesophilic Cheatomiaceae have been demonstrated to be both present and active in the soil. Using Stable Isotope Probing with ^{13}C cellulose substrate, Eichorst and Kuske (2012) showed that species of *Chaetomium* actively decay cellulose added to soil. It is reasonable to believe that thermophilic members of the Chaetomiaceae do the same.

Species in the order Eurotiales are also commonly associated with decaying plant material. For example, the well-known fungus *Aspergillus fumigatus* is cited as one of the most frequent species recovered from composts and other plant debris (Taylor et al. 2015). *A. fumigatus* also shows a pan-global distribution, which Pringle et al. (2005) have suggested may be due to the role of humans in expanding composting processes. Another member of the Eurotiales, the thermophilic fungus *Thermomyces lanuginosus* also shows seemingly ubiquitous distribution. In our present study, it was the most frequently isolated taxon and derived from a variety of substrates and locations. Langarica-Fuentes et al. (2014) also found that along with *Talaromyces thermophilus* (another species in the Eurotiales), *T. lanuginosus* accounted for 65% of sequences obtained via 454' barcoding of the fungal community in the middle and center of an in-vessel compost system. Similarly, it was the top isolate in studies of thermophilic fungi from soils in India (Maheshwari et al. 1987; Rajasekaran and Maheshwari 1993). Still, Rajasekaran and Maheshwari (1993) were unable to detect actively growing *T. lanuginosus* in soil with immunofluorescence assays. However, Hedger and Hudson (1974) reported that *T. lanuginosus* shows commensal interactions with cellulolytic fungal thermophiles (*Chaetomium thermophile* and *Humicola insolens*) and subsists on the sugar byproducts from cellulose decomposition. Thus, it may be that this species performs best in a consortium with cellulolytic thermophiles and requires other fungal partners to grow. If there is adequate cellulose in a given substrate, cellulose degrading fungi can likely support commensal fungi, thus providing a niche in soil for species such as *T. lanuginosus*.

Soils undergo diurnal temperature fluctuations to upwards of 70 °C, so soil is a suitably hot substrate for thermophilic fungi (Powell et al. 2012). Leaf litter and herbivore droppings also experience similar swings in temperature (Fig. 4.1). In addition, thermophilic fungi are more readily isolated from soil after precipitation events, indicating that they are responsive to changes in the soil environment (Powell et al. 2012). Taken together, these factors suggest that thermophilic fungi can inhabit many microhabitats, including soil, provided that they have access to moisture and appropriate temperatures.

4.5   Conclusions

Much remains to be learned about the ecology of thermophilic fungi. Although it has long been known that these fungi can be isolated from soil, herbivore droppings, and other substrates, most studies have focused on composted plant materials in either natural or anthropogenic settings. In contrast, our surveys have shown that nearly all ecosystems provide thermophilic fungi with at least transient access to decomposing plant material, and sufficiently high temperature and moisture (see Fig. 4.1). Our results suggest that such transient microenvironments might be the primary habitats. At the level of resolution provided by ribosomal ITS sequences, there is little evidence for habitat specialization or geographical restrictions among thermophiles. Thermophiles in the Ascomycota are distributed across three orders, with several phylogenetic lineages within each order. We found members of most lineages across wide latitudes, elevations, substrate, and ecosystem types, ranging from desert shrublands and grasslands to montane forests to northern grasslands.

4.6   Future Perspectives

Thermophilic fungi have provided many contributions to science, both in their utility to industry and in the advancement of basic understanding in biology. Information on the distribution of thermophilic fungi, and a better grasp on their natural diversity and roles in the environment, will help further the field of microbial ecology and will aid in bioprospecting new, potentially useful organisms for biotechnology. Although next-generation sequencing methods can detect thermophilic fungi in environmental samples, many thermophiles have close mesophilic relatives, and as a result, the assessment of thermophily often requires evaluation based on growth in the laboratory rather than on sequence analysis alone. Accordingly, it is likely that fungal thermophiles are overlooked in environmental sequencing data. Similarly, culture-based methods of community analysis often employ only temperatures suitable for mesophiles, and temperatures optimal for the growth of thermophiles or psychrophiles are not considered. Moreover, it is possible that certain fungal thermophiles are unculturable and are only detected as DNA in environmental surveys. These circumstances thereby result in a need for a unified, comprehensive approach to appraising and understanding not only the biology of thermophilic fungi, but also the ecology of non-thermophilic microbes that share environments with thermophiles.

Acknowledgements This research was supported in part by a National Science Foundation award to the University of New Mexico (UNM) for the Sevilleta Long-Term Ecological Research program. We acknowledge support for DNA sequencing from the UNM Department of Biology's Molecular Biology Facility. Data analysis was aided by computing resources of the UNM Center for Evolutionary & Theoretical Immunology (CETI) under National Institutes of Health grant P30GM110907, and the UNM Center for Advanced Research Computing, supported in part by the National Science Foundation.

Funding statement Sandia National Laboratories is a multimission laboratory managed and operated by National Technology and Engineering Solutions of Sandia, LLC, a wholly owned subsidiary of Honeywell International, Inc., for the U.S. Department of Energy's National Nuclear Security Administration under contract DE-NA0003525.

References

Abou Alhamed MF, Shebany YM (2012) Endophytic *Chaetomium globosum* enhances maize seedling copper stress tolerance. Plant Biol 14:859–863

Allen PJ, Emerson R (1949) Guayule rubber: microbiological improvement by shrub retting. Ind Eng Chem 41:346–365

Ames LM (1963) A monograph of the Chaetomiaceae. US Army Res Dev Ser No. 2, Washington, DC, pp 9–125

Andrey DO, Kaiser L, Emonet S, Erard V, Chalandon Y, Van Delden C (2017) Cerebral rhizomucor infection treated by posaconazole delayed-release tablets in an allogeneic stem cell transplant recipient. Int J Infect Dis 55:24–26

Beckner M, Ivey ML, Phister TG (2011) Microbial contamination of fuel ethanol fermentations. Lett Appl Microbiol 53:387–394

Berka RM, Grigoriev IV, Otillar R, Salamov A, Grimwood J, Reid I, Ishmael N, John T, Darmond C, Moisan MC, Henrissat B, Coutinho PM, Lombard V, Natvig DO, Lindquist E, Schmutz J, Lucas S, Harris P, Powlowski J, Bellemare A, Taylor D, Butler G, de Vries RP, Allijn IE, van den Brink J, Ushinsky S, Storms R, Powell AJ, Paulsen IT, Elbourne LD, Baker SE, Magnuson J, Laboissiere S, Clutterbuck AJ, Martinez D, Wogulis M, de Leon AL, Rey MW, Tsang A (2011) Comparative genomic analysis of the thermophilic biomass-degrading fungi *Myceliophthora thermophila* and *Thielavia terrestris*. Nat Biotechnol 29:922–929

Blagodatskaya E, Kuzyakov Y (2013) Active microorganisms in soil: critical review of estimation criteria and approaches. Soil Biol Biochem 67:192–211

Bustamante J (2006) Thermophilic fungi on the Sevilleta National Wildlife Refuge. PhD diss., University of New Mexico

Cooney DG, Emerson R (1964) Thermophilic fungi: an account of their biology, activities and classification. W.H. Freeman, San Francisco, CA, p 188

Deacon LJ, Pryce-Miller EJ, Frankland JC, Bainbridge BW, Moore PD, Robinson CH (2006) Diversity and function of decomposer fungi from a grassland soil. Soil Biol Biochem 38:7–20

Edgar RC (2004) MUSCLE: multiple sequence alignment with high accuracy and high throughput. Nucleic Acids Res 32:1792–1797

Edgar RC (2013) UPARSE: highly accurate OTU sequences from microbial amplicon reads. Nat Methods 10:996

Eichorst SA, Kuske CR (2012) Cellulose-responsive bacterial and fungal communities in geographically and edaphically different soils identified using stable isotope probing. Appl Environ Microbiol 78:2316–2327

Fergus CL (1971) The temperature relationships and thermal resistance of a new thermophilic Papulaspora from mushroom compost. Mycologia 63:426–431

Fergus CL, Sinden JW (1969) A new thermophilic fungus from mushroom compost: *Thielavia thermophila* spec. nov. Can J Bot 47:1635–1637

Gardes M, Bruns TD (1993) ITS primers with enhanced specificity for basidiomycetes-application to the identification of mycorrhizae and rusts. Mol Ecol 2:113–118

Hawksworth D, Lücking R (2017) Fungal diversity revisited: 2.2 to 3.8 million species. Microbiol Spectr 5:1–17

Hedger JN, Hudson HJ (1974) Nutritional studies of *Thermomyces lanuginosus* from wheat straw compost. Trans Br Mycol Soc 62:129–143

Herrera J, Khidir HH, Eudy DM, Porras-Alfaro A, Natvig DO, Sinsabaugh RL (2010) Shifting fungal endophyte communities colonize *Bouteloua gracilis*: effect of host tissue and geographical distribution. Mycologia 102:1012–1026

Hirose D, Degawa Y, Inaba S, Tokumasu S (2012) The anamorphic genus *Calcarisporiella* is a new member of the Mucoromycotina. Mycoscience 53:256–260

Hoffmann K, Pawłowska J, Walther G, Wrzosek M, De Hoog GS, Benny GL, Kirk PM, Voigt K (2013) The family structure of the Mucorales: a synoptic revision based on comprehensive multigene-genealogies. Persoonia 30:57–76

Houbraken J, Spierenburg H, Frisvad JC (2012) Rasamsonia, a new genus comprising thermotolerant and thermophilic Talaromyces and Geosmithia species. A Van Leeuw J Microb 101:403c421

Houbraken J, de Vries RP, Samson RA (2014) Modern taxonomy of biotechnologically important Aspergillus and Penicillium species. Adv Appl Microbiol 86:199–249

Houbraken J, Samson RA, Yilmaz N (2016) Taxonomy of Aspergillus, Penicillium and Talaromyces and its significance for biotechnology. Aspergillus and Penicillium in the Post-Genomic Era. Caister Academic Press, Norfolk, pp 1–15

Hutchinson MI, Powell AJ, Tsang A, O'Toole N, Berka RM, Barry K, Grigoriev IV, Natvig DO (2016) Genetics of mating in members of the Chaetomiaceae as revealed by experimental and genomic characterization of reproduction in *Myceliophthora heterothallica*. Fungal Genet Biol 86:9–19

Kane BE, Mullins JT (1973) Thermophilic fungi in a municipal waste compost system. Mycologia 65:1087–1100

Kerekes J, Kaspari M, Stevenson B, Nilsson RH, Hartmann M, Amend A, Bruns TD (2013) Nutrient enrichment increased species richness of leaf litter fungal assemblages in a tropical forest. Mol Ecol 22:2827–2838

Kögel-Knabner I (2002) The macromolecular organic composition of plant and microbial residues as inputs to soil organic matter. Soil Biol Biochem 34:139–162

Koukol O (2016) Myriococcum revisited: a revision of an overlooked fungal genus. Plant Syst Evol 302:957–969

Langarica-Fuentes A, Zafar U, Heyworth A, Brown T, Fox G, Robson GD (2014) Fungal succession in an in-vessel composting system characterized using 454 pyrosequencing. FEMS Microbiol Ecol 88:296–308

Larsson A (2014) AliView: a fast and lightweight alignment viewer and editor for large datasets. Bioinformatics 30:3276–3278

Li W, Cowley A, Uludag M, Gur T, McWilliam H, Squizzato S, Park YM, Buso N, Lopez R (2015) The EMBL-EBI bioinformatics web and programmatic tools framework. Nucleic Acids Res 43:580–W584

Lindt W (1886) Mitteilungen über einige neue pathogene Shimmelpilze. Arch Exp Pathol Pharmakol 21:269–298

López-Mondéjar R, Zühlke D, Becher D, Riedel K, Baldrian P (2016) Cellulose and hemicellulose decomposition by forest soil bacteria proceeds by the action of structurally variable enzymatic systems. Sci Rep 6:25279

Maddison WP, Maddison DR (2010) Mesquite: a modular system for evolutionary analysis. 2011; Version 2.75. http://mesquiteproject.org

Maheshwari R, Kamalam PT, Balasubramanyam PV (1987) The biogeography of thermophilic fungi. Curr Sci 56:151–155

Maheshwari R, Bharadwaj G, Bhat MK (2000) Thermophilic fungi: their physiology and enzymes. Microbiol Mol Biol Rev 64:461–488

Massimo NC, Devan MN, Arendt KR, Wilch MH, Riddle JM, Furr SH, Steen C, U'Ren JM, Sandberg DC, Arnold AE (2015) Fungal endophytes in aboveground tissues of desert plants: infrequent in culture, but highly diverse and distinctive symbionts. Microb Ecol 70:61–76

Mehrotra RS, Aneja KR (1990) An introduction to mycology. New Age International, New Delhi, pp 1–737

Miehe H (1907a) Die selbsterhitzung des Heus. Eine biologische studie. Gustav Fischer, Jena, pp 1–127

Miehe H (1907b) *Thermoidium sulfureum* n.g. n.sp., etin neuer Wärmepilz. Berichte der Deutsch Bot Ges 25:510–515

Miehe H (1930a) Die Wärmebildung von Reinkulturen im Hinblick auf die ätiologie der Selbsterhitzung pflanzlicher Stoffe. Arch Mikrobiol 1:78–118

Miehe H (1930b) Über die Selbsterhitzung des Heues. Arb Dtsch Landwirtsch Gesellsch Berlin 111:76–91

Morgenstern I, Powlowski J, Ishmael N, Darmond C, Marqueteau S, Moisan M, Quenneville G, Tsang A (2012) A molecular phylogeny of thermophilic fungi. Fungal Biol 116:489–502

Mouchacca J (2000a) Thermotolerant fungi erroneously reported in applied research work as possessing thermophilic attributes. World J Microbiol Biotechnol 16:869–880

Mouchacca J (2000b) Thermophilic fungi and applied research: a synopsis of name changes and synonymies. World J Microbiol Biotechnol 16:881–888

Moustafa A-WF, Abdel-Azeem AM (2008) *Thielavia gigaspora*, a new thermotolerant ascomycete from Egypt. Microbiol Res 163:441–444

Natvig DO, Taylor JW, Tsang A, Hutchinson MI, Powell AJ (2015) *Mycothermus thermophilus* gen. et comb. nov., a new home for the itinerant thermophile *Scytalidium thermophilum* (*Torula thermophila*). Mycologia 107:319–327

Neher DA, Weicht TR, Bates ST, Leff JW, Fierer N, Brayton KA (2013) Changes in bacterial and fungal communities across compost recipes, preparation methods, and composting times. PLoS One 8(11):e79512

Oliveira TB, Gomes E, Rodrigues A (2015) Thermophilic fungi in the new age of fungal taxonomy. Extremophiles 19:31–37

Pan WZ, Huang XW, Wei KB, Zhang CM, Yang DM, Ding JM, Zhang KG (2010) Diversity of thermophilic fungi in Tengchong Rehai National Park revealed by ITS nucleotide sequence analyses. J Microbiol 48:146–152

Peixoto-Nogueira SC, Sandrim VC, Guimarães LHS, Jorge JA, Terenzi HF, Polizeli MLTM (2008) Evidence of thermostable amylolytic activity from *Rhizopus microsporus* var. *rhizopodiformis* using wheat bran and corncob as alternative carbon source. Bioprocess Biosyst Eng 31:329–334

Platt AR, Woodhall RW, George AL Jr (2007) Improved DNA sequencing quality and efficiency using an optimized fast cycle sequencing protocol. Biotechniques 43:58–62

Powell AJ, Parchert KJ, Bustamante JM, Ricken JB, Hutchinson MI, Natvig DO (2012) Thermophilic fungi in an aridland ecosystem. Mycologia 104:813–825

Pringle A, Baker DM, Platt JL, Wares JP, Latge JP, Taylor JW (2005) Cryptic speciation in the cosmopolitan and clonal human pathogenic fungus *Aspergillus fumigatus*. Evolution 59:1886–1899

Rajasekaran AK, Maheshwari R (1993) Thermophilic fungi: an assessment of their potential for growth in soil. J Biosci 18:345–354

Richardson MJ (2001) Diversity and occurrence of coprophilous fungi. Mycol Res 105:387–402

Rubin EM (2008) Genomics of cellulosic biofuels. Nature 454:841–845

Salar RK (2018) Thermophilic fungi: basic concepts and biotechnological applications. CRC Press, Boca Raton, FL, pp 1–334

Salar RK, Aneja KR (2007) Thermophilic fungi: taxonomy and biogeography. J Agric Techonol 3:77–107

Sharpton TJ, Stajich JE, Rounsley SD, Gardner MJ, Wortman JR, Jordar VS, Maiti R, Kodira CD, Neafsey DE, Zeng Q, Hung CY (2009) Comparative genomic analyses of the human fungal pathogens *Coccidioides* and their relatives. Genome Res 19:1722–1731

Stamatakis A (2006) RAxML-VI-HPC: maximum likelihood-based phylogenetic analyses with thousands of taxa and mixed models. Bioinformatics 22:2688–2690

St-Germain G, Robert A, Ishak M, Tremblay C, Claveau S (1993) Infection due to *Rhizomucor pusillus*: report of four cases in patients with leukemia and review. Clin Infect Dis 16:640–645

Straatsma G, Samson RA, Olijnsma TW, Den Camp HJO, Gerrits JP, Van Griensven LJ (1994) Ecology of thermophilic fungi in mushroom compost, with emphasis on *Scytalidium thermophilum* and growth stimulation of *Agaricus bisporus* mycelium. Appl Environ Microbiol 60:454–458

Subrahmanyam A (1999) Ecology and distribution. In: Thermophilic moulds in biotechnology. Springer, Dordrecht, pp 13–42

Tansey MR (1971) Isolation of thermophilic fungi from self-heated, industrial wood chip piles. Mycologia 63:537–547

Tansey MR (1973) Isolation of thermophilic fungi from alligator nesting material. Mycologia 65:594–601

Tansey MR (1975) Fungi associated with growing stalagtites. Mycologia 67:171–172

Tansey MR (1977) Enrichment, isolation and assay of growth of thermophilic and thermotolerant fungi in lignin-containing media. Mycologia 69:463–476

Tansey MR, Brock TD (1972) The upper temperature limit for eukaryotic organisms. Proc Natl Acad Sci U S A 69:2426–2428

Tansey MR, Jack MA (1976) Thermophilic fungi in sun-heated soils. Mycologia 68:1061–1075

Taylor JW, Hann-Soden C, Branco S, Sylvain I, Ellison CE (2015) Clonal reproduction in fungi. Proc Natl Acad Sci U S A 112:8901–8908

Tedersoo L, Sánchez-Ramírez S, Kõljalg U, Bahram M, Döring M, Schigel D, May T, Ryberg M, Abarenkov K (2018) High-level classification of the Fungi and a tool for evolutionary ecological analyses. Fungal Divers 90:135–159

Tiquia SM (2005) Microbial community dynamics in manure composts based on 16S and 18S rDNA T-RFLP profiles. Environ Technol 26(10):1104–1114

Tsiklinsky P (1899) Sur les mucédinées thermophiles. Ann Inst Pasteur 13:500–505

van den Brink J, Samson RA, Hagen F, Boekhout T, de Vries RP (2012) Phylogeny of the industrial relevant, thermophilic genera *Myceliophthora* and *Corynascus*. Fungal Divers 52:197–207

van den Brink J, van Muiswinkel GCJ, Theelen B, Hinz SWA, de Vries RP (2013) Efficient plant biomass degradation by thermophilic fungus *Myceliophthora heterothallica*. Appl Environ Microbiol 79:1316–1324

von Klopotek A (1976) *Thielavia heterothallica* spec. nov., die perfekte Form von *Chrysosporium thermophilum*. Arch Microbiol 107:223–224

White TJ, Bruns T, Lee SJ, Taylor JW (1990) Amplification and direct sequencing of fungal ribosomal RNA genes for phylogenetics. PCR Protoc 18:315–322

Winnepenninckx B, Backeljau T, Wachter R (1993) Complete small ribosomal subunit RNA sequence of the chiton (Lischke, 1873) (Mollusca, Polyplacophora). Nucleic Acids Res 21:1670–1670

Chapter 5
Ecology and Biotechnology of Thermophilic Fungi on Crops Under Global Warming

Robert Russell M. Paterson (ID) **and Nelson Lima**

5.1 Introduction

More extreme climates will occur under climate change. These are likely to affect detrimentally the ability to grow crops where temperature increase will cause most problems and the most likely change is high temperatures, as is well known (http://unfccc.int/paris_agreement/items/9485.php) (Paterson and Lima 2010a, 2011, 2017a; Paterson et al. 2013). The changes will affect negatively the diseases of crops with fungal maladies being particularly important (Paterson and Lima 2010a; Paterson and Lima 2011; Paterson et al. 2013; Paterson and Lima 2017a).

Climate change affects the plant disease triangle comprising the environment, plant host and pathogen (Paterson and Lima 2017a). Some fungi can produce mycotoxins on crops (Medina et al. 2015) (Paterson and Lima 2010a; Paterson and Lima 2011; Baranyi et al. 2015) and the infection process is similar to that for fungal plant pathogens not producing mycotoxins. An alternative route for mycotoxin contamination is via stored crops where there may be greater control from, for example, keeping the commodity dry.

Mycotoxins are low-molecular-weight fungal secondary metabolites some of which have statutory limits imposed on crops by trading blocks and nations. These compounds have toxicities and so are biologically active. However, secondary metabolites also include antibiotics and other drugs of which fungi are well-known

R. R. M. Paterson (✉)
Department of Plant Protection, Universiti Putra Malaysia, Serdang, Selangor, Malaysia

CEB—Centre of Biological Engineering, Campus de Gualtar, University of Minho,
Braga, Portugal
e-mail: russellpaterson@upm.edu.my

N. Lima
CEB—Centre of Biological Engineering, Campus de Gualtar, University of Minho,
Braga, Portugal

© Springer Nature Switzerland AG 2019
S. M. Tiquia-Arashiro, M. Grube (eds.), *Fungi in Extreme Environments:
Ecological Role and Biotechnological Significance*,
https://doi.org/10.1007/978-3-030-19030-9_5

producers, for example, penicillin, cephalosporin and statins. Production of myco-toxins is highly influenced by climate change (Magan et al. 2011), affecting the succession of fungi and mycotoxins on crops towards those that grow and are pro-duced respectively at higher temperatures (Paterson and Lima 2010a,2011), where a progression of temperatures is observed. This only occurs with *A. flavus* and afla-toxins, after which there are no conventional fungi that produce well-known myco-toxins. All these fungi do not grow, or not well, above 35 °C. Higher temperatures are obtained readily currently and will become common under global warming (Paterson and Lima2017a). In countries such as Pakistan (Paterson and Lima 2011) and Australia (Guardian, UK newspaper 8 Jan 2018), astonishingly high tempera-tures are recorded at ca. 45–50 °C, or more.

Interestingly, the increase in temperature anticipated from climate change may lead to thermotolerant and thermophilic fungi (TTF) (Salar 2018) with new biotech-nological properties. For example, fungal enzyme activity is responsible for some plant diseases, especially those involved in wilts and rots, and many of the same enzymes have utility in industrial processes, for example, cellulases, ligninases, pectinases and amylases. New high-temperature enzymes may be obtained from novel thermophilic plant pathogenic fungal species.

The distinction between mesophilic, thermotolerant and thermophilic is impre-cise and fungi capable of simply growing above 45 °C have been considered ther-mophilic. If growth up to 45 °C is considered as mesophilic, where do the thermotolerant fungi fit? Morgenstern et al. (2012) usefully consider thermophilic species as those which grow faster at 45 than 34 °C. 'Thermotolerant' can then be applied to fungi which grow at or above 40 °C, distinguishing them from most mesophiles, but with optima below this threshold.

Salar (2018) appears to support the following classifications for fungi: Thermophilic have optima for growth greater or equal to 45 °C; thermotolerant have maxima less than or equal to 50 °C, with a minimum 'considerably' less than 20 °C; A novel category referred to as thermophilus includes all fungi growing at elevated temperatures, hence including TTF; thermoduric are those with reproductive struc-ture surviving ca. 80 °C but have with normal growth temperatures of 22–25 °C; transitional thermophiles are those that grow below 20 °C but withstand near 40 °C. It is unclear how helpful so many definitions are, some of which overlap. Redefinitions may be required as global warming continues and higher tempera-tures become the norm.

The increase in fungal disease of crops because of climate change has been antic-ipated in various studies (Paterson and Lima 2010a, 2011, 2017a; Paterson et al. 2013). The magnified level of maladies could occur from more virulent fungi and/or less resistant crops. Increase in virulence would result from the new conditions favouring the pathogen over the plant. Alternatively, strains of the fungi may be selected for under the novel climate as it is well known that fungi are mutable and are likely to change more than a particular crop could become resistant to the dis-ease within a particular period of time. The fungus could mutate more readily into a virulent strain from, for example, higher levels of UV irradiation from the sun's rays penetrating the atmosphere than from the destruction of the ozone layer (Paterson and Lima 2011; Paterson et al. 2013). Finally, different fungal taxa with higher growth temperatures could become novel pathogen crops in the manner previously described for mycotoxigenic fungi.

The scope of this chapter commences with the premise that fungi are changing from those that grow at low temperatures to those that are thermophilic. This indicates a significant change in biodiversity. These changes will have a large impact on biotechnology in that novel fungi will be increasingly studied with useful properties. However, there will also be losses of biodiversity with a potentially negative biotechnological impact. Naming fungi correctly is essential; otherwise inventories for biotechnology will become misleading. These points are illustrated uniquely using mycotoxigenic fungi and plant pathogens. The field of bioprospecting is also covered. No other work combines these aspects in a unified report.

5.2 Fungal Succession as Illustrated by Mycotoxigenic Fungi

What evidence is there that mycotoxigenic fungi succeed one another when the growing conditions change (Paterson and Lima 2011), which would be an indication of what will occur with other fungi under climate change? In fact, there is a great deal of evidence from crops in Europe where novel mycotoxins are becoming problematic whereas others were predominant previously (Paterson and Lima 2017a).

Conventional mycotoxigenic fungi are mesophilic, although *Aspergillus flavus* and *A. parasiticus* may be thermotolerant marginally, the two most important and dangerous aflatoxin-producing fungi. Temperatures are likely to increase well beyond these ranges and especially in the tropics, although these high temperatures will appear eventually in non-tropical climates if climate change is not addressed. Which fungi and conventional mycotoxins (e.g. aflatoxin, ochratoxin A, deoxynivalenol and fumonisins) will succeed? Paterson and Lima (2010a) suggested that *A. flavus* may become extinct because of climate change, which would be advantageous to crop production as they would have low levels of aflatoxin. Furthermore, there is less relevance in studying conventional fungi and mycotoxins if they become increasingly rare on crops. Paterson and Lima (2011) discuss how mycotoxin contamination is predictable if the same fungi are involved under climate change but becomes unpredictable should novel fungi dominate.

The ecology of fungi under climate change towards high temperatures is interesting. Existing fungi, currently at low concentrations, may dominate when temperature change conforms to the growth optima of these fungi. Alternatively, species from higher temperature regions may be introduced to the more temperate regions as invasive species.

Of greatest significance is that aflatoxigenic fungi will be replaced by TTF. The succession of *A. flavus* and aflatoxin fungi has been reported frequently (Paterson and Lima 2017a), with additional examples from fusaria. The increased frequency of aflatoxin detection indicates an increase in the fungi responsible and represents a biochemical diagnostic for the fungi (Paterson and Kozakiewicz 1997). And vice versa, the decreased detection of the previous mycotoxins acts to indicate fewer producing fungi and a similar principle could be employed with other fungi that produce secondary metabolites on crops. This method could compliment culture-independent PCR methods (Paterson 2012).

Increased aflatoxin B1 contamination in Europe was associated with hot weather, contributing to increased *A. flavus* on crops in Italy (Paterson and Lima 2011; Giorni et al. 2007), and *A. flavus* colonized maize by outcompeting *Fusarium* species during the hot and dry episodes in the same country in 2003 (Giorni et al. 2008). Increased aflatoxin M1 in milk in 2012 related to more aflatoxins than normal in animal feed from southeast Europe (Alfonso and Botana 2015). A survey (EFSA 2007) correlated aflatoxin contamination of crops in southern Europe with a recent subtropical climate, and a model for *A. flavus* growth and aflatoxin B1 production indicated that aflatoxin contamination will increase in maize from climate change, where there was an increase in aflatoxin risk in many countries at plus 2 °C. Furthermore, *A. flavus* colonized ripening maize by outcompeting *Fusarium* species (Magan et al. 2011). Extreme temperatures of >35 °C resulted in a change from *Fusarium verticillioides* and contamination with fumonisins of maize to *A. flavus* and aflatoxins (Giorni et al. 2007). No aflatoxin contamination was detected in Serbian maize in 2009–2011, but prolonged hot and dry weather in 2012 correlated with 69% of samples being contaminated (Kos et al. 2013), with an implied increase in the producing fungi. Increases in aflatoxins in maize in Hungary may be from climate change (Dobolyi et al. 2013), and aflatoxin-producing *A. flavus* isolates in several maize fields were isolated in 2012–2013, whereas before none were found: Aflatoxin B1 levels in cereals and animal feed above the EU limit in 4.8% of samples were observed in Hungary (Borbély et al. 2010), and aflatoxins were also detected in maize kernels in Hungary in 2012 (Baranyi et al. 2015) indicating a climate change event.

This succession has been observed with the fusaria that grow optimally at higher, although mesophilic, temperatures (Paterson and Lima 2017a). The displacement of the formerly predominant species, *Fusarium culmorum* and *Microdochium nivale*, by the more virulent plant pathogen *F. graminearum*, as a result of warm European summers, has been reported (FAO 2008). *F. graminearum* was the most abundant *Fusarium* on wheat in the Netherlands in the early 2000s, whereas *F. culmorum* was dominant in the 1990s (Waalwijk et al. 2004). The fungus has increased on wheat in the UK, while *F. culmorum* is less important (Edwards 2009) and similarly for Germany (Miedaner et al. 2008). *F. poae* was highest in Poland followed by *F. tricinctum*, *F. avenaceum*, *F. culmorum* and *F. graminearum* (Logrieco and Moretti 2008): a significant increase in *F. graminearum* has now been observed (Stępień and Chełkowski 2008). T-2 and HT-2 toxin contamination are more prevalent on oats and barley in the UK related to *F. langsethiae* detection in grains (Edwards 2009) and by implication climate change. In addition, this species has become prevalent on barley in recent years in northern France (Moretti and Logrieco 2015) and produces toxins (Magan et al. 2011). High temperatures favour growth of the fumonisin producer *F. verticillioides* in maize (FAO 2008). A more toxigenic 3-acetylated deoxynivalenol chemotype of *F. graminearum* replaced the 15-acetylated deoxyni-valenol chemotype in Canada (Ward et al. 2008), indicating that lower taxonomic ranks of species are involved. Minnesota, USA, has witnessed the emergence of a novel *Fusarium* taxon called the 'Northland population', which does not produce deoxynivalenol or nivalenol (Paris et al. 2015). Hence, TTF are likely to succeed other mycotoxigenic fungi in the future as temperature increases, and these previously mentioned results are extremely useful for predicting what will happen to fungi in general under climate change.

5.3 Threat Posed by Existing Mycotoxins from Global Warming

Perhaps surprisingly, the only conventional mycotoxin-producing TTF are those that manufacture patulin (Fig. 5.1). There is also a group of potentially patulin-producing TTF. Hence, patulin may become more prevalent in the future through global warming, which requires monitoring. The concentration of patulin in apple products is subject to regulations and is produced in apples by *Penicillium expansum* (Wright 2015). More commodities will require assessing for patulin contamination which are currently only analysed for the other conventional mycotoxins, because of patulin production from TTF and global warming. New commercial markets will be formed in mycotoxin analysis and mycotoxin remediation which may be considered as forms of biotechnology.

It is worthwhile summarizing the toxicity of patulin: The probable primary biochemical lesions and the early cellular events leading to toxic cell injury or cellular deregulation caused by patulin, involve initial non-protein sulfhydryl depletion, leading to (a) an event cascade of altered ion permeability and/or intercellular communication, (b) oxidative stress and (c) cell death, involving inhibition of macromolecular biosynthesis (Paterson and Lima 2010b). The International Agency for Research on Cancer has since 1986 maintained that patulin is not a carcinogen, because of the lack of studies demonstrating its carcinogenicity, but this classification may change in the future as science progresses (Pfenning et al. 2016).

Byssochlamys, *Paecilomyces* anamorphs and *Thermoascus* were considered as patulin-producing fungi or suspected producing fungi. However, from recent nomenclatural changes, *Byssochlamys fulva* is now considered as *Paecilomyces fulvus* and *Byssochlamys nivea* is *Paecilomyces niveus* (Frisvad 2018). Contamination of canned goods can occur with these fungi as they survive high temperatures used in the industry. Species identification is difficult and the taxonomy has been

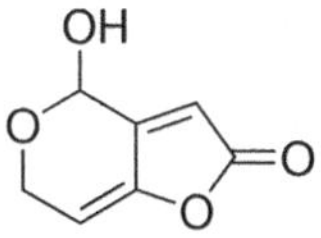

Fig. 5.1 Chemical structure of patulin

subjected to a number of revisions to delineate which taxa produce patulin (Samson et al. 2009), although more work is required particularly on the effect of high temperature on patulin production. Frisvad (2018) considered that a high percentage of fungi listed as producing patulin were designated predominantly from (a) misidentifications of the fungus or toxin and/or (b) carry-over in the analytical equipment, which is remarkable. Can this be ascribed to these reasons in all cases? There are major faults in the procedures and in the literature reporting them, if this is the case. Correct identifications can effectively only be carried out by a small number of taxonomists under current systems, although the need to identify the fungi and mycotoxin production is from a much wider group of scientists. Other interpretations, such as only a few strains of a species having the capacity, or mutation of the fungi from environmental and/or cultural conditions, cannot be dismissed (Paterson and Lima 2015a). Also, subtle changes in growth conditions may allow expression of the patulin pathway in some laboratories but not others and preservation methods may have inhibited expression of the pathway. Contamination of cultures is another possibility. It indicates that the current (chemo) taxonomies cannot be applied universally (Paterson et al. 2017, 2018). A fundamental criterion is that scientific methods should be repeatable by other scientists in other laboratories and so simpler methods are required which can be used in most laboratories and by various taxonomic skill levels. The correct naming of fungi forms the basis of many aspects of biotechnology.

Paecilomyces variotii sensu *lato* is a common fungus in the air, subtropical soil, tropical soil, compost and wood, and is frequently in foods such as rye bread, margarine, peanuts, peanut cake, cereals and heat-treated fruit juices (Houbraken et al. 2006). Strains of *P. variotii* produce patulin, although the species may represent a complex of *P. divaricatus*, *P. formosus*, *P. saturatus* and *P. variotii*. *P. brunneolus* may form another species within the complex, but is very similar to *P. variotii*. However, only *P. saturatus* is listed as producing patulin (Samson et al. 2009). A single representative of *P. brunneolus* grew more at 30 than 37 °C, whereas it is the reverse for *P. variotii*, perhaps indicating that *P. variotii* is a TTF and *P. brunneolus* is not. A chemotaxonomic revision of these fungi is required in relation to patulin production under high temperature which might stimulate patulin production. Furthermore, *P. variotii* is claimed as the anamorph of *Byssochlamys spectabilis*. *P. variotii* appears as the earliest name applied to this species and a new combination into *Byssochlamys* may be required if *B. spectabilis* is not proposed for protection. The case has not been reported upon by an ICTF-recognized working group and remains undecided (Hawksworth 2015). Frisvad (2018) states that *B. variotii* is now named *P. variotii*, which may confuse the non-expert.

B. nivea is a well-recognized name for a patulin-producing species (Samson et al. 2009) although it appears that the designation has been superseded by *P. nivea* (Frisvad 2018). Also, Frisvad (2018) stated that *B. variotii* is now correctly named as *P. variotii* as mentioned previously and the situation appears confusing. At least one strain grew at 46 °C (Zhang et al. 2016) and so is a TTF. The optimum temperature for *P. variotii* growth is 50 °C and the maximum is 55 °C (Table 5.1): this patulin-producing species (Maheshwari et al. 2000) is a valid candidate to succeed

Table 5.1 Patulin production and potential production of some thermophilic and thermotolerant fungi (Maheshwari et al. 2000)

Fungus	Temp. optimum (°C)	Temp. maximum (°C)	Comment
Paecilomyces variotii	50	55	Of this species complex, *P. saturatus* may be the patulin-producing species
Byssochlamys nivea	–	At least 46 (Zhang et al. 2016)	Well-known patulin producer
Thermoascus aurantiacus	49–52	61	IDH positive; patulin to be confirmed
Byssochlamys verrucosa	20–53	–	*B. verrucosa* is more related to *Thermoascus*. IDH positive; patulin to be confirmed

mesophilic mycotoxigenic fungi in nature. *B. nivea* (now *P. nivea*) is also a valid candidate as it is at least thermotolerant and produces patulin. Patulin (Fig. 5.1)-producing fungi mentioned herein (Table 5.1) would be considered thermophilic but also grow within the thermotolerant range and a precise study of the optima and maxima of these fungi is merited. Patulin, produced by TTF, could become more prevalent on crops and even displacing aflatoxin in the future under climate change. Interestingly, this mycotoxin was considered as a potential antibiotic, also indicating how novel biotechnologically relevant compounds could arise under climate change.

The isoepoxydon dehydrogenase (idh) gene of the patulin metabolic pathway was detected in *B. verrucosa*, which is an indicator of the ability to produce patulin in fungi (Paterson et al. 2003), although patulin was not detected (Hosoya et al. 2014). This could reflect that the mycotoxin was produced below the detection limits of an analytical system, or that appropriate growth conditions were not employed, rather than the fungus being incapable of producing the compound. Testing patulin production at high temperatures would be appropriate. The temperature optimum or optima lie within a range that extends to 52 °C, which is thermophilic. Hence this species is a possible candidate for a conventional mycotoxin producer that could replace the mesophilic mycotoxigenic fungi, if patulin production was confirmed. However, *B. verrucosa* is more related to *Thermoascus* than *Byssochlamys* by analyses of the ITS, and partial β-tubulin DNA region (Samson et al. 2009).

Thermoascus aurantiacus is a well-defined species which again possessed the idh gene, although patulin was undetected (Hosoya et al. 2014). *Thermoascus* has been isolated from various foods including maize and could become prominent under climate change. The incidences of spoilage by these fungi are increasing which may be climate change related. The species has a temperature optimum between 49 and 52 °C and a maximum of 61 °C. To reiterate, no other TTF produce other conventional mycotoxins.

5.4   Secondary Metabolites from Thermophilic Fungi That May Contaminate Crops

A list of potential mycotoxins and other secondary metabolites from TTF are given in (Paterson and Lima 2017a) and Table 5.2. Some examples include gliotoxin from *A. fumigatus*; estatin A and B from *Myceliophthora thermophile*; and talathermophilins from *Talaromyces duponti*. Other secondary metabolites will be discovered from TTF which could provide novel pharmaceuticals or become novel mycotoxins. Fungal diseases of crops in general have similarities to mycotoxin contamination by fungi of crops, and this will now be illustrated by a fungal disease of oil palm called basal stem rot.

5.5   *Ganoderma boninense* Oil Palm Disease

Palm oil has many uses including (a) a component of many foods and cosmetics, (b) cooking, (c) pharmaceuticals and (d) biodiesel (Paterson and Lima 2018). It contributes significantly to the economies of many countries and Malaysia and Indonesia particularly, as these are the highest producers by far. However, the production of palm oil contributes to climate change predominantly because of deforestation and burning of peat to grow OP. Equally, the growth of OP is affected negatively by climate change and diseases of the palm will be affected. Levels of disease are likely

Table 5.2 Secondary metabolites from thermophilic fungi (Paterson and Lima 2017a)

Fungus	Temp. optimum (°C)	Temp. maximum (°C)	Secondary metabolites	Habitat
Aspergillus fumigatus	37	65	Gliotoxin, fumigatins, fumigaclavines, fumiquinazolines, fumitremorgins, verruculogens, helvolic acids	Composts, tree bark, crops
Myceliophthora thermophila	45–50	55	Estatin A and B	Wood pulp
Talaromyces duponti	45–50	60	Talathermophilins, thermolides	Guayule shrub, fermented straw, compost
Thermomyces lanuginosus	45–50	60	Thermolides, bacterial-like hybrid macrolactones	Compost, moist oats, cereal grains, mushroom compost, hay, leaf mould peat, garden compost, various plant substances

to increase under climate change (Paterson and Lima 2010a), with novel fungi causing more maladies, or even mutated strains becoming involved (Paterson et al. 2013). The novel fungi may have beneficial properties for biotechnology consistent with the theme of this chapter.

Oil palm and fungal diseases of oil palm have adapted to the tropical climates in which the plant grows. There is considerable concern about the sustainability of oil palm and one of the major issues is an increase in disease from an increase in climate change (Rival 2017; Corley and Tinker 2016; Paterson et al. 2013; Paterson and Lima 2018). The most important disease in Malaysia and Indonesia is basal stem rot caused by *G. boninense* (Fig. 5.2), the growth optimum of which is approximately 32 °C (Paterson et al. 2013), although more stains require testing. The species responsible have not been satisfactorily determined and there may be novel taxa of the genus more adapted to higher temperatures. This will be explained in more detail to indicate which novel species could become a fungal disease.

There are many other wood-rotting fungi within the plantations that could cause the rot disease (Naidu et al. 2015; Paterson et al. 2000; Treu 1998). *G. boninense* has adapted sequentially to jungle/forest trees, coconut/rubber and OP (Flood et al. 2000a). The fungus reproduces sexually, allowing genetic variation to occur: It produces millions of spores for propagation, even from one basidiome, thereby permitting further variation. Evolution in *Ganoderma* and increasing complexity in population structure will occur where interactions with oil palm are longest, when pathogens are evolving and adapting to environmental changes (Merciere et al. 2017). Rapid evolution of taxa adapted to the novel conditions of CC can be

Fig. 5.2 The white rot fungus *Ganoderma* on oil palm (https://www.google.com/search?q=oil+p alm+ganoderma+disease+and+symptom+photo&source=lnms&tbm=isch&sa=X&ved=0ahUKE wjkvMztiazaAhVDLVAKHbQFBa4Q_AUICigB&biw=1094&bih=547&dpr=1.25#im grc=-yc-jvH7ZylSeM)

anticipated. The variation within the genus associated with OP can be appreciated when (a) *G. boninense*, *G. zonatum*, *G. miniatocinctum* and *G. tornatum* (Rashid et al. 2014) and (b) *G. ryvardense* (Corley and Tinker 2016) may cause the disease and these species, or other related taxa, may become dominant in the future. The taxonomy of *Ganoderma* remains unresolved and species concepts are reviewed frequently (Paterson and Lima 2015b; Zhou et al. 2015). The situation is complicated by the plant disease website Plantwise (https://www.plantwise.org/KnowledgeBank/Datasheet.aspx?dsid=24924) where BSR is associated initially with *G. boninense*, but the name *G. orbiforme* is used when the distribution map is consulted: *G. orbiforme* is not considered a synonym of *G. boninense* (Wang et al. 2014) and forms a separate species. Nevertheless, authentic *G. orbiforme* may also be involved in BSR. The indigenous, or introduced, *Ganoderma* in Malaysia and Indonesia in existence with oil palm are capable of adapting to CC more quickly than can the oil palm.

A related malady is the enigmatic upper stem rot which is poorly studied. OP with upper stem rot will tend to remain standing longer than those with BSR as they are more stable (Flood et al. 2000b) making the rot appear less serious: it is important only in deep peat and inland valley soils and appears sporadically in Malaysia and Indonesia. Interestingly, there are conflicting reports of the cause being *Phellinus noxius*, another basidiomycete white rot fungus (Paterson 2007), with *G. boninense* perhaps a secondary infection (Corley and Tinker 2016). Contrary reports suggest that *G. boninense* causes the disease: *Ganoderma* was confirmed in all cases of upper stem rot examined in Sumatra, sometimes with a *Phellinus* sp. *P. noxius* could become a more important disease of OP under CC if conditions favour this fungus.

Thermophilic fungi will be increasingly important as temperature increases under CC (Paterson and Lima 2017a) as mentioned above. *G. collosum* is a thermophilic white rot (Paterson 2007) *Ganoderma* which grows optimally at 40 °C and up to 45 °C. The fungus has been shown to degrade date palm, but has not been tested against OP (Adaskaveg and Gilbertson 1995). Hence, there already exists a palm-degrading *Ganoderma* adapted to high temperatures. Other palm-degrading fungi could also adapt to causing disease in stressed OP from the consequences of CC (Naidu et al. 2017). A question this current chapter asks is what other properties will these TTF have of interest to biotechnology apart from those already discussed.

5.6   Increased Thermophilic Fungi to Impact Biotechnology

One of the main subjects of this chapter is that the climate change conditions may select for fungi with novel biotechnological potential as best illustrated with mycotoxigenic fungi. Increases in plant disease may occur and new mycotoxins may appear which will require novel methods for diagnoses, consequently contributing to the bioscience industries. More and/or different equipment will be required which will boost profits for analytical equipment companies, although there will be lost profits from a reduced need to analyse for fungi or mycotoxins which will no longer be relevant. The toxicity of novel mycotoxins will require assessments which will lead to technological developments implying more resources being expended in these areas.

Improved analytical methods related to biotechnology for patulin will be required if it becomes more prevalent, many of which will be HPLC related. In addition, molecular biology procedures will be employed to identify the fungi isolated from the crops where patulin contamination occurs. Analytical methods will need development based on first principles for each potential toxin in the case of novel mycotoxins from TTF. HPLC, MS and NMR will all be required for these new compounds. Novel toxicity testing and bioassays may be required. New immuno-affinity columns (IAC) may be produced, for example, allowing a new market for diagnostic companies. However, IAC for patulin may not be possible as the molecule is too small to allow the correct immune response in the production animals. This requires weighing against the reduced requirement for equipment to detect the mycotoxins present currently.

When the TTF grow and produce mycotoxins on crops, methods will be required to control these sources of food contamination. Control of current mycotoxins might be possible using biocontrol microorganisms applied to the crop (Chulze et al. 2015). Similarly, novel biocontrol microorganisms may be developed to control TTF as they become problematic. These biocontrol organisms will need to be effective at the new extreme temperatures.

With the possibility of patulin contamination becoming more prevalent, it is useful to consider biotechnological procedures to reduce levels and degrading the molecule may be possible employing patulin-degrading microorganisms (Zhu et al. 2015).

As novel mycotoxins emerge, so might new pharmaceuticals from the TTF, especially if bioprospecting projects are undertaken. Novel fungi with new properties may be found. Refocusing on TTF because of their potential to infect crops may yield promising new (high-temperature) enzymes, organic acids and pharmaceuticals, or fungi that can produce existing ones more efficiently. The search for novel pharmaceuticals, enzymes and organic acids, for example, is often associated with bioprospecting (Paterson and Lima 2017b). The change in climate to thermophilic temperatures implies that novel fungi will almost certainly occur adapted to the novel conditions. In addition, mutations may occur allowing fungi to adapt and thrive under the hot temperatures.

There are few bioprospecting projects that are exploring the current fungi found in particular regions or niches (Stierle and Stierle 2017), but these could usefully be increased. Pingal (2017) discussed a project based in Iwokrama rainforest in Guyana. Novel fungi could be discovered under CC if a new project was established and as proposed in the current chapter. But first a new survey of existing fungi is required, and similarly for the Sarawak rainforest in Malaysia (Shaw 2017) and other biodiverse regions. Unfortunately, useful fungi could become extinct because of CC. These fungi may have yielded new antibiotics, for example, which are especially important in the current era of resistant diseases (Paterson and Lima 2017c) and may represent a missed opportunity.

TTF are often studied for high-temperature enzyme production but are also known to produce secondary metabolites with antimicrobial activities and can be used for composting activities (Singh et al. 2016). Houbraken et al. (2014) investigated the taxonomy of TTF which had biotechnological properties and Maheshwari et al. (2000) discussed TTF enzymes. TTF are seen as having the potential to contribute significantly towards a more sustainable world and represent an understudied group. This may lead to them being investigated for their properties and lead to new enzymes and metabolites. Crops could be modified to withstand higher temperatures from climate change by transforming genes from TTF as (Kumar et al. 2016).

Another negative impact of climate change may be that some useful fungi will become less prevalent or extinct as climate changes. This may provide the impetus for increasing bioprospecting surveys (Paterson and Lima 2017b) before these fungi effectively disappear. The environments that require surveys are fields where crops are grown and which are analysed for mycotoxins and the rainforests and plantation floors where OP are grown or intended to be grown. Thermophilic fungi that cause basal stem rot in oil palm disease under climate change may possess highly active high-temperature ligninolytic enzymes capable of converting plant material into biofuel or animal feed.

5.7 Conclusions

Climate change will create a new environment for biotechnology and for microorganisms such as fungi. Surveys of hot regions (e.g. the tropics) that grow important crops are required to determine which TTF are present and if they can produce any of the important mycotoxins. PCR of metabolic pathway genes may be useful for this purpose (see above). Also, the presence of the TTF on crops may be higher than is currently appreciated, because seldom are high temperatures employed to isolate fungi from foodstuffs. The presence of TTF in currently temperate countries requires investigation to anticipate which fungi will succeed those isolated under mesophilic conditions. When assessing the mycoflora, higher incubation temperatures are recommended of at least 42.5 to as much as 65 °C. The succession of fungi on crops because of global warming will not cease after temperatures increase beyond the mesophilic range. It is likely that patulin will increasingly become a threat under climate warming as discussed herein. The new environmental conditions may allow isolation of novel fungi with useful properties for biotechnological advancement. However, what will we have missed in the mean time?

5.8 Future Perspectives

The anticipated outcomes of climate change are likely severe. However, there may be some positive aspects such as greater availability of fungi able to grow at high temperatures and producing novel (high-temperature) enzymes and pharmaceuticals. Efficient lignin degraders could be discovered capable of degrading by-products from plants much quicker than is possible currently, permitting efficient conversion to biofuels and animal feeds. The most dangerous mycotoxin, aflatoxin, may be less common on crops. Nevertheless, these are oases in an outlook which is bleak, if climate change is not addressed and deaccelerated as outlined in the Paris climate conference. Negative consequences for fungi include the extinction of potentially useful fungi present in the environment under current conditions. The presence of the aflatoxin is likely to increase, and other mycotoxins will be in higher concentrations before the possible reduction of aflatoxin on crops. Patulin may become a major problem. Crop diseases are likely to increase particularly in the tropics. Finally, the effects of climate change on biotechnology will be profound, although it remains unclear if the pros will outweigh the cons. This is not the case concerning the overall consequences of climate change, which remain severe.

Acknowledgments This study was supported by the Portuguese Foundation for Science and Technology (FCT). It was within the strategic funding of UID/BIO/04469/2013 unit, COMPETE 2020 (POCI-01-0145-FEDER-006684) and BioTecNorte operation (NORTE-01-0145-FEDER-000004), which was funded by the European Regional Development Fund within Norte2020—Programa Operacional Regional do Norte. RRMP received gratefully the IOI Professorial Chair in Plant Protection for 2018 at the Universiti Putra Malaysia.

References

Adaskaveg JE, Gilbertson RL (1995) Effects of incubation time and temperature on in vitro selective delignification of silver leaf oak by *Ganoderma colossum*. Appl Environ Microbiol 61:138–144

Alfonso A, Botana LM (2015) Considerations about international mycotoxin legislation, food security, and climate change. In: Botana MJ, Sainz LM (eds) Climate change and mycotoxins. Walter de Gruyter GmbH, Berlin, pp 153–179

Baranyi N, Kocsubé S, Varga J (2015) Aflatoxins: climate change and biodegradation. Curr Opin Food Sci 5:60–66

Borbély M, Sipos P, Pelles F, Gyori Z (2010) Mycotoxin contamination in cereals. J Agroalim Proc Technol 16:96–98

Chulze SN, Palazzini JM, Torres AM, Barros G, Ponsone ML, Geisen R, Schmidt-Heydt M, Köhl J (2015) Biological control as a strategy to reduce the impact of mycotoxins in peanuts, grapes and cereals in Argentina. Food Addit Contam Part A Chem Anal Control Expo Risk Assess 32:471–479

Corley RHV, Tinker PB (2016) The oil palm. Wiley, Chichester

Dobolyi C, Sebők FVJ et al (2013) Occurrence of aflatoxin producing *Aspergillus flavus* isolates in maize kernel in Hungary. Acta Aliment Hung 42:451–459

Edwards SG (2009) *Fusarium* mycotoxin content of UK organic and conventional oats. Food Addit Contam Part A Chem Anal Control Expo Risk Assess 26:1063–1069

EFSA (European Food Safety Authority) (2007) Opinion of the scientific panel on contaminants in the food chain on a request from the commission related to the potential increase of consumer health risk by a possible increase of the existing maximum levels for aflatoxins in almonds, hazelnuts and pistachios and derived products. EFSA J 446:1–127

FAO (Food and Agriculture Organization) (2008) Climate change: implications for food safety. http://www.fao.org/docrep/010/i0195e/i0195e00.HTM. Accessed 14 Feb2017

Flood J, Hasan Y, Turner PD, O'Grady E (2000a) The spread of *Ganoderma* from infective sources in the field and its implications for management of the disease in oil palm. In: Flood J, Bridge PD, Holderness M (eds) *Ganoderma* diseases of perennial crops. CABI Publishing, Wallingford, pp 101–112

Flood J, Bridge PD, Holderness M (2000b) *Ganoderma* diseases of perennial crops. CABI Publishing, Wallingford

Frisvad JC (2018) A critical review of producers of small lactone mycotoxins: patulin, penicillic acid and moniliformin. World Mycotoxin J 11:73–100

Giorni P, Magan N, Pietri A, Bertuzzi T, Battilani P (2007) Studies on *Aspergillus* section Flavi isolated from maize in northern Italy. Int J Food Microbiol 113:330–338

Giorni P, Battilani P, Magan N (2008) Effect of solute and matric potential on in vitro growth and sporulation of strains from a new population of *Aspergillus flavus* isolated in Italy. Fungal Ecol 1:102–106

Hawksworth DL (2015) Naming fungi involved in spoilage of food, drink, and water. Curr Opin Food Sci 5:23–28

Hosoya K, Nakayama M, Tomiyama D, Matsuzawa T, Imanishi Y, Ueda S, Yaguchi T (2014) Risk analysis and rapid detection of the genus *Thermoascus*, food spoilage fungi. Food Control 41:7–12

Houbraken J, Samson RA, Frisvad JC (2006) *Byssochlamys*: significance of heat resistance and mycotoxin production. Adv Exp Med Biol 571:211–224

Houbraken J, Vries RPD, Samson RA (2014) Modern taxonomy of biotechnologically important *Aspergillus* and *Penicillium* species. Adv Appl Microbiol 86:199–249

Kos J, Mastilović J, Janić Hajnal E (2013) Natural occurrence of aflatoxins in maize harvested in Serbia during 2009–2012. Food Control 34:31–34

Kumar V, Chattopadhyay A, Ghosh S, Irfan M, Chakraborty N, Datta A (2016) Improving nutritional quality and fungal tolerance in soya bean and grass pea by expressing an oxalate decarboxylase. Plant Biotechnol 14:1394–1405

Logrieco AF, Moretti A (2008) Between emerging and historical problems: an overview of the main toxigenic fungi and mycotoxin concerns in Europe. In: Leslie JF, Bandyopadhyay R, Visconti A (eds) Mycotoxins: detection methods, management, public health and agricultural trade. CABI, Wallingford, pp 139–153

Magan N, Medina A, Aldred D (2011) Possible climate-change effects on mycotoxin contamination of food crops pre- and postharvest. Plant Pathol 60:150–163

Maheshwari R, Bharadwaj G, Bhat MK (2000) Thermophilic fungi: their physiology and enzymes. Microbiol Mol Biol Rev 64:461–488

Medina A, Rodriguez A, Magan N (2015) Climate change and mycotoxigenic fungi: impacts on mycotoxin production. Curr Opin Food Sci 5:99–104

Merciere M, Boulard R, Carasco-Lacombe C, Klopp C, Lee YP, Tan JS, Syed Alwee SR et al (2017) About *Ganoderma boninense* in oil palm plantations of Sumatra and peninsular Malaysia: ancient population expansion, extensive gene flow and large scale dispersion ability. Fungal Biol 121:529–540

Miedaner T, Cumagun CJR, Chakraborty S (2008) Population genetics of three important head blight pathogens *Fusarium graminearum*, *F. pseudograminearum* and *F. culmorum*. J Phytopathol 156:129–139

Moretti A, Logrieco AF (2015) Climate change effects on the biodiversity of mycotoxigenic fungi and their mycotoxins in preharvest conditions in Europe. In: Botana LM, Sainz LM (eds) Climate change and mycotoxins. Walter de Gruyter GmbH, Berlin, pp 91–108

Morgenstern I, Powlowski J, Ishmael N, Darmond C, Marqueteau S, Moisan MC, Quenneville G, Tsang A (2012) A molecular phylogeny of thermophilic fungi. Fungal Biol 116:489–502

Naidu Y, Idris AS, Nusaibah SA, Norman K, Siddiqui Y (2015) In vitro screening of biocontrol and biodegradation potential of selected hymenomycetes against *Ganoderma boninense* and infected oil palm waste. For Pathol 45:474–483

Naidu Y, Siddiqui Y, Rafii MY, Saudc HM, Idris AS (2017) Investigating the effect of white-rot hymenomycetes biodegradation on basal stem rot infected oil palm wood blocks: biochemical and anatomical characterization. Ind Crop Prod 108:872–882

Paris MPK, Liu YJ, Nahrer K, Binder EM (2015) Climate change impacts on mycotoxin production. In: Botana MJ, Sainz LM (eds) Climate change and mycotoxins. Walter de Gruyter GmbH, Berlin, pp 133–152

Paterson RRM (2007) *Ganoderma* disease of oil palm—a white rot perspective necessary for integrated control. Crop Prot 26:1369–1376

Paterson RRM (2012) *Idh* PCR not only for *Penicillium*. Food Control 25:421

Paterson RRM, Kozakiewicz Z (1997) *Penicillium* and *Aspergillus* mycotoxins - diagnostic characters and quantitative data from commodities and cultures. Cereal Res Commun 25:271–275

Paterson RRM, Lima N (2010a) How will climate change affect mycotoxins in food? Food Res Int 43:1902–1914

Paterson RRM, Lima N (2010b) Toxicology of mycotoxins. EXS 100:31–63

Paterson RRM, Lima N (2011) Further mycotoxin effects from climate change. Food Res Int 44:2555–2566

Paterson RRM, Lima N (2015a) Mutagens affect food and water biodeteriorating fungi. Curr Opin Food Sci 5:8–13

Paterson RRM, Lima N (2015b) Editorial for the special issue on *Ganoderma*. Phytochemistry 114:5–6

Paterson RRM, Lima N (2017a) Thermophilic fungi to dominate aflatoxigenic/mycotoxigenic fungi on food under global warming. Int J Environ Res Public Health 14:199

Paterson RRM, Lima N (2017b) Bioprospecting. Success, potential and constraints. Springer, Cham

Paterson RRM, Lima N (2017c) Bioprospecting insights. In: Paterson RRM, Lima N (eds) Bioprospecting. Success, potential and constraints. Springer, Cham, pp 299–303

Paterson RRM, Lima N (2018) Climate change affecting oil palm agronomy, and oil palm cultivation increasing climate change, require amelioration. Ecol Evol 8:452–461

Paterson RRM, Holderness M, Kelley J, Miller RNG, O'Grady E (2000) *In vitro* biodegradation of oil-palm stem using macroscopic fungi from south-east Asia: a preliminary investigation. In: Flood J, Bridge PD, Holderness M (eds) *Ganoderma* diseases of perennial crops. CABI Publishing, Wallingford, pp 129–138

Paterson RRM, Kozakiewicz Z, Locke T, Brayford D, Jones SCB (2003) Novel use of the iso-epoxydon dehydrogenase gene probe of the patulin metabolic pathway and chromatography to test penicillia isolated from apple production systems for the potential to contaminate apple juice with patulin. Food Microbiol 20:359–364

Paterson RRM, Sariah M, Lima N (2013) How will climate change affect oil palm fungal diseases? Crop Prot 46:113–120

Paterson RRM, Venâncio A, Lima N, Guilloux-Bénatier M, Rousseaux S (2017) Predominant mycotoxins, mycotoxigenic fungi and climate change related to wine. Food Res Int 103:478–491

Paterson RRM, Soares C, Ouhibi S, Lima N (2018) Alternative patulin pathway unproven. Int J Food Microbiol 269:87–88

Pfenning C, Esch HL, Fliege R, Lehmann L (2016) The mycotoxin patulin reacts with DNA bases with and without previous conjugation to GSH: implication for related α, β -unsaturated carbonyl compounds? Arch Toxicol 90:433–448

Pingal R (2017) Iwokrama fungal/plant bioprospecting project 2000–2003 – a model for the future? In: Paterson RRM, Lima N (eds) Bioprospecting. Success, potential and constraints. Springer, Cham, pp 167–196

Rashid M, Rakib M, Bong CJ, Khairulmazmi A, Idris AS (2014) Genetic and morphological diversity of *Ganoderma* species isolated from infected oil palms (*Elaeis guineensis*). Int J Agric Biol 16:691–699

Rival A (2017) Breeding the oil palm (*Elaeis guineensis* Jacq.) for climate change. OCL 24:D107

Salar R (2018) Thermophilic fungi basic concepts and biotechnological applications. CRC Press, Baton Rouge, LA

Samson RA, Houbraken J, Varga J, Frisvad JC (2009) Polyphasic taxonomy of the heat resistant ascomycete genus *Byssochlamys* and its *Paecilomyces* anamorphs. Persoonia 22:14–27

Shaw J (2017) The role of biodiversity centres in bioprospecting: a case study from Sarawak. In: Paterson RRM, Lima N (eds) Bioprospecting. Success, potential and constraints. Springer, Cham, pp 295–298

Singh B, Poças-Fonseca MJ, Johri BN, Satyanarayana T (2016) Thermophilic molds: biology and applications. Crit Rev Microbiol 42:985–1006

Stępień L, Chełkowski J (2008) *Fusarium* head blight of wheat: pathogenic species and their mycotoxins. World Mycotox J 156:129–139

Stierle AA, Stierle DB (2017) Secondary metabolites of mine waste acidophilic fungi. In: Paterson RRM, Lima N (eds) Bioprospecting. Success, potential and constraints. Springer, Cham, pp 213–243

Treu R (1998) Macrofungi in oil palm plantations of South East Asia. Mycologia 12:10–14

Waalwijk C, van der Lee T, de Vries I, Hesselink T, Arts J, Kema GHJ (2004) Synteny in toxigenic *Fusarium* species: the fumonisin gene cluster and the mating type region as examples. Eur J Plant Pathol 110:533–544

Wang DM, Wu SH, Yao YJ (2014) Clarification of the concept of *Ganoderma orbiforme* with high morphological plasticity. PLoS One 9:1–12

Ward T, Clear RM, Rooney AP, O'Donnell K, Gaba D, Patrick S, Starkey DE, Gilbert J, Geiser DM, Nowicki TW (2008) An adaptive evolutionary shift in *Fusarium* head blight pathogen populations is driving the rapid spread of more toxigenic *Fusarium graminearum* in North America. Fungal Genet Biol 45:473–484

Wright SAI (2015) Patulin in food. Curr Opin Food Sci 5:105–109

Zhang X, Guo Y, Ma Y, Chai Y, Li Y (2016) Biodegradation of patulin by a *Byssochlamys nivea* strain. Food Control 64:142–150

Zhou L, Cao Y, Wu S, Vlasák J, Li D, Li M, Dai Y (2015) Global diversity of the *Ganoderma lucidum* complex (Ganodermataceae, Polyporales) inferred from morphology and multilocus phylogeny. Phytochemistry 114:7–15

Zhu R, Feussner K, Wu T, Yan F, Karlovsky P, Zheng X (2015) Detoxification of mycotoxin patulin by the yeast *Rhodosporidium paludigenum*. Food Chem 179:1–5

Chapter 6
Soil Microfungi of Israeli Deserts: Adaptations to Environmental Stress

Isabella Grishkan ⓘD

6.1 Introduction

It is now well established that the development and functioning of soil microbial communities are strongly affected by environmental factors at different geographic scales (Caruso et al. 2011; Fierer et al. 2009; Makhalanyane et al. 2015). For that reason, diversity and distribution of soil fungi in stressful habitats is an important topic to study because it can shed light on the mechanisms of survival and adaptation of microorganisms in extreme environmental conditions. Deserts represent such stressful habitats where severe climate and limited resources greatly influence biota formation (Sterflinger et al. 2012).

Numerous mycological studies have been conducted in desert soils all over the world (Abdullah et al. 1986; Bates et al. 2012; Ciccarone and Rambelli 1998; Conley et al. 2006; Grishkan and Nevo 2010a; Halwagy et al. 1982; Mouchacca 1993; Mulder and El-Hendawy 1999; Ranzoni 1968; Romero-Olivares et al. 2013; Zhang et al. 2016). It has been shown that both taxonomic and functional diversities of soil fungi in the arid zone are highly dependent on water availability, temperature regime, and organic matter content (Zak 2005 and references therein). Because of high spatiotemporal heterogeneity in resource availability, fungal species richness in the desert soils may be greater than expected based solely on consideration of abiotic conditions (Wicklow 1981; Zak et al. 1995).

Deserts cover more than 60% of Israeli territory (Atlas of Israel 1985). Such massive distribution of the desert area in the country offers an excellent opportunity to study soil fungi and their adaptive strategies on a broad environmental scale: from the semiarid to extremely arid regions, with annual rainfall ranging from 300 mm to 25 mm, respectively. The gradient also covers a range of altitude and

I. Grishkan (✉)
Institute of Evolution, University of Haifa, Haifa, Israel
e-mail: bella@evo.haifa.ac.il

© Springer Nature Switzerland AG 2019   97
S. M. Tiquia-Arashiro, M. Grube (eds.), *Fungi in Extreme Environments:*
Ecological Role and Biotechnological Significance,
https://doi.org/10.1007/978-3-030-19030-9_6

vegetation diversity, and such combinations of different climatic, microclimatic, and edaphic factors can principally govern the structure of soil microfungal communities (Grishkan and Nevo 2010b).

6.2   Site Description and Edaphic Characteristics

Our most comprehensive studies devoted to desert soil mycobiota were conducted in three regions of the Negev Desert—Nahal (Wadi) Shaharut (southern Negev), Makhtesh Ramon (central Negev), and Wadi Nizzana (western Negev)—as well as along the altitudinal and latitudinal gradients of the Arava Valley (Fig. 6.1).

Fig. 6.1 Map of the southern part of Israel showing the locations of study sites at the Negev Desert and the Arava Valley

The main Israeli desert, Negev, occupies at least 55% of the country at its southern part. The world "*negev*" means "dryness" in the original Hebrew connoting also "south" (Hillel 1982). In the arid regions of the Negev, annual rainfall ranges from more than 200 mm in the north of the central part to about 30 mm in the extreme south (Bitan and Rubin 1991). The distribution of rainfall within the rainy season (November–April) is highly irregular with wide fluctuations from year to year. Most of the Negev area consists of marine sedimentary rocks, mainly limestones and chalks. The landscape is represented by rocky hills and mountains, wadis, plateaus, coarse sediments, and sands. The typical soil of the Negev is loess—a buff-colored, fine-granulated, wind-borne shallow deposit of desert dust (Singer 2007).

Wadi Shaharut is located in the extremely arid region of southern Negev, with an average annual rainfall less than 40 mm (Bitan and Rubin 1991). The mean daily temperature of the coldest month (January) is 9–11 °C and of the warmest months (July–August) is 36–38 °C. The wadi has two opposite slopes—south facing (SFS, dips about 35°) and north facing (NFS, dips about 30°)—separated by about 150 m at the valley bottom (VB). The slopes consist of limestone and are characterized by various sediments, mainly stony debris (Dan et al. 1976). Very sparse dwarf shrubs (*Zygophyllum dumosum* and *Reaumuria hirtella*) cover the middle and upper parts of the NFS. In the VB, occupied by coarse desert alluvium soil, sparse shrub vegetation is accompanied by *Tamarix* sp. trees.

Makhtesh Ramon (MR) is a unique geological feature located at the southern boundary of the Negev Highlands. This landform is the world's largest erosion cirque or makhtesh (the Hebrew word for "mortar" or "crater") (Garfunkel 1978). The cirque is 40 km long and 2–10 km wide and is extended in north-eastern–southwestern direction, with rims about 900–1000 m above sea level. The climate in the region is extremely arid, with only sporadic rainfall from 85 mm per year on the northern rim to 56 mm per year in the central cirque (Ward et al. 2000). The mean temperatures of the coldest month (January) and warmest month (July) are 8–13 °C and 24–31 °C, respectively (Atlas of Israel 1985). Desert lithosols on sandstone, hard limestone, and chalk are characteristic of the area (Dan et al. 1976); on the north-facing slope of the cirque southern rim (NFS), the sand is enriched by ferric and manganese compounds giving the surface a reddish-brown tincture.

The MR area is considered a transition from the Irano-Turanian phytogeographic region (steppe vegetation) in the north to the Saharo-Arabian region (true desert) in the south (Zohary 1962). Adjacent to the northern part of the cirque steppe zone is covered mainly by the dwarf shrub *Artemisia herba-alba*. The outside southern south exposed desert zone is occupied by sparse shrub vegetation (mainly *Z. dumosum*, *Capparis aegyptia*, and *Haloxylon persicum*) with occasional *Tamarix* trees. A south-facing slope of the northern rim of the cirque (SFS) is covered by very sparse *Z. dumosum* and the geophyte *Bellevalia desertorum*. On the NFS, vegetation is entirely absent.

Wadi Nizzana is located within the Hallamish dune field at western Negev. Long-term mean annual precipitation is about 95 mm, and annual potential evaporation is ~2600 mm (Evenari 1981). The mean temperatures are near 27 °C during the hottest month of July and 12 °C during the coldest month of January (Rosenan and Gilad

1985).Most of the dunes have a mobile crest covered by very sparse vegetation (<5%). The lower flanks of the dunes and the sandy interdunes have 10–20% vegetation cover (Kadmon and Leschner 1994) and are almost entirely covered by biological soil crusts (BSC) represented by four types of cyanobacterial crusts and one type of moss-dominated crust. Significant differences characterized their species composition, biochemical and physical parameters, and wetness duration (Kidron et al. 2010). Within the interdunes, fine-grained flat patches (usually less than 50×50 m) are scattered. These sediments, known as playas, are comprised of high (50–80%) amounts of fines (silt and clay), which leads to low porosity (Blume et al. 1995), and therefore to low water infiltration, promoting runoff generation (e.g., Kidron and Vonshak 2012). Because of low infiltration, the playa surfaces experienced high evaporation rates which results, in turn, in an almost absolute absence of plants and in high accumulation of salts at a relatively low depth of >10 cm (Blume et al. 1995; Yu and Steinberger 2012). The playa surfaces mostly lack BSC (Kidron and Vonshak 2012), and only ~20–25% of the surfaces which have slightly concave contours and therefore receive lower amounts of runoff are covered by 0.5-cm-thick crusts.

Another Israeli desert, Arava, is located in the southeastern part of the country and belongs to the longitudinal Syrian-African Rift Valley. This desert, which is more than 160 km in length, extends from the Dead Sea in the north to the Gulf of Eilat (Gulf of Aqaba) in the south. The elevation of the valley varies from about 400 m below sea level at the Dead Sea area to 210 m a.s.l. in the region of Arvat Yafruq, in the center of the valley, and then decreases southward to sea level at Eilat (the Red Sea) (Goldreich and Karni 2001). The climate of the Arava region is defined as hyperarid (Goldreich and Karni 2001), with an annual precipitation of 15–50 mm (Ginat et al. 2011). Mean annual maximal temperature during the hottest month of July is 39.5 °C, and mean annual minimal temperatures during the coldest month of January are 7.8 °C and 13.6 °C in Sapir (near the center of the valley) and Sedom (near the Dead Sea), respectively (Goldreich and Karni 2001). Annual potential evaporation is ~3200 mm. The major part of the Arava Valley floor consists of stony coarse alluvial soil, mainly on gypseous chalk formations (Singer 2007). The very sparse vegetation of the Arava Valley is represented by dwarf shrubs such as *Nitraria retusa, Traganum nudatum, Haloxylon salicornicum, Anabasis articulata*, and *Z. dumosum* accompanied by rare *Acacia* trees (Danin 1983).

Some edaphic parameters of the studied soils in the central, southern Negev, and the Arava Valley are presented in Table 6.1. They provide insight into hostility and stressfulness of these desert environments for living organisms as a whole and fungi in particular. In the majority of localities, gravimetric moisture content was very low even in winter; in Wadi Shaharut, the driest periods were found not only in the summer, but also in the winter, due to strong windy weather. Summer temperatures in the uppermost soil layers (0–1 cm) of the open sun-exposed localities during the afternoon hours exceeded 50 °C. The soils in the studied areas are slightly alkaline, with pH in most cases lower in shrubby as compared to open localities and with opposite variations in low organic matter content.

Table 6.1 Selected edaphic parameters (mean ± SD) in different sites of the Negev Desert and Arava Valley

Locality	Moisture content, % (n = 6) Winter	Summer	Temperature, °C (n = 6) Winter	Summer	pH (n = 6)	Organic matter, % (n = 3)
Makhtesh Ramon area, central Negev						
Steppe, open	5.9 ± 2.7	1.2 ± 0.1	18.4 ± 1.0[a]	48.6 ± 2.3	8.4 ± 0.2	1.06 ± 0.32
Steppe, shrubs	5.8 ± 3.8	1.4 ± 0.2	16.3 ± 0.9	36.8 ± 2.0	8.1 ± 0.4	1.66 ± 0.32
Cirque, SFS	5.4 ± 1.6	1.4 ± 0.3	27.7 ± 3.8	50.1 ± 2.2	8.7 ± 0.4	0.40 ± 0.01
Cirque, NFS	1.1 ± 0.3	0.7 ± 0.2	21.2 ± 1.8	41.3 ± 1.6	8.5 ± 0.1	0.14 ± 0.06
Desert, open	1.8 ± 0.8	0.8 ± 0.1	22.9 ± 2.6	37.3 ± 2.5	8.0 ± 0.1	0.36 ± 0.25
Desert, shrubs	4.3 ± 2.9	2.1 ± 0.7	17.4 ± 2.1	33.8 ± 3.3	7.9 ± 0.2	4.24 ± 4.03
Wadi Shaharut, southern Negev						
SFS, open	0.5 ± 0.3	0.6 ± 0.1	18.3 ± 0.8	50.3 ± 0.9	7.6 ± 0.3	0.16 ± 0.02
SFS, stones	0.7 ± 0.2	0.6 ± 0.1	15.4 ± 0.9	36.8 ± 1.3	7.6 ± 0.1	0.21 ± 0.02
NFS, open	0.7 ± 0.2	1.1 ± 0.2	14.3 ± 0.4	37.1 ± 3.5	8.2 ± 0.2	0.21 ± 0.04
NFS, shrubs	1.3 ± 0.7	1.4 ± 0.2	12.7 ± 0.3	29.7 ± 0.7	7.2 ± 0.1	2.23 ± 0.33
VB, open	0.4 ± 0.1	0.8 ± 0.1	16.3 ± 0.3	54.4 ± 1.3	8.3 ± 0.1	0.43 ± 0.29
VB, shrubs	0.8 ± 0.2	0.7 ± 0.1	14.1 ± 0.8	35.5 ± 0.5	7.8 ± 0.2	1.68 ± 0.24
Arava Valley						
South, open	0.7 ± 0.3	0.3 ± 0.1	23.8 ± 2.4[b]	48.9 ± 2.9	7.4 ± 0.1	0.16 ± 0.02
South, shrubs	0.8 ± 0.2	0.4 ± 0.1	16.8 ± 1.4	42.0 ± 2.7	7.5 ± 0.06	0.73 ± 0.1
Center, open	1.2 ± 0.2	0.6 ± 0.2	25.5 ± 1.0	52.5 ± 1.4	7.6 ± 0.15	0.37 ± 0.18
Center, shrubs	1.6 ± 0.6	0.7 ± 0.3	22.1 ± 0.7	43.2 ± 2.1	7.8 ± 0.2	1.91 ± 0.96
North, open	0.5 ± 0.1	0.3 ± 0.05	22.8 ± 1.0	53.4 ± 2.3	7.6 ± 0.1	0.37 ± 0.14
North, shrubs	0.4 ± 0.05	0.3 ± 0.1	20.5 ± 0.6	42.3 ± 2.2	7.7 ± 0.2	0.74 ± 0.21

[a]Measured during the hours of 3–4 p.m. (steppe), 1–2 p.m. (cirque, SFS), 11–12 a.m. (cirque, NFS), and 8–9 a.m. (desert) (wintertime and summertime for the respective seasons)
[b]Measured during the hours of 9–10 a.m. (south), 12 a.m.–1 p.m. (center), and 2–3 p.m. (north)

In the crusted sandy profiles at Wadi Nizzana, alkaline pH gradually increased with depth, while at the playa, it gradually decreased through the profiles (Table 6.2). Contrarily to the sandy habitats, substantial increase with depth in the values of electric conductivity characterized the playa profiles attesting for high salinity and limited water infiltration. As for organic matter content (OM), except for the sandy BSC, all other values were very low, less than 0.3%. Notably, a small peak of OM was registered at 10–20 cm of the sandy sediments which is probably a result of root concentration (mainly of annuals) at this depth. Even following medium rain events

Table 6.2 Distribution of selected edaphic parameters through crusted sandy and playa profiles at Wadi Nizzana, western Negev (CBx and CBm, cyanobacterial crusts, xeric and mesic section, respectively; MD, moss-dominated crusts; P-cr and P-nc, crusted and non-crusted playa, respectively)

Depth, cm	Organic matter content, %					Electrical conductivity, mS/cm					pH				
	CBx	CBm	MD	P-cr	P-nc	CBx	CBm	MD	P-cr	P-nc	CBx	CBm	MD	P-cr	P-nc
0–0.2	0.76	1.31	2.5	0.15	0.10	0.59	0.84	1.55	0.98	1.09	8.3	8.4	8.02	9.05	8.89
0.2–1	0.15	0.3	1.85	0.04	0.11	0.41	0.81	1.05	1.02	1.05	8.6	8.7	8.7	8.9	8.49
1–5	0.13	0.24	0.29	0.08	0.08	0.22	0.52	0.34	1.96	3.37	9.05	9.1	9.06	8.95	8.98
5–10	0.20	0.10	0.34	0.13	0.09	0.48	0.46	0.36	4.61	6.5	9.09	9.12	9.17	9.19	8.58
10–20	0.23	0.17	0.44	0.18	0.14	0.46	0.34	0.26	12.5	17.07	9.15	9.25	9.16	8.39	8.34
20–30	0.17	0.13	0.09	0.17	0.13	0.47	0.38	0.28	15.3	17.45	9.02	9.37	9.18	8.19	8.21
30–40	0.15	0.08	0.10	0.20	0.19	0.42	0.39	0.3	15.5	17.31	9.26	9.31	9.25	8.37	8.18
40–50	0.06	0.10	0.09	0.14	0.16	0.42	0.39	0.35	17	17.55	9.28	9.32	9.29	8.2	8.31

(20 mm), water infiltration at non-crusted and crusted playa formations was limited to the upper 20 cm and 30 cm, respectively, whereas in all sandy habitats it reached >30 cm (Grishkan and Kidron 2016).

6.3 Sampling Design and Methodology

In the studied desert areas, the sampling was designed in order to cover maximal regional, local, and seasonal environmental variability. At Wadi Shaharut, the Makhtesh Ramon area, and along the Arava Valley, soil samples were collected from the upper soil layers of 0–2 cm from sun-exposed bare open localities and wherever it was possible—under the nearby shrub canopies. Six samples were taken from each locality at each sampling area during summer and winter; at Wadi Shaharut, samples were taken in each of four seasons.

At Wadi Nizzana, we aimed to study the vertical distribution of microfungal communities through depth in two different soil formations—crusted sandy dunes and playas. Samples were collected in summer in soil profiles from the layers of 0–0.2, 0.2–1, 1–5, 5–10, 10–20, 20–30, 30–40, and 40–50 cm. Four samples, 5 m apart, were taken in each sandy crust type (cyanobacterial and moss-dominated) and the playa (non-crusted and crusted) from each layer.

Soil microfungi were isolated using the soil dilution plate method (Davet and Rouxel 2000). This method, in spite of certain limitations and biases (e.g., Jeewon and Hyde 2007), "is simple and rapid, gives reasonable results, and yields excellent comparative data" (Bills et al. 2004, pp. 284–285), and remains a useful approach in studying the ecology of fungal communities (e.g., Schmit and Lodge 2005).

In the course of the studies, the following characteristics of microfungal communities were examined: (a) general species composition; (b) contribution of major groupings with different life-history strategies, such as mesic *Penicillium* spp., thermotolerant *Aspergillus* spp. and teleomorphic Ascomycota (producing morphologically expressed sexual fruit bodies), and xeric melanin-containing microfungi, to community structure; (c) dominant groups of species in different localities; (d) diversity characteristics—species richness, heterogeneity, and evenness; (e) quantitative parameter—density of microfungal isolates; and (f) the influence of main edaphic factors (soil moisture, temperature, pH, and organic matter content) on microfungal community structure and diversity level. Detailed descriptions of the isolation and identification procedure and data analyses are given in the corresponding articles (Grishkan et al. 2007; Grishkan and Nevo 2010a; Grishkan and Kidron 2016). Besides the analysis of culturable soil mycobiota in the Israeli deserts at community level, this chapter also contains information on the genetic structure of the populations of an ascomycetous fungus, *Aspergillus nidulans* in the soil of Wadi Shaharut (southern Negev) (Hosid et al. 2010a, b).

6.4   Composition and Structure of Mycobiota

The soils of Israeli deserts are inhabited by comparatively rich diversity of cultur-able microfungi—more than 420 identified species representing 135 genera. The data showed once again that, in spite of climatic hostility, desert soils which are characterized by high spatiotemporal heterogeneity in resource availability can maintain diverse microfungal biota (Wicklow 1981; Zak et al. 1995). An analogous pattern was found along the precipitation gradient in Israel for another group of soil microorganisms, bacteria, which the overall taxonomic diversity at the species level in the arid site was high and similar to that found in the humid Mediterranean, Mediterranean, and semiarid sites (Tripathi et al. 2017). The following features characterized the structure of microfungal communities in the soil of Israeli deserts.

6.4.1   Dominance of Melanin-Containing Fungi

Expectedly, melanin-containing species dominated the majority of topsoil micro-fungal communities in the studied desert regions (Fig. 6.2). These species are well-known stress-tolerant microorganisms resistant to solar and UV radiation, high temperature, desiccation, oligotrophic conditions, and chemical and radioactive pollution (Grishkan 2010 and references therein). A dominance of dark-colored microfungi was found in almost all mycologically studied desert soils (e.g., Abdullah et al. 1986; Christensen 1981; Ciccarone and Rambelli 1998; Grishkan et al. 2015; Halwagy et al. 1982; Mulder and El-Hendawy 1999; Ranzoni 1968; Zak 2005).

Importantly, composition of this dominant group in different localities was het-erogeneous. Melanin-containing fungi with large thick-walled and multicellular conidia dominated communities in the most microclimatically and edaphically stressful localities of the MR area (Fig. 6.2a), significantly increasing their contribu-tion in the most climatically stressful summer season, and overwhelmingly pre-vailed in all localities and in all seasons at Wadi Shaharut, southern Negev (Fig. 6.2b). These species were mainly represented by *Alternaria atra, A. alternata*, and *A. phragmospora*, accompanied by *A. chlamydospora* (Wadi Shaharut) and *Monodyctis fluctuata* (the Arava Valley) (Fig. 6.2). Such species (from the genera *Alternaria* and *Ulocladium*) were the most widespread in the soils of the Atacama Desert, which is known as one of the driest locations on Earth (Conley et al. 2006). Likewise, these fungi, belonging to the order Pleosporales, were found to predomi-nate in the biological soil crusts studied by the culture-independent molecular approaches at the Colorado Plateau, in the semiarid grassland in central New Mexico, and in the southwestern deserts of the USA (Bates et al. 2010, 2012; Porras-Alfaro et al. 2011) as well as in the crusted sand of the Tengger Desert, China (Zhang et al. 2016).

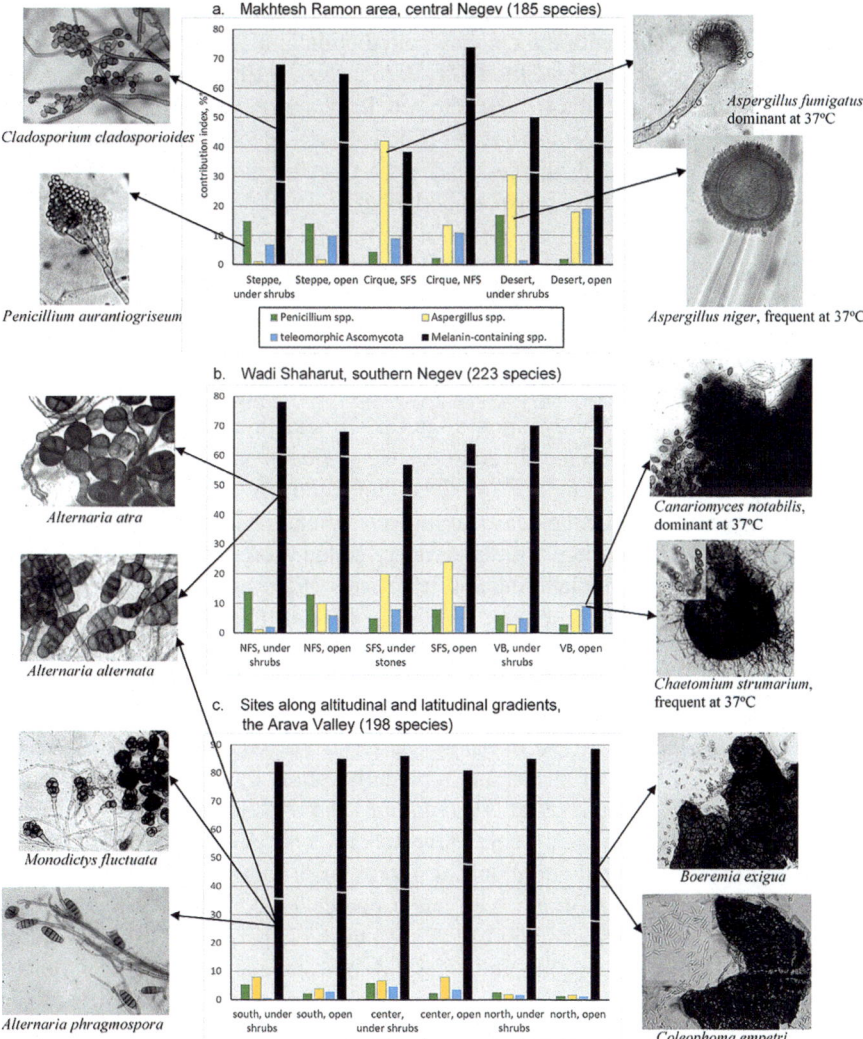

Fig. 6.2 Contribution of main microfungal groupings to local communities in the soil of different sites of the Negev Desert and the Arava Valley, Israel (the area below the white line on the bars of melanin-containing fungi indicates contributions of species with large multicellular spores)

The multicellular spore morphology should be considered an important adaptive feature of desert soil mycobiota and is similar to one of the survival adaptations of annual plants in the desert. Gutterman (2002) reported on two main adaptive strategies in seed dispersal of desert annuals: mass production of tiny dustlike seeds (escape strategy) and production of relatively large lignified seeds in smaller numbers (protection strategy). The main core of soil microfungi in most Negev and Arava localities follows the protection adaptive strategy possessing an ability to successfully survive under extreme temperatures, drought, and UV radiation. In nature, isolated species with dark, many-celled conidia also inhabit stressful and highly fluctuable plant surface (Ellis 1971, 1976; Ellis and Ellis 1997). In laboratory conditions, Durrell and Shields (1960) showed that the survival time of thick-walled multicellular conidia of *Stemphylium ilicis* under the same irradiation was 30-fold longer than for thin-walled one-celled conidia of *Aspergillus niger*. Thus, dark-colored, many-celled conidia carry out both dispersal and resting functions, which are vital in climatically and microclimatically stressful desert habitats.

By contrary, in the shrub localities of the Makhtesh Ramon area (central Negev), species with one-celled conidia prevailed (such as *Cladosporium cladosporioides* in the steppe communities) (Fig. 6.2a). Similar prevalence of dark-colored species with comparatively small one-celled conidia was characteristic of the SFS open localities of the northern Wadi Keziv, Upper Galilei (Grishkan et al. 2003a).

In the soils of Arava Valley, another group of melanin-containing species frequently and abundantly occurred. These species—*Boeremia exigua*, *Coleophoma empetri*, and *Phoma medicaginis*—produce comparatively small (5–15-μm-long) light-colored conidia inside the dense multilayered picnidial fruit bodies (Fig. 6.2c). Similar to the melanized fungi with large many-celled spores, these picnidial fungi are known also as phylloplane-inhabiting species (Ellis and Ellis 1997; Sutton 1980), and thick-walled dark-brown or black spherical fruit bodies provide them with protection against extremely stressful and highly fluctuating environmental conditions.

Picnidial species significantly increased their contribution at the northern part of the Arava Valley located at 190 m below sea level. The increase in abundance of picnidial fungi in the soil of this area is apparently caused by the weakening of abiotic stress (decrease in the level of UV radiation) and the consequential weakening in the competition with melanin-containing species with large many-celled spores dominating the microfungal communities in proximate areas. A similar prevalence of other dark-colored microfungi with small one-celled conidia, *A. niger* and *C. cladosporioides*, was characteristic of the sand of the hypersaline Dead Sea shore (Grishkan et al. 2003b)—an extremely stressful environment but receiving very low UV radiation because of its location at more than 400 m below sea level.

Notably, in soil profiles of Wadi Nizzana (western Negev), melanized fungi with large multicellular conidia prevailing at the uppermost layers in both the crusted sandy and playa habitats substantially contributed also to the communities at 10–50 cm of the non-crusted playa profiles, as well as in the deepest layers of the crusted playa profiles (Fig. 6.3). It indicates that melanin which is known to protect fungal cells from various kinds of stresses, together with many-celled spore morphology, can help microfungal species also to survive, although in very low amounts, under severe stress of strongly limited aeration and high salinity characteristic of playa formations.

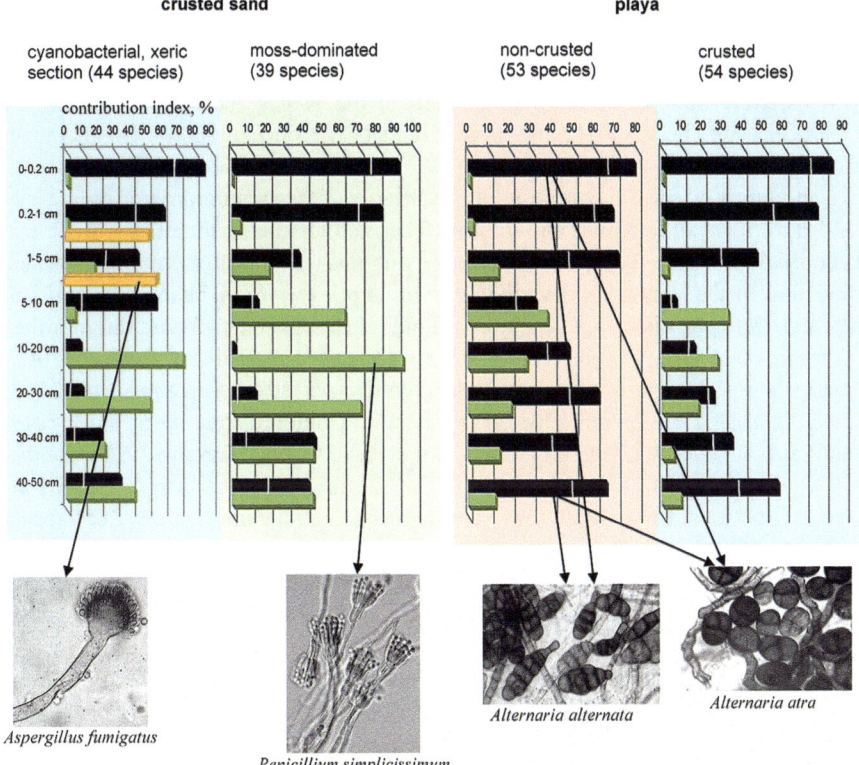

Fig. 6.3 Vertical dynamics of contribution of main microfungal groupings to communities in different sandy and playa profiles at Wadi Nizzana, western Negev (melanin-containing species, *Penicillium* spp., *Aspergillus* spp.; the area left from the white line on the bars of melanin-containing species indicates contribution of species with large multicellular spores)

6.4.2 *Different Distributional Patterns of* Aspergillus *and* Penicillium

According to the number of species, *Penicillium* and *Aspergillus* are the richest genera in Mycota—354 species (Visagie et al. 2014) and 339 species (Samson et al. 2014), correspondingly. They are also known as two main soil genera (e.g., Domsch et al. 2007) and were numerously represented in the soil of Israeli deserts—43 and 39 species of penicillii and aspergilli, respectively. *Penicillium* and *Aspergillus* are taxonomic relatives belonging to the same family—Trichocomaceae—but they display different, even opposite, life-history strategies and geographical trends: mesophilic *Penicillium* is characteristic of cool-temperate mycobiotas, while *Aspergillus*, consisting of many thermotolerant and thermophilic and osmotolerant and osmophilic species, is more widely distributed in warm xeric regions (e.g., Christensen 1981; Klich 2002; Domsch et al. 2007). Likewise, in the soil of Israeli deserts, these genera displayed opposite spatial and seasonal distributional tendencies. *Penicillium* spp. formed a minor group in all topsoil communities studied being remarkably more abundant in the soil under shrubs and correlating significantly and positively with moisture and organic matter content in the Negev localities, whereas for *Aspergillus* spp. these correlations were reverse (Grishkan et al. 2007; Grishkan and Nevo 2010a).

Importantly, aspergilli composed the basic core of thermotolerant mycobiota isolated at 37 °C in all localities and seasons. Thermotolerant communities were dominated by *A. fumigatus*; *A. niger* was the second most abundant *Aspergillus* species prevailing in the winter thermotolerant communities at Wadi Shaharut (Grishkan 2018). *A. fumigatus* is known as one of the most frequent and abundant thermotolerant species in a variety of desert regions (e.g., Abdullah et al. 1986; Bokhary 1998; Christensen 1981; Halwagy et al. 1982; Mouchacca 1993; Oliveira et al. 2013; Powell et al. 2012; Ranzoni 1968). This species also dominated microfungal communities isolated at 45 °C from the soil of the Arava Valley. Moreover, *A. fumigatus* overwhelmingly prevailed in the mesophilic communities of the extremely hot sunexposed SFS localities in Makhtesh Ramon (summer and winter) and Wadi Shaharut (only summer) (Fig. 6.2).

A different picture was observed in the depth-wise distribution of the two main soil genera at Wadi Nizzana (Grishkan and Kidron 2016). *A. fumigatus* prevailed only at 1–10 cm depth of the most xeric section of the cyanobacterial crusts, while the middle layers of all studied sandy crust profiles (10–20 cm) harbored microfungal communities overwhelmingly dominated by one or two *Penicillium* species (Fig. 6.3). The increase in abundance of penicillii might be associated with the penetration of massively produced very small fungal spores during water infiltration and their deposition mainly at 10–20 cm depth. It is likely that at these depths, mesic penicillii met the appropriate abiotic conditions (lower temperatures, a small increase in organic matter content) for successful survival and competition with the stress-selected melanized microfungi prevailing in the topsoil communities.

6.4.3 Sexual Reproduction as an Adaptive Strategy of Desert Mycobiota

In the soils of Israeli deserts, ascomycetes with morphologically expressed sexual stage comprised the one fourth of the species. A significant part of teleomorphic Ascomycota belonged to the genus *Chaetomium*—28 species. This genus is known as cellulolytic (e.g., Domsch et al. 2007), but is also considered characteristic of desert soils (Christensen 1981) and was abundantly recorded in the Dead Sea coastal sand (11 species, Grishkan et al. 2003b). Our study on the relationship between ecological stress and sex evolution in soil microfungi (Grishkan et al. 2003c) showed a highly significant increase in the proportion of sexuals in mycobiota with an increasing salinity/aridity stress southward in Israel. This trend was explained by the high adaptive plasticity of perfect fungi in a highly stressful environment, which is associated with sexual reproduction. Importantly, most of the isolated ascomycetes had thick-walled, dark brown or black perithecia, and almost all of them also produced large (10 μ and more) dark-colored ascospores, thus displaying morphological adaptations to stressful environmental conditions. Among the surveyed sites, the proportion of teleomorphic ascomycetes climaxed in the highly stressful Dead Sea coastal area (33% species, 18.5% isolates—Grishkan et al. 2003b). In the desert soils, the contribution of sexual species was less significant, but together with aspergilli, they formed the main part of the thermotolerant mycobiota—54 teleomorphic species were isolated at 37 °C, and 35 of them only at 37 °C. Apparently, melanin-containing species with protective many-celled spore morphology that dominated mesophilic microfungal communities cannot develop at high temperatures because none of their isolates were grown in the laboratory at 37 °C. At the same time, thermotolerant teleomorphic species led by *Canariomyces notabilis*, *Chaetomium strumarium*, and *Ch. nigricolor* could not only survive but also germinate and be active during a long period of high temperatures in deserts.

6.5 Density of Isolates as a Quantitative Characteristic of Microfungal Communities

Expectedly, the density of microfungal isolates (expressed in colony-forming units—CFU—per gram of dry soil) in all studied locations exposed a highly significant positive correlation with organic matter content displaying maximum values in shrub localities (Fig. 6.4a–c). Seasonally, summer was the least abundant season on microfungal density, except in the NFS locality in Makhtesh Ramon and the open SFS locality in Wadi Shaharut. In these most edaphically stressful environments, the summer increase in CFU number was caused by a more abundant development of xeric melanin-containing *Alternaria* species (Makhtesh Ramon) or thermotolerant *A. fumigatus* (Wadi Shaharut), which comprised, respectively, more than 70% and nearly 45% of all isolates obtained from these localities in this season.

In soil profiles at Wadi Nizzana, density of microfungal isolates abruptly decreased with depth (Fig. 6.4d). Similar reduction in fungal biomass was also found in other parts of the world (e.g., Bissett and Parkinson 1979; Rodriquez et al. 1990; Pandey et al. 1991; Fierer et al. 2003) and was attributed mainly to the depth-wise decline in carbon availability and diminishing aeration. Obviously, a sharp decline in isolate density which expressed a highly positive linear relationship with organic matter (OM) content was expected at the boundary between the OM-enriched crust and the underlying sandy soil. At the playa, the decline in the CFU numbers was apparently associated with the reduction in water availability, as found during periodical moisture measurements.

Depth-wise decrease in isolate density was much more remarkably expressed in the playa soils (and especially in the non-crusted playa) than in the sandy profiles. Likewise, the CO_2 evolution rate and microbial biomass were lower in the playa

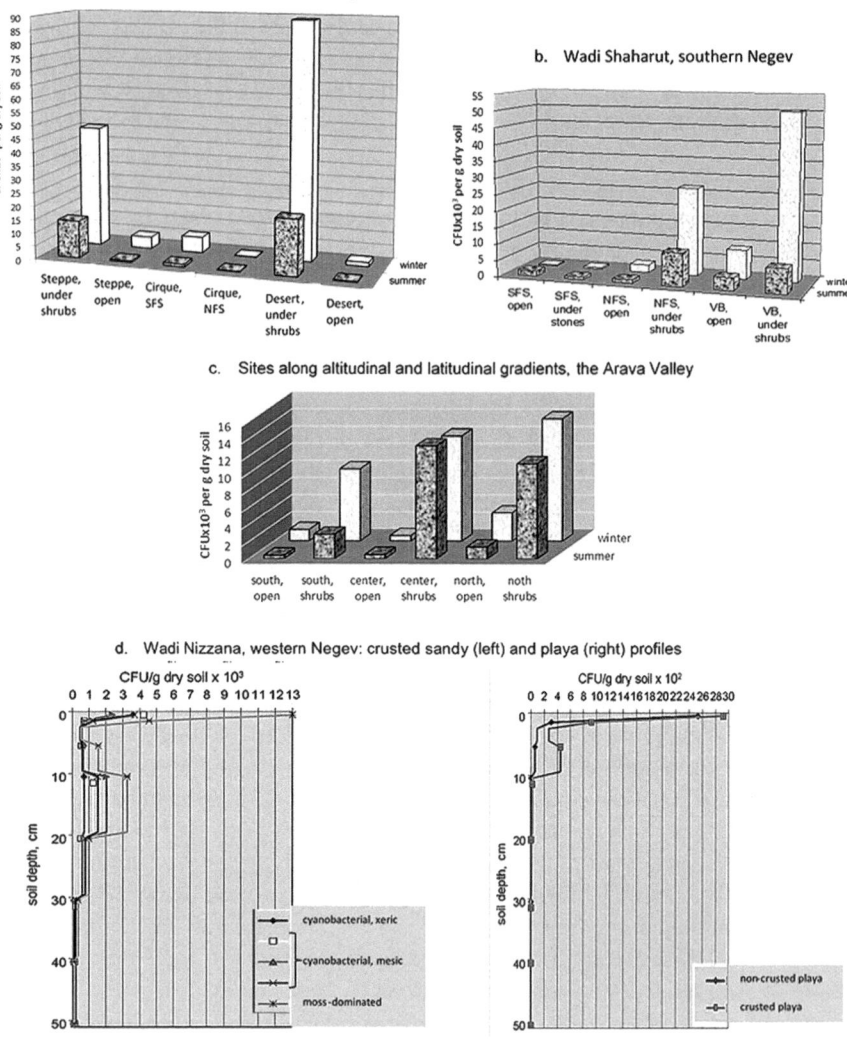

Fig. 6.4 Spatial and seasonal dynamics of density of microfungal isolates (in CFU—colony-forming units) in the soil of different localities at the Negev Desert and the Arava Valley

(Yu and Steinberger 2012). Apparently, diminishing water infiltration and aeration as well as high salt accumulation in the playa soil create harsh environmental conditions for survival and development of fungi as well as for their passive migration by water (Trevors et al. 1990; Breitenbeck et al. 1988; Abu-Ashour et al. 1994). This may explain the fact that the crusted playa profiles, which occupy slightly convex surfaces and therefore facilitate deeper infiltration, harbor 1.5-fold more microfungal isolates at the 5–10-cm layer than the non-crusted playa habitats at the 0.2–1-cm layer (Fig. 6.4d).

Density of microfungal isolates is considered a non-reliable quantitative characteristic because of the impossibility to distinguish between colonies arising from either active mycelium or resting spores (Parkinson 1994). However, sharp spatial and seasonal differences in isolate densities revealed in the studied desert soils cannot be simply ignored. These differences may indirectly indicate significant variations in fungal biomass associated with spatiotemporal variability of edaphic and microclimatic conditions.

6.6 Effect of Different Environmental Aspects on the Characteristics of Microfungal Communities

The unique geographical position of the Arava Valley, with its northern part located below sea level, gave an opportunity to test the influence of different environmental aspects—locality position along altitudinal and latitudinal gradients, locality type (open and under shrubs), and season (summer and winter)—on the characteristics of microfungal communities. The analysis (Table 6.3) showed that locality type

Table 6.3 Data of two-way unbalanced ANOVA analysis (by XLSTAT, http://www.xlstat.com) for the effect of locality type, locality position, season, and interactions between them on different parameters of microfungal communities at Arava Valley

Parameter	Locality type	Locality position	Season	Locality type × locality position	Locality type × season	Locality position × season	Locality type× locality position x season
Species richness	49.8****	8.9***	NS	NS	NS	NS	NS
Shannon index	17.2****	NS	NS	3.2@	NS	NS	NS
Evenness	NS	NS	NS	5.6*	NS	NS	NS
Melanin-containing spp.	NS	NS	NS	NS	NS	7.8*	NS
Melanized spp. with multicellular spores	NS	23.3****	NS	NS	NS	11.3****	NS
Picnidial spp.	35.2****	33.4****	17.3****	5.4*	NS	NS	NS
Penicillium spp.	5.21@	NS	NS	NS	NS	NS	NS
Isolate density	141.5****	12.6****	5.0@	9.6****	NS	NS	NS

@≤0.05; *≤0.01; **≤0.005; ***≤0.001; ****≤0.0001

strongly influenced most measured parameters of microfungal communities followed by locality position along altitudinal and latitudinal gradients. The latter aspect highly significantly affected the abundance of melanin-containing fungi with large multicellular spores which lost their dominant position in the area located 190 m below sea level to the species with picnidial fruit bodies. Importantly, cluster analysis based on the relative abundance of species also revealed that, in the majority of cases, the communities from the same locality type (open or under shrubs) were more similar to each other than the communities from different localities at the same sampling area. A similar pattern was revealed in our study devoted to soil microfungi along the precipitation gradient in northern and central Negev (Grishkan et al. 2006). It indicated that, in most cases, the crusted and shrub localities, separated only by a few meters or less, differed in microfungal community structure much more significantly than crusted or shrub localities at a distance of tens of kilometers. This observation again confirms the fact that microclimatic and edaphic factors play an essential role in the development of soil microfungal communities, and their composition and structure can be a sensitive indicator of changing environmental conditions at a microscale.

6.7 Genetic Divergence and Mode of Reproduction in Populations of *Aspergillus nidulans* from the Soil of Wadi Shaharut, Southern Negev

Aspergillus nidulans is a fungus easily generating in culture with two morphologically distinct kinds of sporulation: sexual (teleomorphic state, producing ascospores) and asexual (anamorphic state, producing conidia). The genetic divergence of *A. nidulans* was studied on regional and local scales using 15 microsatellite (SSR) markers (Hosid et al. 2010a). Three populations of the fungus isolated from the soil of the northern Mediterranean and the desert canyons were found to be genetically distinct. The estimated genetic divergences corresponded to geographical distances and ecological differences between the canyons (Fig. 6.5). On a regional scale, SSR polymorphism tended to increase with severity of ecological conditions being maximal in the desert population of the fungus.

Testing the reproductive structure of the populations of *A. nidulans* indicated the presence of sexuality in the northern populations and predominant clonality in the desert population (Hosid et al. 2010b). The predominantly clonal character of the desert population of the fungus was explained by the assumption that for relevant multilocus systems of a fungus, only several haplotypes could survive in the rather constant, extremely stressful desert conditions. Additionally, the very low density of *A. nidulans* populations in the soil of Wadi Shaharut, which reduced the probability of finding a sexual partner, might favor predominant clonality via selfing.

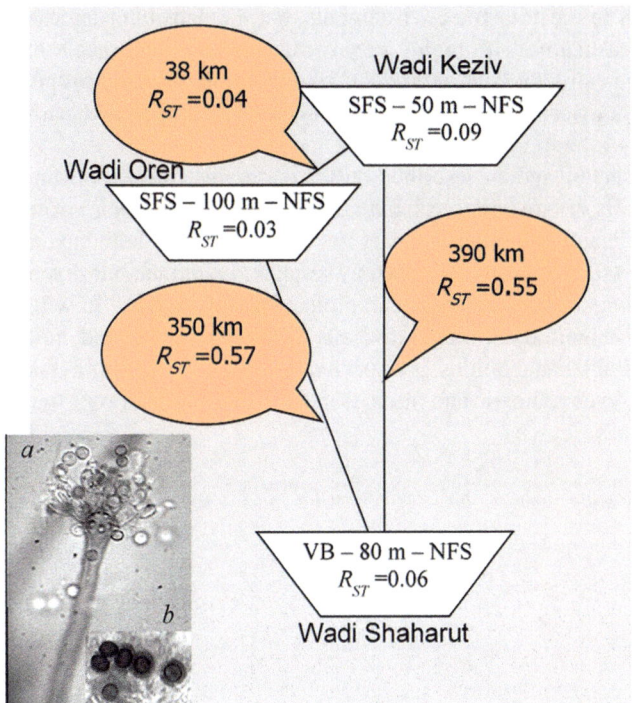

Fig. 6.5 Genetic distances between populations and subpopulations of *Aspergillus nidulans* (*a*, asexual conidial sporulation; *b*, sexual ascospores) from the northern Mediterranean (Wadi Oren and Wadi Keziv) and Israeli desert (Wadi Shaharut) regions (adopted from Hosid et al. 2010a)

6.8   Conclusion

This study showed that comparatively rich soil mycobiota of the Israeli deserts has developed under highly stressful climatic and edaphic conditions. The mycobiota displayed remarkable adaptive strategies to harsh desert stresses reflected in diverse phenotypic and biological traits: (a) melanin-containing fungi with large, thick-walled, and multicellular conidia dominated the majority of topsoil communities and lost their dominant position either to the species with picnidial fruit bodies in the less UV-radiated area located 190 m below sea level or to thermophilic *A. fumigatus* in the extremely hot localities (e.g., SFS of Makhtesh Ramon); (b) melanized species with protective spore morphology prevailed also in the deep layers of bare playa profiles characterized by high salinity and strongly limited water infiltration; (c) mesic *Penicillium* spp. dominated in the middle depths of sandy and playa profiles due to the ability of their abundantly produced small spores to penetrate during water infiltration; (d) aspergilli (mainly *A. fumigatus*) and teleomorphic perithecial ascomycetes comprised a basic part of thermotolerant mycobiota; (e) relatively small spatiotemporal variations characterized the structure of microfungal communities in most hostile environments, but the density of microfungal isolates fluctuated drastically, with highly positive dependence on organic matter content; and (f) genetic structure and reproductive mode of the *A. nidulans* population from southern Negev were substantially distinct as compared to the northern Mediterranean populations of the fungus.

The conducted studies described only a culturable fraction of fungal communities in the soils of Israeli deserts, but nevertheless, this fraction is known to constitute a significant and essential part of soil mycobiota both taxonomically and functionally (see Domsch et al. 2007). We plan to continue our mycological investigations using culture-independent molecular approaches. It will substantially increase the magnitude of fungal diversity in the desert soils and enrich our knowledge about fungal adaptations and survival strategies in such extreme environment with highly severe climatic and microclimatic conditions and very limited availability of nutrient resources.

Acknowledgements I thank the Israeli Ministry of Absorption for financial support of this research.

References

Abdullah SK, Al-Khesraji TO, Al-Edany TY (1986) Soil mycoflora of the Southern Desert of Iraq. Sydowia 39:8–16

Abu-Ashour J, Joy DM, Lee H, Whiteley HR, Zelin S (1994) Transport of microorganisms through soil. Water Air Soil Pollut 75:41–158

Atlas of Israel (1985) 3rd ed. Surveys of Israel, Tel Aviv

Bates ST, Nash THIII, Sweat KG, Garcia-Pichel F (2010) Fungal communities of lichen-dominated biological soil crusts: diversity, relative microbial biomass, and their relationship to disturbance and crust cover. J Arid Environ 74:1192–1199

Bates ST, Nash THIII, Garcia-Pichel F (2012) Patterns of diversity for fungal assemblages of biological soil crusts from the southwestern United States. Mycologia 104:353–366

Bills GF, Christensen M, Powell M, Thorn G (2004) Saprobic soil microfungi. In: Mueller GM et al (eds) Biodiversity of microfungi, Inventory and Monitoring Methods. Elsevier Academic Press, San Diego, CA, pp 271–302

Bissett J, Parkinson D (1979) Fungal community structure in some alpine soils. Can J Bot 57:1630–1641

Bitan A, Rubin S (1991) Climatic atlas of Israel for physical and environmental planning and design. Ramot Publishing, Tel-Aviv

Blume HP, Yair A, Yaalon DH, Berkowicz SM (1995) An initial study of pedogenic features along a transect across longitudinal dunes and interdune areas. Nizzana region, Negev, Israel. Adv Geoecol 29:51–64

Bokhary HA (1998) Mycoflora of desert sand dunes of Riyadh Region, Saudi Arabia. J King Saud Univ 10(1):15–29

Breitenbeck GA, Yang H, Dunigan EP (1988) Water-facilitated dispersal of inoculant *Bradyrhizobia japonicum* in soils. Biol Fertil Soils 7:58–62

Caruso T, Chan Y, Lacap DC, Lau MCY, McKay CP, Pointing SP (2011) Stochastic and deterministic processes interact in the assembly of desert microbial communities on a global scale. ISME J 5:1406–1413

Christensen M (1981) Species diversity and dominance in fungal community. In: Carroll GW, Wicklow DT (eds) The fungal community, Its Organization and Role in the Ecosystem. Marcell Dekker, New York, pp 201–232

Ciccarone C, Rambelli A (1998) A study on micro-fungi in arid areas.notes on stress-tolerant fungi. Plant Biosyst 132(1):17–20

Conley CA, Ishkhanova G, McKay CP, Cullings K (2006) A preliminary survey of non-lichenized fungi cultured from the hyperarid Atacama Desert of Chile. Astrobiology 6:521–526

Dan J, Yaalon DH, Koyumdisky H, Raz H (1976) The soil of Israel. Division of Science Publisher/ TheVolcani Center, Beit Dagan

Danin A (1983) Desert vegetation of Israel and Sinai. Cana Publishing House, Jerusalem

Davet P, Rouxel F (2000) Detection and isolation of soil microfungi. Science Publisher Inc., Enfield (NH)/Plymouth

Domsch KH, Gams W, Anderson TH (2007) Compendium of soil fungi. 2nd revised ed. Academic Press, New York

Durrell LW, Shields LM (1960) Fungi isolated in culture from soils of the Nevada test site. Mycologia 52:636–641

Ellis MB (1971) Dematiaceoushyphomycetes. Commonwealth Mycological Institute, Kew, Surrey

Ellis MB (1976) More Dematiaceous Hyphomycetes. Commonwealth Mycological Institute, Kew, Surrey

Ellis MB, Ellis JP (1997) Microfungi on land plants. An identification handbook. Richmond Publishing Co. Ltd., Slough

Evenari ML (1981) Ecology of the Negev Desert, a critical review of our knowledge. In: Shuval H (ed) Development in arid zone, ecology and environmental quality. Balban, Philadelphia, PA, pp 1–33

Fierer N, Schimel JP, Holden PA (2003) Variations in microbial community composition through two soil depth profiles. Soil Biol Biochem 35:167–176

Fierer N, Strickland MS, Liptzin D, Bradford MA, Cleveland CC (2009) Global patterns in below-ground communities. Ecol Lett 12:1238–1249

Garfunkel Z(1978) The Negev – regional synthesis of sedimentary basins. 10th International Congress of Sedimentology. Guidebook, Jerusalem, pp. 35–110

Ginat H, Shlomi Y, Batarseh S, Vogel J (2011) Reduction in precipitation levels in the Arava Valley (southern Israel and Jordan), 1949-2009. J Dead Sea Arava Res 1:1–7

Goldreich Y, Karni O (2001) Climate and precipitation regime in the Arava Valley, Israel. Isr J Earth Sci 50:53–59

Grishkan I (2010) Ecological stress: Melanization as a response in fungi to radiation. In: Horikoshi K et al (eds) Extremophiles handbook. Springer, Dordrecht, pp 1135–1145

Grishkan I (2018) Thermotolerantmycobiota of Israeli soils. J Basic Microbiol 58:30–40

Grishkan I, Zaady E, Nevo E (2006) Soil crust microfungi along a southward rainfall aridity gradient in the Negev desert, Israel. Eur J Soil Biol 42:33–42

Grishkan I, Kidron GJ (2016) Vertical divergence of microfungal communities through the depth in different soil formations at Nahal Nizzana, western Negev desert, Israel. Geomicrobiol J 7:564–577

Grishkan I, Nevo E (2010a) Spatiotemporal distribution of soil microfungi in the Makhtesh Ramon area, central Negev desert, Israel. Fungal Ecol 3:326–337

Grishkan I, Nevo E (2010b) Soil microfungi of the Negev Desert, Israel – adaptive strategies to climatic stress. In: Veress B, Szigethy J (eds) Horizons in earth science research, vol 1. Nova Science Publishers Inc., New York, pp 313–333

Grishkan I, Nevo E, Wasser SP, Beharav A (2003a) Adaptive spatiotemporal distribution of soil microfungi in "Evolution Canyon" II, Lower Nahal Keziv, western Upper Galilee, Israel. Biol J Linn Soc 79:527–539

Grishkan I, Nevo E, Wasser SP (2003b) Micromycete diversity in the hypersaline Dead Sea coastal area (Israel). Mycol Prog 2(1):19–28

Grishkan I, Korol AB, Nevo E, Wasser SP (2003c) Ecological stress and sex evolution in soil microfungi. Proc R SocLond B 270:13–18

Grishkan I, Beharav A, Kirzhner V, Nevo E (2007) Adaptive spatiotemporal distribution of soil microfungi in "Evolution Canyon" III, Nahal Shaharut, extreme Southern Negev desert, Israel. Biol J Linn Soc 90:263–277

Grishkan I, Jia R-L, Kidron GJ, Li X-R (2015) Cultivable microfungal communities inhabiting biological soil crusts in the Tengger Desert, China. Pedosphere 25:351–363

Gutterman Y (2002) Survival adaptations and strategies of annuals occurring in the Judean and Negev Deserts of Israel. Israel J Plant Sci 50:165–175

Halwagy R, Moustafa AF, Kamel S (1982) Ecology of the soil mycoflora in the desert of Kuwait. J Arid Environ 5:109–125

Hillel D (1982) Negev: land, water, and life in a desert environment. Praeger Publishers, New York

Hosid E, Yusim E, Grishkan I, Frenkel Z, Wasser SP, Nevo E, Korol A (2010a) Microsatellite diversity in natural populations of ascomycetous fungus, *Emericella nidulans,* from different climatic-edaphic conditions in Israel. Israel J Ecol Evol 56:119–134

Hosid E, Yusim E, Grishkan I, Frenkel Z, Wasser SP, Nevo E, Korol A (2010b) Mode of reproduction in natural populations of ascomycetous fungus, *Emericella nidulans,* from Israel. Genet Res 92:83–90

Jeewon R, Hyde KD (2007) Detection and diversity of fungi from environmental samples: traditional versus molecular approaches. In: Varma A, Oelmuller R (eds) Soil biology, Advanced Techniques in Soil Microbiology, vol 11. Springler, Berlin, Heidelberg, pp 1–15

Kadmon R, Leschner H (1994) Ecology of linear dunes: effect of surface stability on the distribution and abundance of annual plants. Adv Geoecol 28:125–143

Kidron GJ, Vonshak A (2012) The effects of heavy winter rains and rare summer rains on biological soil crusts in the Negev Desert. Catena 95:6–11

Kidron GJ, Vonshak A, Dor I, Barinova S, Abeliovich S (2010) Properties and spatial distribution of microbiotic crusts in the Negev Desert, Israel. Catena 82:92–101

Klich MA (2002) Identification of common *Aspergillus* species. Centraalbureauvoor Schimmelcultures, Utrecht

Makhalanyane TP, Valverde A, Gunnigle E, Frossard A, Ramond J-B, Cowan DA (2015) Microbial ecology of hot desert edaphic systems. FEMS Microbiol Rev 39:203–222

Mouchacca J (1993) Thermophilic fungi in desert soils: a neglected extreme environment. In: Allsopp D et al (eds) Microbial diversity and ecosystem function. CAB International, Wallingford, pp 265–288

Mulder JL, El-Hendawy H (1999) Microfungi under stress in Kuwait's coastal saline depressions. Kuwait J Sci Eng 26(1):157–172

Oliveira LG, Cavalcanti MAQ, Fernandes MJS, Lima DMM (2013) Diversity of filamentous fungi isolated from the soil in the semiarid area, Pernambuco. Braz J Arid Environ 95:49–53

Pandey RR, Chaturvedi AP, Dwivedi RS (1991) Ecology of microfungi in soil profiles of guava orchard with reference to edaphic factors. Proc Natl Acad Sci India Sect B Biol Sci 61:97–108

Parkinson D (1994) Filamentous fungi. In: Mickelson SH, Bigham JM (eds) Methods of soil analysis, part 2, microbiological and biochemical properties, SSSA Book Series, vol 5, pp 329–350

Porras-Alfaro A, Herrera J, Navig DO, Lipinski K, Sinsabaugh R (2011) Diversity and distribution of soil fungal communities in a semiarid grassland. Mycologia 103:10–21

Powell AJ, Parchert KJ, Bustamante JM, Ricken JB, Miriam I, Hutchinson MI, Natvig DO (2012) Thermophilic fungi in an arid land ecosystem. Mycologia 104:813–825

Ranzoni FV (1968) Fungi isolated in culture from soils of the Sonoran desert. Mycologia 60:356–371

Rodriquez C, Betucci L, Roquebert MF (1990) Fungal communities of volcanic ash soils along an altitudinal gradient in Mexico. II Vertical distribution. Pedobiologia 34:51–59

Romero-Olivares AL, Baptista-Rosas RC, Escalante AE, Bullock SH, Riquelme M (2013) Distribution patterns of Dikarya in arid and semiarid soils of Baja California, Mexico. Fungal Ecol 6:92–101

Rosenan N, Gilad M (1985) Atlas of Israel. Meteorological data, sheet IV/2. Carta, Jerusalem (Israel)

Samson RA, Visagie CM, Houbraken J, Hong S-B, Hubka V, Klaassen CHW, Perrone G, Seifert KA, Susca A, Tanney JB, Varga J, Kocsubé S, Szigeti G, Yaguchi T, Frisvad JC (2014) Phylogeny, identification and nomenclature of the genus *Aspergillus*. Stud Mycol 78:141–173

Schmit JP, Lodge DJ (2005) Classical methods and modern analysis for studying fungal diversity. In: Dighton J, White JF Jr, Oudemans P (eds) The fungal community, its organization and role in the ecosystem. CRC Press, Boca Raton, FL, pp 193–213

Singer A (2007) The soils of Israel. Springer, Berlin, Heidelberg

Sterflinger K, Tesei D, Zakharova K (2012) Fungi in hot and cold deserts with particular reference to microcolonial fungi. Fungal Ecol 5:453–462

Sutton BC (1980) The coelomycetes. Commonwealth Mycological Institute, Kew, Surrey

Trevors JT, van Elsas JD, van Overbeek LS, Starodub M (1990) Transport of a genetically engineered *Pseudomonas* fluorescent strain through a soil microcosm. Appl Environ Microbiol 56:401–408

Tripathi BM, Moroenyane I, Sherman C, Lee YK, Adams JM, Steinberger Y (2017) Trends in taxonomic and functional composition of soil microbiome along a precipitation gradient in Israel. Microb Ecol 74(1):168–176

Visagie CM, Houbraken J, Frisvad JC, Hong SB, Klaassen CH, Perrone G, Seifert KA, Varga J, Yaguchi T, Samson RA (2014) Identification and nomenclature of the genus *Penicillium*. Stud Mycol 78:343–371

Ward D, Saltz D, Olsvig-Whittaker L (2000) Distinguishing signal from noise: long-term studies of vegetation in Makhtesh Ramon erosion cirque, Negev desert, Israel. Plant Ecol 150:27–36

Wicklow DT (1981) Biogeography and conidial fungi. In: Cole GT, Kendrick B (eds) Biology of conidial fungi, vol 1. Academic Press, New York, pp 417–447

Yu J, Steinberger Y (2012) Spatiotemporal changes in abiotic properties, microbial CO2 evolution, and biomass in playa and crust-covered interdune soils in a sand-dune desert ecosystem. Eur J Soil Biol 50:7–14

Zak J (2005) Fungal communities of desert ecosystems: links to climate change. In: Dighton J, White JF Jr, Oudemans P (eds) The fungal community, its organization and role in the ecosystem. CRC Press, Boca Raton, FL, pp 659–681

Zak J, Sinsabaugh R, MacKay WP (1995) Windows of opportunity in desert ecosystems: their applications to fungal community development. Can J Bot 73:S1407–S1414

Zhang T, Jia R-L, Yu L-Y (2016) Diversity and distribution of soil fungal communities associated with biological soil crusts in the southeastern Tengger Desert (China) as revealed by 454 pyrosequencing. Fungal Ecol 23:156–163

Zohary M (1962) Plant life of palestine. Ronald Press, New York

Chapter 7
Extremotolerant Black Fungi from Rocks and Lichens

Claudio Gennaro Ametrano, Lucia Muggia, and Martin Grube ⓘ

7.1 Introduction

In contrast to the majority of fungi, which live more or less comfortably inside of hosts or substrates, some lineages have specialised to thrive on exposed surfaces. With little competition on nutrient-deprived conditions, such a stressful lifestyle also requires adaptations to fluctuations of hydration, among other challenges. Fungi in these situations also need to shield their cell content from excessive radiation using various kinds of pigments in their cell walls. The symbiotic thalli of lichen-forming fungi are one example on how to cope with conditions of exposed surfaces. Lichen-forming fungi develop diverse forms of characteristically compacted mycelial morphologies to filter light to sheltered photosynthetic algae. Another large group of fungi does not depend on symbiotic partners, but it is characterised by dark pigments, and unlike lichens, these fungi may tolerate much higher levels of air pollution. These 'black' fungi, as they are commonly called, are the focus of this chapter. We provide an overview of the current understanding of diversity and taxonomy of black fungi and review their phenotypic traits. As we previously recognised ecological and evolutionary links of black fungi with the lichen symbiosis, we also discuss first results from co-culture experiments of black fungi with algae. Finally, we also review first insights gained from -omics approaches.

The phenotypes of black fungi have evolved in different lineages of ascomycetes. Under the extremes of abiotic conditions, the typical morphology is usually restricted to vegetative mycelia with insufficient diagnostic characters for species

C. G. Ametrano · L. Muggia
Department of Life Science, University of Trieste, Trieste, Italy
e-mail: claudiogennaro.ametrano@phd.units.it; lmuggia@units.it

M. Grube (✉)
Institute of Biology, University of Graz, Graz, Austria
e-mail: martin.grube@uni-graz.at

© Springer Nature Switzerland AG 2019
S. M. Tiquia-Arashiro, M. Grube (eds.), *Fungi in Extreme Environments: Ecological Role and Biotechnological Significance*,
https://doi.org/10.1007/978-3-030-19030-9_7

recognition. Hence it is practically impossible to identify the species directly in the environment or with a stereomicroscope. Diagnostic phenotypic characters are often expressed under controlled laboratory conditions, even though species commonly lack sexual structures. Analyses of DNA sequence data are the only way to prove the relationship of species with similar morphology and to place these fungi in a phylogenetic framework. Molecular approaches showed both that many lineages exist which have not been named so far while described taxa turned out to be polyphyletic (e.g. Ertz et al. 2014). DNA data are still of limited use to recognise species as little is known about the genetic variation within species, but they clearly showed that black fungi are phylogenetically more diverse than previously thought and that they primarily belong to two lineages of ascomycetes, the early diverging clades of Dothideomycetes and Eurotiomycetes in Ascomycota (Fig. 7.1).

7.2 Phylogenetic Relationships of Black Fungi Within Dothideomycetes and Eurotiomycetes

With more than 19,000 species, Dothideomycetes is the class with the largest number of species in Ascomycota (Kirk et al. 2008; Schoch et al. 2009a, b). Representatives of Dothideomycetes generally pursue a wide diversity of lifestyles, a diversity which is also scattered across many different clades (Egidi et al. 2014; Muggia et al. 2016). Within Dothideomycetes, the order Capnodiales is particularly rich in extremotolerant species. They comprise fungi isolated from rocks of Antarctic dry valleys (Onofri et al. 2007a, b, Selbmann et al. 2005; Egidi et al. 2014), high-altitude rocks of the Alps (Selbmann et al. 2014), hot deserts (Muggia et al. 2015) and black yeasts from marine salters (Gunde-Cimerman et al. 2000). Different life-styles can be found even within genera. For example, the genus *Rachicladosporium* (Capnodiales) includes both rock inhabitants and plant pathogens (Egidi et al. 2014). The environmental versatility of closely related fungi suggests that they keep a shared set of traits, which facilitates adaption to new habitats. Ruibal et al. (2009) and Egidi et al. (2014) show that many of the rock-inhabiting fungi, isolated from both mild and extreme climates, are found in the family Teratosphaeriaceae (Capnodiales), such as *Friedmanniomyces endolithicus*, *Elasticomyces elasticus* and *Recurvomyces mirabilis*. While these species seem to be widespread extremo-tolerant fungi, *Cryomyces antarcticus* seems to be an endemic extremophile confined to Antarctica so far (Ruibal et al. 2009).

Black fungi in Eurotiomycetes are mainly found in the order Chaetothyriales (Gueidan et al. 2011). Chaetothyriales are mostly known as saprophytic and pathogenic fungi—with a wide variety of hosts (Geiser et al. 2006; Teixeira et al. 2017), but they also comprise a relevant number of species living on rocks (Sterflinger et al. 1999; Ruibal et al. 2005), for instance, species from the genera *Knufia*, *Bradymyces*, *Cladophialophora*, *Capronia* and *Strelitziana* (Ruibal et al. 2008; Réblová et al. 2016). It has been suggested that rock-inhabiting species in

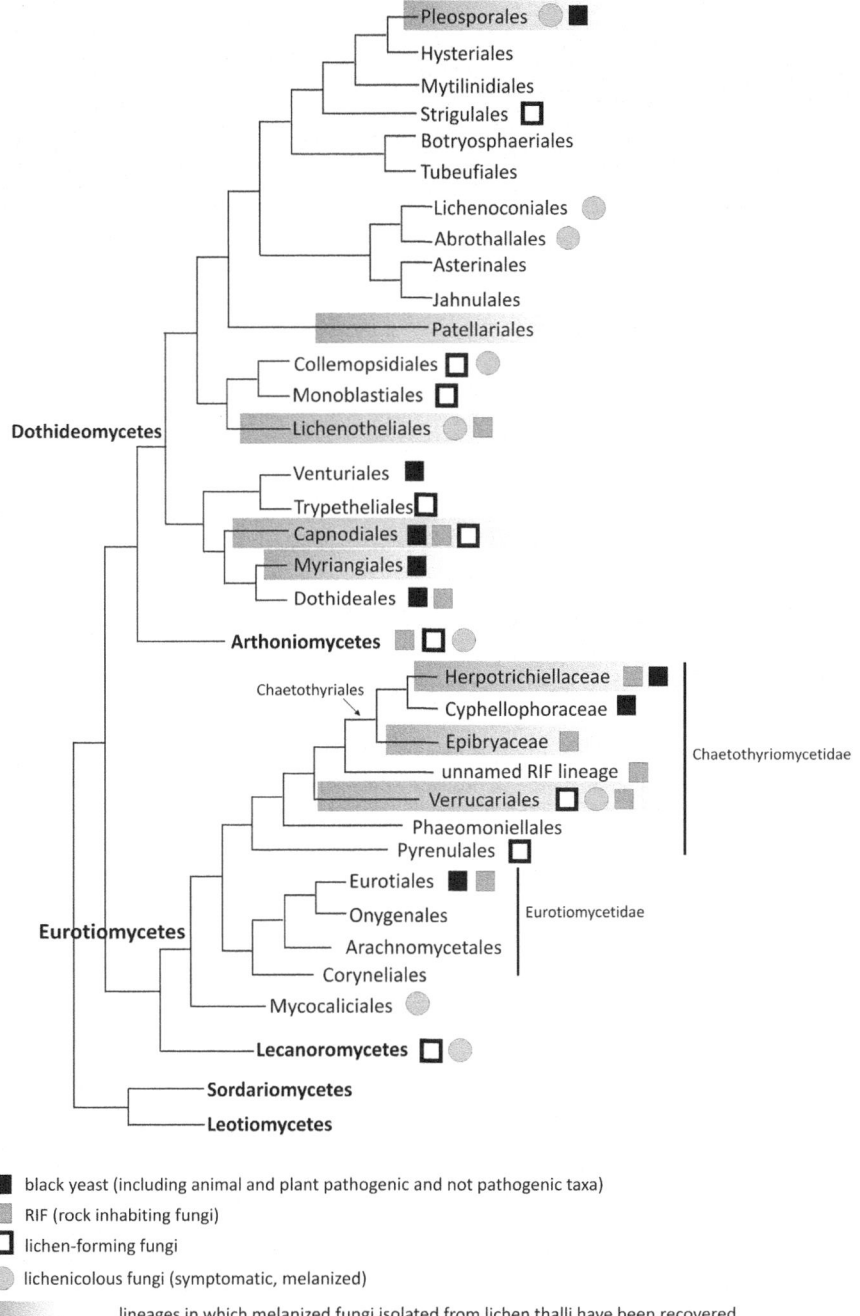

Fig. 7.1 Schematic representation of the major lineages in Dothideomycetes and Eurotiomycetes (Ascomycota) in which black fungi with different lifestyles are found. Lineages in which black fungi have been isolated from lichen thalli are shaded in grey. The phylogeny was graphically reconstructed merging information from most recent phylogenetic studies of Hyde et al. (2013), Gueidan et al. (2014), Chen et al. (2015), Liu et al. (2017) and Teixeira et al. (2017)

Eurotiomycetes are ancestral to present lineages of pathogenic and lichenised fungi in Chaetothyriales and Verrucariales (Gueidan et al. 2008), because they appear particularly diverse in early diverging lineages of these orders.

It is impossible to give a precise number of extremotolerant black fungi. As mentioned above, the characterisation of many black fungi suffers from ambiguous diagnostic characters and the lack of sexual reproductive forms (e.g. Crous et al. 2004; Muggia et al. 2008; Taylor 2011). Molecular phylogenetic studies alone cannot always resolve species delimitation problems of black fungi; thus authors considered the species as complexes containing (so far) cryptic species (Hyde et al. 2013; Egidi et al. 2014; Muggia et al. 2015). An integrative, or polyphasic, taxonomic approach is required to circumscribe closely related lineages. This has been demonstrated with the species complex of *Aureobasidium pullulans*, where a redefinition of its varieties by Zalar et al. (2008) initiated more detailed analyses that resulted in species descriptions for the former varieties (Gostinčar et al. 2014).

7.3 Important Features of Extremotolerant Fungi

7.3.1 Plasticity of Growth

The typical growth forms of polyextremotolerant black fungi and their ability to switch among them is a key adaptation to cope with stressful environmental conditions. Filamentous mycelia are able to explore the surrounding environment for nutrients. Substantially different mycelial organisations can be formed by the same strain on rock surfaces and on culture media. The hyphae of rock-inhabiting mycelia are often short celled, and mycelia may become compact to form microcolonies. These slow-growing forms may also build 'skins' and 'shells' made of extracellular matrix and minerals on the surface of the cell wall (Gorbushina 2007). Apparently, the compact shape of fungal microcolonies also protects black fungi against heat and desiccation, by optimising the volume/surface ratio, according to Gorbushina (2007). Moreover, colonial growth can form extensive biofilms, which are highly resistant to antifungals (Kirchhoff et al. 2017). Many species of black fungi, especially in Dothideales, can also switch between filamentous and yeast growth forms (pleomorphism) depending on the circumstances of the habitat. Transitions exist to completely budding forms, which are often called black yeasts. This ability is termed phenotypic plasticity, which has been recently studied in *Aureobasidium pullulans* by Slepecky and Starmer (2009). In addition, variation known as nongenetic heterogeneity exists between individual cells within genetically uniform mycelia, which grow under the same conditions. When single cells of black fungi become detached from the surface, they commonly develop budding forms (Staley et al. 1982; Gorbushina et al. 1993). Even neighbouring cells in a hyphal thread may differ in vitality according to microscopic analysis, and hyphal outgrowths may occasionally grow internally in neighbouring dead cells of the same hyphal segment

(Grube, unpublished observations). The non-genetic phenotypic heterogeneity results from differences in gene expression, which occur either stochastically or as a result from the relative position of cells in a mycelium (Hewitt et al. 2016). We think that this type of variation is a so far neglected factor of fungal adaptation to stressful environments.

7.3.2   Melanisation

A characteristic phenotypic trait of all black fungi is the presence of melanins in the cell walls. Melanins are a group of polymeric secondary compounds that have been interpreted as 'fungal armour' (Gómez and Nosanchuk 2003; Keller 2015), as they act as protective agents against a wide range of environmental stresses (Sterflinger 2006). The term 'melanins' indicates a black pigment of biological origin but provides little information about the chemical structure of the polymeric molecules. Melanins are produced by a wide range of organisms, including plants, animals and certain species of bacteria (Swan 1974). The polymer structures are still insufficiently characterised because the linkages among the precursors—as building blocks—and the composition are variable. Therefore, melanins are classified according to their precursors and biosynthetic pathways (Eisenman and Casadevall 2012). Eumelanins, the most common type of melanins, are produced during the oxidation of tyrosine (and/or phenylalanine) to 3,4-dihydroxyphenylalanine (DOPA) or dopaquinone which further undergoes cyclisation to 5,6-dihydroxyindole or 5,6-dihydroxyindole-2-carboxylic acid (Butler and Day 1998). Pheomelanins are initially synthesised like eumelanins, but DOPA undergoes cysteinylation into cysteinyl-dopa, which further polymerises into various derivatives of benzothiazines (Plonka and Grabacka 2006). Allomelanins, the least studied group of polymers, are produced through polymerisation of 1,8-dihydroxynaphthalene (DHN). In this pathway, the precursor molecule is acetyl-coenzyme A (acetyl-CoA) or malonyl-CoA. The first step, formation of 1,3,6,8-tetrahydroxynaphthalene, is catalysed by a polyketide synthase (PKS). Subsequently, a sequence of reactions produce the intermediary compounds scytalone, 1,3,8-trihydroxynaphthalene, vermelone and DHN, which is then polymerised to yield melanin (Butler and Day 1998; Langfelder et al. 2003; Plonka and Grabacka 2006). Most melanins characterised from ascomycetes are DHN-melanins (allomelanins) but DOPA-melanins (eumelanins) occur as well (Eisenman and Casadevall 2012); however the latter are more typical for basidiomycetes (Butler and Day 1998).

The best known function of melanins is to protect against UV radiation, which is of particular importance for the rock-inhabiting fungi thriving on bare rock surfaces in open environments. The melanins accumulate in the cell walls of spores and hyphae. Singaravelan et al. (2008) showed by in vitro experiments how physiological stress caused by UV radiation enhanced the synthesis of melanin as an adaptive response in *Aspergillus niger*. A significantly higher concentration of melanin was measured in the conidia, which positively correlated with their germination capacity.

Further, in the phytopathogenic fungus *Bipolaris oryzae* the expression of 1,3,8-tri-hydroxy-naphthalene reductase (THR1) gene—essential for DHN-melanin production—is transcriptionally enhanced by increased doses of UV radiation (Kihara et al. 2004). Melanins accomplish the same function also in basidiomycetes: the pathogenic black yeast *Cryptococcus neoformans* shows a lower susceptibility to UV light damages when the cells are protected by melanins (Wang and Casadevall 1994). Both DOPA and DHN-melanins are efficient protectors also against ionising radiation (Pacelli et al. 2017). The radioprotective properties of fungal melanins derive from a combination of physical shielding and free radical quenching (Dadachova et al. 2008). Ionising radiation alters the oxidative-reduction potential of melanins and is correlated with a faster growth rate in melanised fungi, suggesting that melanins might also function as energy traps (Dadachova et al. 2007; Dadachova and Casadevall 2008).

Melanins accumulate within the cell wall of black fungi as electron-dense granules, which contain various functional groups such as carboxyl, phenolic, hydroxyl and amine. These functional groups provide multiple binding sites for metal ions. The maximum binding capacity of fungal melanins decreases from copper to calcium, magnesium and zinc, respectively (Fogarty and Tobin 1996). Binding mechanisms of fungal melanins have been studied mostly for Cu, which binds mainly at a phenolic hydroxyl group and at a carboxyl group, as in humic acids (Fogarty and Tobin 1996). Though fungal melanins can bind Cu, they show a higher affinity for Fe, if both ions are co-present in a solution: in this case Fe is able to partially substitute Cu (Senesi et al. 1987). The high affinity of fungal melanins to Fe and Cu has also been demonstrated in the lower, melanised cortex of parmelioid lichens and in the melanised apothecia of the lichen *Trapelia involuta*, respectively (Fortuna et al. 2017; McLean et al. 1998). The capacity of melanins to bind ions becomes biologically relevant especially when toxic metal ions are abundant in the environment around the fungal cells: when bound, their decreased concentration allows the fungi to grow also in contaminated environments. Moreover, binding and exposition of cations on the hyphal surface can protect fungi from antagonistic microbes, either reducing the availability of microelements or interfering with the activity of hydrolytic enzymes (Fogarty and Tobin 1996).

Though essential to aerobic life, high electronegativity renders oxygen (O_2) one of the most reactive elements on Earth. During its reduction to water, reactive oxygen species (ROS) are generated as by-products and cause oxidative stress (Turrens 2003). Also other abiotic stresses, such as desiccation, freezing, heavy metals and other xenobiotic compounds, are likely to induce oxidative stress in fungi (Jamieson 1998; Lushchak 2011). ROS (e.g. superoxide radical, hydrogen peroxide and hydroxyl radical) cause severe cell damage, and living organisms have developed both enzymatic and non-enzymatic defence mechanisms. In fungi (both basidiomycetes and ascomycetes), melanins have a relevant redox buffer function and act as non-enzymatic defences against oxidative stress, as known for example in *Inonotus obliquus*, *Phellinus robustus*, *Aspergillus carbonarius*, *Paecilomyces variotii* (Shcherba et al. 2000), *Cryptococcus neoformans* (Jacobson and Tinnell 1993),

Exophiala dermatitidis, Alternaria alternata (Jacobson et al. 1995) and *Aspergillus nidulans* (Goncalves and Pombeiro-Sponchiado 2005). The redox buffering capacity of melanins has been studied mostly in pathogenic fungi identifying melanin as a virulence factor, as one of the most common reaction to pathogens is the production of ROS by leukocytes. Moreover, in pathogenic black fungi melanins have also been claimed to generate appressorium turgor, which is essential to penetrate animal/plant tissues (Sterflinger 2006).

Melanins seem also to confer resistance to osmotic stresses, and fungi isolated on saline media are almost exclusively melanised (Gunde-Cimerman et al. 2000). In this context, the action mechanism of melanins has not been fully elucidated, though Plemenitaš et al. (2008) hypothesised that the dense, shieldlike layer of melanin granules accumulated in the cell wall may reduce loss of osmoprotective substances during salt stress. Therefore, the reduction of cell permeability carried out by melanins would lead to an increased efficiency of the cells to counteract the osmotic stress.

7.3.3 Oligotrophy, Unusual Carbon Sources, Desiccation and Temperature Tolerance

On bare rock surfaces with limited nutrient resources and discontinuous presence of liquid water, fungi must be able to exploit a wide range of carbon sources deposited by dust, water or in the form of volatile organic compounds (VOCs; Prenafeta-Boldú et al. 2001, 2006). There have also been studies using C^{14}-labelled CO_2 and HCO_3^- (Mirocha and DeVay 1971; Palmer and Friedmann 1988), which suggest the capacity of some fungi and black fungi to directly fix carbon dioxide. Yet, these studies still need confirmation and additional work to find out the possible pathways of CO_2 incorporation. Lacking Calvin cycle metabolism, they would need to incorporate carbon via any other potential pathways of carbon uptake (Bar-Even et al. 2012). So far studies mostly focused on the spectrum of organic compounds efficiently usable by these fungi. For example, aerobic metabolism of a large spectrum of L and D forms of monosaccharides, disaccharides and alcohols has been investigated (Sterflinger 2006). Ethanol can be usually degraded, whereas the oxidation of methanol is rare; the use of meso-erythritol is also often reported and several rock-inhabiting fungi are even able to degrade simple and polycyclic aromatic hydrocarbons (Prenafeta-Boldu et al. 2006; Sterflinger 2006; Nai et al. 2013). *Knufia petricola* (Chaetothyriales, Eurotiomycetes), in particular, was proposed as a model organism for further analysis of the physiology of rock-inhabiting black fungi (Nai et al. 2013). This fungus indeed tolerates and grows on media containing monoaromatic compounds, confirming its capacity to utilise recalcitrant carbon sources eventually spurned by other microorganisms (Nai et al. 2013). Further, one of the most striking evidences that fungi are capable of exploiting unusual carbon sources is the black mould *Racodium cellare* (Dothideomycetes, Capnodiales).

Its metabolism seems to benefit volatiles released by wine barrels as carbon source and it is also able to grow using other VOCs (Tribe et al. 2006). Another black fungus associated with alcoholic vapours is *Baudoinia compniacensis*, which is frequently found near distilleries (Scott et al. 2007).

The ability to degrade aromatic compounds appears more typical for black fungi in Chaetothyriales, and in particular to the members of the family Herpotrichiellaceae. This family has been mainly studied for its role in human pathogenesis, but many species have been isolated from hydrocarbon-rich environments as well, such as soil contaminated by hydrocarbon, fuel tanks, washing machines, soap dispensers, indoor moist environments or rotten wood (Prenafeta-Boldu et al. 2006; Badali et al. 2011; Zalar et al. 2011; Isola et al. 2013). These and other recent works highlighted a possible connection between neurotropism (affinity of a pathogen for brain tissues) and the metabolism of aromatic hydrocarbons in the environment (Moreno et al. 2018a).

In this context, hydrocarbon assimilation may represent an additional virulence factor, as the brain contains monoaromatic catecholamine neurotransmitters (e.g. dopamine). Moreover, neurotransmitter catabolism compounds and other substances detected in the human brain are also found as products of lignin degradation (Prenafeta-Boldu et al. 2006). Because some pathogenic fungi have been isolated from environmental sources too, it has been hypothesised that hydrocarbon-rich environments could represent a possible pathogen reservoir. However, the environmental *Cladophialophora psammophila*, congeneric with the notorious human pathogen *C. bantiana*, lacks pathogenicity (Badali et al. 2011). Isola et al. (2013) confirm this finding with isolates of *Exophiala* species, where several pathogens and hydrocarbons associated with black fungi have been found mostly in ecologically divergent lineages of *E. xenobiotica*. With the shared adaptive traits of pathogens and extremotolerant or hydrocarbon-growing black fungi, reciprocal segregation in specific niches is not always complete. Opportunistic pathogens may still have the ability to grow outside the host, while some environmental strains could occasionally become pathogenic in immunocompromised hosts. *E. mesophila*, indeed, is the first reported clinical strain able to grow on alkylbenzenes as well (Blasi et al. 2016), and environmental *Fonsecaea erecta* is able to infect and survive in animal host tissue (Vicente et al. 2017). The ability to thrive in polluted environment and to use aliphatic and aromatic hydrocarbon as energy and carbon source in an otherwise extremely oligotrophic environment is of particular interest for the application in bioremediation of polluted environmental matrices and gas effluent biofilters (Kennes and Veiga 2004; Blasi et al. 2016).

Black fungi also share few other characteristics which make them surviving in extreme environments, including the capacity to suspend their metabolism for long periods until favourable conditions re-establish. The desiccation tolerance is enhanced by the accumulation of the disaccharide trehalose, which stabilises enzymes and cell membranes, avoiding degradation and breakage during dehydration phases (Sterflinger 2006). On the other hand, glycerol is the most abundant compatible solute produced by halotolerant black yeasts, such as *Hortaea werneckii*, to compensate the loss of water from the cells in highly concentrated salt solutions (Plemenitaš et al. 2008). The desiccation tolerance is highly correlated with temperature tolerance. Dried fungal colonies are metabolically inactive and can survive temperature up to 120 °C (relatively common on bare rocks exposed to sun) for short time spans. Otherwise, temperatures between 35 and 75 °C are lethal for hydrated colonies (Sterflinger 1998). Rock fungi from cold environments usually produce also a high amount of extracellular polymeric substances (EPS) to increase their resistance against freeze-thaw damage (Selbmann et al. 2015). The best example of psychrophilic black fungi is *Cryomyces antarcticus*, isolated from rocks from Antarctic dry deserts, which shows a growth optimum below 15 °C and still has a detectable growth near 0 °C (Onofri et al. 2007a). Moreover, this fungus and other black fungi isolated from cryptoendolithic Antarctic communities are able to survive repeated freeze-thaw cycles, outstandingly frequent in Antarctic summer season.

7.4 Lifestyle Versatility of Black Fungi

The above-outlined traits of black fungi facilitated their adaptation in a wide range of niches. Many of them seem to be widespread environmental species, whereas certain lineages comprise specific clinical strains (de Hoog et al. 2013). Many strains are also recurrent endophytic components in plants, where they may symbiotically enhance thermal tolerance of their plant hosts (known as dark-septate endophytes; e.g. Rodriguez et al. 2008). Furthermore their thermal tolerance can facilitate their occurrence as opportunists and pathogens in warm-blooded animals (including humans, as well; de Hoog et al. 2000). In the case of many species of human black yeast pathogens, the pathogenicity seems to be mostly coincidental and suggests that the original niche lies outside the human host. These pathogenic black yeasts lack a specific mechanism to enter the host tissue, suggesting a low-level specialisation. Their pathogenic potential in the animal tissues is attributed to thermal tolerance, pleomorphic growth, melanisation of cell walls, and the ability to degrade complex carbohydrates. The ability to tolerate and degrade a range of toxic aromates may also explain why many black fungi are able to associate with ants (Vasse et al. 2018). In fact, Voglmayr et al. (2011) suggested that the basis for the tolerance factors of ant-associated black fungi could be traced back to the adaptation to the lichen habitat, which is discussed later.

However, the highest diversity of black fungi has been detected in rocky environments, and it has been speculated that these fungi represent the ancestors of those lineages of black fungi, which later evolved other lifestyles (Gueidan et al. 2011; de Hoog et al. 2013), including lichens. Black rock-inhabiting fungi are found in every climatic zone on a wide range of surfaces; they also tolerate polar latitudes and extreme altitudes (Onofri et al. 2007a, Gostinčar et al. 2012; Fig. 7.2). Furthermore they also colonise diverse artificial building materials including plastics

Fig. 7.2 Natural environments where black rock-inhabiting fungi co-occur with lichens: (**a**) Alpine habitat at high elevation (Mt. Rosa, Western Alps, 4500 m a.s.l.); (**b**) rock scree richly colonised by lichens; (**c**) outcrops and walls at low elevation (Taya Tal, The Czech Republic)

and concrete. Concrete surfaces, with their enormous extent, are of particular importance. Fungi on concrete have earlier been studied in the context of biodeterioration and pollution (e.g. Krumbein 2012 and references therein). Strains of potentially allergenic *Alternaria* (causing asthma and chronic rhinosinusitis) and *Epicoccum* are frequently found among isolates from concrete (unpublished data). Thus, concrete surfaces might also be considered in public health as a source of fungi contributing to the aerial mycobiome. Spores and hyphal fragments of *Alternaria*, in particular, are among the most abundant allergens spread in airborne samples (Banchi et al. 2018), and the genus is always reported in the pollen bulletins of air monitoring.

7.5 Links to Lichen Symbiosis

Gorbushina and Broughton (2009) considered the rock surface as a kind of 'symbiotic playground', where they found antibiosis (detrimental interactions between species) to be rare and counterproductive. The authors also mentioned that co-cultivation of the cyanobacterium *Nostoc* with a rock-inhabiting fungus (*Sarcinomyces*) resulted in a specific association, without presenting this association in greater detail. Such associations have been observed previously, e.g. by Turian (1977), in his description of *Coniosporium aeroalgicola*. This species, a dematiaceous hyphomycete, seems to be a ubiquitous component of aereo-algal communities and is able to form some sort of symbiotic structures with algae. Rock-inhabiting microcolonial fungi may develop into lichenoid structures within months when co-cultured with lichen algae (Gorbushina et al. 2005). Brunauer et al. (2007) reported an interesting new black fungal strain (ALr-1) isolated from rock-inhabiting lichen *Lecanora rupicola*. During co-culture with the isolated algae of the host lichen, this black fungus started to cover the algal colonies with a layer-like mycelium recalling a primitive form of a lichen thallus. Through a phylogenetic analysis Brunauer et al. (2007) could then also show that this black fungus is basal to typically lichen-forming lineages in the Chaetothyriomycetidae.

The study of lichen-infecting fungi has a long history, and fungal infections have been observed even before lichens were found to represent a fungal-algal symbiosis. Their propagative structures were used to characterise 2000 so far described species of lichenicolous fungi (Diederich et al. 2018). Microscopic analyses show that there are often additional and dark-coloured fungal hyphae in lichens, which cannot clearly be assigned to known species. Culturing and sequencing approaches have shown more recently that many more fungi have a so far unrecognised association with lichens (e.g. Fernández-Mendoza et al. 2017; Muggia and Grube 2018). Their roles in the lichen symbiosis and phylogenetic relationships still need to be explored. Yet, a significant fraction of fungi isolated from lichens belong to black fungi, which are also known from bare rock habitats. Work of Harutyunyan et al. (2008) suggests that rock-inhabiting lichens in arid environments are particularly rich in black fungi otherwise occurring on rocks.

The shared occurrence of black fungi on rocks and on lichens, as well as the transient ability of certain isolates to form associations with algae, indicates a biological link between extremotolerant, lichenised and lichen-inhabiting lifestyles that are reminiscent of a common phylogenetic past. Potential common ancestries of lichenised fungal lineages and black fungi have already been documented by phylogenetic analyses, which place rock inhabitants basal to the mainly lichenised lineages of Arthoniomycetes and Verrucariales (Gueidan et al. 2008; Ruibal et al. 2009). In these groups we also find both complex morphologies with stratified lichen thalli (and occasionally subsequent loss of thallus formation and sporadic evolution towards the lichen-infecting lifestyle). In comparison, the lichen representatives in Dothideomycetes are scattered among different clades within this huge class (Muggia et al. 2008; Nelsen et al. 2009). These lineages generally do not form complex thallus structures and are more closely related to fungi adapted to other lifestyles, in particular to those growing in oligotrophic rock environments.

Gostinčar et al. (2012) suggested that small protective molecules that are known to accumulate in black fungi as stress-responsive osmolytes could also be linked with potential transition from rock-inhabiting to the lichenised lifestyle. In particular, the polyol metabolism could be involved in both extremotolerance and lichenisation. Ribitol, sorbitol and erythritol (as well as glucose by cyanobacteria) are provided by photoautotrophic symbionts to the fungal partners as 'food molecules', where they are transformed to mannitol (Friedl and Büdel 2008). Efficient osmolyte metabolism, as found in oligotrophic black fungi, could therefore be a preadaptation to facilitate the transition to a lichen symbiotic lifestyle.

Few black fungal species indeed form lichen symbioses with a peculiar morphology. For example, the microfilamentous cushions formed by the genera *Cystocoleus* and *Racodium* consist of algal threads that are enwrapped by fungal hyphae (Muggia et al. 2008). Other (small) lineages in the Dothideales suggest that there is a widespread capacity for evolution of this symbiotic lifestyle (Nelsen et al. 2009), but the lichenised forms never develop a typical thallus morphology as found in the primarily lichenised classes Lecanoromycetes or Arthoniomycetes. To explore the link between rock-colonising and algal-associated lifestyles in more detail, Muggia et al. (2013, 2015) studied the genus *Lichenothelia*. The genus was described by Hawksworth (1981) who already suggested a relationship with Dothideales due to features of the ascomata. Lichenotheliaceae and the order Lichenotheliales (Hyde et al. 2013) are meanwhile confirmed as monophyletic lineage within Dothideomycetes, based on sequences from five species (e.g. *Lichenothelia arida*, *L. calcarea*, *L. convexa*, *L. rugosa* and *L. umbrophila*, and a yet undescribed *Lichenothelia* sp.; Hyde et al. 2013; Muggia et al. 2013, 2015), thus pending the analysis of 23 further accepted species (Henssen 1987; Øvstedal and Smith 2001; Atienza and Hawksworth 2008; Zhurbenko 2008; Etayo 2010; Muggia et al. 2013, 2015; Valadbeigi et al. 2016). *Lichenostigma* was earlier thought to be a closely related genus comprising lichenicolous species, but recent phylogenetic analyses revealed its position at the basis of the primarily lichenised class Arthoniomycetes (Ertz et al. 2014).

Although the mycelium of *Lichenothelia* never builds a typical lichen thallus with an internal algal layer, algae are often found in close contact with the fungal hyphae (Fig. 7.3e, f). So far, in vitro experiments by which both growth rate and structure of mixed culture of *Lichenothelia* with algae were tested (Fig. 7.4b–d, g, h) did not provide clear experimental evidence of symbiosis establishment (Ametrano et al. 2017). This might be attributed to the rich medium conditions, and it thus remains to be tested, whether nutrient-deprived conditions could make the association with algal cells more important for the fungal species. Other black fungi might also suit as experimental models to study lichen-like associations, and particularly, some species of the highly diverse mycobiota associated with lichens may gain benefits from algal products in the host symbiosis (Muggia and Grube 2018). To test potential interactions of lichen-associated and other black fungi with algae, Muggia et al. (2018) introduced a novel method for co-cultivation. They encapsulated fungi in alginate

Fig. 7.3 Habit and lifestyles of black fungi in nature (in parenthesis fungal order and/or class are reported): (**a**) lichenicolous black fungus (Dothideomycetes) spreading hyphae on the apothecium of the lichen *Lecanora polytropa*; (**b**) lichen parasitic fungus *Lichenostigma rouxii* (Arthoniomycetes) developing hyphae and ascomata (arrow) on the thallus areolas of *Pertusaria* sp., sample n. SPO1428; (**c**) *Lichenostigma epirupestre* (Arthoniomycetes) on rock in between thallus areolas of *Pertusaria* sp., sample n. SPO1433; (**d**) lichenicolous and rock-inhabiting *Lichenothelia arida* (Lichenotheliales, Dothideomycetes) on rocks developing abundant acomata (arrow) at thallus centre, sample n. L2162; (**e**) lichenicolous and rock-inhabiting *Lichenothelia scopularia* (Lichenotheliales, Dothideomycetes), thallus in which algae (arrow) are visible and wrapped by the melanised hyphae, sample n. L2181; (**f**) rock-inhabiting *Lichenothelia* sp. growing in proximity of algal colonies in rock crevices, sample n. L1298. Scale bars: A, D = 1 mm; B, C, E, F = 0.5 mm

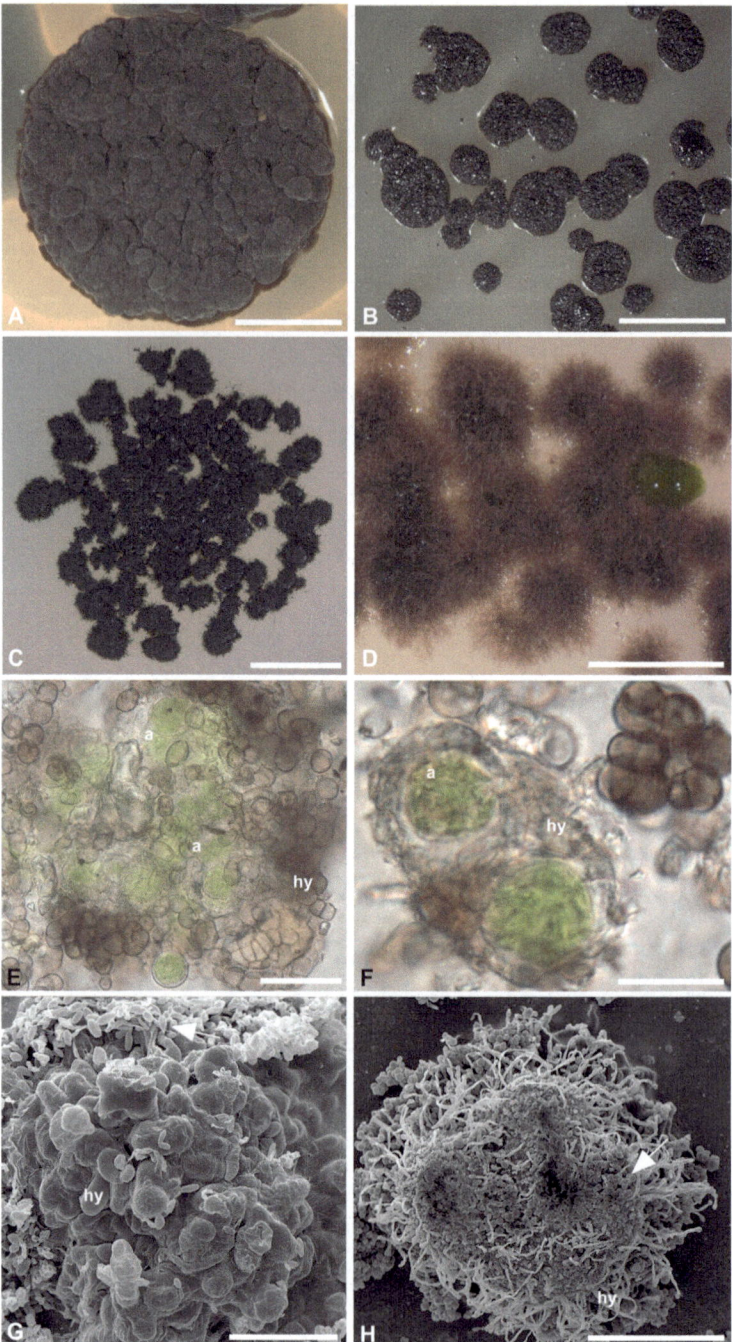

Fig. 7.4 Habit of black fungi in culture and their co-growth with algae. (**a**) Culture isolate of chaeto-thyrealean black fungus strain A564; (**b**) culture isolate of *Saxomyces americanus* (strain L1853); (**c**) culture isolate of *Lichenothelia convexa* (strain L1844); (**d**) *Lichenothelia calcarea* co-cultured with *Coccomyxa* sp. (strain PL2-1); (**e, f**) thallus of environmental sample of *Lichenothelia* sp. (sample L2181) in which fungal hyphae wrap algal cells of *Trebouxia* sp.; (**g**) scanning electron microscopy (SEM) microphotograph of *Saxomyces alpinus* (CCFEE 5470) co-cultured with *Coccomyxa* sp. strain PL2-1; (**h**) SEM microphotograph of *Lichenothelia convexa* (strain L1844) co-cultured with *Trebouxia* sp. Scale bars: **a** = 4 mm; **b, c** = 2 mm; **d** = 1 mm; **e** = 25 μm; **f** = 10 μm; **g, h** = 100 μm

together with algal strains isolated from lichens. Symbiotic interactions that are not normally observed in nature can be artificially enforced with this approach. While intertwined growth of slow-growing black fungal mycelia (*Saxomyces* and *Lichenothelia*) and algae was commonly observed, only the specific lichenicolous fungus *Muellerella* developed a layer-like mycelial growth together with algal colonies. Encapsulation in alginate may be a versatile method to experimentally analyse the specificity of interactions among black fungi and algae under a wide range of parameters. Despite the fact that they are little studied, such lichen-like interactions seem to be surprisingly ubiquitous. Removal of dark matter from concrete in urban environments by using adhesive tapes and subsequent microscopic analysis reveals that black fungi cover extensive surfaces in a biofilm-like manner together with microscopic algae (Grube, unpublished). The associations strikingly resemble the lichenoid interaction structures found in co-culture experiments, with algal cells attached to fungal hyphae or partly enwrapped by fungal microcolonies.

Why can black fungi only form very primitive types of associations with algae, but not a typical lichen thallus equipped with a coherent fungal cortex? Lichenised fungi usually shelter their algal partners beneath a protective peripheral fungal layer, which develops by the conglutination of the outer cell walls of the fungi. We argue that one important step towards the evolution of typical lichen thalli from potential black fungal ancestors was the downregulation of fungal melanin production. Highly cross-linked melanins make fungal cell walls rigid and prevent appropriate responses of cell colonies to mechanic stress imposed by fluctuating hydration conditions in the natural environment. Instead, downregulation of melanin and conglutination of cells in a joint extracellular matrix facilitate the functionality of coherent, hydration-responsive symbiotic structures as required in lichen thalli. Moreover, a strongly melanised fungal upper layer would also prevent the transmission of light to the symbiotic algal partner. Most lichens therefore screen against high doses of light found by other means, such as formation of microscopic crystals that are deposited outside of the cell walls (as extrolites). With a crystal-studded layer, light is also better transmitted to the algae in metabolically active humid stage, whereas it is largely reflected in the shrunk dried stage of thalli when metabolism is suspended. The development of highly complex thallus structures of lichenised fungi, and thus structures of their propagation, requires the presence and functionality of compatible algae. In contrast, most black fungi associate live and propagate independently from optional symbiotic stages, which explains the excessive tolerance for environmental variation and the adaptability to a wide range of ecological conditions.

7.6 Omics Approaches for the Study of Black Fungi

'Omics' approaches, genomics, transcriptomics, proteomics, etc. speeded up the view on the organisation of life in twenty-first-century biology. Large amounts of data are generated by these approaches that require substantial statistical and computational analyses (Zhang et al. 2010), as well as easy access to information for comparative analyses. Fungal comparative genomics have become an affordable and popular endeavour, as exemplified by a project to sequence and analyse 1000 fungal genomes (http://1000.fungalgenomes.org). Yet, we are only beginning to interpret the details of the biology of black fungi within an omics framework.

More than 37 genomes of chaetothyrealean black yeasts are now available in different databases (Moreno et al. 2018b). Black yeast genomes of Chaetothyriomycetidae are similar in size, ranging from 25.8 Mb in *Capronia coronata* CBS 617.96 to 43 Mb in *Cladophialophora immunda* CBS 834.96. These sizes are in the lower average of filamentous ascomycetes, and the length of genomes is not correlated with adaptations to the ecological extremes. However, several protein families in black fungal genomes have undergone extensive gene duplication events (Teixeira et al. 2017). Among the expanded families, cytochromes p450 (CYP), drug efflux pumps, alcohol dehydrogenase (ADH) and aldehyde dehydrogenase (ALDH) seem to be widely distributed across black yeasts. Retention of duplicated genes suggests that the broadening of corresponding metabolic processes and promiscuity for substrates played a role to adapt to their habitats. Based on these findings, Teixeira et al. (2017) suggested that detoxification by black fungi occurs by catabolism of phenolic compounds via phenylacetic acid and homogentisate.

Since its beginning, genome sequencing of Dothideomycetes has advanced much further and, as of March 2018, the Joint Genome Institute site listed 123 sequenced genomes. The great interest in Dothideomycetes is certainly associated with the large number of plant pathogens in that class (Ohm et al. 2012). The number of sequenced extremotolerant fungi is still fairly low though. Sterflinger et al. (2014) used comparative genomics to study the differences between *Cryomyces antarcticus* and the mesophilic fungi *Neurospora crassa*, *Coniosporium apollinis* (a species highly tolerant to UV radiation, desiccation and high temperature), the halotolerant *Hortaea werneckii* and the human pathogens *Exophiala dermatitidis* and *Cladosporium sphaerospermum*. However, as Sterflinger et al. (2014) concluded, the analysis of draft genomes did not reveal any significant deviations of *Cryomyces* genomes and those of mesophilic hyphomycetes. Subsequent analyses revealed duplications of genes potentially associated with stress tolerance. The genome of *Hortaea werneckii* revealed enrichment of metal cation transporters, beside other duplications (Lenassi et al. 2013). Genome and proteome analyses in four varieties of *Aureobasidium pullulans* detected genes possibly associated with degradation of aromatic compounds, in addition to most of the enzyme families involved in decomposition of carbohydrates and many sugar transporters (Gostinčar et al. 2014). All of the components of the high-osmolarity glycerol pathway were present, and the genomes were enriched in putative stress-tolerant genes, e.g. aquaporins and aquaglyceroporins, alkali-metal cation transporters, genes for the synthesis of compatible solutes and melanin, and bacteriorhodopsin-like proteins. The differences in the genomes among the four *Aureobasidium* varieties prompted Gostinčar et al. (2014) to distinguish them into different species, and this raises the question about the amount of variation that has to be expected when further ecologically different strains would be investigated. Additional genome sequences, with limited effort of comparative analysis, have been provided from Antarctic extremotolerant black fungi of the genus *Rachicladosporium*

(Coleine et al. 2017). Even though the results show that extremotolerance in fungi is not reflected in genome size, genes potentially conferring certain metabolic adaptations to stress tolerance seem to be enriched.

Genome sequencing provides important new insights, but it is still not possible to directly infer the involvement of genes in particular functions by merely using the common annotation classifiers. The genome sequences neither provide information about their regulation. Further work is therefore needed to relate the information of genome sequencing with biological roles, particularly of paralogous genes in larger gene families. For a better understanding of biological roles, information obtained from genome sequencing should therefore be complemented with transcriptomics and proteomics. As gene expression might be modulated by environmental factors, transcriptomics may suit better than genomics in testing specific hypotheses about functional aspects of stress tolerance. Within chaetothyrealean black yeasts, most experiments have been conducted on *Exophiala (Wangiella) dermatitidis*, which may be considered a model among pathogenic black fungi. *E. dermatitidis* has been sampled from diverse environments, ranging from glaciers to saunas and dishwashers (Zalar et al. 2011; Zupančič et al. 2016). Therefore, its transcriptomic responses to a wide range of temperatures (1–45 °C) could be analysed (Blasi et al. 2015). The data showed that *E. dermatitidis* usually responds to low temperatures by upregulating genes which modify lipidome composition towards membrane fluidity, whereas there is almost no stress signal in the transcriptome when the fungus is at 45 °C. However, since variations in membrane fluidity patterns among black fungi have been recognised before, any generalisations should be tempered with caution (Turk et al. 2011). Adaptation to ionising radiations and the role of melanin (claimed by Dadachova et al. 2008) have been investigated from a transcriptomic perspective exposing both wild and melanin-deficient mutants of *E. dermatitidis*. According to these results, a high number of genes (3000) are differentially expressed, and an increased growth rate has been observed in both strains (in comparison to the corresponding non-irradiated samples), when they are exposed to a low dose of ionising radiations. However, the majority of regulated genes overlap between wild strain and melanin-deficient mutant; therefore, transcriptional response to the radiation is mainly determined by cellular components other than melanin. The expression of ribosomal biogenesis genes is significantly upregulated only in the wild-type strain (Robertson et al. 2012), suggesting that transcription regulators contribute to both pigment production and general transcriptional response. Comparative genomic and transcriptomic approaches have complemented in the study of Chen et al. (2014), who investigated the *E. dermatitidis* genome and the transcriptional response to low pH. The genome encodes three independent pathways for the melanin synthesis; these were active during pH stress, likely acting as a defence against oxidative damages occurring under stress conditions. The most recent transcriptomics applied to *E. dermatitidis* was an artificial infection of an ex vivo skin model experiment, which aimed at identifying changes in gene expression during infection and potential virulence factors among coding and non-coding RNAs. Even though the yeastlike growth is prevalent during infection, there are evidences of upregulation of genes related to hyphal growth as well. Melanins (often considered as virulence factor) and genes associated to its production are only upregulated in the L-tyrosine pathway, which produces pyomelanin (these melanins are hypothesised to be involved in iron uptake as an indirect factor of pathogenicity). The other melanin pathways are instead not modulated during skin infection. Moreover, *E. dermatitidis* switches to gluconeogenesis pathway in order to respond to decreasing levels of glucose when growing on skin instead of growing in a glucose-rich medium. Genes related to metal cation (Fe, Mg) transporters are also upregulated (Poyntner et al. 2016).

Within Dothideomycetes most scientific attention is focused on economically important crop pathogens, while still little is known about transcriptomics of extremotolerant fungi in this group. Nevertheless, due to its halotolerance and its biotechnological potential, *Hortaea werneckii* is among the better-studied dothidealean fungi to date. Its genome assembly has been recently improved via PacBio sequencing and combined with gene expression analyses to complement the genome annotation (Sinha et al. 2017). These analyses confirmed the previous hypothesis of a recent whole genome duplication event (Lenassi et al. 2013) and identified the presence of some novel high-osmolarity glycerol (HOG) pathway components. These are principally similar in other fungi, such as *Saccharomyces cerevisiae* and *Wallemia ichthyophaga*, but relevant differences, which still need further study, may account for the advanced salt tolerance of *Hortaea* (Plemenitaš et al. 2014).

Proteomics has been suggested as a valid approach to investigate peculiar features of extremophilic black fungi (Marzban et al. 2013) but limited proteomic analyses have so far been carried out. The proteome of *Cryomyces antarcticus* was analysed by Zakharova et al. (2014). Conserved protein families involved in the housekeeping metabolism were found, but a fraction of the protein pool differs significantly from other fungal species and might confer adaptation to extreme conditions. *C. antarcticus* differed by the expression of high levels of heat-shock proteins (HSP), even when it is cultured under optimal temperature conditions. This apparent lack of a stringent regulation of these proteins, which are also present during dehydration (Zakharova et al. 2013), makes *C. antarcticus* perfectly adapted to its rather constant, extreme environment, but less competitive than other mesophilic stress-tolerant fungi. The overall number of proteins detected was low, suggesting that only a limited part of the genome, which has an average size among fungi (Sterflinger et al. 2014), is actually transcribed. This was also found for the proteome of *Friedmanniomyces*.

The alteration of protein expression patterns in rock-inhabiting fungi has been analysed also under temperature variation (Tesei et al. 2012) and Mars-like simulated conditions (a combination of temperature extremes, dryness, low O_2 pressure and high radiations; Zakharova et al. 2014). In these almost anaerobic conditions *Cryomyces antarcticus*, *Knufia perforans* and *Exophiala jeanselmei* still proved to be metabolically active yet exhibiting a significant decrease of expressed proteins during the first 24 h of stress exposure. Although both the extremotolerant (e.g. *Knufia*) and extremophilic (e.g. *Cryomyces*) microcolonial fungi are active under stressful abiotic condition, a main difference has been found: extremophiles are always equipped with a proteome withstanding harsh conditions and just slightly tend to downregulate their activity, and extremotolerants are flexible to change their proteomic profile according to the environmental conditions.

7.7 Conclusion and Future Perspectives

Certain genomic signatures of extremotolerance seem to be widespread in extremotolerant black fungi. Their roles may become clear with more detailed comparisons of genomic data in the future. Also, transcriptomic analyses may find out more details about their differential expression under conditions of stress. It will be particularly interesting to analyse how rapidly the transcriptional responses occur under fluctuating hydration conditions, and how metabolic activities resume after long periods of metabolic suspense. Also, further survival strategies of black fungi should be considered. For example, non-genetic heterogeneity in fungal colonies could indicate a potential self-feeding strategy in the oligotrophic habitats, as dead cells in a colony may be decomposed and used as source of nutrients for surviving cells. We think that a comprehensive understanding of fungal stress tolerance requires an interdisciplinary biophysical and biochemical approach. Hence, we foresee a bright research perspective for black fungi.

Acknowledgments We thank Theodora Gössler for assistance in the lab. Financial support was granted by the Austrian Science Foundation (FWF P24114 to LM).

References

Ametrano CG, Selbmann L, Muggia L (2017) A standardized approach for co-culturing dothidealean rock-inhabiting fungi and lichen photobionts *in vitro*. Symbiosis 73:1–10

Atienza V, Hawksworth DL (2008) *Lichenothelia renobalesiana* sp. nov. (Lichenotheliaceae), for a lichenicolous ascomycete confused with *Polycoccum opulentum* (Dacampiaceae). Lichenologist 40:87–96

Badali H, Prenafeta-Boldu FX, Guarro J, Klaassen CH, Meis JF, De Hoog GS (2011) *Cladophialophora psammophila*, a novel species of Chaetothyriales with a potential use in the bioremediation of volatile aromatic hydrocarbons. Fungal Biol 115:1019–1029

Banchi E, Ametrano CG, Stanković D, Verardo P, Moretti O, Gabrielli F, Lazzarin S, Borney MF, Tassan F, Tretiach M, Pallavicini A, Muggia M (2018) DNA metabarcoding uncovers fungal diversity of mixed airborne samples in Italy. PLoS One 13:0194489

Bar-Even A, Noor E, Milo R (2012) A survey of carbon fixation pathways through a quantitative lens. J Exp Bot 63:2325–2342

Blasi B, Tafer H, Tesei D, Sterflinger K (2015) From glacier to sauna: RNA-seq of the human pathogen black fungus *Exophiala dermatitidis* under varying temperature conditions exhibits common and novel fungal response. PLoS One 10:e0127103

Blasi B, Poyntner C, Rudavsky T, Prenafeta-Boldú FX, De Hoog S, Tafer H, Sterflinger K (2016) Pathogenic yet environmentally friendly? Black fungal candidates for bioremediation of pollutants. Geomicrobiol J 33:308–317

Brunauer G, Blaha J, Hager A, Türk R, Stocker-Wörgötter E, Grube M (2007) An isolated lichenicolous fungus forms lichenoid structures when co-cultured with various coccoid algae. Symbiosis 44:127–136

Butler MJ, Day AW (1998) Fungal melanins: a review. Can J Microbiol 44:1115–1136

Chen Z, Martinez DA, Gujja S, Sykes SM, Zeng Q, Szaniszlo PJ, Wang Z, Cuomo CA (2014) Comparative genomic and transcriptomic analysis of *Wangiella dermatitidis*, a major cause of phaeohyphomycosis and a model black yeast human pathogen. G3 4(4):561–578

Chen KH, Miadlikowska J, Molnár K, Arnold AE, U'Ren JM, Gaya E, Gueidan C, Lutzoni F (2015) Phylogenetic analyses of eurotiomycetous endophytes reveal their close affinities to Chaetothyriales, Eurotiales, and a new order - Phaeomoniellales. Mol Phylogenet Evol 85:117–130

Coleine C, Masonjones S, Selbmann L, Zucconi L, Onofri S, Pacelli C, Stajich JE (2017) Draft genome sequences of the antarctic endolithic fungi *Rachicladosporium antarcticum* CCFEE 5527 and *Rachicladosporium* sp. CCFEE 5018. Genome Announc 5:e00397-17

Crous PW, Groenewald JZ, Mansilla JP, Hunter GC, Wingfield MJ (2004) Phylogenetic reassessment of *Mycosphaerella* spp. and their anamorphs occurring on Eucalyptus. Stud Mycol 50:195–214

Dadachova E, Casadevall A (2008) Ionizing radiation: how fungi cope, adapt, and exploit with the help of melanin. Curr Op Microbiol 11:525–531

Dadachova E, Bryan RA, Huang X, Moadel T, Schweitzer AD, Aisen P, NOsanchuk JD, Casadevall A (2007) Ionizing radiation changes the electronic properties of melanin and enhances the growth of melanized fungi. PLoS One 2:e457

Dadachova E, Bryan RA, Howell RC, Schweitzer AD, Aisen P, Nosanchuk JD, Casadevall A (2008) The radioprotective properties of fungal melanin are a function of its chemical composition, stable radical presence and spatial arrangement. Pigment Cell Melanoma Res 21:192–199

de Hoog GD, Queiroz-Telles F, Haase G, Fernandez-Zeppenfeldt G, Angelis DA, Gerrits van den Ende AHG, Matos T, Peltroche-Llavsahuanga H, Pizzirani-Kleiner AA, Rainer J, Richard-Yegres N, Vicente V, Yegres F (2000) Black fungi: clinical and pathogenic approaches. Med Mycol 38(Suppl1):243–250

de Hoog GS, Vicente VA, Gorbushina AA (2013) The bright future of darkness—the rising power of black fungi: black yeasts, microcolonial fungi, and their relatives. Mycopathologia 175:365–368

Diederich P, Lawrey J, Ertz D (2018) The 2018 classification and checklist of lichenicolous fungi, with 2000 non-lichenized, obligately lichenicolous taxa. Bryologist 12:340–426

Egidi E, de Hoog GS, Isola D, Onofri S, Quaedvlieg W, De Vries M, Verkley GJM, Stielow JB, Zucconi L, Selbmann L (2014) Phylogeny and taxonomy of meristematic rock-inhabiting black fungi in the Dothideomycetes based on multi-locus phylogenies. Fungal Divers 65:127–165

Eisenman HC, Casadevall A (2012) Synthesis and assembly of fungal melanin. Appl Microbiol Biotechnol 93:931–940

Ertz D, Lawrey JD, Common RS, Diederich P (2014) Molecular data resolve a new order of Arthoniomycetes sister to the primarily lichenized Arthoniales and composed of black yeasts, lichenicolous and rock-inhabiting species. Fungal Divers 66:113–137

Etayo J (2010) Líquenes y hongos liquenícolas de Aragón. Guineana 16:1–501

Fernández-Mendoza F, Fleischhacker A, Kopun T, Grube M, Muggia L (2017) ITS 1 metabarcoding highlights low specificity of lichen mycobiomes at a local scale. Mol Ecol 26:4811–4830

Fogarty RV, Tobin JM (1996) Fungal melanins and their interactions with metals. Enzym Microb Technol 19:311–317

Fortuna L, Baracchini E, Adami G, Tretiach M (2017) Melanization affects the content of selected elements in parmelioid lichens. J Chem Ecol 43:1–11

Friedl T, Büdel B (2008) Photobionts. In: Nash TIII (ed) Lichen biology, 2nd edn. Cambridge University Press, Cambridge, pp 9–26

Geiser DM, Gueidan C, Miadlikowska J, Lutzoni F, Kauff F, Hofstetter V, Fraker E, Schoch CL, Tibell L, Untereiner WA, Aptroot A (2006) Eurotiomycetes: Eurotiomycetidae and Chaetothyriomycetidae. Mycologia 98:1053–1064

Gómez BL, Nosanchuk JD (2003) Melanin and fungi. Curr Opin Infect Dis 16:91–96

Goncalves RDCR, Pombeiro-Sponchiado SR (2005) Antioxidant activity of the melanin pigment extracted from *Aspergillus nidulans*. Biol Pharm Bull 28:1129–1131

Gorbushina AA (2007) Life on the rocks. Environ Microbiol 9:1613–1631

Gorbushina AA, Broughton WJ (2009) Microbiology of the atmosphere-rock interface: how biological interactions and physical stresses modulate a sophisticated microbial ecosystem. Annu Rev Microbiol 63:431–450

Gorbushina AA, Krumbein WE, Hamman CH, Panina L, Soukharjevski S, Wollenzien U (1993) Role of black fungi in color change and biodeterioration of antique marbles. Geomicrobiol J 11:205–221

Gorbushina AA, Beck A, Schulte A (2005) Microcolonial rock inhabiting fungi and lichen photobionts: evidence for mutualistic interactions. Mycol Res 109:1288–1296

Gostinčar C, Muggia L, Grube M (2012) Polyextremotolerant black fungi: oligotrophism, adaptive potential, and a link to lichen symbioses. Front Microbiol 3:390

Gostinčar C, Ohm RA, Kogej T, Sonjak S, Turk M, Zajc J, Zalar P, Grube M, Sun H, Han J, Sharma A, Chiniquy J, Ngan CY, Lipzen A, Barry K, Grigoriev IV, Gunde-Cimerman N (2014) Genome sequencing of four *Aureobasidium pullulans* varieties: biotechnological potential, stress tolerance, and description of new species. BMC Genomics 15:549

Gueidan C, Villaseñor CR, De Hoog GS, Gorbushina AA, Untereiner WA, Lutzoni F (2008) A rock-inhabiting ancestor for mutualistic and pathogen-rich fungal lineages. Stud Mycol 61:111–119

Gueidan C, Ruibal C, de Hoog GS, Schneider H (2011) Rock-inhabiting fungi originated during periods of dry climate in the late Devonian and middle Triassic. Fungal Biol 115:987–996

Gueidan C, Aptroot A, da Silva Cáceres ME, Badali H, Stenroos S (2014) A reappraisal of orders and families within the subclass Chaetothyriomycetidae (Eurotiomycetes, Ascomycota). Mycol Prog 13:1027–1039

Gunde-Cimerman N, Zalar P, de Hoog S, Plemenitaš A (2000) Hypersaline waters in salterns–natural ecological niches for halophilic black yeasts. FEMS Microbiol Ecol 32:235–240

Harutyunyan S, Muggia L, Grube M (2008) Black fungi in lichens from seasonally arid habitats. Stud Mycol 61:83–90

Hawksworth DL (1981) *Lichenothelia*, a new genus for the *Microthelia aterrima* group. Lichenologist 13:141–153

Henssen A (1987) *Lichenothelia*, a genus of microfungi on rocks. Biblioth Lichenol 25:257–293

Hewitt SK, Foster DS, Dyer PS, Avery SV (2016) Phenotypic heterogeneity in fungi: importance and methodology. Fungal Biol Rev 30:176–184

Hyde KD, Jones EG, Liu JK et al (2013) Families of Dothideomycetes. Fungal Divers 63:1–313

Isola D, Selbmann L, de Hoog GS, Fenice M, Onofri S, Prenafeta-Boldú FX, Zucconi L (2013) Isolation and screening of black fungi as degraders of volatile aromatic hydrocarbons. Mycopathologia 175:369–379

Jacobson ES, Tinnell SB (1993) Antioxidant function of fungal melanin. J Bacteriol 175:7102–7104

Jacobson ES, Hove E, Emery HS (1995) Antioxidant function of melanin in black fungi. Infect Immun 63:4944–4945

Jamieson DJ (1998) Oxidative stress responses of the yeast *Saccharomyces cerevisiae*. Yeast 14:1511–1527

Keller NP (2015) Translating biosynthetic gene clusters into fungal armor and weaponry. Nat Chem Biol 11:671–677

Kennes C, Veiga MC (2004) Fungal biocatalysts in the biofiltration of VOC-polluted air. J Biotechnol 113:305–319

Kihara J, Moriwaki A, Ito M, Arase S, Honda Y (2004) Expression of THR1, a 1, 3, 8-trihydroxynaphthalene reductase gene involved in melanin biosynthesis in the phytopathogenic fungus *Bipolaris oryzae*, is enhanced by near-ultraviolet radiation. Pigment Cell Melanoma Res 17:15–23

Kirchhoff L, Olsowski M, Zilmans K, Dittmer S, Haase G, Sedlacek L, Steinmann E, Buer J, Rath PM, Steinmann J (2017) Biofilm formation of the black yeast-like fungus *Exophiala dermatitidis* and its susceptibility to antiinfective agents. Sci Rep 7:42886

Kirk PM, Cannon PF, Minter DW, Stalpers JA (2008) Dictionary of the fungi. CABI, Wallingford

Krumbein WE (2012) Microbial interactions with mineral materials. In: Houghton DR, Smith RN, Eggins HO (eds) Biodeterioration 7. Springer Science & Business Media, Dordrecht, pp 78–100

Langfelder K, Streibl M, Jahn B, Haase G, Brakhage AA (2003) Biosynthesis of fungal melanins and their importance for human pathogenic fungi. Fungal Genet Biol 38:143–158

Lenassi M, Gostinčar C, Jackman S, Turk M, Sadowski I, Nislow C, Jones S, Birol I, Cimerman NG, Plemenitaš A (2013) Whole genome duplication and enrichment of metal cation transporters revealed by de novo genome sequencing of extremely halotolerant black yeast *Hortaea werneckii*. PLoS One 8:e71328

Liu JK, Hyde KD, Jeewon R, Phillips AJL, Maharachchikumbura SSN, Ryberg M, Liu ZY, Zhao Q (2017) Ranking higher taxa using divergence times: a case study in Dothideomycetes. Fungal Divers 84:75–99

Lushchak VI (2011) Adaptive response to oxidative stress: bacteria, fungi, plants and animals. Comp Biochem Physiol C Toxicol Pharmacol 153:175–190

Marzban G, Tesei D, Sterflinger K (2013) A review beyond the borders: proteomics of microclonial black fungi and black yeasts. Nat Sci 5:640

McLean J, Purvis OW, Williamson BJ, Bailey EH (1998) Role for lichen melanins in uranium remediation. Nature 391:649

Mirocha CJ, DeVay JE (1971) Growth of fungi on an inorganic medium. Can J Microbiol 17:1373–1378

Moreno LF, Ahmed AA, Brankovics B, Cuomo CA, Menken SBJ, Taj-Aldeen SJ, Faidah H, Stielow JB, Teixeira MM, Prenafeta-Boldú FX, Vicente VA, de Hoog S (2018a) Genomic understanding of an infectious brain disease from the desert. G3 8(3):909–922

Moreno LF, Vicente VA, de Hoog S (2018b) Black yeasts in the omics era: achievements and challenges. Med Mycol 56:S32–S41

Muggia L, Grube M (2018) Fungal diversity in lichens: from extremotolerance to interactions with algae. Life 8:15

Muggia L, Hafellner J, Wirtz N, Hawksworth DL, Grube M (2008) The sterile microfilamentous lichenized fungi *Cystocoleus ebeneus* and *Racodium rupestre* are relatives of plant pathogens and clinically important dothidealean fungi. Mycol Res 112:50–56

Muggia L, Gueidan C, Knudsen K, Perlmutter G, Grube M (2013) The lichen connections of black fungi. Mycopathologia 175:523–535

Muggia L, Kocourková J, Knudsen K (2015) Disentangling the complex of *Lichenothelia* species from rock communities in the desert. Mycologia 107:1233–1253

Muggia L, Fleischhacker A, Kopun T, Grube M (2016) Extremotolerant fungi from alpine rock lichens and their phylogenetic relationships. Fungal Divers 76:119–142

Muggia L, Kraker S, Gößler T, Grube M (2018) Enforced fungal-algal symbioses in alginate spheres. FEMS Microbiol Ecol 365(14):fny11

Nai C, Wong HY, Pannenbecker A, Broughton WJ, Benoit I, de Vries RP, Gueidan C, Gorbushina AA (2013) Nutritional physiology of a rock-inhabiting, model microcolonial fungus from an ancestral lineage of the Chaetothyriales (Ascomycetes). Fungal Genet Biol 56:54–66

Nelsen MP, Lücking R, Grube M, Mbatchou JS, Muggia L, Plata ER, Lumbsch HT (2009) Unravelling the phylogenetic relationships of lichenised fungi in Dothideomyceta. Stud Mycol 64:135–144

Ohm RA, Feau N, Henrissat B, Schoch CL, Horwitz BA, Barry KW et al (2012) Diverse lifestyles and strategies of plant pathogenesis encoded in the genomes of eighteen *Dothideomycetes* fungi. PLoS Pathog 8:e1003037

Onofri S, Selbmann L, De Hoog GS, Grube M, Barreca D, Ruisi S, Zucconi L (2007a) Evolution and adaptation of fungi at boundaries of life. Adv Space Res 40:1657–1664

Onofri S, Zucconi L, Selbmann L, de Hoog S, de los Ríos A, Ruisi S, Grube M (2007b) Fungal associations at the cold edge of life. In: Seckbach J (ed) Algae and cyanobacteria in extreme environments. Springer, Dordrecht, pp 735–757

Øvstedal DO, Smith RL (2001) Lichens of Antarctica and South Georgia: a guide to their identification and ecology. Cambridge University Press, Cambridge

Pacelli C, Bryan RA, Onofri S, Selbmann L, Shuryak I, Dadachova E (2017) Melanin is effective in protecting fast and slow growing fungi from various types of ionizing radiation. Environ Microbiol 19:1612–1624

Palmer RJ, Friedmann EI (1988) Incorporation of inorganic carbon by antarctic cryptoendolithic fungi. Polarforschung 58:189–191

Plemenitaš A, Vaupotič T, Lenassi M, Kogej T, Gunde-Cimerman N (2008) Adaptation of extremely halotolerant black yeast *Hortaea werneckii* to increased osmolarity: a molecular perspective at a glance. Stud Mycol 61:67–75

Plemenitaš A, Lenassi M, Konte T, Kejžar A, Zajc J, Gostinčar C, Gunde-Cimerman N (2014) Adaptation to high salt concentrations in halotolerant/halophilic fungi: a molecular perspective. Front Microbiol 5:199

Plonka PM, Grabacka M (2006) Melanin synthesis in microorganisms-biotechnological and medical aspects. Acta Biochim Pol 53:429–443

Poyntner C, Blasi B, Arcalis E, Mirastschijski U, Sterflinger K, Tafer H (2016) The transcriptome of *Exophiala dermatitidis* during *ex-vivo* skin model infection. Front Cell Infect Microbiol 6:136

Prenafeta-Boldú FX, Kuhn A, Luykx D, Anke H, van Groenestijn JW, de Bont JAM (2001) Isolation and characterisation of fungi growing on volatile aromatic hydrocarbons as their sole carbon and energy source. Mycol Res 105:477–484

Prenafeta-Boldú FX, Summerbell R, de Hoog SG (2006) Fungi growing on aromatic hydrocarbons: biotechnology's unexpected encounter with biohazard? FEMS Microbiol Rev 30:109–130

Réblová M, Hubka V, Thureborn O, Lundberg J, Sallstedt T, Wedin M, Ivarsson M (2016) From the tunnels into the treetops: new lineages of black yeasts from biofilm in the Stockholm metro system and their relatives among ant-associated fungi in the Chaetothyriales. PLoS One 11:e0163396

Robertson KL, Mostaghim A, Cuomo CA, Soto CM, Lebedev N, Bailey RF, Wang Z (2012) Adaptation of the black yeast *Wangiella dermatitidis* to ionizing radiation: molecular and cellular mechanisms. PLoS One 7:e48674

Rodriguez RJ, Henson J, Van Volkenburgh E, Hoy M, Wright L, Beckwith F, Kim YO, Redman RS (2008) Stress tolerance in plants via habitat-adapted symbiosis. ISME J 2:404–416

Ruibal C, Platas G, Bills GF (2005) Isolation and characterization of melanized fungi from limestone formations in Mallorca. Mycol Prog 4:23–38

Ruibal C, Platas G, Bills GF (2008) High diversity and morphological convergence among melanised fungi from rock formations in the Central Mountain System of Spain. Persoonia 21:93–110

Ruibal C, Gueidan C, Selbmann L, Gorbushina AA, Crous PW, Groenewald JZ, Muggia L, Grube M, Isola D, Schoch CL, Staley JT, Lutzoni F, de Hoog GS (2009) Phylogeny of rock-inhabiting fungi related to Dothideomycetes. Stud Mycol 64:123–133

Schoch CL, Crous PW, Groenewald JZ et al (2009a) A class-wide phylogenetic assessment of Dothideomycetes. Stud Mycol 64:1–15

Schoch CL, Sung GH, López-Giráldez F et al (2009b) The Ascomycota tree of life: a phylum-wide phylogeny clarifies the origin and evolution of fundamental reproductive and ecological traits. Syst Biol 58:224–239

Scott JA, Untereiner WA, Ewaze JO, Wong B, Doyle D (2007) *Baudoinia*, a new genus to accommodate *Torula compniacensis*. Mycologia 99:592–601

Selbmann L, de Hoog GS, Mazzaglia A, Friedmann EI, Onofri S (2005) Fungi at the edge of life: cryptoendolithic black fungi from Antarctic deserts. Stud Mycol 51:1–32

Selbmann L, Isola D, Egidi E, Zucconi L, Gueidan C, de Hoog GS, Onofri S (2014) Mountain tips as reservoirs for new rock-fungal entities: *Saxomyces* gen. nov. and four new species from the Alps. Fungal Divers 65:167–182

Selbmann L, Zucconi L, Isola D, Onofri S (2015) Rock black fungi: excellence in the extremes, from the Antarctic to space. Curr Genet 61:335–345

Senesi N, Sposito G, Martin JP (1987) Copper (II) and iron (III) complexation by humic acid-like polymers (melanins) from soil fungi. Sci Tot Environ 62:241–252

Shcherba VV, Babitskaya VG, Kurchenko VP, Ikonnikova NV, Kukulyanskaya TA (2000) Antioxidant properties of fungal melanin pigments. Appl Biochem Microbiol 36:491–495

Singaravelan N, Grishkan I, Beharav A, Wakamatsu K, Ito S, Nevo E (2008) Adaptive melanin response of the soil fungus *Aspergillus niger* to UV radiation stress at "Evolution Canyon", Mount Carmel, Israel. PLoS One 3:e2993

Sinha S, Flibotte S, Neira M, Formby S, Plemenitaš A, Cimerman NG, Lenassi M, Gostinčar C, Stajich JE, Nislow C (2017) Insight into the recent genome duplication of the halophilic yeast *Hortaea werneckii*: combining an improved genome with gene expression and chromatin structure. G3 7(7):2015–2022

Slepecky RA, Starmer WT (2009) Phenotypic plasticity in fungi: a review with observations on *Aureobasidium pullulans*. Mycologia 101:823–832

Staley JT, Palmer F, Adams JB (1982) Microcolonial fungi: common inhabitants on desert rocks? Science 215:1093–1095

Sterflinger K (1998) Ecophysiology of rock inhabiting black yeasts with special reference to temperature and osmotic stress. Antonie Van Leeuwenhoek 74:271–281

Sterflinger K (2006) Black yeasts and meristematic fungi: ecology, diversity and identification. In: Rosa CA, Peter G (eds) Biodiversity and ecophysiology of yeasts. Springer, Berlin, Heidelberg, pp 501–514

Sterflinger K, de Hoog GS, Haase G (1999) Phylogeny and ecology of meristematic ascomycetes. Stud Mycol 43:5–22

Sterflinger K, Lopandic K, Pandey RV, Blasi B, Kriegner A (2014) Nothing special in the specialist? Draft genome sequence of *Cryomyces antarcticus*, the most extremophilic fungus from Antarctica. PLoS One 9:e109908

Swan G (1974) Structure, chemistry, and biosynthesis of the melanins. In: Herz W, Grisebach H, Kirby GW (eds) Progress in the chemistry of organic natural products. Springer, Vienna, pp 521–582

Taylor JW (2011) One fungus = one name: DNA and fungal nomenclature twenty years after PCR. IMA Fungus 2:113–120

Teixeira MM, Moreno LF, Stielow BJ Muszewska A, Hainaut M, Gonzaga L, Abouelleil A, Patané JS, Priest M, Souza R, Young S, Ferreira KS, Zeng Q, da Cunha MM, Gladki A, Barker B, Vicente VA, de Souza EM, Almeida S, Henrissat B, Vasconcelos AT, Deng S, Voglmayr H, Moussa TA, Gorbushina A, Felipe MS, Cuomo CA, de Hoog GS (2017) Exploring the genomic diversity of black yeasts and relatives (Chaetothyriales, Ascomycota). Stud Mycol 86:1–28

Tesei D, Marzban G, Zakharova K, Isola D, Selbmann L, Sterflinger K (2012) Alteration of protein patterns in black rock inhabiting fungi as a response to different temperatures. Fungal Biol 116:932–940

Tribe HT, Thines E, Weber RW (2006) Moulds that should be better known: the wine cellar mould, *Racodium cellare* Persoon. Mycologist 20:171–175

Turian G (1977) *Coniosporium aeroalgicolum* sp. hov., moissisure dèmatièe semi-lichenisante. Ber Schweiz Bot Ges 87:19–24

Turk M, Plemenitas A, Gunde-Cimerman N (2011) Extremophilic yeasts: plasma-membrane fluidity as determinant of stress tolerance. FungalBiol 115:950–959

Turrens JF (2003) Mitochondrial formation of reactive oxygen species. J Physiol 552:335–344

Valadbeigi T, Schultz M, Von Brackel W (2016) Two new species of *Lichenothelia* (Lichenotheliaceae) from Iran. Lichenologist 48:191–199

Vasse M, Voglmayr H, Mayer V et al (2018) A phylogenetic perspective on the association between ants (Hymenoptera: Formicidae) and black yeasts (Ascomycota: Chaetothyriales). Proc R Soc B 284:20162519

Vicente VA, Weiss VA, Bombassaro A, Moreno LF, Costa FF, Raittz RT, Leão AC, Gomes RR, Bocca AL, Fornari G, de Castro RJA, Sun J, Faoro H, Tadra-Sfeir MZ, Baura V, Balsanelli E, Almeida SR, Dos Santos SS, Teixeira MM, Soares Felipe MS, do Nascimento MMF, Pedrosa FO, Steffens MB, Attili-Angelis D, Najafzadeh MJ, Queiroz-Telles F, Souza EM, De Hoog S

(2017) Comparative genomics of sibling species of *Fonsecaea* associated with human chromoblastomycosis. Front Microbiol:8, 1924

Voglmayr H, Mayer V, Maschwitz U, Moog J, Djieto-Lordon C, Blatrix R (2011) The diversity of ant-associated black yeasts: insights into a newly discovered world of symbiotic interactions. Fungal Biol 115:1077–1091

Wang Y, Casadevall A (1994) Decreased susceptibility of melanized *Cryptococcus neoformans* to UV light. Appl Environ Microbiol 60:3864–3866

Zakharova K, Tesei D, Marzban G, Dijksterhuis J, Wyatt T (2013) Microcolonial fungi on rocks: a life in constant drought? Mycopathologia 175:537–547

Zakharova K, Sterflinger K, Razzazi-Fazeli E, Noebauer K, Marzban G (2014) Global proteomics of the extremophile black fungus *Cryomyces antarcticus* using 2D-electrophoresis. Nat Sci 6:978

Zalar P, Gostinčar C, de Hoog GS, Uršič V, Sudhadham M, Gunde-Cimerman N (2008) Redefinition of *Aureobasidium pullulans* and its varieties. Stud Mycol 61:21–38

Zalar P, Novak M, de Hoog GS, Gunde-Cimerman N (2011) Dishwashers—a man-made ecological niche accommodating human opportunistic fungal pathogens. Fungal Biol 115:997–1007

Zhang W, Li F, Nie L (2010) Integrating multiple 'omics' analysis for microbial biology: application and methodologies. Microbiology 156:287–301

Zhurbenko MP (2008) A new species from the genus *Lichenothelia* (Ascomycota) from the Northern Ural. Mikol Fitopatol 42:240–243

Zupančič J, Novak Babič M, Zalar P, Gunde-Cimerman N (2016) The black yeast *Exophiala dermatitidis* and other selected opportunistic human fungal pathogens spread from dishwashers to kitchens. PLoS One 11:e0148166

Chapter 8
Basidiomycetous Yeast of the Genus *Mrakia*

Masaharu Tsuji (ID)**, Sakae Kudoh, Yukiko Tanabe, and Tamotsu Hoshino**

8.1 Introduction

The first fungus recorded from Antarctica was *Scleotium antarcticum* collected at Danco Land, Antarctic Peninsula, on an SY Belgica expedition (Bommer and Rouissean 1900). Over 1000 fungal species from 421 genera have been isolated and recorded from Antarctica (Bridge and Spooner 2012); the list of known species from culturing and collection consists of 68% ascomycetes, 23% basidiomycetes, 5% zygomycetes, and the final 4% made up of various other lineages. Fell et al. (2006) reported that approximately 40% of fungi isolated in Dry Valley, a low-temperature and low-moisture region, are occupied by basidiomycetous yeasts.

Despite several reports indicating that basidiomycetous yeast of the genus *Mrakia* makes up the majority of the mycobiota and is one of the most adaptive fungi in Antarctica (Di Menna 1966b; Tsuji et al. 2013a), there are no reviews about the genus *Mrakia*. In this chapter, we review the history of the genus *Mrakia* and also the biological potential of the genus.

M. Tsuji (✉)
National Institute of Polar Research (NIPR), Tokyo, Japan

S. Kudoh · Y. Tanabe
National Institute of Polar Research (NIPR), Tokyo, Japan

Department of Polar Science, SOKENDAI (The Graduate University for Advanced Studies), Tokyo, Japan

T. Hoshino
Department of Life and Environmental Science, Faculty of Engineering, The Hachinohe Institute of Technology, Hachinohe, Aomori, Japan

© Springer Nature Switzerland AG 2019   145
S. M. Tiquia-Arashiro, M. Grube (eds.), *Fungi in Extreme Environments: Ecological Role and Biotechnological Significance*,
https://doi.org/10.1007/978-3-030-19030-9_8

8.2   Taxonomic History of the Genus *Mrakia*

Three new isolates of *Candida* spp. were collected by di Menna (1966a) from Scott
Base, Ross Island, Antarctica, and were considered to be new species. These three
isolates have several common characteristics, such as the ability to reduce NO_3, and
the production of starch-like extracellular polysaccharide compounds. However,
they also differ in some characteristics, including sugar sources used for carbon
assimilation and fermentation. The three isolates were classified as *C. nivalis*,
C. gelida, and *C. frigida* on the basis of their characteristics. The history of the
genus *Mrakia* perhaps began with this report. Fell et al. (1969) identified a heterobasidiomycetous life cycle in a *Candida* spp. isolated from soil from Antarctica. At
this point, *C. stokesii*, *C. nivalis*, *C. gelida*, and *C. frigida* were redesigned as
Leucosporidium stokesii, *L. nivalis*, *L. gelidum*, and *L. frigidum*, respectively. About
two decades after the publication of the article by Fell and colleagues, the genus
Leucosporidium was reclassified based on the coenzyme Q (CoQ) system. *L. stokesii*, *L. nivalis*, *L. gelidum*, and *L. frigidum* all express CoQ_8, so these species were
again reclassified and transferred to the newly established genus *Mrakia* as *Mrakia
stokesii*, *M. nivalis*, *M. gelida*, and *M. frigida*. In contrast, other *Leucosporidium*
species express CoQ_9 or CoQ_{10} (Yamada and Komagata 1987). The genus *Mrakiella*
was established in order to accommodate species in the *Mrakia* clade for which
sexual cycles had not been observed (Margesin and Fell 2008). Subsequently, the
genus *Mrakiella* was integrated into the genus *Mrakia* (Liu et al. 2015).

Fell and Kurtzman (1990) reported that closely related homothallic species, such
as *Rhodotorula*, *Candida*, and *Mrakia*, can be identified by using 230 base pairs of
the 18S rRNA large subunit. Suh and Sugiyama (1993) used this approach for the
phylogenetic analysis of basidiomycetous yeast using 18S rRNA sequences. Diaz
and Fell (2000) analyzed the phylogenetic relationship of *Mrakia* species based on
sequences of the intergenic spacer (IGS) and internal transcribed spacer (ITS)
regions. In addition, Fell et al. (2000) reported that basidiomycetous yeasts can be
classified to the species level using the D1/D2 domain sequence of the 26S
rRNA. Recently, the phylogenetic analysis of the genus *Mrakia* was performed
using concatenated ITS and D1/D2 region sequences (Tsuji et al. 2016a, 2018).

Currently, the genus *Mrakia* includes a total of 10 species, namely, *M. aquatica*,
M. arctica, *M. blollopis*, *M. cryoconiti*, *M. frigida*, *M. gelida*, *M. hoshino-
nis*, *M. nicombsii*, *M. psychrophila*, and *M. robertii* (Yamada and Komagata 1987;
Xin and Zhou 2007; Thomas-Hall et al. 2010; Liu et al. 2015; Tsuji et al. 2018,
2019). A phylogenetic tree of *Mrakia* species generated using the ITS region and
D1/D2 domain sequences is shown in Fig. 8.1. According to the phylogenetic analysis, *Mrakia curviuscula* is located in a clade far from the genus *Mrakia* (Fig. 8.1).
Fell (2011) reported that this species was isolated from moss from a forest in Central
Russia. In comparison, other *Mrakia* species were isolated from snow, soil, and
glaciers in cold environments. Moreover, *M. curviuscula* grows at 25 °C, while
other *Mrakia* species fail to grow at 25 °C. Fell therefore suggests that *M. curviuscula*

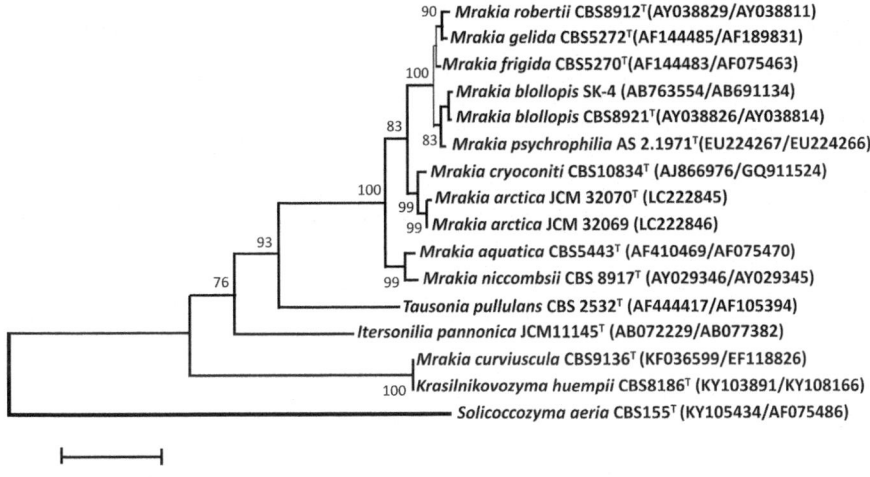

Fig. 8.1 Phylogenetic tree of *Mrakia* spp. and other related species using the ITS region and D1/ D2 domain sequences. Maximum likelihood analysis of the ITS region-LSU D1/D2 domain sequences of genus *Mrakia* and closely related species. The tree was constructed by maximum likelihood analysis with MEGA7. Bootstrap percentages of maximum likelihood analysis over 50% from 1000 bootstrap replicates. Bootstrap percentages from 1000 replications are shown on the branches. *Solicoccozyma aerius* CBS155[T] was used as an out-group. The scale bar represents 0.05 substitutions per nucleotide position

should be removed from the genus *Mrakia*. Currently, this opinion is widely accepted regarding the classification of the genus *Mrakia*.

8.3   Physiological Characteristics

The most important characteristics for classification of yeast taxonomy are the sugars used as sources for the assimilation of carbon and the ability to ferment different sugars, as well as the sequence information for the ITS region and D1/D2 domain. The primary characteristics of *Mrakia* spp. are shown in Table 8.1. For the assimilation of carbon, *Mrakia* spp. commonly can use glucose and sucrose. *M. frigida* is not able to assimilate maltose. *M. gelida* and *M. robertii* cannot assimilate lactose. *M. arctica* and *M. aquatica* are unable to assimilate inositol. Moreover, *M. blollopis*, *M. frigida*, and *M. robertii* can grow in vitamin-free conditions. *Mrakia* spp. therefore have different characteristics for each species regarding the sources used for carbon or nutrient assimilation.

Little is currently known about ethanol fermentation by basidiomycetous yeasts. Species in the basidiomycetous yeast *Mrakia* are known for their ability to ferment sugars. In fact, seven of the nine *Mrakia* species are able to ferment glucose and sucrose (Table 8.1). *M. blollopis* CBS8921[T] and *M. psychrophila* AS2.1971[T] are not

Table 8.1 Comparison of physiological characteristics of the genus *Mrakia*

	M. arctica	M. aquatica	M. blollopis	M. cryoconiti	M. frigida	M. gelida	M. niccombsii	M. psychrophila	M. robertii
Maximum growth temperature	20	20	20	20	19	20	20	18	20
Assimilation of									
Glucose	+	+	+	+	+	+	+	+	+
Sucrose	+	+	+	+	+	+	+	+	+
Maltose	+	+	+	+	–	+	+	+	+
Lactose	+	+	w	+	+	–	w	+	–
Inositol	–	–	w/+	+	+	v	w	v	w/+
Fermentation of									
Glucose	+	–/w	+	–	w	w	–	d	+
Sucrose	+	v	+	–	w	w	nd	d	+
Raffinose	nd	v	–	–	w	w	nd	–	–
Galactose	nd	–	–	–	w	w	nd	–	–
Lactose	nd	–	–	–	–	–	nd	–	D
Maltose	nd	v	–	–	w	d	nd	–	D
Growth on									
50% Glucose	–	–	–	+	+	–	–	–	–
Vitamin-free	–	–	w	–	+	–	–	–	w/+

Main physiology test results for characteristics of the *Mrakia* species. Physiological data were taken from Fell et al. (1969), Xin and Zhou (2007), Thomas-Hall et al. (2010) and Tsuji et al. (2018)

+ positive, w weak, – negative, v variable, d delayed positive, nd no data

Fig. 8.2 Cell morphology of *Mrakia blollopis* SK-4. Micrograph of *Mrakia blollopis* SK-4. The fungi were inoculated on 1.5% (w/v) water agar and incubated for 4 weeks at 4 °C. Bar: 10 µm

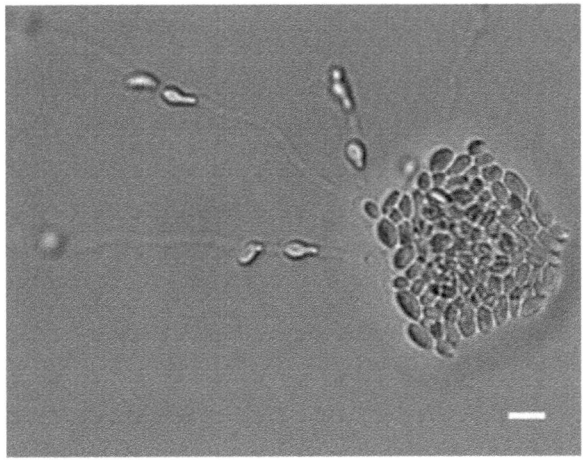

able to ferment raffinose, galactose, lactose, or maltose. *M. frigida* CBS5270[T] and *M. gelida* CBS5272[T] are unable to utilize lactose for fermentation, and *M. robertii* CBS8912[T] is unable to ferment raffinose or galactose.

M. blollopis SK-4 (Fig. 8.2), isolated from an algal matt from Naga-ike in Skarvsnes ice-free area, East Antarctica, shares a high degree of homology of its region and D1/D2 domain sequences with *M. blollopis* CBS8921[T] (>99.6%), and strain SK-4 demonstrates similar characteristics of carbon assimilation with *M. blollopis* CBS8921[T]. Strain SK-4 is clearly able to ferment maltose; however, CBS8921[T] cannot ferment maltose (Table 8.2). Results from phylogenetic analysis indicate that SK-4 should be classified as *M. blollopis*; nevertheless, SK-4 has different characteristics regarding carbon assimilation and fermentative ability for maltose. Therefore, a comparison of the physiological characteristics of strain SK-4 with CBS8921[T] is needed. The maximum growth temperature of *M. blollopis* SK-4 is 22 °C, whereas the maximum growth temperature of CBS8921[T] is 20 °C. *M. blollopis* SK-4 differs from CBS8921[T] in substrate utilization as well. The strain SK-4 thrives on lactose, D-arabinose, and inositol medium, unlike CBS8921[T], which also grows well on vitamin-free medium. A comparison of the fermentation abilities reveals that *M. blollopis* SK-4 can ferment typical sugars such as glucose, sucrose, galactose, maltose, lactose, raffinose, trehalose, and melibiose, while CBS8921[T] is a weaker fermenter, and is limited in the variety of sugars it can use for fermentation (Table 8.2).

8.4 Ecological Role

Mrakia spp. have been isolated from various extreme environments, including the Arctic, Siberia, the Alps, Alaska, Patagonia, and Antarctica (Morgesin et al. 2005; Thomas-Hall et al. 2010; Singh & Singh 2012; de Garcia et al. 2012; Panikov &

Table 8.2 Comparison of physiological characteristics of *Mrakia blollopis* SK-4 and *Mrakia blollopis* CBS8912[T]

Characteristic	*M. blollopis* SK-4	*M. blollopis* CBS8921[T]
Maximum growth temperature	22 °C	20 °C
Assimilation of		
Lactose	+	w
Inositol	+	w/+
D-Arabinose	+	w/−
Ethanol	w/−	+
Growth on 50% glucose	w/−	−
Growth on vitamin-free medium	+	w
Fermentation of		
Galactose	+	−
Lactose	+	−
Raffinose	+	−
Maltose	+	−

Main physiology test results for characteristics of *M. blollopis* SK-4 and type strain of *Mrakia blollopis*. Physiological data were taken from Thomas-Hall et al. (2010) and Tsuji et al. (2013c)
+ positive, *w* weak, − negative

Sizova 2007). Interestingly, this genus has been detected in frozen fish, frozen yogurt, and soil from Hokkaido, Japan (Komagata and Nakase 1965; Moreira et al. 2001; Nakagawa et al. 2004). According to di Menna (1966b), approximately 24% of culturable yeasts in the soil from Ross Island, Antarctica, are *Mrakia* spp. Moreover, about 35% of the culturable fungi in soil surrounding lakes and in lake sediment from Skarvsnes ice-free area, East Antarctica, belong to the genus *Mrakia* (Tsuji et al. 2013a). These results indicate that *Mrakia* spp. are the dominant and most adaptive fungi in Antarctica.

In a glacier retreating area of the Norway High Arctic, the genus *Mrakia* was isolated from the glacier near the terminus position, and likely served a role as a carbon source supplier by decomposing surrounding cell wall and cell membrane remnants with secreted extracellular enzymes (Tsuji et al. 2016b). Pathan et al. (2010) tested a strain of *Mrakia* spp. isolated from an Arctic glacier for extracellular enzyme activity at various temperatures. In that study, *Mrakia* secreted urease, amylase, protease, and lipase, and these enzymes were highly active when *Mrakia* spp. was cultured at 22 °C, compared to when it was cultured at 8 °C. Yeasts isolated from Patagonia, Argentina, are reported by de Garcia et al. (2007) to show higher lipolytic activity at 4 °C than at 20 °C. Moreover, de Garcia et al. (2012) reported that *Mrakia* spp. isolated from the Patagonian Andes in Argentina produce esterase at 5 °C. Turchetti et al. (2008) showed that approximately 77% of the *Mrakia* spp. isolated from alpine glaciers exhibit lipolytic activity at 4 °C. We have previously reported that lipase activity of the *Mrakia* strain isolated from Skarvsnes ice-free area is higher at 4 °C than at 15 °C (Tsuji et al. 2013b, 2014a, 2015b). *M. arctica* was isolated from Canadian Arctic, and the optimum temperature for lipase secretion from this isolate is −3 °C, and the highest level of protein degradation activity by *M. arctica* occurs at 10 °C. Singh et al. (2016) report that *M. blollopis*, isolated from the Norwegian Arctic, shows only weak protease activity, and our previous whole genome analysis of an Antarctic strain of *M. blollopis* strain shows that it lacks the protease K gene and ice-binding gene (Tsuji et al. 2015a). Extracellular enzyme tests indicate that the genus *Mrakia* is able to decompose a variety of organic materials over a wide range of temperatures. The *Mrakia* spp. may therefore play an important role in the biogeochemical cycles of polar ecosystems. However, we have only fragmented information regarding the ecological role of *Mrakia* in polar regions. Further experiments are required in order to elucidate the ecological role of the *Mrakia* spp.

8.5 Biotechnological Applications

Cold environments cover a large portion of the planet, with many ecosystems permanently exposed to temperatures below 5 °C (Feller and Gerday 2003). As microbes adapted to cold environments are able to grow at temperatures below 0 °C, it is expected that these organisms employ unique physiological tools in order to survive, such as cold-adapted enzymes and ice-binding proteins (IBPs) (Buzzini et al. 2012; Tiquia-Arashiro and Rodrigues 2016).

   M. frigida and *M. gelida* have been evaluated for heat shock response. When these microorganisms are incubated at 20 °C and 25 °C for 3 h, both species induce heat-shock proteins, such as hsp70, hsp90, and 110 kDa proteins. When *M. frigida* is incubated at 25 °C for 3 h, this species induces increased levels of the 90 kDa protein (Deegenaars and Watson 1998).

   Margesin et al. (2005) report the presence of a cold-active alkaline pectate lyase in *M. frigida*. This *M. frigida* pectate lyase (*M. frigida* PL) is highly secreted when the fungus is grown at 1–10 °C. The temperature for optimal activity of the enzyme is 25°–30 °C, and the optimal pH is 8.5–9.0. As for the thermal stability of *M. frigida* PL, about 60% of the relative activity is lost when incubated for 15 min at 30 °C. Moreover, when this enzyme is incubated for 48 h at 2 °C, 5–21% of initial activity is lost.

   M. frigida strain 2E00797 possesses a cold-active toxin lethal to the pathogenetic crab yeast *Metschnikowia bicuspidata* (Hua et al. 2010). This killer protein toxin was purified and characterized by Liu et al. (2012). After purification, a 55.6 kDa single band is detectable by sodium dodecyl-polyacrylamide gel electrophoresis (SDS-PAGE). This protein has about 35% sequence homology with a protein kinase, but Liu and colleagues did not determine its activity. The optimal temperature for toxicity of the purified protein was 16 °C, and the enzyme is stable at temperatures below 25 °C, even after 60 min of preincubation. The optimal pH for toxin activity was 4.5, and the enzyme was stable, even after 24-h preincubation at pH 4.5. The highest killing activity of this enzyme was recorded in the presence of 3% (w/v) NaCl.

Milk fat curdle in sewage is one of the most refractory materials to active sludge treatment under low-temperature conditions. In an effort to solve this problem using a bioremediation agent, we previously collected about 75 species of *Mrakia* from soil surrounding a lake and from lake sediments in Skarvsnes ice-free area, East Antarctica. We tested the isolates for their ability to decompose milk fat under low-temperature conditions and evaluated their potential for application to an active sludge system in a cold climate. Fifty-six of the 75 *Mrakia* species exhibited a clear zone, consistent with milk fat decomposition (Shimohara et al. 2012). *M. blollopis* SK-4 demonstrated one of the largest clear zones under low temperature. Therefore, we expected that activated sludge containing the yeast strain SK-4 would have the potential to improve the removal of milk fat biological oxygen demand (BOD_5), and tested it in the removal of milk fat in a low-temperature environment. Consequently, SK-4 showed a BOD removal rate that was 1.25-fold higher than that of the control. The SK-4 lipase was then purified, and the enzyme was found to be quite stable under a wide range of temperatures and pH, even in the presence of various metal ions and organic solvents. The effect of nitrogen concentration on cell morphology, and on the formation of the clear zone around the colonies, which is indicative of lipase activity, was also evaluated on fresh cream agar at various culturing temperatures. When grown on high-nitrogen fresh cream agar, the largest clear zone was formed around colonies at 4 °C, and the cell morphology was a yeast form. The cell morphology of SK-4 during lipase secretion, based on fluorescence in situ hybridization (FISH), was that of a yeast form. These results indicate that SK-4 takes a yeast form when growing in aquatic environments, and in this form it may secrete more lipase than when in a mycelial form (Tsuji et al. 2014a). *Mrakia* sp. YSAR-9, isolated from the Arctic, has a high homology of the ITS region and D1/D2 domain sequences compared with those of *M. blollopis* CBS8921^T (> 99%). YSAR-9 is believed to be the same species as *M. blollopis* SK-4. However, when YSAR-9 is inoculated on lipase assay agar at 8 °C and 22 °C, this yeast demonstrates the largest clear zone at 22 °C (Pathan et al. 2010), while SK-4 forms its largest clear zones at 4 °C and fails to form clear zones at 20 °C (Tsuji et al. 2015b). Strong lipase secretion under low temperature was therefore thought to be a unique characteristic of each strain.

The fermentation abilities of *Mrakia* spp. were previously described in the section of physiological characteristics. In general, the fermentation abilities were uncommon characteristics for basidiomycetous yeast (Fell 2011). However, all *Mrakia* species could ferment glucose and sucrose except *M. aquatica and M. cryoconiti* (Fell and Margesin 2011). *M. frigida, M. blollopis, M. gelida,* and *M. robertii* were evaluated for ethanol fermentation in tests using a home brewing kit. Thomas-Hall et al. (2010) report that all the strains are able to ferment sucrose, but are unable to completely convert sucrose to ethanol, and cell growth is stopped in the presence of ethanol at concentrations greater than 2% (v/v). The ability of strain SK-4 to strongly ferment various sugars, and the findings by Thomas-Hall and colleagues, led us to test SK-4 for its ability to ferment ethanol using various glucose concentrations at 10 °C. Consequently, it was found that SK-4 is able to consume glucose at all the concentrations tested (4%, 6%, and 12%), with 14.4 g/l, 20.4 g/l, and 48.2 g/l ethanol being produced, respectively. Moreover, SK-4 was evaluated for ethanol fermentation using lignocellulosic biomass hydrolysates, such as eucalyptus and Japanese cedar for substrates. This yeast strongly ferments Japanese cedar hydrolysate, but not eucalyptus hydrolysate (Tsuji et al. 2013d). Since SK-4 has both acetic acid and formic acid tolerances during fermentation, the differences in SK-4 fermentability using Japanese cedar and eucalyptus remain unclear. Direct ethanol fermentation (DEF) from a cellulosic biomass was firstly reported by Takagi et al. (1977). In this technique, enzymatic hydrolysis and ethanol fermentation are carried out at the same time. In the presence of a high concentration of glucose, cellulase activity is considerably depressed. However, when yeast is mixed with an enzymatic hydrolysis reactor, glucose is formed from the cellulase activity on the cellulolytic biomass. The glucose is maintained at a low concentration and is rapidly converted to ethanol by the yeast. Moreover, when a lignocellulosic biomass is saccharified and fermented at the same time, major fermentation inhibitors like furfural and 5-methylfolate (5-MHF) are maintained at very low concentrations compared to the concentrations of major fermentation inhibitors in the enzymatic hydrolysate (Thomsen et al. 2009). Therefore, this technique is expected to improve saccharification and ethanol fermentation rates. Since the inhibition of ethanol fermentation is thought to be prevented by DEF, SK-4 was used for direct ethanol fermentation. SK-4 is able to ferment eucalyptus by DEF, although high concentrations of glucose remain in the DEF solutions. Since a nonionic surfactant is thought to combine with lignin (Eriksson et al. 2002), this unproductive fermentation of eucalyptus was thought to be potentially improved by the addition of the nonionic surfactant. In fact, the ethanol concentration was increased 1.6-fold compared to the ethanol concentration in the DEF solution not containing the nonionic surfactant (Tsuji et al. 2014b).

8.6 Conclusions and Future Perspectives

Mrakia spp., especially *M. blollopis* SK-4, have potential for use as bioremediation organisms for degrading milk fat under low-temperature conditions and demonstrate good potential for use in ethanol fermentation. Does this genus truly lack ice-binding proteins? How many lipase isozymes does this genus have? What is the optimal pH and temperature for *Mrakia* spp.-mediated ethanol fermentation? About 50 years ago, di Menna reported that *Mrakia* spp. is the dominant yeast in Antarctic environments. Since then, it has been demonstrated that this genus has several interesting characteristics, such as its fermentative ability, osmotolerance, and secretion of a stable lipase. For biological and physiological characteristics, information regarding *Mrakia* remains fragmentary and limited. Nevertheless, *Mrakia* spp. are thought to have good potential as agents for bioremediation and bioethanol production under conditions of low temperature. Studies must continue on the physiological, morphological, and genomic characteristics of *Mrakia* spp. We believe that the results obtained in previous studies will contribute to the progress of the related research fields and hope that further investigation will offer many opportunities to obtain more valuable knowledge on the Antarctic microbes and their potential uses for human activities. In the near future, this genus will become an important agent in the field of low-temperature microbiology.

Acknowledgment This research was partially supported byan NIPR Research Project (KP-309), a JSPS Grant-in-Aid for Young Scientists (A) to M. Tsuji (No. JP16H06211). Institution for Fermentation, Osaka, for Young Scientists to M. Tsuji (no. Y-2018–004), and the ArCS (Arctic Challenge for Sustainability) provided by the Ministry of Education, Culture, Sports, Science and Technology, Japan. We are deeply grateful to Masaki Uchida (NIPR).

References

Boomer E, Rousseau M (1900) Note préliminaire sur les champignons recueillis par l'Expedition Antarctique Belge. Bull Acad R Sci Belgiq Clas Sci 8:640–646

Bridge PD, Spooner BM (2012) Non-lichenized Antarctic fungi: transient visitors or members of a cryptic ecosystem? Fungal Ecol 5:381–394

Buzzini P, Branda E, Goretti M, Turchetti B (2012) Psychrophilic yeasts from worldwide glacial habitats: diversity, adaptation strategies and biological potential. FEMS Microbiol Ecol 82:217–241

de Garcia V, Brizzio S, Libkind D, Buzzini P, van Broock M (2007) Biodiversity of cold adapted yeasts from glacial meltwater rivers in Patagonia Argentina. FEMS Microbiol Ecol 59:331–341

de Garcia V, Brizzio S, Broock MR (2012) Yeasts from glacial ice of Patagonian Andes, Argentina. FEMS Microbiol Ecol 82:540–550

Deegenaars ML, Watson K (1998) Heat shock response in psychrophilic and psychrotrophic yeast from Antarctica. Extremophiles 2:41–49

di Menna ME (1966a) Three new yeasts from Antarctica soils: *Candida nivalis*, *Candida gelida* and *Candida frigida* spp n. Antonie Van Leeuenwoek 32:25–28

di Menna ME (1966b) Yeasts in Antarctic soil. Antonie Van Leeuwenhoek 32:29–38

Diaz MR, Fell JW (2000) Molecular analyses of the IGS & ITS regions of rDNA of the psychrophilic yeasts in the genus *Mrakia*. Antonie Van Leeuwenhoek 77:7–12

Eriksson T, Börjesson J, Tjerneld F (2002) Mechanism of surfactant effect in enzymatic hydrolysis of lignocellulose. Enzym Microb Technol 31:353–364

Fell JW (2011) *Mrakia* Y. Yamada & Komagata (1987). In: Kurtzman CP, Fell JW, Boekhout T (eds) The yeast, a taxonomic study, 5th edn. Elsevier, Amsterdam, pp 1503–1510

Fell JW, Kurtzman CP (1990) Nucleotide sequence analysis of a variable region of the large subunit rRNA for identification of marine-occurring yeasts. Curr Microbiol 21:295–300

Fell JW, Margesin R (2011) *Mrakiella* Margesin & Fell (2008). In: Kurtzman CP, Fell JW, Boekhout T (eds) The yeast, a taxonomic study, 5th edn. Elsevier, Amsterdam, Netherlands, pp 1847–1852

Fell JW, Statzell AC, Hunter IL, Phaff HJ (1969) *Leucosporidium* gen. n., the heterobasidiomycetous stage of several yeasts of the genus Candida. Antonie Van Leeuwenhoek 35:433–462

Fell JW, Boekhout T, Fonseca A, Scoretti G, Statzelll-Tallman A (2000) Biodiversity and systematics of basidiomycetous yeasts as determined by large-subunit rDNA D1/D2 domain sequence analysis. Int J Syst Evol Microbiol 50:1351–1371

Fell JW, Scorzetti G, Connell L, Craig S (2006) Biodiversity of micro-eucaryotes in Antarctic Dry Valley soil with <5% soil moisture. Soil Biol Biochem 38:3107–3119

Feller G, Gerday C (2003) Psychrophilic enzymes: hot topics in cold adaptation. Nat Rev Microbiol 1:200–208

Hua MX, Chi Z, Liu GL, Buzdar MA, Chi ZM (2010) Production of a novel and cold-active killer toxin by *Mrakia frigida* 2E00797 isolated from sea sediment in Antarctica. Extremophiles 14:515–521

Komagata K, Nakase T (1965) New species of the genus Candida isolated from frozen foods. J Gen Appl Microbiol 11:255–267

Liu GL, Wang K, Hua MX, Buzdar MA, Chi ZM (2012) Purification and characterization of the cold-active killer toxin from the psychrotolerant yeast Mrakia frigida isolated from sea sediments in Antarctica. Process Biochem 47:822–827

Liu XZ, Wang QM, Göker M, Groenewald M, Kachalkin AV, Lumbsch HT, Millanes AM, Wedin M, Yurkov AM, Boekhout T, Bai FY (2015) Towards an integrated phylogenetic classification of the *Tremellomycetes*. Stud Mycol 81:85–147

Margesin R, Fell JW (2008) *Mrakiella cryoconiti* gen. nov., sp. nov., a psychrophilic, anamorphic, basidiomycetous yeast from alpine and arctic habitats. Int J Syst Evol Microbiol 58:2977–2982

Margesin R, Fauster V, Fonteyne PA (2005) Characterization of cold-active pectate lyases from psychrophilic *Mrakia frigida*. Lett Appl Microbiol 40:453–459

Masaharu Tsuji, Yukiko Tanabe, Warwick F. Vincent, Masaki Uchida, (2019) Mrakia hoshinonis sp. nov., a novel psychrophilic yeast isolated from a retreating glacier on Ellesmere Island in the Canadian High Arctic. International Journal of Systematic and Evolutionary Microbiology 69 (4):944–948

Moreira SR, Schwan RF, de Carvalho P, Wheals AE (2001) Isolation and identification of yeasts and filamentous fungi from yoghurts in Brazil. Braz J Microbiol 32:117–122

Nakagawa T, Nagaoka T, Taniguchi S, Miyaji T, Tomizawa N (2004) Isolation and characterization of psychrophilic yeasts producing cold-adapted pectinolytic enzymes. Lett Appl Microbiol 38:383–387

Panikov NS, Sizova M (2007) Growth kinetics of microorganisms isolated from Alaskan soil and permafrost in solid media frozen down to −35°C. FEMS Microbiol Ecol 59:500–512

Pathan AAK, Bhadra B, Begum Z, Shivaji S (2010) Diversity of yeasts from puddles in the vicinity of Midre Lovénbreenglacier, arctic and bioprospecting for enzymes and fatty acids. Curr Microbiol 60:307–314

Shimohara K, Fujiu S, Tsuji M, Kudoh S, Hoshino T, Yokota Y (2012) Lipolytic activities and their thermal dependence of *Mrakia* species, basidiomycetous yeast from Antarctica. J Water Waste (in Japanese) 54:691–696

Sinclair NA, Stokes JL (1965) Obligately psychrophilic yeasts from the polar regions. Can J Microbiol 11:259–269

Singh P, Singh SM (2012) Characterization of yeast and filamentous fungi isolated from cryoconite holes of Svalbard, Arctic. Polar Biol 35:575–583

Singh SM, Tsuji M, Gawas-Sakhalker P, Loonen MJJE, Hoshino T (2016) Bird feather fungi from Svalbard Arctic. Polar Biol 39:523–532

Suh SO, Sugiyama J (1993) Phylogeny among the basidiomycetous yeasts inferred from small subunit ribosomal DNA sequence. J Gen Microbiol 139:1595–1598

Takagi M, Abe S, Suzuki S, Emert GH, Yata A (1977) A method for production of alcohol directly from cellulose using cellulase and yeast. In Bioconversion Symposium Proceedings. IIT, Delhi, pp 551–571

Thomas-Hall SR, Turchetti B, Buzzini P, Branda E, Boekhout T, Threelen B, Watson K (2010) Cold-adapted yeasts from Antarctica and Italian alps-description of three novel species: *Mrakia robertii* sp. nov., *Mrakia blollopis* sp. nov. and *Mrakiella niccombsii* sp. nov. Extremophiles 14:47–59

Thomsen MH, Thygesen A, Thomsen AB (2009) Identification and characterization of fermentation inhibitors formed during hydrothermal treatment and following SSF of wheat straw. Appl Microbiol Biotechnol 83:447–455

Tiquia-Arashiro SM, Rodrigues D (2016) Thermophiles and Psychrophiles in Nanotechnology. In: Extremophiles: applications in nanotechnology. Springer International Publishing, New York, pp 89–127

Tsuji M, Fujiu S, Xiao N, Hanada Y, Kudoh S, Kondo H, Tsuda S, Hoshino T (2013a) Cold adaptation of fungi obtained from soil and lake sediment in the Skarvsnes ice-free area, Antarctic. FEMS Microbiol Lett 346:121–130

Tsuji M, Singh SM, Yokota Y, Kudoh S, Hoshino T (2013b) Influence of initial pH on ethanol production by Antarctic Basidiomycetous yeast *Mrakia blollopis*. Biosci Biotechnol Biochem 77:2483–2485

Tsuji M, Yokota Y, Shimohara K, Kudoh S, Hoshino T (2013c) An application of wastewater treatment in a cold environment and stable lipase production of Antarctic basidiomycetous yeast *Mrakia blollopis*. PLoS One 8:e59376

Tsuji M, Goshima T, Matsushika A, Kudoh S, Hoshino T (2013d) Direct ethanol fermentation from lignocellulosic biomass by Antarctic Basidiomycetous yeast *Mrakia blollopis* under a low temperature condition. Cryobiology 67:241–243

Tsuji M, Yokota Y, Kudoh S, Hoshino T (2014a) Effects of nitrogen concentration and culturing temperature on lipase secretion and morphology of the Antarctic basidiomycetous yeast *Mrakia blollopis*. Int J Res Eng Sci 2:49–54

Tsuji M, Yokota Y, Kudoh S, Hoshino T (2014b) Improvement of direct ethanol fermentation from woody biomasses by Antarctic basidiomycetous yeast *Mrakia blollopis* under a low temperature condition. Cryobiology 68:303–305

Tsuji M, Kudoh S, Hoshino T (2015a) Draft genome sequence of cryophilic basidiomycetous yeast *Mrakia blollopis* SK-4, isolated from an algal mat of Naga-ike Lake in the Skarvsnes ice-free area, East Antarctica. Genome Announc 3:e01454–e01414

Tsuji M, Yokota Y, Kudoh S, Hoshino T (2015b) Comparative analysis of milk fat decomposition activity by *Mrakia* spp. isolated from Skarvsnes ice-free area, East Antarctica. Cryobiology 70:293–296

Tsuji M, Kudoh S, Hoshino T (2016a) Ethanol productivity of cryophilic basidiomycetous yeast *Mrakia* spp. correlates with ethanol tolerance. Mycoscience 57:42–50

Tsuji M, Uetake J, Tanabe Y (2016b) Changes in the fungal community of Austre Brøggerbreen deglaciation area, Ny-Ålesund, Svalbard, High Arctic. Mycoscience 57:448–451

Tsuji M, Tanabe Y, Vincent WF, Uchida M (2018) *Mrakia arctica* sp. nov., a new psychrophilic yeast isolated from an ice island in the Canadian High Arctic. Mycoscience 59:54–58

Turchetti B, Buzzini P, Goretti M, Branda E, Diolaiuti G, D'Agata C, Smiraglia C, Vaughan-Martini A (2008) Psychrophilic yeasts in glacial environments of Alpine glaciers. FEMS Microbiol Ecol 63:73–83

Xin M, Zhou P (2007) *Mrakia psychrophila* sp. nov., a new species isolated from Antarctic soil. J Zhejiang Univ Sci B 8:260–265

Yamada Y, Komagata K (1987) *Mrakia* gen. nov., a heterobasidiomycetous yeast genus for the Q_8-equipped, self-sporulating organisms which produce a unicellular metabasidium, formerly classified in the genus *Leucosporidium*. J Gen Appl Microbiol 33:455–457

Chapter 9
Adaptation Mechanisms and Applications of Psychrophilic Fungi

Muhammad Rafiq, Noor Hassan, Maliha Rehman, and Fariha Hasan (iD)

9.1 Introduction

Frozen environments (cryosphere) represent world's largest share of psychrophilic habitats including snow, ice sheets, ice lake, ice caps, permafrost, glaciers, frozen parts of the ocean, frozen rivers and lakes (Musilova et al. 2015; Kudryashova et al. 2013), in both polar regions (NOAA 2018), glaciers and lakes of nonpolar mountain ranges (Walsh et al. 2016; Salazar and Sunagawa 2017), and man-made freezers (Ahmad et al. 2010) and refrigerators (Flores et al. 2012). Psychrophilic environment is harsh due to low temperature along with at least one of these, i.e., UV rays, low nutrients and water availability, freeze-thaw cycles, and osmotic pressures, and yet these are of ecological and environmental importance. Cold conditions, and other limiting factors, strongly influence survival of organisms in a cold habitat (Margesin and Miteva 2011). Freezing temperature damages cells by disrupting them via ice crystals, stops the activity of enzymes and other proteins, and decreases fluidity of cytoplasm and membranes, thus hindering their normal function in low-temperature environment without proper adaptation tools (Raymond et al. 2007). Cold temperature freezes cell wall and cell membrane that leads to inability to carry out transportation in or out of the cells. Similarly, a frozen cytoplasm is unable to offer favorable environment for enzymes to perform the biochemical processes of a cell. Low temperature affects structure of enzymes which could not achieve their activation energy required to metabolize a reaction (Chandler 2018).

M. Rafiq · N. Hassan · M. Rehman · F. Hasan (✉)
Department of Microbiology, Faculty of Biological Sciences, Quaid-i-Azam University, Islamabad, Pakistan

M. Rehman
Department of Microbiology, Faculty of Life Sciences and Informatics Balochistan
University of Information Technology, Engineering and Management Sciences
(BUITEMS), Quetta, Pakistan

© Springer Nature Switzerland AG 2019
S. M. Tiquia-Arashiro, M. Grube (eds.), *Fungi in Extreme Environments: Ecological Role and Biotechnological Significance*,
https://doi.org/10.1007/978-3-030-19030-9_9

For survival in extreme environments soil fungi compete with microbes of the soil, acquire the intermittent nutrients, and also utilize secondary metabolites for survival (Yogabaanu et al. 2017). Ice veins inside the glaciers and ice sheets represent micro-environment that serves as habitat for existence of microbes (Thomas and Dieckmann 2002). Microorganisms in the ice veins face many physicochemical stresses, i.e., low water activity and pH, lowered solute diffusion rates, and damage to the membranes owing to ice crystal formation. Psychrophiles demonstrate various structural and functional approaches for their survival under reduced liquid water, extremely cold temperature, high solar radiation, and nutrient scarcity (Garcia-Lopez and Cid 2017).

Psychrophiles and psychrotrophs include all three domains of life such as archaea, prokaryotes (e.g., bacteria), and eukaryotes (e.g., fungi) (Margesin and Miteva 2011; Boetius et al. 2015; Hassan et al. 2016); inhabit stressful low-temperature environments; and are dependent on each other for active ecological processes. Fungi are widely distributed in the cryosphere (Hoshino and Matsumoto 2012), and play an important role in nutrient recycling; thus they are termed as "the survivor community." They also decompose organic compounds under subzero temperatures (Tsuji 2016).

Cold environment constitutes extremely diverse cold-adapted fungi including representatives of all phyla (Wang et al. 2017). Cold-tolerant fungi, belonging to phyla Ascomycota, Deuteromycota, Zygomycota and Basidiomycota, including *Mucor, Cladosporium, Alternaria, Aspergillus, Penicillium, Lecanicillium, Botrytis, Geomyces, Monodictys*, and *Rhizopus*, have been reported from Antarctica (Kostadinova et al. 2009). Cold-adapted fungi are varied; dwell as saprobes, symbionts, parasites, and pathogens of plant and animal; and also carry out critical functions in diverse ecosystems. Few fungal species cause diseases in plants and animals in cold regions (Wang et al. 2017) and can have both ecologic as well as economic impact on vegetation or animal life.

Adaptation to low temperature makes fungi an appealing resource for obtaining new enzymes and secondary metabolites for use in biotechnology and pharmaceuticals (Wang et al. 2017). Fungi secrete cellulose, hemicellulose and lignin-degrading enzymes, secondary metabolites, and bioactive compounds, and have great potential in biotechnological applications. In nature, soil fungi decompose dead plants, carry out mineral cycling to maintain soil fertility, and thus have an important role in biogeochemistry (Watkinson 2016), the same role of fungi in low-temperature habitats. Scientists have reviewed cold-adapted fungi, properties of enzymes, biotechnological applications and use of metagenomics to screen for enzymes, cold gene expression systems and enzymes used for washing purpose (Cavicchioli et al. 2011), synthesis of biotechnologically important cold-active enzymes, genome sequences, proteomics and transcriptomics of adaptation mechanisms under cold conditions (Feller 2013; Alcaíno et al. 2015), and influence of climate change on microbes of permafrost and their function (Jansson and Taş 2014). Boetius et al. (2015) explained microbial ecology, composition of frozen waters and biogeochemical activities of the microbial communities, and living strategies and ecological functions of cold-adapted fungi reviewed by Wang et al. (2017). This chapter elaborates strategies used by cold-adapted fungi to survive in cold and avenues to exploit their strategies as potential applications in various industries and biotechnology.

9.2 Adaptation Mechanisms

In cold temperature, fatty acid tails of phospholipids become rigid due to less move-ment, and fluidity of membrane is decreased, thus decreasing permeability to mol-ecules (oxygen and glucose) into the cell. Long exposure to temperatures below-freezing points freezes the liquid inside the cell and forms crystals that dam-ages the membrane, resulting in death of cell (Chandler 2018).

Low temperature affects the cells by impeding chemical reaction rate, denaturing proteins, enhancing water viscosity, limiting activities of microbial enzymes and fluidity of cell membrane (Hassan et al. 2016), and restraining water availability as a solvent for biochemical reactions (Wynn-Williams and Edwards 2000) and fre-quent freeze-thaw cycles (Montiel 2000).

Eukaryotic microorganisms survive in hypersaline environments by accumula-tion of "compatible solutes" in their cytoplasm (Oren 1999) and maintain intracel-lular concentrations of sodium ions below the toxic level (Plemenitaš et al. 2008).

In fungi, melanin provides protection against the undesirable effects of UV radia-tion (Gessler et al. 2014), drying, high amount of salts, heavy metals, and radionuclides. Melanin helps fungi to live under high electromagnetic radiation in higher altitudes and deserts and on plant surfaces (Zhdanova et al. 2005; Dighton et al. 2008; Grishkan 2011).

Radiations from sunlight comprise UV-A and -B radiations with shorter wave-lengths that cause damage to biological systems in glaciers (Cockell and Knowland 1999). To counteract this, organisms have developed repair processes like photore-activation, base excision repair, nucleotide excision repair, and mismatch repair (Rastogi et al. 2010a, b). UV-absorbing pigments are produced by some organisms (Rastogi et al. 2010a, b). Solar UV-A interacts with cellular photosensitizers that generate reactive oxygen species and induce oxidative stress with proteins as the main target for damage. UV-B negatively affects ecology and evolution of biological systems (Cockell and Blaustein 2001).

Various strategies of cold tolerance in fungi include production of antifreeze proteins (AFPs), plasma membrane fluidity, trehalose, compatible solutes, and many other cold-shock proteins and mechanisms (Robinson 2001). Scientists are looking for molecular or genetic basis of adaptations. High expression of unknown or novel genes in *Glaciozyma antarctica* PI12 could have an important role in cold adaptation (Firdaus-Raih et al. 2018).

9.2.1 Plasma Membrane Fluidity Maintenance

Microorganisms living in cold habitats deal with low temperature by changing composition of lipid membrane (Russell 1990) and increasing level of unsaturated fatty acids. Increased unsaturation of lipids is observed at low temperature in *Geomyces pannorum*, with decrease in production of ergosterol in *Mortierella elongate* (Weinstein et al. 2000). *M. elongate* showed increase in production of stearidonic acid, a fatty acid previously reported in psychrotrophic zygomycetes. *Rhodosporidium diobovatum* (psychrotolerant Arctic yeast) demonstrates increased membrane fluidity through unsaturation of fatty acids (Turk et al. 2011).

9.2.2   Compatible Solutes

Compatible solutes are low-molecular-weight osmoregulators that stabilize the cells and provide favorable environment for function of enzymes and other molecules inside cell in cold, heat, drought, and other stress conditions. These solutes have cryoprotective ability and maintain membrane and cytoplasm's structure and function. Different classes of compatible solutes produced by psychrophilic fungi to cope with low temperature include polyols, melanin, mycosporines, trehalose, and betaine (Ruisi et al. 2007). Cold-adapted fungi also adapt to repeated freeze-thaw cycles, low water availability, osmotic stress, desiccation, low nutrient availability, and high UV radiation (Ruisi et al. 2007).

9.2.2.1   Polyols

Polyols are organic compounds which contain more than two hydroxyl functional groups, for example sugar alcohol, including mannitol and glycerol. Synthesis of compatible solutes by enzymatic activities is elicited by induced dehydration and osmotic stress in fungi at low temperature, and glycerol is one of them (Pascual et al. 2003). Fungi use mannitol to store carbon, balance redox, and serve as an antioxidant and stress tolerant (Son et al. 2012). Turgor pressure can be controlled against decline in external water potential by raising mannitol and glycerol concentrations (Grant 2004). It is known that mannitol has protective role in water stress condition and can be used as a protective agent in cryoenvironment (Weinstein et al. 1997). Han and Prade (2002) reported glycerol and erythritol synthesis in *Aspergillus nidulans*, triggered by exposure to high salinity.

9.2.2.2   Trehalose

Increase in trehalose concentration is observed on exposure of fungi (e.g., *Hebeloma* sp., *Humicola marvinii,* and *Mortierella elongate*) to cold environment (Tibbett et al. 1998a; Weinstein et al. 2000).

Lack in ergosterol and increase in trehalose concentration in *Mortierella elongate* at low temperature have been documented by Weinstein et al. (2000). Trehalose accumulates in fungal hyphae and reproductive bodies to protect from adverse effects of low temperature (Robinson 2001).

9.2.2.3   Betaine

Betaine is glycerolipid with a non-phosphorous, polar moiety attached to diacylglycerol through ether linkage. It is found in many lower eukaryotes like bryophytes, algae, protozoa and fungi, and some prokaryotic bacteria. There are three types of betaine: diacylglyceryl-trimethyl-homoserine,   diacylglyceryl-hydroxymethyl-trimethyl-β-alanine, and diacylglyceryl-carboxyhydroxymethylcholine (Murakami et al. 2018). Betaine is soluble in water and protects the cells by two mechanisms:

i) by osmoregulation to adjust osmotic pressure in and outside the cell, and ii) also acting as scavenger of reactive oxygen species. Studies indicated the presence of gene responsible for production of betaine on genome of *Aspergillus fumigatus*. Betaine is produced in a two-step process of oxidation followed by dehydration. Substrate choline is converted to betaine aldehyde (BA) by monooxygenase and BA is transformed to betaine by BA dehydrogenase (Chen and Murata 2011). Hoffmann and Bremer (2011) and Bashir et al. (2014) reported that bacteria can use betaine both as antistress molecule in extreme environment and a source of energy, whereas Lambou et al. (2013) reported fungi to use betaine as a source of carbon and energy.

9.2.2.4 Mycosporines

Mycosporine having oxo-carbonyl chromophores has been found in terrestrial fungi (Shick and Dunlap 2002). Basidiomycetous yeasts, *Rhodotorula minutia* and *R. slooffiae*, produced mycosporine-glutaminol-glucoside (Sommaruga et al. 2004). An Antarctic fungus *Arthrobotrys ferox* produced carotenoid pigments and mycosporines, having a strong role in UV protection (Arcangeli and Cannistraro 2000). Cold-adapted *Dioszegia patagonica* sp. nov, a yeast from Patagonia, accumulated carotenoid and mycosporines (Trochine et al. 2017). Mycosporines are not extensively studied in fungi inhabiting polar and nonpolar regions, but their occurrence in other fungi enables them to shield from UV.

9.2.2.5 Melanin

In mesophilic fungi, melanin plays a role as virulence factor in pathogenesis of fungi, stress protection (e.g., oxidative, UV), attachment, and penetration of appressorium (Yu et al. 2013). All biological kingdoms synthesize melanin (Eisenman and Casadevall 2012) which protects them from UV and ionizing radiation and desiccation.

9.2.3 Cold-Active Enzymes

These are known for sustaining microbial proliferation including fungi, at a very low temperature (Kuddus et al. 2011; Hassan et al. 2017). In cold environment, psychrophiles face low enzyme activity, modified transport systems, reduced membrane fluidity, and protein cold-denaturation among others (D'Amico et al. 2006). Elevated amounts of unsaturated and methyl-branched fatty acids and shorter acyl-chain fatty acids are produced by psychrophiles that increase fluidity of membrane (Chintalapati et al. 2004). Cold-shock proteins are also produced to assist in membrane fluidity or protein folding (Phadtare 2004), and antifreeze proteins hinder growth of ice crystal (Sarmiento et al. 2015). As temperature drops, proteins are denatured due to decrease in water molecule availability (Karan et al. 2012). A number of structural adaptations are known in cold-adapted enzymes that makes these enzymes flexible as compared to mesophilic or thermophilic enzymes. It makes them catalytically active at low temperatures (Siddiqui and Cavicchioli 2006), as well as thermolabile. Psychrophilic enzymes have more flexibility and activity at reduced temperatures: high surface hydrophobicity, reduced core hydrophobicity, decreased ratio of arginine/lysine, increased glycine residues, less proline in loops, with more α-helices, more nonpolar residues on surface of protein, weaker protein interactions, hydrogen bonds and other electrostatic interactions, and less/ weaker metal-binding sites, less disulfide bridges, reduced secondary structures, with increased number and size of loops, and increased conformational entropy of the unfolded protein state (Feller 2010; Cavicchioli et al. 2011). Therefore, rate of reaction in psychrophilic enzymes decreases when temperature decreases (Feller 2013). Interestingly, cold-adapted xylanases are reported more active at low temperatures, and more thermolabile at higher temperatures (Collins et al. 2002). Psychrophilic *Humicola fuscoatra* and *H. marvinii* recovered from Antarctica and solubilized produced phosphatase and extracellular protease at 15 °C (Weinstein et al. 1997). Hassan et al. (2017) reported production of lipases, amylases, phosphatases, proteases, and DNAase from different fungal species isolated from Siachen glacier, Pakistan. He et al. (2017) gave new insights into *Aspergillus oryzae* cold-adapted amylase and application of gene AmyA1 in the food and starch industries. Cold-adapted *Cladosporium herbarum* ER-25 produced extracellular invertase and assisted in removal of toxical dark-brown pigments (melanoidins) along with laccase and manganese peroxidase (Taskin et al. 2016).

9.2.4 Antifreeze Proteins (AFP)

Antifreeze protein is an effective strategy used by psychrophilic organisms, for survival at subzero temperature (Duman 2001). AFP DUF3494-type proteins are present in all domains of life specifically restricted to cold-adapted taxa (Bowman 2017). Ice growth and nucleation are hindered by AFPs and organism stays supercooled until atmospheric temperature is lowered below freezing point.

New fungal AFP has been identified and purified from psychrophilic *Antarctomyces psychrotrophicus* (Ascomycetes) (Xiao et al. 2010). AFP-producing fungi are pathogenic for different plant species (Snider et al. 2000; Hoshino et al. 2003; Hoshino 2005).

9.2.5 Exopolysaccharides (EPS)

Exopolysaccharide production is an adaptive strategy used by fungi to survive in extreme condition by preventing damages in subzero temperature. *Phoma herbarum* CCFEE 5080 from Antarctica was observed for EPS production (Selbmann et al. 2002).

9.3 Applications

Psychrophilic fungi (metabolite or whole cell) can be used as biotechnological product (Fig. 9.1) for production of compounds, and bioremediation in cold regions and their proteins can be used in medical research, molecular biology, biotechnology, detergents or cosmetics, and food or feed technologies (Margesin and Feller 2010; Tiquia-Arashiro and Rodrigues 2016).

9.3.1 Novel Source of Cold-Active Enzymes

Low-temperature-active enzymes represent a striking reserve for biotechnological applications (Santiago et al. 2016; Hamid et al. 2014; Cavicchioli et al. 2011; Tiquia and Mormile 2010), with uses in food processing, textile, detergents, feed stocks, bioremediation, cosmetics, paper, and pharmaceutical industries (Javed and Qazi 2016). Psychrophilic yeasts produce cold-active enzymes, used in fine chemical synthesis, and various domestic and environmental applications (Hamid et al. 2014). They do not require processes requiring heating that hampers the quality, sustainability, and cost-effectiveness of production at industrial level (Santiago et al. 2016), and elimination of heating results in saving substantial energy, efficient function at low temperatures, increased yield, and high stereo-specificity, and avoids the unwanted chemical reactions that occur at high temperatures. Psychrophilic fungi produce various intra- and extracellular enzymes, which enable them to confront

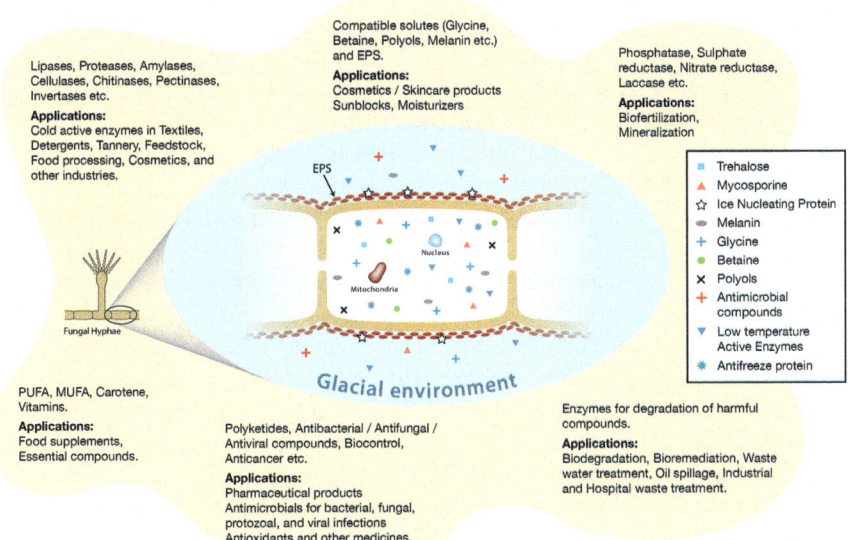

Fig. 9.1 Schematic representation of adaptation mechanisms of psychrophilic fungi that can be used for potential biotechnological purposes

and aid in harsh conditions and in degradation of large molecules and uptake of nutrients (Gerday et al. 2000; Feller and Gerday 2003; Gomes and Steiner 2004; Margesin et al. 2005).

Yeast and fungi from cold habitats deliver usefulness of fermentation procedures feasible at room temperature, that reduce production cost and influence on environment (Perfumo et al. 2018), and are economically important based on their activity at moderate and low temperatures (Allen et al. 2002; Margesin et al. 2002).

Poveda et al. (2018) isolated pectinase producing *Geomyces* sp. strain F09-T3-2 from marine sponges in Antarctica, with probable uses in food and beverage industry. Psychrophilic fungi from Baramulla (Jammu and Kashmir) produced cold-active pectinases (pectin esterase, exo-galacturanase, and endo-galacaturanase) for potential in the wine making and juice industries (Singh et al. 2012). Polygalacturonase from psychrophilic *Sclerotinia borealis* (Takasawa et al. 1997) has applicability in fruit ripening, pollen, and abscission.

Yusof et al. (2017) characterized the sequence of a chitinase produced by psychrophilic yeast, *Glaciozyma antarctica* PI12. Fungi belonging to Ascomycota and Basidiomycota from Antarctic soil and sea samples produced cold-adapted hydrolytic enzymes (e.g., phytase, glucosidase, chitinase, invertase, tannase, pectinase, lipase, protease, α-amylase, cellulase, subtilase, and xylanase) and oxidoreductases (laccase and superoxide dismutase) (Duarte et al. 2018). Cold and pH-tolerant *Penicillium* spp. produced cold-active lipases (Pandey et al. 2016).

Ascomycetes, Deuteromycetes, Basidiomycetes, and white-rot fungi produce laccases that degrade lignin and have been used in petrochemical, pulp, paper, and textile industries; food processing; medical and health care; and designing of biosensors and nanotechnology (Upadhyay et al. 2016).

Cold-active cellulases by psychrophilic microorganisms can hydrolyze biomass at low temperature and convert cellulosic biomass into monomeric sugars for bioethanol production (Tiwari et al. 2015). *Aspergillus niger* SH3 from Himalayan region (India) produced endoglucanase, *β*-glucosidase, FPase, and xylanase and can be a potential candidate for biofuel production (Tiwari et al. 2015). Cellulose decomposing *Cladosporium* (WR-C1) was isolated from a hypothermal litter layer (Da-qing et al. 2016). Cellulases and lipases produced by *M. arctica* reported to be highly active at 3 °C and have significant role in biogeochemical cycle of glacial ecosystems (Tsuji et al. 2018). *Verticillium* sp. *AnsX*1 having enhanced cellulytic activity in cold was recovered from Antarctic and has potential for bioconversion of lignocellulosic biomass into biofuels (Wang et al. 2013).

Efficient activity of endo-1, 4-β-glucanase (endoglucanase) is reported at low temperature from *Cladosporium, Penicillium, Cadophora,* and *Geomyces* by Duncan et al. (2006), and Gawas-Sakhalkar et al. (2012) reported phosphatase activity of *Penicillium citrinum, Aspergillus niger*, whereas *Aspergillus aculeatus* exhibited amylase and pectinase activity.

Psychrophilic enzymes have a great prospective as detergents for cleaning/washing at low temperature (Cavicchioli et al. 2011). Novozymes have developed Celluzyme® and Celluclean® using cellulases from cold-adapted *Humicola insolens* (Adapa et al. 2014). Mukherjee and Singh (2011) reported α-amylase with possible use in the food and textile industries and as additive in detergent for cold washing. They have a great potential of applications in "peeling" of leather at industrial scale, baking and wine industry, food and feed industry, molecular biology, cheese ripening, resizing denim jeans, and paper industry (Petrescu et al. 2000; Mayordomo et al. 2000).

Phytase was produced by *Morchella importuna*, a psychrophilic mushroom which can be used as fish feed additive enzyme (Taskin et al. 2016).

9.3.2 Pharmaceutical Products

Fungi are reported to produce pharmaceutical products (Schulz et al. 2002) but the recovery of such bioactive metabolites from fungi of cold regions is quite rare. *Penicillium lanosum* and *Penicillium soppii* synthesized bioactive secondary metabolites such as cycloaspeptide A and griseofulvin (Frisvad et al. 2006). Psychrophilic *Penicillium jamesonlandense* produced cyclic peptides cycloaspeptide A and D (Frisvad et al. 2006). *Penicillium ribium* was found to synthesize compound, cyclic nitropeptide psychrophilin A (Dalsgaard et al. 2004a; Frisvad et al. 2006), whereas *Penicillium rivulorum* produced communesin G and H and psychrophilin B and C (Dalsgaard et al. 2004b, 2005). *Penicillium algidum* synthesized cycloaspeptide A and D and psychrophilin D (Dalsgaard et al. 2005). These cyclic peptides reported only in fungal isolates from cold habitats showed antimalarial and insecticidal properties (Dalsgaard et al. 2005; Lewer et al. 2006), along with other biological activities.

Polyketides (PKs) have antimicrobial activity and other clinically important applications. PKs promote struggle for nutrients, to demote the potentials of its competitors and to establish chemical interaction with organisms in its vicinity (Mukherjee et al. 2012). Penilactones A and B, the oxygenated polyketides, were produced from *Penicillium crustosum* PRB-2 from deep sea of Antarctic (Wu et al. 2012), and 5 fungal hybrid polyketides, including cladosins, were obtained from deep-sea *Cladosporium sphaerospermum* 2005-01-E3. Cladosin C demonstrated slight activity against influenza A H1N1 virus (Wu et al. 2014). Chloro-trinoreremophilane sesquiterpene, eremophilane sesquiterpenes, and eremofortine recovered from an Antarctic *Penicillium* sp. PR19N-1 showed cytotoxic activity against cancer cell lines (Wu et al. 2013). *Dichotomomyces cejpii* F31-1, a marine fungus, produced polyketide Scequinadoline A showing inhibitory activity against dengue virus serotype 2 production (Wu et al. 2018). Polyketide, anthraquinone-xanthone, from *Engyodontium album* LF069 exhibited inhibition against methicillin-resistant *Staphylococcus aureus* (Wu et al. 2016).

Psychrophilic halophilic *Penicillium chrysogenum* from Vestfold Hills' saline lake produced bis-anthraquinone (rugulosin and skyrin) with possible application as insecticide and medicine (Parker et al. 2000; Sumarah et al. 2005). Some important and potential bioactive secondary metabolites by fungi of Antarctic were documented by Marinelli et al. (2004) and Rojas et al. (2009). Fungi from King George Island, Antarctic, and Svalbard, showed antimicrobial potential against *Bacillus subtilis, Bacillus cereus, Pseudomonas aeruginosa, Enterococcus faecalis,* and *Escherichia coli* (Yogabaanu et al. 2017).

Moghaddam and Soltani (2014) isolated psychrophilic endophytic fungi *Phoma* sp., *P. herbarum,* and *Dothideomycetes* spp., with an ability to synthesize metabolites active against phytopathogenic fungi and antibacterial activity against ice-nucleating *Pseudomonas syringae*. Depsipeptide, chaetomiamide, and diketo-piperazines showing anticancer and cytotoxic activity were recovered from endo-phytic *Chaetomium* sp. (Wang et al. 2017).

9.3.3   Bioremediation Potentials

Psychrophilic microbes are useful for bioremediation of waste water and soil in temperate regions in winter. Bioremediation potential of psychrophilic fungi is not studied well yet; however, it would be quite effective in cold regions.

Mortierella sp. from Antarctica used dodecane as carbon and energy source and can be a good candidate for bioremediation of hydrocarbon spill (Hughes et al. 2007). Antarctic *Aspergillus fumigatus* degraded phenol via production of phenol hydroxy-lase, hydroquinone hydroxylase, and catechol 1,2-dioxygenase (Gerginova et al. 2013).

D'Annibale et al. (2006) reported *Allescheriella* sp. DABAC 1, *Stachybotrys* sp. DABAC 3, and *Phlebia* sp. DABAC 9 to produce laccase and peroxidases, and removed naphthalene, dichloroaniline isomers, o-hydroxybiphenyl, and 1,1-binaph-thalene. *Stachybotrys* sp. DABAC 3 remediated 9,10-anthracenedione and 7H-benz[DE]anthracen-7-one. Dechlorination of polychlorinated biphenyls (PCBs) has been demonstrated by *Phanerochaete chrysosporium* (Bedard et al. 2006). *Candida antarctica* could degrade petroleum compounds (Hua et al. 2004).

9.3.4   Pigment/Lipid Production

Pigments and lipids synthesized by psychrophilic fungi confront low temperatures. Increased amount of lipids like fatty acids and polyunsaturated triglycerides has been found in psychrotolerant and psychrophilic fungi (Weinstein et al. 2000).

Singh et al. (2014) reported pigments (carotenoid) and fatty acids (linoleic, stearic, linolenic, myristic, heptadecanoic, and palmitic acid) from cold-tolerant fungus, *Thelebolus microspores*. Linolenic acid is used as a food supplement for patients of diabetic neuropathy, eczema, and cardiovascular disease. Carotenoid biosynthesis was also reported in *Neurospora crassa* at low temperature (Castrillo et al. 2018).

9.3.5   Exopolysaccharide (EPS) Production

The production of EPS is the response to stress or harsh conditions. Mycelium of fungi surrounded by EPS has high growth rate as compared to unembedded myce-lium in response to repeated exposure to freeze-thaw cycles (Selbmann et al. 2002). *Phoma herbarum* CCFEE 5080, an Antarctic fungal isolate, showed production of exopolysaccharide identified as β 1-3, 1-6 glucan of 7.4 × 10 6 Dalton (Selbmann et al. 2002). Meristematic black fungi isolated from Antarctica were reported by Onofri (1999) and Selbmann et al. (2005) for production of extracellular polymeric substances around their hyphae that surround their multicellular conidia and same is the case found in *Friedmanniomyces endolithicus*.

Endolithic fungus *Cryomyces antarcticus* CCFEE 515 isolated from the most comparable referent for Mars environment present on Earth, McMurdo Dry Valleys of Antarctica. It is used as eukaryotic model for astrobiological studies and in space experiments under UV and ionizing radiation (Selbmann et al. 2018).

Melanized microorganisms are dominant in harsh environments, like soils contaminated with radionuclides (Dadachova et al. 2007). Upregulation of many genes is caused by exposure to radiation, and an inducible microhomology-mediated recombination pathway is expected as a possible mechanism for eukaryotic evolution.

Exopolysaccharide is often used in cryopreservation, e.g., alginate beads containing EPS preserve the sample from freezing damage (Martinez et al. 1999). Psychrophilic Antarctic *Thelebolus* sp. IITKGP-BT12 produced EPS characterized as glucan and showed antiproliferative activity in cancer cells (Mukhopadhyay et al. 2014).

9.3.6 Biofertilization Capabilities

In nature, phosphorus is found in both inorganic and organic states, and it is one of the principal nutrients required for the crop development and increased yield. Soil comprises inorganic phosphates in insoluble form and plants cannot uptake insoluble form, it is useless for plants until solubilized. Solubilization changes the inorganic phosphates into organic soluble state, which the plants can take up.

Microorganisms play a key role in solubilization of phosphates to its organic soluble counterpart via chelation, exchange reaction, and acidification (Narsian and Patel 2000; Reyes et al. 2002). Bacteria, actinomycetes, and fungi involved in phosphate solubilization have been reported (Trivedi and Pandey 2007; Stibal et al. 2009; Nenwani et al. 2010; Singh et al. 2011). Ectomycorrhizal macromycetes (Sharma and Baghel 2010) and ectomycorrhizal *Hebeloma* (Tibbett et al. 1998b) produce phosphatase, whereas *Penicillium* and *Aspergillus niger* from nonpolar cold habitats produced inorganic phosphatase (Goenadi and Sugiarto 2000; Pandey et al. 2008). *Aspergillus niger*-1 and 2, from tundra in Arctic Archipelago of Svalbard, showed an ability for phosphate solubilization. Cold-tolerant *Penicillium citrinum* PG162 produced intracellular acid phosphatase (Gawas-Sakhalkar et al. 2012). Cold-tolerant fungi with an ability to produce phosphatase (Singh et al. 2011; Tibbett et al. 1998a, b; Gawas-Sakhalkar et al. 2012) suggest a good potential of biofertilizers in place of chemical fertilizers with efficient activity and ecofriendly characters.

9.4 Conclusions

Present review gives a detailed account of adaptability processes of cold-adapted fungi and how their strategies could be exploited for applications in biotechnology and industry. Psychrophilic fungi are a splendid resource of new and unique products and can have numerous opportunities in food industry, pharmaceuticals, enzymes, and so on. Unfortunately, these are not studied extensively yet, and therefore hold a promising future. The fungi in low-temperature environments including icy habitats and deep-sea environments are of diverse nature and are in abundance. Their strategies to thrive under extreme conditions make them versatile and their metabolites can be of potential use in many dimensions.

9.5 Future Perspectives

This review provides a baseline or food for thought regarding the exploitation of cold-adapted fungi and their metabolites for biotechnology and industrial uses. Adaptive mechanisms of low-temperature fungi need to be investigated further on molecular and genetic basis. Two of the most important avenues are pharmaceuticals and replacing synthetic compounds with biobased or biologically synthesized metabolites of use in industry and biotechnology. Psychrophilic fungi need to be investigated in practical application for the bioremediation of domestic, industrial, and hospital wastes because they are active at low temperature and can effectively work in winter season all over the globe. Therefore, we strongly recommend bioprospecting for fungal diversity in cold habitats and investigate their processes in detail.

Acknowledgement We would like to thank and acknowledge Simon Powell, University Graphics Officer, School of Earth Sciences, University of Bristol, UK, for illustration.

References

Adapa V, Ramya LN, Pulicherla KK, Rao KR (2014) Cold active pectinases: advancing the food industry to the next generation. Appl Biochem Biotechnol 172:2324–2337

Ahmad B, Javed I, Shah AA, Hameed A, Hasan F (2010) Psychrotrophic bacteria isolated from −20°C freezer. Afr J Biotechnol 9:718–724

Alcaíno J, Cifuentes V, Baeza M (2015) Physiological adaptations of yeasts living in cold environments and their potential applications. World J Microbiol Biotechnol 31:1467–1473. https://doi.org/10.1007/s11274-015-1900-8

Allen D, Huston AL, Weels LE, Deming JW (2002) Biotechnological use of psychrophiles. Encycl Environ Microbiol:1–17

Arcangeli C, Cannistraro S (2000) *In situ* Raman microspectroscopic identification and localization of carotenoids: approach to monitoring of UV-B irradiation stress on Antarctic fungus. Biopolymers 57:179–186

Bashir A, Hoffmann T, Smits SHJ, Bremer E (2014) Dimethylglycine provides salt and temperature stress protection to *Bacillus subtilis*. Appl Environ Microbiol 80(9):2773–2785

Bedard DL, Bailey JJ, Brandon LR, Jerzak GS (2006) Development and characterization of stable sediment-free anaerobic bacterial enrichment cultures that dechlorinate Aroclor 1260. Appl Environ Microbiol 72:2460–2470

Boetius A, Anesio AM, Deming JW, Mikucki JA, Rapp JZ (2015) Microbial ecology of the cryosphere: sea ice and glacial habitats. Nat Rev Microbiol 13:677–690

Bowman JP (2017) Chapter 15: Genomics of psychrophilic bacteria and archaea. In: Margesin R (ed) Psychrophiles: from biodiversity to biotechnology. Springer International Publishing, New York, pp 345–387

Castrillo M, Luque EM, Carmen JPMM, Corrochano LM, Avalos J (2018) Transcriptional basis of enhanced photoinduction of carotenoid biosynthesis at low temperature in the fungus *Neurospora crassa*. Res Microbiol 169: 278–289

Cavicchioli R, Charlton T, Ertan H, Omar SM, Siddiqui K, Williams T (2011) Biotechnological uses of enzymes from psychrophiles. Microb Biotechnol 4:449–460. https://doi.org/10.1111/j.1751-7915.2011.00258.x

Chandler S (2018) The Effect of Temperature on Cell Membranes. Updated March 13. https://sciencing.com/effect-temperature-cell-membranes-5516866.html

Chen TH, Murata N (2011) Glycinebetaine protects plants against abiotic stress: mechanisms and biotechnological applications. Plant Cell Environ 34:1–20

Chintalapati S, Kiran MD, Shivaji S (2004) Role of membrane lipid fatty acids in cold adaptation. Cell Mol Biol (*Noisy-le-Grand*) 50:631–642

Cockell C, Blaustein AR (2001) Ecosystems, evolution, and ultraviolet radiation. Springer, New York

Cockell CS, Knowland J (1999) Ultraviolet radiation screening compounds. Biol Rev 74:311–345

Collins T, Meuwis MA, Stals I, Claeyssens M, Feller G, Gerday C (2002) A novel family 8 xylanase: functional and physico-chemical characterization. J Biol Chem 277:35133–35139

D'Amico S, Collins T, Marx JC, Feller G, Gerday C (2006) Psychrophilic microorganisms: challenges for life. EMBO Rep 7:385–389

D'Annibale A, Rosetto F, Leonardi V, Federici F, Petruccioli M (2006) Role of autochthonous filamentous fungi in bioremediation of a soil historically contaminated with aromatic hydrocarbons. Appl Environ Microbiol 72:28–36

Dadachova E, Bryan RA, Huang X, Moadel T, Schweitzer AD, Aisen P et al (2007) Ionizing radiation changes the electronic properties of melanin and enhances the growth of melanized fungi. PLoS One 2:e457

Dalsgaard PW, Larsen TO, Frydenvang K, Christophersen C (2004a) Psychrophilin A and Cycloaspeptide D, novel cyclic peptides from the psychrotolerant fungus *Penicillium ribeum*. J Nat Prod 67:878–881

Dalsgaard PW, Blunt JW, Munro MH, Larsen TO, Christophersen C (2004b) Psychrophilin B and C: Cyclic nitropeptides from the psychrotolerant fungus *Penicillium rivulum*. J Nat Prod 67:1950–1952

Dalsgaard PW, Larsen TO, Christophersen C (2005) Bioactive cyclic peptides from the psychrotolerant fungus *Penicillium algidum*. J Antibiot 58:141

Da-qing W, Wen-ran J, Tai-peng S, Yu-tian M, Wei Z, Hong-yan W (2016) Screening psychrophilic fungi of cellulose degradation and characteristic of enzyme production. J Northeast Agric Univ 23:20–27

Dighton J, Tugay T, Zhdanova N (2008) Fungi and ionizing radiation from radionuclides. FEMS Microbiol Lett 281:109–120

Duarte AWF, dos Santos JA, Vianna MV, Vieira JMF, Mallagutti VJ, Inforsato FJ, Wentzel LCP, Lario LD, Rodrigues A, Pagnocca FC, Pessoa A Jr, Sette LD (2018) Cold-adapted enzymes produced by fungi from terrestrial and marine Antarctic environments. Crit Rev Biotechnol 38:600–619. https://doi.org/10.1080/07388551.2017.1379468

Duman JG (2001) Antifreeze and ice nucleator proteins in terrestrial arthropods. Annu Rev Physiol 63:327–357

Duncan SM, Farrell RL, Thwaites JM, Held BW, Arenz BE, Jurgens JA, Blanchette RA (2006) Endoglucanase-producing fungi isolated from Cape Evans historic expedition hut on Ross Island, Antarctica. Environ Microbiol 8:1212–1219

Eisenman HC, Casadevall A (2012) Synthesis and assembly of fungal melanin. Appl Microbiol Biotechnol 93:931–940

Feller G (2010) Protein stability and enzyme activity at extreme biological temperatures. J Phys Condens Matter 22:323101. https://doi.org/10.1088/0953-8984/22/32/323101

Feller G (2013) Psychrophilic enzymes: from folding to function and biotechnology. Scientifica 512840. https://doi.org/10.1155/2013/512840

Feller G, Gerday C (2003) Psychrophilic enzymes: hot topics in cold adaptation. Nat Rev Microbiol 1:200

Firdaus-Raih M, Hashim NHF, Bharudin I, Abu Bakar MF, Huang KK, Alias H, Lee BKB, Isa MNM, Mat-Sharani S, Sulaiman S, Tay LJ, Zolkefli R, Noor YM, Law DSN, Rahman SHA, Md-Illias R, Abu Bakar FD, Najimudin N, Murad AMA, Mahadi NM (2018) The *Glaciozyma antarctica* genome reveals an array of systems that provide sustained responses towards temperature variations in a persistently cold habitat. PLOS One 13(1):e0189947. https://doi.org/10.1371/journal.pone.0189947

Flores GE, Bates ST, Caporaso JG, Lauber CL, Leff JW, Knight R, Fierer N (2012) Diversity, distribution and sources of bacteria in residential kitchens. Environ Microbiol 15:588–596

Frisvad JC, Larsen TO, Dalsgaard PW, Seifert KA, Louis-Seize G, Lyhne EK, Jarvis BB, Fettinger JC, Overy DP (2006) Four psychrotolerant species with high chemical diversity consistently producing cycloaspeptide A, *Penicillium jamesonlandense* sp. nov., *Penicillium ribium* sp. nov., *Penicillium soppii* and *Penicillium lanosum*. Int J Syst Evol Microbiol 56:1427–1437

Garcia-Lopez E, Cid C (2017) Glaciers and ice sheets as analog environments of potentially habitable icy worlds. Front Microbiol 8:1407

Gawas-Sakhalkar P, Singh S, Simantini N, Ravindra R (2012) High-temperature optima phosphatases from the cold-tolerant Arctic fungus *Penicillium citrinum*. Polar Res 31. https://doi.org/10.3402/polar.v31i0.11105

Gerday C, Aittaleb M, Bentahir M, Chessa JP, Claverie P, Collins T, D'Amico S, Dumont J, Garsoux G, Georlette D, Hoyoux A (2000) Cold-adapted enzymes: from fundamentals to biotechnology. Trends Biotechnol 18:103–107

Gerginova M, Manasiev J, Yemendzhiev H, Terziyska A, Peneva N, Alexieva Z (2013) Biodegradation of phenol by Antarctic strains of *Aspergillus fumigatus*. Z Naturforsch C 68:384–393. https://doi.org/10.5560/ZNC.2013.68c0384

Gessler NN, Egorova AS, Belozerskaya TA (2014) Melanin pigments of fungi under extreme environmental conditions. Appl Biochem Microbiol 50:105–113

Goenadi DH, Sugiarto Y (2000) Bioactivation of poorly soluble phosphate rocks with a phosphorus-solubilizing fungus. Soil Sci Soc Am J 64:927–932

Gomes J, Steiner W (2004) The biocatalytic potential of extremophiles and extremozymes. Food Technol Biotechnol 42:223–235

Grant WD (2004) Life at low water activity. Philos Trans R Soc B Biol Sci 359:1249–1267

Grishkan I (2011) In: Horikoshi K (ed) Extremophiles handbook. Springer Verlag, Tokyo, pp 1135–1146

Hamid B, Rana RS, Chauhan D, Singh P, Mohiddin FA, Sahay S, Abidi I (2014) Psychrophilic yeasts and their biotechnological applications - a review. Afr J Biotechnol 13:2188–2197

Han KH, Prade RA (2002) Osmotic stress-coupled maintenance of polar growth in *Aspergillus nidulans*. Mol Microbiol 43:1065–1078

Hassan N, Rafiq M, Hayat M, Shah AA, Hasan F (2016) Psychrophilic and psychrotrophic fungi: a comprehensive review. Rev Environ Sci Biotechnol 15:147–172

Hassan N, Rafiq M, Hayat M, Nadeem S, Shah AA, Hasan F (2017) Potential of psychrotrophic fungi isolated from Siachen glacier, Pakistan, to produce antimicrobial metabolites. Appl Ecol Environ Res 15:1157–1171

He L, Mao Y, Zhang L, Wang H, Alias SA, Gao B, Wei D (2017) Functional expression of a novel α-amylase from Antarctic psychrotolerant fungus for baking industry and its magnetic immobilization. BMC Biotechnol 17:22. https://doi.org/10.1186/s12896-017-0343-8

Hoffmann T, Bremer E (2011) Protection of *Bacillus subtilis* against cold stress via compatible-solute acquisition. J Bacteriol 193:1552–1562

Hoshino T (2005) Ecophysiology of snow mold fungi. Curr Top Plant Biol 6:27–35

Hoshino T, Matsumoto N (2012) Cryophilic fungi to denote fungi in the cryosphere. Fungal Biol Rev 26:102–105

Hoshino T, Kiriaki M, Nakajima T (2003) Novel thermal hysteresis proteins from low temperature basidiomycete, *Coprinus psychromorbidus*. Cryo Letters 24:135–142

Hua ZZ, Chen Y, Du GC, Chen J (2004) Effects of biosurfactants produced by *Candida antarctica* on the biodegradation of petroleum compounds. World J Microbiol Biotechnol 20:25–29

Hughes KA, Bridge P, Clark MS (2007) Tolerance of Antarctic soil fungi to hydrocarbons. Sci Total Environ 372:539–548

Jansson J, Taş N (2014) The microbial ecology of permafrost. Nat Rev Microbiol 12. https://doi.org/10.1038/nrmicro3262

Javed A, Qazi JI (2016) Psychrophilic microbial enzymes implications in coming biotechnological processes. Am Scient Res J Eng Technol Sci 23:103–120

Karan R, Capes MD, DasSarma S (2012) Function and biotechnology of extremophilic enzymes in low water activity. Aquat Biosyst 8(1). https://doi.org/10.1186/2046-9063-8-4

Kostadinova M, Krumova E, Tosi S, Pashova, Angelova M (2009) Isolation and identification of filamentous fungi from Island Livingston, Antarctica. Biotech Biotechnol Equip 23:267–270

Kuddus M, Roohi AJ, Ramteke PW (2011) An overview of cold-active microbial α-amylase: adaptation strategies and biotechnological potentials. Biotechnology 10:246–258

Kudryashova EB, Chernousova EY, Suzina NE, Ariskina EV, Gilichinsky DA (2013) Microbial diversity of Late Pleistocene Siberian permafrost samples. Microbiology 82:341–351

Lambou K, Pennati A, Valsecchi I, Tada R, Sherman S, Sato H, Beau R, Gadda G, Latgé JP (2013) Pathway of glycine betaine biosynthesis in *Aspergillus fumigatus*. Eukaryot Cell 12:853–863

Lewer P, Graupner PR, Hahn DR, Karr LL, Duebelbeis DO, Lira JM, Anzeveno PB, Fields SC, Gilbert JR, Pearce C (2006) Discovery, synthesis, and insecticidal activity of cycloaspeptide E. J Nat Prod 69:1506–1510

Margesin R, Miteva V (2011) Diversity and ecology of psychrophilic microorganisms. Res Microbiol 162:346–361

Margesin R, Feller G (2010) Biotechnological applications of psychrophiles. Environ Technol 31:835–844. https://doi.org/10.1080/09593331003663328

Margesin R, Feller G, Gerday C, Russell NJ (2002) In: Bitton G (ed) Encyclopedia of environmental microbiology, vol 2. Wiley, New York, pp 871–885

Margesin R, Fauster V, Fonteyne PA (2005) Characterization of cold-active pectate lyases from psychrophilic *Mrakia frigida*. Lett Appl Microbiol 40:453–459

Marinelli F, Brunati M, Sponga F, Ciciliato I, Losi D, Van Trappen S, Göttlich E, De Hoog S, Rojas JL, Genilloud O (2004) Biotechnological exploitation of heterotrophic bacteria and filamentous fungi isolated from benthic mats of Antarctic lakes. In: Kurtböke I, Swings J (eds) Microbial genetic resources and biodiscovery. Queensland Complete Printing Services, Queensland, pp 163–184

Martínez D, Rosa A-G, Revilla MA (1999) Cryopreservation of in vitro grown shoot-tips of *Olea europaea* L. var. Arbequina. Cryo-Letters 20:29–36

Mayordomo I, Randez-Gil F, Prieto JA (2000) Isolation, purification, and characterization of a cold-active lipase from *Aspergillus nidulans*. J Agric Food Chem 48:105–109

Moghaddam MSH, Soltani J (2014) Psychrophilic endophytic fungi with biological activity inhabit Cupressaceae plant family. Symbiosis 63:79–86

Montiel PO (2000) Soluble carbohydrates (trehalose in particular) and cry protection in polar biota. Cryo Letters 21:83–90

Mukherjee G, Singh SK (2011) Purification and characterization of a new red pigment from *Monascus purpureus* in submerged fermentation. Process Biochem 46:188–192

Mukherjee M, Mukherjee PK, Horwitz B, Zachow C, Berg G, Zeilinger S (2012) Trichoderma–plant–pathogen interactions: advances in genetics of biological control. Ind J Microbiol 52

Mukhopadhyay SK, Chatterjee S, Gauri SS, Das SS, Mishra A, Patra M, Ghosh AK, Das AK, Singh SM, Dey S (2014) Isolation and characterization of extracellular polysaccharide Thelebolan produced by a newly isolated psychrophilic Antarctic fungus *Thelebolus*. Carbohydr Polym 104:204–212

Murakami H, Nobusawa T, Hori K, Shimojima M, Ohta H (2018) Betaine lipid is crucial for adapting to low temperature and phosphate deficiency in Nannochloropsis. Plant Physiol 177(1):181–193

Musilova M, Tranter M, Bennett SA, Wadham J, Anesio AM (2015) Stable microbial community composition on the Greenland ice sheet. Front Microbiol 6:193

Narsian V, Patel HH (2000) *Aspergillus aculeatus* as a rock phosphate solubilizer. Soil Biol Biochem 32:559–565

Nenwani V, Doshi P, Saha T, Rajkumar S (2010) Isolation and characterization of a fungal isolate for phosphate solubilization and plant growth promoting activity. J Yeast Fung Res 1:009–014

NOAA (2018). National Ocean Service, National Oceanographic and Atmospheric Administration, Department of Commerce) as well as non-polar regions. What is the cryosphere? https://ocean-service.noaa.gov/facts/cryosphere.html. Last updated: 06/25/18

Onofri S (1999) Antarctic microfungi. In: Seckbach J (ed) Enigmatic microorganisms and life in extreme environments. Kluwer Academic Publishers, Dordrecht/Boston/London, pp 323–336

Oren A (1999) Bioenergetic aspects of halophilism. Microbiol Mol Biol Rev 63:334–348

Pandey A, Das N, Kumar B, Rinu K, Trivedi P (2008) Phosphate solubilization by *Penicillium* spp. isolated from soil samples of Indian Himalayan region. World J Microbiol Biotechnol 24:97–102

Pandey N, Dhakar K, Jain R, Pandey A (2016) Temperature dependent lipase production from cold and pH tolerant species of *Penicillium*. Mycosphere 7:1533–1545

Parker JC, McPherson RK, Andrews KM, Levy CB, Dubins JS, Chin JE, Perry PV, Hulin B, Perry DA, Inagaki T, Dekker KA (2000) Effects of skyrin, a receptor-selective glucagon antagonist, in rat and human hepatocytes. Diabetes 49:2079–2086

Pascual S, Melgarejo P, Magan N (2003) Water availability affects the growth, accumulation of compatible solutes and the viability of the biocontrol agent *Epicoccum nigrum*. Mycopathologia 156:93–100

Perfumo A, Banat IM, Marchant R (2018) Going green and cold: biosurfactants from low-temperature environments to biotechnology applications. Trends Biotechnol 36:277–289

Petrescu I, Lamotte-Brasseur J, Chessa JP, Ntarima P, Claeyssens M, Devreese B, Marino G, Gerday C (2000) Xylanase from the psychrophilic yeast *Cryptococcus adeliae*. Extremophiles 4:137–144

Phadtare S (2004) Recent developments in bacterial cold-shock response. Curr Issues Mol Biol 6:125–136

Plemenitaš A, Vaupotič T, Lenassi M, Kogej T, Gunde-Cimerman N (2008) Adaptation of extremely halotolerant black yeast *Hortaea werneckii* to increased osmolarity: a molecular perspective at a glance. Stud Mycol 61:67–75

Poveda G, Gil-Durán C, Vaca I, Levicán G, Chávez R (2018) Cold-active pectinolytic activity produced by filamentous fungi associated with Antarctic marine sponges. Biol Res 51:28

Rastogi RP, Richa, Singh SP, Häder D-P, Sinha RP (2010a) Mycosporine-like amino acids profile and their activity under PAR and UVR in a hot-spring cyanobacterium *Scytonema* sp. HKAR-3. Austral J Bot 58:286–293

Rastogi RP, Richa KA, Tyagi MB, Sinha RP (2010b) Molecular mechanisms of ultraviolet radiation-induced DNA damage and repair. J Nucleic Acids 2010:592980

Raymond J, Fritsen C, Shen K (2007) An Ice-binding protein from an Antarctic sea ice bacterium. FEMS Microbiol Ecol 61:214–221

Reyes I, Bernier L, Antoun H (2002) Rock phosphate solubilization and colonization of maize rhizosphere by wild and genetically modified strains of *Penicillium rugulosum*. Microb Ecol 44:39–48

Robinson CH (2001) Cold adaptation in Arctic and Antarctic fungi. New Phytol 151:341–353

Rojas JL, Martín J, Tormo JR, Vicente F, Brunati M, Ciciliato I, Losi D, Van Trappen S, Mergaert J, Swings J, Marinelli F (2009) Bacterial diversity from benthic mats of Antarctic lakes as a source of new bioactive metabolites. Mar Genomics 2:33–41

Ruisi S, Barreca D, Selbmann L, Zucconi L, Onofri S (2007) Fungi in Antarctica. Rev Environ Sci Biotechnol 6:127–141

Russell NJ (1990) Cold adaptation of microorganisms. Philos Trans R Soc B 326:595–611

Salazar G, Sunagawa S (2017) Marine microbial diversity. Curr Biol 27:R489–R494

Santiago M, Ramírez-Sarmiento CA, Zamora RA, Parra LP (2016) Discovery, molecular mechanisms and industrial applications of cold-active enzymes. Front Microbiol 7:1408

Sarmiento F, Peralta R, Blamey JM (2015) Cold and hot extremozymes: Industrial relevance and current trends. Front Bioeng Biotechnol 3:148

Schulz B, Boyle C, Draeger S, Römmert AK, Krohn K (2002) Endophytic fungi: a source of novel biologically active secondary metabolites. Mycol Res 106:996–1004

Selbmann L, Onofri S, Fenice M, Federici F, Petruccioli M (2002) Production and structural characterization of the exopolysaccharide of the Antarctic fungus *Phoma herbaru m* CCFEE 5080. Res Microbiol 153:585–592

Selbmann L, De Hoog GS, Mazzaglia A, Friedmann EI, Onofri S (2005) Fungi at the edge of life: cryptoendolithic black fungi from Antarctic desert. Stud Mycol 51:32

Selbmann L, Pacelli C, Zucconi L, Dadachova E, Moeller R, de Vera JP, Onofri S (2018) Resistance of an Antarctic cryptoendolithic black fungus to radiation gives new insights of astrobiological relevance. Fungal Biol 122:546–554

Sharma R, Baghel RK (2010) Dynamics of acid phosphatase production of the ectomycorrhizal mushroom *Cantharellus tropicalis*. Afr J Microbiol Res 4:2072–2078

Shick JM, Dunlap WC (2002) Mycosporine-like amino acids and related gadusols: biosynthesis, accumulation, and UV-protective functions in aquatic organisms. Annu Rev Physiol 64:223–262

Siddiqui KS, Cavicchioli R (2006) Cold-adapted enzymes. Annu Rev Biochem 75:403–433. https://doi.org/10.1146/annurev.biochem.75.103004.142723

Singh MS, Yadav SL, Singh KS, Singh P, Singh NP, Ravindra R (2011) Phosphate solubilizing ability of two Arctic *Aspergillus niger* strains. Polar Res 30:7283

Singh S, Mandal SK (2012) Optimization of processing parameters for production of pectinolytic enzymes from fermented pineapple residue of mixed *Aspergillus* species. Jordan J Biol Sci 5:307–314

Singh SM, Singh PN, Singh SK, Sharma PK (2014) Pigment, fatty acid and extracellular enzyme analysis of a fungal strain *Thelebolus microsporus* from Larsemann Hills, Antarctica. Polar Rec 50:31–36

Snider CS, Hsiang T, Zhao G, Griffith M (2000) Role of ice nucleation and antifreeze activities in pathogenesis and growth of snow molds. Phytopathology 90:354–361

Sommaruga R, Libkind D, van Broock M, Whitehead K (2004) Mycosporine-glutaminol-glucoside, a UV-absorbing compound of two *Rhodotorula* yeast species. Yeast 21:1077–1081

Son H, Lee J, Lee YW (2012) Mannitol induces the conversion of conidia to chlamydospore-like structures that confer enhanced tolerance to heat, drought, and UV in *Gibberella zeae*. Microbiol Res 167:608–615. https://doi.org/10.1016/j.micres.2012.04.001

Stibal M, Anesio AM, Blues CJD, Tranter M (2009) Phosphatase activity and organic phosphorus turnover on a high Arctic glacier. Biogeosciences 6:913–922

Sumarah MW, Miller JD, Adams GW (2005) Measurement of a rugulosin-producing endophyte in white spruce seedlings. Mycologia 97:770–776

Takasawa T, Sagisaka K, Yagi K, Uchiyama K, Aoki A, Takaoka K, Yamamato K (1997) Polygalacturonase isolated from the culture of the psychrophilic fungus *Sclerotinia borealis*. Can J Microbiol 43:417–424

Tan H, Tang J, Li X, Liu T, Miao R, Huang Z, Wang Y, Gan B, Peng W (2017) Biochemical characterization of a psychrophilic phytase from an artificially cultivable morel *Morchella importuna*. J Microbiol Biotechnol 27. https://doi.org/10.4014/jmb.1708.08007

Taskin M, Ortucu S, Unver Y, Tasar OC, Ozdemir M, Kaymak HC (2016) Invertase production and molasses decolourization by cold-adapted filamentous fungus *Cladosporium herbarum* ER-25 in non-sterile molasses medium. Process Saf Environ Prot 103:136–143

Thomas DN, Dieckmann GS (2002) Antarctic sea ice--a habitat for extremophiles. Science 25:641–644

Tibbett M, Grantham K, Sanders FE, Cairney JWG (1998a) Induction of cold active acid phosphomonoesterase activity at low temperature in psychrotrophic ectomycorrhizal *Hebeloma* spp. Mycol Res 102:1533–1539

Tibbett M, Sanders FE, Cairney JWG (1998b) The effect of temperature and inorganic phosphorus supply on growth and acid phosphatase production in arctic and temperate strains of ectomycorrhizal *Hebeloma* spp. in axenic culture. Mycol Res 102:129–135

Tiquia SM, Mormile M (2010) Extremophiles–A source of innovation for industrial and environmental applications. Environ Technol 31(8-9):823

Tiquia-Arashiro SM, Rodrigues D (2016) Thermophiles and psychrophiles in nanotechnology. In: Extremophiles: applications in nanotechnology. Springer International Publishing, New York, pp 89–127

Tiwari R, Nain PKS, Singh S, Adak A, Saritha M, Rana S, Sharma A, Nain L (2015) Cold active holocellulase cocktail from *Aspergillus niger* SH3: process optimization for production and biomass hydrolysis. J Taiwan Inst Chem Eng 56:57–66

Trivedi P, Pandey A (2007) Low temperature phosphate solubilization and plant growth promotion by psychrotrophic bacteria, isolated from Indian Himalayan region. Res J Microbiol 2:454–461

Trochine A, Turchetti B, Vaz ABM, Brandao L, Rosa LH, Buzzini P, Rosa C, Libkind D (2017) Description of *Dioszegia patagonica* sp. nov., a novel carotenogenic yeast isolated from cold environments. Int J Syst Evol Microbiol 67:4332–4339

Tsuji M (2016) Cold-stress responses in the Antarctic basidiomycetous yeast *Mrakia blollopis*. R Soc Open Sci. 3:160106. https://doi.org/10.1098/rsos.160106

Tsuji M, Tanabe Y, Vincent WF, Uchida M (2018) *Mrakia arctica* sp. nov., a new psychrophilic yeast isolated from an ice island in the Canadian High Arctic. Mycoscience 59:54–58

Turk M, Plemenitaš A, Gunde-Cimerman N (2011) Extremophilic yeasts: plasma-membrane fluidity as determinant of stress tolerance. Fungal Biol 115:950–958

Upadhyay P, Shrivastava R, Agrawal PK (2016) Bioprospecting and biotechnological applications of fungal laccase. 3 Biotech 6:15

Walsh EA, Kirkpatrick JB, Rutherford SD, Smith DC, Sogin M, D'Hondt S (2016) Bacterial diversity and community composition from seasurface to subseafloor. ISME J 10:979–989

Wang N, Zang J, Ming K, Liu Y, Wu Z, Ding H (2013) Production of cold-adapted cellulase by *Verticillium* sp. isolated from Antarctic soils. Electron J Biotechnol 16:10–10

Wang M, Tian J, Xiang M, Liu X (2017) Living strategy of cold-adapted fungi with the reference to several representative species. Mycology 8:178–188

Watkinson SC (2016) The fungi. Chapter 5: Physiology and adaptation. pp 141–187

Weinstein RN, Palm ME, Johnstone K, Wynn-Williams DD (1997) Ecological and physiological characterization of *Humicola marvinii*, a new psychrophilic fungus from fellfield soils in the maritime Antarctic. Mycologia:706–711

Weinstein RN, Montiel PO, Johnstone K (2000) Influence of growth temperature on lipid and soluble carbohydrate synthesis by fungi isolated from fellfield soil in the maritime Antarctic. Mycologia 92(2):222–229

Wu G, Ma H, Zhu T, Li J, Gu Q, Li D (2012) Penilactones A and B, two novel polyketides from Antarctic deep-sea derived fungus *Penicillium crustosum* PRB-2. Tetrahedron 68:9745–9749

Wu G, Lin A, Gu Q, Zhu T, Li D (2013) Four new chloro-eremophilane sesquiterpenes from an antarctic deep-sea derived fungus, *Penicillium* sp. PR19N-1. Mar Drugs 11:1399–1408

Wu G, Sun X, Yu G, Wang W, Zhu T, Gu Q, Li D (2014) Cladosins A–E, hybrid polyketides from a deep-sea-derived fungus, *Cladosporium sphaerospermum*. J Nat Prod 77:270–275

Wu B, Wiese J, Wenzel-Storjohann A, Malien S, Schmaljohann R, Imhoff JF (2016) Engyodontochones, antibiotic polyketides from the marine fungus *Engyodontium album* strain LF069. Chem A Eur J 22:7452–7462

Wu DL, Li HJ, Smith DR, Jaratsittisin J, Xia-Ke-Er XFKT, Ma WZ, Guo YW, Dong J, Shen J, Yang DP, Lan WJ (2018) Polyketides and alkaloids from the marine-derived fungus *Dichotomomyces cejpii* F31-1 and the antiviral activity of Scequinadoline A against Dengue Virus. Mar Drugs 16:229

Wynn-Williams DD, Edwards HGM (2000) Proximal analysis of regolith habitats and protective biomolecules in situ by laser Raman spectroscopy: overview of terrestrial antarctic habitats and Mars analogs. Icarus 144:486–503. https://doi.org/10.1006/icar.1999.6307

Xiao N, Suzuki K, Nishimiya Y, Kondo H, Miura A, Tsuda S, Hoshino T (2010) Comparison of functional properties of two fungal antifreeze proteins from *Antarctomyces psychrotrophicus* and *Typhula ishikariensis*. FEBS J 277:394–403

Yogabaanu U, Weber JFF, Convey P, Rizman-Idid M, Alias SA (2017) Antimicrobial properties and the influence of temperature on secondary metabolite production in cold environment soil fungi. Pol Sci 14:60–67

Yu SM, Ramkumar G, Lee YH (2013) Light quality influences the virulence and physiological responses of *Colletotrichum acutatum* causing anthracnose in pepper plants. J Appl Microbiol 115:509–516

Yusof NY, Firdaus-Raih M, Mahadi NM, Illias RM, Abu Bakar FD, Murad AMA (2017) *In silico* analysis and 3D structure prediction of a chitinase from psychrophilic yeast *Glaciozyma antarctica* PI12. Malays Appl Biol 46:117–123

Zhdanova NN, Zakharchenko VA, Haselwandter K (2005) In: Dighton J, White JF, Oudemans P (eds) The fungal community, its organization and role in the ecosystem. CRC Press, Baton Rouge, LA, pp 759–768

Chapter 10
Melanin as an Energy Transducer and a Radioprotector in Black Fungi

Mackenzie E. Malo and Ekaterina Dadachova ⓘD

10.1 Introduction

Melanins represent a unique and ancient class of pigments that exist through all kingdoms of life with well-studied biological functions, yet despite their ubiquitous nature they have been challenging to define (Solano 2014). Due to their complex composition, function, and distribution melanins can be difficult to classify, and as of yet have eluded all currently available structural analysis. The inability to define melanin structurally impedes the ability to follow the "structure defines function" paradigm, and forces innovation in the field of functional study of the pigment.

Our understanding of the role of melanin in the fungal world has been well characterized yet continues to evolve. Melanized fungi have been observed across all phyla in the kingdom, with some species existing as constitutively melanized (i.e., *Wangiella dermatitidis*), while others only produce melanin under the appropriate conditions (i.e., *Cryptococcus neoformans*) (Cordero and Casadevall 2017). Table 10.1 shows some of the melanized fungi which are considered human pathogens as well as nonpathogenic ones. The presence of melanin corresponds with enhanced survival and improved fitness, imparting an advantage in harsh environments. This is apparent when observing the high incidence of melanized fungi in such extreme locations as the damaged nuclear reactor at Chernobyl (Dighton et al. 2008), the Antarctic rocky deserts (Selbmann et al. 2015), and under simulated Mars-like conditions (Onofri et al. 2008). These three locations deliver varied forms of stress including salinity, aridity, rapid and extreme temperature fluctuations, and little to no nutritional sources, but they also share a unique form of stress: high exposure to ionizing radiation. For the purposes of this chapter we will review the

M. E. Malo · E. Dadachova (✉)
College of Pharmacy and Nutrition, University of Saskatchewan, Saskatoon, Canada
e-mail: ekaterina.dadachova@usask.ca

© Springer Nature Switzerland AG 2019
S. M. Tiquia-Arashiro, M. Grube (eds.), *Fungi in Extreme Environments: Ecological Role and Biotechnological Significance*,
https://doi.org/10.1007/978-3-030-19030-9_10

Table 10.1 Examples of melanized fungi

Known human pathogen	Nonpathogenic in humans
Cryptococcus neoformans	*Cryomyces antarcticus*
Wangiella dermatitidis	*Alternaria alternata*
Histoplasma capsulatum	*Friedmanniomyces endolithicus*
Aspergillus niger	*Cladosporium spherospermum*

properties of melanin that provides resistance to ionizing radiation within the fungal community.

The general term "melanin" refers to a family of chemically different pigments with similar properties including their dark color, insolubility, resistance to acid hydrolysis, and susceptibility to oxidizing agents (Eisenman et al. 2007). There are two common forms of melanin found in fungi including 3,4-dehydroxyphenyalanine (L-DOPA)-melanin, which is synthesized from L-tyrosine or L-DOPA, and dihydroxynaphthalene (DHN)-melanin, which is formed via the polyketide pathway (Butler and Day 1998). During melanin synthesis toxic free radical intermediates are generated, necessitating a system to protect the cell, consequently structures akin to mammalian melanosomes compartmentalize this process and transport the polymer to the extracellular space (Casadevall et al. 2017). Alternatively, some fungi are able to synthesize melanin directly at the cell surface when an extracellular precursor is available, negating the need for self-protection. Location and form of the synthesized melanin then vary depending on the species investigated. In the case of the pathogenic yeast *C. neoformans*, melanin forms concentric layers adjacent to the cell membrane, and internal to the cell wall (Wang et al. 1995). Whereas in other species the polymer is more commonly deposited within the cell wall. Despite variation in deposition locale, melanin interacts with other cellular components within the space, forming a heterogeneous arrangement with lipids, proteins, and carbohydrates. This interaction leads to the formation of a granular microstructure that can arrange to form a more ordered macrostructure that mediates the deposition of newly synthesized melanin, or the movement of other cellular or extracellular components through the cell wall, and ultimately contributing to the unique role that melanin plays (Nosanchuk et al. 2015).

Synthesis of melanin by the cell is a costly endeavor as it can account for a significant portion of the biomass of the cell. In the case of *Agaricus bisporus* it can account for as much as 30% of the dry weight of the spores (Rast and Hollenstein 1977), while for *C. neoformans* the melanized strain has a cell wall approximately double the size of its non-melanized counterpart (Mandal et al. 2007). Considering the presence under the harshest of conditions, and knowing the especially taxing nature of synthesis, melanin must contribute a significant advantage. We will now consider the biological advantage that melanin affords fungi.

Here we will review the recent literature with the goal of not simply providing the examples of the resistance of melanized fungi to ionizing radiation, but with the emphasis on the functional response of melanin to ionizing radiation and the ways this functional response translates into the biological response of melanized fungi to ionizing radiation.

10.2 Biological Response of Melanized Fungi to Ionizing Radiation

The bacterial extremophile *Deinococcus radiodurans*, with an LD_{10} range of 2–15 kGy, is considered the most radiation-resistant microorganism (Sghaier et al. 2008), yet several melanized fungi such as *C. neoformans* and *Histoplasma capsulatum* have been shown to exhibit LD_{10} falling within that same dose range, with many more approaching the 1 kGy range, which is the standard dose for food irradiation in the US (Dadachova and Casadevall 2008). The ability of melanized fungi to survive such high acute doses of ionizing radiation, in addition to the ability to survive the harsh selective pressures of extreme environments with high ionizing radiation, is an example of *radioresistance*. Evidence for the role of melanin in enhanced survival can be found not just in its resistant to ionizing radiation, but also in its response to it.

The phenomenon of *radiotropism*, which refers to the ability to grow towards a radiation source, was observed in a number of strains of melanized fungi such as *Aspergillus versicolor* and *Cladosporium cladosporioides* that were isolated from the damaged Chernobyl reactor (Zhdanova et al. 1991). These fungal strains were able to grow towards radioactive particles, overgrow, and absorb them, and the capacity of accumulation was observed to correlate to the degree of pigmentation. This phenomenon was further characterized when it was observed that the germinating hyphae of these previously exposed spores exhibited directional hyphae growth towards a collimated source of ionizing radiation, independent of a nutritional carbon source, and was promoted by both beta and gamma radiation (Zhdanova et al. 2004). Furthermore, it was found that control fungal strains not previously exposed to radiation did not exhibit radiotrophic qualities, and it was suggested that differences could be due to the increased quantity of melanin in the responsive strains, in addition to some *radioadaptive* response due to previous exposure. Transcriptomic studies exploring the molecular and cellular response to ionizing radiation in *W. dermatitidis* have in fact shown that many transporter genes, as well as a number of genes involved in ribosomal biogenesis are upregulated specifically when melanin is present (Robertson et al. 2012), suggesting a mechanistic means by which the presence of melanin may impart advantage.

The upregulation in response to radiation positions melanized fungi for growth by improving means of nutrient transport and increasing capacity for protein synthesis. It has been shown by several groups that melanized fungal cells do in fact exhibit *radiostimulation*, or improved growth in the presence to radiation. Melanized *C. neoformans* and *W. dermatitidis* both show enhanced growth over non-melanized mutants when exposed to a 0.5 mGy/h radiation with significant increases in colony-forming units (CFU) and ^{14}C-acetate incorporation (Dadachova et al. 2007). Improved spore germination and hyphal growth were also demonstrated in previously exposed fungal cultures relative to radiation-naïve controls, and were found to show selectivity to types of radiation, as well as correlation between response and exposure dose history (Tugay et al. 2006), further suggesting a *radioadaptive* response to ionizing radiation. DNA damage caused by ionizing radiation could promote an adaptive response, and surprisingly, one study showed that cell cycle progression genes were downregulated while a few genes involved in translesion synthesis were upregulated in response to a low dose of ionizing radiation (Robertson et al. 2012), which could provide conditions in which DNA damage could be disregarded enabling the possibility of adaptive mutations. Additionally ionizing radiation has been shown to cause global changes to DNA methylation from bacteria to murine models, presenting an epigenetic mechanism of adaptive response (Miousse et al. 2017).

Metabolic changes in response to exposure to radiation have also shown a correlation when comparing melanized to non-melanized fungal strains. Performing side-by-side metabolic assays with the cell-impermeable tetrazolium salt XTT [2,3-bis-(2-methoxy 4-nitro-5-sulfophenyl)-2H-tetrazolium-5-carboxanilide] versus the cell-permeable salt MTT [3-(4,5-dimethylthiazol-2-yl)-2,5-diphenyltetrazolium bromide] allows to find out where exactly the electron-transfer events are taking place—on the surface of the cell or in the cytoplasm. It was observed in *C. neoformans* that there was an increased reduction of XTT in response to ionizing radiation only in the melanized strain, at doses similar to those at the damaged Chernobyl reactor (Dadachova et al. 2007). This result was not due to the reducing action of melanin alone, as the increase in XTT reduction was only observed in live melanized cells. Conversely, there were no marked changes in MTT reduction following irradiation in the melanized fungi, an additional support that the electron-transfer event is taking place at the cell well where melanin is located. Another experiment looked at the response of melanized *C. neoformans* to three different forms of radiation: light, UV, and gamma radiation. ATP levels were measured as an indicator of cellular energy (Bryan et al. 2011). It was noted that under all conditions, there was a reduction in ATP levels in response to ionizing radiation in the melanized *C. neoformans* relative to the non-melanized control, suggesting a universal melanin-related mechanism resulting in energy expense following irradiation. A more recent study has addressed the differences in metabolic response, as measured using an XTT assay of melanized fungi to protracted low-dose exposure in fast-growing (*C. neoformans*) versus slow-growing (*Cryomyces antarcticus*) species, and found that like the acute doses that are more represented in the literature, the melanized strains displayed increased metabolic activity relative to their non-melanized controls despite the two species having significantly varied metabolism (Pacelli et al. 2018).

Compiling what has been presented thus far it can be suggested that the melanin-related response to irradiation occurs over a broad range of dose rates, under a variety of metabolic models as demonstrated by the various species studied, and in response to different forms of radiation. To assess the mechanism by which melanin contributes to *radioresistance, radiotropism, radiostimulation,* and altered cellular metabolism we will address the functional role of melanin.

10.3 Functional Response of Melanin to Ionizing Radiation

The macrostructure that melanin forms in fungi whereby granules are deposited in the extracellular space as a heterogeneous and rigid structure surrounding the cell or spore suggests a role in *physical shielding*. To assess the role of melanin in *radioresistance* melanized and non-melanized forms of two fungal species were subjected to lethal and sublethal doses of radiation and it was determined that the spherical spatial arrangement of the melanin acted as a shield using the Compton effect to protect the cells from high-energy photons and improving overall survival (Fig. 10.1) (Dadachova et al. 2008). Furthermore, when melanized and non-melanized *C. neoformans* cells were exposed to either sparsely ionizing gamma radiation or densely ionizing alpha particle or deuteron irradiation, the melanized cells showed increased structural stability, as demonstrated by using transmission electron microscopy (TEM) to assess cellular morphology following irradiation (Fig. 10.2) (Malo et al. 2018). It was observed that the presence of a 20–30 nm thick layer of melanin provided an efficient barrier to the ionizing radiation lowering the relative biological effectiveness (RBE) of the densely ionizing radiation. It was also noted that in

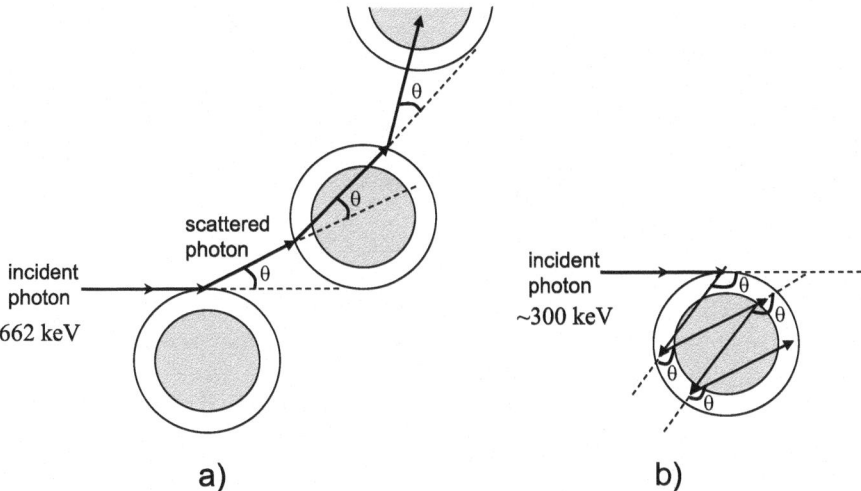

Fig. 10.1 Proposed model of scatter of ionizing radiation by melanin spheres: (**a**) forward scatter of initial high-energy photons incoming at arbitrary angle θ when forward scatter predominates; (**b**) oscillation within melanin spheres of secondary scattered photons which energies fall below 300 keV and both forward and backward scatters take place with backward scatter starting to predominate. *Source* Dadachova et al. (2008). Copyright @ Pigment Cells and Melanoma Research (John Wiley and Sons). Reproduced with permission

Fig. 10.2 Micrographs of melanized and non-melanized *C. neoformans* cells obtained with TEM following alpha particle irradiation. *C. neoformans* cultures were subjected to a sham exposure and to low and high alpha particle irradiation at 0, 150, or 500 Gy (dose rate 72 Gy/min, energy 8.5 MeV). (**a**) H99 strain grown in the presence of L-DOPA; (**b**) H99 strain grown without L-DOPA. Scale bar, 200 nm. *Source* Malo et al. (2018). Copyright @ Fungal Biology (Elsevier). Reproduced with permission

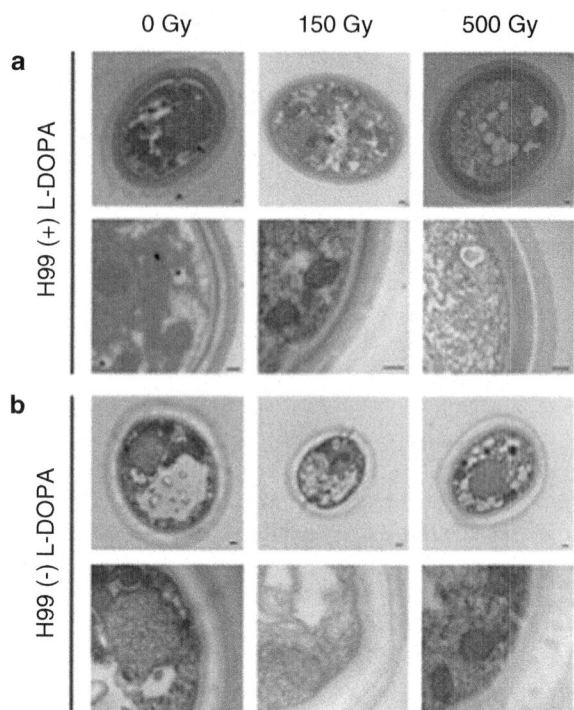

several metabolic studies where MTT activity and ATP levels were measured, the non-melanized controls had higher MTT levels and ATP levels were further depleted following irradiation (Dadachova et al. 2007; Pacelli et al. 2018). It was suggested by the authors that in response to ionizing radiation the cell was upregulating metabolism in order to respond to the radiation assault, while the melanized strains were physically shielded.

Physico-chemical studies have also confirmed the high degree of stability and resistance to ionizing gamma irradiation of melanin occurs first by reducing the energy of the radiation by the Compton effect followed by the capture of the result-ing recoil electron by the stable free radicals in the pigment (Schweitzer et al. 2009). This protective effect has also been demonstrated in a study where mice were fed a melanin-containing black edible mushroom prior to total body irradiation with 9 Gy (Revskaya et al. 2012). Among the mice that were fed with the melanin-containing mushrooms, or white mushrooms supplemented with melanin, 80% survived the dose, while all mice not fed melanin died from gastrointestinal syndrome. The barrier-like *physical shield* that melanin provides clearly protects varied organisms from damage caused by ionizing radiation, but it also appears to provide *chemical shielding* in response to increases in reactive oxygen species (ROS) from resulting free electrons and the radiolysis of water.

The unpaired electrons in melanin readily interact with free radicals and other reactive species providing a melanin-dependent antioxidant system in melanin-containing organisms (Shcherba et al. 2000). In fungi previously reported to present *radioadaptive* properties, it was demonstrated that ionizing radiation induced an upregulation in the expression of melanin almost twofold in addition to increasing activities of several other melanin-regulating enzymes, significantly increasing the antioxidant capacity of the organism (Tugay et al. 2011). More recently, using novel electrochemical reverse engineering methods, two forms of melanin, including a fungal-derived melanin, were shown to have rapid and repeatable scavenging activities (Kim et al. 2017) which would be essential to a radioprotective role in response to ionizing radiation, which generates ROS. In addition to further characterizing melanin's scavenging capabilities this study demonstrated that it was linked to melanin's redox activity. The observation of melanins' redox capability within a physiological range indicates that the pigment could be sensitive to redox content in vivo which suggests a possible role in energy harvesting or redox signaling (Liu et al. 2017).

The concept of *radiosynthesis* is analogous to photosynthesis, in that it proposes a mechanism by which the energy from radiation is converted to chemical energy, and is a phenomenon suggested to explain radiation-induced growth observed in melanized fungi (Dadachova et al. 2007). Melanin is capable of electron transfer, and following irradiation, the velocity of electron transfer increases, and using electron spin resonance spectroscopy (ESR) a stable free radical can be detected in melanin that is altered in response to irradiation. Other studies have also demonstrated the ability for melanin to alter its redox potential resulting in the generation of an electric current following gamma irradiation (Turick et al. 2011).

10.4 Conclusions

The advantage melanin imparts on fungi starts first with its ability to act as a physical shield, reducing the RBE of ionizing radiation and reducing its potential for destruction. It then protects the organism further by scavenging for ROS generated by ionizing radiation. The physical and chemical protective qualities of melanin can explain the increases in cell survival, and the selective growth of melanized species observed in in vitro studies. This is further demonstrated in nature where it was observed that 80% of the fungal species recovered from the damaged nuclear reactor at Chernobyl were melanized (Zhdanova et al. 2000), black rock fungi are some of the only organisms capable of surviving with exposure to high solar radiation in the Antarctic deserts (Selbmann et al. 2015), and various melanized fungal species have survived cosmic radiation while exposed on the Mir Spacecraft (Novikova 2004).

Following shielding, melanin next provides advantage with its electrochemical capacity. In response to exposure to radiation melanin undergoes various electrophysical changes, demonstrates increased capacity for electron transfer, and alters its redox potential. The melanized organism also experiences a melanin-dependent increase in growth and metabolic activity, and in some cases this increased growth carries the organism closer to source the of radiation. The correlation indicates that the changes in melanin are linked to the changes in growth. This suggests that the role that melanin is playing in resistance to ionizing radiation is as a protective agent, and as a player mediating the fungal biological response to ionizing radiation.

10.5   Future Research Directions

The key question yet to be solved is how melanin translates the electrochemical changes that occur in response to ionizing radiation into the biological changes in growth and survival in the organism. Is this due to the ability of melanin to mediate *radiosynthesis*, and if so what is the mechanism, and what pathways are involved? Alternately could it be melanin's role as a redox mediator? Due to its ability to rapidly accept and donate electrons, and its position at the interface of the extracellular and intracellular space, melanin could be positioned for an important role in cellular communication (Liu et al. 2017). If this is the case, what types of signaling events is melanin initiating in response to ionizing radiation?

Improving our understanding of how melanin mediates its effects on biological systems is important for various reasons. By exploiting the shielding properties of melanin we could develop novel methods to protect individuals undergoing radiation treatments, or individuals exposed to radiation following nuclear accidents. Uncovering the capacity for melanin-related energy transduction in fungi will allow us to consider the role that the fungi could contribute as energy generators, which could be of great importance as our global environment changes in response to global warming.

References

Bryan R, Jiang Z, Friedman M, Dadachova E (2011) The effects of gamma radiation, UV and visible light on ATP levels in yeast cells depend on cellular melanization. Fungal Biol 115:945–949

Butler MJ, Day AW (1998) Fungal melanins: a review. Can J Microbiol 44:1115–1136

Casadevall A, Cordero RJB, Bryan R, Nosanchuk J, Dadachova E (2017) Melanin, radiation, and energy transduction in fungi. Microbiol Spectr 5:1–6

Cordero RJB, Casadevall A (2017) Functions of fungal melanin beyond virulence. Fungal Biol Rev 31:99–112

Dadachova E, Casadevall A (2008) Ionizing radiation: how fungi cope, adapt, and exploit with the help of melanin. Curr Opin Microbiol 11:525–531

Dadachova E, Bryan RA, Huang X, Moadel T, Schweitzer AD, Aisen P, Nosanchuk JD, Casadevall A (2007) Ionizing radiation changes the electronic properties of melanin and enhances the growth of melanized fungi. PLoS One 2:e457

Dadachova E, Bryan RA, Howell RC, Schweitzer AD, Aisen P, Nosanchuk JD, Casadevall A (2008) The radioprotective properties of fungal melanin are a function of its chemical composition, stable radical presence and spatial arrangement. Pigment Cell Melanoma Res 21:192–199

Dighton J, Tugay T, Zhdanova N (2008) Fungi and ionizing radiation from radionuclides. FEMS Microbiol Lett 281:109–120

Eisenman HC, Mues M, Weber SE, Frases S, Chaskes S, Gerfen G, Casadevall A (2007) *Cryptococcus neoformans* laccase catalyses melanin synthesis from both D- and L-DOPA. Microbiology 153:3954–3962

Kim E, Kang M, Tschirhart T, Malo M, Dadachova E, Cao G, Yin JJ, Bentley WE, Wang Z, Payne GF (2017) Spectroelectrochemical reverse engineering DemonstratesThat Melanin's redox and radical scavenging activities are linked. Biomacromolecules 18(12):4084–4098

Liu Y, Kim E, Li J, Kang M, Bentley WE, Payne GF (2017) Electrochemistry for bio-device molecular communication: the potential to characterize, analyze and actuate biological systems. Nano Commun Networks 11:76–89

Malo ME, Bryan RA, Shuryak I, Dadachova E (2018) Morphological changes in melanized and non-melanized *Cryptococcus neoformans* cells post exposure to sparsely and densely ionizing radiation demonstrate protective effect of melanin. Fungal Biol 122(6):449–456

Mandal P, Roy TS, Das TK, Banerjee U, Xess I, Nosanchuk JD (2007) Differences in the cell wall architecture of melanin lacking and melanin producing *Cryptococcus neoformans* clinical isolates from India: an electron microscopic study. Braz J Microbiol 38:662–666

Miousse IR, Kutanzi KR, Koturbash I (2017) Effects of ionizing radiation on DNA methylation: from experimental biology to clinical applications. Int J Radiat Biol 93:457–469

Nosanchuk JD, Stark RE, Casadevall A (2015) Fungal melanin: what do we know about structure? Front Microbiol 6:1463

Novikova ND (2004) Review of the knowledge of microbial contamination of the Russian manned spacecraft. Microb Ecol 47(2):127–132

Onofri S, Barreca D, Selbmann L, Isola D, Rabbow E, Horneck G, De Vera JP, Hatton J, Zucconi L (2008) Resistance of Antarctic black fungi and cryptoendolithic communities to simulated space and Martian conditions. Stud Mycol 61:99–109

Pacelli C, Bryan RA, Onofri S, Selbmann L, Zucconi L, Shuryak I, Dadachova E (2018) The effect of protracted X-ray exposure on cell survival and metabolic activity of fast and slow growing fungi capable of melanogenesis. Environ Microbiol Rep 10(3):255–263

Rast D, Hollenstein GO (1977) Architecture of the Agaricus bisporus spore wall. Can J Bot 55:2251–2262

Revskaya E, Chu P, Howell RC, Schweitzer AD, Bryan RA, Harris M, Gerfen G, Jiang Z, Jandl T, Kim K, Ting LM, Sellers RS, Dadachova E, Casadevall A (2012) Compton scattering by internal shields based on melanin-containing mushrooms provides protection of gastrointestinal tract from ionizing radiation. Cancer Biother Radiopharm 27(9):570–576

Robertson KL, Mostaghim A, Cuomo CA, Soto CM, Lebedev N, Bailey RF, Wang Z (2012) Adaptation of the black yeast Wangiella dermatitidis to ionizing radiation: molecular and cellular mechanisms. PLoS One 7(11):e48674

Selbmann L, Zucconi L, Isola D, Onofri S (2015) Rock black fungi: excellence in the extremes, from the Antarctic to space. Curr Genet 61:335–345

Sghaier H, Ghedira K, Benkahla A, Barkallah I (2008) Basal DNA repair machinery is subject to positive selection in ionizing-radiation-resistant bacteria. BMC Genomics 9:297

Shcherba VV, Babitskaia VG, Kurchenko VP, Ikonnikova NV, Kukulianskaia TA (2000) Antioxidant features of fungal melanin pigments. Prikl Biokhim Mikrobiol 36:569–574

Solano F (2014) Melanins: skin pigments and much more—types, structural models, biological functions, and formation routes. New J Sci 2014:1–28

Schweitzer AD, Howell RC, Jiang Z, Bryan RA, Gerfen G, Chen CC, Mah D, Cahill S, Casadevall A, Dadachova E (2009) Physico-chemical evaluation of rationally designed melanins as novel nature-inspired radioprotectors. PLoS One 4(9):e7229

Tugay T, Zhdanova NN, Zheltonozhsky V, Sadovnikov L, Dighton J (2006) The influence of ionizing radiation on spore germination and emergent hyphal growth response reactions of microfungi. Mycologia 98(4):521–527

Tugay TI, Zheltonozhskaya MV, Sadovnikov LV, Tugay AV, Farfan EB (2011) Effects of ionizing radiation on the antioxidant system of microscopic fungi with radioadaptive properties found in the Chernobyl exclusion zone. Health Phys 101:375–382

Turick CE, Ekechukwu AA, Milliken CE, Casadevall A, Dadachova E (2011) Gamma radiation interacts with melanin to alter its oxidation-reduction potential and results in electric current production. Bioelectrochemistry 82(1):69–73

Wang Y, Aisen P, Casadevall A (1995) *Cryptococcus neoformans* melanin and virulence: mechanism of action. Infect Immun 63:3131–3136

Zhdanova NN, Lashko TN, Redchits TI, Vasilevskaia AI, Borisiuk LG, Siniavskaia OI, Gavriliuk VI, Muzalev PN (1991) The interaction of soil micromycetes with "hot" particles in a model system. Mikrobiol Zh 53:9–17

Zhdanova NN, Tugay T, Dighton J, Zheltonozhsky V, Mcdermott P (2004) Ionizing radiation attracts soil fungi. Mycol Res 108:1089–1096

Zhdanova NN, Zakharchenko VA, Vember VV, Nakonechnaya LT (2000) Fungi from Chernobyl: mycobiota of the inner regions of the containment structures of the damaged nuclear reactor. Mycol Res 104(12):1421–1426

Chapter 11
Fungi in Biofilms of Highly Acidic Soils

Martina Hujslová [iD] and Milan Gryndler

11.1 Introduction

This chapter summarizes current knowledge about the diversity of acidophilic fungi and proposes new insight into their ecology. In this regard, the list of highly acidic localities (pH <3) along with the list of methodological approaches applied for studying fungal diversity and inventories of the main fungal groups and/or species connected with the studied habitats is useful in many aspects of studies of extremophilic fungi. The text of our contribution further tracts the concept of acidophilic behavior in fungi in connection with basic knowledge about the evolution of acidophilic fungi, possible strategy used for colonization of highly acidic sites, and main fungal roles within acidophilic communities. Some aspects of the above information suggest the linkage between acidophilic behavior and biofilm life strategy as an ecological phenomenon of adaptation to life in highly acidic environments including soils.

There are several review articles dealing with the diversity and functioning of acidophilic microorganisms and the whole communities. However, most of them aim mainly on prokaryotic organisms and where eukaryotic assemblage is under review acidophilic fungi are mentioned only marginally (e.g., Johnson 1998, 2012; Aguilera 2013; Aguilera et al. 2016; Quatrini and Johnson 2018). Some of the review articles focused exclusively on microbial communities inhabiting highly acidic habitats like acid mine drainage systems (Baker and Banfield 2003; Das et al. 2009; Méndez-García et al. 2015) or corroding concrete sewer environment (Li et al. 2017). The first paper summarizing particularly fungal data has been published

M. Hujslová (✉)
Laboratory of Fungal Biology, Institute of Microbiology ASCR, Prague 4, Czech Republic

M. Gryndler
Faculty of Science, Department of Biology, Jan Evangelista Purkyně University in Ústí nad Labem, Ústí nad Labem, Czech Republic

© Springer Nature Switzerland AG 2019
S. M. Tiquia-Arashiro, M. Grube (eds.), *Fungi in Extreme Environments: Ecological Role and Biotechnological Significance*,
https://doi.org/10.1007/978-3-030-19030-9_11

by Gross and Robbins (2000). The authors reviewed all fungi and yeasts isolated from habitats having pH up to 4 along with their morphological and habitat characteristics and provided a guide for identification of 81 fungal species. This chapter focuses exclusively on fungi forming the core of the fungal assemblage inhabiting highly acidic habitats (pH < 3). This is the first comprehensive view on acidophilic fungi, their diversity, functioning, and ecology.

11.2 The Concept of Acidophilic/Acidotolerant Fungi

Fungi are common inhabitants of the soil and are, as a group of organisms, highly taxonomically and functionally diversified. Fungal species can be divided into three ecological groups according to their abilities to inhabit extreme environments: mesophiles, generalists, and specialists (Gostinčar et al. 2010). Mesophiles do not tolerate extreme conditions and only their resting structures may survive under harsh conditions for a considerable time. Generalists tolerate various stressful environments but not the most extreme ones. Although they successfully populate stressful habitats in the nature, in the laboratory they grow best under moderate conditions. Specialists can tolerate or even prefer extreme environmental conditions. The specialists with some preference for extreme conditions are called extremophilic species (extremophiles). Gostinčar et al. (2010) distinguish two groups of extremophiles: adaptive and obligate. Adaptive extremophiles have much broader ecological amplitude compared to obligate ones. Though the obligate behavior is typical for prokaryotes, it is occasionally encountered also in fungi. For instance an obligate halophilic fungus *Wallemia ichthyophaga* has been isolated and described exclusively from hypersaline environments (Gostinčar et al. 2010).

Acidophiles are organisms (mainly microorganisms) capable of growing in the environments with low pH (Johnson 2012). Cavicchioli and Thomas (2000) define acidophiles as organisms that have a pH optimum for growth close or below 3. Magan (1997) delimits acidophiles as organisms showing active growth at pH values below 4. According to generally accepted consensus, the organisms inhabiting low pH habitats can be divided into three categories based on their growth optima. Extreme acidophiles have pH optima for growth lower than 3, moderate acidophiles grow best at pHs 3–5, and acid-tolerant organisms have pH growth optima at values higher than 5 but are also metabolically active in acidic environments (Johnson 2012).

Almost all fungal species known to date from highly acidic environments are able to grow at pH 1 or even lower (*Acidomyces acidophilus*) (Starkey and Waksman 1943; Gould et al. 1974; Hölker et al. 2004; Yamazaki et al. 2010; Hujslová et al. 2010, 2013, 2014; Vázquez-Campos et al. 2014). At the same time, they do not require low pH values to survive, since they can grow in a broad range of pH. With respect to the concepts mentioned above, currently known fungal species inhabiting highly acidic sites can be regarded as extreme or moderate acidophiles with broad ecological amplitude. No obligate acidophilic fungus has been described to date (Hujslová et al. 2017).

11.3 The Origin of Acidophilic Life Strategy

Extreme acidophiles are widely distributed throughout the three domains of life (Johnson 2009) and, besides prokaryotic microorganisms, they include eukaryotes like fungi, algae, protozoa, and rotifera. It appears that the ability to grow at low pH has arisen independently several times and recent acidophilic organisms thus do not have a single acidophilic common ancestor (Johnson 2012).

The evolutionary distances between acidophilic species that have been detected in Tinto River and their neutrophilic relatives showed that adaptations associated with the transition from a neutral to an acidic environment must develop relatively rapidly (Amaral-Zettler et al. 2002). This ability to adapt to environments with the broad pH range has been observed in very different evolutionary lineages (Amaral-Zettler 2013). Close phylogenetic relations between acidophilic and neutrophilic fungi (Amaral-Zettler et al. 2002, 2003; Baker et al. 2004, 2009; Hujslová et al. 2017) support the notion of rapid development of the adaptation to acidic environments.

Baker et al. (2004) suggest that acidophilic fungi have a selective advantage in surviving in acid environments that other closely related fungi do not have. It seems that highly acidic sites represent a vacant niche with limited competition. Acidophilic fungi, as weak competitors, then take their opportunity and colonize these sites (Hujslová et al. 2017). This strategy is typical for generalist species that successfully populate stressful habitats where competition with mesophiles is limited. Consequently, the generalists can be taken as a reservoir of potential candidates for the evolution of taxa specialized for living under extreme conditions because they are able to persist across a range of environments due to their "robust" genotypes (Gostinčar et al. 2010). This mechanism of colonization is probably commonly used across various extreme environments (Gostinčar et al. 2010) and extremophilic fungal communities consisting of both generalists and specialists represent the typical pattern encountered in a variety of extreme habitats, e.g., acidic, alkaline, and hypersaline (Hujslová et al. 2017).

11.4 The Main Reasons for Studying Soil Acidophilic Fungi

Fungi from extreme environments, including acidophilic ones, belong to biotechnologically most attractive organisms. They are studied as a source of enzymes and metabolites with uncommon properties and may actively participate within bioremediation processes (Johnson 1998; Baker and Banfield 2003). The review focusing on biotechnological potential of acidophilic fungi is in preparation.

The second reason is more essential. There is a notion that environments with extreme conditions were probably far more widespread during the early life of our planet. Therefore, some extremophilic organisms are representatives of archaic forms of life (Johnson 1998). However, eukaryotic extremophiles including fungi, as noted above, are able to develop their secondary ability to grow under extreme conditions relatively quickly. These events thus probably occurred many times throughout different geological periods. As a result, the group of recent extremophilic fungi probably represents a complex mix of organisms with very different phylogenesis, ecological traits, and taxonomic position. The research on acidophilic fungal communities thus may bring radically new information about various mechanisms of adaptation and is highly desirable. Acidic soils represent complex environments with diversified mycoflora which can serve as particularly rich source of interesting fungi.

11.5   Chemical and Biological Characteristics of Extremely Acidic Environments

Highly acidic environments (pH <3) are of both natural and anthropogenic origin and their genesis is closely connected with microbial processes such as dissimilatory oxidation of elemental sulfur, reduced sulfur compounds, and ferrous iron. The first one is typically encountered in geothermal and volcanic areas where elemental sulfur is oxidized by autotrophic and heterotrophic microorganisms to sulfuric acid, which can result in the lowering of the pH (Johnson 1998).

The second and the third processes are involved in oxidation of sulfide minerals. In locations where metal sulfide-rich rocks occur, microbial oxidation together with exposure to air and water contributes to the generation of highly acidic solutions referred to as acid rock drainage (ARD) or acid mine drainage (AMD). Since sulfide minerals largely encompass metals such as Au, Ag, Cu, Zn, and Pb, the waters acidified by products of oxidation of these minerals are often rich in these elements (Baker and Banfield 2003; Tiquia-Arashiro and Rodrigues 2016) and the types and concentrations of heavy metals present in the environment are determined by local geochemistry. Extremely acidic environments may be enriched also in soluble metalloid elements, of which the most important one is arsenic (Johnson 1998). Besides high concentrations of hydrogen ions and heavy metals, high concentrations of dissolved sulfates are also typical in extremely acidic habitats (Baker and Banfield 2003; Zak and Wildman 2004). At the same time, high temperature represents the feature often encountered in these places (Brock 1978).

Highly acidic environments may be classified as oligotrophic since they contain relatively low concentrations of dissolved organic carbon that can be exploited by saprotrophic organisms as a source of nutrition. Dissolved organic carbon is derived from biomass of primary producers. In sites where sunlight is missing, like abandoned deep mines, the dissolved organic carbon is produced exclusively by chemolitho-autotrophic organisms and, as such, is of strictly prokaryotic origin (Johnson 1998). Illuminated sites mainly depend on photoautotrophic acidophiles including eukaryotic microalgae (Johnson 1998; Amaral-Zettler et al. 2002). Heterotrophic microorganisms including fungi depend on carbon nutrition derived from biomass of primary producers (Johnson 1998, 2012). Potential extraneous sources of organic matter like bat guano or wooden components of mine roof supports in subterranean environments as well as terrestrial organic carbon sources such as leaf litter accumulated on the surface of acidic soils can be encountered as sources of organic matter potentially available for saprotrophic acidophiles (Das et al. 2009; Johnson 2012).

11.6   Acidophilic Fungi in Biofilms

Since acid mine drainage systems represent the global environmental problem, these sites have been extensively studied. The most abundant biological structures encountered in these places were the microbial communities forming biofilms. This is not surprising if the ability of the organisms to live in biofilms represents an adaptation to extreme environmental conditions, including acidity. In abandoned pyrite mine Cae Coch in North Wales, streamer and slime growths have been estimated to exceed 100 m³ in volume. It is the largest accumulation of macroscopic acidophilic biomass yet described (Johnson 1998). Macroscopic structures like streamers, slimes, mats, snottites (stalactite-like structures), and drapes were found to cover 30% of the surfaces occurring in the acidic mine drainage environments (Méndez-García et al. 2015). These microhabitats differ in physicochemical factors like temperature, pH, and ionic strength, and thus harbor communities formed by different microorganisms (Baker and Banfield 2003).

Fungi have been described as an abundant element of the acidophilic biofilm and the biomass of fungal hyphae represents a significant portion of the total biomass in biofilm communities (Baker and Banfield 2003). This finding has been confirmed by microscopic observation (Baker et al. 2004; López-Archilla et al. 2004a; Zirnstein et al. 2012) as well as culture-independent approaches like fluorescent in situ hybridization (FISH) and cloning and sequencing of SSU rDNA gene (Bond et al. 2000; Amaral-Zettler et al. 2002; Baker et al. 2004, 2009; Zirnstein et al. 2012). High amounts of the fungal biomass have also been confirmed by analysis of intact polar lipids and total fatty acids (Bühring et al. 2012). In addition, Zirnstein et al. (2012) have found out that fungal hyphae made up between 10 and 20% of the total cell biomass in stalactite biofilms compared to approximately 5% portion in acid streamer biofilms. Higher amounts of fungal filaments were observed in flowing acid mine drainage solutions than in pools where the water was stagnant (Baker et al. 2004). Fungi belong to the most abundant groups in mature biofilms (Wilmes et al. 2009; Zirnstein et al. 2012).

It is thus obvious that fungi provide the backbone for three-dimensional structure of biofilms, anchor them to physical surfaces, and serve as secondary surfaces for other organisms during the biofilm development (Baker and Banfield 2003). Filamentous character of hyphae of mycelial fungi may contribute to mechanical stability of biofilm structures and may speed up the transport of metabolites within the volume of macroscopic biofilms. This "mechanical" function of biofilm fungi may be of great ecological importance. Wilmes et al. (2009) described that fungal filaments in mature biofilms provide structural support for pellicles subjected to shear stress in flowing streams. Robbins et al. (2000) have mentioned that fungi significantly help to form drip (stalactite) structures. Another example of significant contribution of fungi to formation of biofilms has been provided by Amaral-Zettler et al. (2003) in highly acidic Tinto River.

Because the majority of soil microorganisms are living as members of the biofilm, it is probable that significant analogies in the ecology of acidophilic organisms inhabiting the soil and organisms inhabiting biofilms in other acidic environments (slimes, mats, and snottites mentioned above) do exist. Unfortunately, the function of filamentous fungi in biofilms inhabiting acidic soils was hitherto almost neglected in spite of the fact that the contribution of fungi to the formation of three-dimensional cohesive biofilms may stabilize soil structure.

11.7   Diversity of Acidophilic Fungi

Since prokaryotic members of acidophilic microbial communities actively participate on the genesis of extremely acidic places, the majority of investigations have been focused on bacterial communities. Nevertheless, the evidence of eukaryotic life, including the fungal one, in extremely acidic habitats has been gathered as well. One of the first observations of fungi in a highly acidic substrate has been presented by Starkey and Waksman (1943) who obtained two fungal isolates from acid solutions containing copper sulfate having pH values between 0.2 and 0.7. Later Sletten and Skinner (1948) isolated two fungi from a sulfuric acid-rich solution, and these two isolates were identical to the one of the fungi mentioned by Starkey and Waksman. In 1953, a report of microbiological assemblage inhabiting acid mine waters in West Virginia and Pennsylvania was presented by Joseph (Joseph 1953). Seven fungal genera have been detected during this study. Later on, altogether 186 fungal species have been described from acid streamers and surrounding soils from areas in Ohio and West Virginia by Cooke (1976). In 2000, Gross and Robbins summarized the data on fungi isolated from habitats characterized by pH values below 4, primarily soils, and published the list of 81 species with their morphological features along with literature sources.

During the last decades, the interest in studying acidophilic microbial communities has increased and fungi have been detected in various extremely acidic substrates including acid mine waters and sediments (López-Archilla and Amils 1999; López-Archilla et al. 2001, 2004b; Gadanho and Sampaio 2006; Oggerin et al. 2016), mine process water (Vázquez-Campos et al. 2014), acid mine biofilms (Bond et al. 2000; Amaral-Zettler et al. 2002; López-Archilla et al. 2004a; Baker et al. 2004, 2009; Macalady et al. 2007; Zirnstein et al. 2012), corroding concrete sewer environments (summarized in Li et al. 2017), oil shale by-products (de Goes et al. 2017), and soils (Hujslová et al. 2010, 2017). Some of these studies have proposed just the evidence of the fungi and their potential role within the studied substrate. Some of them have also provided the data on fungal diversity (see Table 11.1). It is apparent that the core fungal assemblage mostly consists of species inhabiting exclusively highly acidic habitats as well as of taxa known from less acidic but otherwise extreme environments (Hujslová et al. 2017).

The first acidophilic fungi have been described in the beginning of this century. New fungal species *Hortaea acidophila* (current name *Neohortaea acidophila*) has been described from an extract of brown coal containing humic and fulvic acids at pH 0.6 (Hölker et al. 2004). Selbmann et al. (2008) have completed the data on fungal isolates published by Starkey and Waksman (1943), Sletten and Skinner (1948), and later Gould et al. (1974), Sigler and Carmichael (1974), and Gimmler et al. (2001) and concluded that the dark pigmented isolates studied in these papers are identical. The authors described them as a new acidophilic fungus *Acidomyces acidophilus*. The first record of acidophilic fungus detected in sexual state has been reported by Yamazaki et al. (2010) as *Teratosphaeria acidotherma*. Later on, this fungus has been ascribed to the genus *Acidomyces* (Hujslová et al. 2013). New acidophilic genus *Acidiella* with two new species *Acidiella bohemica* (Hujslová et al. 2013) and *Acidiella uranophila* (Vázquez-Campos et al. 2014; Kolařík et al. 2015) has been described. There is however a possibility that *Acidiella bohemica* and *A. uranophila* are conspecific (Hujslová et al. 2017). Recently, four new acidophilic fungal species *Acidea extrema*, *Acidothrix acidophila*, *Soosiella minima* (Hujslová et al. 2014) and *Coniochaeta fodinicola* (Vázquez-Campos et al. 2014) have been described.

The inventories of mycoflora of acid sites indicate wide geographic distribution of different species of acidophilic fungi. With exception of *Neohortaea acidophila* and *Soosiella minima*, all currently known acidophilic species have been reported from more than one highly acidic site (see Table 11.1). *Acidomyces acidophilus* is a typical member of the acidophilic fungal assemblage known from distinct highly acidic places (Selbmann et al. 2008; Hujslová et al. 2013). *Acidomyces acidothermus*, another extremophilic species, must be taken as a ubiquitous inhabitant of acidic environments as well, being detected in the USA, the Czech Republic, Iceland, Japan, China, and Australia (see Table 11.1).

Further, at least some species of acidophilic fungi are not strictly specialized to a particular substrate/environment. For example, *Acidiella bohemica*, originally described from highly acidic soil (Fig. 11.1), has also been detected as the most abundant species in acidic oil shale by-products (de Goes et al. 2017). In addition, Amaral-Zettler et al. (2002) probably detected this species in Tinto River by molecular tools (Hujslová et al. 2017, see Table 11.1). The capnodialean fungal isolate obtained by Oggerin et al. (2016) from Tinto River seems to be very similar to *Acidiella uranophila* described from mine waters (see Table 11.1). An acidophilic soil fungus *Acidothrix acidophila* has been detected as one of the four members of a consortium attained during laboratory cultivation of archaeal Richmond mine acidophilic nanoorganisms (Krause et al. 2017) as well as from acidic soil (see Table 11.1 and Fig. 11.2). *Coniochaeta fodinicola* belongs to one of the most abundant species at highly acidic soil of Cihelna v Bažantnici site studied by Hujslová et al. (2017) as well as in uranium mine water (Vázquez-Campos et al. 2014). The genus *Coniochaeta* has also been noticed among the most abundant fungal groups

Table 11.1 The list of highly acidic localities (pH <3) from which fungal diversity data are available

Locality	The main physicochemical characteristics	Studied samples	Methods for studying diversity	The main fungal groups/species connected with highly acidic habitats	References
Richmond mine, Iron Mountains, California, USA (subterranean)	pH 0.8–1.38 Temperature 30–50 °C metal-rich (Fe, Zn, As, Cu) High levels of sulfates	Biofilm	**Combination of culture dependent and culture independent** (18S rRNA and β-tubulin cloning and sequencing, fluorescent in situ hybridization)	Dothideomycetes ***Acidomyces acidothermus*** Eurotiomycetes	Baker et al. (2004)
		Biofilm	**Culture independent** (18S rRNA cloning and sequencing, fluorescent in situ hybridization)	Dothideomycetes ***Acidomyces acidothermus*** Basidiomycota	Baker et al. (2009)
		Biofilm	**Culture independent** (Illumina sequencing of 18S and 28S rRNA)	Dothideomycetes (Capnodiales) Eurotiales	Aliaga-Goltsman et al. (2015)
Tinto River, Spain, IPB	pH 2 metal rich (Fe, Zn, Cu) High levels of sulfates	Water	**Culture dependent** Identification based on phenotype	Dematiaceous fungi	López-Archilla et al. (2001)
		Sediment Water	**Culture dependent** Identification based on phenotype	*Acremonium* Dematiaceous fungi *Lecythophora* (=current name *Coniochaeta*) *Mortierella*	López-Archilla et al. (2004b)
		Biofilm	**Culture independent** (18S rRNA cloning and sequencing)	Six fungal clones: RT3n2—***Acidiella bohemica?***, RT3n5, RT5in6—***Acidea extrema?***, RT5iin23, RT5iin1, RT5iin3	Amaral-Zettler et al. (2002)
		Sediment Water	**Culture dependent** Identification based on phenotype and molecular marker (ITS rDNA)	Dothideomycetes ***Acidiella uranophila?*** Eurotiomycetes *Aspergillus* sp. *Penicillium* spp. *Talaromyces* spp. Sordariomycetes Diaporthales Hypocreales Sordariales	Oggerin et al. (2016)

Site	Conditions	Sample	Method	Fungi	Reference
Tinto River, Spain, IPB		Water	**Culture dependent** Identification based on phenotype	*Acremonium*-like fungi Dematiaceous fungi *Penicillium* spp.	López-Archilla and Amills (1999)
Odiel River, Spain, IPB	Less extreme than Tinto River			*Aspergillus* spp. *Penicillium* spp. *Trichoderma* spp.	
Abandoned mines of Tinto River, Spain, IPB	pH 2.4–2.5 metal rich (Fe, Zn, Cu)	Water	**Culture independent** (18S rRNA cloning and sequencing and TGGE—eukaryotic and fungus-specific primers were used)	Ascomycota (7 clones, 6 TGGE records) Clone C573–*Acidea extrema?* Basidiomycota (2 clones, 4 TGGE records) Zygomycota (1 clone, 6 TGGE records)	Gadanho and Sampaio (2006)
Sao Domingos, Portugal, IPB (several subterranean sites)	pH 2.5–2.9			Ascomycota (1 clone, 3 TGGE records) Basidiomycota (1 clone, 5 TGGE records) Zygomycota (5 TGGE records)	
Tinto River, Spain, IPB	pH 2	Biofilm Sediment	**Culture independent** (pyrosequencing of 18S rRNA)	Leotiomycetes Sordariomycetes	Amaral-Zettler (2013)
Davis mine, USA	pH 2.7	Biofilm Sediment		Agaricomycetes Leotiomycetes Sordariomycetes	
Kiesberg mine, Banat Mountains, Romania	pH 2.8 High levels of sulfates	Biofilm	**Culture dependent** Identification based on phenotype	*Aspergillus* spp. Dematiaceous fungi *Penicillium* spp.	Gherman et al. (2007a)
	pH 1–1.5 High levels of sulfates			*Penicillium* spp.	Gherman et al. (2007b)

(continued)

Table 11.1 (continued)

Locality	The main physicochemical characteristics	Studied samples	Methods for studying diversity	The main fungal groups/species connected with highly acidic habitats	References
Sainokawara hot spring, Japan	pH 1.5 Temperature 96 °C	Tree branches Biofilm	**Culture dependent** Identification based on phenotype and molecular markers (SSU, ITS, LSU rDNA)	*Acidomyces acidothermus*	Yamazaki et al. (2010)
Xiang Mountain sulfide mine, Anhui Province, China	pH 3 High levels of Fe and sulfates	Water	**Culture independent** (18S cloning and sequencing)	*Penicillium* sp.	Hao et al. (2010)
Königstein uranium mine, Germany (subterranean)	Temperature 13–15 °C pH 2.5–2.7 Metal rich (Fe, U) High levels of sulfates	Biofilm	**Combination of** microscopic observations and 18S rRNA cloning and sequencing	Ascomycota Chytridiomycota Zygomycota	Zirnstein et al. (2012)
Ranger uranium mine, Australia	pH 1.7–1.8 High levels of dissolved colloidal salts	Water	**Culture dependent** Identification based on phenotype and molecular markers (ITS rDNA)	Seven fungal isolates: *Acidomyces acidothermus* *Acidiella uranophila* *Coniochaeta fodinicola*	Vazquéz-Campos et al. (2014)

	Soil	**Culture dependent** Identification based on phenotype and molecular markers (18S, 28S, ITS rDNA, β-tubulin)	
Soos National Nature Reserve, The Czech Republic	Area lacks vegetation pH 1.6–2.2 High levels of sulfate salts	*Acidomyces acidophilus* *Acidothrix acidophila* *Hypholoma fasciculare* *Penicillium* sp. 4	Hujslová et al. (2010)
		Acidomyces acidophilus *Acidea extrema* *Chaetomium* sp. *Penicillium spinulosum* **Soosiella minima** Unidentified dark mycelia	Hujslová et al. (2017)
	Zone between the bare soil and vegetation pH 1.9–2.7 High levels of sulfate salts	*Acidomyces acidophilus* *Penicillium* spp. *Mucor* sp.	Hujslová et al. (2010)
Mirová active kaolin quarry, The Czech Republic	Area lacks vegetation pH 1.5–2.5 Sulfur-rich brown coal layers Humic acids	*Acidea extrema* *Acidiella bohemica* *Acidomyces acidophilus* *Acidomyces acidothermus* *Acidothrix acidophila* *Penicillium* spp. Unidentified dark/hyaline mycelia	Hujslová et al. (2017)
Jimlíkov active kaolin quarry, The Czech Republic	Area lacks vegetation pH 2–4 Sulfur-rich brown coal layers	*Acidea extrema* *Acidiella bohemica* *Acidomyces acidophilus* *Acidomyces acidothermus* *Penicillium simpl.* s.l. Unidentified dark/hyaline mycelia	
Cihelna v Bažantnici National Monument, The Czech Republic	Area lacks vegetation pH 1–2 Claystone layers enrich by brown coal	*Acidea extrema* *Acidiella bohemica* *Acidomyces acidophilus* *Coniochaeta fodinicola* *Hypholoma fasciculare* Unidentified dark/hyaline mycelia	

(continued)

Table 11.1 (continued)

Locality	The main physicochemical characteristics	Studied samples	Methods for studying diversity	The main fungal groups/species connected with highly acidic habitats	References
Oil shale beds, Irati Formation in São Mateus do Sul, Brazil	Oil shale by-products—fine shale particles pH 2.4–3.6		**Culture dependent** Identification based on phenotype and molecular marker (ITS rDNA)	Six fungal genera *Acidiella* **Acidiella bohemica** *Aspergillus* *Cladosporium* *Ochroconis* *Penicillium* *Talaromyces*	de Goes et al. (2017)
Los Rueldos abandoned mercury underground mine, NW Spain	pH ~ 2	Biofilm Water	**Culture independent** (pyrosequencing of 18S rRNA, 18S cloning and sequencing, DGGE)	Ascomycota—Helotiales Basidiomycota Chytridiomycota	Mesa et al. (2017)

The main physicochemical characteristics of the studied sites, types of collected samples, methodological approaches applied for studying, and main fungal groups and/or species connected with highly acidic habitats are included. Current names are used for all taxa. Acidophilic fungal species are in bold

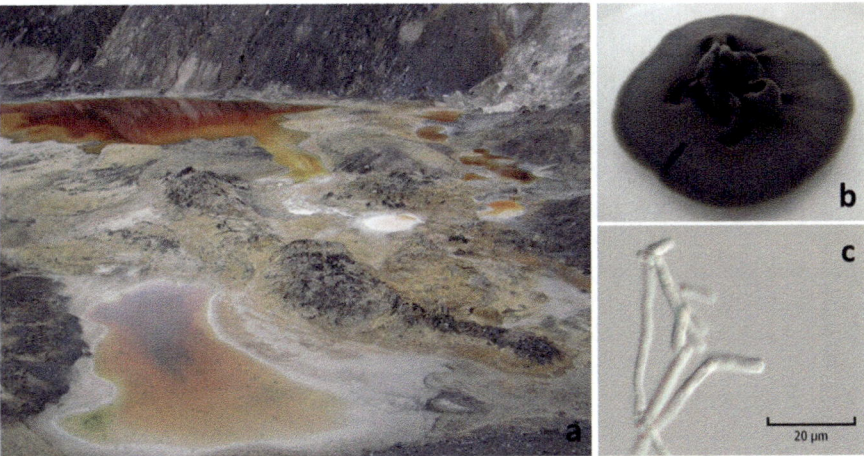

Fig. 11.1 (**a**) General view of Mírová kaolin quarry; (**b**) *Acidiella bohemica*. Colony on MEA at 24 °C, 28 days; (**c**) Thin-walled mycelium fragmented to arthroconidia. Microscopic photo was made by Miroslav Kolařík

Fig. 11.2 (**a**) General view of Soos National Nature Reserve; (**b**) *Acidothrix acidophila*. Colony on MEA (pH 2) at 24 °C, 21 days; (**c**) conidiophores and conidia. Microscopic photo was made by Miroslav Kolařík

detected in Tinto River by López-Archilla et al. (2004b) (see Table 11.1). Another fungus dominating in highly acidic soil, *Acidea extrema*, probably occurs in Tinto River locality since two eukaryotic rDNA clones RT3n5 and RT5in6 found here by Amaral-Zettler et al. (2002) and a clone C573 detected at the same locality by Gadanho and Sampaio (2006) seem to be similar to this species (see Table 11.1).

Nevertheless, compared to other acidophilic fungi, *Acidea extrema* is not exclusively associated with highly acidic sites as it has been reported from slightly alkaline Antarctic soil with pH 8 as well (Hujslová et al. 2014).

Several other fungi, such as *Penicillium* species, *Aspergillus fumigatus*, *Hypholoma fasciculare, Talaromyces helicus* var. *major*, and *Trichoderma harzianum*, are present in highly acidic soils (Hujslová et al. 2017). Indeed, the list of hitherto known acidophilic fungi is not complete because many highly acidic environments, especially soils, were still not studied in this regard and the overview of the taxa presented in Table 11.1 must be taken as provisional.

Insufficient identification of the fungal taxa to the species level in many cases hampers the precise comparison of the fungal communities in different acidic environments. It is, however, apparent that acidophilic fungal assemblage is worldwide similar and it is much more diversified than it has been supposed so far (Hujslová et al. 2017). Considerable portion of the above-cited observations of extremophilic fungal taxa in acidic soils and other acidic substrates strongly suggests that it is the acidity of the environment and not its type (i.e., soil vs. acidic waters) what determines the community of the inhabiting fungi. Thus, some similarities in composition and behavior of acidophilic mycoflora can be supposed across different acidic substrates/environments.

11.8 The Main Functions of Fungi in Acidophilic Communities

Just as in other habitats acidophilic fungi actively participate in carbon cycling as decomposers (e.g., Méndez-García et al. 2015; Mesa et al. 2017). They keep organic carbon levels low and produce dissolved carbonate ions, which are likely important for the growth of chemolitho-autotrophic acidophilic prokaryotes (Baker and Banfield 2003). Baker et al. (2004) proposed that fungi may impact the community structure and function by the consumption of organic waste products and the production of organic polymers and other compounds, possibly including antibiotics.

Recently, the application of omics-based technologies helps to reveal genetic background of fungal role in metabolism of acidophilic biofilm community (Mosier et al. 2013, 2016). For example, the fundamental role of the melanized fungus *Acidomyces acidothermus* in metabolism of taurine, the compound that protects the organisms against osmotic stress, was noted. In addition, this fungus actively participates in biosynthesis of phosphatidylethanolamine lipids that may prevent the excessive uptake of Fe cations (Mosier et al. 2013).

The upregulated transcripts produced by *Acidomyces acidothermus*, involved in denitrification and in degradation of complex carbon sources, were detected in floating biofilms. On the contrary in streamer biofilms transcripts linked to central carbon metabolism and stress alleviation are upregulated (Mosier et al. 2016).

Another natural process in which acidophilic fungi also actively participate is biomineralization, biologically assisted formation of minerals (Oggerin et al. 2016). An example of biomineralization is the formation of jarosite, a basic hydrous sulfate of potassium and iron, associated with acid mine drainage and acid sulfate soil environments. Formation of this mineral is assisted by acidophilic filamentous fungus *Purpureocillium lilacinum* which has been isolated from the banks of Rio Tinto (Oggerin et al. 2013).

11.9 Methodology Used in Studies of Acidophilic Fungi

Already the first studies using traditional cultivation approaches provided the evidence of fungi inhabiting highly acidic habitats and the basic knowledge dealing with their diversity and ecology. The identification of the obtained isolates was based exclusively on morphological features. Later on, the application of molecular markers such as 18S, 28S, ITS rDNA, or β-tubulin gene contributed to the more accurate identification of the isolated strains and first acidophilic fungi have been reliably distinguished as taxonomical units (e.g., Selbmann et al. 2008; Hujslová et al. 2013, 2014; Vázquez-Campos et al. 2014). The use of culture-independent approaches helped to reveal the portion of fungal diversity that escaped attention using classical cultivation methods. All methods used for studying fungal diversity in highly acidic sites are summarized in Table 11.1.

Advanced molecular technologies have been used to elucidate genetic background of the functioning of acid mine drainage biofilm community of Richmond Mine. Within the frame of this study, the reconstruction of partial genome of *Acidomyces acidothermus* has been performed, providing the insight into the main metabolic pathways in which this fungus actively participates (Mosier et al. 2013, 2016). It is probable that similar molecular approaches will be used more extensively in future to further increase our understanding of life in highly acidic habitats.

The investigations focusing on isolation of fungi inhabiting extremely acidic soils applied traditional cultivation techniques that are commonly used for isolation of soil fungi. These methods were based on direct inoculation and suspension plating method. Classical cultivation media like malt extract agar or soil extract agar (pH 5.5) as well as special media with low pH value (pH 2) were used to cover the demands of various fungi from those preferring soil with moderate pH to acidotolerant/acidophilic ones (Hujslová et al. 2010, 2017).

If the extreme values of pH are to be simulated in isolation media for acidophilic fungi, high concentrations of agar have to be used to reach satisfactory solidity because the solidification of agar is difficult under acidic conditions (Hujslová et al. 2013). High concentration of agar is an unnatural attribute of the cultivation medium which probably leads to stimulation of growth of some organisms and suppression of the growth of others (Harris 1985).

Still other methodological aspect of the isolation of fungi from acidic biofilms may be important: the purification of isolates. As the fungi are typically present in acidic environments (soils) as members of complex communities and, at the same time, are typically slow growing, they have to be mechanically separated from other community members before cultivation attempts. The failure in this step may constitute a serious obstacle for the obtaining of pure isolates. The separation of microbial cells is not always an easy task, mainly if the soil biofilm is used as a source of isolates (M. Hujslová and M. Gryndler, unpublished observation). In these cases, the isolation efficiency might be potentially increased by using selective antibiotics added to the cultivation medium.

The above text suggests that contemporary isolation techniques may be insufficient to cover a representative portion of the acidophilic mycoflora diversity. The obtaining of cultures of many important acidophilic organisms including fungi is probably not possible which, in turn, makes taxonomic and ecological studies very difficult. Any future progress in isolation technique development thus may bring new perspectives for the research of acidophilic mycoflora of biofilms in general and soil biofilms in particular.

11.10 Conclusions

A considerable portion of acidophilic fungi that inhabit acidic waters and biofilm accumulations can also be found in soils. At the same time, the formation of biofilms in acidic environments, including soils, may represent the adaptation to extreme acidity. This strongly suggests that acidity is the most important factor that determines microbial community behavior and composition at the site, probably regardless of the type of the environment. Further, hitherto performed inventories of mycoflora of acid sites show wide geographic distribution of different species of acidophilic fungi.

The above facts suggest that acidophily in fungi, connected with their life strategy as biofilm inhabitants, represents a general ecological phenomenon that merits serious scientific study. More detailed knowledge in taxonomy, physiology, and diversity is needed for deeper understanding processes and adaptations used by acidophilic fungi in soils. Specific research techniques directed to fungi inhabiting acidophilic biofilms should be developed.

Acknowledgements Footnote: This text has been created within the frame of the project 17-09946S supported by the Czech Science Foundation. We would like to thank Dr. Miroslav Kolařík (Institute of Microbiology ASCR, The Czech Republic) for providing microscopic photos.

References

Aguilera A (2013) Eukaryotic organisms in extreme acidic environments, the Río Tinto Case. Life 3:363–374

Aguilera A, Olsson S, Puerte-Sánchez F (2016) Physiological and phylogenetic diversity of acidophilic eukaryotes. In: Quatrini R, Johnson DB (eds) Acidophiles. Life in extremely acidic environment. Caister Academic Press, Norfolk, pp 107–118

Aliaga-Goltsman DS, Comolli LR, Thomas BC, Banfield JF (2015) Community transcriptomics reveals unexpected high microbial diversity in acidophilic biofilm communities. ISME J 9:1014–1023

Amaral-Zettler LA (2013) Eukaryotic diversity at pH extremes. Front Microbiol 3:1–17

Amaral-Zettler LA, Gomez F, Zettler E, Keenan BG, Amils R, Sogin ML (2002) Eukaryotic diversity in Spain's river of fire. Nature 417:137

Amaral-Zettler LA, Messerli MA, Laatsch AD, Smith PJS, Sorgin ML (2003) From genes to genomes: beyond biodiversity in Spain's Rio Tinto. Biol Bull 204:205–209

Baker BJ, Banfield JF (2003) Microbial communities in acid mine drainage. FEMS Microbiol Ecol 44:139–152

Baker BJ, Lutz MA, Dawson SC, Bond PL, Banfield JF (2004) Metabolically active eukaryotic communities in extremely acidic mine drainage. Appl Environ Microbiol 70(10):6264–6271

Baker BJ, Tyson GW, Goosherst L, Banfield JF (2009) Insights into the diversity of eukaryotes in acid mine drainage biofilm communities. Appl Environ Microbiol 75(7):2192–2199

Bond PL, Druschel GK, Banfield JF (2000) Comparison of acid mine drainage microbial communities in physically and geochemically distinct ecosystems. Appl Environ Microbiol 66(11):4962–4971

Brock TD (1978) Thermophilic micro-organisms and life at high temperatures. Springer-Verlag, New York, p 465

Bühring SI, Schubotz F, Harms C, Lipp JS, Amils R, Hinrichs K-U (2012) Lipid signatures of acidophilic microbial communities in an extreme acidic environment – Río Tinto, Spain. Org Geochem 47:66–77

Cavicchioli R, Thomas T (2000) Extremophiles. In: Lederberg J, Alexander M, Bloom BR, Hopwood D, Hull R, Iglewski BH, Laskin AI, Oliver SG, Schaechter M, Summers WC (eds) Encyclopedia of microbiology. Academic Press, San Diego, CA, pp 317–337

Cooke WB (1976) Fungi in and near streams carrying acid mine-drainage. Ohio J Sci 76(5):231

de Goes KCGP, da Silva JJ, Lovato GM, Iamanaka BT, Massi FP, Andrade DS (2017) Talaromyces sayulitensis, Acidiella bohemica and Penicillium citrinum in Brazilian oil shale by-products. Antonie Van Leeuwenhoek 110(12):1637–1646

Das BK, Roy A, Koschorreck M, Mandal SM, Wendt-Potthoff K, Bhattacharya J (2009) Occurrence and role of algae and fungi in acid mine drainage environment with special reference to metals and sulfate immobilization. Water Res 43:883–894

Gadanho M, Sampaio JP (2006) Microeukaryotic diversity in the extreme environments of the Iberian Pyrite Belt: a comparison between universal and fungi-specific primer sets, temperature gradient gel electrophoresis and cloning. FEMS Microbiol Ecol 57:139–148

Gherman VD, Bréheret J-G, Bularda M-D (2007a) Microorganism populations within gelatinous formations from Kiesberg mine in Banat Mountains, Romania. Studia Universitates Babeş-Bollyai Biologia LII 2:109–118

Gherman VD, Bréheret J-G, Bularda M-D (2007b) Microorganism associations within a thin acid solution film from an old mine in Banat Mountains. Studia Universitates Babeş-Bollyai Biologia LII 2:119–127

Gimmler H, de Jesus J, Greiser A (2001) Heavy metal resistance of the extreme acidotolerant filamentous fungus *Bispora* sp. Microb Ecol 42:87–98

Gostinčar C, Grube M, de Hoog S, Zalar P, Gunde-Cimerman G (2010) Extremotolerance in fungi: evolution on the edge. FEMS Microbiol Ecol 71:2–11

Gould WD, Fujikawa JI, Cook FD (1974) A soil fungus tolerant to extreme acidity and high salt concentrations. Can J Microbiol 20:1023–1027

Gross S, Robbins EI (2000) Acidophilic and acid-tolerant fungi and yeasts. Hydrobiologia 433:91–109

Hao C, Wang L, Gao Y, Zhang L, Dong H (2010) Microbial diversity in acid mine drainage of Xiang Mountain sulfide mine, Anhui Province, China. Extremophiles 14:465–474

Harris JE (1985) Gelrite as an agar substitute for the cultivation of mesophilic *Methanobacter* and *Methanobrevibacter* species. Appl Environ Microbiol 50:1107–1109

Hölker U, Bend J, Pracht R, Tetsch L, Müller T, Höfer M, de Hoog GS (2004) *Hortaea acidophila*, a new acid-tolerant black yeast from lignite. Antonie Van Leeuwenhoek 86:287–294

Hujslová M, Kubátová A, Chudíčková M, Kolařík M (2010) Diversity of fungal communities in saline and acidic soils in the Soos National Natural Reserve, Czech Republic. Mycol Prog 9:1–15

Hujslová M, Kubátová A, Kostovčík M, Kolařík M (2013) *Acidiella bohemica* gen. et sp. nov. and *Acidomyces* spp. (Teratosphaeriaceae), the indigenous inhabitants of extremely acidic soils in Europe. Fungal Divers 58:33–45

Hujslová M, Kubátová A, Kostovčík M, Blanchette RA, de Beer ZW, Chudíčková M, Kolařík M (2014) Three new genera of fungi from extremely acidic soils. Mycol Prog 13:819–831

Hujslová M, Kubátová A, Bukovská P, Chudíčková M, Kolařík M (2017) Extremely acidic soils are dominated by species-poor and highly specific fungal communities. Microb Ecol 73:321–337

Johnson DB (1998) Biodiversity and ecology of acidophilic microorganisms. FEMS Microbiol Ecol 27:307–317

Johnson DB (2009) Extremophiles: acid environments. In: Schaechter M (ed) Encyclopedia of microbiology. Academic Press, Oxford, pp 107–126

Johnson DB (2012) Geomicrobiology of extremely acidic subsurface environments. FEMS Microbiol Ecol 81:2–12

Joseph JM (1953) Microbiological study of acid mine waters: preliminary study. Ohio J Sci 53(2):123

Kolařík M, Hujslová M, Vázquez-Campos X (2015) Acidotolerant genus *Fodinomyces* (Ascomycota: Capnodiales) is a synonym of *Acidiella*. Czech Mycol 67:37–38

Krause S, Bremges A, Münch PC, McHardy AC, Gescher J (2017) Characterisation of a stable laboratory co-culture of acidophilic nanoorganisms. Sci Rep 7:3289. https://doi.org/10.1038/s41598-017-03315-6

Li X, Kappler U, Jiang G, Bond PL (2017) The ecology of acidophilic microorganisms in the corroding concrete sewer environment. Front Microbiol 8:683. https://doi.org/10.3389/fmicb.2017.00683

López-Archilla AI, Amils R (1999) A comparative ecological study of two acidic rivers in southwestern Spain. Microb Ecol 38:146–156

López-Archilla AI, Marin I, Amils R (2001) Microbial community composition and ecology of an acidic aquatic environment: the Tinto River, Spain. Microb Ecol 41:20–35

López-Archilla AI, Gérard E, Moreira D, López-García P (2004a) Macrofilamentous microbial communities in the metal-rich and acidic River Tinto, Spain. FEMS Microbiol Lett 235:221–228

López-Archilla AI, González AE, Terrón MC, Amils R (2004b) Ecological study of the fungal populations of the acidic Tinto River in southwestern Spain. Can J Microbiol 50:923–934

Macalady JL, Jones DS, Lyon EH (2007) Extremely acidic, pendulous cave wall biofilms from the Frasassi cave system, Italy. Environ Microbiol 9(6):1402–1414

Magan N (1997) Fungi in extreme environments. In: Wicklow DT, Soderstrom BE (eds) The Mycota IV. Environmental and microbial relationships. Springer-Verlag, Berlin, pp 99–114

Méndez-García C, Peláez AI, Mesa V, Sánchez J, Golyshina OV, Ferrer M (2015) Microbial diversity and metabolic networks in acid mine drainage habitats. Front Microbiol 6:475. https://doi.org/10.3389/fmicb.2015.00475

Mesa V, Gallego JLR, Gonzáles-Gil R, Lauga B, Sánchez J, Méndez-García PAI (2017) Bacterial, archaeal, and eukaryotic diversity across distinct microhabitats in an acid mine drainage. Front Microbiol 8:1756. https://doi.org/10.3389/fmicb.2017.01756

Mosier AC, Justice NB, Bowen BP, Baran R, Thomas BC, Northen TR, Banfield JF (2013) Metabolites associated with adaptation of microorganisms to an acidophilic, metal-rich environment identified by stable-isotope-enabled metabolomics. mBio 4(2):00484–00412. https://doi.org/10.1128/mBio.00484-12

Mosier AC, Miller CS, Frischkorn KR, Ohm RA, Li Z, LaButti K, Lapidus A, Lipzen A, Chen C, Johnson J, Lindquist EA, Pan C, Hettich RL, Grigoriev IV, Singer SW, Banfield JF (2016) Fungi contribute critical but spatially varying roles in nitrogen and carbon cycling in acid mine drainage. Front Microbiol 7:238

Oggerin M, Tornos F, Rodríguez N, del Moral C, Sánchez-Román M, Amils R (2013) Specific jarosite biomineralization by *Purpureocillium lilacinum*, an acidophilic fungi isolated from Río Tinto. Environ Microbiol 15(8):2228–2237

Oggerin M, Tornos F, Rodriguez N, Pascual L, Amils R (2016) Fungal iron biomineralization in Rio Tinto. Fortschr Mineral 6(2):37. https://doi.org/10.3390/min6020037

Quatrini R, Johnson DB (2018) Microbiomes in extremely acidic environments: functionalities and interactions that allow survival and growth of prokaryotes at low pH. Curr Opin Microbiol 43:139–147

Robbins EI, Rodgers TM, Alpers CN, Nordstrom DK (2000) Ecogeochemistry of the surface food web at pH 0-2.5 in Iron Mountain, California, USA. Hydrobiologia 433:15–23

Selbmann L, Hoog GS, de Zucconi L, Isola D, Ruisi S, Gerrits van den Ende AHG, Ruibal C, De Leo F, Urzì C, Onofri S (2008) Drought meets acid: three new genera in a dothidealean clade of extremotolerant fungi. Stud Mycol 61:1–20

Sigler L, Carmichael JW (1974) A new acidophilic *Scytalidium*. Can J Microbiol 20:267–268

Sletten O, Skinner CE (1948) Fungi capable of growing in strongly acid media and in concentrated copper sulfate solutions. J Bacteriol 56:679–681

Starkey RL, Waksman SA (1943) Fungi tolerant to extreme acidity and high concentrations of copper sulfate. J Bacteriol 45:509–519

Tiquia-Arashiro SM, Rodrigues D (2016) Alkaliphiles and Acidophiles in Nanotechnology. In: Extremophiles: applications in nanotechnology. Springer International Publishing, New York, pp 129–162

Vázquez-Campos X, Kinsela AS, Waite TD, Collins RN, Neilan BA (2014) *Fodinomyces uranophilus* gen. nov. sp. nov. and *Coniochaeta fodinicola* sp. nov., two uranium mine-inhabiting Ascomycota fungi from northern Australia. Mycologia 106(6):1073–1089

Wilmes P, Remis JP, Hwang M, Auer M, Thelen MP, Banfield JF (2009) Natural acidophilic biofilm communities reflect distinct organismal and functional organization. ISME J 3:266–270

Yamazaki A, Toyama K, Nakagiri A (2010) A new acidophilic fungus *Teratosphaeria acidotherma* (Capnodiales, Ascomycota) from a hot spring. Mycoscience 51:443–455

Zak JC, Wildman HG (2004) Fungi in stressful environments. In: Mueller GM, Bills GF, Foster MS (eds) Biodiversity of fungi, inventory and monitoring methods. Elsevier Academic Press, London, pp 303–315

Zirnstein I, Arnold T, Krawczyk-Bärsch E, Jenk U, Bernhard G, Röske I (2012) Eukaryotic life in biofilms formed in a uranium mine. Microbiologyopen 1(2):83–94

Chapter 12
Global Proteomics of Extremophilic Fungi: Mission Accomplished?

Donatella Tesei ⓘ**, Katja Sterflinger, and Gorji Marzban**

12.1 Introduction

12.1.1 *Extremophilic and Extremotolerant Fungi as Model Organisms for Stress Resistance*

Extremophiles are organisms living under conditions that extend far beyond those considered optimal for human life and therefore grow under extremes of temperature, pH, salinity, pressure, radiation, or other limiting constraints (Burg et al. 2011). Organisms, which tolerate and thrive at the edges of the extreme conditions and can only grow optimally under more human environmental conditions, are instead known as extremotolerants (Cavicchioli 2002).

That of extreme is a rather relative concept, as it has been shaped based on the ambient parameters supporting mammalian life (Selbmann et al. 2013). Nevertheless, it has become clear that life under extreme environmental conditions is mostly a prerogative of microorganisms (Gunde-Cimerman et al. 2003). An increasing number of investigations carried out during the last decades showed the ability of fungi to sustain and even to thrive at different environmental extremes such as permafrost, snow and glacier ice (Abyzov et al. 1998; Nienow and Friedmann 1993), cold and hot deserts (Selbmann et al. 2005), ocean depths (López-García et al. 2002), hypersaline waters (Gunde-Cimerman et al. 2000), acidic environments (Selbmann et al. 2008), and areas contaminated by pollution and radioactivity (Blasi et al. 2017; Dadachova and Casadevall 2008).

The organism life expectancy seems to be connected to either existing or extraordinary fast adaptive cellular or metabolic characteristics (Magan 2007). Eukaryotic

D. Tesei (✉) · K. Sterflinger · G. Marzban
Department of Biotechnology, University of Natural Resources and Life Sciences,
Vienna, Austria
e-mail: donatella.tesei@boku.ac.at

© Springer Nature Switzerland AG 2019
S. M. Tiquia-Arashiro, M. Grube (eds.), *Fungi in Extreme Environments:*
Ecological Role and Biotechnological Significance,
https://doi.org/10.1007/978-3-030-19030-9_12

cells have evolved over millions of years tailored cellular response to both biotic and abiotic stresses in the natural ambiance, which manage several aspects of cell machinery like gene expression, metabolism/catabolism, cell cycle, cytoskeletal arrangement and homeostasis, as well as modification of enzymatic activity (Mafart et al. 2001). Such stress tolerance reactions can produce both immediate and long-term adaptations—the latter called S-selected strategies—which are especially crucial for the survival of an organism under continuous exposure to extreme environmental parameters (Magan 2007).

Due to their fascinating lifestyle, extremotolerant and extremophilic organisms may serve as model organisms for studies of the stress resistance aiming at the elucidation of the molecular basis of survival. Moreover, proteins and metabolites from extremotolerant and extremophilic producers offer some advantages over novel proteins and bioactive components from less tolerant microorganisms. Enzymes produced by both thermophiles and psychrophiles have for instance received particular attention for their commercial value (Tiquia-Arashiro 2014; Tiquia-Arashiro and Rodrigues 2016). The search for new products and compounds has additionally promoted a renaissance for unusual and interesting organisms as emerging model systems, which also include black fungi, a morphological group of ascomycetes, whose physiological features include a high stress resilience (Onofri et al. 2012; Pacelli et al. 2016; Selbmann et al. 2014; Sterflinger et al. 2012). Previous studies suggested that this group of fungi, which is specified by melanin generation and absence of sporulation, has adapted survival strategies quite different from those of the majority of fungal species (Blasi et al. 2017; Tesei et al. 2017). Hence, an extended knowledge about the systems biology of extremophiles is of crucial impact.

12.1.2 Black Fungi

Microcolonial fungi (MCF) and black yeasts are among the most stress-resistant eukaryotes known up to now (de Hoog and Grube 2008). Grouped under the general name black fungi, they represent a heterogeneous taxonomic group having polyphyletic origins within the orders of *Ascomycota Chaetothyriales, Dothideales, Capnodiales, Pleosporales, Xylariales, Myriangiales, Mycocaliciales,* and *Hysteriales,* among others (Selbmann et al. 2008). While MCF prevalently live on rock substrates, several black yeast species have a life cycle in association with plant, animal, and human hosts (Fig. 12.1). Black fungi's high stress tolerance is demonstrated by the wide geographical distribution, which includes the most extreme habitats on the planet—ranging from mountain peaks (Branda et al. 2010) to nuclear power plants (Selbmann et al. 2014)—where the life of the majority of organisms is at risk. The role of black fungi as human opportunists and pathogens (Woo et al. 2013), as degraders of volatile compounds (Prenafeta-Boldú et al. 2012), and as a dominant member of the epi- and endolithic microbial populations (Sterflinger and Krumbein 1997) has additionally emerged (Sert et al. 2007).

Fig. 12.1 Examples of the three main ecotypes of black fungi and black yeasts grown on 2% malt extract agar (MEA). (**a**) Microcolonial fungi from extreme environments: *Cryomyces minteri* MA6029, Antarctica; (**b**) thermotolerant microcolonial fungi from lower Alpine and Mediterranean areas: *Knufia perforans* MA3307, Vienna, Austria; (**c**) black yeasts, *Cladophialophora minutissima* MA6077, Yosemite National Park, California. MA Nr.: strain number in the Austrian Center of Biological Resources and Applied Mycology culture collection (ACBR). Photos: Christian Voitl

Survival tests demonstrated how these organisms can cope with extreme physico-chemical stress factors far beyond those in their natural habitats. Additionally some species showed resistance against high doses of ionizing radiation, by conversion of β and γ radiation into chemical energy used for growth (Dadachova et al. 2007; Pacelli et al. 2016; Robertson et al. 2012). On these premises, pathogenicity has been reconsidered as the result of the combination of extremotolerance and assimilative abilities of aromatic compounds, together enhancing the nature of black fungi, which is primarily opportunistic (Teixeira et al. 2017). Further investigations demonstrated the ability of black fungi, especially MCF, to endure outer space as well as simulated Martian conditions (Onofri et al. 2012; Zakharova et al. 2014a), thereby giving support to the hypothesis of lithopanspermia and interplanetary transfer of life (Onofri et al. 2008).

Microcolonial growth morphology; strong melanization of the multilayered cell wall; synthesis of carotenoids, exopolysaccharides (EPS), trehalose, and polyols; and lack of sexual reproductive structures are adaptations to the extreme environments spread across the black fungi group, despite its polyphyletic origins (Mafart et al. 2001; Magan 2007). While melanin plays a role in UV protection, pathogenicity, and improving the desiccation and radiation tolerance, polyols and trehalose act as osmoprotectants and as stabilizer of enzymes and membranes, respectively. The absence of spores or conidia results in the use of each single vegetative cell as both a survival and a dispersal state and represents a crucial energy-saving mechanism under unfavorable life conditions (Pacelli et al. 2016).

Multi-OMICS analyses recently performed on black fungi species indicate that these organisms have a rather different strategy to cope with nonoptimal conditions than other fungal species. The presence of "stress-resistant" proteins and the thermostability of the basic set of proteins have been suggested to be crucial aspects of the ecological plasticity that characterizes black fungi (Tesei et al. 2015b). The potential of these organisms however does not solely lie in the protein pool. Transcriptomics studies recently carried out showed how black fungi recur to further mechanisms of tolerance involving diverse *RNA* species. *CircRNAs, ncRNAs,* and fusion transcripts seem to enable the synthesis of proteins under temperature stress, thus being at the base of both survival and pathogenicity (Blasi et al. 2015; Poyntner et al. 2016).

The scientific results obtained over the last few years of research demonstrated how black fungi offer potential as natural resource of novel proteins and compounds, the latter acting as stabilizer of proteins and cellular structures (Armengaud et al. 2014; Gostinčar et al. 2012; Moreno et al. 2018; Seyedmousavi et al. 2014; Sterflinger et al. 2012; Tesei et al. 2017; Zakharova et al. 2014a). This combination of properties supports the adaptation not only to temperature changes but also to a multitude of other biotic and abiotic factors like osmotic, UV and oxidative stress, desiccation, water supply, and nutrient availability (Vember and Zhdanova 2001) and might prove to be of great biotechnological significance. The study of black fungi ecophysiology and adaptations could additionally provide tools to shed light on the actual limits for life and the existence of life beyond planet Earth.

The investigation of the proteome—as a pool of all synthesized proteins, their abundance, variations, and modifications as well as their interactions and networks (Aebersold and Mann 2003; Geisow 1998)—in part already had and will aid to improve the understanding of the ecology, physiology, and system biology of these organisms. Analyzing protein expression profiles and their modulation under a given set of conditions that are out of the organisms' growth range or beyond those considered optimal for human life could thereby provide invaluable information about the cellular mechanisms of stress resistance.

12.1.3   Discovery of the Stress Adaptation in Extremophilic and Extremotolerant Fungi by Proteomics

Investigations of the proteome—as a global collection of all expressed proteins in response to one single physiological state (Wasinger et al. 1995)—can play an important role in the elucidation of the ecophysiology of the adaptive behavior toward different kinds of stresses. As the product of transcription and translation, protein processing, and turnover (Burg et al. 2011), a proteome contains the whole information about the proteins that are synthesized, their abundance, variations, modifications, interactions, and networks (Aebersold and Mann 2003; Geisow 1998; Pandey and Mann 2000). Proteomic measurements are achieved through a combination of highly sensitive instrumentation and the application of powerful computational methods to produce high-throughput qualitative (i.e., protein identity) and quantitative (i.e., protein abundance) data and therefore simplify the clarification of cellular pathways and processes in a given cell as well as a community. On this premise, proteomics of extremophilic and extremotolerant fungi allow to understand the modifications at the base of structure and stability of proteins, biologically active under extreme conditions (Evilia 2018). Discovery of adaptive mechanisms helps, therefore, to develop strategies at gene and protein level toward novel therapeutics and biotechnological innovations.

Despite the extremophilic character of fungi is known, there is still very little information regarding the molecular mechanisms, which lie behind their stress resistance. The majority of studies about microbial adaptation mechanisms have been indeed so far focused on the stress response of a low number of widely investigated mesophilic model organism such as *Candida*, *Penicillium*, and *Saccharomyces*, among others. During the last decade the number of "omics" studies concentrated on stress-resistant fungi has increased significantly and reached over less known species, thereby raising our understanding of the system biology of those microorganisms to a higher level (Moreno et al. 2018). However, the knowledge at the functional proteomic level is unfortunately still poor, mainly due to the challenging application of protein techniques to extremophiles, whose major bottleneck is the sample preparation. The lack of proteomics data for most of extremophiles indeed indicates major analytical and methodological challenges (Marzban et al. 2013). As a consequence of the adaptation to extreme conditions, proteins have evolved a range of specific properties (Siddiqui and Thomas 2008), which actually hamper the extraction and separation procedures. Generally, thermostable and chemical denaturant-resistant proteins hardly undergo denaturation, while proteins with altered surface charges—e.g., from acidophiles or alkaliphiles—usually cause migration troubles in the isoelectric focusing (IEF) (Burg et al. 2011). Similarly, less water-soluble proteins—from organisms inhabiting dry environments—are susceptible to irreversible precipitation.

In the case of black fungi, the rigid cell wall and the high content of melanin, pigments, polysaccharides, and lipids represent major obstacles to cell disruption and to protein determination and separation, respectively. Moreover, the fungal protein solubility is quite low and tends toward spontaneous precipitation (Marzban et al. 2013). Therefore, the low protein yield and the high content of impurities represent the major issues toward the proteomics of extremophilic fungi. The analysis of their protein profiles thereby requires optimization and a case-by-case assessment of the protein preparation and separation procedures. Although the molecular basis of both extremotolerance and extremophilia is far from being clarified, crucial indications have been obtained. Comparative proteomic profiling methodologies such as classical 2D-E and 2D-DIGE and the more recent shotgun proteomics techniques have been applied to the investigation of the response mechanisms toward a number of stress like supra- and sub-optimal temperature values, salinity, pH, desiccation, Mars-like conditions, and host-pathogen interaction (Magan 2007; Onofri et al. 2012; Seyedmousavi et al. 2013; Wasinger et al. 2013). The analysis of the repertoire and abundance of proteins for both the rock inhabitant and pathogenic species under temperature stress reported striking changes in the proteome. Interestingly, rock-associated species with different ecology and distribution all react to temperatures far above their growth optimum by decreasing the number of expressed proteins. The opposite trend is instead observed as a result of the exposure to temperatures far below the optimum (Tesei et al. 2012). A similar reaction was detected in the thermophilic and pathogenic species *E. dermatitidis*; however high temperature (45 °C) seemed not to induce any stress response, likely reflecting the high level of adaptation of the fungus to elevated temperature. Long-term exposure to the cold (1 °C) reduced metabolic activities of carbon and pyruvate metabolism, along with triggering rearrangements of the cell wall and cell membrane (Tesei et al. 2015a). Downregulation of genes involved in metabolic processes, including metabolism of amino acids, carbohydrates, and glycolysis, is consistent with and complementary to earlier reports in E. *dermatitidis* transcriptomics (Blasi et al. 2015). A downregulation of the metabolism seems to represent the typical response of black fungi also upon desiccation and the exposure to Mars-like conditions (CO_2 and pressure) (Zakharova et al. 2014a). While this phenomenon is mostly observed in extremophilic species, extremotolerant mesophilic black fungi react to water loss by expressing additional proteins (Zakharova et al. 2013). These results mostly indicate that black fungi, independently on their ecotype, habitat, and degree of specialization, play similar strategies while coping with nonoptimal environmental conditions. Such strategies additionally appear to be quite different from the stress response mechanisms exhibited by other fungal species like mesophilic hyphomycetes (Jami et al. 2010a) and human fungal pathogens (Enjalbert and Whiteway 2003; Jami et al. 2010a), where a HSR is generally observed. A direct connection between the extremophilic nature of the organisms and the minimization of changes at the protein pattern level has been further detected and can be explained with the recourse of black fungi to energy-saving mechanisms. On this basis, a fine-tuning regulation of the protein expression upon unfavorable growth conditions—mostly involving housekeeping proteins—was suggested. Temperature and desiccation-resistant proteins might constitute a basic set of proteins commonly present in black fungi and supporting the tolerance of temperature and desiccation stress. If morpho- and physiological characters—spread across the black fungi group—were initially considered as the main factors responsible for the stress resistance, these studies demonstrated that a fine modulation of the proteome indeed plays a major role in the survival of these model organisms.

The combination of different proteomics workflows and bioinformatics analyses will allow in the near future to gain information about protein identities and functionalities as well as cellular pathways and processes, involved in the survival of these organisms. The availability of whole-genome sequences and annotation in particular has already had a key role in accelerating the study on mechanisms of adaptation. Comparative genomics has proved that extremophiles in general possess the unique set of genes and proteins that empower them with biochemical machinery necessary to thrive in extreme environments (Kumar et al. 2018). One such example of adaption is the initiation of the unfolded protein response (UPR) in psychrophiles. It is known that the rate of protein folding is adversely affected by low temperature (Piette et al. 2010). The activation of UPR and the simultaneous upregulation of the proteasome and induced expression of genes involved in protein folding aid in maintaining protein-folding homeostasis of the ER under temperature stress (Su et al. 2016). In organisms thriving in hypersaline environments, both genome alterations and protein structural modifications were reported (Plemenitaš et al. 2014). The genome of the obligate halophile and basidiomycetous fungus

Wallemia ichthyophaga displays a significant expansion of several protein families, including the hydrophobins, small proteins ($\leq$20 kDa) also present in the cell wall of filamentous fungi and involved in an array of processes of cellular growth and development (Wösten 2001). A number of genes encoding for hydrophobins undergo differential expression in the presence of varying salt concentration (Zajc et al. 2013). Similarly to what is observed in halophilic archaea, these proteins also display a higher proportion of acidic amino acids compared to homologs from other fungi, which might play a role in water and salt binding to prevent conformational change and loss of activity (Siglioccolo et al. 2011). Molecular biology data have shown how the same regulatory pathways and proteins can be of importance for survival under a variety of stress conditions. Proton/Na + symport switch as well as the Rim101 transcription factor and calcineurin-dependent regulatory are for instance both involved in high salt and alkaline conditions (Lambert et al. 1997). Studies of *Yarrowia lipolytica*, the only known ascomycete to grow on alkaline media and salt at near-saturating point, elucidated a key role for proteins such as Hsp12 (Champer et al. 2012). The stress protein Hsp12, that under normal conditions provides a launch of emergency responses to stress allowing only a short-term survival in all studied yeasts, promotes rearranging and repairing of the membrane compartments under prolonged stress conditions (Epova et al. 2012).

Proteomics has also provided the tools to improve the understanding of the mechanisms at the base of pathogenicity. Until now proteomic methods have particularly been used to screen for specific biomarkers of infections and virulence factors (Champer et al. 2012; Rodrigues et al. 2014), to characterize the response to antifungal agents (Kniemeyer 2011), and to identify fungal opportunists and clinical species (e.g., *Aspergillus* sp.) (Lau et al. 2013). Vesicle proteomics in the opportunistic yeasts *Cryptococcus neoformans*, *Histoplasma capsulatum*, and *Paracoccidioides brasiliensis* has supported an important role for the extracellular vesicles (EV) as transport vehicles of microorganism modulators to distant sites inside the host, through the detection of proteins (e.g., laccase, urease, phosphatase) with both immunological and pathogenic activities (Rodrigues et al. 2014; Vallejo et al. 2012). As pathogens undergo stress exposure themselves during infection, pathogenicity and stress resilience are strictly connected matters. Increased temperature is likely the first stressor that opportunistic and pathogenic fungi encounter after gaining entry to the host and proteins such as the mitochondrial manganese superoxide dismutase (SOD) have been suggested to augment adaptation to human host body temperature through the rapid degradation of superoxide into hydrogen peroxide and O_2 (Brown et al. 2007). Proteomics reports documented the role of several other proteins involved in pathogenicity: Cu, Zn superoxide dismutase, peroxidases, and the glutathione system, all being involved in the resistance to oxygen radicals and laccases, important for the production of the pigment melanin, which is a free radical scavenger and thus plays a protective role in stress resistance (Garcia-Rivera et al. 2005).

Based on these premises, proteomics appears to be a very powerful tool in understanding how extremophiles persist at extreme ecological conditions. Through the elucidation of what proteins are present under given conditions, their levels, and posttranslational modifications, proteomics also plays a significant role in the field of biotechnology and can directly contribute to the screening for biotechnologically relevant enzymes.

The first aim of the work presented in this chapter is thus to perform a systematic review of the state of the art of proteomics workflows applied to the investigation of extremotolerant and extremophilic fungal species, with a special focus on black fungi. To this end, we have collected and described methods for protein identification and quantitative analysis as well as for the preparation of different types of samples aiming at a critical evaluation of their advantages and pitfalls. While doing this, we wish to introduce the reader to diverse approaches for managing proteomics of black fungi, including strains' cultivation, protein extraction, and bioinformatics tools for meaningful data interpretation. By presenting examples of strategies to successfully overcome sample-related challenges and by providing a first comprehensive overview of the current status of proteome research in the black fungi field, we wish to ultimately promote the application of protein science to the study of emerging model organisms and to thereby contribute to make proteomics of black fungi an established field from an emerging one. If proteomics has been for long time considered as not yet entered at full right in black fungi research, the progresses and achievement in method optimization have finally created the conditions for more studies to come in the future.

12.2 Methodologies and Approaches

12.2.1 *Fungal Cultivation In Vitro and Sample Collection*

The cultivation of black fungi under laboratory conditions usually aims at the simulation of their natural habitat or of stress conditions. According to the research targets, fungal cultivation can be performed on solid or in liquid media. Cultivation in medium broth within flask, vials, or in microtiter plates is particularly suited to experiments, where the effects of fungal exposure to pollutants in liquid or aerosol form (e.g., toluene, phenol) are to be tested (Blasi et al. 2016, 2017). Chemicals in solid or powdered form can also be added to liquid media, in order to maximize the chances of contact with the fungus. Incubation in liquid media generally results in enhanced growth rates and in a consequent production of more biomass in shorter time ranges, as compared with the growth in petri dishes. However, biomass collection from liquid media is inevitably laborious and time consuming, as it requires filtration or centrifugation to separate the mycelium from the culture supernatant.

On the other hand, fungal biomass can be easily harvested from the surface of solid media by means of a scalpel or a spatula. Growth on solid media is also compatible with the use of neutral matrix of cellulose sheet, which eases biomass collection as well as transfer and is especially useful in workflows, where separation of the biomass from the media is required before treatment (Zakharova et al. 2013).

Due to a particularly slow growth rate, which has evolved hand in hand with extremotolerance, obtaining an amount of biomass sufficient for protein analyses might require weeks. Among black fungi, black yeasts usually display shorter generation times while rock-associated species grow at a slow pace. An accurate estimation of the growth rate is therefore paramount in view of a wise planning of the experimental work.

The inoculation of solid media can be carried out by direct transfer of small amounts of fungal biomass on top of the agar by means of inoculation needles or loops (the latter, to be used in the case of yeasts). Alternatively, drops of a cell suspension can be transferred to the solidified medium to be thereafter distributed using a spreader. Slightly different methods are used for the preparation of liquid cultures. As for yeasts, a cell suspension is generally used as inoculum. Alternative procedures need to instead be applied for those species of black fungi exhibiting microcolonial growth and lacking both sporulation and budding. Biomass gentle disruption with a homogenizer can be performed in order to break down hyphae or cell agglomerates into single cells, whose concentration can thereafter be assessed by cell count or OD measurement (Nai 2014). Accidental cell breakage is unfortunately a drawback of this method; therefore, cell survivability and viability must be evaluated each time. Otherwise, pellet or biomass can be directly added to the liquid media. If cell concentration cannot be determined, it is recommended to resort to biomass dry weight.

Experimental conditions are set based on the research aim. Extreme or more moderate values of temperature, humidity, salinity, UV, radiation, CO_2 and ozone concentration, etc. can be applied alone or in combination, to simulate abiotic stress. Similarly, biotic stress can be mimicked. As widely reported in literature, adaptation and survival to both abiotic and biotic stressors is often associated with melanization (Blasi et al. 2016, 2017; Cordero et al. 2018; Poyntner et al. 2018; Selbmann et al. 2018; Sterflinger and Krumbein 1995). The exposure of melanotic fungi to stress conditions thereby most often results in an enhanced production of melanin, one of the major contaminants interfering with proteomic analyses. In order to counteract or to reduce melanin synthesis to a minimum, the growth and incubation of cultures of black fungi and black yeasts in the dark are suggested.

12.2.2 Sample Preparation

Due to the protein-specific properties evolved in conditions of abiotic extremes, the preparation of protein samples is the most crucial and laborious of all aspects of a proteomics workflow and is the very step that requires sample-designed optimization when working with extremotolerant and extremophilic fungi (Siddiqui and Thomas 2008). Thermostable proteins and proteins particularly resistant to denaturation, hardy cell walls, polyphenolic compounds, polyols, and exopolysaccharides are some of the aspects potentially hindering the progress of proteomic analyses. Additionally, a slow growth rate often observed in species from the extremes results

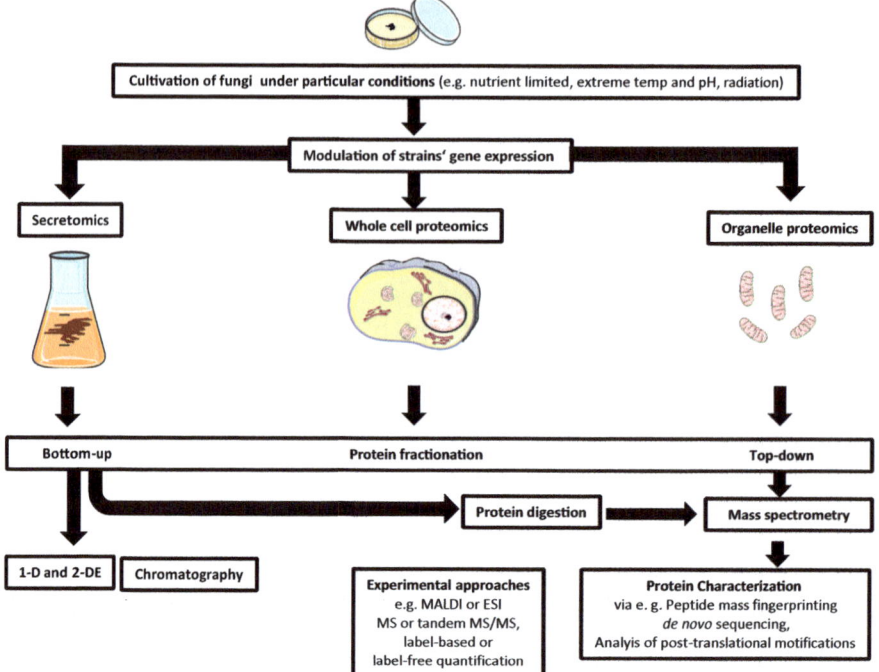

Fig. 12.2 General workflow for fungal proteomics. Both bottom-up and top-down approaches are indicated. As compared to top-down, a bottom-up approach allows, however, a wider methodological analysis of the samples by independent tools and repeats

in very poor amounts of mycelium as starting material for protein extraction. Besides the obstacles posed by the sample, the choice of reagents and techniques is also a crucial aspect in proteomics workflows which needs to be assessed case by case (Fig. 12.2). According to the downstream analyses, the preparation of protein samples can thereby be performed using a number of different methodologies.

Different approaches are additionally chosen to achieve extraction of whole-cell proteome or of protein subfractions. While the first amounts to the whole-sample extract, the subcellular proteome comprises three main fractions: the secretome, consisting of the culture supernatant and containing extracellular proteins; the insoluble fraction, consisting of membranes, membrane proteins, and large complexes; and the soluble fraction, which primarily contains cytoplasmic proteins (Burg et al. 2011). An integrated analysis of the different fractions potentially provides the most complete overview about protein function and compartmentalization within the cell. In the following sections, proteomics approaches in use, as well as the adaptation and optimization of these approaches to the use with extremophiles, are described.

12.2.2.1 Whole-Cell Proteome

The definition of "whole-cell proteome" includes the sum of all cytoplasmic, sub-cellular, and membrane proteins. Prerequisite to the isolation of the whole-cell protein content for all proteomic analyses is the protein extraction, whose aim is to lead to the purest obtainable protein mixture, free from interfering compounds such as polysaccharides, lipids, nucleic acids, phenols, and lower molecular weight components of cell matrix. Fungal proteins are especially effortful to extract, due to the chitin content of the fungal cell wall, and, moreover in the case of black fungi, strong pigmentation by melanin that additionally disturbs all consequent steps from isolation to quantification procedures (Isola et al. 2011; Özhak-Baysan et al. 2015; Uranga et al. 2017). We are for instance often confronted with very complex samples having a cell matrix, which is specialized in the protection of the cell content (e.g., proteins and biomolecules) from physicochemical environmental harms. All the strategies to disintegrate the cells and reach out for the molecules of interest can thereby potentially be outsmarted by millions of years of evolution. However, a combination of different approaches using mechanical as well as biochemical interventions showed success and paved the way for extensive investigation of fungal proteome (Marzban et al. 2013).

The disruption of the mycelium can be performed using mechanical and enzymatic methodologies or a combination of both (Barreiro et al. 1991). The rigid cell wall of black fungi makes the effective cell disruption a crucial step for consequent separation methodologies (gel-based or gel-free approaches). Glass and metal beads in beating mills are often used to achieve the disruption of fungal biomass. As an extensive milling is generally required to successfully break the cells, performing this step at 4 °C is crucial, in order to minimize heat generation as well as protein lysis by intracellular proteases or protein fragmentation and oligomerization (Coumans et al. 2010; Fernández-Acero et al. 2006; Jami et al. 2010b; Yildirim et al. 2011). The milling can be also performed in a mortar with pestle under liquid nitrogen (Kniemeyer 2011); however the reachable particle size is incomparable with that obtained with milling procedures. Moreover, this technique requires a bigger effort not to damage the material. Freeze-drying or lyophilization of the biomass can additionally be used before the milling procedure to aid cell breakage (Uranga et al. 2017).

Mechanical cell disruption can also be combined with homogenization buffers with different concentrations of detergents and chaotropes, to enhance the disintegration of cell walls and inhibit the activity of proteases (Barreiro et al. 1991). Urea at different molarities (7–9 M) and 2 M thiourea are often used to solubilize and re-solubilize proteins in filamentous fungi (Vödisch et al. 2011). However, care must be taken when overcoming the saturation limit of chaotropes in order to prevent crystallization at temperatures other than room temperature and higher. High concentrations of urea and thiourea can also interfere with different protein determination procedures later on. The addition of NaOH at 0.1–1 M can further enhance the protein solubility and destabilize the cell wall tremendously (Suh et al. 2012). Protein solubilization is also achieved by adding nonionic detergents as CHAPS to the cell homogenization buffers. CHAPS is generally used at concentrations in the range of 1–4%, depending on the sample type and the protein yield (Barreiro et al. 1991). DTT and ß-mercaptoethanol are added to both increase protein solubility and achieve protein reduction (Longo et al. 2014; Lu et al. 2010; Oh et al. 2010). Ampholytes at a concentration of 0.5–2% can also be used to aid protein solubility; nevertheless their interference with downstream sample labeling procedures as well as the heat formation and high currents during the isoelectric focusing step must be taken into consideration (Sørensen et al. 2009). Tris–HCl or phosphate buffers at 10–30 mM are additionally added to the sample to keep the pH constant and/or to be compatible with protein labeling (Kubitschek-Barreira et al. 2013). Protease inhibitor cocktails with or without (for MS applications) EDTA can be added to the mycelium upon sample collection or to the homogenization buffer to reduce lysis of the proteins to a minimum (Onofri et al. 2012). As protease inhibitors can potentially disturb the isoelectric focusing, urea and thiourea can be used instead in the homogenization buffer. Protease inhibitors can otherwise be removed from the protein sample through buffer exchange procedures.

Enzymatic and chemical cell wall lysis was also used in fungi, however with less success (Barreiro et al. 1991). Multilayered formations and the deposition of melanin at the cell wall level in black fungi make the extraction and solubilization of proteins more challenging than in filamentous fungi types. Therefore, the selection of an appropriate method resulting in a good protein yield must be optimized for each species. Quite recently, a protoplast-based system was established for genetic transformation of black fungi (Noack-Schönmann et al. 2014). Such workflow, which applies a combination of fungal enzymes to produce protoplasts, could possibly serve as a starting point in proteomics application to enhance extraction of intracellular proteins.

The removal of interfering compounds is another critical step on the way to obtaining a high-quality protein extract. Pigments, exopolysaccharides, nucleic acids, and lipids are examples of the compounds mostly hampering protein extraction in fungi. Melanins and carotenoids are long chain polymers with several functions in cell protection (Cordero et al. 2017; Flieger et al. 2018). In melanotic fungi, multiple layers of melanin granules from phenolic or indolic precursors (Nosanchuk and Casadevall 2003) accumulate at the cell wall level contributing to a thick and rough cell surface (Eisenman and Casadevall 2012). Carotenoids are usually masked by melanin, however are also located in the cell walls, and are displayed in melanin mutant strains (Nai 2014). Despite efforts, the elimination of pigments remains still challenging [87]. Bleaching using chlorine or hydrogen peroxide prior or after biomass disruption can be attempted however not without risking to affect the proteins disulfide bonds. The benefit is that both components are labile and do not remain in the sample as additional residues (*personal communication*). However, their influence on the mass spec analysis must be studied. The use of charcoal powder by direct addition to the homogenization buffer has also proved to be quite useful to reduce the abundance of melanin in the final extract, however not without drawbacks (personal communication). The removal of melanin goes hand in hand with a low protein yield, probably due to the loss of cell wall proteins as well as melanin-bound cellular proteins (Jacobson 2000). Attention needs to be paid also when using phenol-based methods for protein extraction, followed by precipitation steps. As polyphenolic compounds, melanins are soluble in organic solvents such as phenol (Amin et al. 2018), thus resulting in the co-extraction of melanin and proteins. Furthermore, melanin and proteins coprecipitate in the presence of organic solvents, thus resulting in darkly pigmented protein pellets. Methods for the successful isolation of melanin are available and include multiple precipitation steps induced by shifts in the pH conditions of the medium as well as boiling at high temperature (Amin et al. 2018; Pinto et al. 2018). Parameters—temperature and pH—however influence protein stability during extraction and are therefore not compatible with protein downstream analysis.

Precipitation steps are used to selectively purify proteins from several types of contaminants ranging from salts to polysaccharides and fatty acids, which can interfere with both gel-based (Crichton et al. 2013) and gel-free analyses. Precipitation with trichloroacetic acid (TCA) and acetone or a combination of the two is often used for fungal protein extracts (Bhadauria et al. 2009). Nevertheless the precipitation in ammonium acetate in methanol has also been described for black fungi (Isola et al. 2011; Tesei et al. 2015b; Vödisch et al. 2011; Zakharova et al. 2014b) as a way to bypass protein re-solubilization problems often occurring when using TCA. Precipitation steps are sometimes not enough to achieve the efficient elimination of interfering compounds, and washes of the protein pellet with organic solvents as methanol and acetone at different concentrations are therefore additionally performed. Nevertheless, these steps are in the case of black fungi often only contributing to a reduction in the amounts of contaminants, whose complete removal is extremely arduous. Polysaccharides, for instance, which make up the core of the cell wall, represent a substantial part of the cell lysate. A wide array of different proteins are anchored in various ways to cell wall polysaccharides like glucans and chitin, which makes both the isolation of the proteins and the removal of polysaccharides difficult (Pitarch et al. 2008). Most species of black fungi are also well known to produce massive amounts of extracellular polysaccharides (EPS) forming matrices or capsules around the cell surface and having roles in stress protection and infection (Breitenbach et al. 2018; Rodrigues et al. 2014). As these polymers are associated with the mycelia, they are hardly removed during the extraction procedures, thereby hindering proteomic protocols and interfering with spectrophotometric methods (Bianco and Perrotta 2015; Marzban et al. 2013). A number of strategies to enhance the removal of polysaccharides, fatty acids, and gelatinous material in general have been reported in literature for fungal species and could be applied to black fungi. These strategies most often involve high-speed centrifugation and/or filtration as well as sample dialysis before or after precipitation (Adav et al. 2010; Fragner et al. 2009).

The abovementioned procedures for the preparation of whole-cell protein extract can be applied to both gel-based and gel-free approaches, however not without the need for adjustments and optimization, which have to be assessed case by case. The requirements can indeed be different among gel-based and gel-free methods for protein separation. For instance, if maintaining the protein integrity is crucial in gel-based techniques—especially native gel proteomics—protein fragmentation is less of an issue in gel-free proteomics, where proteins necessarily undergo tryptic digestion prior to chromatographic separation (Fernández et al. 2014). Keeping the amount of detergents, salts, and organic solvents in the sample at a minimum is paramount when using all kinds of separation procedures. Despite aiding the extraction and improving the solubility of hydrophobic proteins, detergents interfere with mass spectrometry (MS) analysis and must therefore be removed from protein samples (Yeung et al. 2008). Salt interference significantly affects the quality of the first-dimensional electrophoresis, isoelectric focusing (IEF) (Wu et al. 2010). Similarly, organic solvents, used in the extraction procedures for removal of distinct interfering compounds such as phenolic compounds and pigments, often contribute to the hardening of the protein pellet, thereby hindering protein solubilization (Parkhey et al. 2015). In all cases where the use of chemicals and reagents potentially interfering with MS cannot be avoided as this would result in a very poor protein yield, methods such as dialysis or buffer exchange are applied. Along with commercially available kits for sample cleanup, which often result in considerable protein loss (Marzban et al. 2013), the use of filter units has become in recent years more popular (Wiśniewski et al. 2009). Filter-aided sample preparation (FASP) or generally methods carrying out detergent removal, buffer exchange, chemical modification, and protein digestion into centrifugal filters have proved to be effective also for the processing of recalcitrant fungal samples (Adav et al. 2010; Wiśniewski et al. 2009; Zhong et al. 2018; Zoglowek et al. 2018).

A very low number of studies report protein extraction workflows, which only make use of MS-compatible reagents and were successfully applied to the analysis of proteins from extremophilic and extremotolerant fungal species. Romsdahl and colleagues adapted 100 mM triethylammonium bicarbonate (TEAB) extraction buffer containing protease inhibitors in combination with mechanical homogenization of mycelia at 4 °C to extract total proteins from melanotic fungal strains isolated from or exposed to the International Space Station (ISS) (Romsdahl et al. 2018). The same protocol was also used for both mycelial protein and secretome extraction in black rock-inhabiting fungi (Tesei et al. in preparation). After debris elimination through high-speed centrifugation, despite no effort was put in trying to remove contaminants, coloration was not observed in the supernatant. The persistence of a slight pigmentation due to melanin and carotenoids was instead observed in secretome samples, as expected.

12.2.2.2 Secretomics of Extremophiles and Extremotolerant Fungi

As the first fungal genome was sequenced, researchers employed comparative proteome analyses to discover how fungi could adapt to the occupation of a wide variety of ecological niches. The secretome or extracellular proteome explains a repertory of protein entities released by fungal cells with crucial importance for the organism to acquire nutrients and communicate with the environment (Krijger et al. 2014).

Secreted proteins are necessary for the exchange of information with the environment, especially in sessile organisms, which are precluded from actively seeking out nutrients, such as several species of extremophilic fungi. Fungi exhibit a wide diversity of nutritional lifestyles, ranging from strict saprobe—feeding on dead or decaying organic matter—to pathogenic or parasitic, whose life cycle is in strict association with a living host organism. Obviously fungal secretome plays important roles in the degradation of organic material, and in managing directly or indirectly symbiotic and pathogenic lifestyle with their hosts (Krijger et al. 2014). The secretome of extremophiles additionally holds particular interest for identifying enzymes of potential biotechnological value, such as proteases and cellulases (Blumer-Schuette et al. 2008; Miyazaki et al. 2005; Sanchez-Pulido and Andrade-Navarro 2007).

Secretomics of extremophiles use different strategies and employ various techniques like IPG-shotgun, SDS-PAGE-MS/MS (or GeLC-MS/MS), and 2-DE-MS/MS (Ellen et al. 2009; Muddiman et al. 2010; Vincent et al. 2012). Taking into consideration that there are no methodological boundaries existing for application of large-scale and high-throughput tools, like LC/LC-MS/MS (Vincent et al. 2012; Williams et al. 2010), the gel-based secretomics have been ever a favorite approach (Vincent et al. 2012). The sample preparation by means of fractionation is hitherto to be considered as the most critical step in the analysis of the secretome. Therefore, additional strategies are necessary to guarantee that the fractionation is robust enough (e.g., including test systems for cytoplasmic markers), so that the obtained data can be annotated with high confidence (Williams et al. 2010). The main challenge is to differentiate secretome from proteins, which are not originating from the extracellular repertory (Barreiro et al. 2012). The secretory proteins are generally collected from the nutrient media after filtration or centrifugation to separate the mycelia tissue from the culture supernatant. Proteins are thereafter isolated through precipitation. Secretome fractions may also contain membrane or outer membrane proteins that are released by endopeptidases during cultivation or by cell death (Ellen et al. 2009); this can be avoided by culture collection at early stages of growth, although early harvesting downturns the protein yield. Other types of contaminants such as pigments and polysaccharides often populate the secretome of melanotic and extremophilic fungi, and protein separation by means of precipitation is sometimes not sufficient to guarantee their removal. However, downstream procedures for the cleanup of secretomics samples prior to MS such as FASP demonstrated their efficacy in the elimination of contaminants from the extracts. Despite the fact that robustness against impurities is not among the strengths of in-solution digestion (Wiśniewski et al. 2009), this method proved a particular effectivity in separation of melanin from fungal protein extracts (Tesei et al. in preparation). While peptides elute in a colorless buffer, melanin is retained on top of the membrane filter.

Hydrophilic proteins can be separated from the insoluble hydrophobic fraction by phase extraction and centrifugation (e.g., chloroform extraction); however this method has its limitations concerning MS (e.g., detergents). Obstacles with gel electrophoresis caused by hydrophobic proteins including post-solubilization precipitation can occur during the IEF and lead to vertical streaking lines in the gels (Rabilloud et al. 2008). Methanol as proven solvent has been often used for non-soluble proteins accompanied with thermal denaturation and insolvent digestion with higher success than detergents and more compatibility for MS (Mitra et al. 2007; Zhang et al. 2007). Dimethyl sulfoxide (DMSO), an additive to animal cell culture media, can also be used to increase the protein solubility of membrane-based proteins (*personal communication*).

Studying the insoluble proteome of extremophiles helps to explore the membrane and surface proteins, which are associated with environmental interactions. For example, the halophilic alkalithermophile *Natranaerobius thermophilus* is assumed to accommodate to multifactorial environmental extremes by huge depot of Na + (K+)/H+ antiporters (Mesbah et al. 2009). Proteomic studies of non-soluble fractions discovered proteins, which are associated with signal transduction/transport of different substances in the psychrophilic methanogens *Methanococcoides burtonii* (Burg et al. 2010; Williams et al. 2010).

In order to identify those proteins different strategies are suggested, which envisage the development of specific protocols in direction of sample collection and post-identification protein analysis (Barreiro et al. 2012; Jami et al. 2010b). The multifunctional proteins and proteins, which are rarely ever identified from secretomics samples, represent the main challenge. Whereas the post-identification protein analyses are focused on the cultivation and harvesting time ranges (detection of proteins at different time points of cultivation) to justify if a protein is secreted or released by cell lysis, the rare proteins can only apply the available proteomics data (Barreiro et al. 2012; Bendtsen et al. 2004). Fractionation of proteins and/or tryptic peptides prior to MS represents a good strategy for the enrichment and the detection of less abundant peptides and protein species (Ly and Wasinger 2011). An intensive research in the literature for possible routes of secretion or lysis is therefore strongly recommendable. Signal peptides can help discriminate intracellular proteins from proteins that are ultimately destined to the secretory pathway or to the cell membrane (Kapp et al. 2009). Although the signal sequence is usually removed in the mature protein, a number of bioinformatics tools are available to aid the prediction of the presence and location of signal peptide cleavage sites in amino acid sequences from different organisms. In the near future, we expect that the developments in the area of data storage and annotations deliver cutting-edge progresses toward secretome knowledge levels, which are deciding for the understanding of fungi survival evolution, pathogenicity, and extracellular communication.

12.2.2.3 Subcellular, Membrane, and Vesicle Proteomics of Extremophilic Fungi

Most of the proteomics research work hitherto carried out in fungi has been focused on the whole-cell proteome and on the secretome. Several crucial biological processes however occur specifically within cell compartments (Kim et al. 2007), therefore shedding light on the proteome profile of organelles, and vesicles would give additional insight not only into the cell basic metabolism but also into stress survival and pathogenesis. The complexity of protein extraction from extremophilic and extremotolerant fungi however reflects the low number of subproteomics studies.

The nature of the fungal cell wall, especially in melanotic species, requires resorting to harsh treatments for cell disruption, which can on the other hand compromise the organelle integrity (De Oliveira and de Graaff 2011). Bead milling-mediated breakage of the biomass is preferentially performed at low temperatures and is coupled with the use of lysis buffers added with protease inhibitors (Grinyer et al. 2004). More gentle methods for the mechanical breakage of the mycelium have been suggested, such as the French pressure and mortar instead of the bead beating. Sonication can be additionally applied for the lysis of organelles. Similarly, biomass boiling as well as the use of detergents should be avoided (Nandakumar and Marten 2002). Enzymatic methods like cell wall digestion and protoplast formation could represent a valuable alternative, however not without possible drawbacks. Protoplast formation protocols extend for hours and have effects on the physiology of the fungus. Decreased reproducibility of the method is instead experienced when using different batches of enzyme cocktails (De Oliveira and de Graaff 2011). Organelle isolation is made also difficult by the presence of microtubules supporting the hyphal development, which increases the clustering of different organelles, thus making their separation more difficult (Bianco and Perrotta 2015). An additional obstacle to extraction of proteins from organelles is represented by proteases, which are abundantly secreted by fungi, thereby enhancing the common problem of protein degradation common to other eukaryotes.

Subfractionation procedures are applied following cell disruption and usually involve low-speed centrifugation to allow a first separation of the organelles based on size and weight. Ultracentrifugation steps are additionally performed to obtain the distribution of organelles in different fractions. Along with centrifugation, antibody-based methods are also used for organelle separation. Immunomagnetic separation (IMS), among others, proved to be quite efficient; however its application is strictly dependent on the availability of antibodies as well as on the antibody specificity. One method increasingly used to enrich organelles is free-flow electrophoresis (FFE). In FFE, the organelle mixture moves along carrier medium between two slanted plates (Karkowska-Kuleta and Kozik 2015).

If the characterization of organelle proteins in extremophilic and extremotolerant species of fungi is yet to be accomplished and optimization work needs to be done in order to adapt preexisting methods to the study of these organisms, cell wall proteomics recently witnessed the fast development of a wide variety of techniques. The fungal cell wall acts as a protective layer sheltering the cell from the outer environment and represents the place of cell-to-cell first contact and communication. The cell surface is actively involved also in phenomena such as pathogenicity and infection; thus cell wall-accessible molecules are targets for host immunity as well as for new drugs (Karkowska-Kuleta and Kozik 2015). With the cell wall playing a critical role in the biology and the ecophysiology of fungal organisms, investigating the structure and function of cell wall proteins is of great importance, also in view of a full understanding of yet-unresolved issues such as cell wall morphogenesis and host-pathogen interaction.

Many proteins are located at the cell wall level, most of them displaying modification with N- and O-linked glycans. A number of these proteins originate as cell membrane glycosylphosphatidylinositol (GPI)-linked polypeptides and end up in the cell wall, where they are covalently linked to ß-1,6-glycans (Gow et al. 2017). Others are instead associated to the wall via non-covalent interactions, or disulfide bonds to polypeptides that are themselves covalently bound to structural glycans (Agustinho et al. 2018).

The analysis of proteinaceous components of the fungal cell wall generally involves the separation of the wall from the protoplast and can be carried out either with or without cell disruption, however avoiding cell breakage and the consequent release of cell content. To this purpose, usually vortex or gentle agitation of the fungal cells with vertical turntable is chosen over more invasive methods (Klis et al. 2007). Protocols involving mechanical disruption of the cells include washing steps with buffers with ionic strength for the removal of cytoplasmic and membranous contaminants (Pitarch et al. 2008). The use of a DTT-based homogenization buffer for cell incubation has been shown to help protein extraction in black yeasts without affecting membrane integrity (Longo et al. 2014). Short enzymatic treatments can otherwise be performed (Vialás et al. 2012). As in the procedures described earlier for the extraction of whole-cell proteome, detergents and denaturing and reducing agents are used, however here mostly for the isolation of proteins non-covalently incorporated into the wall. Additional steps are needed to extract proteins associated with cell wall polysaccharides, and they usually involve hydrolytic enzymes and chemicals (Karkowska-Kuleta and Kozik 2015). Several classes of commercially available enzymes—e.g., β-glucanases and chitinases—have been applied to the cleavage of specific proteins based on the nature of their linkage to the wall (i.e., alkali-sensitive linkage (ASL), GPI linkage, chitin-bound proteins). Alkali and acids serve a similar purpose. Trifluoromethanesulfonic acid (TFMS), among others, is often used to isolate and simultaneously deglycosilate proteins, in order to ease protein gel-based separation (Maddi et al. 2009). The biomass incubation in cell wall digestion cocktails can also be performed in the presence of protease inhibitor cocktails to diminish protein degradation (Champer et al. 2016). Similarly, live intact cells can be incubated with proteolytic enzymes—e.g., trypsin, chymotrypsin, and proteinase K (Olaya-Abril et al. 2014)—to promote protein cleavage and the release of peptides, which can thereafter be identified by mass spectrometry. Such procedure is better known as "cell shaving" (Yin et al. 2008). All these strategies firstly set up and developed in bacteria and classical yeasts are increasingly being applied to the study of species of black fungi, especially the opportunistic and pathogenic ones, to investigate the role of surface-exposed proteins during infection.

Because of its impact on fungal pathogenesis, the proteome of extracellular vesicles (EVs) has additionally gained increasing interest in recent years. The production and secretion of EVs—whose size ranges from 50 nm to 400 nm in diameter (Oliveira et al. 2013)—have been observed in a number of species of melanotic fungi and yeasts (Joffe et al. 2016; Rodrigues et al. 2014, 2015; Vallejo et al. 2012); nevertheless many aspects related to EVs remain unknown. EV biogenesis and release are yet to be fully elucidated; however a number of studies confirmed the diversity in EV composition (Rodrigues et al. 2015). EVs are characterized by a conserved set of molecules across species—e.g., lipid raft molecules, membrane trafficking molecules, MHC class I molecules, and heat-shock proteins (Furi et al. 2017)—suggesting life-preserving functions (De Toro et al. 2015). The comparison of EV protein profiles from opportunistic species further revealed overlapping in protein composition, thus suggesting the presence of signature proteins with clear roles during infection (Vallejo et al. 2012). Remarkably, most of the proteins found in fungal vesicular fractions lack the characteristic signal peptides required for conventional secretion (Oliveira et al. 2010; Rodrigues et al. 2008); thereby the origin of EV from unconventional or still unknown pathways of secretion can be hypothesized.

The analysis of EVs by proteomic-based approaches requires the purification of the vesicles from cell-free culture supernatants. In order to aid the recovery of EVs, the culture supernatant is initially concentrated by membrane ultrafiltration using membrane with high molecular weight cutoff (MWCO, 100 kDa). A number of separation techniques can be used individually or combined, to remove aggregates and obtain a vesicles fraction. These methods include sequential centrifugation and ultracentrifugation, density gradient centrifugation, filtration (0.22 μm or 0.1 μm) polymer-based precipitation, and immuno-affinity (Simpson et al. 2008; Taylor and Shah 2015). Density gradient separation combines sucrose gradients with ultracentrifugation; polymer-based precipitation uses polyethylene glycol; immunoselection employs antibodies binding to vesicle surface markers. Ultracentrifugation ($100,000 \times g$) appears to be the method most frequently used for the isolation of EVs from black fungi, however not without limitations. Low recovery of EVs and of the rations of exosomal proteins and non-vesicular macromolecule contamination have been observed independently on the cell type and on the organism (Furi et al. 2017; Taylor et al. 2011). Further purification steps in HPLC can additionally be performed to aid in the removal of contaminating non-exosomal material (Rodrigues et al. 2008). Along with vesicle enrichment, melanin contamination represents the major obstacle to proteomic analyses of EVs from melanotic fungi, as it affects both protein determination and separation. Melanin is often found in liquid media during incubation of black fungi; thus concentrating the culture supernatant prior to vesicle isolation also results in the concentration of this pigment. Notably, melanin is additionally found in and exported extracellularly by EVs (Eisenman et al. 2011; Joffe et al. 2016; Rodrigues et al. 2014). Filter-aided procedures for protein sample cleanup such as buffer exchange could support the removal of pigments. Due to melanins' large size (Langfelder et al. 2003), multiple ultrafiltration steps on membranes with different cutoffs could be necessary to separate melanins from the protein fraction.

The thorough isolation of pure fractions and the integrity of the sample (e.g., membrane integrity) are both crucial aspects having an impact on downstream analyses. Protein extraction from purified vesicles is generally achieved without mechanical disruption. Proteins can be isolated using the TRIzol extraction procedures for RNA and protein analyses or recovered by centrifugation after precipitation (e.g., TCA) on ice and afterward be lyophilized or resuspended in a buffer

(Taylor et al. 2011; Vallejo et al. 2012). Alternatively, they can be solubilized by incubation in chaotrope-based homogenization buffers and thereafter be processed for reduction, alkylation, and finally digestion (Rodrigues et al. 2008). In the majority of studies the characterization of protein fractions from EVs has been carried out resorting to gel-free techniques as liquid chromatography-tandem mass spectrometry (LC-MS/MS) (Rodrigues et al. 2014).

12.2.3 Protein Separation and Identification

Two major proteomic technologies have been applied to resolve the fungal proteome: gel-based separation techniques coupled to mass spectrometry (MS) or tandem mass spectrometry (MS/MS) and shotgun (gel-free) methods based on LC-MS/MS.

The different methodologies for protein separation via SDS-PAGE acrylamide gels have been very popular for decades through their simplicity and robustness. SDS-PAGE in one- and two-dimensional feature has also been for long time the most often used approach applied to separate fungal proteins. In particular, SDS-PAGE has been proven to be the most favorite tool to study hydrophobic proteins with lowered solubility, such as membrane- and cell wall-embedded proteins. The separation and visualization of proteins from MCF and black yeasts were indeed mostly performed by using one- and two-dimensional electrophoresis (1-D and 2-DE). These methodologies showed to be capable of visualization of a high number of proteins and to also visualize the alteration of protein patterns (Bhadauria et al. 2009; Isola et al. 2011; Tesei et al. 2012).

One-dimensional electrophoresis is a useful technique for the rapid fingerprinting of protein samples as well as to assess the quality of protein extract (e.g., protein degradation); it is often used for protein separation prior to MS/MS and additionally represents an essential step of immunoblotting techniques (González-Fernández et al. 2014; Guimarães et al. 2011; Loginov and Šebela 2016). Despite the simplicity of this technique and its low resolution, 1-DE led to the identification of proteins as malate dehydrogenase and peptidyl-prolyl cis-trans isomerase, contributing to important steps forward in fungal proteomics (Dodds et al. 2009; Jorrín-Novo et al. 2009; Marzban et al. 2013). Two-dimensional gel electrophoresis (2-DE) aids in the characterization of protein profiles and their alterations in response to different experimental conditions, and eases the detection of protein isoforms as well as of posttranslational modifications (Bhadauria et al. 2007; De Oliveira and de Graaff 2011; Tesei et al. 2015b).

Despite the major contribution given by gel-based techniques to the investigation of the fungal proteome in species from extreme habitats, the number of applications has decreased over the last years in favor of gel-free techniques, which allow more in-depth proteomic analyses (Loginov and Šebela 2016). Complete and annotated sequences of a growing number of fungal genomes in combination with the development of more sophisticated and sensitive high-throughput LC-MS instrumentation have indeed resulted in the identification of increased number of proteins. The gel-free LC-MS-based methodology like isobaric tag for relative and absolute quantitation (iTRAQ) was for instance reported to be the most effective technique for the identification of membrane-associated proteins in filamentous fungi (Barreiro et al. 2012; Ouyang et al. 2010).

Nevertheless, the success of gel-based and gel-free approaches is limited both within the boundaries of the sample preparation and protein yields. Therefore, the choice between classic gel-based techniques—or more exactly their updated version—and gel-free ones strictly depends on the research goal and on the nature of the sample.

In the next section, the functionality and the applications of gel-based and gel-free techniques to the analysis of proteins from extremophilic and extremotolerant fungi will be described, not without reviewing the need for workflow optimization as required by difficult samples.

12.2.3.1 Gel-Based Protein Separation: One-Dimensional and Two-Dimensional SDS-PAGE

In fungal proteomics, 1-D SDS-PAGE has been extensively employed to investigate hydrophobic proteins, such as membrane- and cell wall-anchored proteins. Indeed, one of the first intracellular proteomic studies in filamentous fungi biology was carried out on white/red *P. chrysosporium* and *Lentinula edodes* (De Oliveira and de Graaff 2011). Iron-binding plasma membrane proteins were analyzed by using only 1-D SDS-PAGE, since the authors failed to detect them by 2-DE. Another example is a similar approach to explore mitochondrial proteins of *Neurospora crassa*, a filamentous fungus secreting enzyme capable of complete digest of plant cellulose (Schmitt et al. 2006).

One-dimensional SDS-PAGE is to be seen as the first routine step in the analysis of fungi proteomics and occupies a fix position in every established workflow despite its disadvantages, e.g., low resolution and absolute protein amounts that can be applied to the gel (Fig. 12.3a). It can be used in native feature as the only and sole tool for the separation of membrane and insoluble proteins in combination with mass spectrometry (Nguyen et al. 2005). This variation of SDS-PAGE, termed Blue Native PAGE (BN-PAGE), showed extensive progress in the investigation of insoluble or native large protein complexes, which are embedded in fungal membranes or mycelium and secretome (Crichton et al. 2013).

1-D SDS-PAGE remains a qualitative methodology and the protein separation is not optimal and results in more proteins of identical molecular weight falling into a single and thereby heterogeneous band. However, it remains versatile and flexible in the case of comparative analysis with large number of samples. Through the rapid fingerprinting of fungal extracts, 1-D SDS-PAGE allows differentiating among various genotypes as well as between wild-type strains and mutants, different cultivation modes, or environmental stresses.

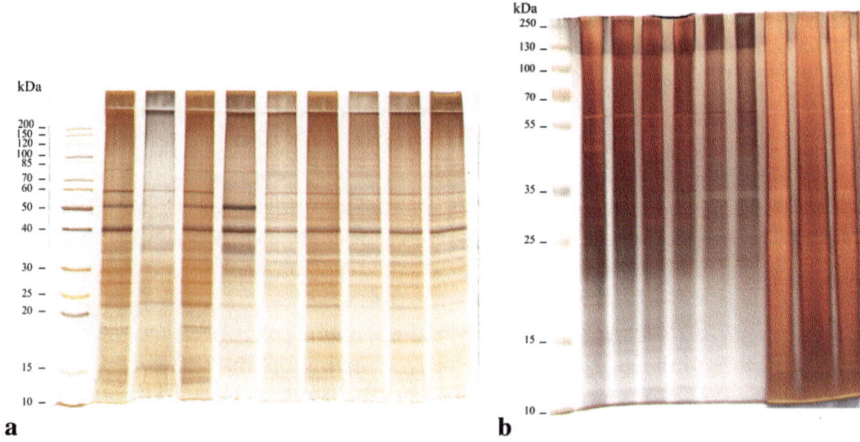

Fig. 12.3 SDS-PAGE 1-D electrophoresis gel. (**a**) Total protein extracts of the rock-associated fungus *Knufia chersonesos* wild-type MA5789 and mutant MA5790 using different extraction methods. Protein amount: 4 μg. (**b**) Secretomes of *K. chersonesos* MA5789. Interference of melanin, present in the samples, is observed in the gel. As silver nitrate binds to both melanin and proteins, a dark background is generated on the gel after silver staining, thereby masking the protein bands. Protein amount: 2 μg

It is quite often applied particularly for the separation of membrane proteins and for the detection of immunogenic proteins and fungal allergens through antibody-based immunoblotting techniques (Supek et al. 2011; Westwood et al. 2005). Despite the borders of denaturing and native SDS-PAGE effectivity, it can provide a first but, in some cases, the only technique to have a prompt glimpse inside the proteome of fungal species.

Two-dimensional quantitative electrophoresis (2-DE) in combination with silver staining or DIGE technology and a wide spectrum of mass spectrometry techniques have a proven history in the investigation of extremotolerant and extremophilic black fungi (Tesei et al. 2012, 2015b; Zakharova et al. 2013, 2014a, b) (Fig. 12.4). Since the first and most prominent systematic mapping of lignocellulolytic *Trichoderma harzianum* has been performed by MS-MALDI-TOF and 2-DE as reported by Bianco and Perrotta (2015), a series of different researchers started to explore the extremophile fungi by various criteria. The response to zinc in pathogenic *Paracoccidioides* sp. yeast cells could be clarified by comparative 2-DE during zinc starvation. The results showed the physiological rearrangement of *Paracoccidioides* sp. to the probable oxidative stress induced during zinc withdrawal (de Arruda Grossklaus et al. 2013; Parente et al. 2013). Earlier mycelial proteins were separated by 2-DE, analyzed by peptide mass fingerprinting (PMF) and tandem MS and successfully identified (Jami et al. 2010a). Accordingly, protein maps for different filamentous fungi could be established using 2-DE and MS (Lu et al. 2010; Ravalason et al. 2008; Yildirim et al. 2011). Similarly to 1-DE, the 2-DE technique has also provided the unique tool for immunomics in fungi (Barreiro et al. 2012). By the use of high-resolution 2-DE in combination with IgE immunoblot-

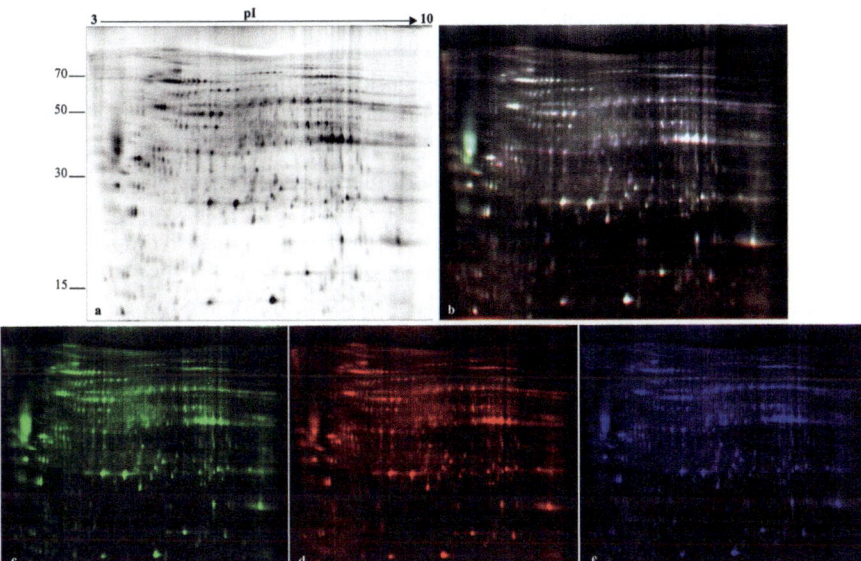

Fig. 12.4 SDS-PAGE 2-D electrophoresis gel of total protein extracts of the opportunistic black yeast *E. dermatitidis* CBS 525.76. (**a**) Classical 2-D, protein spots are visualized by silver staining; (**b**) 2-D DIGE, overlay of the three images obtained by separating on the same gel three samples differentially labeled with fluorescent dyes; (**c–e**) three separate images obtained using three different lysine-binding fluorescent dyes

ting, the allergenic proteins of fungal spores of *Aspergillus versicolor* could be identified and a component-resolved allergen testing could be obtained using patients' serum IgEs (Benndorf et al. 2008).

The invention of quantitative differential electrophoresis (difference gel electrophoresis, DIGE) revolutionized the comparative analysis and provided a reliable approach by 2-DE technique. Before the development of DIGE, the major obstacle using 2-DE was the irreproducibility of the gels running with the identical repeats. The fluorescent dye-based labeling and the labeling chemistry allow the multiplexing of the samples applied into one single IEF strip (Marouga et al. 2005). A maximum of three different samples can be run in the same gel, along with a reference sample, created by pooling all samples involved in the experiment. This achievement catapulted 2-DE into a new era and comparative mapping could thereby be extended to quantitative proteomics and be directly coupled with MS pipelines for protein identification and characterization, a milestone which allowed the establishment of universal protocols and the standardization of the running conditions. In the meanwhile, two or three different applications for DIGE protein labeling have been developed. Together with the minimal dyes (lysine-based) and saturation (cysteine-based) dyes for the quantitative comparison of protein expression even in samples with as little as 3–5 μg protein, dyes for the visualization of complex oxidative responses of the proteome are nowadays also available (Kondo and Hirohashi 2007; Kratochwill et al. 2015; Strohkamp et al. 2016).

Although the general proteomics workflow remains identical for any analyzed sample, the first attempt to establish a workflow for black fungi is not older than one decade (Isola et al. 2011). High melanin and low soluble protein content slowed down the process of isolation of cell components not only at the proteome but also at the transcriptomic and genomic level. The developments at the protein analytical level were promising and the investigation of the proteome modulation in extremophilic and extremotolerant fungi became soon a reality. The first step was the mapping of the proteins in the black yeast *Exophiala jeanselmei* and the meristematic fungus *Coniosporium perforans* (Isola et al. 2011). The first 2-D protein maps opened the doors for extensive studies and analyses of protein alterations under different environmental conditions (Tesei et al. 2012; Zakharova et al. 2013, 2014a, b). However, the full optimization of a working protocol needed more time and experimental genius. First deep investigation which involved a complex experimental design and a combination of 2-D DIGE and mass spectrometry needed over 2 years for step-by-step optimization and the generation of bioinformatics data about thermal stress tolerance of the opportunistic black yeast *Exophiala dermatitidis* (Tesei et al. 2015a).

The improved proteomics workflows could be recruited for experiments with similar samples, with adjustments needing to be assessed case by case. Despite the optimization work, melanin still represents the main bottleneck and melanin-contaminated extracts aggravate stubbornly all proteomics approaches with black fungi (Moreno et al. 2018) (Fig. 12.3b). Besides the sample-related drawbacks, 2-DE remains itself a very time-consuming and laborious technique. The available image analysis delivers many mismatches, which are to be manually improved. The resolution of 2-DE is a point of concern, since there are spots, which contain multiple proteins. Despite typical disadvantages and limitations, 2-DE however represents a unique methodology due to the capability to generate visible protein maps and images after separation of proteins. Images, which can serve proteome comparative analysis and maps, can be directly used for spot picking and can therefore undergo mass spectrometric analyses.

12.2.3.2 Gel-Free Approaches: SHOTGUN Proteomics

As compared to gel-based approaches, mass spectrometry (MS)-based proteomics such as shotgun proteomics allows a comprehensive study of the proteome involving a multitude of samples in complex experimental designs and combines the qualitative and quantitative study of the proteome with an enhanced efficiency of separation and sensitivity of protein identification (Fernández et al. 2014; Marcotte 2007). Such properties can potentially serve different applications within the study of extremotolerant and extremophilic species, including, among others, the investigation of the stress survival and of the evolution in extreme environments as well as the elucidation of the mechanisms involved in fungal infection. If shotgun proteomics often entails a quicker workflow for sample processing due to the possibility of simultaneous analysis of multiple samples, it however requires complex downstream analyses aiming at the biological interpretation of the results (Sinitcyn et al. 2018). Statistics, computational technologies, and bioinformatics are indeed paramount to proteomics data mining, especially since large amount of data are involved. Moreover, the integration of proteomics data with other OMICS data has been increasingly gaining importance (Kohl et al. 2014).

Peptide-based shotgun proteomics, also called bottom-up proteomics, is most often the method of choice for the analysis of the fungal proteome. Bottom-up proteomics, in combination with either protein labeling or label-free techniques (Huang et al. 2015), happens to be also the format currently most used in proteomics (Sinitcyn et al. 2018). Sample preparation for bottom-up techniques involves reduction alkylation steps followed by in-solution enzymatic cleavage (by a trypsin and LysC protease mix) of the protein extract into peptides, which are then subjected to liquid chromatography (LC)-based separation and MS/MS analysis. Proteolytic peptides can be analyzed in a data-dependent or a data-independent manner. In data-dependent acquisition (DDA) mode, only the peaks (i.e., peptides) with higher intensities at a given chromatographic elution time are chosen for fragmentation and subsequently identified through protein database search (Huang et al. 2015). In data-independent acquisition (DIA) experiments, all peptides within a certain m/z range undergo fragmentation and fragment ion spectral libraries, established beforehand, are used to mine the complete fragmentation maps (Gillet et al. 2012). The actual number of identified proteins depends on the sample complexity and the proteins' dynamic concentration range—namely the diverse abundance of different classes of proteins—as well as the availability of sequenced and annotated genomes (Rohrbough et al. 2007).

The earliest research works to apply bottom-up proteomics in black fungi and black yeasts aimed at getting an insight into the extracellular proteome, especially of opportunistic and pathogenic species. Albuquerque et al. (2008) used reversed-phase liquid chromatography-tandem mass spectrometry (LC-MS/MS) to identify proteins from purified extracellular vesicles from the clinically important melanotic dimorphic fungus *Histoplasma capsulatum*. Two hundred and six out of 283 proteins could be identified by sequence analysis, revealing their involvement on the onset of pathogenesis and host immune responses along with cell architecture and cell growth. Many of these proteins were also described in the proteome of vesicles from *Cryptococcus neoformans*, the causative agent of cryptococcosis, where a similar protocol for protein fractionation and identification by LC-MS/MS was applied (Rodrigues et al. 2008). Interestingly, the rate of protein detection among different strains varied considerably according to the presence of capsular structures surrounding the cell, suggesting that polysaccharides in extracellular fractions might interfere with protein digestion. The secretome of *Paracoccidioides brasiliensis*, the dimorphic ascomycete, and the causative agent of paracoccidioidomycosis were additionally inspected. The overall characterization of extracellular proteins in the supernatants of the yeast phase of *P. brasiliensis* was achieved by LC-MS/MS. The analysis of vesicle and vesicle-free fractions resulted in the identification of 85 and 140 proteins, respectively, with 35% of the sequences—most likely vesicle-free proteins, which bound to vesicles during sample preparation steps—overlapping both fractions. Also in this case, several vesicle-associated sequences had orthologues in other fungal extracellular vesicles.

Along with protein identification, bottom-up proteomics allows the comparative quantitative analysis of complex protein samples by means of a great variety of approaches and instrumentations based on the methodology strength for a specific research aim. Relative protein abundance can be measured using stable isotopes or label-free methods. The label-based techniques isobaric tagging like iTRAQ (isobaric tags for relative and absolute quantification) and TMT (tandem mass tag) are among the most popular techniques in quantitative proteomics. Both iTRAQ and TMT use chemical labels, which bind to protein samples after proteolysis in a peptide-specific manner without altering peptides' chromatographic and ionization properties (Bianco and Perrotta 2015; Chahrour et al. 2015). In iTRAQ, isobaric mass labels are covalently linked to peptides and yield "signature" or reporter ions following peptide fragmentation that can be used to identify and quantify individual members within a multiplex set of samples using tandem mass spectrometry (Wiese et al. 2007). Similarly to iTRAQ, TMT reagents use isotopomer labels and release "daughter ions" (Chahrour et al. 2015). iTRAQ is available in 4-plex and 8-plex formats, while TMT is available in 2-plex, 6-plex, and 10-plex formats. Together with the samples, a reference sample, which is a pool of all samples included in the experiment, is also used. Protein samples are pooled, then separated in a first dimension by strong cation exchange (SCX)—separation by charge—or reversed-phase chromatography (RP)—separation by hydrophobicity—and thereafter analyzed by MS/MS (Rohrbough et al. 2007). The combination of the two separation techniques in a single chromatographic column is known as multidimensional protein identification technology (MudPIT) and is applied to the separation of complex biological mixtures.

Although all mentioned techniques show high potentials, their successful application in the study of black fungi is still in the early stages, mostly due to the difficulties posed by the sample. First studies of the fungal proteome resorting to iTRAQ were carried out in species of *Aspergillus* and *Fusarium*, to investigate alterations of the total proteome in response to antifungal drugs and mycotoxins (Cagas et al. 2011; Taylor et al. 2008). Similarly, iTRAQ found applications also in the characterization of the secretome of biotechnologically relevant strains such as *A. niger* and *Trichoderma reesei* (Adav et al. 2010, 2012). More recently, iTRAQ labeling and strong cation exchange in combination with ESI-Q-TOF MS/MS were reported for the study of the aluminum stress in the extremotolerant yeast *Cryptococcus humicola* (Zhang et al. 2015). In order to assess changes in total protein abundance between control and aluminum-treated samples, protein relative concentrations were calculated based on the reporter peak area. TMT labeling coupled with reversed-phase LC-tandem MS (MS/MS) was applied to the molecular characterization of the melanotic hyphomycete and human opportunist *A. niger* JSC, isolated from the International Space Station (ISS; Romsdahl et al. 2018). The comparison with a terrestrial reference strain allowed to get an insight into the adaptive mechanisms to space travel conditions at the protein level (total protein). Protein differential abundance especially affected proteins involved in the stress resistance, starvation response, and nutrient acquisition (Romsdahl et al. 2018). Through TMT approach, the existence of a distinct strain of *A. niger* onboard the ISS also showing a higher melanin content than its ground counterparts was revealed.

Label-free methods for quantitative proteomics have been also applied in fungi aiming at the discovery of processes and molecules governing the stress response. Although isotope labeling approaches are more accurate than label-free methods, the latter are relatively inexpensive and unaffected by labeling process-related technical bias (Bianco and Perrotta 2015; Huang et al. 2015). Moreover, there is no limit to the number of samples that can be compared. In a typical label-free experiment, samples are individually analyzed and compared after independent analyses, usually using spectral counting or peak intensity measurement, which approximate protein abundance (Wasinger et al. 2013). Spectral counting is a relatively simple procedure based on the foundation that a higher number of peptides can be identified from a more abundant protein (Washburn et al. 2001). The number of mass spectra recorded for a peptide in a sample thereby linearly correlates with its molar amount; therefore, the sum of spectral counts for peptides associated with a particular protein can be used to estimate protein amount (Bianco and Perrotta 2015).

Label-free quantitative proteomics based on sequential window acquisition of all theoretical fragment ion spectra (SWATH)-MS is currently the best known workflow for label-free proteomics. Quantitative proteomics using SWATH-MS provides good reproducibility, accuracy, and precision in quantification of proteins, and is suitable for detecting negligible protein differentiation (less than twofold) (Li et al. 2017). SWATH was reported for the study of the mechanism of formation and differentiation of sclerotia, a dormant form displayed by several fungal species in relation to extreme environmental conditions. Proteomes of both sclerotia and hyphae at different developmental stages were analyzed in the medicinal fungus *Polyporus umbellatus* (Li et al. 2017). A total of 1234 proteins were identified and quantified by LC-MS using SWATH in combination with DDA mode. The results of protein differential abundance analysis revealed the role of oxidative stress in triggering sclerotia formation from hypha and further highlighted the importance of antioxidant defensing for sclerotia growth. SWATH was used to achieve a comprehensive protein identification and regulation in the mycotoxin producer *Penicillium verrucosum* at different time points of the growth curve (Nöbauer et al. 2017). Notably, as no protein information was previously available for the organism, an "ab initio" translated database from the sequenced genome of *P. verrucosum* was applied for the identification of 3488 proteins. A SWATH setup was also applied for comparative quantitative analysis of *A. fumigatus* proteome aiming at the characterization of conidia's adaptation at different culture ages (Anjo et al. 2017). Information-dependent acquisition (IDA) was used in combination with SWATH to generate peptide fragmentation spectra for library creation. A time course evaluation of the proteome revealed how the metabolic state of conidia switched from very active—at the beginning of the cultivation—to dormant, as the availability of nutrients decreased. The increase levels of hydrolytic enzymes toward the end of cultivation in a 30-day-old conidia cell went hand in hand with cell autolysis.

Nano ultra-performance liquid chromatography coupled with tandem mass spectrometry (UPLC-MS^E) is another non-labeling approach for the quantitative analysis of proteins. UPLC-MS^E is a relatively new MS/MS technique for the simultaneous acquisition of precursor ion information and fragment ion data at low and high collision energy in one analytical run, and has been thereby shown to improve protein and proteome coverage compared to the conventional LC-MS/MS approach (Murad et al. 2011). In this technique, abundance values of proteins are assessed based on the average intensity value of the top three ionized tryptic peptides from a constitutive protein, detected in all replicates and with a low variance coefficient (Rodrigues et al. 2016). This internal standard protein is thereby used to normalize the protein expression levels in the samples. Informatics tools are additionally used for proper quantitative comparisons (Parente-Rocha et al. 2015). UPLC-MS^E has served a thorough characterization of *Paracoccidioides* sp. response to environmental stresses trying to get an insight into the mechanisms at the base of pathogenicity. De Silva Rodrigues et al. (2016) applied UPLC-MS^E to identify differentially expressed proteins during osmotic shock in *P. lutzii*. The results of this study suggest that the response to osmotic stress could help the fungus to cope with the host environment during dissemination to organs and tissues. Proteomic responses to carbon starvation were additionally investigated in *Paracoccidioides* sp. to simulate adaptation to host during infection. A total of 421 proteins were found to be differentially regulated by 1.5-fold change, suggesting a metabolism reprogramming during carbon starvation in favor of gluconeogenesis and ethanol production (Lima Pde et al. 2014). Further proteomics investigations of *P. lutzii* and *P. brasiliensis* were carried out to shed light on the survival mechanisms of human pathogenic fungi toward nitrosative stress, as the main defensive strategies applied by immune cells during infection (Parente et al. 2015). Among the 66 downregulated proteins, proteins related to carbohydrate energy conservation and the mitochondrial electron transport chain were found. Proteins showing increased expression (i.e., 76) during nitrosative stress induced by adding GSNO (S-nitrosoglutathione) included those related to lipids and branched-chain amino acid metabolism as well as to the oxidative stress response, such as superoxide dismutase (SOD) and cytochrome c peroxidase (CCP). Further aspects of the infection were examined in *P. brasiliensis* (Parente-Rocha et al. 2015). The proteomic response of the dimorphic fungus to macrophage internalization was analyzed and, similarly to what is observed for nitrosative stress defense, the results of protein differential analysis revealed proteins involved in the oxidative stress response—i.e., thioredoxins along with SOD and CCP—to be upregulated. The rest of the 139 proteins found to be positively regulated out of the 7845 peptides identified were related to amino acid catabolism, cell rescue, defense, and virulence, suggesting that *P. brasiliensis* adapts to the macrophage milieu by reprogramming its metabolism to produce glucose and inhibiting protein synthesis.

The rapid advancement in bottom-up MS technologies has paved the way for faster analyses of larger number of proteins resulting in high amounts of data regarding protein IDs and quantitative peptide information. By relying on protein digestion, bottom-up techniques, however, can leave behind important information regarding posttranslational modifications (PTMs) or sequence variants like those giving rise to protein isoforms. Moreover, alterations or sequence variations may occur on peptides, causing their relation to one another to be lost following digestion (Catherman et al. 2014). Top-down approaches were developed to compensate for the limits of bottom-up proteomics, by using MS to characterize intact proteins whose intact and fragment ion masses are both measured. Top-down is a promising tool for the detection of proteoforms originating from PTMs, alternative splicing, or genetic variations. Nevertheless, the technical effort of proteome-wide investigation at the intact protein level has caused top-down proteomics to lag behind bottom-up in terms of proteome coverage, sensitivity, and throughput (Catherman et al. 2014). Similarly, software for the interpretation of top-down has also been slower to develop and the protocols for accurate quantification are still under development (Collier and Muddiman 2012). The application of top-down proteomics in fungal analysis is therefore still very limited. Up to now, intact protein studies were carried out in filamentous fungi of agricultural and medical interest. In the work from Collier et al. (2008), whole-organism SILAC labeling was applied for qualitative and quantitative top-down proteomics in *A. flavus* to assess the biological effects of several growth parameters and antifungal agents.

Each of the methods here mentioned has its own merits and drawbacks; the choice of the optimal method should therefore be evaluated in accordance with the experimental design and downstream needs. Being microcolonial fungi and black yeasts at the same time emerging model organisms and challenging samples for biochemical analysis, label-based and label-free quantitative proteomics are hitherto far from being fully exploited in black fungi proteomic research, as revealed by the limited number of papers published in this field up to now.

12.2.3.3   Use of MALDI-TOF MS for the Identification of Fungal Species

Based on the UV irradiation of the sample while embedded in a matrix, MALDI-TOF MS (matrix-assisted laser desorption/ionization time-of-flight mass spectrometry) has several applications, ranging from the analysis of intact proteins to diagnostics. Over the last years, multiple studies have reported MALDI as a diagnostic technique in medical centers worldwide for the identification of bacteria, yeasts, and fungi (Panda et al. 2015; Paul et al. 2017, 2018; Putignani et al. 2011). The discrimination between microorganisms is based on highly specific peptide profiling (i.e., ribosomal fingerprinting) and has proven great precision, low costs, and speed of analysis (turnaround time of approximately 10 min) (Özhak-Baysan et al. 2015). MALDI has thus become a valid alternative to other identification techniques such as sequencing methodologies targeting ribosomal DNA (rDNA) genes (e.g., internal transcribed spacers, ITS), especially for the analysis of large numbers of isolates. Putignani et al. (2011) performed MALDI-TOF proteome phenotype profiling on more than 300 clinically relevant samples including several *Candida* species, by achieving a high analytical performance of yeasts and yeast-like identifications (IDs). Along with *Candida*, members of many fungal genera such as

Aspergillus, Fusarium, Penicillium, or *Trichoderma* have been successfully identified by MALDI-TOF MS (Chalupová et al. 2014; Del Chierico et al. 2012; Ranque et al. 2014).

MS-based species identification is especially relevant in black yeasts and black fungi, in view of the lack of reproductive cells, which hinders definitive identification through classical morphological, biochemical, and physiological methods (Özhak-Baysan et al. 2015). Nevertheless, a small number of studies have hitherto reported a MALDI-based identification of melanotic fungi, mostly due to the lack of a standardized protocol for protein extraction and a partial database availability (Borman et al. 2017; Kondori et al. 2015; Özhak-Baysan et al. 2015). More specifically, signal suppression was observed in darkly pigmented fungi due to the presence of melanin, resulting in poor fingerprint mass spectra containing few peaks of low relative abundance yield (Buskirk et al. 2011). The majority of the studies focus on the identification of black yeasts of the genus *Exophiala*, which includes agents of severe potentially fatal infections in both immunocompromised and immunocompetent individuals (Chalkias et al. 2014; Jeong et al. 2010; Kusenbach et al. 1992). To address the problems regarding the sample preparation and the lack of standardization, recently Paul et al. (2018) embraced an optimization work, which encompassed protein extraction as well as the creation of an in-house database for the rapid identification of melanized fungal isolates. Earlier, a broad in-house library was generated by Becker et al. (2014) by comparing MALDI-TOF MS with the classical identification methods, however only for filamentous fungi.

Based on these premises, MALDI-TOF MS will increasingly be applied in the future for the rapid diagnostic and detection of emerging clinically important melanized fungi.

12.3   Bioinformatics and Functional Analyses of Fungal Proteins

In the past, the analytical methodologies recruited for proteomics were based on conventional tools for protein characterization. The technical challenges comprised the overall coverage of physicochemical properties for a few thousands of proteins. Next-generation sequencing technology has represented the true milestones in science and paved the way for an upcoming challenge. In parallel proteomics involved interdisciplinary tools to evaluate the genes at the protein expression level.

Currently, proteomic analyses of extremophilic and extremotolerant fungi are challenged by various aspects, which have much to do with protein-related features such as protein identification, quantification, posttranslational modifications (PTMs), structure and function, and exploration of interactions and networks. The ultimate progress in fungal proteomics will be, however to achieve both protein identification and the understanding of protein function in the cell physiology.

In the last decade, we witnessed the generation of an extraordinary amount of proteomics and genomics data; however, the functional annotation leaves us waiting. The great ability of different proteomic separation tools such as gel-based and gel-free methodologies combined with high-resolution MS features makes data analysis using bioinformatics approaches and novel mathematical algorithms the major bottleneck for mass interpretation of obtained information.

The data generated in the field of black fungi and black yeasts are now increasing (Marzban et al. 2013). The genomic sequence data is progressively completing and the odyssey of homology searches in several available proteomic databases like UniProt (http://www.uniprot.org), Munich Information Center of Protein Sequences (MIPS) (http://www.helmholtz-muenchen.de/en/ibis) for plant pathogenic fungi, the Sanger Institute Fungal Sequencing (http://sanger.ac.uk/Projects/Fungi/), Central *Aspergillus* Data Repository (CADRE, http://www.cadre-genomes.org.uk/), Fungal Genome (http://fungalgenomes.org/), Fungal BLAST (https://www.yeastgenome.org/blast-fungal), FungiDB (http://fungidb.org/fungidb/), MycoBank (http://www.mycobank.org/), Fungal Annotation Project (https://www.uniprot.org/program/Fungi/), and Q Bank (http://www.q-bank.eu/fungi/) is becoming an exciting task. The global databases are gaining more efficiency for the homology searches (e.g., ExPASy (http://world-2dpage.expasy.org/repository/), PRoteomics IDEntifications database (PRIDE, www.ebi.ac.uk/pride/), MASCOT (http://www.matrixscience.com), and UNITE (https://unite.ut.ee/)). Additionally, different advanced platforms make the identification of fungal secretomics possible based on prediction of signal peptides, e.g., comparative fungal genomics platform (CFGP, http://cfgp.riceblast.snu.ac.kr/main.php) and SignalP 4.1 (http://www.cbs.dtu.dk/services/SignalP/). In both cases the correct genomic sequence data is the prerequisite for the prediction of secretory protein. Whereas CFGP encompasses secretome, transcription factor, and mitochondrial genome databases, among others, SignalP aids the prediction of signal peptide cleavage sites. Furthermore, the Fungal Secretome KnowledgeBase (FunSecKB and FunSecKB2 (Meinken et al. 2014); http://bioinformatics.ysu.edu/secretomes/fungi.php) and the Fungal Secretome Database (FSD; http://fsd.snu.ac.kr/) have been established to collect, manage, and use the data. FunSecKB collects secretomes from all available fungal protein data in the NCBI RefSeq database. The FSD is at present time the most accurate platform for putative secretory proteins, since it uses different databases like SigCleave, SigPred, and RPSP to screen those proteins not considered positive by SignalP (Choi 2010). By increasing volume of data, the integration of more platforms using efficient algorithms must be developed to minimize the false-positive and false-negative results.

In the last years, a number of tools for the localization of intracellular proteins in fungi have been developed (e.g., WoLF PSORT, MultiLoc2, SherLoc2, MSLoc-DT, SCLpred) (Blum et al. 2009; Briesemeister et al. 2010; Horton et al. 2007; Mooney et al. 2011; Zhang et al. 2014). This aspect holds particular importance in black fungi where, as described earlier, the characterization of the subcellular proteome can be especially cumbersome. Along with pre-fractionation techniques, protein localization can be effectively demonstrated by means of fluorescent microscopy.

However, this requires fluorescent dyes, recombinant reporter proteins, or protein-specific antibodies, which might not be available for proteins not yet characterized as well as for unknown proteins. Bioinformatics tools can instead provide predictions of subcellular localization for a large number and a great variety of proteins, solely based on the amino acidic sequence. As protein localization and function are somewhat related, software-based predictions can additionally unravel protein potential functions and roles in biological processes.

In the example of black microcolonial fungi and black yeasts, however, we must still wait for the completion of annotation of genomics data, which will simplify the functional exploration of these still unknown living entities. Protein analysis is to a certain extent possible even when RNA-seq-based genome annotations are unavailable; however it requires resorting to a number of bioinformatics tools for the generation, in primis, of a database of predicted proteins. Currently AUGUSTUS is the best known tool for the "ab initio" translation of genome sequences based on training databases (i.e., annotated genome sequences from closely related species) (Hoff and Stanke 2013). Homology searches and PFAM (http://pfam.xfam.org/) analysis are then performed in order to obtain information about the biological function of the proteins (i.e., PFAM clan) (Nöbauer et al. 2017). By allowing the identification of proteins in yet not annotated organisms, "ab initio" translated databases ease the investigation of emerging model organisms and non-model-model organisms, thereby serving as basis for further and more detailed investigations.

The major challenge, however, will be the exploration of interactomics, biological networks, and PTMs by the means of epiproteomics, which are the key process regulators of extremotolerance and resistance.

12.4　Conclusions

The discovery of organisms like extremophilic and extremotolerant fungi and in particular black fungi influenced our knowledge about pathways and strategies that provide a biological systemic stubbornness against climate changes or sudden catastrophic environmental events. Primary efforts involved genomics and proteomics and more recently metabolomics and aimed at getting a more clear insight into the cell biology system as well as to obtain a comprehensive overview of the cellular processes under stress. However, the first ever step of the workflow, the sample preparation, has proven to be the real bottleneck for all of the analytical downstream approaches at present time. The standardization of extraction protocols will therefore be a foremost test for all future studies, in order for the extraction of fungal macromolecules and metabolites to be performed routinely or even be tailored for MS or DIGE, as it is already for plant or animal tissue.

The successful analyses of proteomic samples from extremophilic and extremotolerant fungal species carried out up to date have relied on the optimization of sample preparation workflows as well as on the availability of progressively improving mass spectrometry technologies and labeling protocols. Such approaches showed to be adequate to deliver results and to characterize novel proteins or to explore already known pathways further. Despite the growing amounts of fungal functional data, genome annotation remains the main hindrance to the understanding of biological pathways leading to stress resistance. The application of pipelines for gene prediction, functional annotation, and comparative analysis as well as RNA-seq-based annotations to the study of extremophilic and extremotolerant fungi will therefore make the final breakthrough in the OMICS of these emerging model organisms.

Additional optimization work will be necessary in the field of fungal secretomics, which has hitherto proven to be the most challenging area in the investigation of adaptive capabilities of fungi in general and which still is in its early stages.

12.5 Future Perspective

A closer look to the habitats of the extremophilic fungi reveals a rare and desirable bioeconomy. The growing interest in the investigation of extremophilic fungi seems to have its roots in our time and in the challenges for our living space. The understanding of the evolutionary-based strategies of fungi to cope with harsh living conditions is of special interest for us.

We are currently witnessing climate change and global warming is undeniably underway. Both these aspects have had an influence on our own habitat, also having devastating effects on crop plants, which are essential for our existence. The urgency to counteract this state of things has sparked interest in alternative sources of energy, recycling of plastic materials, and pollution prevention, among others, consistently with the concept of sustainable living. The development of new ideas and the search for new products and biotechnological applications have at this point gained added value.

Fungi interactomics will therefore be in future the main challenge on the way to a full understanding of the system biology in a multidimensional network of genome, transcriptome, secretome, proteome, and metabolome. It seems clear that a cooperative functioning of all these levels motorizes the machinery of resistance to extreme environmental condition, while the organisms at the same time try to utilize and recycle the nutrients available as dead organic material. The future of extremophilic interactomics research looks mesmerizing and opens windows to technologies and strategies ranging from climate-resistant organisms and products on Earth to rapid transformation of Martian environment into a living space for humans.

Acknowledgements The authors would like to acknowledge the Austrian Science Fund (FWF-der Wissenschaftsfonds, Hertha-Firnberg Project T872-B22) and the BOKU Equipment GmbH as the founder of the VIBT-Extremophile Center for the technical and financial support of our research, whose outcomes widely contributed to this book chapter.

We also thank FWF for the project FWF-P24206, which allowed the initiation and the pioneering proteomics work on extremophilic black fungi.

The authors apologize to all researchers whose work could not be cited in this book chapter due to limits in manuscript dimensions.

References

Abyzov SS, Mitskevicha IN, Poglazovaa MN, Narciss I, Lipenkov VY, Bobiflc NE, Koudryashovc BB, Pashkevichc VM (1998) Long-term conservation of viable microorganisms in ice sheet of Central Antarctica. SPIE Conf. Instruments, Methods Mission. Astrobiology 3441:75–84

Adav SS, Li AA, Manavalan A, Punt P, Sze SK (2010) Quantitative iTRAQ secretome analysis of *Aspergillus niger* reveals novel hydrolytic enzymes. J Proteome Res 9:3932–3940

Adav SS, Chao LT, Sze SK (2012) Quantitative secretome of Trichoderma reesei strains reveals enzymatic composition for lignocellulosic biomass degradation. Mol Cell Proteomics 65:1–46

Aebersold R, Mann M (2003) Mass spectrometry-based proteomics. Nature 422:198–207

Agustinho DP, Miller LC, Li LX, Doering TL (2018) Peeling the onion: the outer layers of *Cryptococcus neoformans*. Mem Inst Oswaldo Cruz 113:1–8

Albuquerque PC, Nakayasu ES, Rodrigues ML, Frases S, Casadevall A, Zancope-Oliveira RM, Almeida IC, Nosanchuk JD (2008) Vesicular transport in *Histoplasma capsulatum*: an effective mechanism for trans-cell wall transfer of proteins and lipids in ascomycetes. Cell Microbiol 10:1695–1710

Amin S, Rastogi RP, Sonani RR, Ray A, Sharma R, Madamwar D (2018) Bioproduction and characterization of extracellular melanin-like pigment from industrially polluted metagenomic library equipped *Escherichia coli*. Sci Total Environ 635:323–332

Anjo SI, Figueiredo F, Fernandes R, Manadas B, Oliveira M (2017) A proteomic and ultrastructural characterization of *Aspergillus fumigatus*' conidia adaptation at different culture ages. J Proteomics 161:47–56

Armengaud J, Trapp J, Pible O, Geffard O, Chaumot A, Hartmann EM (2014) Non-model organisms, a species endangered by proteogenomics. J Proteomics 105:5–18

de Arruda Grossklaus D, Bailão AM, Vieira Rezende TC, Borges CL, de Oliveira MA, Parente JA, de Almeida Soares CM (2013) Response to oxidative stress in Paracoccidioides yeast cells as determined by proteomic analysis. Microbes Infect 15:347–364

Barreiro C, García-estrada C, Martín JF (1991) Proteomics methodology applied to the analysis of filamentous fungi - new trends for an impressive diverse group of organisms. In: Prasain J, Harn G (eds) Tandem mass spectrometry- applications and principles. Intech-Open Access Publisher, Rijeka, Croatia, pp 127–160

Barreiro C, Martín JF, García-Estrada C (2012) Proteomics shows new faces for the old penicillin producer *Penicillium chrysogenum*. J Biomed Biotechnol 2012:105109

Becker PT, De Bel A, Martiny D, Ranque S, Piarroux R, Cassagne C, Detandt M, Hendrickx M (2014) Identification of filamentous fungi isolates by MALDI-TOF mass spectrometry: clinical evaluation of an extended reference spectra library. Med Mycol 52:826–834

Bendtsen JD, Nielsen H, von Heijne G, Brunak S (2004) Improved prediction of signal peptides: SignalP 3.0. J Mol Biol 340:783–795

Benndorf D, Müller A, Bock K, Manuwald O, Herbarth O, Von Bergen M (2008) Identification of spore allergens from the indoor mould *Aspergillus versicolor*. Allergy Eur J Allergy Clin Immunol 63:454–460

Bhadauria V, Zhao W, Wang L, Zhang Y, Liu J, Yang J, Kong LA, Peng YL (2007) Advances in fungal proteomics. Microbiol Res 162:193–200

Bhadauria V, Banniza S, Wang L-X, Wei Y-D, Peng Y-L (2009) Proteomic studies of phytopathogenic fungi, oomycetes and their interactions with hosts. Eur J Plant Pathol 126:81–95

Bianco L, Perrotta G (2015) Methodologies and perspectives of proteomics applied to filamentous fungi: from sample preparation to secretome analysis. Int J Mol Sci 16:5803–5829

Blasi B, Tafer H, Tesei D, Sterflinger K (2015) From glacier to sauna: RNA-seq of the human pathogen black fungus *Exophiala dermatitidis* under varying temperature conditions exhibits common and novel fungal response. PLoS One 10:e0127103

Blasi B, Poyntner C, Rudavsky T, Prenafeta-Boldú FX, de Hoog GS, Tafer H, Sterflinger K (2016) Pathogenic yet environmentally friendly? Black fungal candidates for bioremediation of pollutants. Geomicrobiol J 33:308–317

Blasi B, Tafer H, Kustor C, Poyntner C, Lopandic K, Sterflinger K (2017) Genomic and transcriptomic analysis of the toluene degrading black yeast *Cladophialophora immunda*. Sci Rep 7:1–13

Blum T, Briesemeister S, Kohlbacher O (2009) MultiLoc2: integrating phylogeny and Gene Ontology terms improves subcellular protein localization prediction. BMC Bioinformatics 10:274

Blumer-Schuette SE, Kataeva I, Westpheling J, Adams MW, Kelly RM (2008) Extremely thermophilic microorganisms for biomass conversion: status and prospects. Curr Opin Biotechnol 19:210–217

Borman AM, Fraser M, Szekely A, Larcombe DE, Johnson EM (2017) Rapid identification of clinically relevant members of the genus exophiala by matrix-assisted laser desorption ionization-time of flight mass spectrometry and description of two novel species, *Exophiala campbellii* and *Exophiala lavatrina*. J Clin Microbiol 55:1162–1176

Branda E, Turchetti B, Diolaiuti G, Pecci M, Smiraglia C, Buzzini P (2010) Yeast and yeast-like diversity in the southernmost glacier of Europe (Calderone Glacier, Apennines, Italy). FEMS Microbiol Ecol 72:354–369

Breitenbach R, Silbernagl D, Toepel J, Sturm H, Broughton WJ, Sassaki GL, Gorbushina AA (2018) Corrosive extracellular polysaccharides of the rock-inhabiting model fungus *Knufia petricola*. Extremophiles 22:165–175

Briesemeister S, Rahnenführer J, Kohlbacher O (2010) Going from where to why—interpretable prediction of protein subcellular localization. Bioinformatics 26:1232–1238

Brown SM, Campbell LT, Lodge JK (2007) *Cryptococcus neoformans*, a fungus under stress. Curr Opin Microbiol 10:320–325

Burg DW, Lauro FM, Williams TJ, Raftery MJ, Guilhaus M, Cavicchioli R (2010) Analyzing the hydrophobic proteome of the Antarctic archaeon *Methanococcoides burtonii* using differential solubility fractionation. J Proteome Res 9:664–676

Burg D, Ng C, Ting L, Cavicchioli R (2011) Proteomics of extremophiles. Environ Microbiol 13:1934–1955

Buskirk AD, Hettick JM, Chipinda I, Law BF, Siegel PD, Slaven JE, Green BJ, Beezhold DH (2011) Fungal pigments inhibit the matrix-assisted laser desorption/ionization time-of-flight mass spectrometry analysis of darkly pigmented fungi. Anal Biochem 411:122–128

Cagas SE, Jain MR, Li H, Perlin DS (2011) Profiling the *Aspergillus fumigatus* proteome in response to caspofungin. Antimicrob Agents Chemother 55:146–154

Catherman AD, Skinner OS, Kelleher NL (2014) Top Down proteomics: facts and perspectives. Biochem Biophys Res Commun 445:683–693

Cavicchioli R (2002) Extremophiles and the Search for Extraterrestrial Life. Astrobiology 2:281–292

Chahrour O, Cobice D, Malone J (2015) Stable isotope labelling methods in mass spectrometry-based quantitative proteomics. J Pharm Biomed Anal 113:2–20

Chalkias S, Alonso CD, Levine JD, Wong MT (2014) Emerging pathogen in immunocompromised hosts: *Exophiala dermatitidis* mycosis in graft-versus-host disease. Transpl Infect Dis 16:616–620

Chalupová J, Raus M, Sedlářová M, Šebela M, Sebela M (2014) Identification of fungal microorganisms by MALDI-TOF mass spectrometry. Biotechnol Adv 32:230–241

Champer J, Diaz-Arevalo D, Champer M, Hong TB, Wong M, Shannahoff M, Ito JI, Clemons KV, Stevens DA, Kalkum M (2012) Protein targets for broad-spectrum mycosis vaccines: quantitative proteomic analysis of *Aspergillus* and *Coccidioides* and comparisons with other fungal pathogens. Ann N Y Acad Sci 1273:44–51

Champer J, Ito J, Clemons K, Stevens D, Kalkum M (2016) Proteomic analysis of pathogenic fungi reveals highly expressed conserved cell wall proteins. J Fungi 2:6

Choi J, Park J, Kim D, Jung K, Kang S, Lee Y-H (2010) Fungal Secretome Database: Integrated platform for annotation of fungal secretomes. BMC Genomics 11:105

Collier TS, Muddiman DC (2012) Analytical strategies for the global quantification of intact proteins. Amino Acids 43:1109–1117

Collier TS, Hawkridge AM, Georgianna DR, Payne GA, Muddiman DC (2008) Top-down identification and quantification of stable isotope labeled proteins from *Aspergillus* flavus using online nano-flow reversed-phase liquid chromatography coupled to a LTQ-FTICR mass spectrometer. Anal Chem 80:4994–5001

Cordero RJB, Vij R, Casadevall A (2017) Microbial melanins for radioprotection and bioremediation. J Microbial Biotechnol 10:1186–1190

Cordero RJB, Robert V, Cardinali G, Arinze ES, Thon SM, Casadevall A (2018) Impact of yeast pigmentation on heat capture and latitudinal distribution. Curr Biol 28:2657–2664.e3

Coumans JVF, Moens PDJ, Poljak A, Al-Jaaidi S, Pereg L, Raftery MJ (2010) Plant-extract-induced changes in the proteome of the soil-borne pathogenic fungus *Thielaviopsis basicola*. Proteomics 10:1573–1591

Crichton PG, Harding M, Ruprecht JJ, Lee Y, Kunji ERS (2013) Lipid, detergent, and coomassie blue G-250 affect the migration of small membrane proteins in blue native gels: mitochondrial carriers migrate as monomers not dimers. J Biol Chem 288:22163–22173

Dadachova E, Casadevall A (2008) Ionizing radiation: how fungi cope, adapt, and exploit with the help of melanin. Curr Opin Microbiol 11:525–531

Dadachova E, Bryan RA, Huang X, Moadel T, Schweitzer AD, Aisen P, Nosanchuk JD, Casadevall A (2007) Ionizing radiation changes the electronic properties of melanin and enhances the growth of melanized fungi. PLoS One 2(5):e457

De Toro J, Herschlik L, Waldner C, Mongini C (2015) Emerging roles of exosomes in normal and pathological conditions: new insights for diagnosis and therapeutic applications. Front Immunol 6:1–12

Del Chierico F, Masotti A, Onori M, Fiscarelli E, Mancinelli L, Ricciotti G, Alghisi F, Dimiziani L, Manetti C, Urbani A, Muraca M, Putignani L (2012) MALDI-TOF MS proteomic phenotyping of filamentous and other fungi from clinical origin. J Proteomics 75:3314–3330

Dodds PN, Rafiqi M, Gan PHP, Hardham AR, Jones DA, Ellis JG (2009) Effectors of biotrophic fungi and oomycetes: pathogenicity factors and triggers of host resistance. New Phytol 183:993–1000

Eisenman HC, Casadevall A (2012) Synthesis and assembly of fungal melanin. Appl Microbiol Biotechnol 93:931–940

Eisenman HC, Chow SK, Tsé KK, McClelland EE, Casadevall A (2011) The effect of L-DOPA on *Cryptococcus neoformans* growth and gene expression. Virulence 2:329–336

Ellen AF, Albers SV, Huibers W, Pitcher A, Hobel CFV, Schwarz H, Folea M, Schouten S, Boekema EJ, Poolman B, Driessen AJM (2009) Proteomic analysis of secreted membrane vesicles of archaeal *Sulfolobus* species reveals the presence of endosome sorting complex components. Extremophiles 13:67–79

Enjalbert B, Whiteway M (2003) Stress-induced gene expression in *Candida albicans*: absence of a general stress response. Mol Biol Cell 14:1460–1467

Epova E, Guseva M, Kovalyov L, Isakova E, Deryabina Y, Belyakova A, Zylkova M, Shevelev A (2012) Identification of proteins involved in pH adaptation in extremophile yeast *Yarrowia lipolytica*. In: Heazlewood J (ed) Proteomic applications in biology. In Tech Open, Rijeka, Croatia, pp 209–224

Evilia C (2018) Understanding protein adaptations can help us solve real problems. Semin Cell Dev Biol:9–10

Fernández-Acero FJ, Jorge I, Calvo E, Vallejo I, Carbú M, Camafeita E, López JA, Cantoral JM, Jorrín J (2006) Two-dimensional electrophoresis protein profile of the phytopathogenic fungus Botrytis cinerea. Proteomics 6:S88–S96

Fernández RG, Redondo I, Jorrin-Novo JV (2014) Making a Protein Extract from Plant Pathogenic Fungi for Gel- and LC-Based Proteomics. In: Jorrin-Novo J., Komatsu S, Weckwerth W, Wienkoop S (eds) Plant Proteomics. Methods in Molecular Biology (Methods and Protocols), vol 1072. Humana Press, Totowa, NJ

Flieger K, Knabe N, Toepel J (2018) Development of an improved carotenoid extraction method and characterisation of the carotenoid composition under oxidative and cold stress in the rock inhabiting fungus *Knufia petricola* A95. J Fungi 4:1–10

Fragner D, Zomorrodi M, Kües U, Majcherczyk A (2009) Optimized protocol for the 2-DE of extracellular proteins from higher basidiomycetes inhabiting lignocellulose. Electrophoresis 30:2431–2441

Furi I, Momen-Heravi F, Szabo G (2017) Extracellular vesicle isolation: present and future. Ann Transl Med 5:263–263

Garcia-Rivera J, Tucker S, Feldmesser M, Williamson P, Casadevall A (2005) Laccase expression in murine pulmonary *Cryptococcus neoformans* infection. Infect Immun 73:3124–3127

Geisow MJ (1998) Proteomics: one small step for a digital computer, one giant leap for humankind. Nat Biotechnol 16:206

Gillet LC, Navarro P, Tate S, Röst H, Selevsek N, Reiter L, Bonner R, Aebersold R (2012) Targeted data extraction of the MS/MS spectra generated by data-independent acquisition: a new concept for consistent and accurate proteome analysis. Mol Cell Proteomics 11:O111.016717

González-Fernández R, Redondo I, Jorrin-Novo JV (2014) Making a protein extract from plant pathogenic fungi for Gel- and LC-based proteomics. In: Jorrin-Novo JV, Komatsu S, Weckwerth W, Wienkoop S (eds) Plant proteomics, Methods and Protocols. Humana Press, New York, NY, pp 93–109

Gostinčar C, Muggia L, Grube M (2012) Polyextremotolerant black fungi: oligotrophism, adaptive potential, and a link to lichen symbioses. Front Microbiol 3:390

Gow NAR, Latge J, Munro CA (2017) The fungal cell wall: structure, biosynthesis and function. J Microbiol Spectr 5:1–25

Grinyer J, McKay M, Herbert B, Nevalainen H (2004) Fungal proteomics: mapping the mitochodrial proteins of a *Trichoderma harzianum* strain applied for biological control. Curr Genet 45:170–175

Guimarães AJ, Nakayasu ES, Sobreira TJP, Cordero RJB, Nimrichter L, Almeida IC, Nosanchuk JD (2011) *Histoplasma capsulatum* heat-shock 60 orchestrates the adaptation of the fungus to temperature stress. PLoS One 6:e14660

Gunde-Cimerman N, Zalar P, de Hoog GS, Plemenitaš A (2000) Hypersaline waters in saltern - natural ecological niches for halophilic black yeasts. FEMS Microbiol Ecol 32:235–240

Gunde-Cimerman N, Sonjak S, Zalar P, Frisvad JCC, Diderichsen B, Plemenitaš A (2003) Extremophilic fungi in arctic ice: a relationship between adaptation to low temperature and water activity. Phys Chem Earth 28:1273–1278

Hoff KJ, Stanke M (2013) WebAUGUSTUS—a web service for training AUGUSTUS and predicting genes in eukaryotes. Nucleic Acids Res 41:123–128

de Hoog GS, Grube M (2008) Black fungal extremes. Stud Mycol 61

Horton P, Park KJ, Obayashi T, Fujita N, Harada H, Adams-Collier CJ, Nakai K (2007) WoLF PSORT: protein localization predictor. Nucleic Acids Res 35:W585–W587

Huang Q, Yang L, Luo J, Guo L, Wang Z, Yang X, Jin W, Fang Y, Ye J, Shan B, Zhang Y (2015) SWATH enables precise label-free quantification on proteome scale. Proteomics 15:1215–1223

Isola D, Marzban G, Selbmann L, Onofri S, Laimer M, Sterflinger K (2011) Sample preparation and 2-DE procedure for protein expression profiling of black microcolonial fungi. Fungal Biol 115:971–977

Jacobson ES (2000) Pathogenic roles for fungal melanins. Clin Microbiol Rev 13:708–717

Jami M-S, Barreiro C, García-Estrada C, Martín J-F (2010a) Proteome analysis of the penicillin producer *Penicillium chrysogenum*: characterization of protein changes during the industrial strain improvement. Mol Cell Proteomics 9:1182–1198

Jami M-S, García-Estrada C, Barreiro C, Cuadrado A-A, Salehi-Najafabadi Z, Martín J-F (2010b) The *Penicillium chrysogenum* extracellular proteome. Conversion from a food-rotting strain to a versatile cell factory for white biotechnology. Mol Cell Proteomics 9:2729–2744

Jeong ES, Shin JH, Shin MG, Suh SP, Ryang DW (2010) Fungemia due to *Exophiala dermatitidis*. Korean J Clin Microbiol 13:135

Joffe LS, Nimrichter L, Rodrigues ML, Del Poeta M (2016) Potential roles of fungal extracellular vesicles during infection. mSphere 1:e00099-16

Jorrín-Novo JV, Maldonado AM, Echevarría-Zomeño S, Valledor L, Castillejo MA, Curto M, Valero J, Sghaier B, Donoso G, Redondo I (2009) Plant proteomics update (2007-2008): Second-generation proteomic techniques, an appropriate experimental design, and data analysis to fulfill MIAPE standards, increase plant proteome coverage and expand biological knowledge. J Proteomics 72:285–314

Kapp K, Schrempf S, Lemberg MK, Dobberstein B (2009) Post-targeting functions of signal peptides. In: Madame Curie Bioscience Database [Internet]. Landes Bioscience, Austin, TX, pp 2000–2013

Karkowska-Kuleta J, Kozik A (2015) Cell wall proteome of pathogenic fungi. Acta Biochim Pol 62:339–351

Kim Y, Nandakumar MP, Marten MR (2007) Proteomics of filamentous fungi. Trends Biotechnol 25:395–400

Klis FM, de Jong M, Brul S, de Groot PWJ (2007) Extraction of cell surface-associated proteins from living yeast cells. Yeast 24:253–258

Kniemeyer O (2011) Proteomics of eukaryotic microorganisms: the medically and biotechnologically important fungal genus *Aspergillus*. Proteomics 11:3232–3243

Kohl M, Megger D a, Trippler M, Meckel H, Ahrens M, Bracht T, Weber F, Hoffmann AC, Baba H a, Sitek B, Schlaak JF, Meyer HE, Stephan C, Eisenacher M (2014) A practical data processing workflow for multi-OMICS projects. Biochim Biophys Acta Proteins Proteomics 1844:52–62

Kondo T, Hirohashi S (2007) Application of highly sensitive fluorescent dyes (Cydye dige fluor saturation dyes) to laser microdissection and two-dimensional difference gel electrophoresis (2d-Dige) for cancer proteomics. Nat Protoc 1:2940–2956

Kondori N, Erhard M, Welinder-Olsson C, Groenewald M, Verkley G, Moore ERB (2015) Analyses of black fungi by matrix-assisted laser desorption/ionization time-of-flight mass spectrometry (MALDI-TOF MS): species-level identification of clinical isolates of *Exophiala dermatitidis*. FEMS Microbiol Lett 362:1–6

Kratochwill K, Bender TO, Lichtenauer AM, Herzog R, Tarantino S, Bialas K, Jörres A, Aufricht C (2015) Cross-omics comparison of stress responses in mesothelial cells exposed to heat- versus filter-sterilized peritoneal dialysis fluids. Biomed Res Int 2015(628158):1–12

Krijger JJ, Thon MR, Deising HB, Wirsel SGR (2014) Compositions of fungal secretomes indicate a greater impact of phylogenetic history than lifestyle adaptation. BMC Genomics 15:1–18

Kubitschek-Barreira PH, Curty N, Neves GWP, Gil C, Lopes-Bezerra LM (2013) Differential proteomic analysis of *Aspergillus fumigatus* morphotypes reveals putative drug targets. J Proteomics 78:522–534

Kumar A, Alam A, Tripathi D, Rani M, Khatoon H, Pandey S, Ehtesham NZ, Hasnain SE (2018) Protein adaptations in extremophiles: an insight into extremophilic connection of mycobacterial proteome. Semin Cell Dev Biol 84:147–157

Kusenbach G, Skopnik H, Haase G, Friedrichs F, Döhmen H (1992) *Exophiala dermatitidis* pneumonia in cystic fibrosis. Eur J Pediatr 151:344–346

Lambert M, Blanchin-Roland S, Le Louedec F, Lepingle A, Gaillardin C (1997) Genetic analysis of regulatory mutants affecting synthesis of extracellular proteinases in the yeast *Yarrowia lipolytica*: identification of a RIM101/pacC homolog. Mol Cell Biol 17:3966–3976

Langfelder K, Streibel M, Jahn B, Haase G, Brakhage A (2003) Biosynthesis of fungal melanins and their importance for human pathogenic fungi. Fungal Genet Biol 38:143–158

Lau AF, Drake SK, Calhoun LB, Henderson CM, Zelazny AM (2013) Development of a clinically comprehensive database and a simple procedure for identification of molds from solid media by matrix-assisted laser desorption ionization-Time of flight mass spectrometry. J Clin Microbiol 51:828–834

Li B, Tian X, Wang C, Zeng X, Xing Y, Ling H, Yin W, Tian L, Meng Z, Zhang J, Guo S (2017) SWATH label-free proteomics analyses revealed the roles of oxidative stress and antioxidant defensing system in sclerotia formation of Polyporus umbellatus. Sci Rep 7:1–13

Lima PS, Casaletti L, Bailão AM, de Vasconcelos AT, Fernandes Gda R, Soares CM (2014) Transcriptional and proteomic responses to carbon starvation in Paracoccidioides. PLoS Negl Trop Dis 8:e2855

Loginov D, Šebela M (2016) Proteomics of survival structures of fungal pathogens. N Biotechnol 33:655–665

Longo LVG, da Cunha JPC, Sobreira TJP, Puccia R (2014) Proteome of cell wall-extracts from pathogenic Paracoccidioides brasiliensis: comparison among morphological phases, isolates, and reported fungal extracellular vesicle proteins. EuPA Open Proteom 3:216–228

López-García P, Rodrguez-Valera F, Pedrós-Alió C, Moreira D (2002) Unexpected diversity of smal eukaryotes in Deep-Sea Antarctic Plankton. Nature 409:603–606

Lu X, Sun J, Nimtz M, Wissing J, Zeng A-P, Rinas U (2010) The intra- and extracellular proteome of Aspergillus niger growing on defined medium with xylose or maltose as carbon substrate. Microb Cell Fact 9:1–13

Ly L, Wasinger VC (2011) Protein and peptide fractionation, enrichment and depletion: tools for the complex proteome. Proteomics 11:513–534

Maddi A, Bowman SM, Free SJ (2009) Trifluoromethanesulfonic acid-based proteomic analysis of cell wall and secreted proteins of the ascomycetous fungi Neurospora crassa and Candida albicans. Fungal Genet Biol 46:768–781

Mafart P, Couvert O, Leguérinel I (2001) Effect of pH on the heat resistance of spores. Int J Food Microbiol 63:51–56

Magan N (2007) Fungi in extreme environments. In: Kubicek C, Druzhinina I (eds) Environmental and microbial relationships, The Mycota, vol 4. Springer, Berlin/Heidelberg, pp 85–103

Marcotte EM (2007) How do shotgun proteomics algorithms identify proteins? Nat Biotechnol 25:755–757

Marouga R, David S, Hawkins E (2005) The development of the DIGE system: 2D fluorescence difference gel analysis technology. Anal Bioanal Chem 382:669–678

Marzban G, Tesei D, Sterflinger K (2013) A review beyond the borders: proteomics of microclonial black fungi and black yeasts. Nat Sci 05:640–645

Meinken J, Asch DK, Neizer-Ashun KA, Chang G-H, Cooper CR, Min XJ (2014) FunSecKB2: a fungal protein subcellular location knowledgebase. Comput Mol Biol 4:1–17

Mesbah NM, Cook GM, Wiegel J (2009) The halophilic alkalithermophile Natranaerobius thermophilus adapts to multiple environmental extremes using a large repertoire of Na +(K+)/H+ antiporters. Mol Microbiol 74:270–281

Mitra SK, Gantt JA, Ruby JF, Clouse SD, Goshe MB (2007) Membrane proteomic analysis of Arabidopsis thaliana using alternative solubilization techniques. J Proteome Res 6:1933–1950

Miyazaki K, Hirase T, Kojima Y, Flint HJ (2005) Medium- to large-sized xylo-oligosaccharides are responsible for xylanase induction in Prevotella bryantii B14. Microbiology 151:4121–4125

Mooney C, Wang YH, Pollastri G (2011) SCLpred: protein subcellular localization prediction by N-to-1 neural networks. Bioinformatics 27:2812–2819

Moreno LF, Vicente VA, de Hoog S (2018) Black yeasts in the omics era: achievements and challenges. Med Mycol 56:32–41

Muddiman D, Andrews G, Lewis D, Notey J, Kelly R (2010) Part II: defining and quantifying individual and co-cultured intracellular proteomes of two thermophilic microorganisms by GeLC-MS2 and spectral counting. Anal Bioanal Chem 398:391–404

Murad AM, Souza GHMF, Garcia JS, Rech EL (2011) Detection and expression analysis of recombinant proteins in plant-derived complex mixtures using nanoUPLC-MSE. J Sep Sci 34:2618–2630

Nai C (2014) Rock-inhabiting fungi studied with the aid of the model black fungus *Knufia petricola* A95 and other related strains

Nandakumar MP, Marten MR (2002) Comparison of lysis methods and preparation protocols for one- and two-dimensional electrophoresis of *Aspergillus oryzae* intracellular proteins. Electrophoresis 23:2216–2222

Nguyen CH, Tsurumizu R, Sato T, Takeuchi M (2005) Taka-amylase A in the conidia of *Aspergillus oryzae* RIB40. Biosci Biotechnol Biochem 69:2035–2041

Nienow J, Friedmann IE (1993) In: Friedmann EI (ed) Terrestrial lithophytic (rock) communities. Wiley-Liss, New York, NY, pp 343–412

Noack-Schönmann S, Bus T, Banasiak R, Knabe N, Broughton WJ, Den Dulk-Ras H, Hooykaas PJJ, Gorbushina AA (2014) Genetic transformation of *Knufia petricola* A95 - a model organism for biofilm-material interactions. AMB Express 4:80

Nöbauer K, Hummel K, Mayrhofer C, Ahrens M, Setyabudi FMC, Schmidt-Heydt M, Eisenacher M, Razzazi-Fazeli E (2017) Comprehensive proteomic analysis of *Penicillium verrucosum*. Proteomics 17:1–5

Nosanchuk JD, Casadevall A (2003) The contribution of melanin to microbial pathogenesis. Microreview 5:203–223

Oh YT, Ahn CS, Kim JG, Ro HS, Lee CW, Kim JW (2010) Proteomic analysis of early phase of conidia germination in *Aspergillus nidulans*. Fungal Genet Biol 47:246–253

Olaya-Abril A, Jiménez-Munguía I, Gómez-Gascón L, Rodríguez-Ortega MJ (2014) Surfomics: shaving live organisms for a fast proteomic identification of surface proteins. J Proteomics 97:164–176

de Oliveira JM, de Graaff LH (2011) Proteomics of industrial fungi: trends and insights for biotechnology. Appl Microbiol Biotechnol 89:225–237

Oliveira DL, Nakayasu ES, Joffe LS, Guimarães AJ, Sobreira TJP, Nosanchuk JD, Cordero RJB, Frases S, Casadevall A, Almeida IC, Nimrichter L, Rodrigues ML (2010) Biogenesis of extracellular vesicles in yeast. Commun Integr Biol 3:533–535

Oliveira DL, Rizzo J, Joffe LS, Godinho RMC, Rodrigues ML (2013) Where do they come from and where do they go: candidates for regulating extracellular vesicle formation in fungi. Int J Mol Sci 14:9581–9603

Onofri S, Barreca D, Selbmann L, Isola D, Rabbow E, Horneck G, de Vera JPP, Hatton J, Zucconi L (2008) Resistance of Antarctic black fungi and cryptoendolithic communities to simulated space and Martian conditions. Stud Mycol 61:99–109

Onofri S, de la Torre R, de Vera J-P, Ott S, Zucconi L, Selbmann L, Scalzi G, Venkateswaran KJ, Rabbow E, Sánchez Iñigo FJ, Horneck G (2012) Survival of rock-colonizing organisms after 1.5 years in outer space. Astrobiology 12:508–516

Ouyang H, Luo Y, Zhang L, Li Y, Jin C (2010) Proteome analysis of *Aspergillus* fumigatus total membrane proteins identifies proteins associated with the glycoconjugates and cell wall biosynthesis using 2D LC-MS/MS. Mol Biotechnol 44:177–189

Özhak-Baysan B, Ögünc D, Dögen A, Ilkit M, De Hoog GS (2015) MALDI-TOF MS-based identification of black yeasts of the genus *Exophiala*. Med Mycol 53:347–352

Pacelli C, Selbmann L, Zucconi L, De Vera J-P, Rabbow E, Horneck G, de la Torre R, Onofri S (2016) BIOMEX experiment: ultrastructural alterations, molecular damage and survival of the fungus *Cryomyces antarcticus* after the experiment verification tests. Orig Life Evol Biosph 47:187–202

Panda A, Ghosh AK, Mirdha BR, Xess I, Paul S, Samantaray JC, Srinivasan A, Khalil S, Rastogi N, Dabas Y (2015) MALDI-TOF mass spectrometry for rapid identification of clinical fungal isolates based on ribosomal protein biomarkers. J Microbiol Methods 109:93–105

Pandey A, Mann M (2000) Proteomics to study genes and genomes. Nature 405:837–846

Parente AFA, de Rezende TCV, de Castro KP, Bailão AM, Parente JA, Borges CL, Silva LP, Soares CM (2013) A proteomic view of the response of *Paracoccidioides* yeast cells to zinc deprivation. Fungal Biol 117:399–410

Parente AFA, Naves PEC, Pigosso LL, Casaletti L, McEwen JG, Parente-Rocha JA, Soares CMA (2015) The response of *Paracoccidioides* spp. to nitrosative stress. Microbes Infect 17:575–585

Parente-Rocha JA, Parente AFA, Baeza LC, Bonfim SMRC, Hernandez O, McEwen JG, Bailão AM, Taborda CP, Borges CL, De Almeida Soares CM (2015) Macrophage interaction with paracoccidioides brasiliensis yeast cells modulates fungal metabolism and generates a response to oxidative stress. PLoS One 10:1–18

Parkhey S, Chandrakar V, Naithani SC, Keshavkant S (2015) Efficient extraction of proteins from recalcitrant plant tissue for subsequent analysis by two-dimensional gel electrophoresis. J Sep Sci 38:3622–3628

Paul S, Singh P, Rudramurthy SM, Chakrabarti A, Ghosh AK (2017) Matrix-assisted laser desorption/ionization–time of flight mass spectrometry: protocol standardization and database expansion for rapid identification of clinically important molds. Future Microbiol 12:1457–1466

Paul S, Singh P, Sharma S, Prasad GS, Rudramurthy SM, Chakrabarti A, Ghosh AK (2018) MALDI-TOF MS-based identification of melanized fungi is faster and reliable after the expansion of in-house database. Proteomics Clin Appl 1800070:1–8

Piette F, D'Amico S, Struvay C, Mazzucchelli G, Renaut J, Tutino ML, Danchin A, Leprince P, Feller G (2010) Proteomics of life at low temperatures: trigger factor is the primary chaperone in the Antarctic bacterium *Pseudoalteromonas haloplanktis* TAC125. Mol Microbiol 76:120–132

Pinto L, Granja LFZ, Almeida MA, Alviano DS, Silva MHD, Ejzemberg R, Rozental S, Alviano CS (2018) Melanin particles isolated from the fungus *Fonsecaea pedrosoi* activates the human complement system. Mem Inst Oswaldo Cruz 113:e180120

Pitarch A, Nombela C, Gil C (2008) Cell wall fractionation for yeast and fungal proteomics. Methods Mol Biol 425:217–239

Plemenitaš A, Lenassi T, Konte T, Kejžar A, Zajc J, Gostinčar C, Gunde-Cimerman N (2014) Adaptation to high salt concentrations in halotolerant/halophilic fungi: a molecular perspective. Front Microbiol 5:1–12

Poyntner C, Blasi B, Arcalis E, Mirastschijski U, Sterflinger K, Tafer H (2016) The transcriptome of *Exophiala dermatitidis* during ex-vivo skin model infection. Front Cell Infect Microbiol 6:1–19

Poyntner C, Mirastschijski U, Sterflinger K, Tafer H (2018) Transcriptome study of an *Exophiala dermatitidis* PKS1 mutant on an ex vivo skin model: is melanin important for infection? Front Microbiol 9:1–13

Prenafeta-Boldú FX, Guivernau M, Gallastegui G, Viñas M, de Hoog GS, Elías A (2012) Fungal/bacterial interactions during the biodegradation of TEX hydrocarbons (toluene, ethylbenzene and p-xylene) in gas biofilters operated under xerophilic conditions. FEMS Microbiol Ecol 80:722–734

Putignani L, Del Chierico F, Onori M, Mancinelli L, Argentieri M, Bernaschi P, Coltella L, Lucignano B, Pansani L, Ranno S, Russo C, Urbani A, Federici G, Menichella D (2011) MALDI-TOF mass spectrometry proteomic phenotyping of clinically relevant fungi. Mol Biosyst 7:620–629

Rabilloud T, Chevallet M, Luche S, Lelong C (2008) Fully denaturing two-dimensional electrophoresis of membrane proteins: a critical update. Proteomics 8:3965–3973

Ranque S, Normand AC, Cassagne C, Murat JB, Bourgeois N, Dalle F, Gari-Toussaint M, Fourquet P, Hendrickx M, Piarroux R (2014) MALDI-TOF mass spectrometry identification of filamentous fungi in the clinical laboratory. Mycoses 57:135–140

Ravalason H, Jan G, Mollé D, Pasco M, Coutinho PM, Lapierre C, Pollet B, Bertaud F, Petit-Conil M, Grisel S, Sigoillot J-C, Asther M, Herpoël-Gimbert I (2008) Secretome analysis of *Phanerochaete chrysosporium* strain CIRM-BRFM41 grown on softwood. Appl Microbiol Biotechnol 80:719

Robertson KL, Mostaghim A, Cuomo CA, Soto CM, Lebedev N, Bailey RF, Wang Z (2012) Adaptation of the black yeast *Wangiella dermatitidis* to ionizing radiation: molecular and cellular mechanisms. PLoS One 7:e48674

Rodrigues ML, Nakayasu ES, Oliveira DL, Nimrichter L, Nosanchuk JD, Almeida IC, Casadevall A (2008) Extracellular vesicles produced by *Cryptococcus neoformans* contain protein components associated with virulence. Eukaryot Cell 7:58–67

Rodrigues ML, Nakayasu ES, Almeida IC, Nimrichter L (2014) The impact of proteomics on the understanding of functions and biogenesis of fungal extracellular vesicles. J Proteomics 97:177–186

Rodrigues ML, Godinho RMC, Zamith-Miranda D, Nimrichter L (2015) Traveling into outer space: unanswered questions about fungal extracellular vesicles. PLoS Pathog 11:1–6

Rodrigues LNDS, Brito WA, Parente AFA, Weber SS, Bailão AM, Casaletti L, Borges CL, Soares CMA (2016) Osmotic stress adaptation of *Paracoccidioides lutzii*, Pb01, monitored by proteomics. Fungal Genet Biol 95:13–23

Rohrbough JG, Galgiani JN, Wysocki VH (2007) The application of proteomic techniques to fungal protein identification and quantification. Ann N Y Acad Sci 1111:133–146

Romsdahl J, Blachowicz A, Chiang A, Singh NK, Stajich JE, Kalkum M, Venkateswaran KJ, Wang C (2018) Characterization of *Aspergillus niger* isolated from the international space station. mSystems 3:1–13

Sanchez-Pulido L, Andrade-Navarro MA (2007) The FTO (fat mass and obesity associated) gene codes for a novel member of the non-heme dioxygenase superfamily. BMC Biochem 8:23

Schmitt S, Prokisch H, Schlunck T, Camp DG, Ahting U, Waizenegger T, Scharfe C, Meitinger T, Imhof A, Neupert W, Oefner PJ, Rapaport D (2006) Proteome analysis of mitochondrial outer membrane from *Neurospora crassa*. Proteomics 6:72–80

Selbmann L, Hoog GS, De Mazzaglia A, Friedmann EI, Onofri S (2005) Fungi at the edge of life: cryptoendolithic black fungi from Antarctic desert. Stud Mycol 51:1–32

Selbmann L, de Hoog GS, Zucconi L, Isola D, Ruisi S, van den Ende AH, Ruibal C, De Leo F, Urzì C, Onofri S (2008) Drought meets acid: three new genera in a dothidealean clade of extremotolerant fungi. Stud Mycol 61:1–20

Selbmann L, Egidi E, Isola D, Onofri S, Zucconi L, de Hoog GS, Chinaglia S, Testa L, Tosi S, Balestrazzi A, Lantieri A, Compagno R, Tigini V, Varese GC (2013) Biodiversity, evolution and adaptation of fungi in extreme environments. Plant Biosyst 147:237–246

Selbmann L, Zucconi L, Isola D, Onofri S (2014) Rock black fungi: excellence in the extremes, from the Antarctic to space. Curr Genet 61:335–345

Selbmann L, Pacelli C, Zucconi L, Dadachova E, Moeller R, de Vera JP, Onofri S (2018) Resistance of an Antarctic cryptoendolithic black fungus to radiation gives new insights of astrobiological relevance. Fungal Biol 122:546–554

Sert H, Sümbül H, Sterflinger K (2007) Microcolonial fungi from antique marbles in Perge/Side/Termessos (Antalya/Turkey). Antonie van Leeuwenhoek. Int J Gen Mol Microbiol 91:217–227

Seyedmousavi S, Guillot J, de Hoog GS (2013) Phaeohyphomycoses, emerging opportunistic diseases in animals. Clin Microbiol Rev 26:19–35

Seyedmousavi S, Netea MG, Mouton JW, Melchers WJG, Verweij PE, de Hoog GS (2014) Black yeasts and their filamentous relatives: principles of pathogenesis and host defense. Clin Microbiol Rev 27:527–542

Siddiqui KS, Thomas T (2008) Protein adaptation in extremophiles. Nova Science Publishers Inc., Hauppauge, NY

Siglioccolo A, Paiardini A, Piscitelli M, Pascarella S (2011) Structural adaptation of extreme halophilic proteins through decrease of conserved hydrophobic contact surface. BMC Struct Biol 11:50

Simpson RJ, Jensen SS, Lim JWE (2008) Proteomic profiling of exosomes: current perspectives. Proteomics 8:4083–4099

Sinitcyn P, Daniel Rudolph J, Cox J (2018) Computational methods for understanding mass spectrometry–based shotgun proteomics data. Annu Rev Biomed Data Sci 1:207–234

Sørensen LM, Lametsch R, Andersen MR, Nielsen PV, Frisvad JC (2009) Proteome analysis of *Aspergillus niger*: lactate added in starch-containing medium can increase production of the mycotoxin fumonisin B2 by modifying acetyl-CoA metabolism. BMC Microbiol 9:255

Sterflinger K, Krumbein WE (1995) Multiple stress factors affecting growth of rock-inhabiting black fungi. Bot Acta 108:490–496

Sterflinger K, Krumbein WE (1997) Dematiaceous fungi as a major agent of biopitting for Mediterranean marbles and limestones. Geomicrobiol J 14:219–230

Sterflinger K, Tesei D, Zakharova K (2012) Fungi in hot and cold deserts with particular reference to microcolonial fungi. Fungal Ecol 5:453–462

Strohkamp S, Gemoll T, Habermann JK (2016) Possibilities and limitations of 2DE-based analyses for identifying low-abundant tumor markers in human serum and plasma. Proteomics 16:2519–2532

Su Y, Jiang X, Wu W, Wang M, Hamid MI, Xiang M, Liu X (2016) Genomic, transcriptomic and proteomic analysis provide insights into the cold adaptation mechanism of the obligate psychrophilic fungus *Mrakia psychrophila*. G3 6:3603–3613

Suh M-J, Fedorova ND, Cagas SE, Hastings S, Fleischmann RD, Peterson SN, Perlin DS, Nierman WC, Pieper R, Momany M (2012) Development stage-specific proteomic profiling uncovers small, lineage specific proteins most abundant in the *Aspergillus* Fumigatus conidial proteome. Proteome Sci 10:30

Supek F, Bošnjak M, Škunca N, Šmuc T (2011) REVIGO summarizes and visualizes long lists of gene ontology terms. PLoS One 6:e21800

Taylor D, Shah S (2015) Methods of isolating extracellular vesicles impact down-stream analyses of their cargoes. Methods 87:3–10

Taylor RD, Saparno A, Blackwell B, Anoop V, Gleddie S, Tinker NA, Harris LJ (2008) Proteomic analyses of *Fusarium graminearum* grown under mycotoxin-inducing conditions. Proteomics 8:2256–2265

Taylor D, Zacharias W, Taylor C (2011) Exosome isolation for proteomics analyses and RNA profiling. In: Simpson RJ, Greening DW (eds) Serum/plasma proteomics, Methods and Protocols. Humana Press, New York, USA, pp 235–246

Teixeira MM, Moreno LF, Stielow BJ, Muszewska A, Hainaut M, Gonzaga L, Abouelleil A, Patané JSL, Priest M, Souza R, Young S, Ferreira KS, Zeng Q, da Cunha MML, Gladki A, Barker B, Vicente VA, de Souza EM, Almeida S, Henrissat B, Vasconcelos ATR, Deng S, Voglmayr H, Moussa TAA, Gorbushina A, Felipe MSS, Cuomo CA, de Hoog GS (2017) Exploring the genomic diversity of black yeasts and relatives (*Chaetothyriales, Ascomycota*). Stud Mycol 86:1–28

Tesei D, Marzban G, Zakharova K, Isola D, Selbmann L, Sterflinger K, Rosling A (2012) Alteration of protein patterns in black rock inhabiting fungi as a response to different temperatures. Fungal Biol 116:932–940

Tesei D, Marzban G, Marchetti-Deschmann M, Tafer H, Arcalis E, Sterflinger K (2015a) Proteome of tolerance fine-tuning in the human pathogen black yeast *Exophiala dermatitidis*. J Proteomics 128:39–57

Tesei D, Marzban G, Marchetti-Deschmann M, Tafer H, Arcalis E, Sterflinger K (2015b) Protein functional analysis data in support of comparative proteomics of the pathogenic black yeast *Exophiala dermatitidis* under different temperature conditions. Data Brief 5:372–375

Tesei D, Tafer H, Poyntner C, Piñar G, Lopandic K, Sterflinger K (2017) Draft genome sequences of the black rock fungus *Knufia petricola* and its spontaneous nonmelanized mutant. Genome Announc 5:1–2

Tiquia-Arashiro SM (2014) Thermophilic carboxydotrophs and their applications in biotechnology. In: Tiquia-Arashiro SM, Mormile M (eds) Springer briefs in microbiology, extremophilic bacteria. Springer, New York, NY, p 131

Tiquia-Arashiro SM, Rodrigues DF (2016) Thermophiles and psychrophiles in nanobiotechnology. In: Tiquia-Arashiro SM, Rodrigues DF (eds) Extremophiles: applications in nanotechnology. Springer, Cham/New York, pp 89–127

Uranga CC, Ghassemian M, Hernández-Martínez R (2017) Novel proteins from proteomic analysis of the trunk disease fungus *Lasiodiplodia theobromae* (Botryosphaeriaceae). Biochim Open 4:88–98

Vallejo M, Nakayasu E, Matsuo A, Sobreira TP, Longo LG, Ganiko L, Almeida I, Puccia R (2012) Vesicle and vesicle-free extracellular proteome of *Paracoccidioides brasiliensis*: comparative analysis with other pathogenic fungi. J Proteome Res 11:1676–1685

Vember VV, Zhdanova NN (2001) Peculiarities of linear growth of the melanin containing fungi *Cladosporium sphaerospermum* Perz. and *Alternaria alternata* (Fr.) Keissler. Mikrobiol Zh 63:3–12

Vialás V, Perumal P, Gutierrez D, Ximénez-Embún P, Nombela C, Gil C, Chaffin WL (2012) Cell surface shaving of *Candida albicans* biofilms, hyphae, and yeast form cells. Proteomics 12:2331–2339

Vincent D, Kohler A, Claverol S, Solier E, Joets J, Gibon J, Lebrun M, Plomion C, Martin F (2012) Secretome of the free-living mycelium from the ectomycorrhizal basidiomycete *Laccaria bicolor*. J Proteome Res 11:157–171

Vödisch M, Scherlach K, Winkler R, Hertweck C, Braun HP, Roth M, Haas H, Werner ER, Brakhage AA, Kniemeyer O (2011) Analysis of the *Aspergillus fumigatus* proteome reveals metabolic changes and the activation of the pseurotin A biosynthesis gene cluster in response to hypoxia. J Proteome Res 10:2508–2524

Washburn MP, Wolters D, Yates JR III (2001) Large-scale analysis of the yeast proteome by multi-dimensional protein identification technology. Nat Biotechnol 19:242–247

Wasinger VC, Cordwell SJ, Cerpa-Poljak A, Yan JX, Gooley AA, Wilkins MR, Duncan MW, Harris R, Williams KL, Humphery-Smith I (1995) Progress with gene-product mapping of the mollicutes: *Mycoplasma genitalium*. Electrophoresis 16:1090–1094

Wasinger VC, Zeng M, Yau Y (2013) Current status and advances in quantitative proteomic mass spectrometry. Int J Proteomics 2013:1–12

Westwood GS, Huang SW, Keyhani NO (2005) Allergens of the entomopathogenic fungus *Beauveria bassiana*. Clin Mol Allergy 3:1

Wiese S, Reidegeld KA, Meyer HE, Warscheid B (2007) Protein labeling by iTRAQ: a new tool for quantitative mass spectrometry in proteome research. Proteomics 7:340–350

Williams TJ, Burg DW, Ertan H, Raftery MJ, Poljak A, Guilhaus M, Cavicchioli R (2010) Global proteomic analysis of the insoluble, soluble, and supernatant fractions of the psychrophilic archaeon *Methanococcoides burtonii* part II: the effect of different methylated growth substrates. J Proteome Res 9:653–663

Wiśniewski JR, Zougman A, Nagaraj N, Mann M (2009) Universal sample preparation method for proteome analysis. Nat Methods 6:359–362

Woo PCY, Ngan AHY, Tsang CCC, Ling IWH, Chan JFW, Leung S-Y, Yuen K-Y, Lau SKP (2013) Clinical spectrum of *Exophiala* infections and a novel *Exophiala* species, *Exophiala hongkongensis*. J Clin Microbiol 51:260–267

Wösten HAB (2001) Hydrophobins: multipurpose proteins. Annu Rev Microbiol 55:625–646

Wu HC, Chen TN, Kao SH, Shui HA, Chen WJ, Lin HJ, Chen HM (2010) Isoelectric focusing management: an investigation for salt interference and an algorithm for optimization. J Proteome Res 9:5542–5556

Yeung YG, Nieves E, Angeletti RH, Stanley ER (2008) Removal of detergents from protein digests for mass spectrometry analysis. Anal Biochem 382:135–137

Yildirim V, Özcan S, Becher D, Büttner K, Özcengiz G (2011) Characterization of proteome alterations in *Phanerochaete chrysosporium* in response to lead exposure. Proteome Sci 9:1–15

Yin QY, de Groot PWJ, de Koster CG, Klis FM (2008) Mass spectrometry-based proteomics of fungal wall glycoproteins. Trends Microbiol 16:20–26

Zajc J, Liu Y, Dai W, Yang Z, Hu J, Gostin C (2013) Genome and transcriptome sequencing of the halophilic fungus *Wallemia ichthyophaga*: haloadaptations present and absent. BMC Genomics 14:1–20

Zakharova K, Tesei D, Marzban G, Dijksterhuis J, Wyatt T, Sterflinger K (2013) Microcolonial fungi on rocks: a life in constant drought? Mycopathologia 175:537–547

Zakharova K, Marzban G, de Vera J-P, Lorek A, Sterflinger K (2014a) Protein patterns of black fungi under simulated mars-like conditions. Sci Rep 4:1–7

Zakharova K, Sterflinger K, Razzazi-fazeli E, Noebauer K, Marzban G (2014b) Global proteomics of the extremophile black fungus *Cryomyces antarcticus* using 2D-electrophoresis. Nat Sci 6:978–995

Zhang N, Chen R, Young N, Wishart D, Winter P, Weiner JH, Li L (2007) Comparison of SDS- and methanol-assisted protein solubilization and digestion methods for *Escherichia coli* membrane proteome analysis by 2-D LC-MS/MS. Proteomics 7:484–493

Zhang SW, Liu YF, Yu Y, Zhang TH, Fan XN (2014) MSLoc-DT: a new method for predicting the protein subcellular location of multispecies based on decision templates. Anal Biochem 449:164–171

Zhang J, Zhang L, Qiu J, Nian H (2015) Isobaric tags for relative and absolute quantitation (iTRAQ)-based proteomic analysis of *Cryptococcus humicola* response to aluminum stress. J Biosci Bioeng 120:359–363

Zhong Z, Li N, Liu L, He B, Igarashi Y, Luo F (2018) Label-free differentially proteomic analysis of interspecific interaction between white-rot fungi highlights oxidative stress response and high metabolic activity. Fungal Biol 122:774–784

Zoglowek M, Brewer H, Norbeck A (2018) Discovery of novel cellulases using proteomic strategies. In: Lübeck M (ed) Cellulases: methods and protocols. Springer, New York, NY, pp 103–113

Part II
Biotechnological Applications
of Extremophilic Fungi

Chapter 13
Yeast Thriving in Cold Terrestrial Habitats: Biodiversity and Industrial/ Biotechnological Applications

Marcelo Baeza ⓘ, Oriana Flores, Jennifer Alcaíno, and Víctor Cifuentes

13.1 Introduction

Environments with constant temperatures at 5 °C or below represent more than 80% of our planet, including polar regions, high mountains, glaciers, and deep oceans (Russell 1990). In spite of the fact that microorganisms are considered the primary organic matter recyclers in these cold environments, our current knowledge about them is still scarce, representing a pending challenge to microbiologists. Although yeasts are regarded as important nutrient recyclers due to their heterotrophic metabolism and ability to degrade organic macromolecules, their ecological role remains mainly unknown (Buzzini et al. 2012; Antony et al. 2016). Yeasts have evolved physiological adaptations to survive, grow, and successfully proliferate under the presence of several stress factors in cold environments, in addition to the low temperatures, nutrient deprivation, desiccation, and high UV radiation (D'Amico et al. 2006; Margesin and Miteva 2011; Wilkins et al. 2013; De Maayer et al. 2014). In addition to the ecological interest on cold-adapted microorganisms, the study of these microorganisms, especially of yeasts, has attracted the attention of researchers due to their potential applications in diverse industrial/biotechnological fields (Thomas-Hall et al. 2010; Tiquia and Mormile 2010; Buzzini et al. 2012; de Garcia et al. 2012; Buzzini and Margesin 2014a; Zalar and Gunde-Cimerman 2014; Tiquia-Arashiro and Rodrigues 2016). In this chapter, an actualized analysis of cold-adapted yeasts is presented that reveals the cold regions from which more studies and yeast species have been described. The identification of yeasts in the original works is contrasted to their current taxonomical classification, and according to the latter the yeast species that could be considered more ubiquitous in cold environments around

M. Baeza (✉) · O. Flores · J. Alcaíno · V. Cifuentes
Facultad de Ciencias, Departamento de Ciencias Ecológicas, Universidad de Chile, Santiago, Chile
e-mail: mbaeza@uchile.cl

© Springer Nature Switzerland AG 2019
S. M. Tiquia-Arashiro, M. Grube (eds.), *Fungi in Extreme Environments: Ecological Role and Biotechnological Significance*,
https://doi.org/10.1007/978-3-030-19030-9_13

the earth is suggested. Furthermore, the incipient data referring to yeasts obtained in culture-independent approaches of cold environments and the novel or less reviewed elsewhere applied potential in diverse productive areas of cold-adapted yeasts is discussed.

13.2   Yeasts from Cold Environments

The study of cold regions is challenging because of their difficult accessibility and the need for substantial funds to perform accurate scientific research. For these reasons, the published works about isolation and identification of yeasts from cold environments are still limited and concentrated in areas of better accessibility of the earth, as can be observed in Fig. 13.1. Yeast species have been described from approximately 84 cold localities around the world, which according to the number of works reported from them and their geographic location can be mainly grouped in those from Shetland South Archipelago and Antarctic Peninsula, Patagonia, Alps, Apennines, and Italian glaciers, and sites close to Davis base and South Victoria Land at Antarctica. According to the current number of works reported, the most studied locations have been King George Island and South Victoria Land, 14 and 8 works, respectively. Concerning the yeast species identified from cold environments, in literature it is commonly stated that the vast majority of them belong to the genus *Cryptococcus*, an aspect that had changed profoundly in last years. The percentages of different genera of cold-adapted yeast species reported are shown in Fig. 13.2. The top five yeast genera described from cold environments, considering the reported identification in the original works (Fig. 13.2a), are by far dominated by *Cryptococcus* followed by *Rhodotorula*, and appearing in a minor proportion the genera *Candida*, *Mrakia*, and *Dioszegia*. However, the top five genera drastically change when the current taxonomical classification of yeasts is considered (Fig. 13.2b). In this case, *Mrakia* and *Naganishia* are the main genera, followed by *Rhodotorula*, *Candida*, and *Leucosporidium*, but all five species in similar proportions. As it can be observed, the most drastic change of genera when comparing the results from both taxonomical classifications is the proportion of the genus *Cryptococcus* that decreases from a 27% to only a 3%, and to a lesser extent in *Rhodotorula* that decreases from 15% to 7%. According to the current taxonomic classification of yeasts, the locations in which the greatest number of different yeast species have been reported (see Fig. 13.1) are King George Island (81), followed by du Geant-Miage Glacier (49), Kandalaksha (42), Nahuel Huapi Lake (34), Calderone Glacier (32), Kongsfjorden (31), South Victoria Land (29), Olsztyn (25), Nahuel Huapi Park (24), Ross Sea region (22), and Mount Tronador (20). The cold-adapted yeast species that have been isolated and identified from more cold environments around the world are *Debaryomyces hansenii*, *Rhodotorula mucilaginosa*, *Cystobasidium laryngis*, *Vishniacozyma victoriae*, and *Papiliotrema laurentii* (shown in triangles in Fig. 13.1). On the other hand, there is a high number of cold-adapted yeasts that have been isolated and identified from only one location, corresponding to 55% at the species level and 48% at the genus level of all the yeasts that have been isolated and identified from cold environments.

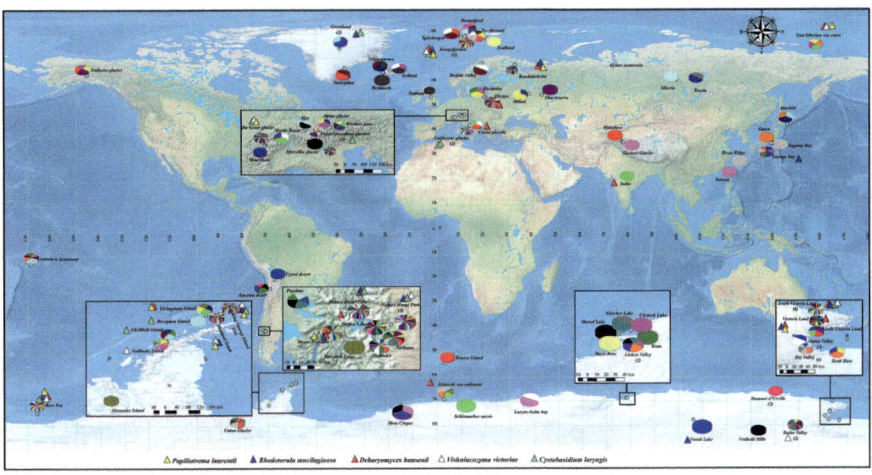

Fig. 13.1 Identified yeasts isolated from different cold environments on earth. The pie chart at each location represents the fraction of different species described. The number of yeast isolation/identification reports from each locality from which at least two works have been published is indicated in parenthesis. Yeast species that have been isolated from at least ten locations are indicated in triangles. Data considered only works where molecular taxonomical markers were used for yeast identification, such as the ITS or D1/D2 region of the rDNA, and was collected from literature revisions performed until January 2018 (di Menna 1966; Goto et al. 1969; Baharaeen and Vishniac 1982; Gounot 1986; Baublis et al. 1991; Ray et al. 1992; Babjeva and Reshetova 1998; Golubev 1998; Ma et al. 1999; Montes et al. 1999; Petrescu et al. 2000; Scorzetti et al. 2000; Nagahama et al. 2001; Bab'eva et al. 2002; Thomas-Hall 2002; Vishniac 2002, 2006; Birgisson et al. 2003; Libkind et al. 2003, 2009a, b; Guffogg et al. 2004; Nakagawa et al. 2004; Bergauer et al. 2005; Gilichinsky et al. 2005; Starmer et al. 2005; Arenz et al. 2006; Fell et al. 2006; Butinar et al. 2007, 2011; Chen et al. 2007; de García et al. 2007; Margesin et al. 2007; Sansone et al. 2007; Xin and Zhou 2007; Connell et al. 2008, 2009, 2010; Turchetti et al. 2008, 2011, 2013; Zalar et al. 2008; Bridge and Newsham 2009; D'Elia et al. 2009; Pavlova et al. 2009; Branda et al. 2010; Kachalkin 2010; Konishi et al. 2010; Pathan et al. 2010; Thomas-Hall et al. 2010; Brandão et al. 2011, 2017; Vaz et al. 2011; Carrasco et al. 2012; Singh and Singh 2012; Uetake et al. 2012; Zhang et al. 2012, 2013, 2017; Duarte et al. 2013, 2016; Godinho et al. 2013; Laich et al. 2013, 2014; Rovati et al. 2013; Tsuji et al. 2013a; Vaca et al. 2013; Ejdys et al. 2014; Furbino et al. 2014; Mestre et al. 2014; Selbmann et al. 2014a, b; Vasileva-Tonkova et al. 2014; Coleine et al. 2015; Jacques et al. 2015; Pulschen et al. 2015; Barahona et al. 2016; Martinez et al. 2016; Filipowicz et al. 2017; Hassan 2017; Martorell et al. 2017; Trochine et al. 2017)

13.3 Identification of Cold-Adapted Yeasts Using Metagenomic Approaches

Most of our current knowledge about the microbial diversity of cold environments is due to studies based on traditional culture-dependent methodologies. However, it is known that only a small proportion of viable microorganisms are recovered from diverse environmental samples using this strategy, estimated to be only a 17% in the case of fungi (Holdgate 1977). The environmental microbial ecology has been

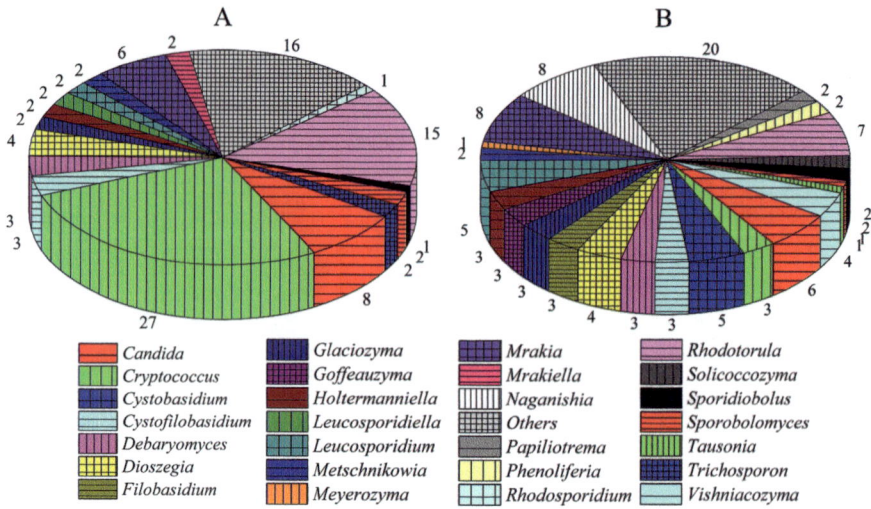

Fig. 13.2 Proportion of cold-adapted yeasts described at the genera level. The percentages of cold-adapted yeast species belonging to different genera are shown according to the description in the original publication (**a**) and to current taxonomical classification (**b**). Source: data as in Fig. 13.1

revolutionized by the development of culture-independent methodologies, especially by the use of metagenomics approaches that allow a more exhaustive exploration of the taxonomic composition and metabolic profiles of communities of diverse environments (Rondon et al. 2000). Initially, metagenomic libraries were constructed using total DNA purified directly from the environmental samples for sequencing analysis followed by functional characterization by heterologous expression methods (Rondon et al. 2000; Voget et al. 2003; Delmont et al. 2011). The advances of high-throughput DNA sequencing technologies, bioinformatic tools, and databases allowed a progressive discovery of novel genes, physiological pathways, and species from diverse environments (Simon et al. 2009; Berlemont et al. 2011; Pearce et al. 2012; Choudhari et al. 2014b; Culligan et al. 2014; Alves Junior et al. 2015; Rivkina et al. 2015), and also it has helped to compare the microbial composition in different environments (Varin et al. 2012; Choudhari et al. 2014a; Lee et al. 2015; Johnston et al. 2016). In metagenomic studies where eukaryotes have been identified, it has been observed that they are present in a lower proportion (less than a 10%) than prokaryotes. Taking into account microbial eukaryotes, they are mostly represented by ascomycetous and basidiomycetous fungi (Simon et al. 2009; Rogers et al. 2013, 2013; Baeza et al. 2017b).

The identification of microorganisms based on metagenomic data generated from metagenome analysis needs robust bioinformatic tools and databases; the latter has been more developed for prokaryotes (Kim et al. 2013). A fungi database was recently published from a metagenomic study on Antarctic soil samples, which was initially performed to search bacteria capable of degrading diesel oil

(Donovan et al. 2018). Another molecular approach used to identify microbial species is the "metabarcoding," which is based on the amplification by the polymerase chain reaction (PCR) of species-discriminating barcode genes or sequences followed by next-generation sequencing (Kim et al. 2013). This methodology has been successfully used in the identification of fungal species from Antarctic regions, using the internal transcribed spacers (ITS) and the D1/D2 domain of the large ribosomal DNA as barcode sequences (Kurtzman et al. 2015; Antony et al. 2016).

Metagenomic studies detecting fungi from cold environments are currently very limited and even more scarce for yeasts. In a study of permafrost in the high Canadian Arctic, it was found that the proportion of sequences belonging to fungi was 200- to 10,000-fold lower than the proportion of prokaryotic DNA sequences (Yergeau et al. 2010); a lower proportion of fungal sequences was also observed in samples obtained from ice core sections from Lake Vostok (Rogers et al. 2013). In a study of permafrost from Muot da Barba Peider (Eastern Swiss Alps), operational taxonomic units (OTUs) including lichenized fungi of the genera *Lecidea*, *Acarospora*, or *Umbilicaria* and yeasts of the genera *Rhodotorula*, *Cryptococcus*, *Mrakia*, and *Leucosporidium* were found (Frey et al. 2016). In an amplicon-metagenomic analysis of soils from islands of the South Shetland archipelago and Antarctic Peninsula, and from Union Glacier at Antarctica based on the D1/D2 rDNA region, OTUs corresponding to fungi were distributed in 87 genera and 123 species (Baeza et al. 2017b). Regarding yeasts, the highest fraction of them corresponded to basidiomycetes, which is according to data obtained from viable yeast studies. Also, 37 yeast genera were found that were not previously cultivated from Antarctic samples.

13.4 Applications of Cold-Adapted Yeasts

The best known potential of cold-adapted yeasts that can be applied in several fields lies in their hydrolytic enzymes, such as lipases, amylases, and proteases, which are valuable for industrial processes or for the generation of products that require enzymes highly active at low temperatures. These have been revised in previous reviews and book chapters (Gerday et al. 2000; Tutino et al. 2010; Cavicchioli et al. 2011; Buzzini et al. 2012; Feller 2013; Gerday 2013, 2014; Joshi and Satyanarayana 2013; Białkowska and Turkiewicz 2014; Buzzini and Margesin 2014a, b; Alcaíno et al. 2015; Sarmiento et al. 2015; Baeza et al. 2017a). Here, some newfangled or less revised applied potentialities of cold-adapted yeasts are briefly discussed.

The multiple health benefits of fermented foods are known (Marco et al. 2017), and currently low-temperature fermentations are valued to improve flavor, for example, in the production of dairy products, bread, and alcoholic beverages (Kanellaki et al. 2014). Another example is the wine industry, which is continuously evolving according to the consumer preferences. Currently, there is high demand for wines having a lower alcohol content (Schmidtke et al. 2012; Varela et al. 2012; van Bussel et al. 2018) as they have less negative impacts on flavor and aroma perception, and on human health (Golddner et al. 2009; Kutyna et al. 2010). Even though there are physical-chemical techniques to reduce the alcohol content in wine, an alternative and promising tool to produce this kind of wines is the use of cold-adapted *Saccharomyces* and non-*Saccharomyces* yeasts to perform the must fermentation at lower temperatures. The psychrotolerant yeast *Candida sake* was tested in the fermentation of concentrated natural must from Spanish wineries (Tempranillo 2012) at 12 °C. By this approach, the typical cold-associated *S. cerevisiae* lag phase in the fermentation was avoided, and the obtained wine had a significantly lower ethanol content (a 30% fold reduction) and glycerol (50% fold reduction, but sorbitol was produced as a compatible osmolyte), and different aromatic features (Ballester-Tomás et al. 2017). A psychrophilic strain of *S. cerevisiae* immobilized onto apple cuts was successfully used in continuous wine fermentation at 5–15 °C. Under these conditions, the microbiological contamination was reduced, and the obtained wine was similar to dry wines considering total and volatile acids, but it had an overall improved quality (Kourkoutas et al. 2002). In beer brewing at 5 °C using domestic refrigerators, a freeze-dried immobilized format of a cold-adapted strain of *S. cerevisiae* (AXAZ-1) showed a good performance, obtaining beer of good clarity, with an acceptable amount of volatile compounds, and lower diacetyl and polyphenol contents than commercial beers (Gialleli et al. 2017).

Microorganisms are the primary organic matter recyclers in cold environments, displaying a wide range of metabolic activities, including the degradation of a broad range of hydrocarbons (Margesin 2007). For example, the use of phenol, n-hexadecane, and methanol as a carbon source was described in cold-adapted yeasts isolated from Antarctica (Fernández et al. 2017). Yeasts isolated from the Alps, including *Cryptococcus terreus* and species of *Rhodotorula*, also showed the ability to degrade phenol and phenol-related mono-aromatic compounds, in some cases using immobilized yeast cells on zeolite at temperatures as low as 10 °C and phenol concentration from 10 to 12.5 mM (Margesin et al. 2003; Bergauer et al. 2005; Krallish et al. 2006). For the treatment of other contaminant/refractory materials as wastewater containing cow's milk, an Antarctic yeast identified as *Mrakia blollopis* showed promissory results. This yeast showed a high biochemical oxygen demand (BOD) removal rate on a wastewater model at low temperature, an ability that was attributed to the production of cold-active lipase (Tsuji et al. 2013b). These examples, and surely many other similar ongoing works, support the potential of cold-adapted yeasts to be applied in bioremediation of contaminated soils and water.

In the field of postharvest fruits, vegetables, and grains, the biological control of pests is gaining space as an alternative to reduce the current chemical treatments (Liu et al. 2013; Sangorrín et al. 2014). Yeasts of different genera have been used for the biocontrol of fungal phytopathogens, in which the reported control mechanisms include competition for nutrients and space, production of cell wall-degrading enzymes and antifungal compounds, mycoparasitism, and induction of host resistance (El-Tarabily and Sivasithamparam 2006). Cold-adapted yeasts represent good candidates for postharvest biocontrol, especially for refrigerated products. Epiphytic isolates of *Aureobasidium pullulans* and *Rhodotorula mucilaginosa* obtained from healthy pears from two Patagonian cold-storage packing houses were able to reduce to a 33% the fruit decay produced by *Penicillium expansum*, and the lesion diameter was reduced by an 88% after 60 days of incubation in cold (Robiglio et al. 2011). Similarly, cold-adapted yeasts isolated from soils of Tibet showed a high potential to be used in biocontrol. Isolates identified as *R. mucilaginosa* and *Cryptococcus laurentii* (now *Papiliotrema laurentii*) reduced the incidence of decay on cherry tomatoes due to *P. expansum* and *Botrytis cinerea* about a 65% and a 52%, respectively (Hu et al. 2015, 2017). A *Leucosporidium scottii* isolate from Antarctic soils showed to be a good biocontrol agent for blue and gray mold on two apple cultivars; moreover, the isolate was resistant to commonly used postharvest fungicides, which allows its use in combination to low-dose fungicides in an integrated management practice (Vero et al. 2013).

Another potential field for application of cold-adapted yeasts is in the development of alternative combustibles to fossil fuel, for example, through the fermentation of raw materials such as starch and lignocellulose (Tiquia-Arashiro and Mormile 2013). Advances in this area have included methods for the conversion of these feedstocks to fermentable sugars (ozonolysis, acid or alkaline hydrolysis), design of bioreactors, growth condition optimization, and immobilization of yeasts. However, a significant challenge is to simplify and reduce the cost of bioethanol production by the simultaneous saccharification and fermentation of raw sources rich in starch or cellulose (Petrovič 2015) by supplementing to the *S. cerevisiae* fermentation process amylases and/or cellulases. The enzymes currently available in the market are active at temperatures $\geq 50\ °C$; therefore, enzymes having high activity at lower temperatures to make the process more cost effective are needed. In this way, the amylases and cellulases produced by cold-adapted yeasts and fungi such as *Tetracladium* sp., *Cystofilobasidium capitatum*, *Rhodotorula glacialis*, and *M rakia blollopis* (Hamid 2015; Carrasco et al. 2016, 2017; Daskaya-Dikmen et al. 2018) are attractive candidates to be applied in the bioethanol industry. Furthermore, the Antarctic yeast *M. blollopis* has a unique capacity to ferment cellulosic biomass, reaching up to 12 g L^{-1} of alcohol from Japanese cedar and Eucalyptus pulp (Tsuji et al. 2013c). Another attractive eco-friendly fuel is biodiesel, which is currently produced from plant oils or waste cooking oils/fats. However, these sources of biodiesel have the inconvenience of their inconsistent composition and quality, and a considerable formation of C_1–C_4 hydrocarbons, which limits the production capacity (Bateni et al. 2017; Mishra and Goswami 2018). Good alternatives are oleaginous yeasts such as *Yarrowia lipolytica* that produce lipids from various economical substrates by fermentation under aerobic conditions (Xie 2017). Oleaginous microorganisms with the potential to be used as biodiesel sources have been isolated from cold environments. Among yeasts isolated from Tibetan Plateau, *Cryptococcus* species accumulated more than 30% of lipid content (Li et al. 2012), and isolates of *Rhodotorula glacialis* a 68% lipid/biomass, reaching a lipid/glucose yield of 16% (Amaretti et al. 2010). Antarctic isolates of *Rhodotorula glutinis* and *R. glacialis* showed high lipid production and accumulation (5–7 g L^{-1}) and accumulated large amounts of lipids per gram of biomass (47–77% w/w) (Viñarta et al. 2016).

13.5 Conclusions and Future Perspectives

It is clear that our knowledge about yeasts living in cold environments is very limited because the majority of studies are based on culture-dependent methodologies and concentrated in some locations, in spite of the majority of earth's biosphere being cold. It is desirable to broaden the research to more cold regions, for example glacial areas more easily accessible. Along with that, the application of culture-independent methodologies is essential, since although efforts are being made to develop cultivation methodologies to emulate environmental conditions, this is a tough task because of the unknown macro- and micronutrient requirements and in many cases the yeasts grow only in association with other organisms. Culture-independent methodologies have advanced greatly, including the purification of nucleic acid directly from diverse environmental materials and high-throughput sequencing. However, a pending task in this direction is the availability of a robust sequence database of fungi/yeasts for correct identification, including the detection of genes. This will be an essential input to knowing the metabolic potential of fungal communities and to finding novel genes desirable for application in diverse areas.

References

Alcaíno J, Cifuentes V, Baeza M (2015) Physiological adaptations of yeasts living in cold environments and their potential applications. World J Microbiol Biotechnol 31:1467–1473

Alves Junior N, Meirelles PM, de Oliveira SE, Dutilh B, Silva GG, Paranhos R, Cabral AS, Rezende C, Iida T, de Moura RL, Kruger RH, Pereira RC, Valle R, Sawabe T, Thompson C, Thompson F (2015) Microbial community diversity and physical-chemical features of the Southwestern Atlantic Ocean. Arch Microbiol 197:165–179

Amaretti A, Raimondi S, Sala M, Roncaglia L, De Lucia M, Leonardi A, Rossi M (2010) Single cell oils of the cold-adapted oleaginous yeast *Rhodotorula glacialis* DBVPG 4785. Microb Cell Fact 9:73

Antony R, Sanyal A, Kapse N, Dhakephalkar PK, Thamban M, Nair S (2016) Microbial communities associated with Antarctic snow pack and their biogeochemical implications. Microbiol Res 192:192–202

Arenz BE, Held BW, Jurgens JA, Farrell RL, Blanchette RA (2006) Fungal diversity in soils and historic wood from the Ross Sea Region of Antarctica. Soil Biol Biochem 38:3057–3064

Bab'eva IP, Lisichkina GA, Reshetova IS, Danilevich VN (2002) *Mrakia curviuscula* sp. Nov.: a new species of psychrophilic yeasts from forest substrates. Microbiology 71:526–532

Babjeva I, Reshetova I (1998) Yeast resources in natural habitats at polar circle latitude. Food Bioech 36:1–5

Baeza M, Alcaíno J, Cifuentes V, Turchetti B, Buzzini P (2017a) Cold-active enzymes from cold-adapted yeasts. In: Sibirny AA (ed) Biotechnology of yeasts and filamentous fungi. Springer International Publishing, Cham, pp 297–324

Baeza M, Barahona S, Alcaíno J, Cifuentes V (2017b) Amplicon-metagenomic analysis of fungi from Antarctic terrestrial habitats. Front Microbiol 8:35

Baharaeen S, Vishniac HS (1982) *Cryptococcus lupi* sp. nov., an Antarctic basidioblastomycete. Int J Syst Bacteriol 32:229–232

Ballester-Tomás L, Prieto JA, Gil JV, Baeza M, Randez-Gil F (2017) The Antarctic yeast *Candida sake*: understanding cold metabolism impact on wine. Int J Food Microbiol 245:59–65

Barahona S, Yuivar Y, Socias G, Alcaíno J, Cifuentes V, Baeza M (2016) Identification and characterization of yeasts isolated from sedimentary rocks of union glacier at the Antarctica. Extremophiles 20:479–491

Bateni H, Saraeian A, Able C (2017) A comprehensive review on biodiesel purification and upgrading. Biofuel Res J 4:668–690

Baublis JA, Wharton RA Jr, Volz PA (1991) Diversity of micro-fungi in an Antarctic dry valley. J Basic Microbiol 31:1–12

Bergauer P, Fonteyne PA, Nolard N, Schinner F, Margesin R (2005) Biodegradation of phenol and phenol-related compounds by psychrophilic and cold-tolerant alpine yeasts. Chemosphere 59:909–918

Berlemont R, Pipers D, Delsaute M, Angiono F, Feller G, Galleni M, Power P (2011) Exploring the Antarctic soil metagenome as a source of novel cold-adapted enzymes and genetic mobile elements. Rev Argent Microbiol 43:94–103

Białkowska A, Turkiewicz M (2014) Miscellaneous cold-active yeast enzymes of industrial importance. In: Buzzini P, Margesin R (eds) Cold-adapted yeasts: biodiversity, adaptation strategies and biotechnological significance. Springer, Berlin, Heidelberg, pp 377–395

Birgisson H, Delgado O, García Arroyo L, Hatti-Kaul R, Mattiasson B (2003) Cold-adapted yeasts as producers of cold-active polygalacturonases. Extremophiles 7:185–193

Branda E, Turchetti B, Diolaiuti G, Pecci M, Smiraglia C, Buzzini P (2010) Yeast and yeast-like diversity in the southernmost glacier of Europe (Calderone Glacier, Apennines, Italy). FEMS Microbiol Ecol 72:354–369

Brandão LR, Libkind D, Vaz AB, Espírito Santo LC, Moliné M, de García V, van Broock M, Rosa CA (2011) Yeasts from an oligotrophic lake in Patagonia (Argentina): diversity, distribution and synthesis of photoprotective compounds and extracellular enzymes. FEMS Microbiol Ecol 76:1–13

Brandão LR, Vaz ABM, Espírito Santo LC, Pimenta RS, Morais PB, Libkind D, Rosa LH, Rosa CA (2017) Diversity and biogeographical patterns of yeast communities in Antarctic, Patagonian and tropical lakes. Fungal Ecol 28:33–43

Bridge PD, Newsham KK (2009) Soil fungal community composition at Mars Oasis, a southern maritime Antarctic site, assessed by PCR amplification and cloning. Fungal Ecol 2:66–74

Butinar L, Spencer-Martins I, Gunde-Cimerman N (2007) Yeasts in high Arctic glaciers: the discovery of a new habitat for eukaryotic microorganisms. Antonie Van Leeuwenhoek 91:277–289

Butinar L, Strmole T, Gunde-Cimerman N (2011) Relative incidence of ascomycetous yeasts in arctic coastal environments. Microb Ecol 61:832–843

Buzzini P, Margesin R (2014a) Cold-adapted yeasts. Biodiversity, Adaptation Strategies and Biotechnological Significance

Buzzini P, Margesin R (2014b) Cold-adapted yeasts: a lesson from the cold and a challenge for the XXI century. In: Buzzini P, Margesin R (eds) Cold-adapted yeasts. Springer, Berlin, Heidelberg, pp 3–22

Buzzini P, Branda E, Goretti M, Turchetti B (2012) Psychrophilic yeasts from worldwide glacial habitats: diversity, adaptation strategies and biotechnological potential. FEMS Microbiol Ecol 82:217–241

Carrasco M, Rozas JM, Barahona S, Alcaíno J, Cifuentes V, Baeza M (2012) Diversity and extracellular enzymatic activities of yeasts isolated from King George Island, the sub-Antarctic region. BMC Microbiol 12:251

Carrasco M, Villarreal P, Barahona S, Alcaíno J, Cifuentes V, Baeza M (2016) Screening and characterization of amylase and cellulase activities in psychrotolerant yeasts. BMC Microbiol 16:21

Carrasco M, Alcaíno J, Cifuentes V, Baeza M (2017) Purification and characterization of a novel cold adapted fungal glucoamylase. Microb Cell Fact 16:75

Cavicchioli R, Charlton T, Ertan H, Mohd Omar S, Siddiqui KS, Williams TJ (2011) Biotechnological uses of enzymes from psychrophiles. J Microbial Biotechnol 4:449–460

Chen SC, Chen YC, Kwang J, Manopo I, Wang PC, Chaung HC, Liaw LL, Chiu SH (2007) *Metschnikowia bicuspidata* dominates in Taiwanese cold-weather yeast infections of *Macrobrachium rosenbergii*. Dis Aquat Organ 75:191–199

Choudhari S, Lohia R, Grigoriev A (2014a) Comparative metagenome analysis of an Alaskan glacier. J Bioinform Comput Biol 12:1441003

Choudhari S, Dial RJ, Kumar D, Shain DH, Grigoriev A (2014b) Sequence composition diversity in Alaskan glacier and other metagenomes

Coleine C, Selbmann L, Ventura S, D'Acqui LP, Onofri S, Zucconi L (2015) Fungal biodiversity in the Alpine Tarfala Valley. Microorganisms 3:612–624

Connell L, Redman R, Craig S, Scorzetti G, Iszard M, Rodriguez R (2008) Diversity of soil yeasts isolated from South Victoria Land, Antarctica. Microb Ecol 56:448–459

Connell L, Barrett A, Templeton A, Staudigel H (2009) Fungal diversity associated with an active Deep Sea Volcano: Vailulu'u Seamount, Samoa. Geomicrobiol J 26:597–605

Connell LB, Redman R, Rodriguez R, Barrett A, Iszard M, Fonseca A (2010) *Dioszegia antarctica* sp. nov. and *Dioszegia cryoxerica* sp. nov., psychrophilic basidiomycetous yeasts from polar desert soils in Antarctica. Int J Syst Evol Microbiol 60:1466–1472

Culligan EP, Sleator RD, Marchesi JR, Hill C (2014) Metagenomics and novel gene discovery: promise and potential for novel therapeutics. Virulence 5:399–412

D'Amico S, Collins T, Marx JC, Feller G, Gerday C (2006) Psychrophilic microorganisms: challenges for life. EMBO Rep 7:385–389

D'Elia T, Veerapaneni R, Theraisnathan V, Rogers SO (2009) Isolation of fungi from Lake Vostok accretion ice. Mycologia 101:751–763

Daskaya-Dikmen C, Karbancioglu-Guler F, Ozcelik B (2018) Cold active pectinase, amylase and protease production by yeast isolates obtained from environmental samples. Extremophiles 22:599–606

de García V, Brizzio S, Libkind D, Buzzini P, van Broock M (2007) Biodiversity of cold-adapted yeasts from glacial meltwater rivers in Patagonia, Argentina. FEMS Microbiol Ecol 59:331–341

de Garcia V, Brizzio S, van Broock MR (2012) Yeasts from glacial ice of Patagonian Andes, Argentina. FEMS Microbiol Ecol 82:540–550

De Maayer P, Anderson D, Cary C, Cowan DA (2014) Some like it cold: understanding the survival strategies of psychrophiles. EMBO Rep 15:508–517

Delmont TO, Robe P, Cecillon S, Clark IM, Constancias F, Simonet P, Hirsch PR, Vogel TM (2011) Accessing the soil metagenome for studies of microbial diversity. Appl Environ Microbiol 77:1315–1324

di Menna ME (1966) Yeasts in Antarctic soil. Antonie Van Leeuwenhoek 33:25–28

Donovan PD, Gonzalez G, Higgins DG, Butler G, Ito K (2018) Identification of fungi in shotgun metagenomics datasets. PLoS One 13:e0192898

Duarte AW, Dayo-Owoyemi I, Nobre FS, Pagnocca FC, Chaud LC, Pessoa A, Felipe MG, Sette LD (2013) Taxonomic assessment and enzymes production by yeasts isolated from marine and terrestrial Antarctic samples. Extremophiles 17:1023–1035

Duarte AW, Passarini MR, Delforno TP, Pellizzari FM, Cipro CV, Montone RC, Petry MV, Putzke J, Rosa LH, Sette LD (2016) Yeasts from macroalgae and lichens that inhabit the South Shetland Islands, Antarctica. Environ Microbiol Rep 8(5):874–885

Ejdys E, Biedunkiewicz A, Dynowska M, Sucharzewska E (2014) Snow in the city as a spore bank of potentially pathogenic fungi. Sci Total Environ 470-471:646–650

El-Tarabily KA, Sivasithamparam K (2006) Potential of yeasts as biocontrol agents of soil-borne fungal plant pathogens and as plant growth promoters. Mycoscience 47:25–35

Fell JW, Scorzetti G, Connell L, Craig S (2006) Biodiversity of micro-eukaryotes in Antarctic Dry Valley soils with <5% soil moisture. Soil Biol Biochem 38:3107–3119

Feller G (2013) Psychrophilic enzymes: from folding to function and biotechnology. Scientifica (Cairo) 2013:512840

Fernández PM, Martorell MM, Blaser MG, Ruberto LAM, de Figueroa LIC, Mac Cormack WP (2017) Phenol degradation and heavy metal tolerance of Antarctic yeasts. Extremophiles 21:445–457

Filipowicz N, Momotko M, Boczkaj G, Pawlikowski T, Wanarska M, Cieśliński H (2017) Isolation and characterization of phenol-degrading psychrotolerant yeasts. Water Air Soil Pollut 228:210

Frey B, Rime T, Phillips M, Stierli B, Hajdas I, Widmer F, Hartmann M (2016) Microbial diversity in European alpine permafrost and active layers. FEMS Microbiol Ecol 92:fiw018

Furbino LE, Godinho VM, Santiago IF, Pellizari FM, Alves TM, Zani CL, Junior PA, Romanha AJ, Carvalho AG, Gil LH, Rosa CA, Minnis AM, Rosa LH (2014) Diversity patterns, ecology and biological activities of fungal communities associated with the endemic macroalgae across the Antarctic peninsula. Microb Ecol 67:775–787

Gerday C (2013) Psychrophily and catalysis. Biology (Basel) 2:719–741

Gerday C (2014) Fundamentals of cold-active enzymes. In: Buzzini P, Margesin R (eds) Cold-adapted yeasts: biodiversity, adaptation strategies and biotechnological significance. Springer, Berlin, Heidelberg, pp 325–350

Gerday C, Aittaleb M, Bentahir M, Chessa JP, Claverie P, Collins T, D'Amico S, Dumont J, Garsoux G, Georlette D, Hoyoux A, Lonhienne T, Meuwis MA, Feller G (2000) Cold-adapted enzymes: from fundamentals to biotechnology. Trends Biotechnol 18:103–107

Gialleli AI, Ganatsios V, Terpou A, Kanellaki M, Bekatorou A, Koutinas AA, Dimitrellou D (2017) Technological development of brewing in domestic refrigerator using freeze-dried raw materials. Food Technol Biotechnol 55:325–332

Gilichinsky D, Rivkina E, Bakermans C, Shcherbakova V, Petrovskaya L, Ozerskaya S, Ivanushkina N, Kochkina G, Laurinavichuis K, Pecheritsina S, Fattakhova R, Tiedje JM (2005) Biodiversity of cryopegs in permafrost. FEMS Microbiol Ecol 53:117–128

Godinho VM, Furbino LE, Santiago IF, Pellizzari FM, Yokoya NS, Pupo D, Alves TM, Junior PA, Romanha AJ, Zani CL, Cantrell CL, Rosa CA, Rosa LH (2013) Diversity and bioprospecting of fungal communities associated with endemic and cold-adapted macroalgae in Antarctica. ISME J 7:1434–1451

Golddner MC, Zamora MC, Di Leo P, Gianninoto H, Bandoni A (2009) Effect of ethanol level in the perception of aroma attributes and the detection of volatile compounds in red wine. J Sens Stud 24:243–257

Golubev WI (1998) New species of basidiomycetous yeasts, *Rhodotorula creatinovora* and *R. yakutica,* isolated from permafrost soils of Eastern-Siberian Arctic. Mykologiya I Phytopathologiya 32:8–13

Goto S, Sugiyama J, Iizuka H (1969) A taxonomic study of Antarctic yeasts. Mycologia 61:748–774

Gounot A-M (1986) Psychrophilic and psychrotrophic microorganisms. Cell Mol Life Sci 42:1192–1197

Guffogg SP, Thomas-Hall S, Holloway P, Watson K (2004) A novel psychrotolerant member of the hymenomycetous yeasts from Antarctica: *Cryptococcus watticus* sp. nov. Int J Syst Evol Microbiol 54:275–277

Hamid B (2015) Cold-active α-amylase from psychrophilic and psychrotolerant yeast. J Global Biosci 4:2670–2677

Hassan N (2017) Potential of psychrotrophic fungi isolated from Siachen glacier, Pakistan, to produce antimicrobial metabolites. Appl Ecol Env Res 15:1157–1171

Holdgate MW (1977) Terrestrial ecosystems in the Antarctic. Philos Trans R Soc Lond B Biol Sci 279:5–25

Hu H, Yan F, Wilson C, Shen Q, Zheng X (2015) The ability of a cold-adapted *Rhodotorula mucilaginosa* strain from Tibet to control blue mold in pear fruit. Antonie Van Leeuwenhoek 108:1391–1404

Hu H, Wisniewski ME, Abdelfattah A, Zheng X (2017) Biocontrol activity of a cold-adapted yeast from Tibet against gray mold in cherry tomato and its action mechanism. Extremophiles 21:789–803

Jacques N, Zenouche A, Gunde-Cimerman N, Casaregola S (2015) Increased diversity in the genus *Debaryomyces* from Arctic glacier samples. Antonie Van Leeuwenhoek 107:487–501

Johnston ER, Rodriguez-R LM, Luo C, Yuan MM, Wu L, He Z, Schuur EA, Luo Y, Tiedje JM, Zhou J, Konstantinidis KT (2016) Metagenomics reveals pervasive bacterial populations and reduced community diversity across the Alaska tundra ecosystem. Front Microbiol 7:579

Joshi S, Satyanarayana T (2013) Biotechnology of cold-active proteases. Biology (Basel) 2:755–783

Kachalkin AV (2010) New data on the distribution of some psychrophilic yeasts in the Moscow Region. Microbiology 79:843–847

Kanellaki M, Bekatorou A, Koutinas AA (2014) Low-temperature production of wine, beer, and distillates using cold-adapted yeasts. In: Buzzini P, Margesin R (eds) Cold-adapted yeasts. Springer, Berlin, Heidelberg, pp 417–439

Kim M, Lee KH, Yoon SW, Kim BS, Chun J, Yi H (2013) Analytical tools and databases for metagenomics in the next-generation sequencing era. Genomics Inform 11:102–113

Konishi M, Fukuoka T, Nagahama T, Morita T, Imura T, Kitamoto D, Hatada Y (2010) Biosurfactant-producing yeast isolated from *Calyptogena soyoae* (deep-sea cold-seep clam) in the deep sea. J Biosci Bioeng 110:169–175

Kourkoutas Y, Koutinas AA, Kanellaki M, Banat IM, Marchant R (2002) Continuous wine fermentation using a psychrophilic yeast immobilized on apple cuts at different temperatures. Food Microbiol 19:127–134

Krallish I, Gonta S, Savenkova L, Bergauer P, Margesin R (2006) Phenol degradation by immobilized cold-adapted yeast strains of *Cryptococcus terreus* and *Rhodotorula creatinivora*. Extremophiles 10:441–449

Kurtzman CP, Mateo RQ, Kolecka A, Theelen B, Robert V, Boekhout T (2015) Advances in yeast systematics and phylogeny and their use as predictors of biotechnologically important metabolic pathways. FEMS Yeast Res 15:fov050

Kutyna DR, Varela C, Henschke PA, Chambers PJ, Stanley GA (2010) Microbiological approaches to lowering ethanol concentration in wine. Trends Food Sci Technol 21:293–302

Laich F, Vaca I, Chávez R (2013) *Rhodotorula portillonensis* sp. nov., a basidiomycetous yeast isolated from Antarctic shallow-water marine sediment. Int J Syst Evol Microbiol 63:3884–3891

Laich F, Chávez R, Vaca I (2014) *Leucosporidium escuderoi* f.a., sp. nov., a basidiomycetous yeast associated with an Antarctic marine sponge. Antonie Van Leeuwenhoek 105:593–601

Lee J, Lee HT, Hong WY, Jang E, Kim J (2015) FCMM: a comparative metagenomic approach for functional characterization of multiple metagenome samples. J Microbiol Methods 115:121–128

Li SL, Lin Q, Li XR, Xu H, Yang YX, Qiao DR, Cao Y (2012) Biodiversity of the oleaginous microorganisms in Tibetan Plateau. Braz J Microbiol 43:627–634

Libkind D, Brizzio S, Ruffini A, Gadanho M, van Broock M, Paulo Sampaio J (2003) Molecular characterization of carotenogenic yeasts from aquatic environments in Patagonia, Argentina. Antonie Van Leeuwenhoek 84:313–322

Libkind D, Gadanho M, van Broock M, Sampaio JP (2009a) *Cystofilobasidium lacus-mascardii* sp. nov., a basidiomycetous yeast species isolated from aquatic environments of the Patagonian Andes, and *Cystofilobasidium macerans* sp. nov., the sexual stage of *Cryptococcus macerans*. Int J Syst Evol Microbiol 59:622–630

Libkind D, Moliné M, Sampaio JP, van Broock M (2009b) Yeasts from high-altitude lakes: influence of UV radiation. FEMS Microbiol Ecol 69:353–362

Liu J, Sui Y, Wisniewski M, Droby S, Liu Y (2013) Review: utilization of antagonistic yeasts to manage postharvest fungal diseases of fruit. Int J Food Microbiol 167:153–160

Ma LJ, Catranis CM, Starmer WT, Rogers SO (1999) Revival and characterization of fungi from ancient polar ice. Mycologist 13:70–73

Marco ML, Heeney D, Binda S, Cifelli CJ, Cotter PD, Foligné B, Gänzle M, Kort R, Pasin G, Pihlanto A, Smid EJ, Hutkins R (2017) Health benefits of fermented foods: microbiota and beyond. Curr Opin Biotechnol 44:94–102

Margesin R (2007) Alpine microorganisms: useful tools for low-temperature bioremediation. J Microbiol 45:281–285

Margesin R, Miteva V (2011) Diversity and ecology of psychrophilic microorganisms. Res Microbiol 162:346–361

Margesin R, Gander S, Zacke G, Gounot AM, Schinner F (2003) Hydrocarbon degradation and enzyme activities of cold-adapted bacteria and yeasts. Extremophiles 7:451–458

Margesin R, Fonteyne PA, Schinner F, Sampaio JP (2007) *Rhodotorula psychrophila* sp. nov., *Rhodotorula psychrophenolica* sp. nov. and *Rhodotorula glacialis* sp. nov., novel psychrophilic basidiomycetous yeast species isolated from alpine environments. Int J Syst Evol Microbiol 57:2179–2184

Martinez A, Cavello I, Garmendia G, Rufo C, Cavalitto S, Vero S (2016) Yeasts from sub-Antarctic region: biodiversity, enzymatic activities and their potential as oleaginous microorganisms. Extremophiles 20:759–769

Martorell MM, Ruberto LAM, Fernández PM, Castellanos de Figueroa LI, Mac Cormack WP (2017) Bioprospection of cold-adapted yeasts with biotechnological potential from Antarctica. J Basic Microbiol 57:504–516

Mestre MC, Fontenla S, Rosa CA (2014) Ecology of cultivable yeasts in pristine forests in northern Patagonia (Argentina) influenced by different environmental factors. Can J Microbiol 60:371–382

Mishra VK, Goswami R (2018) A review of production, properties and advantages of biodiesel. Biofuels 9:273–289

Montes MJ, Belloch C, Galiana M, Garcia MD, Andrés C, Ferrer S, Torres-Rodriguez JM, Guinea J (1999) Polyphasic taxonomy of a novel yeast isolated from Antarctic environment; description of *Cryptococcus victoriae* sp. nov. Syst Appl Microbiol 22:97–105

Nagahama T, Hamamoto M, Nakase T, Takami H, Horikoshi K (2001) Distribution and identification of red yeasts in deep-sea environments around the northwest Pacific Ocean. Antonie Van Leeuwenhoek 80:101–110

Nakagawa T, Nagaoka T, Taniguchi S, Miyaji T, Tomizuka N (2004) Isolation and characterization of psychrophilic yeasts producing cold-adapted pectinolytic enzymes. Lett Appl Microbiol 38:383–387

Pathan AA, Bhadra B, Begum Z, Shivaji S (2010) Diversity of yeasts from puddles in the vicinity of midre lovénbreen glacier, arctic and bioprospecting for enzymes and fatty acids. Curr Microbiol 60:307–314

Pavlova K, Panchev I, Krachanova M, Gocheva M (2009) Production of an exopolysaccharide by Antarctic yeast. Folia Microbiol (Praha) 54:343–348

Pearce DA, Newsham KK, Thorne MA, Calvo-Bado L, Krsek M, Laskaris P, Hodson A, Wellington EM (2012) Metagenomic analysis of a southern maritime Antarctic soil. Front Microbiol 3:403

Petrescu I, Lamotte-Brasseur J, Chessa JP, Ntarima P, Claeyssens M, Devreese B, Marino G, Gerday C (2000) Xylanase from the psychrophilic yeast *Cryptococcus adeliae*. Extremophiles 4:137–144

Petrovič U (2015) Next-generation biofuels: a new challenge for yeast. Yeast 32:583–593

Pulschen AA, Rodrigues F, Duarte RT, Araujo GG, Santiago IF, Paulino-Lima IG, Rosa CA, Kato MJ, Pellizari VH, Galante D (2015) UV-resistant yeasts isolated from a high-altitude volcanic area on the Atacama Desert as eukaryotic models for astrobiology. Microbiology 4:574–588

Ray MK, Devi KU, Kumar GS, Shivaji S (1992) Extracellular protease from the Antarctic yeast *Candida humicola*. Appl Environ Microbiol 58:1918–1923

Rivkina E, Petrovskaya L, Vishnivetskaya T, Krivushin K, Shmakova L, Tutukina M, Meyers A, Kondrashov F (2015) Metagenomic analyses of the late Pleistocene permafrost – additional tools for reconstruction of environmental conditions. Biogeosci Discuss 12:12091–12119

Robiglio A, Sosa MC, Lutz MC, Lopes CA, Sangorrín MP (2011) Yeast biocontrol of fungal spoilage of pears stored at low temperature. Int J Food Microbiol 147:211–216

Rogers SO, Shtarkman YM, Koçer ZA, Edgar R, Veerapaneni R, D'Elia T (2013) Ecology of subglacial lake Vostok (Antarctica), based on metagenomic/metatranscriptomic analyses of accretion ice. Biology (Basel) 2:629–650

Rondon MR, August PR, Bettermann AD, Brady SF, Grossman TH, Liles MR, Loiacono KA, Lynch BA, MacNeil IA, Minor C, Tiong CL, Gilman M, Osburne MS, Clardy J, Handelsman J, Goodman RM (2000) Cloning the soil metagenome: a strategy for accessing the genetic and functional diversity of uncultured microorganisms. Appl Environ Microbiol 66:2541–2547

Rovati JI, Pajot HF, Ruberto L, Mac Cormack W, Figueroa LI (2013) Polyphenolic substrates and dyes degradation by yeasts from 25 de Mayo/King George Island (Antarctica). Yeast 30:459–470

Russell NJ (1990) Cold adaptation of microorganisms. Philos Trans R Soc Lond B Biol Sci 326:595–608; discussion 608

Sangorrín MP, Lopes CA, Vero S, Wisniewski M (2014) Cold-adapted yeasts as biocontrol agents: biodiversity, adaptation strategies and biocontrol potential. In: Buzzini P, Margesin R (eds) Cold-adapted yeasts. Springer, Berlin, Heidelberg, pp 441–464

Sansone C, Rita Massardo D, Pontieri P, Maddaluno L, De Stefano M, Maurizio Tredici S, Talà A, Alifano P, Del Giudice L (2007) Isolation of a psychrotolerant strain from fermented tea plant leaves. J Plant Interact 2:169–174

Sarmiento F, Peralta R, Blamey JM (2015) Cold and hot extremozymes: industrial relevance and current trends. Front Bioeng Biotechnol 3:148

Schmidtke LM, Blackman JW, Agboola SO (2012) Production technologies for reduced alcoholic wines. J Food Sci 77:R25–R41

Scorzetti G, Petrescu I, Yarrow D, Fell JW (2000) *Cryptococcus adeliensis* sp. nov., a xylanase producing basidiomycetous yeast from Antarctica. Antonie Van Leeuwenhoek 77:153–157

Selbmann L, Turchetti B, Yurkov A, Cecchini C, Zucconi L, Isola D, Buzzini P, Onofri S (2014a) Description of *Taphrina antarctica* f.a. sp. nov., a new anamorphic ascomycetous yeast species associated with Antarctic endolithic microbial communities and transfer of four *Lalaria* species in the genus *Taphrina*. Extremophiles 18:707–721

Selbmann L, Zucconi L, Onofri S, Cecchini C, Isola D, Turchetti B, Buzzini P (2014b) Taxonomic and phenotypic characterization of yeasts isolated from worldwide cold rock-associated habitats. Fungal Biol 118:61–71

Simon C, Wiezer A, Strittmatter AW, Daniel R (2009) Phylogenetic diversity and metabolic potential revealed in a glacier ice metagenome. Appl Environ Microbiol 75:7519–7526

Singh P, Singh SM (2012) Characterization of yeast and filamentous fungi isolated from cryoconite holes of Svalbard, Arctic. Polar Biol 35:575–583

Starmer W, Fell J, Catranis C, Aberdeen V, Ma L, Zhou S, Rogers S (2005) Yeasts in the genus *Rhodotorula* recovered from the Greenland ice sheet. In: Castello JD, Rogers SO (eds) Life in ancient ice. Princeton University Press, Princeton, NJ, pp 181–195

Thomas-Hall S (2002) *Cryptococcus nyarrowii* sp. nov., a basidiomycetous yeast from Antarctica. Int J Syst Evol Microbiol 52:1033–1038

Thomas-Hall SR, Turchetti B, Buzzini P, Branda E, Boekhout T, Theelen B, Watson K (2010) Cold-adapted yeasts from Antarctica and the Italian Alps-description of three novel species: *Mrakia robertii* sp. nov., *Mrakia blollopis* sp. nov. and *Mrakiella niccombsii* sp. nov. Extremophiles 14:47–59

Tiquia SM, Mormile M (2010) Extremophiles-a source of innovation for industrial and environmental applications. Environ Technol 31(8–9):823

Tiquia-Arashiro SM, Mormile M (2013) Sustainable technologies: bioenergy and biofuel from biowaste and biomass. Environ Technol 34(13):1637–1805

Tiquia-Arashiro SM, Rodrigues D (2016) Thermophiles and psychrophiles in nanotechnology. In: Extremophiles: applications in nanotechnology. Springer International Publishing, New York, pp 89–127

Trochine A, Turchetti B, Vaz ABM, Brandao L, Rosa LH, Buzzini P, Rosa C, Libkind D (2017) Description of *Dioszegia patagonica* sp. nov., a novel carotenogenic yeast isolated from cold environments. Int J Syst Evol Microbiol 67:4332–4339

Tsuji M, Fujiu S, Xiao N, Hanada Y, Kudoh S, Kondo H, Tsuda S, Hoshino T (2013a) Cold adaptation of fungi obtained from soil and lake sediment in the Skarvsnes ice-free area, Antarctica. FEMS Microbiol Lett 346:121–130

Tsuji M, Yokota Y, Shimohara K, Kudoh S, Hoshino T (2013b) An application of wastewater treatment in a cold environment and stable lipase production of Antarctic basidiomycetous yeast *Mrakia blollopis*. PLoS One 8:e59376

Tsuji M, Goshima T, Matsushika A, Kudoh S, Hoshino T (2013c) Direct ethanol fermentation from lignocellulosic biomass by Antarctic basidiomycetous yeast *Mrakia blollopis* under a low temperature condition. Cryobiology 67:241–243

Turchetti B, Buzzini P, Goretti M, Branda E, Diolaiuti G, D'Agata C, Smiraglia C, Vaughan-Martini A (2008) Psychrophilic yeasts in glacial environments of Alpine glaciers. FEMS Microbiol Ecol 63:73–83

Turchetti B, Thomas Hall SR, Connell LB, Branda E, Buzzini P, Theelen B, Müller WH, Boekhout T (2011) Psychrophilic yeasts from Antarctica and European glaciers: description of *Glaciozyma* gen. nov., *Glaciozyma martinii* sp. nov. and *Glaciozyma watsonii* sp. nov. Extremophiles 15:573–586

Turchetti B, Goretti M, Branda E, Diolaiuti G, D'Agata C, Smiraglia C, Onofri A, Buzzini P (2013) Influence of abiotic variables on culturable yeast diversity in two distinct Alpine glaciers. FEMS Microbiol Ecol 86:327–340

Tutino ML, Parrilli E, De Santi C, Giuliani M, Marino G, de Pascale D (2010) Cold-adapted esterases and lipases: a biodiversity still under-exploited. Curr Chem Biol 4:74–83

Uetake J, Yoshimura Y, Nagatsuka N, Kanda H (2012) Isolation of oligotrophic yeasts from supra-glacial environments of different altitude on the Gulkana Glacier (Alaska). FEMS Microbiol Ecol 82:279–286

Vaca I, Faúndez C, Maza F, Paillavil B, Hernández V, Acosta F, Levicán G, Martínez C, Chávez R (2013) Cultivable psychrotolerant yeasts associated with Antarctic marine sponges. World J Microbiol Biotechnol 29:183–189

van Bussel BCT, Henry RMA, Schalkwijk CG, Dekker JM, Nijpels G, Feskens EJM, Stehouwer CDA (2018) Alcohol and red wine consumption, but not fruit, vegetables, fish or dairy products, are associated with less endothelial dysfunction and less low-grade inflammation: the Hoorn Study. Eur J Nutr 57:1409–1419

Varela C, Kutyna DR, Solomon MR, Black CA, Borneman A, Henschke PA, Pretorius IS, Chambers PJ (2012) Evaluation of gene modification strategies for the development of low-alcohol-wine yeasts. Appl Environ Microbiol 78:6068–6077

Varin T, Lovejoy C, Jungblut AD, Vincent WF, Corbeil J (2012) Metagenomic analysis of stress genes in microbial mat communities from Antarctica and the High Arctic. Appl Environ Microbiol 78:549–559

Vasileva-Tonkova E, Romanovskaya V, Gladka G, Gouliamova D, Tomova I, Stoilova-Disheva M, Tashyrev O (2014) Ecophysiological properties of cultivable heterotrophic bacteria and yeasts dominating in phytocenoses of Galindez Island, maritime Antarctica. World J Microbiol Biotechnol 30:1387–1398

Vaz AB, Rosa LH, Vieira ML, de Garcia V, Brandão LR, Teixeira LC, Moliné M, Libkind D, van Broock M, Rosa CA (2011) The diversity, extracellular enzymatic activities and photoprotective compounds of yeasts isolated in Antarctica. Braz J Microbiol 42:937–947

Vero S, Garmendia G, González MB, Bentancur O, Wisniewski M (2013) Evaluation of yeasts obtained from Antarctic soil samples as biocontrol agents for the management of postharvest diseases of apple (Malus × domestica). FEMS Yeast Res 13:189–199

Viñarta SC, Angelicola MV, Barros JM, Fernández PM, Mac Cormak W, Aybar MJ, de Figueroa LI (2016) Oleaginous yeasts from Antarctica: screening and preliminary approach on lipid accumulation. J Basic Microbiol 56:1360–1368

Vishniac HS (2002) *Cryptococcus tephrensis*, sp. nov., and *Cryptococcus heimaeyensis*, sp. nov.; new anamorphic basidiomycetous yeast species from Iceland. Can J Microbiol 48:463–467

Vishniac HS (2006) Yeast biodiversity in the Antarctic. In: Rosa CA, Péter G (eds) Biodiversity and ecophysiology of yeasts. Springer, Berlin, pp 419–440

Voget S, Leggewie C, Uesbeck A, Raasch C, Jaeger KE, Streit WR (2003) Prospecting for novel biocatalysts in a soil metagenome. Appl Environ Microbiol 69:6235–6242

Wilkins D, Yau S, Williams TJ, Allen MA, Brown MV, DeMaere MZ, Lauro FM, Cavicchioli R (2013) Key microbial drivers in Antarctic aquatic environments. FEMS Microbiol Rev 37:303–335

Xie D (2017) Integrating cellular and bioprocess engineering in the non-conventional yeast *Yarrowia lipolytica* for biodiesel production: a review. Front Bioeng Biotechnol 5:65

Xin MX, Zhou PJ (2007) *Mrakia psychrophila* sp. nov., a new species isolated from Antarctic soil. J Zhejiang Univ Sci B 8:260–265

Yergeau E, Hogues H, Whyte LG, Greer CW (2010) The functional potential of high Arctic permafrost revealed by metagenomic sequencing, qPCR and microarray analyses. ISME J 4:1206–1214

Zalar P, Gunde-Cimerman N (2014) Cold-adapted yeasts in Arctic habitats. In: Buzzini P, Margesin R (eds) Cold-adapted yeasts. Springer, Berlin, Heidelberg, pp 49–74

Zalar P, Gostincar C, de Hoog GS, Ursic V, Sudhadham M, Gunde-Cimerman N (2008) Redefinition of *Aureobasidium pullulans* and its varieties. Stud Mycol 61:21–38

Zhang X, Hua M, Song C, Chi Z (2012) Occurrence and diversity of marine yeasts in Antarctica environments. J Ocean Univ China 11:70–74

Zhang T, Zhang YQ, Liu HY, Wei YZ, Li HL, Su J, Zhao LX, Yu LY (2013) Diversity and cold adaptation of culturable endophytic fungi from bryophytes in the Fildes Region, King George Island, maritime Antarctica. FEMS Microbiol Lett 341:52–61

Zhang T, Zhao L, Yu C, Wei T, Yu L (2017) Diversity and bioactivity of cultured aquatic fungi from the High Arctic region. Adv Polar Sci 28:2942

Chapter 14
Pharmaceutical Applications of Thermophilic Fungi

Gurram Shyam Prasad ⓘ

14.1 Introduction

Microbial life is not only restricted to specific environmental conditions like moderate temperature and neutral pH, where salinity, hydrostatic pressure, and ionizing radiations are low. Large group of microbial communities were also found in most diverse environments including extremes of temperature, pressure, salinity, and pH. The microorganisms thriving optimally under one or several of these diverse conditions for their growth are termed as extremophiles which include acidophiles, alkalophiles, halophiles, psychrophiles, thermophiles, hyperthermophiles, radioresistant microbes, barophiles, and endoliths. The term extremophile was used for the first time by MacElroy in 1974 (Gomes and Steiner 2004). The extremophiles that have been identified to date belong to the domain of the archaea. However, these have also been identified in eubacterial and eukaryotic organisms (Burg 2003). Among extremophiles, thermophilic fungi are a small group of mycota with an exceptional mechanism of growing at an elevated temperature of at or above 50 °C. Thermophily in these fungi is not as extreme as in eubacteria or archaea which are able to grow near or above 100 °C in thermal springs, solfatara fields, or hydrothermal vents (Brock 1995; Blohl et al. 1997). The fungi which grow at or above 20 °C and attain maximum growth at or above 50 °C are thermophilic while thermotolerant forms opt 20–55 °C for growth.

Thermophilic fungi form a diverse group of organisms reported from various natural habitats, viz. soils and in habitats where plant material decomposition takes place which includes compost, wood chip piles, nesting material of birds, and

G. S. Prasad (✉)
Department of Microbiology, Vaagdevi Degree and Post Graduate College,
Warangal, Telangana State, India

© Springer Nature Switzerland AG 2019
S. M. Tiquia-Arashiro, M. Grube (eds.), *Fungi in Extreme Environments:
Ecological Role and Biotechnological Significance*,
https://doi.org/10.1007/978-3-030-19030-9_14

animal and municipal refuse (Prasad et al. 2011). These are also reported from stored groundnuts (Ogundero 1981), forest soils (Sandhu and Singh 1985), mushroom composts (Ross and Harris 1983; Miller et al. 1990; Weigant 1992; Stratsma et al. 1994; Johri and Rajani 1999), desert soils (Mouchacca 1995), vermicomposts (Anastasi et al. 2005), hot springs (Lin et al. 2005), and alkalescent thermal springs (Pan et al. 2010). Tubaki et al. (1974) have also recorded thermophilic fungi even from aquatic environments. These fungi survive the stress such as oxygen and desiccation (Mahajan et al. 1986). Majority of the fungi were isolated from herbivore dung and bird nest materials (Mouchacca 1997, 1999). Thermophilic fungi are the principal components of the microflora that develops in heaped stacks of plant material, piles of agricultural and forestry products, and other accumulations of organic matter where the warm, humid, and aerobic environment provides the ideal conditions for their growth (Allen and Emerson 1949). In these habitats, thermophiles may occur as resting propagules or as active mycelium depending on the availability of nutrients and favorable environmental conditions (Khushaldas 2009). The exothermic reactions of the saprophytic mesophilic microflora raise the temperature of the substratum to 40 °C, resulting in a warm environment favoring the germination of spores of thermophilic microflora, and eventually the latter outgrows the mesophilic microorganisms. The thermophilic fungi constitute a heterogeneous physiological group of various genera in the Mastigomycotina, Ascomycotina, Deuteromycotina, and Mycelia Sterilia (Mouchacca 1997). Most significant natural habitats for saprophytic thermophilic fungi are the decomposing organic materials in which thermogenic conditions resulted in the activity of mesophilic microorganisms (Eggins and Coursey 1964). Thermophilic fungi are worldwide in distribution and most species do not show any geographical restrictions and their gene pool is still uncertain.

The first of the known thermophilic fungi, *Mucor pusillus*, was isolated from bread over a century ago by Lindt (1886). A little later, Tsiklinskaya discovered another thermophilic fungus, *Thermomyces lanuginosus*, growing on potato which had been inoculated with garden soil (Tsiklinskaya 1899). Several workers including Cooney and Emerson (1964), Chang (1967), Eggins and Malik (1969), Evans (1971), Crisan (1973), Ofosu-Asiedu and Smith (1973), Tansey and Brock (1978), Kuthubutheen (1983), Ito et al. (1992), Johri and Satyanarayana (1986), Sharma and Johri (1992), Maheshwari (1997), Niehaus et al. (1999), Kohilu et al. (2001), Cordova et al. (2003), Moloney et al. (2004), Fulleringer et al. (2005), Salar and Aneja (2007), Khushaldas (2009), Pan et al. (2010), and Sreelatha et al. (2018) have isolated these fungi from different ecological niches.

Though thermophilic fungi are recognized for wide biotechnological applications, their application in the pharmaceutical industry is unexplored. In this review, potential pharmaceutical applications of thermophilic fungi are discussed.

14.2 Pharmaceutical Applications of Thermophilic Fungi

14.2.1 Biotransformation of Organic Compounds for Bioactive Compound Synthesis

Biotransformation is the method by which an organism or its enzyme brings out minor chemical alterations on compounds that are not part of their metabolism and result in a novel or more useful product than its parent compound which is difficult or impossible to obtain by conventional chemical methods (Prasad et al. 2010). Almost all types of chemical reactions are made achievable by microbial transformations. Further, this process is hazard free, and minimizes the problems of isomerization, racemization, epimerization, and rearrangement that generally occur during the chemical process. Though biotransformation reactions employing mesophilic microbial cultures are well documented, such reactions employing thermophilic organisms are limited. Thermophilic enzymes show thermostability and offer many major advantages over mesophiles (Nguyen et al. 2013; Tiquia-Arashiro 2014). The main advantages of performing biocatalytic reactions using thermophiles are reduced risk of mesophilic microbial contamination, lower viscosity of the medium, improved transfer rates, and improved solubility of substrates (Bruce et al. 1991; Lasa and Berenguer 1994). They are also known to withstand denaturants of extreme acidic and alkaline conditions (Tiquia-Arashiro and Rodrigues 2016a). Thermophilic enzymes, when cloned and expressed in mesophilic hosts, are easy to purify by heat treatment. Their thermostability is also associated with higher resistance to chemical solvents (Pomaranski and Tiquia-Arashiro 2016; Tiquia-Arashiro and Rodrigues 2016b). These reactions are the subject of increasing interest in the pharmaceutical industry because of the demand for enantiomerically pure compounds (Schulze and Wubbolts 1999). Some of the examples of biotransformation reactions employing thermophilic fungi are reported in this section.

14.2.1.1 Biotransformation of Albendazole to Albendazole Sulfoxide by *Rhizomucor pusillus*

Albendazole is a benzimidazole carbamate with a broad-spectrum anti-helminthic activity. It is marketed as a prodrug and after administration; it is rapidly metabolized by oxidation in the liver to form its active metabolite albendazole sulfoxide (Prasad et al. 2008). Synthesis of albendazole sulfoxide chemically is difficult as it is site-specific reaction where oxidation has to take place at sulfur group while the process also leads to environmental pollution. Biotransformation is the best alternative to synthesize albendazole sulfoxide. The mesophilic fungal culture *Cunninghamella blakesleeana* NCIM 687 was reported to transform albendazole to albendazole sulfoxide in a single-step reaction with a yield of 16% at 27 °C (Prasad et al. 2008) while thermophilic fungus *Rhizomucor pusillus* NRRL 28626 could transform albendazole to albendazole sulfoxide with a yield of 60% at an incubation temperature of 45 °C which is far more superior to mesophilic fungus (Prasad et al. 2011).

14.2.1.2 Biotransformation of Ferulic Acid to Guaiacol by *Sporotrichum thermophile*

Vanillic acid which is used as the starting material in chemical synthesis of oxygenated aromatic chemicals such as vanillin, one of the most universally used aromatic molecules in the pharmaceutical, food, and cosmetic industries, was produced from ferulic acid employing thermophilic fungal culture *Sporotrichum thermophile* at a temperature of 50 °C by propenoic chain degradation via an intermediate compound 4-vinyl guaiacol with very high levels of 4798 mg/L with a molar yield of 35% (Topakas et al. 2003). Initially, ferulic acid is decarboxylated to 4-vinyl guaiacol which is further converted to vanillic acid which later undergoes a nonoxidative decarboxylation to guaiacol. The thermophilic fungus *Sporotrichum thermophile* has a considerable potential in performing decarboxylation reaction and synthesis of high yields of vanillic acid compared to mesophilic fungal cultures *Paecilomyces variotii* 3.2 mg/L (Rahouti et al. 1989), *Aspergillus niger* 920 mg/l (Meessen et al. 1996), *Aspergillus niger* 357 mg/L (Meessen et al. 1999), *Schizophyllum commune* 0.13 mM (Ghosh et al. 2005), *Paecilomyces variotii* 115 mg/L (Ghosh et al. 2006), and *Aspergillus niger* K8 116mg/L (Motedayan et al. 2013).

14.2.1.3 Biotransformation of Steroids by Thermophilic Fungi

The most important group of pharmaceuticals are steroid drugs which are effective as antiphlogistics, progestational, male and female sex hormones, blood pressure-regulating agents (Sayanarayana and Chavant 1987), and anabolic, antitumor, sedative, as well as oral contraceptives (Zohri and Abdel-Galil 1999). They are also effective in allergic, dermatologic, and ocular diseases and in cardiovascular therapy (Zohri and Abdel-Galil 1999). Some of them are also used in veterinary medicine. Commercial production of steroids is very much needed whose chemical synthesis requires multistep reactions and is a costly affair. For example cortisone steroid can be synthesized chemically from deoxycholic acid which requires 37 steps and must be carried out under extreme condition of temperature and pressure resulting in high production cost. But, with microbial transformation, the number of steps both chemical and microbial was reduced to two with reduced production cost. Not only mesophilic microorganisms, but also thermophilic fungi are potential enough to synthesize steroid drugs in eco-friendly and economical manner with many more advantages. Cholestenone, which represents an important group of pharmaceuticals, is used against obesity, liver disease, and keratinization and also serves as a precursor for the synthesis of other drug intermediates like androst-4-ene-3,17-dione and androsta-1,4-diene-3,17-dione; it also serves as a major starting material for production of anabolic drugs and contraceptive hormones (Wu et al. 2015). Chemical synthesis of cholestenone is a difficult task as the molecule consists of asymmetric centers and requires harmful chemical solvents. Sayanarayana and Chavant (1987) could efficiently transform cholesterol in a single step employing thermophilic fungi *Acremonium alabamensis* and *Talaromyces emersonii* to cholestenone in an eco-friendly and cost-effective way. Similarly, different biotransformation products of steroids produced by using thermophilic fungi are shown in Table 14.1.

14.2.2 Thermophilic Fungi in Predicting Mammalian Drug Metabolism

Biotransformation is the principal route of elimination of drugs from the body which is metabolized upon administration by a series of enzymatic conversions leading to chemical alteration to more polar and hydrophilic metabolites for easy excretion (Rowland and Tozer 1995). The products of the metabolism are called metabolites which may be inactive or they may have a similar or different degree of therapeutic activity or toxicity than the original drug. The liver is the principal site of drug metabolism and significant levels of drug-metabolizing enzymes also occur in other organs. The biotransformation reactions occur in the liver in two stages classified as phase I and phase II. Phase I reactions occur in microsomes and are catalyzed by a group of enzymes known as cytochrome P450 system that plays a significant role in drug metabolism. The common chemical reactions involved in phase I are aromatic hydroxylations, oxidative *N*-dealkylation, S-oxidation, reduction, and hydrolysis. Phase II reactions occur in liver cells where the parent or the metabolite from phase I gets conjugated by glucuronidation, sulfation, amino acid conjugation, acylation, methylation, or glutathione conjugation to facilitate elimination (Gunaratna 2000).

In the course of drug innovation, understanding drug metabolism and metabolite toxicity is very crucial. Prior to consent to use in humans, a wide range of studies to establish safety and efficacy are mandatory which can be known by drug metabolism studies. Different animal models, microsomal preparations, perfused organ systems, etc. are presently accessible for drug metabolism studies but are associated with some disadvantages. Further, for evaluating pharmacological and toxicological studies large quantities of metabolites are required which is difficult with animal or with in vitro models while chemical synthesis of metabolites in a laboratory is also a tedious and costly process. Microorganisms, especially most of the fungi, were found to possess cytochrome p450 enzyme (CYP450) system and oxidize organic compounds in the same way as mammalian hepatic CYP450 enzymes. Many mammalian phase I (introduction of a functional group) and phase II metabolic reactions (conjugation with endogenous compounds) also occur in microbial models (Abourashed et al. 1999). The filamentous fungi, especially of the

Table 14.1 Some of the examples of biotransformation reactions mediated by thermophilic fungi

Thermophilic fungi	Substrate	Transformation products	Reference
Acremonium alabamensis	Cholesterol Stigmasterol Sitosterol	Cholestenone Stigmastadienone Stigmastadienone	Sayanarayana and Chavant (1987)
Talaromyces emersonii	Cholesterol	Cholestenone	
Humicola fuscoatra	Progesterone	Androst-4-ene-3,17-dione Testosterone Testololactone	Zohri and Abdel-Galil (1999)
H. grisea	Progesterone	Androst-4-ene-3, 17-dione Testosterone Testololactone	Zohri and Abdel-Galil (1999)
H. hyalothermophila	Progesterone	11α-Hydroxyprogesterone 11β-Hydroxyprogesterone 17α -Hydroxyprogesterone 21-Hydroxyprogesterone 11α,17α-Dihydroxy progesterone 11α,17α,21-Trihydroxyprogesterone (epicortisol) 1β,17α,21-Trihydroxyprogesterone (cortisol)	Zohri and Abdel-Galil (1999)
Rhizomucor tauricus	Progesterone	6β-Hydroxyprogesterone 6β-11α-Dihydroxyprogesterone 6β-11α-Diacetoxyprogesterone	Hunter et al. (2007)
	Testosterone	6β-Hydroxy-testosterone 12β-Hydroxy-testosterone	
	Pregnenolone	3β,7β,12β-Trihydroxypregn-5en-20one	
	Androst-4-ene-3, 17-dione	6α-Hydroxy-androst-4-ene-3,17-dione 7α-Hydroxy-androst-4-ene-3,17-dione 6β-Hydroxy-androst-4-ene-3,17-dione 6β-11α-Dihydroxy-androst-4-ene-3,17-dione	
	Dehydroepiandrosterone	3β,7α-Dihydroxy-androst-5en-17one 3β,7β-Dihydroxy-androst-5en-17-one	

genus *Cunninghamella*, are well documented as a model of mammalian biotransformation (Sun et al. 2004). The use of fungi as microbial models in predicting mammalian models is associated with many advantages like low cost, ease of handling, scale-up capacity, and potential to reduce the use of animals (Zhang et al. 2006). Further, fungi can also be used as metabolic factories to produce huge quantities of metabolites which is not possible with other available models. This mode of drug metabolism studies with microbial models and producing metabolites in large quantities using fungi is very convenient and a preparative method for otherwise difficult-to-obtain ones particularly when the structure of metabolite is complex. This approach has been successfully employed by many researchers comparing drug metabolism using mesophilic fungal and mammalian systems (Sun et al. 2004; Zhong et al. 2003; Cha et al. 2001; Moody et al. 1999, 2000; Hezari and Devis 1993) but the potential thermophilic fungi are unexplored in this area. Very few researchers (Prasad et al. 2011, 2018; Sreelatha et al. 2018) studied drug metabolism employing thermophilic fungi.

14.2.2.1 Metabolic Studies of Losartan Using *Rhizomucor pusillus* NRRL 28626

Losartan is the first of new class of antihypertensive drugs (Siegl 1993) and substrate for CYP 2C9 and CYP 3A4 enzymes. It is metabolized in humans by oxidation of C5-hydroxy methyl to the active metabolite carboxylic acid by CYP450 3A4 and by CYP450 2C9 (Lee et al. 2003). The aldehyde metabolite is also observed in human as an intermediate in the oxidation of losartan to carboxylic acid metabolite (Stearns et al. 1995), which is excreted in urine and feces as a conjugate (Boris and Bernahard 2003; Yun et al. 1995). The other routes of metabolism include C-1′, C-3′ hydroxylations and *N*-2 tetrazole glucuronidation. Similarly, the thermophilic fungus *Rhizomucor pusillus* NRRL 28626 biotransformed losartan to five metabolites (Fig.14.1), viz. glucuronic conjugate of losartan (M1), 3-hydroxy-*N*-acetyl losartan (M2); the metabolite M3 was produced by oxidation and acylation of demethylated and dechlorinated parent compound losartan; the other metabolite of losartan produced by *Rhizomucor pusillus* was found to be *N*-acetylated carboxylic acid compound (M4); and the metabolite M5 of losartan produced by the fungus was by decarboxylation of a carboxylic acid metabolite of losartan (Prasad and Srisailam 2018). The metabolite M1 produced by *Rhizomucor pusillus* was also detected by Huskey et al. (1993) using liver microsomes of rat, monkeys, and humans while Krieter et al. (1995) reported in rat intestine. This metabolite was reported to catalyze by UGT superfamily of enzymes in animals (Prasad and Srisailam 2018). The metabolite M2 produced by *Rhizomucor pusillus* is by *N*-acetylation reaction catalyzed by *N*-acetyl transferase in mammals which is a well-known reaction (Prasad et al. 2018). The other metabolite M4 produced by thermophilic fungus was also recorded by Yun et al. (1995) in the mammalian metabolic pathway of losartan catalyzed by CYP3A4.

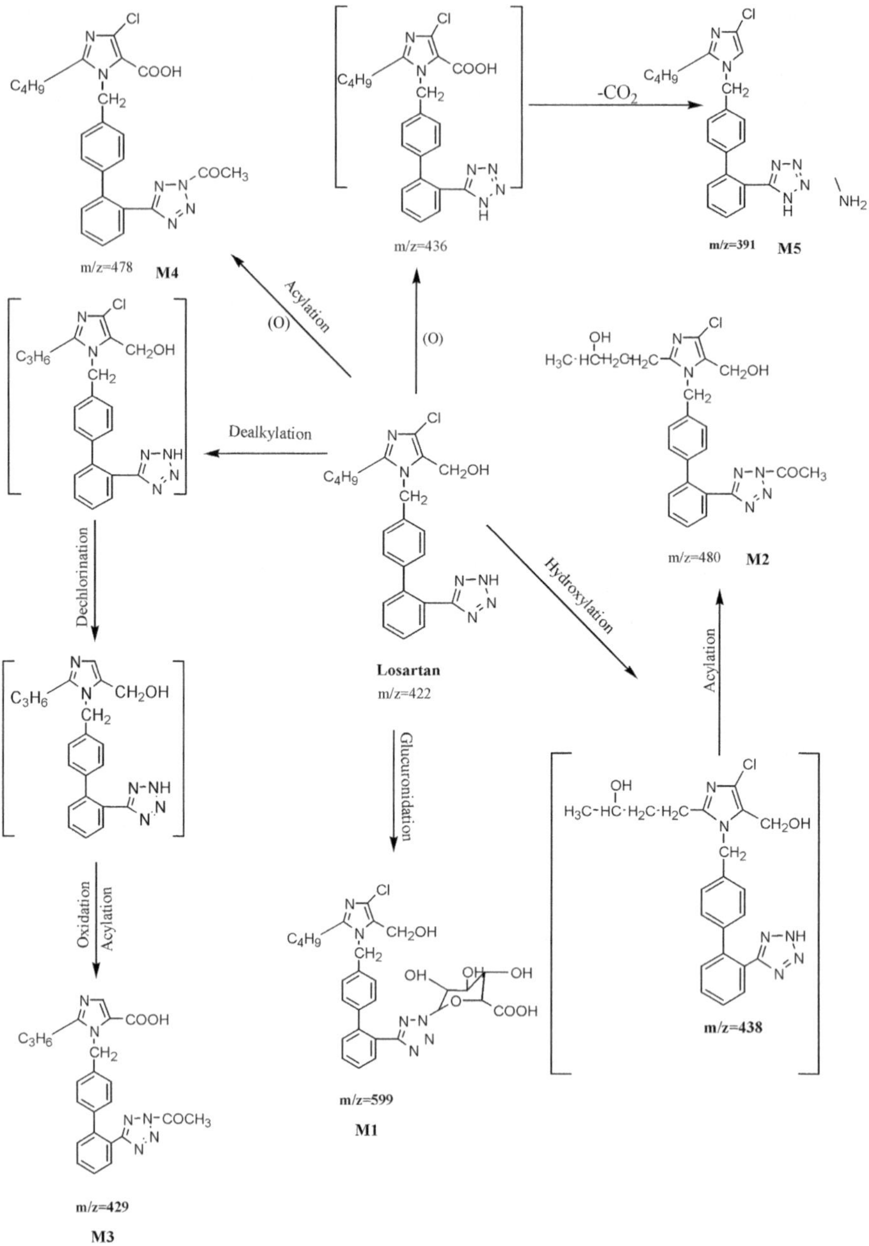

Fig. 14.1 Metabolic pathway of losartan by *Rhizomucor pusillus*. *Source* Prasad and Srisailam (2018).Copyright © International Research Journal of Natural and Applied Sciences (Associated Asia Research Foundation) Reproduced with permission

The fungus *R. pusillus* transformed losartan to five metabolites by oxidation, hydroxylation, acylation, dechlorination, dealkylation, and glucuronidation of losartan which clearly states that this thermophilic fungus *R. pusillus* NRRL 28626 has the ability to catalyze diverse reactions compared to mesophilic fungi. Similarly, the metabolites of losartan produced by the fungus are similar to metabolites of losartan reported in mammals which clearly states that similar type of enzyme system exists in mammals and this fungus.

14.2.2.2 Metabolism of Albendazole by *Rhizomucor pusillus* NRRL 28626

Benzimidazole anti-helminthics with a sulfide group are the most active against intestinal nematodes in humans, as well as in animals. Albendazole is a benzimidazole carbamate with a broad antiparasitic spectrum. The metabolic studies of albendazole have been shown to follow similar pathways in various mammals. These metabolic conversions included oxidation at sulfur alkyl and aromatic hydroxylation, methylation at both nitrogen and sulfur, and carbamate hydrolysis (Gyurik et al. 1981). Albendazole sulfoxide and albendazole sulfone were identified in plasma after oral administration in several species, viz., rat, human, porcine, ovine, bovine, caprine, and chicken (Penicaut et al. 1983; El Amri et al. 1987; Benchaoui et al. 1993; Lanusse et al. 1993; McKellar et al. 1993; Moroni et al. 1995; Csiko et al. 1996). The flavin-containing monooxygenases and cytochrome P-450 (CYP, mainly CYP3A in rat) appear to mediate the conversion of albendazole to albendazole sulfoxide, whereas the biotransformation of albendazole sulfoxide to albendazole sulfone is influenced by CYP only (CYP1A in rat) (Prasad et al. 2011). The drug is rapidly metabolized by oxidation in the liver to form its sulfoxide and sulfone. In animals after parenteral, oral, or intraruminal administration, albendazole is rapidly oxidized to the sulfoxide (Marriner and Bogan 1980); later, albendazole sulfoxide undergoes bioconversion to albendazole sulfone which is pharmacologically inactive (Lacey 1990). Similarly, the thermophilic fungus *Rhizomucor pusillus* NRRL 28626 was reported (Prasad et al. 2011) to biotransform albendazole to albendazole sulfoxide (M1), albendazole sulfone (M2), the major mammalian metabolites of albendazole reported previously and an *N*-methyl metabolite of albendazole sulfoxide (M3), and other metabolite (M4) which clearly states that metabolic pattern of mammals and the thermophilic fungus *Rhizomucor pusillus* NRRL 28626 are similar (Fig.14.2). Hence, this fungus can be used for studying mammalian drug metabolism and large quantities of metabolites can be produced for different pharmacological studies which are not possible with mammals.

14.2.2.3 Metabolism of Spironolactone by *Thermomyces lanuginosus* NCIM-1934

Spironolactone, a mineralocorticoid receptor antagonist and a potassium-sparing diuretic (Brunton et al. 2008), is used for the treatment of congestive heart failure, edema, and ascites in cirrhosis and primary hyperaldosteronism. It is reported to

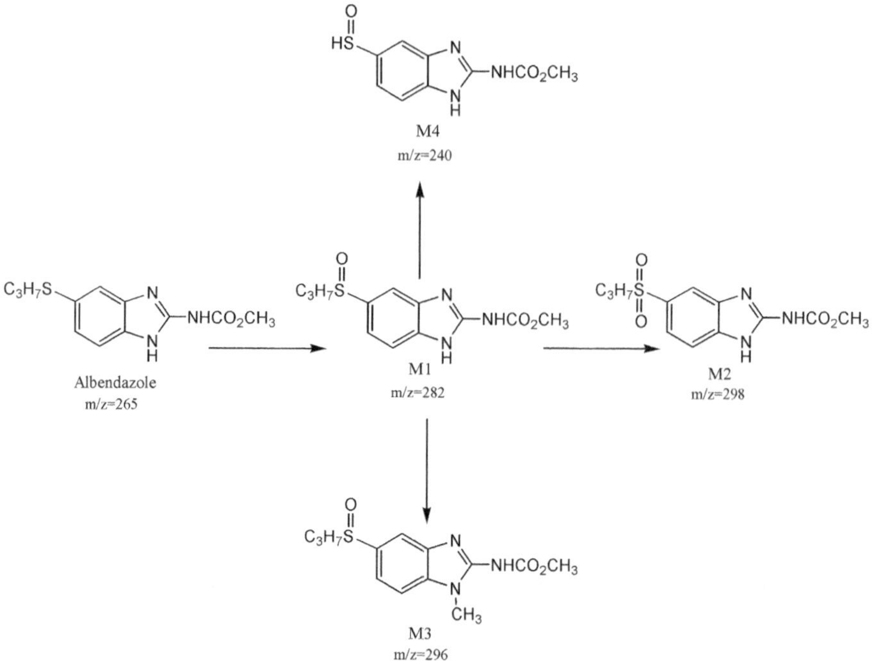

Fig. 14.2 Metabolic pathway of albendazole by *Rhizomucor pusillus*. *Source* Prasad et al. (2011). Copyright © Applied Biochemistry and Biotechnology (Springer Science + Business Media, LLC 2011) Reproduced with permission

biotransform in humans into 7α-thiomethylspironolactone, 6β-hydroxy-7α-thiomethyl spironolactone, and canrenone (IARC Monograph). On the other hand, Sreelatha et al. (2018) in their biotransformation studies employing thermophilic fungus *Thermomyces lanuginosus* reported four metabolites (Fig.14.3) of spirono-lactone, viz. 7-α thiospironolactone (M1), canrenone (M2), 7-α thiomethyl spirono-lactone (M3), 6β-OH-7α-thiomethyl spironolactone (M4), the major mammalian metabolites reported previously in different mammalian, and in vitro models. The metabolite (M1) was recorded earlier by Sherry et al. (1981) in microsomal prepara-tions of guinea pig liver, adrenals, kidneys, and testes. However, Overdiek and Merkus (1987) recorded this metabolite (M1) in humans and Los et al. (1993) in plasma and organs of guinea pigs. The metabolite M2 produced by *Thermomyces lanuginosus* was also recorded earlier by many researchers in different models (Overdiek and Merkus 1987; Gardiner et al. 1989; LaCagnin et al. 1987; Albidy et al. 1997). The metabolite M3 was recorded by Overdiek and Merkus (1987) as a major metabolite of spironolactone in humans (Gardiner et al. 1989). LaCagnin et al. (1987), Los et al. (1993), and Albidy et al. (1997) also reported this metabolite (M3) of spironolactone in different systems. The metabolite 6β-OH-7α-thiomethyl spironolactone (M4) produced by *Thermomyces lanuginosus* was also detected by many researchers in various models (Overdiek and Merkus 1987; Karim et al. 1976;

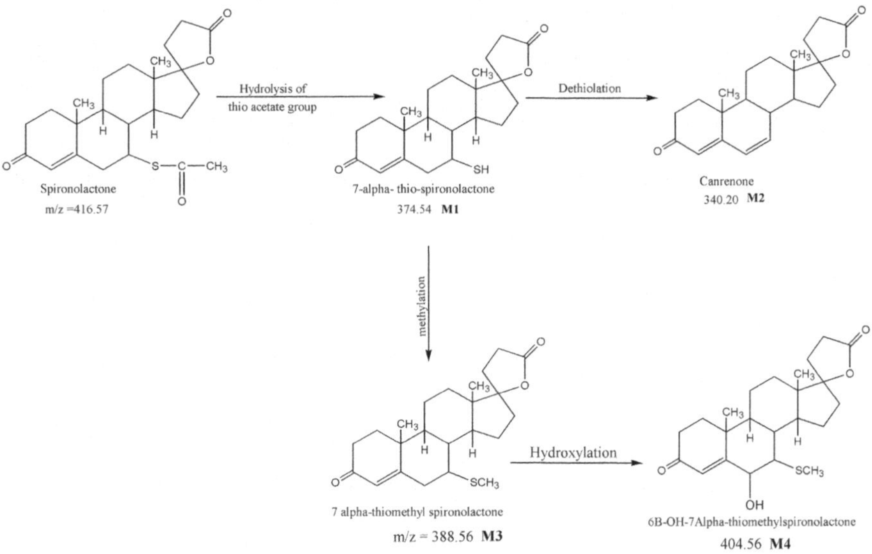

Fig. 14.3 Metabolic pathway of spironolactone by *Thermomyces lanuginosus*. *Source* Sreelatha et al. (2018).Copyright ©Steroids (Elsevier) Reproduced with permission

Albidy et al. 1997; Gardiner et al. 1989). The thermophilic fungi *T. lanuginosus* could generate four metabolites of spironolactone by hydrolysis, dethiolation, methylation, and hydroxylation, reactions which are also reported in mammals, while the mesophilic fungus *Cunninghamella elegans* could transform spironolactone into two hydroxylated derivatives (Marsheck and Karim 1973). On the other hand, *Chaetomium cochloides* could also produce three oxygenated metabolites of spironolactone (Mei et al. 2014). This clearly indicates that the metabolic pattern of *T. lanuginosus* is similar to mammals in contrast to mesophilic fungi reported previously in the biotransformation of spironolactone (Marsheck and Karim 1973; Mei et al. 2014).

14.2.3   Thermophilic Fungi in Novel Drug Discovery

Drug development is a huge challenging task from the laboratory scale to the market. The synthesized compound has to cross different biological assays, in vitro and in vivo drug metabolism, pharmacokinetic studies, and pharmacological and toxicological studies. After performing all the studies, compounds are to be ranked for clinical studies and those which satisfy all the chemical and biological assays will be further tested. Synthesis of compounds similar to that of the parent drug will have added advantage for the drug discovery program (Ravindran et al. 2012). In the case of failure, compounds with next best properties will be tested (Ravindran et al. 2012).

Hence, synthesis of compounds with structure and properties similar to that of the parent compound is crucial. Microbial transformation especially employing thermophilic fungi is a better alternative to generate compounds whose structure and properties resemble the parent compound or for the synthesis of compounds with novel structures similar to parent compounds with enhanced activity or with different therapeutic effects. Some of the examples of metabolites with novel biological activity generated by biotransformation employing thermophilic fungi are described below.

14.2.3.1 Novel Metabolites of Losartan as Human Peroxisome Proliferator Activated Receptor-Gamma (PPAR-γ) and Human Angiotensin Receptor (AT1R) Binders

Peroxisome proliferator-activated receptors (PPARs) are members of the nuclear hormone receptor superfamily of ligand-activated transcription factors which are related to retinoid, steroid, and thyroid hormone receptors (Murphy and Holder 2000). Among three PPAR isotypes PPAR-γ constitutes a prime target for the development of drug candidates to treat type II diabetes and PPAR-γ full agonists may even induce cell growth arrest, apoptosis, and terminal differentiation in various human malignant tumors (Guasch et al. 2012). Hypertension is a chronic disease affecting one-third of adult population worldwide and causing about half of the total mortalities, mainly due to stroke and heart problems. The main PPAR-γ synthetic full agonists studied to date are the thiazolidinedione (TZD) insulin-sensitizing drugs (e.g., rosiglitazone and pioglitazone) which were withdrawn from the market due to their pharmacovigilance and identified undesired adverse effects such as weight gain, edema, bone loss, and congestive heart failure (Ahmadian et al. 2013). Hence, there is a need for producing drugs with both hypertension and insulin resistance to treat simultaneously with the same pharmaceutical agent. Prasad et al. (2018), in his preliminary studies, reported human peroxisome proliferator-activated receptor-gamma (PPAR-γ) and human angiotensin receptor (AT1R) binding activity of novel metabolites of losartan, viz. M3 and M5 (Fig.14.1), produced by biotransformation process employing thermophilic fungi *Rhizomucor pusillus* NRRL 28626. Hence, thermophilic fungi are potential enough to produce novel value-added metabolites.

14.2.3.2 Hepatitis C Virus RNA-Dependent RNA Polymerase NS5B Inhibition Potentials of Albendazole and Its Biotransformed Metabolites

Hepatitis C virus (HCV) infection is a major public health problem, with nearly 3% of the world's population persistently infected with this virus (Fan et al. 2007) causing liver failure and responsible for the majority of liver transplants. The RNA-dependent RNA polymerase NS5B, in particular, has been the subject of intense research for developing new drugs in the past decade because of its essential role in viral replication (Wei et al. 2016). Anti-helminthic drug albendazole and four of its metabolites (Fig.14.2), viz. albendazole sulfoxide (M1), albendazole sulfone (M2), and the two novel metabolites (M3 and M4), produced by albendazole biotransformation employing thermophilic fungus *Rhizomucor pusillus* (Prasad et al. 2011) when studied for hepatitis C virus RNA-dependent RNA polymerase NS5B inhibition in silico showed strong inhibition potentials. Albendazole sulfone (M2) which is an inactive metabolite of albendazole against helminths proved to be active against hepatitis C virus RNA-dependent RNA polymerase NS5B followed by the novel metabolite M3. Albendazole, the anti-helminthic drug and albendazole sulfoxide (M1), the active metabolite of albendazole, showed intermediate degree of inhibition against hepatitis C virus RNA-dependent RNA polymerase NS5B. However, the least inhibition was shown by the metabolite M4. However, inhibition potential of albendazole and four of its metabolites against hepatitis C virus RNA-dependent RNA polymerase NS5B was higher compared to the standard drug sofosbuvir (Prasad and Shravan 2018).

14.2.4 Thermophilic Fungi in Studying the Drug-Drug Interactions

The drug-drug interaction occurs when two or more drugs are administered where one drug affects the activity of the other by increasing or decreasing its activity. These interactions are a major public health concern estimating that approximately 5% of hospital admissions are affected by adverse drug interactions; other estimates are between 3 and 28% (Hutzler et al. 2011), causing unexpected side effects. The probability of interactions increases with other drugs taken (Cascorbi 2012). These interactions are perceptible in patients suffering from chronic ailments such as congestive heart failure, cancer, hypertension, rheumatic diseases, and human immunodeficiency which require multiple drug therapy and may result in unpleasant drug interactions manifest as a loss in drug efficacy (Doucet et al. 2002). Cytochrome enzymes play a major role in metabolizing drugs and the activity of this group of enzymes or a single CYP can determine patient's response to drug therapy. Therefore, modulation of the activity of CYPs by a given drug is a critical issue for assessing the safety and efficacy of a drug. Inhibition of CYP can increase systemic exposure leading to severe toxic side effects of the drug or another concomitantly given medication that is metabolized by the respective CYP(s) (Romet et al. 1994; Wandel et al. 1998). Progress in CYP enzymology and biochemistry in the recent past suggests that the drug interactions are based on enzyme inhibition.

Life-threatening ventricular arrhythmia was recorded when ketoconazole and terfenadine were co-administered (Manahan et al. 1990). Similarly, an interaction between sorivudine and fluorouracil also resulted in fatal toxicity (Watabe 1996; Sokuda et al. 1997). Astemizole and cisapride were withdrawn from the market for causing drug-drug interactions. Mibefradil which is a calcium channel blocker caused rhabdomyolysis when combined with lovastatin and nephrotoxicity in combination with cyclosporine or tacrolimus (Hutzler et al. 2011). The best example of resulting of cardiac toxicity is when anti-histamine terfenadine and the antifungal ketoconazole or the antibiotic erythromycin are co-administered whereby inhibition of CYP 3A4 results in elevated terfenadine levels, resulting in prolongation of the QTc interval (VenkataKrishna et al. 2000). Similarly, on warfarin therapy, the increased bleeding in patients has been attributed to inhibition of its metabolism (Prasad et al. 2016). Many high-profile drugs were withdrawn from the market because of drug-drug interactions. Hence, addressing this issue is crucial. Drug-drug interactions are investigated in vitro with microsomes, expressed enzymes, or cell systems (Prasad et al. 2016). An understanding of the role of drug-metabolizing enzymes in the clearance of drugs and of drug-drug interactions caused by co-administered medications is a vital part of the drug discovery process and its therapeutics. To assess the potential of a drug in inhibiting different P450 enzymes in vitro, a variety of tools currently are available which include human liver tissue, cDNA-expressed P450 enzymes, and specific probe substrates where human liver microsomal preparations are of choice. For inhibition of enzyme selective studies, the utility of tissues from individual donors is limited by the adequacy of catalytic activity present in the tissue. Alternately, recombinant P450s can be used when a specific enzyme is to be investigated. Human liver microsomes and recombinant p450 enzymes are the most preferred test systems as they are readily available than human hepatocytes and p450 kinetic measurements are not confounded with other metabolic processes or cellular uptake. A major disadvantage of these test systems is that they do not represent the true physiological environment (e.g., not all phase II enzymes are present) if that is of interest to the study. Moreover, the process with these systems is a costly affair.

The best alternative would be microorganisms whose enzymes have proved to be versatile biocatalysts and are involved in the biotransformation of complex organic compounds. Most of the fungal cultures were reported to metabolize drugs similar to mammals; especially the fungus of the genus *Cunninghamella* is very well familiar as a model of mammalian biotransformation (Prasad et al. 2016) and as a microbial model in studying drug-drug interactions (Prasad et al. 2016). Thermophilic fungi especially *Rhizomucor pusillus* (Prasad et al. 2011, 2018) and *Thermomyces lanuginosus* (Sreelatha et al. 2018) are reported to possess drug-metabolizing enzymes. This process with the fungi is simple, hazard free, efficient, economical, and eco-friendly. Prasad et al. (2016) studied the metabolic inhibition of meloxicam using CYP2C9 inhibitors using mesophilic fungus *Cunninghamella blakesleeana* NCIM 687. Similarly, Srisailam et al. (2010) studied the prediction of drug interaction of clopidogrel on microbial metabolism of diclofenac. Though thermophilic fungi are reported to possess different drug-metabolizing enzymes (Prasad et al. 2011, 2018; Sreelatha et al. 2018) their use in drug-drug interaction studies remained unexplored. The use of thermophilic fungi in drug-drug interaction studies will be more advantageous compared to currently available in vitro test system and mesophilic fungi.

14.2.5 Thermophilic Fungi in Antibiotic Production

Antibiotic resistance is a global trouble with increased prevalence which showcases the need for drugs designed to overcome this epidemic. With increasing rates of bacterial drug resistance, the number of antibiotic unresponsive infectious diseases increased and the development of new antibiotics has become a crucial focus of the medical community. However, despite a push for new antibiotic therapies, there has been a continued decline in the number of newly approved drugs. Antibiotic resistance, therefore, poses a significant problem. As long as bacteria continue to develop resistance to the antibiotics, the continued isolation, screening, and evaluation of microorganisms from different habitats and their secondary metabolites can only be of benefit to all higher life forms on earth. Hence, search for new antibiotics effective against resistant pathogenic bacteria is currently required. Though thermophilic fungi are potential enough to produce novel antibiotics very few reports on antimicrobial agents are available (Kluepfel et al. 1971; Saito et al. 1979; Chiung et al. 1993). A new crystalline antifungal compound myriocin which is effective against *Candida* species and *Trichophyton granulosum*, *Microsporum gypseum* was isolated from thermophilic ascomycete *Myriococcum albomyce* (Kluepfel et al. 1971). Similarly, a thermophilic fungus *Malbranchea pulchella* var. sulfurea was also reported to produce an antibiotic Tf-26Vx (Saito et al. 1979) which was highly active against Gram-positive and obligate anaerobic Gram-negative bacteria. Chiung et al. (1993) also reported a novel quinine antibiotic named malbranicin from thermophilic fungus *Malbranchea cinnamomea* TAIM 13 T54 which exhibited toxicity against *Staphylococcus aureus* and *Bacillus subtilis*. In the same way, thermophilic fungi are also reported to produce other antibiotics like penicillin G, 6-aminopenicillanic acid, sillucin, miehein, and vioxanthin which are active against both Gram-positive and Gran-negative bacteria. Thermozymocidin was the other antifungal substance produced by thermophilic fungi (Mehrotra 1985; Satyanarayana et al. 1992). Hence, intensive research in this direction with thermophilic fungi is needed.

14.3 Conclusion and Future Prospects

Though thermophilic fungi are potential enough, they have not received required attention. They can be exploited commercially in the pharmaceutical industry and their application can be amplified by technologies such as immobilization which improve the stability (longevity, reusability) of biocatalysts and render continuous production process possible, recombinant DNA techniques which can be used to increase the production of the enzyme responsible for the desired biotransformation, and protein engineering (site-directed mutagenesis) which can help to increase the stability and/or improve the catalytic properties of the enzyme in question, or even to tailor an enzyme for specific purpose. Skillful application of these techniques in combination with conventional methods will certainly help in cost reduction and further make industrial bioprocess feasible and attractive, from an economical point of view. Further, increasing pressure of environmental constraints will favor this process.

In view of nearly unlimited reservoir of thermophilic fungi existing in nature and the exciting achievements of modern biotechnology, there is still an enormous potential awaiting for further progress in pharmaceutical applications in drug discovery, drug-drug interaction, and drug metabolism studies. It is envisaged that biotransformations employing thermophilic fungi will be increasingly exploited as a useful and often unique tool in the pharmaceutical industry.

Acknowledgement The author is thankful to Dr. G. Snithik, Dr. Akshaya Priya, and G. Preethi for extending their support in writing this review.

References

Abourashed EA, Clark AM, Hufford CD (1999) Microbial models of mammalian metabolism of xenobiotics, an updated review. Curr Med Chem 6:359–374

Ahmadian M, Suh JM, Hah N et al (2013) PPAR γ signaling and metabolism: the good, the bad and the future. Nat Med 19:557–566

Albidy AZA, York P, Wong V, Losowsky MS, Chrystyn H (1997) Improved bioavailability and clinical response in patients with chronic liver disease following the administration of a spironolactone: beta-cyclodextrin complex. Br J Clin Pharmacol 44:35–39

Allen PJ, Emerson R (1949) Guayule rubber, microbiological improvement by shrub retting. Ind Eng Chem 41:346–365

Anastasi A, Varese GC, Marchisio VF (2005) Isolation and identification of fungal communities in compost and vermicompost. Mycolagia 94:33–44

Benchaoui HA, Scott EW, Mc Kellar QA (1993) Pharmacokinetics of albendazole, albendazole sulfoxide and netobimin in goats. J Vet Pharmcol Ther 2:237–240

Blochl E, Rachel R, Burggraf S, Hafenbradl D, Jannasch HW, Stetter KO (1997) Pyrolobus fumarii, gen. and sp. nov., represents a novel group of archaea, extending the upper temperature limit for life to 113 degrees C. Extremophiles 1:14–21

Boris S, Bernahard S (2003) Angiotensin II AT1 receptor antagonists. Clinical implications of active metabolites. J Med Chem 46:2261–2270

Brock TD (1995) The road to Yellowstone and beyond. Annu Rev Microbiol 49:1–28

Bruce LZ, Henrik KN, Robert LS (1991) Thermostable enzymes for industrial applications. J Ind Microbiol Biotechnol 8:71–81

Brunton L, Keith P, Blumenthal D, Buxton I (2008) Goodman and Gilman's manual of pharmacology and therapeutics. McGraw-Hill, New York

Burg BV (2003) Extremophiles as a source for novel enzymes. Curr Opin Microbiol 6:213–218

Cascorbi I (2012) Drug interactions-principles, examples and clinical consequences. Dtsch Arztebl Int 109:33–34

Cha CJ, Doerge DR, Cerniglia CE (2001) Biotransformation of malachite green by the fungus *Cunninghamella elegans*. Appl Environ Microbiol 67:4358–4360

Chang Y (1967) The fungi of wheat straw compost. II. Biochemical and physiological studies. Trans Br Mycol Soc 50:667–677

Chiung M, Fujita T, Nakagawa M, Nozaki H, Chen GY, Chen ZC, Nakayama M (1993) A novel Quinone antibiotic from *Malbranchea cinnamomea* TAIM 13T54. J Antibiot 46:1819–1826

Cooney DG, Emerson R (1964) Thermophilic fungi. An account of their biology, activities and classification. W.H. Freeman and Company, San Francisco, CA/London

Cordova J, Roussos S, Raratti J, Nungaray J, Loera O (2003) Identification of Mexican thermophilic and thermotolerant fungal isolates. Micol Appl Int 15:337–344

Crisan EV (1973) Current concepts of thermophilism and thermophilic fungi. Mycologia 65:1973–1978

Csiko GY, Banhidi GY, Semjen G et al (1996) Metabolism and Pharmacokinetics of albendazole after oral administration to chickens. J Vet Pharmacol Ther 19:322–325

Doucet JJA, Noel D, Geffroy CE, Capet C, Coquard A, Couffin E, Fauchais AL, Chassagne P, Schleifer M (2002) Preventable and non-preventable risk factors for adverse drug events related to hospital admissions in the elderly: a prospective study. Clin Drug Investig 22:385–392

Eggins HOW, Coursey DG (1964) Thermophilic fungi associated with Nigerian oil palm produce. Nature 203:1081–1084

Eggins HOW, Malik KA (1969) The occurrence of thermophilic and cellulolytic fungi in pasture land soil. Antonie Van Leewenhoek 35:178–184

El Amri HS, Fargetton X, Delatour P, Batt MA (1987) Sulphoxidation of albendazole by the FAD-containing & cytochrome P-450 dependent mono-oxygenases from pig liver microsomes. Xenobiotica 10:1159–1168

Evans HC (1971) Thermophilic fungi of coal spoil tips II. Occurrence and temperature relations. Trans Br Mycol Soc 57:255–266

Fan Z, Huang XL, Kalinski P, Young S, Rinaldo CR (2007) Dendritic cell function during chronic hepatitis C virus and human immunodeficiency virus type 1 infection. Clin Vaccine Immunol 14:1127–1137

Fulleringer SL, Seguin D, Bexille A, Desterque C, Arne P, Cherette R, Bretagne S, Guillot J (2005) Evolution of the environmental contamination by thermophilic fungi in a Turkey confinement house in France. Poult Sci 85:1875–1880

Gardiner P, Schrode K, Quinlan D, Martin BK, Boreham DR, Rogers MS, Smith SM, Karim A (1989) Spironolactone metabolism: steady state serum levels of the sulfur containing metabolites. J Clin Pharmacol 29:342–347

Ghosh S, Sachan A, Mitra A (2005) Degradation of ferulic acid by a basidiomycetes Schizophyllum commune. World J Microbiol Biotechnol 21:385–388

Ghosh S, Sachan A, Mitra A (2006) Formation of vanillic acid from ferulic acid by *Paecilomyces variotii* MTCC 6581. Curr Sci 90:825–829

Gomes J, Steiner W (2004) The biocatalytic potential of extremophiles and extremozymes. Food Technol Biotechnol 42:223–235

Guasch L, Sala E, Valls C, Mulero M, Pujadas G, Vallve SG (2012) Development of docking-based D-QSAR models for PPAR gamma Full agonists. J Mol Graph Model 36:1–9

Gunaratna C (2000) Drug metabolism and pharmacokinetics in drug discovery: a primer for bioanalytical chemists part-1. Curr Sep 19:1

Gyurik RJ, Chow AW, Zaber B, Brunner EL, Miller JA, Villani AJ et al (1981) Metabolism of albendazole in cattle, sheep, rats and mice. Drug Metab Dispos 9:503–508

Hezari M, Devis PJ (1993) Microbial models of mammalian metabolism. Furosemide glucoside formation using the fungus *Cunninghamella elegans*. Drug Metab Dispos 21:259–267

Hunter AC, Mills PW, Dedi C, Dodd HT (2007) Predominant allylic hydroxylation at carbons 6 and 7 of 4 and 5-ene functionalized steroids by the thermophilic fungus *Rhizomucor tauricus* IMI23312. J Steroid Biochem Mol Biol 108:155–163

Huskey SE, Miller RR, Chiu SH (1993) *N*-Glucuronidation reactions. I. Tetrazole *N*- glucuronidation of selected angiotensin II receptor antagonists in hepatic microsomes from rats, dogs, monkeys, and humans. Drug Metab Dispos 21:792–799

Hutzler JM, Cook J, Fleishaker JC (2011) Drug-drug interactions: designing development programmes and appropriate product labeling. In: Bonate PL, Howard DR (eds) Pharmacokinetics in Drug Development. American Association of Pharmaceutical Scientists

Ito K, Iwashita K, Iwano K (1992) Cloning and sequencing of the xyn C gene encoding acid xylanase of *Aspergillus kawachii*. Biosci Biotechnol Biochem 56:1338–1340

Johri BN, Rajani (1999) Mushroom compost: microbiology and application. In: Bagyaraj BJ, Varma A, Khanna K, Kheri HK (eds) Modern approaches and innovations in soil management. Rastogi Publications Meerut, Uttar Pradesh, pp 345–358

Johri BN, Satyanarayana (1986) Thermophilic moulds; Perspectives in basic and applied research. Indian Rev Lifes Sci 6:75–100

Karim K, Zagarella BA, Hribar J, Dooley M (1976) Spironolactone. I Disposition and metabolism. Clin Pharmacol Ther 19:158–169

Khushaldas MB (2009) Eco-Physiological studies on thermophilic fungi from diverse habitats in Vidarbha region, PhD thesis, R.T.M. Nagpur University, Nagpur

Kluepfel D, Bagli J, Baker H, Charest MP, Kudelski A, Sehgal SN, Vezina C (1971) Myricin, A new antifungal antibiotic from *Myriococcum albomyces*. J Antibiot 25:109–115

Kohilu U, Nigam P, Singh D, Chaudhary K (2001) Thermostable, alkaliphilic and cellulose free xylanase production by *Thermoactinomyces thalophilus* subgroup C. Enzyme Microb Technol 28:606–610

Krieter PA, Colletti AE, Miller RR, Stearns RA (1995) Absorption and glucuronidation of the angiotensin II receptor antagonist losartan by the rat intestine. J Pharmacol Exp Ther 273:816–822

Kuthubutheen AJ (1983) Growth and sporulation of thermophilous fungi on agar containing various carbon and nitrogen compounds. Mycopathologia 82:45–48

Lacagnin LB, Lutsie P, Colby HD (1987) Conversion of spironolactone to 7α-thiomethylspironolactone by hepatic and renal microsomes. Biochem Pharmacol 36:3439–3444

Lacey E (1990) Mode of action of benzimidazoles. Parasitol Today 6:112–115

Lanusse CE, Gascon LH, Prichard PK (1993) Gastrointestinal distribution of albendazole metabolites following netobimin administration to cattle; relationship with plasma disposition kinetics. J Vet Pharmacol Ther 1:38–47

Lasa I, Berenguer J (1994) Thermophilic enzymes and their biotechnological potential. Microbiologia 9:77–89

Lee CR et al (2003) Tolbutamide, Flurbiprofen and Losartan as probes of CYP 2C9 activity in humans. J Clin Pharmacol 43:84–91

Lin L, Zhang J, Wei Y, Chen C, Peng Q (2005) Phylogenetic analysis of several Thermus strains from Rehai of Tengchong. Yunnan, China. Can J Microbiol 51:881–886

Lindt W (1886) Mitteilungen uber einige neue pathogene Schimmelpilze. Arch Exp Pathol Pharmakol 21:269–298

Los LE, Coddington AB, Ramjit HG, Colby HD (1993) Identification of spironolactone metabolites in plasma and target organs of guinea pigs. Drug Metab Dispos 21:1086–1090

Mahajan MK, Johri BN, Guptha RK (1986) Influence of desiccation stress in xerophylic thermophile *Humicola* species. Curr Sci 56:928–930

Maheshwari R (1997) The ecology of thermophilic fungi. In: Janardhana KK, Rajendran C, Natarajan K, Hawksworth DL (eds) Tropical mycology. Oxford and IBH Publishers, Delhi, pp 277–289

Manahan BP, Ferguson CL, Killeavy ES, Lioyd BK, Troy J, Cantilena LR (1990) Torsades de pointes occurring in association with terfenadine use. JAMA 264:2788–2790

Marriner SE, Bogan JA (1980) Pharmacokinetics of albendazole in sheep. Am J Vet Res 7:1126–1129

Marsheck WJ, Karim A (1973) Preparation of metabolites of spironolactone by microbial oxygenation. Appl Environ Microbiol 25:647–649

McKellar QA, Jackson F, Coop RL, Baggot JD (1993) Plasma profiles of albendazole metabolites after administration of netobimin and albendazole in sheep; effects of parasitism & age. Br Vet J 1:101–113

Meessen LL, Delattre M, Haon M, Thibault JF, Ceccaldi BC, Brunerie P, Asther M (1996) A two-step bioconversion process for vanillin production from ferulic acid combining *Aspergillus niger* and *Pycnoporus cinnabarinus*. J Biotechnol 50:107–113

Meessen LL, Delattre M, Haon M, Asther M (1999) Methods for bioconversion of ferulic acid to vanillic acid or vanillin and for the bioconversion of vanillic acid to vanillin using filamentous fungi, US Patent no 5866380

Mehrotra BS (1985) Thermophilic fungi-Biological enigma and tools for the biotechnologist and biologist. Indian Phytopathol 38:211–229

Mei J, Wang L, Wang S, Zhan J (2014) Synthesis of two new hydroxylated derivatives of spirono-lactone by microbial transformation. Bioorg Med Chem Lett 24:3023–3025

Miller FC, Harper ER, Macauly BJ, Gulliever A (1990) Composting based on moderately thermo-philic and aerobic conditions for the production of commercial mushroom growing compost. Aust J Exp Agric 30:287–296

Moloney AP, Callan SM, Murray PG, Tuohy MG (2004) Mitochondrial malate dehydrogenase from the thermophilic, filamentous fungus *Talaromyces emersonii*; Purification of the native enzyme, cloning and over expression of the corresponding gene. Eur J Biochem 271:3115–3126

Moody JD, Freeman JP, Cerniglia CE (1999) Biotransformation of doxepin by *Cunninghamella elegans*. Drug Metab Dispos 27:1157–1164

Moody JD, Heinze TM, Hansen EB, Cerniglia CE (2000) Metabolism of the ethanol amine type antihistamine diphenhydramine (Benadryl) by the fungus *Cunninghamella elegans*. Appl Microbiol Biotechnol 53:310–315

Moroni P, Bouronfosse T, Sauvageon CL, Delatour P, Benoit E (1995) Chiral sulfoxidation of albendazole by the flavin adenine dinucleotide-containing & cytochrome P450-dependent monooxygenases from rat liver microsomes. Drug Metabol Dispos 2:160–165

Motedayan N, Ismail MB, Nazarpour F (2013) Bioconversion of ferulic acid to vanillin by com-bined action of *Aspergillus niger* K8 and *Phanerochaete chrysosporium* ATCC24725. Afr J Biotechnol 12:6618–6624

Mouchacca J (1995) Thermophilic fungi in desert soils: a neglected extreme environment. In: Allsopp D, ColWell RR, Hawksworth DL (eds) Microbial diversity and eco-system function. C.A.B. International, Wallingford, pp 265–288

Mouchacca J (1997) Thermophilic fungi: biodiversity and taxonomic status. Crypt Mycol 18:19–69

Mouchacca J (1999) Thermophilic fungi; present taxonomic concepts. In: Johri BN, Satyanarayana T, Olsen J (eds) Thermophilic moulds in biotechnology. Springer, Dordrecht, pp 43–83

Murphy GJ, Holder JC (2000) PPAR-γ agonists: therapeutic role in diabetes, inflammation and cancer. Trends Pharmacol Sci 21:469–474

Nguyen S, Ala F, Cardwell C, Cai D, McKindles KM, Lotvola A, Hodges S, Deng Y, Tiquia-Arashiro SM (2013) Isolation and screening of carboxydotrophs isolated from composts and their potential for butanol synthesis. Environ Technol 34:1995–2007

Niehaus F, Bertoldo C, Kahler M, Antranikian G (1999) Extremophiles as a source of novel enzymes for industrial applications. Appl Microbiol Biotechnol 51:711–729

Ofosu-Asiedu A, Smith RS (1973) Some factors affecting wood degradation by thermophilic and thermotolerant fungi. Mycologia 65:87–98

Ogundero VW (1981) Isolation of thermophilic and thermotolerant fungi from stored groundnuts in Nigeria and determination of their lipolytic activity. Int Biodeterior Bull 17:51–55

Overdiek HW, Merkus FW (1987) The metabolism and biopharmaceutics of spironolactone in man. Rev Drug Metab Drug Interac 5:273–302

Pan WZ, Huang XW, Wei KB, Zhang CM, Yang DM, Ding JM, Zhang KQ (2010) Diversity of thermohic fungi in Tengchong Rehai National Park revealed by ITS nucleotide sequence analy-ses. J Microbiol 48:146–152

Penicaut B, Maugein P, Maisonneuve H (1983) Pharmacokinetics and urinary metabolism of albendazole in man. Bull Soc Pathol Exot Filiales 76:698–708

Pomaranski E, Tiquia-Arashiro SM (2016) Butanol tolerance of carboxydotrophic bacteria iso-lated from manure composts. Environ Technol 37(15):1970–1982

Prasad GS, Shravan GK (2018) Hepatitis C virus RNA-dependent RNA polymerase NS5B inhibi-tion potentials of anti-helminthic drug albendazole and its biotransformed metabolites: an in silico study. IJPBS 8:341–348

Prasad GS, Srisailam K (2018) Metabolites generation via oxidation, hydroxylation, acyla-tion, dechlorination, dealkylation and glucuronidation of losartan by thermophilic fungus *Rhizomucor pusillus* NRRL 28626. IRJNAS 5:98–115

Prasad GS, Girisham S, Reddy SM, Srisailam K (2008) Biotransformation of albendazole by fungi. World J Microbiol Biotechnol 24:1565–1571

Prasad GS, Girisham S, Reddy SM (2010) Microbial transformation of albendazole. Indian J Exp Biol 48:415–420

Prasad G, Girisham S, Reddy SM (2011) Potential of thermophilic fungus *Rhizomucor pusillus* in biotransformation of antihelminthic drug albendazole. Appl Biochem Biotechnol 165:1120–1128

Prasad GS, Srisailam K, Sashidhar RB (2016) Metabolic inhibition of meloxicam by specific CYP2C9 inhibitors in *Cunninghamella blakesleeana* NCIM 687: in silico and in vitro studies. Springer Plus 5:166

Prasad GS, Shravan Kumar G, Srisailam K (2018) Novel metabolites of losartan as human peroxisome proliferator activated receptor gamma (PPAR γ) and human angiotensin receptor (AT1R) binders: an *in silico* study. IJPBS 8:330–337

Rahouti M, Seigle-Murandi F, Steiman R, Eriksson KE (1989) Metabolism of ferulic acid by Paecilomyces variotii and Pestalotia palmarum. Appl Environ Microbiol 55:2391–2398

Ravindran S, Basu S, Surve P, Lonsane G, Sloka N (2012) Significance of biotransformation in drug discovery and development. J Biotechnol Biomater S13:005

Romet JM, Crawford K, Cyr T, Inaba T (1994) Terfenadine metabolism in human liver: in vitro inhibition by macrolide antibiotics and azole antifungals. Drug Metab Dispos 22:849–857

Ross RC, Harris PJ (1983) The significance of thermophilic fungi in mushroom compost preparation. Sci Hortic AMST 20:61–70

Rowland M, Tozer TN (1995) Clinical pharmacokinetics: concepts and applications, 3rd edition. Section 1:11–17

Saito M, Matsuura I, Okazaki H (1979) Tf26Vx, an antibiotic produced by a thermophilic fungus. J Antibiot 32:1210–1212

Salar RK, Aneja KR (2007) Thermophilic fungi: taxonomy and biogeography. J Agric Technol 3:77–107

Sandhu DK, Singh S (1985) Air-borne thermophilous fungi at Amritsar. India Trans Br Mycol Soc 84:41–46

Satyanarayana T, Johri BN, Klein (1992) Biotechnological potential of thermophilic fungi. In: Arora DK, Elander RP, Mukharji KG (eds) Hand book of applied mycology, vol 4. Marcel Dekker Inc., New York, pp 729–761

Sayanarayana T, Chavant L (1987) Bioconversion and Binding of sterols by thermophilic moulds. Folia Microbiol 32:354–359

Schulze B, Wubbolts MG (1999) Biocatalysis for industrial production of fine chemicals. Curr Opin Biochem 10:609–615

Sharma HSS, Johri BN (1992) The role of thermophilic fungi in agriculture. In: Handbook of applied mycology, vol 4. Marcel Dekker, New York, pp 707–728

Sherry JH, Donnell JPO, Colby HD (1981) Conversion of spironolactone to an active metabolite in target tissues: formation of 7α-thiospironolactone by microsomal preparations from guinea pig liver, adrenals, kidneys and testes. Life Sci 29:2727–2736

Siegl PKS (1993) Discovery of losartan: the first specific non-peptide angiotensin II receptor antagonist. J Hypertension 11:19–22

Sokuda H, Nishiyama T, Ogura K, Nagayama S, Ikeda K, Yamaguchi S, Nakamura Y, Kawaguchi Y, Watable T (1997) Lethal drug interactions of sorivudine, a new antiviral drug, with oral 5-fluorouracil prodrugs. Drug Metab Dispos 25:270–273

Sreelatha B, Prasad GS, RaoV K, Girisham S (2018) Microbial synthesis of mammalian metabolites of Spironolactone by thermophilic fungus *Thermomyces lanuginosus*. Steroids 136:1–7

Srisailam K, Rajkumar V, Veeresham C (2010) Predicting drug interaction of clopidogrel on microbial metabolism of clopidogrel. Appl Biochem Biotechnol 160:1508–1516

Stearns RA, Chakravarty PK, Chen R, Chiu SH (1995) Biotransformation of losartan to its active carboxylic acid metabolite in human liver microsomes. Role of cytochrome p4502C and 3A subfamily members. Drug Metab Dispos 23:207–215

Stratsma G, Samson RA, Olijusma TW, Gerrits JPG, Opden Camp HJM, Gerrits JPG, Griensven LJLD (1994) Ecology of thermophilic fungi in mushroom compost with emphasis on

Scytalidium thermophilum and growth stimulation of *Agaricus bisporus*. Appl Environ Microbial 60:454–455

Sun L, Huang HH, Liu H, Zhang DF (2004) Transformation of verapamil by *Cunninghamella blakesleeana*. Appl Environ Microbiol 70:2722–2727

Tansey MR, Brock TD (1978) Microbial life at high temperatures: ecological aspects. In: Kushner DJ (ed) Microbial life in extreme environments. Academic Press, London, pp 159–195

Tiquia-Arashiro SM (2014) Thermophilic carboxydotrophs and their biotechnological applications. Springerbriefs in microbiology: extremophilic microorganisms. Springer International Publishing, New York, p 131

Tiquia-Arashiro SM, Rodrigues D (2016a) Alkaliphiles and acidophiles in nanotechnology. In: Extremophiles: applications in nanotechnology. Springer International Publishing, New York, pp 129–162

Tiquia-Arashiro SM, Rodrigues D (2016b) Thermophiles and psychrophiles in nanotechnology. In: Extremophiles: applications in nanotechnology. Springer International Publishing, New York, pp 89–127

Topakas E, Kalogeris E, Kekos D, Macris BJ, Christakopoulos P (2003) Bioconversion of ferulic acid into vanillic acid by the thermophilic fungus Sporotrichum thermophile. LWT-Food Sci Technol 36:561–565

Tsiklinskaya P (1899) Sur les muce'dine'es thermophils. Ann Inst Pasteur 13:500–515

Tubaki K, Ito T, Natsuda Y (1974) Aquatic sediment as habitat of thermophilic fungi. Ann Microbiol 24:199–207

Venkatakrishna K, Moltke VLL, Greenblatt DJ (2000) Effects of the antifungal agents on oxidative drug metabolism: clinical relevance. Clin Pharmacokinet 38:111–180

Wandel C, Lang CC, Cowart DC, Girard AF, Bramer S, Flockhart DA, Wood AJ (1998) Effect of CYP3A inhibition on vesnarinone metabolism in humans. Clin Pharmacol Ther 63:506–511

Watabe T (1996) Strategic proposals for predicting drug-drug interactions during new drug development: based on sixteen deaths caused by interactions of the new antiviral sorivudine with 5-fluorouracil prodrugs. J Toxicol Sci 21:299–300

Wei Y, Li J, Qing J, Huang M, Wu M, Gao F, Li D, Hong Z, Kong L, Huang W, Lin J (2016) Discovery of novel hepatitis C virus NS5B polymerase inhibitors by combining random forest, multiple e-pharmacophore modeling and docking. PLoS One 11:e0148181

Weigant WM (1992) Growth characteristics of *Scytalidium thermophilum* in relation to the production mushroom compost. Appl Environ Microbiol 58:1301–1307

Wu K, Song J, Li T (2015) Production purification and identification of Cholest-en-3-one produced by cholesterol oxidase from Rhodococcus sp.in aqueous/organic biphase system. Biochem Insights 8:1–8

Yun CH, Lee HS, Lee H, Rho JK, Jeong HG, Guengerich FP (1995) Oxidation of the angiotensin II receptor antagonist Losartan (DUP 753) in human liver microsomes. Roles of cytochrome P4503A (4) in formation of the active metabolite EXP3174. Drug Metab Dispos 23:285–289

Zhang D, Freeman JP, Sutherland JB, Walker AE, Yang Y, Crniglia C (2006) Biotransformation of chlorpromazine and methdilazine by *Cunninghamella elegans*. Appl Environ Microbiol 62:798–803

Zhong DF, Sun L, Liu L, Huang HH (2003) Microbial transformation of naproxen by *Cunninghamella* species. Acta Pharmacol Sin 24:442–447

Zohri AA, Abdel-Galil MSM (1999) Progesterone transformation by three species of Humicola. Folia Microbiol 44:277–282

Chapter 15
Biotechnological Applications of Halophilic Fungi: Past, Present, and Future

Imran Ali ⓘ**, Samira Khaliq, Sumbal Sajid, and Ali Akbar**

15.1 Introduction

Diverse hypersaline habitats are present globally such as in the form of saline soil, saline water, and salted foods. However, there is diversity in hypersaline soils and heterogeneity in their nature and composition comprising a wide range of minerals present at various depths (Gostinčar et al. 2011). Likewise, water from salterns, salt lakes, sea, oceanic water, and brackish water are all considered as saline but have diverse compositions; for example the concentration of salt in Dead Sea is 78% whereas that of the Great Salt Lake is 33% (Khan et al. 2017). Similarly, salted foods can be formed by the addition of different salts for the purpose of flavors and preservation or for the addition of minerals in the food, in varying concentrations.

The microbial community surviving in hypersaline habitats comes as a blessing in disguise due to their adaptability in these extreme conditions and production of extreme metabolites, which are in high demand in current changing environments and the challenges faced by biotechnological applications (Gunde-Cimerman et al. 2004). There are very few genetic and/or genomic studies on identification and relative abundance of various halophilic species isolated from different sources.

I. Ali (✉)
School of Life Sciences and Engineering, Southwest University of Science and Technology, Mianyang, Sichuan, China

Plant Biomass Utilization Research Unit, Botany Department, Chulalongkorn University, Bangkok, Thailand

Institute of Biochemistry, University of Balochistan, Quetta, Pakistan

S. Khaliq · S. Sajid
Institute of Biochemistry, University of Balochistan, Quetta, Pakistan

A. Akbar
Department of Microbiology, University of Balochistan, Quetta, Pakistan

© Springer Nature Switzerland AG 2019
S. M. Tiquia-Arashiro, M. Grube (eds.), *Fungi in Extreme Environments: Ecological Role and Biotechnological Significance*,
https://doi.org/10.1007/978-3-030-19030-9_15

However, the few investigations that have been conducted are PCR-based strategies, targeting 16S small subunit ribosomal ribonucleic acid (16S rRNA) genes (Baati et al. 2010). The results indicate that some of the most commonly isolated and studied halophilic species may in fact not be significant in situ; for example genus *Haloarcula* only makes less than 0.1% of the in situ community; however, it usually appears in isolation studies (Bakke, et al. 2009). In other recent studies where comparative genomic and proteomic analyses were carried out, it is identified that halophiles have distinct molecular signatures coinciding with environmental adaptations. Using *Sulfolobus solfataricus* as a model, it is identified that halophiles rearrange their extrachromosomal replicons, which are rich in IS elements, at high frequency. However, their chromosomes are quite stable in contrast to non-halophiles and in fact contain a smaller number of transposable IS elements (Berquist et al. 2006). Characterization of the proteome of halophiles indicates low hydrophobicity, repetition of acidic residues, lower occurrence of cysteine, lower inclination for helix formation, and higher propensities for coil structure (Paul et al. 2008). It is often found that the core of most proteins isolated from halophiles is less hydrophobic; for example DHFR protein from halophiles has narrower β-strands as compared to similar proteins isolated from non-halophiles (Miyashita et al. 2017; Zusman et al. 1989). In a similar fashion, it is also observed that halophiles show distinct dinucleotide and codon usage (Paul et al. 2008).

Halophiles are mostly prokaryotes while some eukaryotes also make it to the list. Prokaryotic halophiles can further be divided into two main classes: (1) *Archaea* and (2) *Halobacteria*, which include the extreme halophiles like *Halomonas elongata sp.*, phototrophic halophiles such as *Saliiococcus hispanicus* sp., and methanogenic halophiles like *Methanosarcinales* sp. Likewise, eukaryotic halophiles also include photosynthetic halophiles such as green algae *Dunaliella salina* and heterotrophic halophiles like *Wallemia ichthyophaga* (Gunde-Cimerman et al. 2009). Depending on their ability to thrive on salt concentrations, halophiles may also be classified as slight, moderate, or extreme, and/or as obligate or facultative halophiles (Tiquia et al. 2007). On the contrary, halotolerant species are the ones which do not require salt but can survive in changes pertaining to external salt concentrations (DasSarma and DasSarma 2012). Slight halophiles prefer 0.3–0.8 M (1.7–4.8%—seawater is 0.6 M or 3.5%), moderate halophiles 0.8–3.4 M (4.7–20%), and extreme halophiles 3.4–5.1 M (20–30%) salt content (Sarwar et al. 2015).

Many halophiles accumulate compatible solutes in cells to balance the osmotic stress in their environment (Oren 2013). The aerobic halophilic archaea are famous for "salt in" strategy in which they accumulate high concentrations of K+ and Na+ ions in the cytoplasm by the help of ionic pumps and protein transportation at the expense of energy (Schafer et al. 1999). In "low salt in" strategy, the microorganisms maintain a low concentration of salts in the cytoplasm and instead survive in hypersaline conditions by the use of compatible solutes (Oren 2013). Some halophiles produce acidic proteins that function in high salinity by increasing solvation and preventing protein aggregation, precipitation, and denaturation (Talon et al. 2014). The high salt survival mechanism in prokaryotes is different from eukaryotes, e.g., extremely halophilic archaea accumulate potassium, as high as molar levels, when exposed to high external salinity (Sarwar et al. 2015). In contrast, eukaryotic microorganisms such as fungi cannot tolerate such high intracellular ion concentrations and in such organisms the maintenance of positive turgor pressure at high salinity is mainly due to an increased production and accumulation of glycerol, trehalose, and other organic compatible solutes (Oren 2013). Moreover, metabolites from fungi, mostly being extracellular, are easily extracted as compared to bacteria (Gostinčar et al. 2011) and show better performance in terms of quality and quantity of biotechnological applications (Ali et al. 2016). Hence, investigating biotechnological potential of only prokaryotic halophiles does not do justice, and therefore this chapter is intended to highlight the importance of halophilic fungi in biotechnology which is aimed at attracting industry and academia to further explore the potentials of halophilic fungi for use in biotechnological applications.

15.2 Halophilic Fungi

Halophilic fungi were first reported in 2000 to be active inhabitants of hypersaline environments, when they were found in man-made solar salterns in Slovenia (Gunde-Cimerman et al. 2000). Since then, there are numerous reports of halophilic fungi around the globe. Recently, a criterion has been set for their consideration that the ones isolated from hypersaline habitats of 1.7 M salt concentration and can grow in vitro at or above 3 M concentration of salt should be considered as halophilic fungi (Gunde-Cimerman et al. 2009). Location, time of sampling, dissolved oxygen, water activity, and available organic and inorganic nutrients are found to be the important factors in the geographical distribution, growth, and viability of these fungi (Butinar et al. 2005a, b). The adaptation of halophilic fungi to hypersaline habitats is independent of salt concentrations and they can be found inhabiting any range of salt present, such as in hypersaline waters, and they can be found from freshwater to saturated natural or man-made salterns (Oren 2013). Halophilic fungi from hypersaline habitats have been recognized as either new species, previous ones which may not be recognized as halophilic earlier, or the ones having natural mutation(s), making them new strains. Phylogenetic analyses of these strains show interesting far distant relationships with the fellow species or genus (Gunde-Cimerman et al. 2009). The dominant representatives are different species of black yeast-like and related melanized fungi of the genus *Cladosporium*, different species within the anamorphic *Aspergillus* and *Penicillium*, the teleomorphic *Emericella* and *Eurotium*, certain species of non-melanized yeasts, and *Wallemia* spp. (Butinar et al. 2005a, b).

15.3 Genomes of Halophilic Fungi

Halophilic microorganisms have helped us understand the basics of life and survival in extreme conditions and are useful participants in major biogeochemical cycles of sulfur, nitrogen, carbon, and phosphorous operating in extreme conditions (Oren 2008). They also provide a better comparative understanding of the interactions amongst living organisms in simple versus extreme ecosystem, and contribute to the knowledge of survival in salt stress and selection of best suitable genes for white biotechnology (Gunde-Cimerman et al. 2009). For example, discovery and comparative analysis of the extremely halotolerant but adaptable fungus *Hortaea werneckii* (Lenassi et al. 2013) and the obligate halophile *Wallemia ichthyophaga* (Zajc et al. 2013) revealed novel molecular mechanisms used in combating high salt concentrations in distinct ways. In both of these fungi, the key signaling components are conserved; however, there are structural and regulatory differences.

It is interesting to note that there is a large genetic redundancy in *H. werneckii* and the genes coding for metal cation transporters have increased in number. Surprisingly, studies have revealed that it has also undergone a recent whole-genome duplication (Lenassi et al. 2013). In comparison, *W. ichthyophaga* has a very compact genome of 4884 protein-coding genes, which make up almost three-fourth of the sequence. Amongst this, a significant increase in the hydrophobin cell-wall proteins with multiple cellular functions is observed (Zajc et al. 2013).

Genomic analysis of 26.2 Mbp of the fungus *Eurotium rubrum* (Eurotiomycetes) isolated from Dead Sea reinstates the fact that there is gain in gene families related to stress response and losses with regard to transport processes (Kis-Papo et al. 2014).

15.4 Biotechnological Reports on Halophilic Fungi

Most of the research on halophilic fungi from year 2000 has been focused on morphological and molecular adaptations of these fungi in hypersaline environments (De Hoog et al. 2005; Gostinčar et al. 2011; Gunde-Cimerman et al. 2004, 2009; Oren 2013; Plemenitaš and Gunde-Cimerman 2005; Plemenitaš et al. 2008). However, there is no substantial information on biotechnological applications on halophilic fungi as compared to halophilic bacteria (Oren 2010; Tiquia 2010; Tiquia and Mormile 2010; Tiquia-Arashiro and Rodrigues 2016a, b); therefore, we have tried to summarize all reported applications of halophilic fungi herein.

15.4.1 Production of Bioactive Compounds by Halophilic Fungi

Sepcic et al. (2011) reported a total of 43 fungal species, isolated from various environments. These fungal species were tested for their ability to either metabolize or produce compounds with selected biological activities such as hemolysis, antibacterials, and acetylcholinesterase inhibition. Results indicate that the halophilic fungal species synthesize specific bioactive metabolites under conditions that represent stress for non-adapted species.

It was observed that increased salt concentrations resulted in higher hemolytic activity. This was reasoned to be due to the production of only organic metabolites, which can be dissolved in organic solvents, and hence these halophilic fungi do not produce proteins which can hemolyze erythrocytes. However, low water activity and colder conditions increase the hemolytic activity of most halophilic fungi suggesting the stress immune response.

The appearance of antibacterial potential under stress conditions was seen in a similar pattern for the fungal species with regard to hemolysis. The active extracts exclusively affected the growth of the Gram-positive bacterium tested, *Bacillus subtilis*. None of the extracts tested showed inhibition of acetyl cholinesterase activity. Hence, species such as *Aureobasidium sp.*, *A. pullulans*, *Var. melanogenum*, *H. werneckii*, *T. salinum*, and *Wallemia spp.* perform best for their hemolytic and antibacterial activities and these can be exploited for commercial applications.

Ravindran et al. (2012) highlighted the role of antioxidants and its related enzymes on adaption to salt stress by a halophilic fungus isolated from seawater collected from Dona Paula beach, Goa, India. The fungus was morphologically and molecularly identified as *Phialosimplex sp.* (though in publication authors only used the name of fungus as halophilic fungus, but the accession number at NCBI shows *Phialosimplex sp.*, which could be a later modification). The growth characterization showed that this fungus preferred to grow at 15% of NaCl concentration. The aqueous extracts of *Phialosimplex sp.* exhibited different levels of antioxidant activity in all the in vitro tests performed, such as α,α-diphenyl-β-picrylhydrazyl (DPPH), hydroxyl radical scavenging assay (HRSA), metal chelating assay, and β-carotene–linoleic acid model system and it was concluded that increasing the salt concentration increases the antioxidant capacity of this fungi. Antioxidant enzyme assays such as superoxide dismutase assay, catalase assay, guaiacol peroxidase assay, and glutathione S transferase assay with extracellular and intracellular samples were separately analyzed for activity and it was concluded that the antioxidant enzyme activity was best at 15% of salt concentration. Hence, this species can be subjugated for the production of antioxidants in the presence of high salt concentrations for commercial viability.

Xiao et al. (2013) isolated the moderately halophilic fungal strain *Aspergillus sp.* nov. F1 from a solar saltern in Weihai, Shandong, China. The fungus was tested for its secondary metabolites (cytotoxic compounds) in the presence of salt concentrations. It was found that the increase in salt concentration increased the production of cytotoxic compounds. Three compounds with cytotoxicity were isolated from the ethyl acetate extract of the whole broth and mycelia of *Aspergillus sp.* nov. F1, and identified as ergosterol, rosellichalasin, and cytochalasin E, respectively. The structure elucidation of isolated compounds was performed by 1H and 13C NMR spectral. In terms of quantity, cytochalasin E was the most isolated compound (985 mg), followed by rosellichalasin (712 mg) and ergosterol (346 mg). Bioassay of crude and purified cytotoxic compounds showed anti-cancerous potentials against many tumors. Crude extract was found effective against A549, Hela, BEL-7402, and RKO (data was not provided). The purified compounds showed high toxicity to human tumor cell lines A549, Hela, BEL-7402, and RKO. Ergosterol was found very potent against human colon cancer cell line RKO.

The antibacterial potentials of *Aspergillus flavus*, *Aspergillus gracilis*, and *Aspergillus penicillioides* were checked by plate screening method and by spectrophotometric analyses, against Gram-positive *Bacillus subtilis* and Gram-negative *Escherichia coli* (Ali et al. 2014c). The results showed that halophilic fungal strains were active in producing antibacterial compounds against both Gram-positive and Gram-negative bacteria tested. These results were different from earlier report of Sepcic et al. (2011), in which the halophilic fungi were not effective against *E. coli*. Antioxidant potentials of these halophilic fungi were checked by using thin-layer chromatography and total phenolic content assay. All obligate halophilic fungi showed positive antioxidant potential with *Aspergillus penicillioides* (sp. 2) showing most antioxidant capacity. Some hydrolases (amylase, cellulase, lipase, protease, and xylanase) were screened by using obligate halophilic fungi through plate screening studies and by enzyme assays. Except for *Aspergillus penicillioides* (sp. 2), all of the screened enzymes were found positive at least by one obligate halophile and vice versa. All crude fungal filtrates were obtained at 10% of NaCl concentration.

Zambelli et al. (2015) reported a brief study by taking lyophilized *Cladosporium cladosporioides*, which was previously isolated for fructooligosaccharide production from sucrose. Fructofuranosidase assays were performed by using sucrose as a substrate and dinitrosalicylic acid (reagent) was used to study the calculated amount of fructose obtained. The crude enzyme residual activity was studied at different temperatures which revealed that enzyme was heat stable with an increase in enzyme activity at temperatures from 50 °C to 60 °C and the highest activity was found at 50 °C. Substrate concentration of 600 g/L and the lyophilized mycelium of 40 g/L (at 50 °C) were found to have the highest fructooligosaccharide production (344 g/L after 72 h). By high-performance liquid chromatography, the fructooligosaccharide composition was found as 1-fructofuranosylnystose 22 g/L, 1-kestose 184 g/L, and 1-nystose 98 g/L. A nonconventional disaccharide, namely maltose, was also recovered (30 g/L). Since Fructooligosaccharides have many commercial applications such as their use for curing constipation, traveler's diarrhea, and high cholesterol levels as well as their use as prebiotics and artificial sweeteners; it can be easily deduced that these compounds isolated from the halophilic fungi can be of great commercial use.

Jančič et al. (2016) collected 30 strains of *Wallemia spp.* obtained from different culture banks isolated from various hypersaline environments. The strains were maintained and grown for production of secondary metabolites according to the need of required media and salt concentrations (NaCl and MgCl$_2$) and sugar (glucose). High-performance liquid chromatography-diode array detection was used for the detection of approximately 100 different compounds selected from overall 200 extracts of *Wallemia spp.* The machine learning analysis revealed that NaCl was the most influenced solute amongst all solutes tested. Mass spectroscopic results in this study showed that NaCl significantly affects the biological activity of some compounds. There was an increase in the production of toxic metabolites (wallimidione, walleminol, and walleminone) from strains when the NaCl concentration was increased from 5% to 15%. Since these toxic compounds are known to cause respiratory conditions like asthma, hypersensitivity pneumonitis, rhinosinusitis, bronchitis, and respiratory infections, and *Wallemia spp.* is commonly found in household dust and as a food contaminant, it would be interesting to further our knowledge in this regard.

15.4.2   Production of Enzymes

15.4.2.1   Proteases

Annapurna et al. (2012) collected soil samples from the coast of Mumbai and from the Sambhar Lake in Rajasthan, India. The samples were focused on screening for protease-producing species which were later identified. *Aspergillus flavus* was claimed to be a halophilic fungus (no details were presented for species apart from the addition of 10% NaCl in the production medium of protease) after molecular identification. During the characterization studies the protease from *Aspergillus flavus* was found to be slightly acidic (having optimum activity at pH 6), thermophilic (having highest activity at 57 °C but inconclusive, as the graph just carried an upward peak only), and showing inhibition by the use of HgCl$_2$ and activation by the use of CaCl$_2$. Further work in this direction could help isolate/develop a protease which is stable at higher salt concentrations and could be used in the food and leather industry.

15.4.2.2   Amylases

Ali et al. (2014a) reported an α-amylase from halophilic *Engyodontium album*. The crude enzyme was purified by column chromatography and the molecular mass of enzyme was found to be approximately 50 kDa by SDS-PAGE. The specific activity was found as 132.17 U/mg by enzyme assay. Through enzyme kinetic studies, the Vmax and Km values of 15.36 U/mg and 6.28 mg/mL were found, respectively. Enzyme characterization studies showed polyextremophilic properties of α-amylase from *Engyodontium album*. The amylase was found alkalophilic by showing a steady increase up to pH 9 and even retaining over 95% of its percentage relative activity at pH 10. Thermophilic nature of enzyme was found by a steady increase in enzyme activity at higher temperature up to 60 °C. Enzyme activity was increased by the addition of salt in substrate. The optimum halophilic character was found at 30% (which is a saturation point of NaCl solution), but even at 35% of salt concentration (oversaturation point of NaCl solution) the enzyme retained almost 90% of the activity. $HgCl_2$, $CaCl_2$, $MgCl_2$, and $BaCl_2$ improved enzyme activity, while $FeCl_2$, $ZnCl_2$, EDTA, and β-mercaptoethanol were found to decrease the amylase activity.

Ali et al. (2014b) reported another amylase isolated from *Aspergillus gracilis* and purified by column chromatography. The molecular mass of enzyme was found to be approximately 35 kDa by SDS-PAGE. The specific activity was found as 131.02 U/mg by enzyme assay. Through enzyme kinetic studies, the Vmax and Km values of 8.36 U/mg and 6.33 mg/mL were found, respectively. Enzyme characterization studies showed polyextremophilic properties of α-amylase from *Aspergillus gracilis*. The amylase was found acidophilic by showing an optimum activity at pH 5 and even retaining over 95% of its percentage relative activity at pH 4. Thermophilic nature of enzyme was found by an increase in enzyme activity at higher temperature up to 60 °C. However, there was a sharp decline in enzyme activity after 60 °C. Amylase activity was increased by the addition of salt in substrate. The optimum halophilic character was found at 30%. There was a steady increase from 5% to 25% of salt concentrations, but there was a sharp increase in enzyme activity from 25% to 30% of salt concentration. The amylase was found retaining over 85% of percentage relative activity up to last tested concentration of 40% salt which is a supersaturation point of salt concentration. Unlike the amylase from *Engyodontium album* (Ali et al. 2014a), the α-amylase from *Aspergillus gracilis* was not found much affected by the addition of metallic salts. Only $FeCl_2$ was slightly found to inhibit the enzyme activity. The amylase from *Aspergillus gracilis* was tested for its saline wastewater remediation compared with the commercial-grade amylase from a normal fungus. Synthetic wastewater was made and added with 0–25% of NaCl concentrations. The enzymes were added, incubated, and compared for their performance by dissolved oxygen parameter. The percentage relative activity of amylase from *Aspergillus gracilis* was taken as control. The waste remediation ability of commercial amylase subsequently decreased by the addition of salt in the substrate.

In another study Ali et al. (2016) selected the obligate halophilic fungus *Aspergillus penicillioides* for the purification and characterization of α-amylase. The crude enzyme was purified by column chromatography. The molecular mass of enzyme was found to be approximately 42 kDa by SDS-PAGE. The specific activity was found as 118.42 U/mg by enzyme assay. Through enzyme kinetic studies, the Vmax and Km values of 1.05 mol/min·mg and 5.41 mg/mL were found, respectively. Enzyme characterization studies showed polyextremophilic properties of α-amylase from *Aspergillus penicillioides*. The amylase was found being alkalophilic by showing an optimum activity at pH 9, though the activity declined rapidly after crossing pH 9. Amylase was found to have an increase in enzyme activity at higher temperature up to 80 °C (this by far is the highest temperature of α-amylase ever reported yet from halophilic fungi). Though there was a sharp decline in enzyme activity after 80 °C still the amylase was found to have an activity over 80% at 90 °C. Increase in salinity increased the amylase activity. The optimum halophilic character was found same at 30%. More likely as the amylase from *Engyodontium album* (Ali et al. 2014b), the α-amylase from *Aspergillus penicillioides* was affected by the addition of metallic salts. Only $CaCl_2$ was found slightly activating the enzyme activity, which suggests that this enzyme could be metalloenzyme being activated more by the divalent ions of Ca^{2+}. Enzyme was greatly inhibited by $ZnCl_2$, followed by moderate inhibition from EDTA and $FeCl_2$. Negligible inhibition was observed by the addition of 2 mM concentrations of $BaCl_2$, $HgCl_2$, $MgCl_2$, and β-mercaptoethanol. The compatibility of α-amylase from *Aspergillus penicillioides* was checked by the addition of three commercial detergents (A, B, and C, obtained from local market in Thailand) and incubated for 1 h at 40 °C. Residual enzyme activity was calculated. It was found that enzyme retained at least more than 80% of residual activity in comparison to the control in any tested detergent. The α-amylase from *Aspergillus penicillioides* was also tested for its performance in increasing salt concentrations (from 0% to 5% NaCl) with abovementioned commercial detergents as well as with commercial amylase from a normal fungus. The percentage relative activity results showed gradual decrease in the activities of commercial amylase and detergents in comparison to α-amylase of *Aspergillus penicillioides*, which was taken as control.

Hence these reported amylases which are much stable at higher salt concentrations as well as being thermostable could find their commercial applications in bioremediation of wastewater as well as food-, pharmaceutical-, and fermentation-based industries.

15.4.2.3   Cellulases

Gunny et al. (2014) isolated *Aspergillus terreus* and *Penicillium sp.* The fungal strains were tested for halotolerance by supplementing salt from 0% to 30% in growth medium (no results were provided). Both fungi were tested for their crude cellulases in increasing salt concentrations from 0% to 30%. *Aspergillus terreus* was found of having better hydrolysis capacity at higher salt concentration so was chosen for further studies. The crude cellulases from *Aspergillus terreus* were characterized for salinity, temperature, and stability in ionic liquids. Halostability of the enzyme was tested by incubating the enzyme for either 1 h or 24 h from 0% to 20% of salt concentrations. The enzyme stability was increased from 0% to 15% of salt concentration and was decreased further in both allotted times. The heat stability of enzyme was tested at different temperatures (0–80 °C) at various salt concentrations (from 0 to 3 M NaCl). There was an increase in enzyme stability at any temperature by the increase of salt concentration. At any salt concentration, the enzyme stability was almost found best until 40 °C, after which there was a gradual decrease in enzyme activity. This step suggested that the increase in salt was increasing the thermostability of the enzymes. The cellulase activity was determined in the presence of different ionic liquids ([BMIM][Ac], [EMIM][Ac], and [BMIM] [Cl]). The enzyme activity was found increased from 0% to 10% concentration of ionic liquids after which it declined gradually.

Gunny et al. (2015) carried same cellulases from *Aspergillus terreus*, which was found halophilic and halostable. In this study, production of halophilic cellulases from *Aspergillus terreus* was found positively influenced by the substrate (carboxymethylcellulose, i.e. CMC), salts ($FeSO_4 \cdot 7H_2O$, NaCl, and $MgSO_4 \cdot 7H_2O$), and peptone and physical factors such as size of inoculum and agitation speed. Contrarily, components like yeast extract, KH_2PO_4, KOH, and temperature were found negatively affecting the production of enzyme. Face-centered central composite design was applied on the most positively influenced components (CMC, $FeSO_4 \cdot 7H_2O$, and NaCl) to find the exact amount of optimization. The results showed that the fungus preferred the higher concentration of NaCl and CMC for the cellulase production. The overall optimization studies showed almost double-fold of increase in cellulase production from 0.029 U/mL to 0.0625 U/mL, which shows that by statistical approaches and good experiment designs the production cost of enzyme production can be minimized.

Since cellulases isolated from halophiles are much stable at high salt concentrations, they could be used in the textile, paper, and food industry which require a salt-stable cellulose for better results.

15.4.3 Application of Halophilic Fungi in Bioremediation

15.4.3.1 Phenol Degradation

Jiang et al. (2016) collected activated sludge from the pharmaceutical factory in Wuhan, Hubei, China. The fungus was selected from the microbial mixture on the basis of phenol tolerance. The growth media and all tests in this research were carried by supplementing 5% of NaCl concentration. The obtained colonies were morphologically and molecularly identified into *Debaryomyces sp.*, showing most close resemblance with *Debaryomyces hansenii* and *Debaryomyces subglobosus*. Phenol degradation ability was assessed by supplementing 100–1300 mg/L of phenol to the growth medium. The cultures were incubated at 30 °C at 160 rpm. The readings were obtained up to 72 h by spectrophotometric analysis through cell cultures and residual phenol concentration. The results show that lesser initial concentrations most relevantly up to 500 mg/L took shorter time for phenol degradation as compared to the higher concentrations. Over 900 mg/L of initial phenol concentration took 4 and more days to degrade all phenol in the media. Similarly there was a decrease in biomass of the organism due to increase in phenol concentration. The heavy metal tolerance along with phenol degradation was checked by adding 5 mM salts ($CoCl_2 \cdot 6H_2O$, $MnCl_2 \cdot 4H_2O$, $NiCl_2 \cdot 6H_2O$, and

ZnSO$_4$·7H$_2$O) of metal ions with 500 mg/L of phenol for 48 h at 30 °C and 160 rpm. Control was taken with no heavy metal salt added. Presence of Mn and Zn hardly affected the phenol degradation. Phenol degradation was greatly inhibited by the presence of Co and Ni. The effects of different physical factors (pH, dissolved oxygen, and salinities) on phenol degradation were observed. The pH was checked from 3 to 11 ranges. Neutral pH was found to facilitate the growth of fungus. The optimum pH for phenol degradation was found to be 6.0 at 36 h. For dissolved oxygen the variation (from 50 rpm to 200 rpm) of shaking speed was tested for 48 h at 30 °C. The higher shaking speeds favored the phenol degradation where the most amount of phenol was degraded at 200 rpm. Salinity was checked from 0% to 17% of NaCl concentrations. The most amount of fungal growth and degradation of phenol were found best at 1% of NaCl concentration.

15.4.3.2 Remediation of Halite on Sandstones by Halophilic Fungi

Mansour (2017) collected sandstone samples from the Medamoud, Egypt, and halophilic fungi (*Aspergillus nidulans, Aureobasidium pullulans, Cladosporium sphaerospermum* and *Wallemia sebi*) were obtained from a local culture bank. Salinity tolerance was estimated by growing these fungi at 0% to 25% of NaCl concentration supplemented on potato dextrose agar. Results showed that 5% of NaCl was the best suited growth supplement for halophilic fungi. The fungal strains were also tested on liquid media where they showed better tolerance to salt as compared to the solid media. However, 25% of NaCl concentration was found to inhibit any fungal growth. Stones were cut into small pieces and the fungal media prepared was poured over them filling all the gaps on the surface. The analytical determination of resulted samples showed that the salt concentrations from the rocks treated by halophilic fungi were lower than those of untreated rocks. *Wallemia sebi*, which showed best halotolerance in these experiments, was found best in remediation test too. Hence, *Wallemia sebi* could be used for treatment and bioremediation of hypersaline soils as well as salt damage caused by dampness in building materials such as sandstone.

15.4.3.3 Removal of Heavy Metals by Obligate Halophilic Fungi

Bano et al. (2018) took *Aspergillus flavus, Aspergillus gracilis, Aspergillus penicillioides* (2 strains as sp. 1 and sp. 2), *Aspergillus restrictus*, and *Sterigmatomyces halophilus* and incubated them in 50 mL of potato dextrose broth supplemented with 10% of NaCl concentration and 1000 ppm of heavy metal salts [CdCl$_2$·H$_2$O, CuCl$_2$·2H$_2$O, Fe(NH$_4$)$_2$SO$_4$·6H$_2$O, MnCl$_2$, Pb(NO$_3$)$_2$, and Zn(NO$_3$)$_2$]. The filtrate and biomass obtained after 14 days were undertaken for acid digestion method. Both components were separately analyzed in atomic absorption spectroscopy. Proper controls and blanks were used for comparison of data. The results showed that on average at least 67% of all metals were removed by each obligate halophilic fungus.

Most of the metals were removed by *Aspergillus flavus* (85.6%), narrowly followed by *Sterigmatomyces halophilus* (83.3%). The least activity of metal adsorption (67%) was exhibited by *Aspergillus penicillioides* (sp. 1). In terms of overall metal adsorption at least 63% of each metal was absorbed by all obligate halophilic fungi. Fe and Zn were mostly removed metals of approximate 84% adsorption. Copper was least absorbed with 63.2% of absorption.

15.5 Past and Present Trends of Research in Halophilic Fungal Biotechnology

Considering that it is only very recently that fungi have been recognized as halophilic microorganisms, the amount of reports in this new field of research is considerably high but still a lot of further research in general biotechnology and microbiology of these strains could be carried out since most of the research on halophilic microorganisms has been focused on either halophilic bacteria or, in case of halophilic fungi, adaptation studies. This can be signified by the fact that the total number of publications reported till date is 18 whereas in consecutive years 2014 and 2015 four publications are reported per year. Hence, on average, a trend of approximately two publications per year is observed, regarding biotechnology from halophilic fungi. Most of the work (39%) on halophilic fungal biotechnology has been focused on enzymes, followed by multicovered studies (28%), bioremediation research (22%), and 11% of metabolite reports. It could be postulated that an interest on halophilic fungal enzymes could be due to the polyextremophilic nature of these enzymes. Some of the reports (Esawy et al. 2016; Geoffry and Achur 2017; Lenka et al. 2016; Liu et al. 2017; Sinha et al. 2017) have somewhat touched the coverage of halophilic fungal biotechnology but they do not completely fulfill the criteria due to either focus on other microorganisms or other than proper coverage of biotechnology.

15.6 Conclusions and Future Perspectives

Climate change and global warming are creating challenges for the humanity. Approximately, 25% of irrigated lands are comprised of saline soils and subsequently millions of hectares of land are getting unsuitable for agricultural purposes. Hence, research on halophilic fungi will no longer be a luxury but it will be a need of future biotechnology. One of the reasons of fewer reports on halophilic fungal biotechnology could be the lack of awareness amongst researchers about the tremendous potentials these fungi hold in terms of their metabolites. For example, the amylases from obligate halophilic fungi have been found to consistently provide polyextremophilic characteristics. Especially, the salinity in which the optimum amylase activities are found is far above than the ones reported from halophilic bacteria (Ali et al. 2016). The nature of these amylases makes them perfect candidates to be applied in white biotechnology as most of the industrial operations are carried at extremes of pH, temperature, and low water activity. Similarly, as shown by reports these fungi can work in solving several environmental issues such as bioremediation (Bano et al. 2018; Jiang et al. 2015, 2016). Screening studies show that there is still a lot to be explored about biotechnology from halophilic fungi. The genes present in these fungi can be incorporated into food crops for making them utilize less amount of water. The human expedition on other planets such as Mars may need the aid from these fungi due to their adaptations and metabolite productions in extreme environments. This comprehensive coverage of biotechnology from halophilic fungi is expected to promote the research in this field.

References

Ali I, Akbar A, Anwar M, Yanwisetpakdee B, Prasongsuk S, Lotrakul P, Punnapayak H (2014a) Purification and characterization of extracellular, polyextremophilic α-amylase obtained from halophilic *Engyodontium album*. Iran J Biotech 12:35–40. https://doi.org/10.15171/ijb.1155

Ali I, Akbar A, Yanwisetpakdee B, Prasongsuk S, Lotrakul P, Punnapayak H (2014b) Purification, characterization, and potential of saline waste water remediation of a polyextremophilic α-amylase from an obligate halophilic *Aspergillus gracilis*. Biomed Res Int 2014:7. https://doi.org/10.1155/2014/106937

Ali I, Siwarungson N, Punnapayak H, Lotrakul P, Prasongsuk S, Bankeeree W, Rakshit SK (2014c) Screening of potential biotechnological applications from obligate halophilic fungi, isolated from a man-made solar saltern located in Phetchaburi province. Thail Pak J Bot 46:983–988

Ali I, Akbar A, Anwar M, Prasongsuk S, Lotrakul P, Punnapayak H (2016) Purification and characterization of a polyextremophilic α-amylase from an obligate halophilic *Aspergillus penicillioides* isolate and its potential for souse with detergents. Biomed Res Int 2015:8. https://doi.org/10.1155/2015/245649

Annapurna SA, Singh A, Garg S, Kumar A, Kumar H (2012) Screening, isolation and characterisation of protease producing moderately halophilic microorganisms. Asian J Microbiol Biotechnol Environ Sci 14:603–612

Baati H, Guermazi S, Gharsallah N, Sghir A, Ammar E (2010) Microbial community of salt crystals processed from Mediterranean seawater based on 16S rRNA analysis. Can J Microbiol 56:44–51. https://doi.org/10.1139/W09-102

Bakke P, Carney N, Deloache W, Gearing M, Ingvorsen K, Lotz M, McNair J, Penumetcha P, Simpson S, Voss L, Win M, Heyer LJ, Campbell AM (2009) Evaluation of three automated genome annotations for *Halorhabdus utahensis*. PLoS One 4:6291. https://doi.org/10.1371/journal.pone.0006291

Bano A, Hussain J, Akbar A, Mehmood K, Anwar M, Hasni MS, Ullah S, Sajid S, Ali I (2018) Biosorption of heavy metals by obligate halophilic fungi. Chemosphere 199:218–222. https://doi.org/10.1016/j.chemosphere.2018.02.043

Berquist BR, Müller JA, DasSarma S (2006) 27 genetic systems for Halophilic Archaea. In: Methods in microbiology, vol 35. Academic Press, Cambridge, MA, pp 649–680. https://doi.org/10.1016/S0580-9517(08)70030-8

Butinar L, Sonjak S, Zalar P, Plemenitaš A, Gunde-Cimerman N (2005a) Melanized halophilic fungi are eukaryotic members of microbial communities in hypersaline waters of solar salterns. Bot Mar 48:73–79. https://doi.org/10.1515/BOT.2005.007

Butinar L, Zalar P, Frisvad JC, Gunde-Cimerman N (2005b) The genus Eurotium - members of indigenous fungal community in hyper-saline waters of salterns. FEMS Microbiol Ecol 51:155–166

DasSarma S, DasSarma P (2012) Halophiles. In eLS, (Ed.). doi:https://doi. org/10.1002/9780470015902.a0000394.pub3

De Hoog S, Zalar P, Van Den Ende BG, Gunde-Cimerman N (2005) Relation of halotolerance to human-pathogenicity in the fungal tree of life: an overview of ecology and evolution under stress. In: Adaptation to life at high salt concentrations in Archaea, Bacteria, and Eukarya. Springer, Netherlands, pp 371–395

Esawy MA, Awad GEA, Wahab WAA, Elnashar MMM, El-Diwany A, Easa SMH, El-beih FM (2016) Immobilization of halophilic Aspergillus awamori EM66 exochitinase on grafted k-carrageenan-alginate beads. 3 Biotech 6:29. https://doi.org/10.1007/s13205-015-0333-2

Geoffry K, Achur RN (2017) A novel halophilic extracellular lipase with both hydrolytic and synthetic activities. Biocat Agric Biotechnol 12:125–130. https://doi.org/10.1016/j. bcab.2017.09.012

Gostinčar C, Lenassi M, Gunde-Cimerman N, Plemenitaš A (2011) Chapter 3 - Fungal adaptation to extremely high salt concentrations. In: Laskin AI, Sariaslani S, Gadd GM (eds) Advances in applied microbiology, vol 77. Academic Press, Cambridge, MA, pp 71–96. https://doi. org/10.1016/B978-0-12-387044-5.00003-0

Gunde-Cimerman N, Zalar P, de Hoog S, Plemenitaš A (2000) Hypersaline waters in salterns: natural ecological niches for halophilic black yeasts. FEMS Microbiol Ecol 32:235–240

Gunde-Cimerman N, Zalar P, Petrovič U, Turk M, Kogej T, de Hoog GS, Plemenitaš A (2004) Fungi in salterns. In: Ventosa A (ed) Halophilic microorganisms. Springer, Berlin, pp 103–113. https://doi.org/10.1007/978-3-662-07656-9_7

Gunde-Cimerman N, Ramos J, Plemenitaš A (2009) Halotolerant and halophilic fungi. Mycol Res 113:1231–1241. https://doi.org/10.1016/j.mycres.2009.09.002

Gunny AAN, Arbain D, Edwin Gumba R, Jong BC, Jamal P (2014) Potential halophilic cellulases for in situ enzymatic saccharification of ionic liquids pretreated lignocelluloses. Bioresour Technol 155:177–181. https://doi.org/10.1016/j.biortech.2013.12.101

Gunny AAN, Arbain D, Jamal P, Gumba RE (2015) Improvement of halophilic cellulase production from locally isolated fungal strain. Saudi J Biol Sci 22:476–483. https://doi.org/10.1016/j. sjbs.2014.11.021

Jančič S, Frisvad JC, Kocev D, Gostinčar C, Džeroski S, Gunde-Cimerman N (2016) Production of secondary metabolites in extreme environments: food and airborne Wallemia spp. produce toxic metabolites at hypersaline conditions. PLoS One 11(12):e0169116. https://doi.org/10.1371/ journal.pone.0169116

Jiang Y, Yang K, Wang H, Shang Y, Yang X (2015) Characteristics of phenol degradation in saline conditions of a halophilic strain JS3 isolated from industrial activated sludge. Mar Pollut Bull 99:230–234. https://doi.org/10.1016/j.marpolbul.2015.07.021

Jiang Y, Shang Y, Yang K, Wang H (2016) Phenol degradation by halophilic fungal isolate JS4 and evaluation of its tolerance of heavy metals. Appl Microbiol Biotechnol 100:1883–1890. https:// doi.org/10.1007/s00253-015-7180-2

Khan MN, Lin H, Li M, Wang J, Mirani ZA, Khan SI, Buzdar MA, Ali I, Jamil K (2017) Identification and growth optimization of a Marine Bacillus DK1-SA11 having potential of producing broad spectrum antimicrobial compounds. Pak J Pharm Sci 30:839–853

Kis-Papo T et al (2014) Genomic adaptations of the halophilic Dead Sea filamentous fungus Eurotium rubrum. Nat Commun 5:3745. https://doi.org/10.1038/ncomms4745

Lenassi M et al (2013) Whole genome duplication and enrichment of metal cation transporters revealed by De Novo Genome Sequencing of extremely halotolerant black yeast Hortaea werneckii. PLoS One 8:e71328. https://doi.org/10.1371/journal.pone.0071328

Lenka J, Maharana AK, Priyadarshini S, Mohanty D, Ray P (2016) Isolation and characterization of marine micro-flora from coast of Paradip and screening for potent enzymes. RJPBCS 7:1187–1196

Liu KH, Ding SW, Narsing RMP, Zhang B, Zhang YG, Liu FH, Liu BB, Xio M, Li WJ (2017) Morphological and transcriptomic analysis reveals the osmoadaptive response of endophytic fungus *Aspergillus montevidensis* ZYD4 to high salt stress. Front Microbiol 8:1789. https://doi.org/10.3389/fmicb.2017.01789

Mansour MMA (2017) Effects of the halophilic fungi *Cladosporium sphaerospermum*, *Wallemia sebi*, *Aureobasidium pullulans* and *Aspergillus nidulans* on halite formed on sandstone surface. Int Biodeter Biodegr 117:289–298. https://doi.org/10.1016/j.ibiod.2017.01.016

Miyashita Y, Ohmae E, Ikura T, Nakasone K, Katayanagi K (2017) Halophilic mechanism of the enzymatic function of a moderately halophilic dihydrofolate reductase from *Haloarcula japonica* strain TR-1. Extremophiles 21:591–602. https://doi.org/10.1007/s00792-017-0928-0

Oren A (2008) Microbial life at high salt concentrations: phylogenetic and metabolic diversity. Saline Sys 4:2. https://doi.org/10.1186/1746-1448-4-2

Oren A (2010) Industrial and environmental applications of halophilic microorganisms. Environ Technol 31:825–834. https://doi.org/10.1080/09593330903370026

Oren A (2013) Life at high salt concentrations. In: Rosenberg E, DeLong EF, Lory S, Stackebrandt E, Thompson F (eds) The prokaryotes: prokaryotic communities and ecophysiology. Springer, Berlin, Heidelberg, pp 421–440. https://doi.org/10.1007/978-3-642-30123-0_57

Paul S, Bag SK, Das S, Harvill ET, Dutta C (2008) Molecular signature of hypersaline adaptation: insights from genome and proteome composition of halophilic prokaryotes. Genome Biol 9:R70. https://doi.org/10.1186/gb-2008-9-4-r70

Plemenitaš A, Gunde-Cimerman N (2005) Cellular responses in the halophilic black yeast *Hortaea Werneckii* to high environmental salinity. In: Adaptation to life at high salt concentrations in Archaea, Bacteria, and Eukarya. Springer, Netherlands, pp 453–470

Plemenitaš A, Vaupotič T, Lenassi M, Kogej T, Gunde-Cimerman N (2008) Adaptation of extremely halotolerant black yeast *Hortaea werneckii* to increased osmolarity: a molecular perspective at a glance. Stud Mycol 61:67–75. https://doi.org/10.3114/sim.2008.61.06

Ravindran C, Varatharajan GR, Rajasabapathy R, Vijayakanth S, Kumar AH, Meena RM (2012) A role for antioxidants in acclimation of marine derived pathogenic fungus (NIOCC 1) to salt stress. Microb Pathog 53:168–179. https://doi.org/10.1016/j.micpath.2012.07.004

Sarwar KM, Azam I, Iqbal T (2015) Biology and applications of halophilic bacteria and Archaea: a review. Electron J Biol 11:98–103

Schafer G, Engelhard M, Muller V (1999) Bioenergetics of the Archaea. Microbiol Mol Biol Rev 63:570–620

Sepcic K, Zalar P, Gunde-Cimerman N (2011) Low water activity induces the production of bioactive metabolites in halophilic and halotolerant fungi. Mar Drugs 9:43

Sinha SK, Goswami S, Das S, Datta S (2017) Exploiting non-conserved residues to improve activity and stability of *Halothermothrix orenii* β-glucosidase. Appl Microbiol Biotechnol 101:1455–1463. https://doi.org/10.1007/s00253-016-7904-y

Talon R, Coquelle N, Madern D, Girard E (2014) An experimental point of view on hydration/solvation in halophilic proteins. Front Microbiol 5:66. https://doi.org/10.3389/fmicb.2014.00066

Tiquia SM (2010) Salt-adapted bacteria isolated from the Rouge River and potential for degradation of contaminants and biotechnological applications. Environ Technol 31:967–978

Tiquia SM, Mormile M (2010) Extremophiles–a source of innovation for industrial and environmental applications. Environ Technol 31(8–9):823

Tiquia SM, Davis D, Hadid H, Kasparian S, Ismail M, Sahly R, Shim J, Singh S, Murray KS (2007) Halophilic and halotolerant bacteria from river waters and shallow groundwater along the Rouge River of Southeastern Michigan. Environ Technol 28:297–307

Tiquia-Arashiro SM, Rodrigues D (2016a) Extremophiles: applications in nanotechnology. Springerbriefs in microbiology: extremophilic microorganisms. Springer International Publishing, New York, p 193

Tiquia-Arashiro SM, Rodrigues D (2016b) Halophiles in nanotechnology. In: Extremophiles: applications in nanotechnology. Springer International Publishing, New York, pp 53–58

Xiao L, Liu H, Wu N, Liu M, Wei J, Zhang Y, Lin X (2013) Characterization of the high cytochalasin E and rosellichalasin producing *Aspergillus sp. nov.* F1 isolated from marine solar saltern in China. World J Microbiol Biotechnol 29:11–17. https://doi.org/10.1007/s11274-012-1152-9

Zajc J, Liu Y, Dai W, Yang Z, Hu J, Gostinčar C, Gunde-Cimerman N (2013) Genome and transcriptome sequencing of the halophilic fungus *Wallemia ichthyophaga*: haloadaptations present and absent. BMC Genomics 14:617. https://doi.org/10.1186/1471-2164-14-617

Zambelli P, Serra I, Arrojo LF, Plou FJ, Tamborini L, Conti P, Contente ML, Molinari F, Romano D (2015) Sweet-and-salty biocatalysis: fructooligosaccharides production using *Cladosporium cladosporioides* in seawater. Process Biochem 50:1086–1090. https://doi.org/10.1016/j.procbio.2015.04.006

Zusman T, Rosenshine I, Boehm G, Jaenicke R, Leskiw B, Mevarech M (1989) Dihydrofolate reductase of the extremely halophilic archaebacterium *Halobacterium volcanii*. J Biol Chem 264:18878–18883

Chapter 16
Sporotrichum thermophile Xylanases and Their Biotechnological Applications

Ayesha Sadaf, Syeda Warisul Fatima, and Sunil K. Khare (iD)

16.1 Introduction

Thermostable enzymes are highly desirable commodities for industrial bioprocesses (Karnaouri et al. 2016). The thermophilic fungi like *Thermotoga, Thermoascus, Scytalidium*, etc. which produce thermostable enzymes have been majorly exploited in industrial bioprocesses (Patel and Savanth 2015; Zeldes et al. 2015). *Sporotrichum thermophile* (syn. *Myceliophthora thermophile*) is a potential thermophilic mould belonging to the class ascomycetes (Singh et al. 2016). It has been extensively utilised as a platform for the production of large number of enzymes mainly cellulases, xylanases, phytases, esterases and mannoses (Bala and Singh 2016).

Lignocellulosic biomass is the most plentiful and rich source of renewable energy; hence its utilisation for the production of second-generation biofuels and platform chemicals has been perceived as an alternate strategy to meet the current alarm of fossil fuel depletion (Cherubini 2010; Tiquia-Arashiro and Mormile 2013; Gupta and Verma 2015). Lignocellulosics are mainly composed of three major components, viz. cellulose (40–55%), hemicellulose (20–30%) and lignin (15–20%) (Menon and Rao 2012; Tadesse and Luque 2011). For complete utilisation of biomass the most important process remains the removal of lignin so that the hemicellulose and cellulose portion are saccharified by hydrolytic enzymes like cellulases and xylanases (Ravindran and Jaiswal 2016). Thermostable cellulases and xylanases find great advantage since their hydrolytic rates are higher as compared to their mesophilic counterparts (Plecha et al. 2013; Watanabe et al. 2016). In this context, *S. thermophile* xylanase holds great potential to be used for efficient biomass utilisation. The xylanase from *S. thermophile* belongs to GH10 and GH11 families and has the ability to catalyse the hydrolysis of different types of substituted and

A. Sadaf · S. W. Fatima · S. K. Khare (✉)
Department of Chemistry, Indian Institute of Technology Delhi, Hauz Khas, New Delhi, India
e-mail: skkhare@chemistry.iitd.ac.in

© Springer Nature Switzerland AG 2019
S. M. Tiquia-Arashiro, M. Grube (eds.), *Fungi in Extreme Environments: Ecological Role and Biotechnological Significance*,
https://doi.org/10.1007/978-3-030-19030-9_16

non-substituted xylans, respectively (Van Gool et al. 2012, 2013; Vardakou et al. 2003). The xylanase has been widely studied with a pH optima in the range of 5.0–7.0 and a temperature optima between 50 and 70 °C along with other desirable properties such as thermal, pH and ionic liquid stability (Bala and Singh 2016; Katapodis et al. 2006; Sadaf et al. 2016; Topakas et al. 2003). This xylanase has been produced by both submerged and solid-state mode of fermentation by employing a number of substrates like cotton seed cake, wheat straw and corn cobs (Bala and Singh 2017; Katapodis et al. 2006; Vafiadi et al. 2010). Xylanase as well as cellulase from *S. thermophile* have also been successfully used for biomass conversion processes. For example, xylanase and cellulase from *S. thermophile* namely CMCase and endo- and exo-β-1,4-glucanase have been used for hydrolysis of rice straw and waste tea cup paper yielding sugars of 578.12 and 421.79 mg/g substrate (Bala and Singh 2016). Similarly a thermophilic enzymatic cocktail containing cellobiohydrolase, endoglucanase, mannanase and xylanase activities was used for the saccharification of wheat straw, birch and spruce biomass (Karnaouri et al. 2016).

Hence *S. thermophile* xylanase is viewed as a potent accessory enzyme and holds great significance in the biofuel industry. To date there are few reports which have summarised the research update on *S. thermophile* as well as the different types of enzymes produced by it along with their industrial applications. This chapter encompasses for the first time the detailed description of *S. thermophile* xylanase, its biochemical characteristics and its production levels by different modes of fermentation and by utilising different lignocellulosic substrates. The chapter also sheds some light on the genetic organisation of *S. thermophile*. The proteomic and transcriptomic profiles have also been discussed with a view to gain an insight into the expression level of various lignocellulolytic enzymes.

The last part of the chapter deals with the latest research being conducted on *S. thermophile*. This includes gene editing protocols by CRISPR-Cas system and the ionic liquid stability of *S. thermophile* xylanase.

16.2 Xylanases

Xylan (β-1, 4-D-xylose polymer) is the major form of the polymer hemicellulose and contains wide linkages and branching points (Scheller and Ulvskov 2010). Xylan is basically a heteropolysaccharide consisting of arabinosyl, *O*-acetyl and 4-*O*-methyl-D-glucuronic acid substituents, ferulic and coumaric acids (Collins et al. 2005). Their distribution and composition vary among various plant genera and species. The xylans from dicots are acetylated to various degrees whereas xylans from monocotyledons like grasses, sorghum species and eucalyptus comprise ferulic acid esters attached to arabinofuranosyl and galactopyranosyl residues, respectively (Scheller and Ulvskov 2010). The functional properties of xylans like solubility, interaction with various polymeric substances present in the cell wall as well as degradability by enzymes are influenced by their substitution (Ebringerova and Heinze 2000). Due to the complex structure of the xylan and its various substituents, its enzymatic hydrolysis becomes an important area of study. The hydrolases are required to cleave the glycosidic linkage of xylan as well as its branches. These enzymes include endoxylanase (EC 3.2.1.8), β-xylosidases (EC 3.2.1.37), α-arabinofuranosidases (EC 3.2.1.55), α-glucuronidase (EC 3.2.1.139), feruloyl esterases (EC 3.1.1.73) and xylan esterases (EC 3.1.1.72) (Manju and Singh Chadha 2011).

Xylanases fall under the category of xylanolytic enzymes with 4-xylanohydrolase, endo-1,4-xylanase, endo-1,4-β-xylanase, β-1,4-xylanase, endo-1,4-β-D-xylanase and 1,4-β-xylan xylanohydrolase activities (Goncalves et al. 2015). Major producers of xylanase are the fungal sources while bacteria, plants and actinomycetes are also good producers. For example xylanases from *Trichoderma reesei*, *Trichoderma longibrachiatum* and *Aspergillus niger* are being used commercially due to their high activity (Dhiman et al. 2008; Kumar et al. 2016).

Xylanases have been classified, based on their structure as well as their mechanism of action, into different glycoside hydrolase families, namely GH 5, 7, 8, 9, 10, 11, 12, 16, 26, 30, 43, 44, 51 and 62 (Moreira 2016). Among these the glycoside hydrolase families 10 and 11 are the most widely studied. GH 10 family consists of endo-1,4-β-xylanases (EC 3.2.1.8), endo-1,3-β-xylanases (EC 3.2.1.32) and cellobiohydrolases (EC 3.2.1.91) (Takahashi et al. 2013) whereas GH 11 family consists of 'true xylanases' as they are routinely referred, as these act on mainly D-xylose-containing substrates (Bai et al. 2015b). The endo-xylanase forms the most important class of xylanases as they have the ability to break the glycosidic linkages liberating xylooligosaccharides and hence find extensive applications in the food and feed industry (Harris and Ramalingam 2010). The other important applications of xylanases include pulp and paper processing, clarification of fruit juices, in the textile industry as well as in lignocellulosic biomass processing.

16.2.1 Xylanase from Extremophilic Sources

Extremophiles are microorganisms which possess the unique ability to thrive under extremes of temperature, pH, pressure, salts, etc. and hence the enzymes produced by these 'classified as extremozymes' are also able to function under the above-mentioned stringent conditions (Dalmaso et al. 2015). Alkaliphilic xylanases are required in the pulp and paper industry since the chemical pulping leads to increase in the pH values and hence alkali-stable xylanases are required in this regard (Lin et al. 2013; Weerachavangkul et al. 2012). As a representative example, xylanase from *Alkalibacterium* sp. SL3 was found to be stable in the alkaline pH range of 7.0–12.0 with optima at pH 9.0. The xylanase was also found to be halotolerant by maintaining 60% of its activity at 3 M NaCl. These features made it suitable for biobleaching of paper pulp as well as in the production of xylooligosaccharides (Wang et al. 2017). Similarly an alkalitolerant xylanase from *Bacillus* sp. SN5 has also been reported in a recent study (Bai et al. 2015a). Acidophilic xylanases find application in the food industry for the clarification of juices, baking as well as biobleaching (Sharma et al. 2016; Yegin 2017). The acidic medium enhances the catalytic mechanism of xylan hydrolysis and also prepares the substrate for cleavage. For example, a novel acidophilic and thermostable xylanase has been isolated from the fungal strain, *Thermoascus aurantiacus* M-2 (Ping et al. 2017). The xylanase exhibited a pH and temperature optima of 5.0 and 75 °C and was stable in the broad pH range of 2.0–10.0. Apart from the above described xylanases, cold-adapted xylanases also have immense potential in the food industries since they function at low temperatures which minimises the chances for microbial contamination (Butt et al. 2008). The cold-adapted xylanase from *Penicillium chrysogenum* showed high activity between 2 and 15 °C and between pH 3.0 and 9.0. However, most demanded xylanases are those from thermophiles due to their thermostability at higher temperatures which result in better product as well as substrate solubility (Watanabe et al. 2016). The reactions taking place at higher temperatures reduce the time for hydrolysis as well as lower the viscosity of the medium. The details about thermophilic xylanases are discussed in the next section.

16.2.2   Thermophilic Xylanases

Thermophilic xylanases are produced by a variety of microorganisms including bacteria, fungi and yeasts though fungi have been the most potent producers of thermophilic xylanases (Kumar et al. 2016; Polizeli et al. 2005). Among the fungal sources, the most common producers include *Thermomyces lanuginosus*, *Thermoascus aurantiacus*, *Paecilomyces thermophile*, *Malbranchea cinnamomea*, *Chaetomium thermophilum*, *Sporotrichum thermophile*, etc. (Bala and Singh 2018). These thermostable xylanases find special applications in pulp-prebleaching, bioethanol industry as well as other biomass conversion processes. Table 16.1 enlists some representative thermostable xylanases and their industrial applications.

16.2.3   Structure of Thermophilic Xylanases

To gain an insight into the structure of thermophilic xylanases, their crystal structures have been resolved and studied in detail. An interesting study on structural features of a GH 10 xylanase from thermophilic *Caldicellulosiruptor bescii* revealed stable interactions between the loops of the catalytic cleft region by H-bonding (Zhang et al. 2016). This H-bonding network was found to be mediated by Arg[314] and Lys[306] residues. Moreover, the presence of an aromatic cluster of amino acids like Tyr[17] and Phe[20], and Phe[21] and Phe[337], was thought to increase the interaction between the N- and C-terminus. All these interactions helped in providing stability to the enzyme at high temperatures. Similarly the structure of another well-known xylanase from *Thermoascus aurantiacus* has also been determined (de Souza et al. 2016). It was found that the protein had a predominance of salt bridges. It also

Table 16.1 Thermostable xylanases and their industrial applications

Microorganism	Source	pH and temperature optima	Heat stability	Application	References
Rhizomucor miehei NRRL 3169	Fungus	pH and temperature optima between 5.56–6.0 and 65 °C	Completely stable between 70 and 75 °C for 60 min	Pulp bleaching	Fawzi (2011)
Actinomadura sp	Bacteria	pH and temperature optima of 10.0 and 80 °C, respectively	Completely stable for 5 days at 60 °C	Pulp and paper industry	Taibi et al. (2012)
Penicillium oxalicum	Fungus	pH and temperature optima of 5.0 and 50 °C, respectively	Retained 63% activity at 50 °C for 30 min	Production of xylooligosaccharides	Wang et al. (2013)
Remersonia thermophila CBS	Fungus	pH and temperature optima of 6.0 and 65 °C, respectively	Retained 50% activity at 50 °C for 30 min	Bread making	McPhillips et al. (2014)
Caldicoprobacter algeriensis strain TH7C1	Bacteria	pH and temperature optima of 11.0 and 70 °C, respectively	Retained 50% activity at 80 °C for 4 h	Pulp bleaching	Bouacem et al. (2014)
Malbranchea cinnamomea	Fungus	pH and temperature optima of 6.5 and 80 °C, respectively	Retained 50% activity at 70 °C for 76 min	Production of xylooligosaccharides	Fan et al. (2014)
Thermoascus aurantiacus KKU-PN-I2-1	Fungus	pH and temperature optima of 9.0 and 60 °C, respectively	Retained 70% activity at 50 °C for 90 min	Pulp and paper industry	Chanwicha et al. (2015)
Scytalidium thermophilum ATCC no. 16454	Fungus	pH and temperature optima of 9.0 and 60 °C, respectively	Retained 85% activity at 50 °C for 120 min	Biomass hydrolysis	Kocabas et al. (2015)

(continued)

Table 16.1 (continued)

Microorganism	Source	pH and temperature optima	Heat stability	Application	References
Planococcus sp. SL4	Bacteria	pH and temperature optima of 7.0 and 70 °C, respectively	Completely stable at 55 °C for more than 60 min, retained 50% activity at 60 and 70 °C for 15 min and 2 min, respectively	–	Huang et al. (2015)
Geobacillus sp.	Bacteria	pH and temperature optima of 7.0 and 65 °C, respectively	Completely stable at 65 °C for 60 min at pH 10.0	Pulp and paper industry	Mitra et al. (2015)
Thielavia terrestris Co3Bag1	Fungus	pH and temperature optima of 5.5 and 85 °C	pH stability in the range of 4.5–10.0. Maintained 70–80% of its activity in the temperature range of 65–80 °C	Biofuel production	García-Huante et al. (2017)
Pleurotus ostreatus HAUCC 162 and *Irpex lacteus* CD2	Fungus	pH and temperature optima of 5.0 and 55 °C, respectively. Stable in the presence of Ca^{2+}, Cr^{3+}, Zn^{2+}, Na^+ and $Al3^+$. Inhibited by Li^+, K^+, Cu^{2+}, Fe^{3+}, Ni^{2+}, Mg^{2+}, Cd^{2+}	–	Lignocellulosic hydrolysis	Zhuo et al. (2018)

consisted of a compact hydrophobic core and proline residues at the N-terminal which were thought to provide thermostability to the xylanase. Similarly the thermophilic xylanase from the ruminal fungus, *Neocallimastix patriciarum*, have also been studied structurally (Cheng et al. 2014). A uniquely present N-terminal region (NTR) comprising of 11 amino acid residues was found in this particular protein.

This N-terminal region was linked to the catalytic core by H-bonds, stacking forces and a disulphide bond. This NTR was confirmed to contribute significantly to the thermostability of the xylanase after a series of mutations in this particular region led to a loss in the thermophilicity of the enzyme. The structure of xylanase from the fungus *Thermomyces lanuginosus* showed the presence of disulphide linkages as well as many charged residues leading to the formation of a compact globular structure (Wang et al. 2012). The crystal structure of two thermophilic xylanases from *Chaetomium thermophilum* and *Nonomuraea flexuosa* also provided some interesting thermophilic features (Hakulinen et al. 2003). Both the xylanases were found to contain twisted β-sheets forming a cleft. The *C. thermophilum* xylanase showed the presence of sulphate and calcium ions. The sulphate ions were attached to the arginine residues whereas the calcium ions interacted with the threonine residues. The resultant packing of the enzyme molecule led to the appearance of a compact tetrameric structure. The *N. flexuosa* structure consisted of a large number of side chain-polar interactions as well as several salt bridges. Similarly the structural elucidation of the xylanase from marine bacterium *Geobacillus stearothermophilus* revealed the presence of glutamate residues in the catalytic core of the protein. Homology studies with the local *G. stearothermophilus* showed the presence of five important amino acid residues. These were Thr/Ala, Asn/Asp, Lys/Asn, Iso/Meth and Ser/Thr. These residues were analysed to be responsible for imparting thermostability to the enzyme (Saksono and Sukmarini 2010).

16.3   *Sporotrichum thermophile*

Sporotrichum thermophile (syn *Myceliophthora thermophila*) is a major thermophilic fungus which is known to produce a variety of enzymes having immense industrial potential. Various enzymes produced by this thermophilic mould are xylanases, cellulases, esterases, phytases, mannanases and glucosidases (Gopalan et al. 2015; Kumari et al. 2016). The fungus was classified under ascomycetes and has optimal growth temperature between 40 and 50 °C (L Bergquist et al. 2014). Recently the genome of *S. thermophile* has been sequenced and has provided some pertinent details about the lignocellulolytic and other enzyme machinery.

16.3.1   Sporotrichum thermophile *Xylanase*

S. thermophile produces major endo-1,4-β-xylanolytic activities. These types of xylanases majorly fall under two prominent xylanase families, i.e. GH10 and GH11. While the GH10 xylanases are efficient for the hydrolysis of various xylans as well as their substituted forms, GH11 is more specific and does not act on substituted xylans (Van Gool et al. 2012). In one study the efficiency of two GH10 xylanases of *S. thermophile* was investigated on different types of xylans. The degradation pattern on the basis of substrate specificity of high and low-substituted xylans gave a possible indication about the difference in the protein structure of both xylanases (Van Gool et al. 2012). Further study revealed intra-family differences in the GH10 type of xylanases (Van Gool et al. 2013). The protein structure of both the xylanases showed specific variation in the amino acid sequence. On similar lines, a xylanase with a molecular mass of 25 kDa was purified from *S. thermophile* ATCC 34628 (Katapodis et al. 2003). The enzyme had a pH and temperature optima of 5.0 and 70 °C, respectively. By studying the degradation pattern of different types of xylans as well as the inhibition profile by ω-epoxyalkyl glycosides of d-xylopyranose, the xylanase was concluded to be part of the GH11 family. However a study conducted a little later showed the purification and characterisation of two types of xylanases from *S. thermophile* with a molecular mass of 24 kDa and 48 kDa, respectively. Both the xylanases had a pH and temperature optima of 5.0 and 60 °C and the peptide mass sequencing helped in identifying the xylanases to belong to GH10 and GH11 families, respectively (Vafiadi et al. 2010). A recent study on *S. thermophile* xylanases provided a clear insight into their structural parameters (Basit et al. 2018). Both the xylanases closely resembled a right-handed helix with minor differences occurring in the β-strand. Their main chain structures matched with those of GH11 xylanases of *Acremonium cellulolyticus* and *Trichoderma longibrachiatum*.

16.3.2 Production of Sporotrichum thermophile *Xylanase*

The xylanase from *S. thermophile* possesses various advantageous properties like thermal and pH stability which are of industrial significance; hence different methods to enhance its production level have been attempted by several research groups in the recent past. A representative study utilised corn cobs as carbon source for xylanase production. By using central composite design, 2.7% (w/v) corn cob and 0.7% (w/v) ammonium phosphate were found to yield maximum xylanase activity levels of 56 U/mL (Katapodis et al. 2006). In another study factors such as moisture content and carbon sources (wheat straw and bran) were varied to evaluate the xylanase production levels under solid-state fermentation (SSF). Quite high xylanase activity of 320 U/g was attained under finally optimised conditions (Topakas et al. 2003). Badhan et al. (2007) reported the production of xylanase by using rice straw as the substrate under SSF. A similar study reported xylanase activity levels under SSF by employing wheat bran and citrus pectin. Maximal xylanase activity (1900 U/g) was achieved after 4 days of fermentation at pH 7.0 and 45 °C with wheat bran and citrus pectin in a ratio of 1:1(Kaur and Satyanarayana 2004). Xylanase production has also been optimised by using *Jatropha curcas* seed cake (Sadaf and Khare 2014). Under the final conditions of pH, temperature, moisture level, carbon supplementation and inducers, high titres of xylanase (1025 U/g seed cake) were obtained. This study was advantageous in terms of utilisation of the seed cake which is otherwise underutilised as it is toxic even to animal feed.

16.3.3 Genomics of Sporotrichum thermophile *Xylanase*

Sporotrichum thermophile has emerged out to be a useful and well-characterised thermophilic strain while hunting for thermostable xylanases from filamentous fungi (Liu et al. 2017; Margaritis et al. 1986). The coupled cellulolytic activity has made it even more suitable for studying it genomic features (Wojtczak et al. 1987).

While deciphering the genomic makeup of this strain, it was noticed that genome size extended to 38.74 Mb containing seven telomere-telomere chromosomes comprising of TTAGGG repeats (Singh 2016). With most of the genes residing on chromosomes 2 and 4, main translocation events occurred extensively between chromosomes 1 and 6. The core of protein coding in the genome corresponded to 9110 genes, which were too small as compared to other closely related class of fungi, Sordariomycetes (Berka et al. 2011). It also shared similarity in organisation with *Thielavia terrestris*. The gene model statistics and assembly exhibited G + C content of 51.4% and the gene length of 1733 nt with two exons per gene. Higher GC content at the coding regions pointed towards greater thermal stability. Based on above facts, *T. terrestris* and *S. thermophile* have been placed under the family Chaetomiaceae as they share syntenic relationships (with over 6000 genes in a block between the pair ending with repetitive stretch of AT-rich contents) and orthologues (6279 three-way) (Berka et al. 2011). The genome of *S. thermophile* is further constituted of gene families of large transporters such as AAA, sugar, ABC and MFS along with signalling proteins like WD40 and protein kinases. Several other protein domains like Pfam and glycoside hydrolases of families GH61 and GH11 appeared to be expanded in the genome. Attempts have been made to clone xylanase genes MYCTH_56237 (672 bp) and MYCTH_49824 (693 bp) encoding proteins of 223–230 amino acids (Verma et al. 2013). The bioinformatics analysis matched with endo-β-1,4-xylanases which belong to GH11 family of glycoside hydrolases. Further sequence alignment of these xylanases showed high index of similarities in their primary and secondary structures. The insights of the tertiary structures also provided a clear picture for the mechanism of substrate binding.

Other relevant enzymes of the strain have also been retrieved by genomic analysis. These included polysaccharide lyases (PLs like pectate lyases), glycosyl transferases (GT), carbohydrate esterases, carbohydrate-binding modules and carbohydrate-active proteins (CAZymes): glycoside hydrolases (GHs-polygalacturonases) (Harris et al. 2010; Henrissat and Davies 1997). The diverse array of enzymes produced by *S. thermophile* having differential role eventually helps the fungus to effectively hydrolyse various types of biomass.

16.3.4 Transcriptomic Profile of Sporotrichum thermophile Grown on Different Biomass

Transcriptomic profile of *S. thermophile* was also studied for checking the decomposition pattern of polysaccharides. The growth of the fungus was analysed on glucose, starch, flax, canola, barley (monocot) and alfalfa straw (dicot) (Kolbusz et al. 2014; Xu et al. 2018). These substrates represent major differences in terms of cell wall constituents (cellulose, hemicellulose with a negligible amount of pectin in monocots; 15–20% pectin and xylan in dicots) (Dien et al. 2006; Pahkala et al. 2007). Based on significant differences in cell wall, the transcriptional profiles reflected the secretion of different classes of carbohydrate-active enzymes. The substrate composition influenced the secretion and expression of various enzymes. In case of barley, cellulolytic and xylanolytic enzymes were highly up-regulated, followed by arabinanases, mannanases and pectinases. However, in case of alfalfa there was down-regulated expression of xylanolytic enzymes (Berka et al. 2011). To conclude, the link between expression profiles and orthologues extended to many core lignocellulolytic proteins with the exception for pectinolytic enzymes.

16.4 Applications of *Sporotrichum thermophile* Xylanase

The *S. thermophile* xylanase finds large applications, in various food, biofuel and pulp and paper processes, apart from other uses. Since the *S. thermophile* xylanase is thermostable it can be used efficiently in the biomass conversion of lignocellulosics into fermentable sugars. It has been successfully employed in the saccharification of agrowaste like wheat straw, corn cobs, birch and spruce biomass. The applications of *S. thermophile* xylanases are summarised in Table 16.2.

16.5 Other Enzymes from *Sporotrichum thermophile* and Their Applications

S. thermophile is a thermophilic mould and hence various enzymes secreted by it are thermostable and advantageous from the industrial point of view especially the cocktail of lignocellulolytic enzymes. Apart from xylanase, cellulases (Bajaj et al. 2014), glycosyl hydrolases (Ye et al. 2014), phytases (Kumari et al. 2016), feruloyl esterases (Topakas et al. 2004) and polygalacturonases (Kaur et al. 2004) have also been worked out. *S. thermophile* has been reported to produce high activity of cellulase (42 U/g) by SSF using wheat bran and citrus pectin as substrates (Kaur and Satyanarayana 2004). Two cellobiohydrolases have been purified and characterised from this strain (Gusakov et al. 2007). One of the cellobiohydrolases was found to

Table 16.2 Industrial applications of *S. thermophile* xylanases

Mode of xylanase production	Process conditions	Xylanase characteristics	Application	References
SSF	Corn cob at pH 5.0, 50 °C for 4 days	K_m and V_{max} of 7.1 nm and 3.54 gL^{-1} h^{-1}, respectively	Xylan hydrolysis	Vardakou et al. (2003)
SSF	Corn cob at pH 5.0, 50 °C for 4 days	Molecular mass and pI of 25 kDa and 6.7, respectively. The pH and temperature optima of 5.0 and 70 °C, respectively	Hydrolysis of polysaccharides	Katapodis et al. (2003)
SSF	Rice straw, wheat straw, bagasse, corn cob, wheat bran each incubated for 5 days at 45 °C, pH 7.5	n.s.	Xylan hydrolysis	Badhan et al. (2004)
SSF	Wheat bran and citrus pectin in 1:1 ratio, pH 7.0 at 45 °C with moisture ratio of 1:2. for 4 days	n.s.	Treatment of fruit pulp for clarification of juices	Kaur and Satyanarayana (2004)
Submerged	Corn cobs at 2.7% (w/v) and ammonium phosphate at 0.7% (v/v) concentration for 5 days	pH and temperature optima of 5.0 and 70 °C, respectively	–	Katapodis et al. (2006)
Submerged	2.7% (w/v) wheat straw at pH 5.0, 50 °C for 7 days	pH and temperature optima of *St*Xyn1 *St*Xyn2 were 5.0 and 60 °C, respectively, for both the xylanases $t_{1/2}$ of 60 and 115 min for *St*Xyn1 and *St*Xyn2, respectively	Hydrolysis of xylan	Vafiadi et al. (2010)
SSF	De-oiled *Jatropha curcas* seed cake with a moisture ratio of 1:1.5, pH 9.5 at 35 °C with 1% (w/v) birchwood xylan as inducer	$t_{1/2}$ of 4 h at 45 °C and pH stability in the range of 7–11.0. K_m and V_{max} of 12.54 mg/mL and 454.5 U/mL/min, respectively	Xylooligosaccharide production	Sadaf and Khare (2014)

(continued)

Table 16.2 (continued)

Mode of xylanase production	Process conditions	Xylanase characteristics	Application	References
Submerged	Cane molasses at 8% (v/v) supplemented with ammonium sulphate at 0.5% (v/v) at pH 5.0	pH and temperature optima of 5.0 and 60 °C, respectively	Saccharification of biomass	Bala and Singh (2016)
SSF	Rice bran and rice straw (1:2 w/w), pH 8.9 at 44.16 °C for 7 days of incubation	pH and temperature optima of 12.0 and 50 °C, respectively	Xylooligosaccharide production	Boonrung et al. (2016)
Submerged	Emerson's medium containing soluble starch 15, yeast extract 4, K_2HPO_4 1.5, $MgSO_4$, 0.5 in g/L, at pH 9.0, 40 °C for 48 h	Stable at 50% (v/v) concentrations of [EMIM][OAc] and [BMIM][MeSO$_4$] for 72 h	In situ pretreatment and saccharification of wheat straw	Sadaf et al. (2016)
SSF	Cottonseed cake and wheat straw in 1:1 ratio, pH 5.0 at 45 °C for 96 h at a moisture level of 1:2.5	n.s	Xylooligosaccharide production	Bala and Singh (2017)
Submerged	Buffered minimal methanol medium	pH optima of MYCTH_56237 and MYCTH_49824 were 6.0 and 7.0, respectively, temperature optima of 60 °C for both the xylanases. The V_{max} and K_m for MYCTH_56237 were 2380 U/mg and 8.80 mg/mL, respectively, V_{max} and K_m of MYCTH_49824 were 1750 U/mg and 5.67 mg/mL, respectively	Saccharification of biomass	Basit et al. (2018)

n.s. not specified

Table 16.3 Other enzymes produced by *S. thermophile*

Enzyme	Method of production	Characteristics	Application	References
Polygalacturonase	Solid-state fermentation using wheat bran and citrus pectin in the ratio of 1:1	Temperature and pH optima of 55 °C and 7.0, respectively. $t_{1/2}$ of 4 h at 65 °C. K_m and V_{max} of 0.416 mg/mL and 0.52 µM/mg/min, respectively	Treatment of fruit pulps for better recovery of juice	Kaur et al. (2004)
Feruloyl esterase	Solid-state fermentation using wheat straw	Temperature and pH optima of 60 °C and 6.0, respectively. Molecular weight of 57 kDa	Production of ferulic acids from destarched wheat bran	Topakas et al. (2004)
Laccase	Submerged fermentation	Temperature and pH optima of 60 °C and 3.0, respectively. Stable in organic solvents like DMSO and ethanol	Decolorisation of six synthetic dyes	Kunamneni et al. (2008)
Phytase	Submerged fermentation	Temperature and pH optima of 60 °C and 5.0, respectively. $t_{1/2}$ of 16 h at 60 °C. Molecular weight of 90 kDa		Singh and Satyanarayana (2009)
Aldonolactonase	Submerged fermentation	Temperature and pH optima of 25 °C and 5.0, respectively. Molecular weight of 48 kDa	–	Beeson et al. (2011)
β-Mannosidase	Submerged fermentation	Temperature and pH optima of 40 °C and 5.3, respectively. Molecular weight of 97 kDa	Polysaccharide hydrolysis	Dotsenko et al. (2012)
Cellulase	Submerged fermentation	Temperature and pH optima of 65 °C and 8.0, respectively	Bioethanol production	Dimarogona et al. (2012)
β-Mannanase	Submerged fermentation	Temperature and pH optima of 69 °C and 5.2, respectively. Molecular weight of 48 kDa	Polysaccharide hydrolysis	Dotsenko et al. (2012)
GH7 endoglucanase	Submerged fermentation	Temperature and pH optima of 60 °C and 5.0, respectively. Molecular weight of 65 kDa	Enzymatic liquefaction of biomass	Karnaouri et al. (2014b)

(continued)

Table 16.3 (continued)

Enzyme	Method of production	Characteristics	Application	References
Protease	Submerged fermentation	Temperature and pH optima of 45 °C and 6.5, respectively. Molecular weight of 36.2 kDa	Food industry, leather, detergents and bioremediation	Neto et al. (2015)

be very active on Avicel and cotton. Table 16.3 summarises the other important enzymes from *S. thermophile*.

16.6   Current Trends in *Sporotrichum* Research

16.6.1   Genome Editing

With the advent of versatile technology such as genome editing, the revolutionary outlook has been shifted towards multifarious biotechnological applications (Liu et al. 2017; Singh 2016). The booming CRISPR/Cas9 system has manifested the underlying mechanism of metabolic pathways leading to enhanced thermostability (Karnaouri et al. 2014a). The genome editing/engineering has led to the efficient hydrolysis/degradation of lignocellulosic biomass (Viikari et al. 2007).

Previously, the CRISPR/Cas9 system for gene editing has been well established in yeasts and fungi like *Aspergillus* (Katayama et al. 2016; Nodvig et al. 2015), *Penicillium chrysogenum* (Pohl et al. 2016), *Trichoderma reesei* (Liu et al. 2015) and *Magnaporthe oryzae* (Arazoe et al. 2015). Using CRISPR-Cas interface via genomic engineering, a metabolic pathway of cellulase which in turn has been linked with xylanases and other extracellular hydrolytic enzymes has been edited in *S. thermophile* for higher production of enzymes. Consequently, a genome-wide engineering system for thermophilic fungi has been established based on multiplex-locus editing with the CRISPR/Cas9 technique, the main target locus being cre-1.

Similarly, gene editing-mediated enhanced production of enzymes by targeted mutations on desired gene via NHEJ-mediated events in a one-step transformation into *Myceliophthora heterothallica* strain CBS203 was successfully achieved (Hutchinson et al. 2016; van den Brink et al. 2012). The targeted gene of interest located at loci (amdS gene along with four other regions/constructs *cre-1, res-1, gh1–1* and *alp-1*) was mutated. These multigene disruptions were at different loci and selections of mutants were based on neomycin selection marker. The mutated strain exhibited hypersecretions of desired enzymes nearly 5 to 13-folds more than wild type (Liu et al. 2017).

16.6.2   *Ionic Liquid Stability of* Sporotrichum thermophile

The use of ionic liquids as green solvents for biomass pretreatment and saccharification has gained sufficient interest in the recent years (da Costa Lopes and Bogel-Łukasik 2015; Mahmood et al. 2016). Ionic liquids (ILs) are salts composed of an anion and cation moiety and hence are tuneable for desired applications (Marsh et al. 2004). This property of ILs makes them useful for biomass saccharification for enhanced bioethanol generation (Badgujar and Bhanage 2015). However, the sensitivity of biomass-saccharifying enzymes towards ILs limits the use of these solvents (Sadaf et al. 2018). ILs have been shown to exert inhibitory effect on the activity of many cellulases, xylanases and other hydrolytic enzymes (Thuy Pham et al. 2010; Turner et al. 2003). Xylanases are important for hydrolysis of hemicellulose portion of lignocellulosic biomass. The studies on the IL stability of xylanases are comparatively few as compared to those on cellulases. For example xylanase from *Thermoascus aurantiacus* retained its activity in the presence of 25% [EMIM] [OAc] (Chawachart et al. 2014). Similarly xylanase from *Volvariella volvacea* retained only 50% activity in the presence of 20% [EMIM][DMP] (Thomas et al. 2011). *Amycolatopsis* sp. GDS was slightly more stable in the presence of 10% (v/v) HEMA (2-hydroxyethyl methacrylate) and retained 80% activity (Kshirsagar et al. 2016).

Detail study on IL stability of *S. thermophile* xylanase has been recently described by Sadaf et al. (2016). The crude xylanase was stable and active in the presence of 50% [EMIM][OAc] for as long as 72 h. This is a remarkable stability for any xylanases reported so far. The xylanase was exploited to hydrolyse IL-pretreated wheat straw without the removal of residual IL. The whole process led to the generation of 281 mg reducing sugars per gram wheat straw. The IL-stable xylanase thus enabled the development of one-pot IL pretreatment and simultaneous saccharification process.

16.6.3   *Proteomics of* Sporotrichum thermophile

The studies on secretomes and exo-proteomes of *S. thermophile* provided insights into its cellular, physiological and metabolic functioning (Ghimire and Jin 2017). In addition to extracellular CAZymes involved in the digestion of polysaccharide nutrients, the genomes of *S. thermophile* and *T. terrestris* encoded an assortment of hydrolytic and oxidative enzymes that possibly enhanced their ability to forage non-carbohydrate substrates. The secretome of *S. thermophile* was comprised of ~683 proteins. Out of these, 569 were predicted to be homologues consisting of 180 CAZymes, >65 oxidoreductases, 40 peptidases and >230 proteins with unknown functions (Berka et al. 2011). Based on this data more studies would be needed to decipher the role of these secreted proteins in lignocellulose degradation.

16.7 Conclusions and Future Perspectives

S. thermophile, a thermophilic mould, produces an array of enzymes like cellulases, xylanases, pectinases, mannanases and phytases. The battery of interesting enzymes hold promise for various bioprocesses in food, feed, textile, biofuel, etc. *S. thermophile* xylanase is an interesting enzyme with many novel characteristics such as broad pH, heat and ionic liquid stability. Owing to the above features, the mould holds promise in effective biomass saccharification for biofuel production. So far there are only few studies which have studied the extracellular proteome and transcriptome of *S. thermophile* when grown on different types of substrates giving an indication of the expression levels of various enzymes involved in lignocellulosic biomass hydrolysis. Hence in future intracellular proteomic studies are also required in order to gain a better understanding of the functioning of major metabolic pathways under various growth conditions of the fungus. The IL stability aspect of *S. thermophile* xylanase has been only reported by a single study; therefore further studies in this regard will help in the development of an efficient cost-effective enzyme system. The genome editing of this fungus has also led to the enhanced production of hydrolytic enzymes by targeted mutations. Therefore more genetic manipulations of this kind will assist in enhancing the enzyme levels to a larger industrial scale. Also, bioinformatics studies of *S. thermophile* xylanase are necessary in deciphering the structural integrity of the xylanase under various extreme conditions and will therefore benefit in designing potent and robust enzymes.

References

Arazoe T, Miyoshi K, Yamato T, Ogawa T, Ohsato S, Arie T, Kuwata S (2015) Tailor-made CRISPR/Cas system for highly efficient targeted gene replacement in the rice blast fungus. Biotechnol Bioeng 112:2543–2549

Badgujar KC, Bhanage BM (2015) Factors governing dissolution process of lignocellulosic biomass in ionic liquid: current status, overview and challenges. Bioresour Technol 178:2–18

Badhan AK, Chadha BS, Sonia KG, Saini HS, Bhat MK (2004) Functionally diverse multiple xylanases of thermophilic fungus *Myceliophthora* sp. IMI 387099. Enzym Microb Technol 35:460–466

Badhan A, Chadha B, Kaur J, Saini H, Bhat M (2007) Production of multiple xylanolytic and cellulolytic enzymes by thermophilic fungus *Myceliophthora* sp. IMI 387099. Bioresour Technol 98:504–510

Bai W, Xue Y, Zhou C, Ma Y (2015a) Cloning, expression, and characterization of a novel alkalitolerant xylanase from alkaliphilic *Bacillus* sp. SN5. Biotechnol Appl Biochem 62:208–217

Bai W, Zhou C, Zhao Y, Wang Q, Ma Y (2015b) Structural insight into and mutational analysis of family 11 xylanases: implications for mechanisms of higher pH catalytic adaptation. PLoS One 10:e0132834

Bajaj BK, Sharma M, Rao RS (2014) Agricultural residues for production of cellulase from *Sporotrichum thermophile* LAR5 and its application for saccharification of rice straw. J Mater Environ Sci 5:1454–1460

Bala A, Singh B (2016) Cost-effective production of biotechnologically important hydrolytic enzymes by *Sporotrichum thermophile*. Bioprocess Biosyst Eng 39:181–191

Bala A, Singh B (2017) Concomitant production of cellulase and xylanase by thermophilic mould *Sporotrichum thermophile* in solid state fermentation and their applicability in bread making. World J Microbiol Biotechnol 33:1–10

Bala A, Singh B (2018) Cellulolytic and xylanolytic enzymes of thermophiles for the production of renewable biofuels. Renew Energy. https://doi.org/10.1016/j.renene.2018.09.100

Basit A, Liu J, Miao T, Zheng F, Rahim K, Lou H, Jiang W (2018) Characterization of two endo-β-1, 4-xylanases from *Myceliophthora thermophila* and their saccharification efficiencies, synergistic with commercial cellulase. Front Microbiol 9:1–11

Beeson WT, Iavarone AT, Hausmann CD, Cate JHD, Marletta MA (2011) Extracellular Aldonolactonase from *Myceliophthora thermophila*. Appl Environ Microbiol 77:650–656

Bergquist PL, Morgan HW, Saul D (2014) Selected enzymes from extreme thermophiles with applications in biotechnology. Curr Biotechnol 3:45–59

Berka RM, Grigoriev IV, Otillar R, Salamov A, Grimwood J, Reid I, Ishmael N, John T, Darmond C, Moisan M-C (2011) Comparative genomic analysis of the thermophilic biomass-degrading fungi *Myceliophthora thermophila* and *Thielavia terrestris*. Nat Biotechnol 29:922–927

Boonrung S, Katekaew S, Mongkolthanaruk W, Aimi T, Boonlue S (2016) Purification and characterization of low molecular weight extreme alkaline xylanase from the thermophilic fungus *Myceliophthora thermophila* BF1-7. Mycoscience 57:408–416

Bouacem K, Bouanane-Darenfed A, Boucherba N, Joseph M, Gagaoua M, Hania WB, Kecha M, Benallaoua S, Hacene H, Ollivier B (2014) Partial characterization of xylanase produced by *Caldicoprobacter algeriensis*, a new thermophilic anaerobic bacterium isolated from an Algerian hot spring. Appl Biochem Biotechnol 174:1969–1981

Butt MS, Tahir-Nadeem M, Ahmad Z, Sultan MT (2008) Xylanases and their applications in baking industry. Food Technol Biotechnol 46:22–31

Chanwicha N, Katekaew S, Aimi T, Boonlue S (2015) Purification and characterization of alkaline xylanase from *Thermoascus aurantiacus* var. *levisporus* KKU-PN-I2-1 cultivated by solid-state fermentation. Mycoscience 56:309–318

Chawachart N, Anbarasan S, Turunen S, Li H, Khanongnuch C, Hummel M, Sixta H, Granström T, Lumyong S, Turunen O (2014) Thermal behaviour and tolerance to ionic liquid [emim] OAc in GH10 xylanase from *Thermoascus aurantiacus* SL16W. Extremophiles 18:1023–1034

Cheng YS, Chen CC, Huang CH, Ko TP, Luo W, Huang JW, Liu JR, Guo RT (2014) Structural analysis of a glycoside hydrolase family 11 xylanase from *Neocallimastix patriciarum*: insights into the molecular basis of a thermophilic enzyme. J Biol Chem 289:11020–11028

Cherubini F (2010) The biorefinery concept: using biomass instead of oil for producing energy and chemicals. Energy Convers Manag 51:1412–1421

Collins T, Gerday C, Feller G (2005) Xylanases, xylanase families and extremophilic xylanases. FEMS Microbiol Rev 29:3–23

da Costa Lopes AM, Bogel-Lukasik R (2015) Acidic ionic liquids as sustainable approach of cellulose and lignocellulosic biomass conversion without additional catalysts. ChemSusChem 8:947–965

Dalmaso GZL, Ferreira D, Vermelho AB (2015) Marine extremophiles: a source of hydrolases for biotechnological applications. Mar Drugs 13:1925–1965

de Souza AR, de Araujo GC, Zanphorlin LM, Ruller R, Franco FC, Torres FA, Mertens JA, Bowman MJ, Gomes E, Da Silva R (2016) Engineering increased thermostability in the GH-10 endo-1, 4-β-xylanase from *Thermoascus aurantiacus* CBMAI 756. Int J Biol Macromol 93:20–26

Dhiman SS, Sharma J, Battan B (2008) Industrial applications and future prospects of microbial xylanases: a review. Bioresources 3:1377–1402

Dien BS, Jung H-JG, Vogel KP, Casler MD, Lamb JF, Iten L, Mitchell RB, Sarath G (2006) Chemical composition and response to dilute-acid pretreatment and enzymatic saccharification of alfalfa, reed canarygrass, and switchgrass. Biomass Bioenergy 30:880–891

Dimarogona M, Topakas E, Olsson L, Christakopoulos P (2012) Lignin boosts the cellulase performance of a GH-61 enzyme from *Sporotrichum thermophile*. Bioresour Technol 110:480–487

Dotsenko GS, Semenova MV, Sinitsyna OA, Hinz SWA, Wery J, Zorov IN, Kondratieva EG, Sinitsyn AP (2012) Cloning, purification, and characterization of galactomannan-degrading enzymes from *Myceliophthora thermophila*. Biochem Mosc 77:1303–1311

Ebringerova A, Heinze T (2000) Xylan and xylan derivatives–biopolymers with valuable properties, 1. Naturally occurring xylans structures, isolation procedures and properties. Macromol Rapid Commun 21:542–556

Fan G, Yang S, Yan Q, Guo Y, Li Y, Jiang Z (2014) Characterization of a highly thermostable glycoside hydrolase family 10 xylanase from *Malbranchea cinnamomea*. Int J Biol Macomol 170:482–489

Fawzi E (2011) Highly thermostable xylanase purified from *Rhizomucor miehei* NRL 3169. Acta Biol Hung 62:85–94

Garcia-Huante Y, Cayetano-Cruz M, Santiago-Hernandez A, Cano-Ramirez C, Marsch-Moreno R, Campos JE, Aguilar-Osorio G, Benitez-Cardoza CG, Trejo-Estrada S, Hidalgo-Lara ME (2017) The thermophilic biomass-degrading fungus *Thielavia terrestris* Co3Bag1 produces a hyperthermophilic and thermostable β-1, 4-xylanase with exo-and endo-activity. Extremophiles 21:175–186

Ghimire PS, Jin C (2017) Genetics, molecular, and proteomics advances in filamentous fungi. Curr Microbiol 74:1226–1236

Goncalves GA, Takasugi Y, Jia L, Mori Y, Noda S, Tanaka T, Ichinose H, Kamiya N (2015) Synergistic effect and application of xylanases as accessory enzymes to enhance the hydrolysis of pretreated bagasse. Enzym Microb Technol 72:16–24

Gopalan N, Rodriguez-Duran L, Saucedo-Castaneda G, Nampoothiri KM (2015) Review on technological and scientific aspects of feruloyl esterases: a versatile enzyme for biorefining of biomass. Bioresour Technol 193:534–544

Gupta A, Verma JP (2015) Sustainable bio-ethanol production from agro-residues: a review. Renew Sust Energ Rev 41:550–567

Gusakov AV, Salanovich TN, Antonov AI, Ustinov BB, Okunev ON, Burlingame R, Emalfarb M, Baez M, Sinitsyn AP (2007) Design of highly efficient cellulase mixtures for enzymatic hydrolysis of cellulose. Biotechnol Bioeng 97:1028–1038

Hakulinen N, Turunen O, Janis J, Leisola M, Rouvinen J (2003) Three-dimensional structures of thermophilic β-1, 4-xylanases from *Chaetomium thermophilum* and *Nonomuraea flexuosa*: comparison of twelve xylanases in relation to their thermal stability. Eur J Biochem 270:1399–1412

Harris AD, Ramalingam C (2010) Xylanases and its application in food industry: a review. J Exp Sci 1:1–11

Harris PV, Welner D, McFarland K, Re E, Navarro Poulsen JC, Brown K, Salbo R, Ding H, Vlasenko E, Merino S (2010) Stimulation of lignocellulosic biomass hydrolysis by proteins of glycoside hydrolase family 61: structure and function of a large, enigmatic family. Biochemistry 49:3305–3316

Henrissat B, Davies G (1997) Structural and sequence-based classification of glycoside hydrolases. Curr Opin Struct Biol 7:637–644

Huang X, Lin J, Ye X, Wang G (2015) Molecular characterization of a thermophilic and salt-and alkaline-tolerant xylanase from *Planococcus* sp. SL4, a strain isolated from the sediment of a soda lake. J Microbiol Biotechnol 25:662–671

Hutchinson MI, Powell AJ, Tsang A, O'Toole N, Berka RM, Barry K, Grigoriev IV, Natvig DO (2016) Genetics of mating in members of the Chaetomiaceae as revealed by experimental and genomic characterization of reproduction in *Myceliophthora heterothallica*. Fungal Genet Biol 86:9–19

Karnaouri A, Topakas E, Antonopoulou I, Christakopoulos P (2014a) Genomic insights into the fungal lignocellulolytic system of *Myceliophthora thermophila*. Front Microbiol 5:1–22

Karnaouri AC, Topakas E, Christakopoulos P (2014b) Cloning, expression, and characterization of a thermostable GH7 endoglucanase from *Myceliophthora thermophila* capable of high-consistency enzymatic liquefaction. Appl Microbiol Biotechnol 98:231–242

Karnaouri A, Matsakas L, Topakas E, Rova U, Christakopoulos P (2016) Development of thermophilic tailor-made enzyme mixtures for the bioconversion of agricultural and forest residues. Front Microbiol 7:1–14

Katapodis P, Vrsanska M, Kekos D, Nerinckx W, Biely P, Claeyssens M, Macris BJ, Christakopoulos P (2003) Biochemical and catalytic properties of an endoxylanase purified from the culture filtrate of *Sporotrichum thermophile*. Carbohydr Res 338:1881–1890

Katapodis P, Christakopoulou V, Christakopoulos P (2006) Optimization of xylanase production by *Sporotrichum thermophile* using corn cobs and response surface methodology. Eng Life Sci 6:410–415

Katayama T, Tanaka Y, Okabe T, Nakamura H, Fujii W, Kitamoto K, Maruyama J-I (2016) Development of a genome editing technique using the CRISPR/Cas9 system in the industrial filamentous fungus *Aspergillus oryzae*. Biotechnol Lett 38:637–642

Kaur G, Satyanarayana T (2004) Production of extracellular pectinolytic, cellulolytic and xylanolytic enzymes by thermophilic mould *Sporotrichum thermophile* Apinis in solid state fermentation. Indian J Biotechnol 3:552–557

Kaur G, Kumar S, Satyanarayana T (2004) Production, characterization and application of a thermostable polygalacturonase of a thermophilic mould *Sporotrichum thermophile*. Apinis. Bioresour Technol 94:239–243

Kocabas DS, Güder S, Ozben N (2015) Purification strategies and properties of a low-molecular weight xylanase and its application in agricultural waste biomass hydrolysis. J Mol Catal B Enzym 115:66–75

Kolbusz MA, Di Falco M, Ishmael N, Marqueteau S, Moisan M-C, Baptista CDS, Powlowski J, Tsang A (2014) Transcriptome and exoproteome analysis of utilization of plant-derived biomass by *Myceliophthora thermophila*. Fungal Genet Biol 72:10–20

Kshirsagar SD, Saratale GD, Saratale RG, Govindwar SP, Oh MK (2016) An isolated *Amycolatopsis* sp. GDS for cellulase and xylanase production using agricultural waste biomass. J Appl Microbiol 120:112–125

Kumar V, Marin-Navarro J, Shukla P (2016) Thermostable microbial xylanases for pulp and paper industries: trends, applications and further perspectives. World J Microbiol Biotechnol 32:1–10

Kumari A, Satyanarayana T, Singh B (2016) Mixed substrate fermentation for enhanced phytase production by thermophilic mould *Sporotrichum thermophile* and its application in beneficiation of poultry feed. Appl Biochem Biotechnol 178:197–210

Kunamneni A, Ghazi I, Camarero S, Ballesteros A, Plou FJ, Alcalde M (2008) Decolorization of synthetic dyes by laccase immobilized on epoxy-activated carriers. Process Biochem 43:169–178

Lin XQ, Han S, Zhang N, Hu H, Zheng SP, Ye YR, Lin Y (2013) Bleach boosting effect of xylanase A from *Bacillus halodurans* C-125 in ECF bleaching of wheat straw pulp. Enzym Microb Technol 52:91–98

Liu R, Chen L, Jiang Y, Zhou Z, Zou G (2015) Efficient genome editing in filamentous fungus *Trichoderma reesei* using the CRISPR/Cas9 system. Cell Discov 1:1–11

Liu Q, Gao R, Li J, Lin L, Zhao J, Sun W, Tian C (2017) Development of a genome-editing CRISPR/Cas9 system in thermophilic fungal *Myceliophthora species* and its application to hyper-cellulase production strain engineering. Biotechnol Biofuels 10:1–14

Mahmood H, Moniruzzaman M, Yusup S, Akil HM (2016) Pretreatment of oil palm biomass with ionic liquids: a new approach for fabrication of green composite board. J Clean Prod 126:677–685

Manju S, Singh Chadha B (2011) Production of hemicellulolytic enzymes for hydrolysis of lignocellulosic biomass. In: Pandey A, Larroche C, Ricke SC, Dussap CG, Gnansounou E (eds) Biofuels-alternative feedstocks and conversion processes. Academic Press, Amsterdam, pp 203–228

Margaritis A, Merchant RF, Yaguchi M (1986) Thermostable cellulases from thermophilic microorganisms. Crit Rev Biotechnol 4:327–367

Marsh K, Boxall J, Lichtenthaler R (2004) Room temperature ionic liquids and their mixtures—a review. Fluid Phase Equilib 219:93–98

McPhillips K, Waters DM, Parlet C, Walsh DJ, Arendt EK, Murray PG (2014) Purification and characterisation of a β-1, 4-xylanase from *Remersonia thermophila* CBS 540.69 and its application in bread making. Appl Biochem Biotechnol 172:1747–1762

Menon V, Rao M (2012) Trends in bioconversion of lignocellulose: biofuels, platform chemicals & biorefinery concept. Prog Energy Combust Sci 38:522–550

Mitra S, Mukhopadhyay BC, Mandal AR, Arukha AP, Chakrabarty K, Das GK, Chakrabartty PK, Biswas SR (2015) Cloning, overexpression, and characterization of a novel alkali-thermostable xylanase from *Geobacillus sp*. WBI J Basic Microbiol 55:527–537

Moreira L (2016) Insights into the mechanism of enzymatic hydrolysis of xylan. Appl Microbiol Biotechnol 100:5205–5214

Neto YA, de Oliveira LC, de Oliveira AH, Rosa JC, Juliano MA, Juliano L, Rodrigues A, Cabral H (2015) Determination of specificity and biochemical characteristics of neutral protease isolated from *Myceliophthora thermophila*. Protein Pept Lett 22:972–982

Nodvig CS, Nielsen JB, Kogle ME, Mortensen UH (2015) A CRISPR-Cas9 system for genetic engineering of filamentous fungi. PLoS One 10:e0133085

Pahkala K, Kontturi M, Kallioinen A, Myllymaki O, Uusitalo J, Siika-Aho M, von Weymarn N (2007) Production of bio-ethanol from barley straw and reed canary grass: a raw material study. Paper presented at the 15th European biomass conference and exhibition, Berlin, Germany, 7–11 May

Patel SJ, Savanth VD (2015) Review on fungal xylanases and their applications. Int J Adv Res 3:311–315

Ping L, Wang M, Yuan X, Cui F, Huang D, Sun W, Zou B, Huo S, Wang H (2017) Production and characterization of a novel acidophilic and thermostable xylanase from *Thermoascus aurantiacus*. Int J Biol Macomol 109:1270–1279

Plecha S, Hall D, Tiquia-Arashiro SM (2013) Screening and characterization of soil microbes capable of degrading cellulose from switchgrass (*Panicum virgatum* L.). Environ Technol 34:1895–1904

Pohl C, Kiel J, Driessen A, Bovenberg R, Nygard Y (2016) CRISPR/Cas9 based genome editing of *Penicillium chrysogenum*. ACS Synth Biol 5:754–764

Polizeli M, Rizzatti A, Monti R, Terenzi H, Jorge JA, Amorim D (2005) Xylanases from fungi: properties and industrial applications. Appl Microbiol Biotechnol 67:577–591

Ravindran R, Jaiswal AK (2016) A comprehensive review on pre-treatment strategy for lignocellulosic food industry waste: challenges and opportunities. Bioresour Technol 199:92–102

Sadaf A, Khare SK (2014) Production of *Sporotrichum thermophile* xylanase by solid state fermentation utilizing deoiled *Jatropha curcas* seed cake and its application in xylooligosachharide synthesis. Bioresour Technol 153:126–130

Sadaf A, Morya VK, Khare S (2016) Applicability of *Sporotrichum thermophile* xylanase in the in situ saccharification of wheat straw pre-treated with ionic liquids. Process Biochem 51:2090–2096

Sadaf A, Grewal J, Khare SK (2018) Ionic liquid stable cellulases and hemicellulases: application in biobased production of biofuels. In: Bhaskar T, Pandey A, Mohan SV, Lee DJ, Khanal SK (eds) Waste Biorefinery. Elsevier, Amsterdam, pp 505–532

Saksono B, Sukmarini L (2010) Structural analysis of xylanase from marine thermophilic *Geobacillus stearothermophilus* in Tanjung Api, Poso, Indonesia. Hayati J Biosci 17:189–195

Scheller HV, Ulvskov P (2010) Hemicelluloses. Annu Rev Plant Biol 61:263–289

Sharma A, Parashar D, Satyanarayana T (2016) Acidophilic microbes: biology and applications. In: Rampelotto P (ed) Biotechnology of extremophiles: grand challenges in biology and biotechnology, vol 1. Springer, Cham, pp 215–241

Singh B (2016) *Myceliophthora thermophila* syn. *Sporotrichum thermophile*: a thermophilic mould of biotechnological potential. Crit Rev Biotechnol 36:59–69

Singh B, Satyanarayana T (2009) Characterization of a HAP–phytase from a thermophilic mould *Sporotrichum thermophile*. Bioresour Technol 100:2046–2051

Singh B, Pocas-Fonseca MJ, Johri B, Satyanarayana T (2016) Thermophilic molds: biology and applications. Crit Rev Microbiol 42:985–1006

Tadesse H, Luque R (2011) Advances on biomass pretreatment using ionic liquids: an overview. Energy Environ Sci 4:3913–3929

Taibi Z, Saoudi B, Boudelaa M, Trigui H, Belghith H, Gargouri A, Ladjama A (2012) Purification and biochemical characterization of a highly thermostable xylanase from *Actinomadura sp.* strain Cpt20 isolated from poultry compost. Appl Biochem Biotechnol 166:663–679

Takahashi Y, Kawabata H, Murakami S (2013) Analysis of functional xylanases in xylan degradation by *Aspergillus niger* E-1 and characterization of the GH family 10 xylanase XynVII. Springerplus 2:1–11

Thomas MF, Li LL, Handley-Pendleton JM, Van der Lelie D, Dunn JJ, Wishart JF (2011) Enzyme activity in dialkyl phosphate ionic liquids. Bioresour Technol 102:11200–11203

Thuy Pham TP, Cho CW, Yun YS (2010) Environmental fate and toxicity of ionic liquids: a review. Water Res 44:352–372

Tiquia-Arashiro SM, Mormile M (2013) Sustainable technologies: bioenergy and biofuel from biowaste and biomass. Environ Technol 34(13):1637–1805

Topakas E, Katapodis P, Kekos D, Macris BJ, Christakopoulos P (2003) Production and partial characterization of xylanase by *Sporotrichum thermophile* under solid-state fermentation. World J Microbiol Biotechnol 19:195–198

Topakas E, Stamatis H, Biely P, Christakopoulos P (2004) Purification and characterization of a type B feruloyl esterase (StFAE-A) from the thermophilic fungus *Sporotrichum thermophile*. Appl Microbiol Biotechnol 63:686–690

Turner MB, Spear SK, Huddleston JG, Holbrey JD, Rogers RD (2003) Ionic liquid salt-induced inactivation and unfolding of cellulase from *Trichoderma reesei*. Green Chem 5:443–447

Vafiadi C, Christakopoulos P, Topakas E (2010) Purification, characterization and mass spectrometric identification of two thermophilic xylanases from *Sporotrichum thermophile*. Process Biochem 45:419–424

van den Brink J, van Muiswinkel GC, Theelen B, Hinz SW, de Vries RP (2012) Efficient plant biomass degradation by the thermophilic fungus *Myceliophthora heterothallica*. Appl Environ Microbiol 79:1316–1324

Van Gool M, Van Muiswinkel G, Hinz S, Schols H, Sinitsyn A, Gruppen H (2012) Two GH10 endo-xylanases from *Myceliophthora thermophila* C1 with and without cellulose binding module act differently towards soluble and insoluble xylans. Bioresour Technol 119:123–132

Van Gool MP, Van Muiswinkel GCJ, Hinz SWA, Schols HA, Sinitsyn AP, Gruppen H (2013) Two novel GH11 endo-xylanases from *Myceliophthora thermophila* C1 act differently toward soluble and insoluble xylans. Enzym Microb Technol 53:25–32

Vardakou M, Katapodis P, Samiotaki M, Kekos D, Panayotou G, Christakopoulos P (2003) Mode of action of family 10 and 11 endoxylanases on water-unextractable arabinoxylan. Int J Biol Macomol 33:129–134

Verma D, Kawarabayasi Y, Miyazaki K, Satyanarayana T (2013) Cloning, expression and characteristics of a novel alkalistable and thermostable xylanase encoding gene (Mxyl) retrieved from compost-soil metagenome. PLoS One 8:e52459

Viikari L, Alapuranen M, Puranen T, Vehmaanpera J, Siika-Aho M (2007) Thermostable enzymes in lignocellulose hydrolysis. In: Olsson L (ed) Biofuels: advances in biochemical engineering/biotechnology, vol 108. Springer, Berlin, Heidelberg, pp 121–145

Wang Y, Fu Z, Huang H, Zhang H, Yao B, Xiong H, Turunen O (2012) Improved thermal performance of *Thermomyces lanuginosus* GH11 xylanase by engineering of an N-terminal disulfide bridge. Bioresour Technol 112:275–279

Wang J, Mai G, Liu G, Yu S (2013) Molecular cloning and heterologous expression of an acid-stable endoxylanase gene from *Penicillium oxalicum* in *Trichoderma reesei*. J Microbiol Biotechnol 23:251–259

Wang G, Wu J, Yan R, Lin J, Ye X (2017) A novel multi-domain high molecular, salt-stable alkaline xylanase from *Alkalibacterium sp.* SL3. Front Microbiol 7:2120

Watanabe M, Fukada H, Ishikawa K (2016) Construction of thermophilic xylanase and its structural analysis. Biochemistry 55:4399–4409

Weerachavangkul C, Laothanachareon T, Boonyapakron K, Wongwilaiwalin S, Nimchua T, Eurwilaichitr L, Pootanakit K, Igarashi Y, Champreda V (2012) Alkaliphilic endoxylanase from lignocellulolytic microbial consortium metagenome for biobleaching of eucalyptus pulp. J Microbiol Biotechnol 22:1636–1643

Wojtczak G, Breuil C, Yamada J, Saddler J (1987) A comparison of the thermostability of cellulases from various thermophilic fungi. Appl Microbiol Biotechnol 27:82–87

Xu G, Li J, Liu Q, Sun W, Jiang M, Tian C (2018) Transcriptional analysis of *Myceliophthora thermophila* on soluble starch and role of regulator AmyR on polysaccharide degradation. Bioresour Technol 265:558–562

Ye Z, Zheng Y, Li B, Borrusch MS, Storms R, Walton JD (2014) Enhancement of synthetic Trichoderma-based enzyme mixtures for biomass conversion with an alternative family 5 glycosyl hydrolase from *Sporotrichum thermophile*. PLoS One 9:e109885

Yegin S (2017) Single-step purification and characterization of an extreme halophilic, ethanol tolerant and acidophilic xylanase from *Aureobasidium pullulans* NRRL Y-2311-1 with application potential in the food industry. Food Chem 221:67–75

Zeldes BM, Keller MW, Loder AJ, Straub CT, Adams MW, Kelly RM (2015) Extremely thermophilic microorganisms as metabolic engineering platforms for production of fuels and industrial chemicals. Front Microbiol 6:1–17

Zhang Y, An J, Yang G, Zhang X, Xie Y, Chen L, Feng Y (2016) Structure features of GH10 xylanase from *Caldicellulosiruptor bescii*: implication for its thermophilic adaption and substrate binding preference. Acta Biochim Biophys Sin 48:948–957

Zhuo R, Yu H, Qin X, Ni H, Jiang Z, Ma F, Zhang X (2018) Heterologous expression and characterization of a xylanase and xylosidase from white rot fungi and their application in synergistic hydrolysis of lignocellulose. Chemosphere 212:24–33

Part III
Biosynthesis of Novel Biomolecules and Extremozymes

Chapter 17
Deep-Sea Fungi: Diversity, Enzymes, and Bioactive Metabolites

Muhammad Zain Ul Arifeen, Ya-Rong Xue, and Chang-Hong Liu ⓘ

17.1 Introduction

Fungi are an important component of all ecosystems and are ubiquitous in nature because of their highly versatile physiological behavior (Gostinčar et al. 2009; Tedersoo et al. 2014). Fungi that inhibit deep sea and its sediments at a depth of more than 1000 m below the surface of the sea are called deep-sea fungi (Swathi et al. 2013). The deep-sea environment is one of the extreme environments for living organisms (Tiquia-Arashiro 2012). Absence of sunlight, usually lower temperature (some places like hydrothermal vents have an extremely high temperature up to 400 °C), extreme pH, and high hydrostatic pressure (up to 110 MPa) are the conditions which make deep sea an extreme environment (Damare et al. 2006; Burgaud et al. 2010; Nagano and Nagahama 2012). It has been documented that many environmental factors such as salinity, temperature, oxygen availability, hydrostatic pressure, substrate specificity, light, and availability of substrata significantly affect the abundance, diversity, activity, and distribution of fungi in the natural habitats (Nagano and Nagahama 2012; Batista-García et al. 2017). Despite these extreme living conditions, microorganisms, especially fungi, are found in abundance in deep-sea environments (Gadanho and Sampaio 2005; Nagahama et al. 2006; Wang et al. 2015; Xu et al. 2016; Nagano et al. 2017). It is believed that these fungi came from terrestrial environments, and adapted well to the deep-sea environment through evolution (Redou et al. 2015; Hassan et al. 2016). A recent estimate suggests that global fungal species are approximately 2.2–3.8 million (Hawksworth and Lücking 2017), around 120,000 species have been described by taxonomists (Mueller and Schmit 2006), and only a few species have been identified from the deep-sea

M. Z. U. Arifeen · Y.-R. Xue · C.-H. Liu (✉)
State Key Laboratory of Pharmaceutical Biotechnology, School of Life Sciences
Nanjing University, Nanjing, People's Republic of China
e-mail: chliu@nju.edu.cn

© Springer Nature Switzerland AG 2019
S. M. Tiquia-Arashiro, M. Grube (eds.), *Fungi in Extreme Environments:*
Ecological Role and Biotechnological Significance,
https://doi.org/10.1007/978-3-030-19030-9_17

environments. The biodiversity and their application in biotechnology of deep-sea derived fungal species are not fully understood and need more research to explore their potential.

A wide range of bioactive molecules have been isolated from deep-sea fungal communities, which have the potential to perform different activities such as antibacterial, antiviral, antidiabetics, anti-inflammatory, and antitumor, and some can be used as in important enzymes (Tiquia and Mormile 2010; Shang et al. 2012; Mayer et al. 2013; Wang et al. 2015, 2017a).

The aim of this review is to explain the current research of deep-sea fungal diversity, and the tools and equipment used in the sampling, isolation, and identification of the deep-sea fungi. This review also discusses novel and industrially important enzymes and bioactive molecules recently isolated from deep-sea fungi.

17.2   Biodiversity of Deep-Sea Fungi

Although the deep-sea environment provides a habitat for a vast number of microbial lives, however, the origin, diversity, and distribution of deep-sea fungal community remain largely undiscovered. Höhnk (1969) for the first time reported about deep-sea fungi when he isolated fungi from shells collected at 4610 m depth below the sea. Roth et al. (1964) isolated fungi from surface to a depth of 4500 m from the subtropical Atlantic Ocean. However, these fungi were not able to culture in the lab. Until 1992, Raghukumar et al. (1992) firstly isolated various filamentous fungal strains from calcareous fragments collected from the Bay of Bengal at a depth of 300–860 m using culture-dependent method. After that, several numbers of fungal species have been isolated and identified from many deep-sea environmental samples, such as from the Mariana Trench at a depth of 11,500 m (Takami et al. 1997), the Chagos Trench 5500 m (Raghukumar et al. 2004), the Central Indian Basin 5000 m (Singh et al. 2011), Gulf of Mexico sediment 2400 m (Thaler et al. 2012), South China Sea 2400–4000 m (Zhang et al. 2013), East India Ocean 4000 m (Zhang et al. 2014), Canterbury Basin (New Zealand) 4–1884 mbsf (below the seafloor) (Redou et al. 2015), Okinawa 1190–1589 m (Zhang et al. 2016a), and the pacific ocean off the Shimokita Peninsula, Japan, 1289–2466 mbsf (Liu et al. 2017).

So far more than 120 fungal species have been isolated and identified from the deep sea, using culture-dependent technique (Gadanho and Sampaio 2005; Nagano and Nagahama 2012; Zhang et al. 2013) while culture-independent technique has also been useful for the identification of some strange and unknown fungi. It has been reported that Ascomycota was the dominant species in deep-sea environment representing 78% followed by Basidiomycota (17.3%), Zygomycota (1.5%), and Chytridiomycota (0.8%) while 2.4% was unidentified fungal isolates (Redou et al. 2015; Zhang et al. 2016a). The most diverse and common fungal species were reported to be *Aspergillus* sp., *Penicillium* sp., and *Simplicillium obclavatum*, while

Alternaria alternata, Aureobasidium pullulans, Cryptococcus liquefaciens, Exophiala dermatitidis, Epicoccum nigrum, and *Neosetophoma samarorum* were the rarest fungal species documented in the deep-sea environments (Zhang et al. 2014). Recently, Liu et al. (2017) obtained 69 fungal isolates belonging to 27 species from deep coal-associated sediment samples collected at depths ranging from 1289 to 2457 mbsf in the pacific ocean off the Shimokita Peninsula, Japan. Most of the isolated strains (88% to the total strains) belonged to Ascomycota dominated by *Penicillium* and *Aspergillus,* and only 12% of the strains belong to Basidiomycota.

Besides the richness of biodiversity, many novel fungal species were also reported recently. Six novel phylotypes of genera *Ajellomyces, Podosordaria, Torula,* and *Xylaria* were reported by Zhang et al. (2013) based on their investigation and identification of the obtained cultural fungal species from the sediment of the South China Sea at a depth of 2400–4000 m. One novel fungal phylotype, DSF-Group1, was discovered in deep-sea sediments at depths ranging from 1200 to 10,000 m by using three fungal specific primer sets, targeting the ITS1-5.8S–ITS2-28S rRNA regions (Nagano et al. 2010) and another environmental clade KML11, which strongly supported group of the parasitic genus *Rozella* in *Cryptomycota,* was also identified by using eukaryotic-specific primers EK-82F and EK-1492R (Lara et al. 2010). Similarly, the BCGI clade by using fungus-specific primers nu-SSU-0817 and nu-SSU-1536-39 was also reported (Nagahama et al. 2011). Since most fungal species isolated from deep-sea environments showed similar morphological characteristics to their terrestrial counterparts, molecular phylogenetic analysis played important role in the discovery and identification of the unknown fungi.

The biodiversity of deep-sea fungal communities has largely been dependent on the samples where they obtained, and the methods used for fungal detection. The great depths and high hydrostatic pressure in the deep-sea are the major factors severely constraining sample collection from those extreme environments. Several samplers such as box corer, gravity corer, piston corer, and a grab sampler have been used to collect samples from deep sea and sediments (Nagahama and Nagano 2012). A sampler is used according to the nature of the sample. For example, a grab sampler is used to collect samples from the surface of deep-sea floor which are soft and silt in nature; box corer is suitable for mud- and silt-natured samples but not for sand; the gravity corer and piston corer are used for sampling at various depths from the deep-sea floor. Another powerful and most advanced drilling instruments are used with deep-sea drilling vessel, CHIKYU, able to collect sediments samples from 7000 m below the deep-sea floor (Nagahama and Nagano 2012). Unfortunately, all these samplers cannot absolutely protect the samples from contamination during sampling process Therefore, with the development of novel techniques used in samples from deep-sea environments, more and more fungi will be discovered from the extreme deep-sea habitats in the future.

17.3   Enzymes Derived from Deep-Sea Fungi

The enzymes produced by deep-sea fungi are either unique protein molecules which are not isolated from any terrestrial organisms or it may be a known protein previously isolated from terrestrial source but with novel properties. However, so far only few enzymes with unique properties have been reported from the deep-sea fungal community. Barotolerant proteases were reported from two deep-sea strains of *Aspergillus ustus* (Raghukumar and Raghukumar 1998); alkaline and cold-tolerant proteases were isolated from the deep-sea fungi, *Aspergillus ustus*, were active at various temperatures, pH, and pressure, with an optimum activity at pH 9, and showed stability under high concentration of detergents and salinity (Raghukumar and Raghukumar 1998; Damare et al. 2006), thus allowing laundry under cold water and having the potential to reduce the energy requirement for a chemical reaction by increasing the hydrostatic pressure and decreasing the temperature. Enzymes like this would be very useful in future to lower the energy cost. Polygalacturonases (PGase) are enzymes usually used in the food industry for clarification of fruit juices. Two unique PGase were isolated from deep-sea (4500–6500 m) yeast strains, and showed activity at low temperature (0–10 °C) and high hydrostatic pressure (100 MPa) (Miura et al. 2001; Abe et al. 2006). These enzymes were also tolerant at high (50 mM) concentration of $CuSO_4$ and showed high activity of superoxide dismutase (superoxide radical scavenger) (Abe et al. 2001, 2006). Batista-García et al. (2017) reported three lignocellulolytic-halotolerant fungi (*Cadophora* sp. TS2, *Emericellopsis* sp. TS11, and *Pseudogymnoascus* sp. TS12) from deep-sea sponge samples, which displayed high CMCase and xylanase activities at an optimal temperature of 50–70 °C and optimal pH of 5–8; thus it could be possible to use these enzymes in future for the conversion of lignocellulosic biomass materials (Plecha et al. 2013). Dimethyl phthalate ester is an important environmental pollutant belonging to phthalate esters group which can cause toxicity in the endocrine system, is widely distributed in nature, and is commonly used in plastic preparation. A deep-sea fungus, *Aspergillus versicolor* IR-M4, can degrade dimethyl phthalate ester with the help of phthalate esterase (Wang et al. 2017b).

17.4   Bioactive Compounds from Deep-Sea Fungi

Deep-sea fungi survived in the extreme conditions of the deep-sea and are adapted well through evolution; therefore, scientists considered this group of fungi a novel and new source of medicinally important metabolites for new drug discovery. According to previous data, more than 200 new bioactive materials have been isolated so far from deep-sea fungi (Daletos et al. 2018). Table 17.1 shows some interesting bioactive compounds recently isolated from deep-sea fungi.

Table 17.1 Important bioactive compounds isolated from deep-sea fungi after 2015

Fungal species	Place	Depth (m)	Metabolite	Activity	Chemical formula	Reference
Aspergillus versicolor	South China Sea	2326	Aspergilol I	Anti-HSV-1/antioxidant/antifouling	$C_{34}H_{30}O_{10}$	Huang et al. (2017)
			Aspergilol H		$C_{20}H_{20}O_8$	
			Coccoquinone A		$C_{22}H_{20}O_9$	
Aspergillus versicolor	South China Sea	–	N1 (polysaccharide)	Antioxidant	–	Yan et al. (2016)
Aspergillus wentii	South China Sea	2038	Wentinoids A	Antifungal	$C_{21}H_{32}O_4$	Li et al. (2017)
			Asperolides D	Cytotoxic/antibacterial	$C_{16}H_{16}O_6$	Li et al. (2016a)
			Asperolides E		$C_{16}H_{18}O_5$	
			Aspewentin D	Antibacterial	$C_{19}H_{26}O_3$	Li et al. (2016b)
			Asperether A	Cytotoxic	$C_{19}H_{26}O_4$	Li et al. (2016c)
			Asperether B		$C_{19}H_{26}O_3$	
			Asperether C		$C_{19}H_{26}O_4$	
			Asperether D		$C_{21}H_{28}O_5$	
			Asperether E		$C_{19}H_{26}O_4$	
Aspergillus versicolor	Indian Ocean	3927	Versicoloids A	Antifungal	$C_{19}H_{25}N_3O_4$	Wang et al. (2016)
			Versicoloids B		$C_{19}H_{25}N_3O_5$	
Acaromyces ingoldii	South China Sea	3415	Acaromycin A	Growth inhibitory	$C_{14}H_{12}O_6$	Gao et al. (2016)
Alternaria sp.	South China Sea	3927	Perylenequinone	Bromodomain (BRD4) protein inhibitor	$C_{20}H_{14}O_6$	Ding et al. (2017)
Acremonium sp.	South Atlantic Ocean	2869	Acremeremophilane B	Anti-inflammatory	$C_{21}H_{26}O_5$	Cheng et al. (2016)
			Acremeremophilane E		$C_{21}H_{26}O_5$	
Biscogniauxia mediterranea	Mediterranean Sea	2800	Isopyrrolonaphthoquinone	Enzyme inhibition	$C_{12}H_7NO_4$	Wu et al. (2016)

(continued)

Table 17.1 (continued)

Fungal species	Place	Depth (m)	Metabolite	Activity	Chemical formula	Reference
Chaetomium sp.	-	-	Chaetoviridide A	Antibacterial/cytotoxic	$C_{30}H_{31}N_2O_7Cl$	Wang et al. (2018a)
			Chaetoviridide B		$C_{27}H_{34}NO_7Cl$	
Chaetomium globosum	Indian Ocean	-	Chaetoglobosin E	Antiproliferative	$C_{32}H_{38}O_5N_2$	Zhang et al. (2016b)
Dichotomomyces cejpii	South China Sea	3941	Dichotocejpins A	α-Glucosidase inhibition	$C_{14}H_{14}N_2O_3S$	Fan et al. (2016)
Eutypella sp.	South Atlantic Ocean	5610	Eutypellazine P	Antibacterial	$C_{19}H_{18}N_2O_4S$	Niu et al. (2017a)
			Eutypellazine Q		$C_{19}H_{20}N_2O_5S_3$	
			Eutypellazine R		$C_{20}H_{20}N_2O_3S_2$	
Sarcopodium sp.	Kagoshima Bay	200	Sarcopodinols A	Cytotoxic	$C_{16}H_{34}O_5$	Matsuo et al. (2018)
			Sarcopodinols B		$C_{16}H_{34}O_4$	
Stachybotrys sp.	Atlantic Ocean	2807	Stachybotrin H	Cytotoxic	$C_{25}H_{33}NO_6$	Ma et al. (2019)
			Stachybotrysin H		$C_{27}H_{36}O_8$	
Graphostroma sp.	Atlantic Ocean	2721	Reticulol	Antiallergic	$C_{11}H_{10}O_5$	Niu et al. (2018b)
			Khusinol B	Anti-inflammatory/antiallergic	$C_{15}H_{26}O_2$	Niu et al. (2017c)
			Graphostromanes F	Anti-inflammatory	$C_{15}H_{26}O_3$	Niu et al. (2018a)
Oidiodendron griseum	-	765	Dihydrosecofuscin	Antibacterial	$C_{15}H_{18}O_5$	Navarri et al. (2017)
			Secofuscin		$C_{15}H_{16}O_5$	
Penicillium sp.	South China Sea	1300	Penicilliumin B	Kidney fibrogenic inhibition	$C_{21}H_{30}O_3$	Lin et al. (2017)
Penicillium chrysogenum	Indian Ocean	-	Chrysamides C	Inhibition of proinflammatory cytokine IL-17 production	$C_{26}H_{26}N_4O_{10}$	Chen et al. (2016)

Species	Location	No.	Compound	Activity	Formula	Reference
Penicillium chrysogenum	Indian Ocean	3386	Bipenicilisorin	Cytotoxic	$C_{24}H_{19}O_{12}$	Chen et al. (2017)
			Chrysines B	α-Glucosidase inhibition	$C_{19}H_{18}Cl_2O_8$	Wang et al. (2018c)
			Chrysines C		$C_{19}H_{19}ClO_8$	
			Chrysoxanthone		$C_{16}H_{10}Cl_2O_7$	
			Dichloro-orcinol		$C_7H_5Cl_2O_2$	
			2,4-Dichloroasterric acid		–	
			Methyl chloroasterrate		$C_{18}H_{17}ClO_8$	
			Mono-chlorosulochrin		$C_{17}H_{15}ClO_7$	
			Geodin		$C_{17}H_{12}Cl_2O_7$	
			Chrysoxanthone		$C_{31}H_{26}O_{10}$	
Penicillium brevicompactum	South China Sea	3928	Brevianamide C	Cytotoxic	$C_{21}H_{23}N_3O_3$	Xu et al. (2017)
Penicillium granulatum	Prydz Bay of Antarctica	2284	Roquefortine J	Cytotoxic	$C_{22}H_{21}N_5O_2$	Niu et al. (2018c)
			Spirograterpene A	Antiallergic activity	$C_{20}H_{30}O_3$	Niu et al. (2017b)
Penicillium sumatrense	Indian Ocean	–	Dehydrocurvularin	Inhibition of NO production induced by LPS	$C_{16}H_{18}O_5$	Wu et al. (2017)
Spiromastix sp.	South Atlantic Ocean	2869	Spiromastilactone D	Inhibition of influenza A and B	$C_{20}H_{19}Cl_3O_6$	Niu et al. (2016)
Simplicillium obclavatum	East Indian Ocean	4571	Simplicilliumtide D	Antifouling	$C_{24}H_{37}N_3O_6$	Liang et al. (2016)
Trichobotrys effuse	South China Sea	2918	Trichobotryside A	Antifouling	$C_{20}H_{32}O_8$	Sun et al. (2016)
Williamsia sp.	Southwestern Indian Ocean	1654	CDMW-3, 5,15	Antiallergic	–	Gao et al. (2017)

17.4.1 Antimicrobial Activity

Antimicrobial compounds, asperolides D and E, isolated from the deep-sea fungus *Aspergillus wentii* SD-310 showed moderate antimicrobial activity against *Edwardsiella tarda*, each with an MIC value of 16 µg/mL (Li et al. 2016a). Wang et al. (2016) isolated oxepine-containing alkaloids and xanthones, from the deep-sea-derived fungus *Aspergillus versicolor* SCSIO 05879, which showed antifungal activities against other plant pathogenic fungi especially versicoloids A and B, and exhibited significant antifungal activity against *Colletotrichum acutatum* with an MIC value of 1.6 µg/mL; thus they could be used as a potent candidate for antifungal-agrochemicals. Similarly another antibacterial compound, aspewentin D (Li et al. 2016b), and antifungal compound, wentinoids A (Li et al. 2017), have also been isolated from fungus *Aspergillus wentii* SD-310. Niu et al. (2017a) isolated antibacterial compounds (eutypellazine P, eutypellazine Q, eutypellazine R) from deep-sea fungus, *Eutypella* sp. MCCC 3A00281, which showed inhibitory effects against *Staphylococcus aureus* ATCC 25923 and vancomycin-resistant enterococci (VRE), with an MIC value of 16–32 µM. Chaetoviridides A–C, from deep-sea fungus, *Chaetomium* sp. NA-S01-R1, showed antimicrobial activities against strains of *Vibrio vulnificus* (MIC = 30.5, 7.4, and 15.7 µg/mL), *Vibrio rotiferianus* (MIC = 7.3, 31.3, and 15.3 µg/mL), and *Vibrio campbellii* (MIC = 32.7, 32.3, and NA µg/mL) (Wang et al. 2018a). Another novel compound, anthraquinone, 2-(dimethoxymethyl)-1-hydroxyanthracene-9,10-dione, isolated from the deep-sea fungus *Aspergillus versicolor*, showed strong inhibition against MRSA ATCC 43300 (MIC = 3.9 µg/mL) and MRSA CGMCC 1.12409 (MIC = 7.8 µg/mL) and moderate inhibitory activities against Vibrio strains (Wang et al. 2018b).

17.4.2 Anti-inflammatory Activity

Two new anti-inflammatory compounds (acremeremophilane B and acremeremophilane E) were isolated from deep-sea-derived fungal species, *Acremonium* TVG-S004-0211. These compounds showed inhibition activity against NO production in LPS-activated RAW 264.7 macrophages, with IC_{50} values of 8 µM and 15 µM, respectively, suggesting them to be new anti-inflammatory compounds (Cheng et al. 2016). A nitrophenyl trans-epoxyamides derivative compound, chrysamides C, isolated from the deep-sea fungus, *Penicillium chrysogenum* SCSIO41001, showed a suppressive effect against the production of pro-inflammatory cytokine interleukin-17 with the inhibition rate of 40% at 1 µM (Chen et al. 2016). Several bioactive compounds have been isolated from the deep-sea fungus, *Penicillium granulatum* MCCC 3A00475. Spirograterpene A, a compound isolated from this fungus, showed antiallergic activity against immunoglobulin E (IgE)-mediated rat mast RBL-2H3 cells with 18% inhibition at 20 µg/mL as compared to control one (Loratadine) which showed 35% inhibition at the same concentration (Niu et al. 2017b).

An anti-inflammatory compound called khusinol B isolated and purified from the culture broth of fungus, *Graphostroma* sp. MCCC 3A00421, showed significant anti-inflammatory effect with IC_{50} value of 17 mM; this compound also showed weak antiallergic activity with IC_{50} value of 150 mM and thus could be a good anti-inflammatory candidate (Niu et al. 2017c). Similarly, a new anti-inflammatory compound, graphostromanes F, recently isolated from the same fungi, *Graphostromanes F*, exhibited stronger anti-inflammatory response against lipopolysaccharide-induced nitric oxide in macrophages with IC_{50} value of 14.2 µM, proving to be a potent candidate for anti-inflammatory drugs (Niu et al. 2018a). Gao et al. (2017) isolated three antiallergic compounds (CDMW-3, CDMW-5, and CDMW-15) from a deep-sea fungal species, *Williamsia* sp. MCCC 1A11233 (CDMW), showing potent anti-allergic activity through induction of apoptosis in mast cells. The deep-sea fungi were also investigated for antifood allergic drugs. A polyketide compound reticulol was isolated from the deep-sea-derived fungus *Graphostroma* sp. MCCC 3A00421, which showed antifood allergic and anti-inflammatory effect. Reticulol showed significant inhibition of degranulation with an IC_{50} value of 13.5 µM (Niu et al. 2018b).

17.4.3 Antioxidant Activity

The deep-sea-derived microorganisms produced many extracellular polysaccharides with novel and unique properties and are considered as most promising and potent group of antioxidant compounds (Arena et al. 2009; Sun et al. 2009; Le Costaouëc et al. 2012). Currently used antioxidants are synthetic and taught to be carcinogenic and may cause damage to liver (Laurienzo 2010); therefore, it is important to find and utilize natural products as an antioxidant source. Yan et al. (2016) extracted N1 compound, a novel extracellular polysaccharide, from the deep-sea isolated fungus *Aspergillus versicolor* N2bc. Proper investigation of N1 compounds revealed that it is a potent antioxidant and could be used in future.

17.4.4 Antiviral Activity

The most unique and interesting compound isolated from the deep-sea fungi is the influenza virus inhibitor. A compound called spiromastilactone D can be a potent inhibitor of influenza A and B virus. This compound has the ability to bind to hemagglutinin (HA) protein which usually attaches to the host cell surface receptor called sialic acid (SA) making an HA–SA complex which is essential for the attachment and entry of influenza virus to the host (human) cell. In the presence of this compound, HA cannot attach to SA and thus prevents influenza virus from being entered into the host cell. This compound can also inhibit viral genome replication process by targeting viral ribonucleoprotein complex (Niu et al. 2016); through both these properties this compound could be used in future for the antiviral application.

Another compound called cladosin C, isolated from the deep-sea fungus *Cladosporium sphaerospermum* 2005-01-E3, also showed antiviral activity against influenza A virus (Wu et al. 2014). Anti-HSV1 bioactive compounds (Aspergilol H, Aspergilol I, and Coccoquinone A) have also been isolated from another fungal strain, *Aspergillus versicolor* SCSIO 41502, which showed potent antiviral activity against HSV-1 with EC_{50} values of 4.68 μM, 6.25 μM , and 3.12 μM, respectively (Huang et al. 2017).

17.4.5   Cytotoxic Activity

A cytotoxic compound, asperolides E, isolated from the deep-sea fungus *Aspergillus wentii* SD-310, showed significant cytotoxic effect against HeLa, MCF-7, and NCI-H446 cell lines, with IC_{50} values of 10 μM, 11 μM, and 16 μM, respectively (Li et al. 2016a). Another five cytotoxic compounds (Asperethers A–E) were also isolated from this fungus. All these compounds showed potent cytotoxic effects against A549 cell line with IC_{50} values of 20 μM, 16 μM, 19 μM, 17 μM, and 20 μM, respectively (Li et al. 2016c). Two new secondary metabolites acaromycin A and acaromyester A have been isolated from deep-sea fungal strain *Acaromyces ingoldii* FS121, which showed cell growth inhibitory activities against MCF-7, NCI-H460, SF-268, and HepG-2 with IC_{50} values less than 10 μM (Gao et al. 2016). An antiproliferative compound, chaetoglobosin E, isolated from deep-sea fungus *Chaetomium globosum*, showed good cytotoxic and antitumor activity against two cell lines LNCaP and B16F10, with IC_{50} values of 0.62 μM and 2.78 μM, respectively (Zhang et al. 2016b). Another cytotoxic compound, brevianamide C, extracted form a fugal strain, *Penicillium brevicompactum* DFFSCS025, showed cytotoxic activity against human colon cancer cell line (HCT116) with IC_{50} value of 15.6 μM (Xu et al. 2017). A cytotoxic compound, dehydrocurvularin, isolated from deep-sea fungus, *Penicillium sumatrense* MCCC 3A00612, showed potent inhibition effect against LPS-induced NO production in RAW 264.7 macrophages with IC_{50} value of 0.91 mM as comparable to that of control L-NMMA with IC_{50} vale of 41.91 mM (Wu et al. 2017). New cytotoxic metabolites, sarcopodinols A and B, isolated from deep-sea fungus *Sarcopodium* sp. FKJ-0025, showed cytotoxic activity against human cell lines. Sarcopodinols A exhibited cytotoxic effect against Jurkat cells with an IC_{50} value of 47 μg/mL, while Sarcopodinols B showed cytotoxic activity against HL-60 cells with IC_{50} value of 37 μg/mL, Jurkat cells with IC_{50} value of 47 μg/mL, and Panc1 cells with IC_{50} value of 66 μg/mL (Matsuo et al. 2018). A compound phenylspirodrimane, stachybotrin H, isolated from the deep-sea fungal strain, *Stachybotrys* sp. MCCC 3A00409, showed weak cytotoxic activity against human leukemia (K562), cervical adenocarcinoma (Hela), and promyelocytic leukemia (HL60) cell lines (Ma et al. 2019). Four novel chlorinated azaphilone compounds chaephilone, chaetoviridides A–C, were isolated from deep-sea-derived fungus *Chaetomium* sp. NA-S01-R1, and among these, chaetoviridides A and B showed cytotoxic activities against A549, HeLa, and Hep G2 cell lines, with IC_{50}

values of 15.2 μM, 12.3 μM, and 3.9 μM, and 16.3 μM, 5.6 μM, and 18.2 μM, respectively (Wang et al. 2018a). Another dimeric isocoumarin compound, bipenicilisorin, has also been isolated from fungus, *Penicillium chrysogenum* SCSIO41001, which showed strong cytotoxic activities against various cell lines such as K562, A549, and Huh-7 with IC_{50} values of 6.78 μM, 6.94 μM, and 2.59 μM, respectively (Chen et al. 2017). Another compound, roquefortine J, isolated from the fungus, *Penicillium granulatum* MCCC 3A00475, showed cytotoxic activity against HepG2 tumor cells with an IC_{50} value of 19.5 μM (Niu et al. 2018c).

17.4.6 Bioactive Compounds with Other Activities

A new isopyrrolonaphthoquinone compound isolated from deep-sea fungi, *Biscogniauxia mediterranea* LF657, showed the potential of inhibiting glycogen synthase kinase (GSK-3β) with an IC_{50} value of 8.04 μM as compared to positive control (4-benzyl-2-methyl-1,2,4-thiadiazolidine-3,5-dione) with IC_{50} value of 0.26 μM (Wu et al. 2016). Perylenequinone, a protein inhibitor compound, isolated from the deep-sea fungi, *Alternaria* sp. NH-F6, has the ability to inhibit bromodomain-containing proteins such as BRD4, which is an epigenetic code reader protein, with an inhibition rate of more than 80% at a concentration of 10 μM. Thus it was suggested that this compound has potent antitumoral, antiviral, and anti-inflammatory properties (Ding et al. 2017). A unique compound sesquiterpene methylcyclopentenedione named penicilliumin B was reported for the first time in the fermentation broth of the fungus *Penicillium* sp. F00120. This compound was examined for its reno-protective activities and showed substantial potential to inhibit renal fibrosis in vitro, through oxidative stress disruption mechanism, thus presenting a new type of promising reno-protective agent (Lin et al. 2017). Recently, two new chlorinated diphenyl ethers, chrysines B and C, dichlorinated xanthone, chrysoxanthone, dichloroorcinol, 2,4-dichloroasterric acid, methyl chloroasterrate, mono-chlorosulochrin, and geodin were isolated from deep-sea fungal strain, *Penicillium chrysogenum* SCSIO 41001. All these compounds showed good potential of inhibiting α-glucosidase with IC_{50} values of 0.35 μM, 0.20 μM, 0.04 μM, 0.16 μM, 0.15 μM, 0.09 μM, 0.14 μM, 0.14 μM, and 0.12 mM, respectively, in comparison to positive control (acarbose) with IC_{50} value of 0.28 mM (Wang et al. 2018c). Similarly, another α-glucosidase inhibitor compound called dichotocejpins A, isolated from fungus *Dichotomomyces cejpii* FS110, showed significant α-glucosidase inhibition activity with IC_{50} values of 138 μM (Fan et al. 2016). Liang et al. (2016) investigated deep-sea fungal strain *Simplicillium obclavatum* EIODSF 020, for peptide production and eight new linear peptides, simplicilliumtides A–H, were successfully isolated and purified from the culture broth. Among them, simplicilliumtides D showed best antifouling activity against *Bugula neritina* larva with the EC_{50} value of 7.8 mg/mL and LC_{50}/EC_{50} >100, indicating that simplicilliumtides D has the potential to be a natural antifouling candidate. Another antifouling compound, trichobotryside A, has also been isolated from another deep-sea fungus,

Trichobotrys effuse DFFSCS021, which showed significant antifouling activities against *Bugula neritina* and *Balanus amphitrite* larvae with EC_{50} values of 7.3, 2.5 μg/mL and LC_{50}/EC_{50} > 40.5, 37.4, respectively (Sun et al. 2016).

17.5 Conclusions and Future Perspectives

Even though deep-sea fungi have been extensively studied in recent researches, our understanding and scientific knowledge of deep-sea fungi is still very limited. So far the major focus was on deep-sea bacteria and regardless of the physiological adaptability to low temperature, elevated hydro pressure, and playing important roles in the ecosystem, deep-sea fungal taxa have not been discovered that much. The current known diversity of deep-sea fungi has demonstrated that it is just the tip of the iceberg and a vast extent of the unknown fungal community is yet to be discovered. Deep-sea fungi have revealed much promise in terms of interesting enzymes with novel properties, unique metabolics, and secondary metabolites. Many biotechnological important enzymes have been produced by deep-sea fungal community isolated from a variety of habitats. This diversity of deep-sea fungal products could be due to their genetic diversity based on taxonomy and adaptability to various extreme environmental conditions. Both cultural-dependent and cultural-independent techniques are essential for the investigation of deep-sea fungal diversity. However, cultural-independent techniques proved to be very useful in the discovery of various unknown fungal lineages in the deep sea, which were not able to grow under normal cultural conditions. There is still a very big scope to examine these fungi for other interesting and useful products and these fungi could be the source for novel products like extracellular polysaccharides, enzymes, and other secondary metabolites (Pettit 2011). Future studies should be focused on marine fungal biology to reveal interesting biochemical and physiological features useful to various new biotechnological processes. For example, the finding of new techniques to study uncultured and rare marine-derived fungi and knowing about its physiology and biochemistry will definitely pave the way for future deep-sea mycology.

Acknowledgements This work was supported in part by NSFC (41773083, 31471810, and 31272081) and the National Key R&D Program of China (2017YFD0800705 and 2017YFC 0506005).

References

Abe F, Miura T, Nagahama T, Inoue A, Usami R, Horikoshi K (2001) Isolation of a highly copper-tolerant yeast, *Cryptococcus* sp., from the Japan Trench and the induction of superoxide dismutase activity by Cu^{2+}. Biotechnol Lett 23(24):2027–2034

Abe F, Minegishi H, Miura T, Nagahama T, Usami R, Horikoshi K (2006) Characterization of cold-and high-pressure-active polygalacturonases from a deep-sea yeast, *Cryptococcus liquefaciens* strain N6. Biosci Biotechnol Biochem 70(1):296–299

Arena A, Gugliandolo C, Stassi G, Pavone B, Iannello D, Bisignano G, Maugeri TL (2009) An exopolysaccharide produced by *Geobacillus thermodenitrificans* strain B3-72: antiviral activity on immunocompetent cells. Immunol Lett 123(2):132–137

Batista-García RA, Sutton T, Jackson SA, Tovar-Herrera OE, Balcázar-López E, del Rayo Sánchez-Carbente M, Sánchez-Reyes A, Dobson AD, Folch-Mallol JL (2017) Characterization of lignocellulolytic activities from fungi isolated from the deep-sea sponge *Stelletta normani*. PLoS One 12(3):e0173750

Burgaud G, Arzur D, Durand L, Cambon-Bonavita MA, Barbier G (2010) Marine culturable yeasts in deep-sea hydrothermal vents: species richness and association with fauna. FEMS Microbiol Ecol 73(1):121–133

Chen S, Wang J, Lin X, Zhao B, Wei X, Li G, Kaliaperumal K, Liao S, Yang B, Zhou X, Liu J, Xu S, Liu Y (2016) Chrysamides A-C, three dimeric nitrophenyl trans-epoxyamides produced by the deep-sea-derived fungus *Penicillium chrysogenum* SCSIO41001. Org Lett 18(15):3650–3653

Chen S, Wang J, Wang Z, Lin X, Zhao B, Kaliaperumal K, Liao X, Tu Z, Li J, Xu S, Liu Y (2017) Structurally diverse secondary metabolites from a deep-sea-derived fungus *Penicillium chrysogenum* SCSIO 41001 and their biological evaluation. Fitoterapia 117:71–78

Cheng Z, Zhao J, Liu D, Proksch P, Zhao Z, Lin W (2016) Eremophilane-type sesquiterpenoids from an *Acremonium* sp. fungus isolated from deep-sea sediments. J Nat Prod 79(4):1035–1047

Daletos G, Ebrahim W, Ancheeva E, El-Neketi M, Song W, Lin W, Proksch P (2018) Natural products from deep-sea-derived fungi—a new source of novel bioactive compounds? Curr Med Chem 25(2):186–207

Damare S, Raghukumar C, Muraleedharan UD, Raghukumar S (2006) Deep-sea fungi as a source of alkaline and cold-tolerant proteases. Enzym Microb Technol 39(2):172–181

Ding H, Zhang D, Zhou B, Ma Z (2017) Inhibitors of BRD4 protein from a marine-derived fungus *Alternaria* sp. NH-F6. Mar Drugs 15(3):76

Fan Z, Sun ZH, Liu Z, Chen YC, Liu HX, Li HH, Zhang WM (2016) Dichotocejpins A–C: new diketopiperazines from a deep-sea-derived fungus *Dichotomomyces cejpii* FS110. Mar Drugs 14(9):164

Gadanho M, Sampaio JP (2005) Occurrence and diversity of yeasts in the mid-Atlantic ridge hydrothermal fields near the Azores Archipelago. Microb Ecol 50(3):408–417

Gao XW, Liu HX, Sun ZH, Chen YC, Tan YZ, Zhang WM (2016) Secondary metabolites from the deep-sea derived fungus *Acaromyces ingoldii* FS121. Molecules 21(4):371

Gao YY, Liu QM, Liu B, Xie CI, Cao MJ, Yang XW, Liu GM (2017) Inhibitory activities of compounds from the marine Actinomycete *Williamsia* sp. MCCC 1A11233 variant on IgE-mediated mast cells and passive cutaneous anaphylaxis. J Agric Food Chem 65(49):10749–10756

Gostinčar C, Grube M, De Hoog S, Zalar P, Gunde-Cimerman N (2009) Extremotolerance in fungi: evolution on the edge. FEMS Microbiol Ecol 71(1):2–11

Hassan N, Rafiq M, Hayat M, Aamer AS, Fariha H (2016) Psychrophilic and psychrotrophic fungi: a comprehensive review. Rev Environ Sci Biotechnol 15(2):147–172

Hawksworth DL, Lücking R (2017) Fungal diversity revisited: 2.2 to 3.8 million species. Microbiol Spectr 5(4):FUNK-0052-2016

Höhnk W (1969) Über den pilzlichen Befall kalkiger Hartteile von Meerestieren. Bericht Deutsche Wissenschaftliche Kommission für Meeresforschung 20:129–140

Huang Z, Nong X, Ren Z, Wang J, Zhang X, Qi S (2017) Anti-HSV-1, antioxidant and antifouling phenolic compounds from the deep-sea-derived fungus *Aspergillus versicolor* SCSIO 41502. Bioorg Med Chem Lett 27(4):787–791

Lara E, Moreira D, López-García P (2010) The environmental clade LKM11 and *Rozella* form the deepest branching clade of fungi. Protist 161(1):116–121

Laurienzo P (2010) Marine polysaccharides in pharmaceutical applications: an overview. Mar Drugs 8(9):2435–2465

Le Costaouëc T, Cérantola S, Ropartz D, Ratiskol J, Sinquin C, Colliec-Jouault S, Boisset C (2012) Structural data on a bacterial exopolysaccharide produced by a deep-sea *Alteromonas macleodii* strain. Carbohydr Polym 90(1):49–59

Li XD, Li X, Li XM, Xu GM, Zhang P, Meng LH, Wang BG (2016a) Tetranorlabdane diterpenoids from the deep sea sediment-derived fungus *Aspergillus wentii* SD-310. Planta Med 82(9–10):877–881

Li XD, Li XM, Li X, Xu GM, Liu Y, Wang GB (2016b) Aspewentins D-H, 20-nor-isopimarane derivatives from the deep sea sediment-derived fungus *Aspergillus wentii* SD-310. J Nat Prod 79(5):1347–1353

Li X, Li XM, Li XD, Xu GM, Liu Y, Wang BG (2016c) 20-Nor-isopimarane cycloethers from the deep-sea sediment-derived fungus *Aspergillus wentii* SD-310. RSC Adv 6(79): 75981–75987

Li X, Li XD, Li HM, Xu GM, Liu Y, Wang BG (2017) Wentinoids A-F, six new isopimarane diterpenoids from *Aspergillus wentii* SD-310, a deep-sea sediment derived fungus. RSC Adv 7(8):4387–4394

Liang X, Zhang XY, Nong XH, Wang J, Huang ZH, Qi SH (2016) Eight linear peptides from the deep-sea-derived fungus *Simplicillium obclavatum* EIODSF 020. Tetrahedron 72(22):3092–3097

Lin X, Wu Q, Yu Y, Liang Z, Liu Y, Zhou L, Tang L, Zhou X (2017) Penicilliumin B, a novel sesquiterpene methylcyclopentenedione from a deep sea-derived *Penicillium* strain with renoprotective activities. Sci Rep 7(1):10757

Liu CH, Xin H, Tian-Ning X, Duan N, Ya-Rong X, Tan-Xi Z, Mark L, Kai-Uwe H, Fumio I (2017) Exploration of cultivable fungal communities in deep coal-bearing sediments from 1.3 to 2.5 km below the ocean floor. Environ Microbiol 19(2):803–818

Ma XH, Zheng WM, Sun KH, Gu XF, Zeng XM, Zhang HT, Zhong TH, Shao ZZ, Zhang YH (2019) Two new phenylspirodrimanes from the deep-sea derived fungus *Stachybotrys* sp. MCCC 3A00409. Nat Prod Res 33:386–392

Matsuo H, Nonaka K, Nagano Y, Yabuki A, Fujikura K, Takahashi Y, Ōmura S, Nakashima T (2018) New metabolites, sarcopodinols A and B, isolated from deep-sea derived fungal strain *Sarcopodium* sp. FKJ-0025. Biosci Biotechnol Biochem 82:1323–1326. https://doi.org/10.10 80/09168451.2018.1467264

Mayer A, Rodríguez AD, Taglialatela-Scafati O, Fusetani N (2013) Marine pharmacology in 2009-2011: marine compounds with antibacterial, antidiabetic, antifungal, anti-inflammatory, antiprotozoal, antituberculosis, and antiviral activities; affecting the immune and nervous systems, and other miscellaneous mechanisms of action. Mar Drugs 11(7):2510–2573

Miura T, Abe F, Inoue A, Usami R, Horikoshi K (2001) Purification and characterization of novel extracellular endopolygalacturonases from a deep-sea yeast, *Cryptococcus* sp. N6, isolated from the Japan Trench. Biotechnol Lett 23(21):1735–1739

Mueller G, Schmit J (2006) Fungal biodiversity: what do we know? What can we predict? Biodivers Conserv 16:1–5

Nagahama T, Nagano Y (2012) Cultured and uncultured fungal diversity in deep-sea environments. In: Raghukumar C (ed) Biology of marine fungi. Springer, Berlin, pp 173–187

Nagahama T, Hamamoto M, Horikoshi K (2006) Rhodotorula pacifica sp. nov., a novel yeast species from sediment collected on the deep-sea floor of the north-west Pacific Ocean. Int J Syst Evol Microbiol 56(1):295–299

Nagahama T, Takahashi E, Nagano Y, Abdel-Wahab MA, Miyazaki M (2011) Molecular evidence that deep-branching fungi are major fungal components in deep-sea methane cold-seep sediments. Environ Microbiol 13(8):2359–2370

Nagano Y, Nagahama T (2012) Fungal diversity in deep-sea extreme environments. Fungal Ecol 5(4):463–471

Nagano Y, Nagahama T, Hatada Y, Nunoura T, Takami H, Miyazaki J, Takai K, Horikoshi K (2010) Fungal diversity in deep-sea sediments—the presence of novel fungal groups. Fungal Ecol 3(4):316–325

Nagano Y, Miura T, Nishi S, Lima AO, Nakayama C, Pellizari VH, Fujikura K (2017) Fungal diversity in deep-sea sediments associated with asphalt seeps at the Sao Paulo Plateau. Deep Sea Res II 146:59–67

Navarri M, Jégou C, Bondon A, Pottier S, Bach S, Baratte B, Ruchaud S, Barbier G, Burgaud G, Fleury Y (2017) Bioactive metabolites from the feep subseafloor fungus *Oidiodendron griseum* UBOCC-A-114129. Mar Drugs 15(4):111

Niu S, Si L, Liu D, Zhou A, Zhang Z, Shao Z, Wang S, Zhang L, Zhou D, Lin W (2016) Spiromastilactones: a new class of influenza virus inhibitors from deep-sea fungus. Eur J Med Chem 108:229–244

Niu S, Liu D, Shao Z, Proksch P, Lin W (2017a) Eutypellazines N–S, new thiodiketopiperazines from a deep sea sediment derived fungus *Eutypella* sp. with anti-VRE activities. Tetrahedron Lett 58(38):3695–3699

Niu S, Fan ZW, Xie CL, Liu Q, Luo ZH, Liu G, Yang XW (2017b) Spirograterpene A, a tetracyclic spiro-diterpene with a fused 5/5/5/5 ring system from the deep-sea-derived fungus *Penicillium granulatum* MCCC 3A00475. J Nat Prod 80(7):2174–2177

Niu S, Xie CL, Zhong T, Xu W, Luo ZH, Shao Z, Yang XW (2017c) Sesquiterpenes from a deep-sea-derived fungus *Graphostroma* sp. MCCC 3A00421. Tetrahedron 73(52):7267–7273

Niu S, Xie CL, Xia JM, Luo ZH, Shao Z, Yang XW (2018a) New anti-inflammatory guaianes from the Atlantic hydrotherm-derived fungus *Graphostroma* sp. MCCC 3A00421. Sci Rep 8(1):530

Niu S, Liu Q, Xia JM, Xie CL, Luo ZH, Shao Z, Liu GM, Yang XW (2018b) Polyketides from the deep-sea-derived fungus *Graphostroma* sp. MCCC 3A00421 showed potent anti-food allergic activities. J Agric Food Chem 66(6):1369–1376

Niu S, Wang N, Xie CL, Fan Z, Luo Z, Chen HF, Yang XW (2018c) Roquefortine J, a novel roquefortine alkaloid, from the deep-sea-derived fungus *Penicillium granulatum* MCCC 3A00475. J Antibiot 71:658–661. https://doi.org/10.1038/s41429-018-0046-y

Pettit RK (2011) Culturability and secondary metabolite diversity of extreme microbes: expanding contribution of deep sea and deep-sea vent microbes to natural product discovery. Mar Biotechnol 13(1):1–11

Plecha S, Hall D, Tiquia-Arashiro SM (2013) Screening and characterization of soil microbes capable of degrading cellulose from switchgrass (*Panicum virgatum* L.). Environ Technol 34:1895–1904

Raghukumar C, Raghukumar S (1998) Barotolerance of fungi isolated from deep-sea sediments of the Indian Ocean. Aquat Microb Ecol 15(2):153–163

Raghukumar C, Nagarkar S, Raghukumar S (1992) Association of thraustochytrids and fungi with living marine algae. Mycol Res 96(7):542–546

Raghukumar C, Raghukumar S, Sheelu G, Gupta S, Nath BN, Rao B (2004) Buried in time: culturable fungi in a deep-sea sediment core from the Chagos Trench, Indian Ocean. Deep Sea Res I 51(11):1759–1768

Redou V, Navarri M, Meslet-Cladiere L, Barbier G, Burgaud G (2015) Species richness and adaptation of marine fungi from deep-subseafloor sediments. Appl Environ Microbiol 81(10):3571–3583

Roth JF, Orpurt P, Ahearn DG (1964) Occurrence and distribution of fungi in a subtropical marine environment. Can J Bot 42(4):375–383

Shang Z, Li X, Meng L, Li C, Gao S, Huang C, Wang B (2012) Chemical profile of the secondary metabolites produced by a deep-sea sediment-derived fungus *Penicillium commune* SD-118. Chin J Oceanol Limnol 30(2):305–314

Singh P, Raghukumar C, Verma P, Shouche Y (2011) Fungal community analysis in the deep-sea sediments of the Central Indian Basin by culture-independent approach. Microb Ecol 61(3):507–517

Sun HH, Mao WJ, Chen Y, Guo SD, Li HY, Qi XH, Chen YL, Xu J (2009) Isolation, chemical characteristics and antioxidant properties of the polysaccharides from marine fungus *Penicillium* sp. F23-2. Carbohydr Polym 78(1):117–124

Sun YL, Zhang XY, Nong XH, Xu XY, Qi SH (2016) New antifouling macrodiolides from the deep-sea-derived fungus *Trichobotrys effuse* DFFSCS021. Tetrahedron Lett 57(3):366–370

Swathi J, Narendra K, Sowjanya KM, Satya AK (2013) Evaluation of biologically active molecules isolated from obligate marine fungi. Mintage J Pharm Med Sci 2:45–47

Takami H, Inoue A, Fuji F, Horikoshi K (1997) Microbial flora in the deepest sea mud of the Mariana Trench. FEMS Microbiol Lett 152(2):279–285

Tedersoo L, Bahram M, Põlme S, Kõljalg U, Yorou NS, Wijesundera R, Ruiz LV, Vasco-Palacios AM, Thu PQ, Suija A (2014) Global diversity and geography of soil fungi. Science 346(6213):1256688

Thaler AD, Van Dover CL, Vilgalys R (2012) Ascomycete phylotypes recovered from a Gulf of Mexico methane seep are identical to an uncultured deep-sea fungal clade from the Pacific. Fungal Ecol 5(2):270–273

Tiquia SM, Mormile M (2010) Extremophiles—a source of innovation for industrial and environmental applications. Environ Technol 31(8–9):823

Tiquia-Arashiro SM (2012) Molecular biological technologies for ocean sensing. Humana Press, Totowa, NJ, 295 p

Wang YT, Xue YR, Liu CH (2015) A brief review of bioactive metabolites derived from deep-sea fungi. Mar Drugs 13(8):4594–4616

Wang J, He W, Huang X, Tian X, Liao S, Yang B, Wang F, Zhou X, Liu Y (2016) Antifungal new oxepine-containing alkaloids and xanthones from the deep-sea-derived fungus Aspergillus versicolor SCSIO 05879. J Agric Food Chem 64(14):2910–2916

Wang W, Li S, Chen Z, Li Z, Liao Y, Chen J (2017a) Secondary metabolites produced by the deep-sea-derived fungus Engyodontium album. Chem Nat Compd 53(2):224–226

Wang JW, Xu W, Zhong TH, He GY, Luo ZH (2017b) Degradation of dimethyl phthalate esters by a filamentous fungus Aspergillus versicolor isolated from deep-sea sediments. Bot Mar 60:351

Wang W, Liao Y, Chen R, Hou Y, Ke W, Zhang B, Gao M, Shao Z, Chen J, Li F (2018a) Chlorinated azaphilone pigments with antimicrobial and cytotoxic activities isolated from the deep sea derived fungus Chaetomium sp. NA-S01-R1. Mar Drugs 16:61

Wang W, Chen R, Luo Z, Wang W, Chen R (2018b) Antimicrobial activity and molecular docking studies of a novel anthraquinone from a marine-derived fungus Aspergillus versicolor. Nat Prod Res 32(5):558–563

Wang JF, Zhou LM, Chen ST, Yang B, Liao SR, Kong FD, Lin XP, Wang FZ, Zhou XF, Liu YH (2018c) New chlorinated diphenyl ethers and xanthones from a deep-sea-derived fungus Penicillium chrysogenum SCSIO 41001. Fitoterapia 125:49–54

Wu G, Sun X, Yu G, Wang W, Zhu T, Gu Q, Li D (2014) Cladosins A–E, hybrid polyketides from a deep-sea-derived fungus, Cladosporium sphaerospermum. J Nat Prod 77(2):270–275

Wu B, Wiese J, Schmaljohann R, Imhoff JF (2016) Biscogniauxone, a new isopyrrolonaphthoquinone compound from the fungus Biscogniauxia mediterranea isolated from deep-sea sediments. Mar Drugs 14(11):204

Wu YH, Zhang ZH, Zhong Y, Huang JJ, Li XX, Jiang JY, Deng YY, Zhang LH, He F (2017) Sumalactones A-D, four new curvularin-type macrolides from a marine deep sea fungus Penicillium sumatrense. RSC Adv 7(63):40015–40019

Xu W, Luo ZH, Guo S, Pang KL (2016) Fungal community analysis in the deep-sea sediments of the Pacific Ocean assessed by comparison of ITS, 18S and 28S ribosomal DNA regions. Deep Sea Res I 109:51–60

Xu X, Zhang X, Nong X, Wang J, Qi S (2017) Brevianamides and mycophenolic acid derivatives from the deep-sea-derived fungus Penicillium brevicompactum DFFSCS025. Mar Drugs 15(2):43

Yan MX, Wao WJ, Liu X, Wang SY, Xia Z, Cao SJ, Li J, Qin L, Xian HL (2016) Extracellular polysaccharide with novel structure and antioxidant property produced by the deep-sea fungus Aspergillus versicolor N2bc. Carbohydr Polym 147:272–281

Zhang XY, Zhang Y, Xu XY, Qi SH (2013) Diverse deep-sea fungi from the South China Sea and their antimicrobial activity. Curr Microbiol 67(5):525–530

Zhang XY, Tang GI, Xu XY, Nong XH, Qi SH (2014) Insights into deep-sea sediment fungal communities from the East Indian Ocean using targeted environmental sequencing combined with traditional cultivation. PLoS One 9(10):e109118

Zhang XY, Wang GH, Xu XY, Nong XH, Wang J, Amin M, Qi SH (2016a) Exploring fungal diversity in deep-sea sediments from Okinawa Trough using high-throughput Illumina sequencing. Deep-Sea Res I 116:99–105

Zhang Z, Min X, Huang J, Zhong Y, Wu Y, Li X, Deng Y, Jiang Z, Shao Z, Zhang L, He F (2016b) Cytoglobosins H and I, new antiproliferative cytochalasans from deep-sea-derived fungus Chaetomium globosum. Mar Drugs 14(12):233

Chapter 18
Bioactive Compounds from Extremophilic Marine Fungi

Lesley-Ann Giddings ⓘ and David J. Newman

18.1 Introduction

Natural products are small molecules produced by organisms that are not required for development, growth, or propagation of a species. These secondary metabolites have evolved over billions of years to have high affinities for specific biological targets, making them a prolific source of structurally unique, bioactive compounds for industrial purposes and human health. In the search for new secondary metabolites, natural product chemists traditionally explored easily accessible terrestrial environments. However, by the 1990s we realized that all of the "low-hanging fruit" had been picked and natural product rediscovery rates increased. As such, underexplored environments, such as extreme ecosystems, are now being investigated for new bioactive compounds.

Chemical novelty is often found in underexplored environments, as their unique ecology impacts the availability of nutrients, such as heavy metals or salts, as well as the genes expressed in an organism at a given point in time. Many extreme ecosystems remain unexplored due to the costs and tremendous effort required to obtain samples. These environments are typically characterized as being "non-mesophilic" (either hot or cold), oligotrophic, or subject to osmotic stress, high salt concentrations, extreme pH, limited oxygen and light, severe radiation, high metal concentrations, and high pressures. To inhabit these ecosystems, organisms have mechanisms to regulate temperature, intracellular pH values, solute composition, biochemical redox reactions, and production of other biomolecules, as well as repair DNA,

L.-A. Giddings (✉)
Middlebury College, Middlebury, VT, USA
e-mail: LGiddings@middlebury.edu

D. J. Newman
Newman Consulting LLC, Wayne, PA, USA

© Springer Nature Switzerland AG 2019
S. M. Tiquia-Arashiro, M. Grube (eds.), *Fungi in Extreme Environments: Ecological Role and Biotechnological Significance*,
https://doi.org/10.1007/978-3-030-19030-9_18

protein, and lipid damage. Thus, there is potential to find new molecular structures and bioactivity in extreme environments.

The marine world represents some of the most underexplored extreme environments, as roughly 71% of the Earth's surface is covered with water, mainly oceans. For the purpose of this chapter, we define marine environments as oceans, seas, estuaries, polar sea ice, brine, as well as shallow, deep, and tidal water collections with salt concentrations equivalent to that of sea water (3% w/v). These ecosystems are quite diverse with variable temperatures (~-1.5 °C in sea ice to 400 °C in deep sea hydrothermal vents), pressure (1–1000 atm), light (complete darkness to photic zones), nutrients (nutrient-rich to -limited), and species. Marine environments harbor a wide range of organisms, such as marine (in)vertebrates, plants, algae and their associated microorganisms, as well as sediment-derived microorganisms (Tiquia-Arashiro 2012). Oceans represent the greatest biodiversity with 34 of 36 phyla represented (Donia and Hamann 2003). Such genetic diversity renders significant chemical diversity with new carbon skeletons, high levels of halogenation, and most importantly, novel bioactivity.

It took a long time for researchers to become interested in studying marine environments due to difficulty accessing these sites. Studying oceanic environments requires specialized sampling techniques (Mckindles and Tiquia-Arashiro 2012; Tiquia-Arashiro 2012) and equipment and let us not forget that <1% of these microorganisms are actually culturable on traditional nutrient agar. In the past when new compounds were reported from marine environments, they typically came from easily cultivable organisms. Until the invention of the reliable self-contained underwater breathing apparatus (SCUBA), the isolation of marine natural products lagged behind that of metabolites from the terrestrial world. However, in the 1960s and 1970s, SCUBA diving became an essential tool for collecting marine organisms from ocean depths ranging from 3 to 40 m (Cragg and Newman 2001) and although expensive submersibles and remotely operated vehicles were also developed for deep sea sampling. Since the 1970s, the number of "extremophiles" reported has increased and new sampling and culturing methodologies, journals, databases such as MarinLit, and international symposia are dedicated to the study of these organisms. Now there is a renewed interest in marine natural products, and each year, increasing numbers of new biologically active marine natural products are reported (some from fungal sources) and several of these compounds have entered the preclinical and clinical pipelines. Table 18.1 includes the names of drugs (1–7), either directly from the sea or with structural modifications, which have been approved for clinical use by the FDA and/or the EMA. Their "nominal sources," structure numbers, bioactivities in brief, and relevant reference numbers are also provided. This will be the format used throughout this chapter, so readers interested in specific metabolites may consult the appropriate reference(s).

At least five clinically approved marine drugs are produced by microbial symbionts, which is not surprising as microbes primarily occupy 70% of marine biomass. As such, there is significant interest in bioprospecting marine microbes for natural

Table 18.1   Biological activities of compounds **1–38**

Name	Source	#	Bioactivity
Cytarabine	Sponge/bacterium?	1	Antitumor
Vidarabine	Sponge/bacterium?	2	Antiviral
Ziconotide	Cone snail/Mollusk	3	Severe pain
Eribulin	Sponge/synthetic derivative	4	Antitumor
Omega-3-acids; ethyl esters	Fish/microalgae	5	Lipid metabolism
Trabectedin	Tunicate/uncultured bacterium	6	Antitumor
Brentuximab vedotin	Mollusk/cyanobacterium	7	Antitumor
Cephalosporin C	*C. acremonium*	8	Antibiotic; low activity
Cephalosporin N (now Penicillin N)	*C. acremonium*	9	Antibiotic; low activity
Awajanoran	*Acremonium* sp. AWA16-1	10	Antitumor/antibiotic
Awajanomycin	*Acremonium* sp. AWA16-1	11	A-549 active
Awajanomycin (hydrolyzed ester derivative)	*Acremonium* sp. AWA16-1	12	None reported
Awajanomycin (reduced derivative)	*Acremonium* sp. AWA16-1	13	A-549 active
Cordyheptapeptide C	*Acremonium persicinum* SCSIO 115	14	SF-268; MCF-7 and H460 active
Cordyheptapeptide D	*Acremonium persicinum* SCSIO 115	15	None reported
Cordyheptapeptide E	*Acremonium persicinum* SCSIO 115	16	SF-268; MCF-7 and H460 active
Efrapeptin J	*Tolypocladium* sp.	17	GRP78 downregulator
Efrapeptin F	*Tolypocladium* sp.	18	GRP78 downregulator
Efrapeptin G	*Tolypocladium* sp.	19	GRP78 downregulator
Efrapeptin A (terrestrial source)	*Tolypocladium* sp.		Multiple bioactivities
Efrapeptin B (terrestrial source)	*Tolypocladium* sp.		Multiple bioactivities
Efrapeptin C (terrestrial source)	*Tolypocladium* sp.		Multiple bioactivities
Efrapeptin D (terrestrial source)	*Tolypocladium* sp.		Multiple bioactivities
Efrapeptin E (terrestrial source)	*Tolypocladium* sp.		Multiple bioactivities
Efrapeptin H (terrestrial source)	*Tolypocladium* sp.		Multiple bioactivities
Efrapeptin I (terrestrial source)	*Tolypocladium* sp.		Multiple bioactivities
Ligerin	*Penicillium* sp. MMS351	20	Murine osteosarcoma
Triprostatin A	*A. fumigatus BM939*	21	Cell cycle inhibitor at G2/M
Triprostatin B	*A. fumigatus BM939*	22	Cell cycle inhibitor at M
ds2-try B	Synthesis	23	cytotoxic against multiple lines
Cycloprostratin A	*A. fumigatus BM939*	24	Cell cycle inhibitor at G2/M
Cycloprostratin B	*A. fumigatus BM939*	25	Cell cycle inhibitor at G2/M
Cycloprostratin C	*A. fumigatus BM939*	26	Cell cycle inhibitor at G2/M

(continued)

Table 18.1 (continued)

Name	Source	#	Bioactivity
Cycloprostratin D	*A. fumigatus* BM939	27	Cell cycle inhibitor at G2/M
Cycloprostratin E	*A. sydowii* SCSIO 0035	28	Not cytotoxic in cell lines tested
Demethoxy-fumitremorgin C	*A. fumigatus* BM939	29	M phase inhibitor
Spirotryprostatin A	*A. fumigatus* BM939	30	M phase inhibitor
Spirotryprostatin B	*A. fumigatus* BM939	31	M phase inhibitor;
Prenylcyclotryprostatin B	*A. fumigatus* YK-7	32	Cytotoxic against U937
20-Hydroxycyclotryprostatin B	*A. fumigatus* YK-7	33	No cell line activity
9-Hydroxyfumitremorgin C	*A. fumigatus* YK-7	34	Cytotoxic against U937
6-Hydroxytryprostatin B	*A. fumigatus* YK-7	35	No cell line activity
Spirogliotoxin	*A. fumigatus* YK-7	36	No cell line activity
Zofimarin	*Zopfiella marina* SANK21274	37	Anti-*Candida*
R-135853	Synthesis	38	Antifungal

products, especially because in time, it may be possible to use these, or surrogate hosts of the biogenetic clusters (BGCs), to sustainably produce large quantities of bioactive metabolites. While most microbial secondary metabolites reported from the marine world have been isolated from prokaryotes, fungi are rich sources of natural products, with at least 30–40 BGCs coding for secondary metabolites in a single genome. Fungi are metabolically, morphologically, and phylogenetically diverse. The chemo-diversity of marine fungi from underexplored environments is of even more interest as these environments have a range of ecological factors that impact natural product biosynthesis. In this chapter, we discuss marine-derived fungi from different aspects of marine environments, including shallow-sea isolates, deep sea environments, including sediments and hydrothermal vents, which are unique settings for life due to the increased pressures, temperatures, and presence of toxic elements, then finishing by covering the "Polar Seas." In these areas, and their functional subdivisions, we cover metabolites produced by fungi associated with marine invertebrates, predominantly sponges, and those associated with algae and cyanobacteria, noting that in the latter two cases, these are almost certainly epiphytic organisms in contrast to those in sponges. Chemical structures (**123**) are shown in detail in five figures, and in order to save space, we have used five tables to report compound names, the producing fungus, and any up-to-date biological details available in the reference(s). This review differs from others by our use of geographic and physical extremes to discuss fungal chemical diversity, corresponding bioactivity, and strategies to identify bioactive metabolites from marine environments. We also highlight the occasional occurrence of the same chemistry produced by (micro)organisms from terrestrial and marine environments.

18.2 Definition of Marine Fungi

Since the 1980s, more researchers have published the structures and bioactivities of a number of marine fungal metabolites as can be seen on inspection of the yearly review on marine natural products published in Natural Product Reports (Blunt et al. 2018). Why the delayed interest in marine fungi? To begin to understand why, one must recognize the fact that the mere definition of a marine fungus continues to be debated in publications and at international meetings. In fact, it took marine mycologists almost two decades after the first published study on marine fungi (Barghoorn and Linder 1944) to provide the first comprehensive definition of a marine fungus, Johnson and Sparrow (1961) followed by Kohlmeyer and Kohlmeyer (1979). The latter paper defined obligate marine fungi as growing and sporulating in marine or estuarine environments, whereas facultative marine fungi grow and possibly sporulate in marine, freshwater, or terrestrial habitats. Yet, different definitions are still used to classify fungi within the marine mycology and natural products communities (Overy et al. 2014).

Many marine mycologists only use microscopic techniques to study obligate marine fungi. These investigators consider facultative marine fungi to be (1) dormant spores that may be blown or rained into the ocean and (2) ubiquitous fungi metabolically active in marine environments. Others disagree with the treatment of facultative fungi and continue to advocate for these microorganisms to be included in the definition of marine fungi. These nominally marine facultative fungi elaborate different secondary metabolites from their terrestrial counterparts; thus natural product chemists often use the term "marine-derived" to classify fungi collected from marine environments, as the emphasis is on the bioactive metabolites, rather than the "correct" identification and characterization of the fungal producer. For the purpose of this chapter, we define marine fungi as microorganisms that can (1) reproducibly grow or sporulate in marine environments; (2) be found as symbionts of other marine organisms; or (3) adapt and evolve to thrive in marine environments.

18.2.1 Marine Fungal Sources

Fungi occupy various marine environments, such as the deep sea, hydrothermal vents, polar systems, sediment, sand, and symbiotic environments found in driftwood, algae, marine (in)vertebrates, or plants (i.e., mangroves). Even oceanic air harbors a diverse fungal population (Fröhlich-Nowoisky et al. 2012). Early studies on marine fungi focused on those isolated from macroalgae or seaweeds, submerged wood, and seagrass, followed in the 1970s by more published studies on fungal morphology and taxonomy due to significant technological advances made in microscopy. Three decades later, high-throughput genomic sequencing methods were developed and are now frequently used to classify taxa from marine sources. Distinct and highly divergent phylotypes have been identified in these studies, suggesting that there is a great potential to find more chemical diversity in these environments (Singh 2012). Furthermore, metatranscriptomic sequencing is now revealing some of the following important roles fungi play in marine communities: biogeochemical cycling; decomposition/mineralization of organic matter; functioning as parasites, pathogens, mutualists, and endobionts in host organisms; and playing roles in chemical defense or as attractants.

As of 2015, a total of 1112 documented marine fungi were reported (Jones et al. 2015), but this number may be misleading due to confusion around whether an organism is even a marine fungus or a member of new taxa. Nevertheless, with increasing numbers of marine fungi reported, the number of publications on new bioactive secondary metabolites has increased. Culture-dependent and -independent techniques have been used to identify and characterize new bioactive compounds from these fungal isolates. Here, we will discuss representative bioactive metabolites produced by culturing fungi isolated from specific marine environments, such as the deep sea, hydrothermal vents, driftwood, and polar systems.

18.3 Culture-Based Approaches

18.3.1 Serendipitous Discovery of the First Natural Product Discovered from a "Marine-Derived" Fungus

The clinical use of penicillin marked the beginning of the "Golden Age of Antibiotics" in the 1940s, resulting in the extensive investigation of microbes as sources of new therapeutics. While searching for new antibiotic-producing organisms in 1945, Brotzu examined the microbial flora of sea water near a sewage outlet in Cagliari, Sardinia and isolated a fungus that inhibited the growth of Gram-positive and -negative bacteria. This fungus was identified as *Cephalosporium acremonium* in a 1948 report (Brotzu 1948). That year, a culture was sent to Oxford University for further investigation and over the next 20 years, active components, such as cephalosporins C (**8**) and N (now penicillin N) (**9**), were identified. Cephalosporin C (**8**) had a wide range of bioactivities against a number of penicillin-resistant and -sensitive strains of *Staphylococcus aureus* (Newton and Abraham 1955). Although its bioactivities were weak, cephalosporin C (**8**) was not susceptible to β-lactamases and therefore had potential bioactivity against penicillin-resistant microbes (Abraham and Newton 1956; Newton and Abraham 1956). These results encouraged Lilly to expand upon the 7-aminocephalosporanic acid scaffold to produce many cephalosporins to treat penicillin-resistant infections, and today there are five generations of these agents.

The genus *Cephalosporium* was later renamed *Acremonium* and several marine species have been reported to produce bioactive compounds. Jang et al. reported the new dibenzofuran derivative, awajanoran (**10**), from *Acremonium* sp. AWA16-1, isolated from sea mud at Awajishima Island, Japan (Jang et al. 2006a). The same strain was later shown to produce the γ-lactone-δ-lactam, awajanomycin (**11**), two derivatives, a hydrolyzed ester (**12**) and a reduced carboxyl group (**13**) (Jang et al. 2006b). From *A. persicinum* SCSIO 115 three new cycloheptapeptides, cordyhepta-peptides C–E (**14–16**), were isolated from a marine sediment sample collected in the South China Sea (Chen et al. 2012). A number of other bioactive compounds have been reported from predominately symbiotic *Acremomium* species of marine invertebrates and algae.

Several cytotoxic metabolites have been isolated from marine fungi. Hayakawa et al. reported efrapeptin J (**17**), a new downregulator of GRP78, from a *Tolypocladium* sp. isolated from sea mud in Japan (Hayakawa et al. 2008). Such downregulators exhibit antitumor activity against solid tumors (Fernandez and Tabbara 2000). Efrapeptin J (**17**), a linear pentadecapeptide with a hexahydropyr-rolo-[1,2-*a*]-pyridinium moiety, and other efrapeptins (efrapeptins F–G, **18–19**) with a one- or two-amino acid difference have been isolated from marine-derived *Tolypocladium* sp. with similar bioactivities. Efrapeptins (efrapeptins A–I; struc-tures not shown) are known from terrestrial species of *Tolypocladium* (Gupta et al. 1992; Krasnoff and Gupta 1991, 1992). Thus, osmotic stress may result in differen-tial expression in this genus, and unpurified extracts of efrapeptins exhibit insecti-cidal, anti-parasitic, and antimicrobial activities (Bandani et al. 2000).

In 2013, Vansteelandt et al. isolated a new fumagillin analogue, ligerin (**20**), from *Penicillium* sp. MMS351 collected from seawater (Vansteelandt et al. 2013). Despite the abundance of chlorine in marine environments, there are very few secondary metabolites from marine-derived *Penicillium* strains containing this atom in their chemical structures. The chlorohydrin and C-6 substituents are essential for cyto-toxic activity (Blanchet et al. 2014). Although of scientific interest, no antitumor clinical trials of any fumagillin derivative have proceeded beyond phase II.

18.3.2 Bioactive Compounds from the Deep Sea

The deep sea is a unique marine ecosystem that harbors some of the most diverse fungi and potentially unique molecules/structures. Deep sea environments are defined to be water at depths beyond the euphotic zone (200–300 m) (van Dover et al. 2002). These represent some of the most extensive and remote ecosystems on the planet, as 95% of the Earth's oceans are greater than 1000 m deep (Skropeta and Wei 2014) with hydrostatic pressures increasing by 1 atm with every 10 m increase in depth, to as high as 1100 atm in the Challenger Deep (10,897 m) (Takami et al. 1997). In addition, these environments have temperatures down to ~2 °C, darkness at depths greater than 250 m, sparse nutrients, and low oxygen levels (Skropeta and Wei 2014). The deep-sea microbes mainly inhabit the sediment substratum and obtain nutrients from the sediment or remote surface waters. Yet less than 5% of the deep sea has been explored and only 0.01% of the deep-sea floor has been sampled (Ramirez-Llodra et al. 2010).

Major limitations to exploiting the biodiversity of the deep sea have been developing effective sampling methods and access to this ecosystem. ZoBell coined the term "barophile" (now known as piezophile) to describe microorganisms that thrive under high-pressure conditions (ZoBell and Johnson 1949). Fifteen years later, Roth and coworkers published the first report of deep sea fungal piezophiles isolated from the Atlantic Ocean at a depth of 4450 m (Roth et al. 1964). With the advent of manned submersibles, it was now feasible to identify two new species of organisms a month at depths > 10,000 m (Fisher et al. 2007), and inexpensive collection methods, such as the "mud missile" were developed by Fenical et al. at Scripps Institution of Oceanography (Fenical et al. 2013). Techniques are now available to cultivate deep sea fungi using conditions that mimic the in situ natural environment (Rédou et al. 2015). Using these specialized tools and methods, microbes that live greater than 5000 m below sea level, irrespective of their classifications, can sometimes be cultured.

While some of the deep sea fungal phylotypes are novel, a number of them have "relatives isolated from terrestrial, fresh water, and salt water environments" (Nagano et al. 2010). Thus, it is not surprising that a number of bioactive compounds have been isolated from more cosmopolitan and universal fungi. There are numerous reports on diketopiperazine alkaloids produced from the *Aspergillus fumigatus* BM929 strain isolated from sea sediment 760 m deep. The two tryprostatins A (**21**) and B (**22**) isolated from this strain have different bioactivity profiles (Cui et al. 1995, 1996a, b, 1997; Usui et al. 1998). These compounds also reversed the resistance phenotype in selected cell lines (Woehlecke et al. 2003). Other tryprostatin A and B enantiomers and stereoisomers were evaluated for cytotoxic activity against human carcinoma cell lines, with the diastereomer of tryprostatin B (i.e., ds2-try B, **23**) being more potent than etopside (Zhao et al. 2002). Further synthetic work modeled on tryprostatins yielded multiple compounds with increased cellular cytotoxicity, but none entered clinical trials (Jain et al. 2008).

In 1997, Cui et al. reported the diketopiperazine derivatives, cyclotryprostatins A–D (**24–27**), from the same *A. fumigatus* strain (Cui et al. 1997). These compounds are pentacyclic with structural differences shown in the relevant figures. All inhibited the cell cycle progression of mouse tsFT210 cells at the G2/M phase, with cyclotryprostatin A (**24**) being the most potent, suggesting that the methoxy group and C-12 stereochemistry are important. A variant, cyclotryprostatin E (**28**), with an isopropyl alcohol group at the C-21 position sans the isoprene unit, was isolated from the mycelia of *A. sydowii* SCSIO 0035 (symbiont of the gorgorian coral *Verrucella umbraculum*) but lacked cytotoxic activity in the tested cell lines (He et al. 2012).

Cui et al. had earlier reported the production of a new diketopiperazine, demethoxy fumitremorgin C (**29**), in a companion paper to the initial report on the tryprostatins (Cui et al. 1996b). During a scale-up fermentation of strain BM939, they isolated the bioactive spirotryprostatins A (**30**) and B (**31**) (Cui et al. 1996c, d). They contain a spiro ring system with (**30**) having a C-6 methoxy group and lacking the C-8, C-9 double bond found in (**31**). Other related compounds, prenylcyclotryprostatin B (**32**),

20-hydroxycyclotryprostatin B (**33**), 9-hydroxyfumitremorgin C (**34**), 6-hydroxy-tryprostatin B (**35**), and spirogliotoxin (**36**), were reported from the *A. fumigatus* YK-7 strain isolated from the intertidal zone in Yingkou, China, but only prenylcy-clotryprostatin B (**32**) and 9-hydroxyfumitremorgin (**34**), both with C-6 methoxy groups and C-18 prenyl groups, were active (Wang et al. 2012). Thus, the tryprostatin scaffold appears to be a valuable pharmacophore and several computational and synthetic studies have been published since this study to find other mammalian cell cycle inhibitors (Fani et al. 2015; Yamakawa et al. 2011).

Deep sea fungi have produced compounds that have entered early (pre)clinical trials. For example, in 1985, Japanese researchers at Sankyo laboratories identified the antifungal agent zofimarin (**37**) in the culture broth of *Zopfiella marina* SANK21274, isolated from mud at 120 m in the East China Sea (Ogita et al. 1987; Sato et al. 1985). Zofimarin (**37**) is a sordarin derivative, and the sordarins are a desired class of antifungal agents because they selectively inhibit fungal protein synthesis by inhibiting elongation factor 2 (Domínguez and Martín 1998). Though it had in vitro activity against various fungi and little toxicity in mice (Biabani and Laatsch 1998), it failed in in vivo studies. In 2002, two reports were published on derivatives, with R-135853 (**38**) being selected for further preclinical evaluation (Kaneko et al. 2002a, b). Although active against resistant *C. albicans* strains, and demonstrating in vivo efficacy in various experimental murine models of candidiasis, it was inactive against *C. parapsilosis*, *C. krusei*, and *Aspergillus* spp. (Kamai et al. 2005). Currently, analogs and sordarin-inspired scaffolds are still being investigated for in vivo efficacy against a spectrum of susceptible species (Chakraborty et al. 2016; Wu and Dockendorff 2018).

While several terrestrial strains of *Zopfiella* have been isolated, none produced zofimarin (**37**), highlighting the important impact of the *Z. marina* marine habitat on metabolism. However, it was identified in cultures of terrestrial fungi belonging to the following families: *Microascaceae* (*Graphium putredinis* F12210), soil in the UK (Kennedy et al. 1998); *Xylariaceae* (strain F-064,188), a plant epiphyte from Mauritius (Vicente et al. 2009); *Xylaria* sp. Acra L38, an endophyte isolated from the Thai medicinal plant *Aquilaria crassna* (Chaichanan et al. 2014; Wetwitaklung et al. 2009); *Diatrypaceae* (likely *Eutypa tetragona*), isolated from the fruitbody of *Phellodon melaleucus* in Spain (strain F-081,165); and *Sarothamnus scoparius* from France (strain F-247,493; CBS 284.87) (Vicente et al. 2009). It is still unclear how evolution leads to different species from distinct environments to produce the same natural product. Yet, this phenomenon has been observed with metabolites from sponges and tunicates that are also produced by terrestrial bacteria (Newman 2016). Specifically, unculturable microbes tend to be the true producers of several natural products from marine invertebrates, suggesting that the common denominator is microbes that share similar BGCs. So, many unanswered questions remain. Why are microbes producing the same metabolites in completely different environments? How do these BGCs evolve? A deeper understanding of (1) the ecological roles of secondary metabolites and (2) the influence the environment has on their production will provide more insight into these questions (Fig. 18.1).

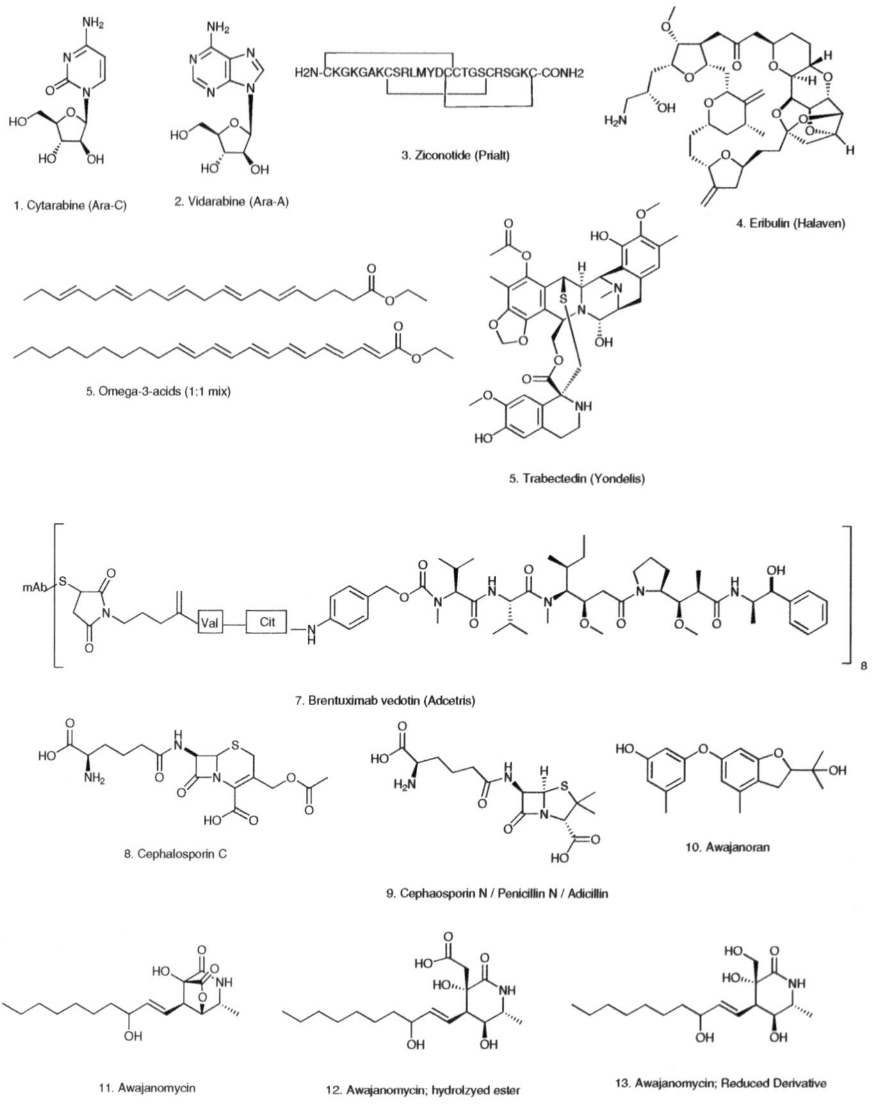

Fig. 18.1 Compounds 1–38

18.3.3   Bioactive Compounds from the Hydrothermal Vents

Within the deep sea are hydrothermal vents, which are metal- and sulfur-rich, high-pressure and -temperature environments that form along continuous mid-ocean ridges (~60,000 km-long). The hot fluids (400 °C) that exit from deep-sea vents are enriched with transition metals, silica, sulfide, and dissolved gases, and the mixing

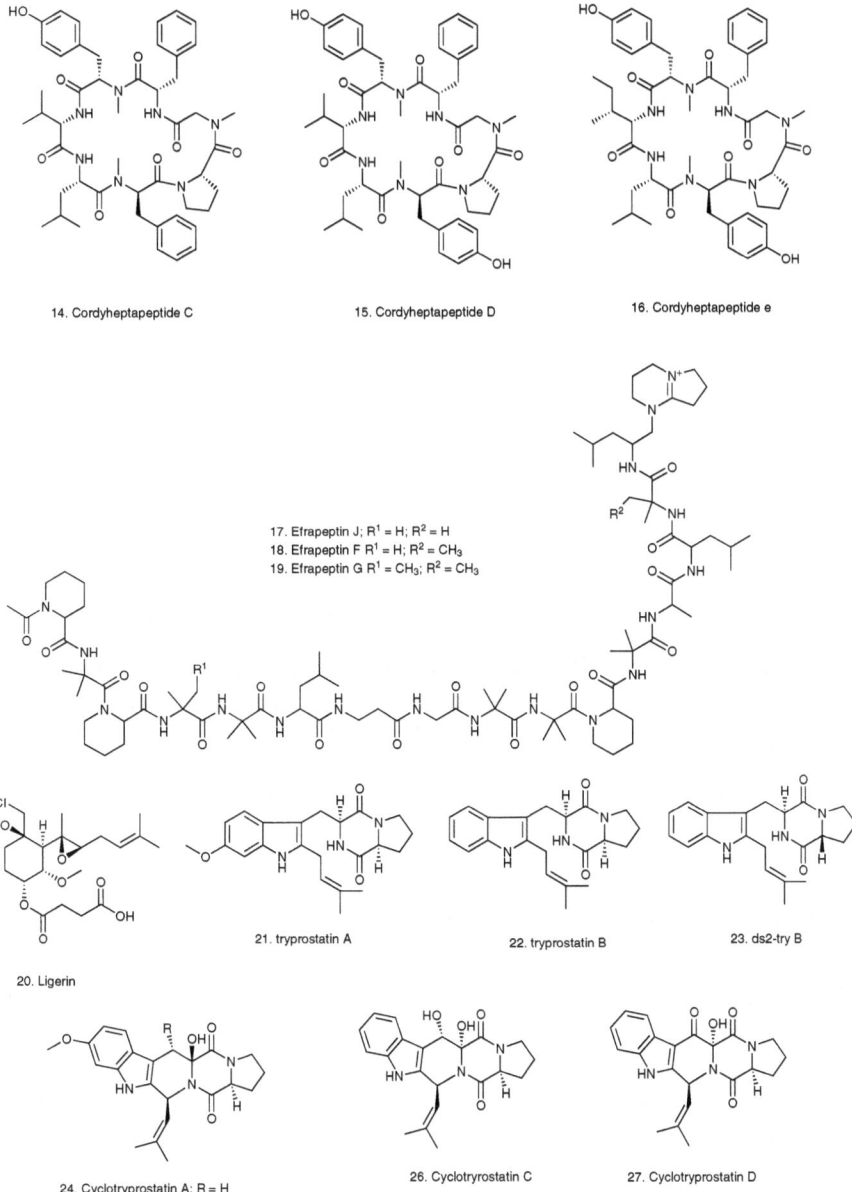

Fig. 18.1 (continued)

of these fluids with the cold ocean water on the sea floor changes the pH and temperature as well as precipitates metal sulfides and minerals. When Weiss et al. reported the first plumes and attendant faunas from the Galapagos Rift in 1977 (Weiss et al. 1977), scientists were surprised because no one expected deep sea hot

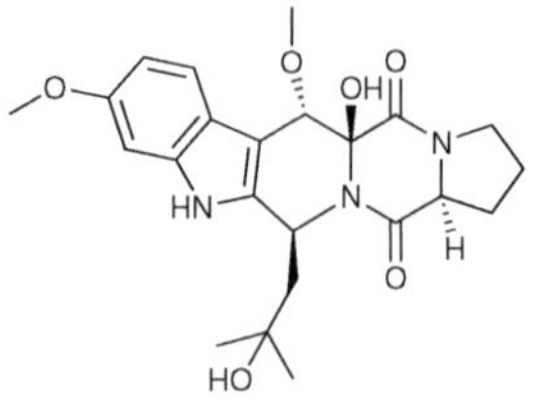

28. Cycloprostatin E

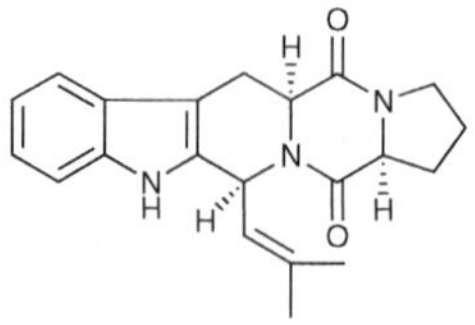

29. Demethoxyfumitremorgin C

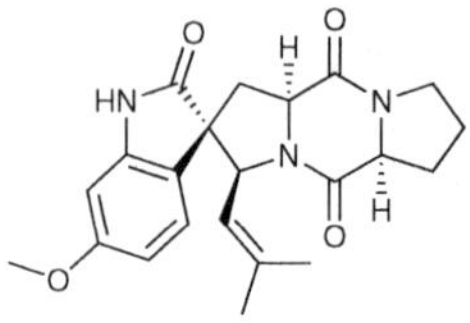

30. Spirotryprostatin A

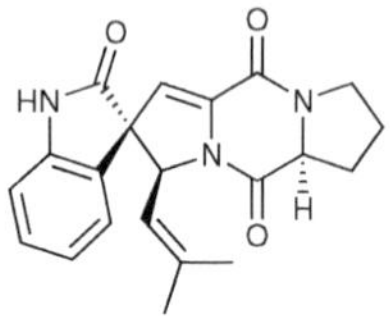

31. Spirotryprostatin B

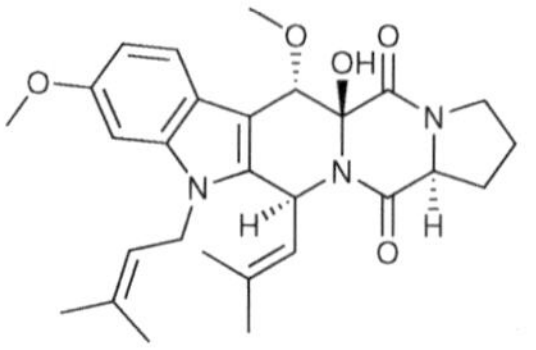

32. Prenylcyclotryprostatin B

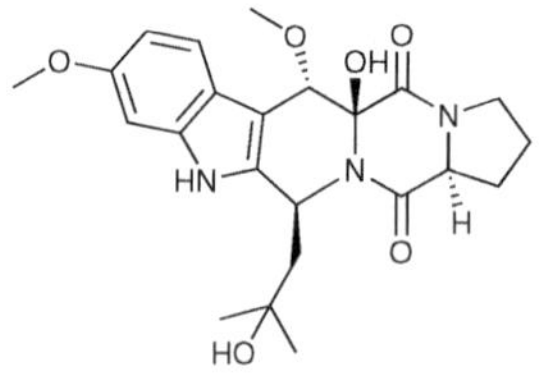

33. 20-Hydroxycycloprostratin B

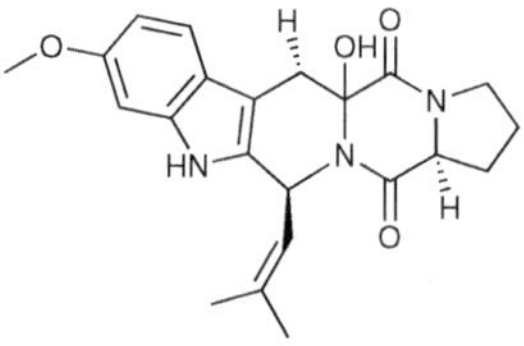

34. 9-Hydroxyfumitremorgin C

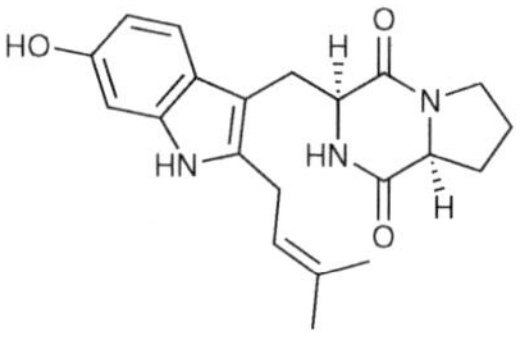

35. 6-Hydroxytryprostatin B

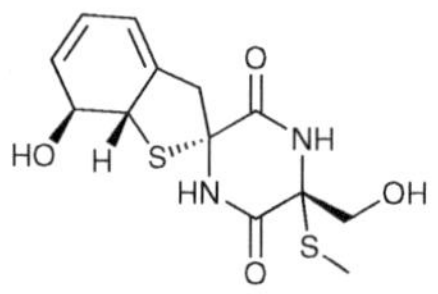

36. Spirogliotoxin

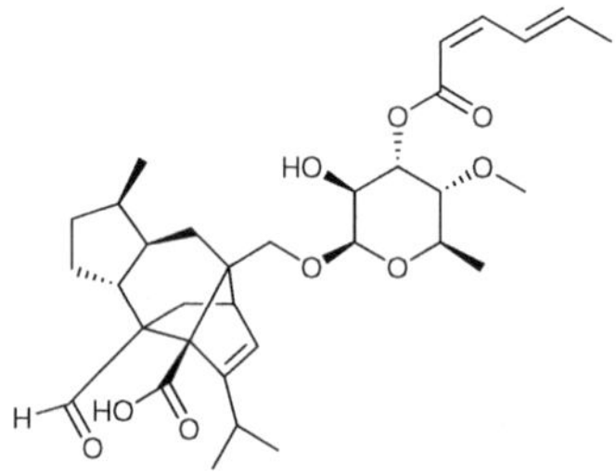

37. Zofimarin

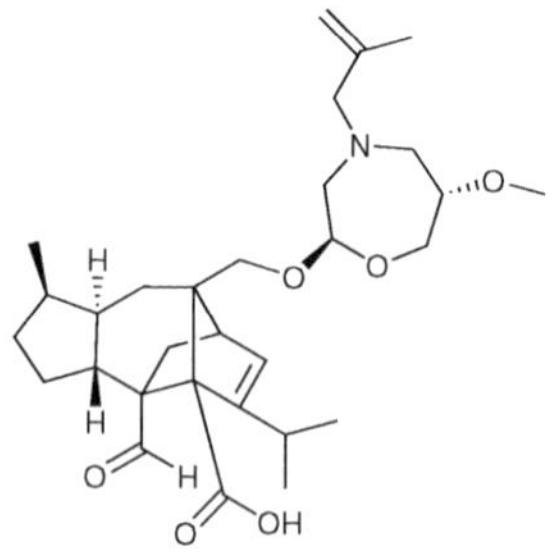

38. R-135853

Fig. 18.1 (continued)

water springs to have abundant communities of benthic organisms. A number of large and unusual marine invertebrates, such as bivalve mollusks and crabs, were photographed by the manned submersible Alvin near these plumes. From 1979 to 1982, expeditions to hydrothermal vents led to the discovery of chemoautotrophs and symbiotic sulfide-oxidizing bacteria in invertebrates. These microbes play important roles in hydrothermal vents by being the primary producers driving the food chain (van Dover 2000; van Dover et al. 2002). Furthermore, the steep physical and chemical gradients of these remarkable ecosystems influence the biogeographical patterns of species distribution. Several complex fungal communities in hydrothermal vents have been characterized, with the sediment being more phylogenetically diverse than the overlaying seawater (López-García et al. 2003). Molecular surveys of the small subunit ribosomal RNA in autochthonous eukaryotic communities have revealed that vent sediment is comprised of anaerobic fungi as well as those from known eukaryotic lineages (Zhang et al. 2014).

Marine fungi have been isolated from a variety of endemic animals, such as crabs, in hydrothermal vents and reported to produce a number of structurally diverse, bioactive metabolites. Pan et al. reported versicomides A–C (**39–41**), an oxepin containing quinazoline (versicomide D, **42**) and four cyclopenin derivatives (7-methoxycyclopeptin, 7-methoxydehydrocyclopeptin, 7-methoxycyclopenin, and 9-hydroxy-3-methoxyviridicatin, [**43–46**]) produced by *A. versicolor* XZ-4 isolated from a Taiwan Kueishantao hydrothermal vent crab (Pan et al. 2017). With the exception of (**44**), (**43, 45,** and **46**) were active against *Escherichia coli*, suggesting that these structural scaffolds may be worth investigating to fine tune their antibacterial activity. The same group also reported a new verrucosidin derivative, methyl isoverrucosidinol (**47**), isolated from *Penicillium* sp. Y-50-10 from the same vent that inhibited *Bacillus subtilis* (Pan et al. 2016). Compound **47** is an unusual isomer of the potent neurotoxin verrucosidin (**48**) (Burka et al. 1983; Whang et al. 1990), based on its conformational isomerization between C-8, C-9 and the C-10, C-11 double bonds (Fig. 18.2; Table 18.2).

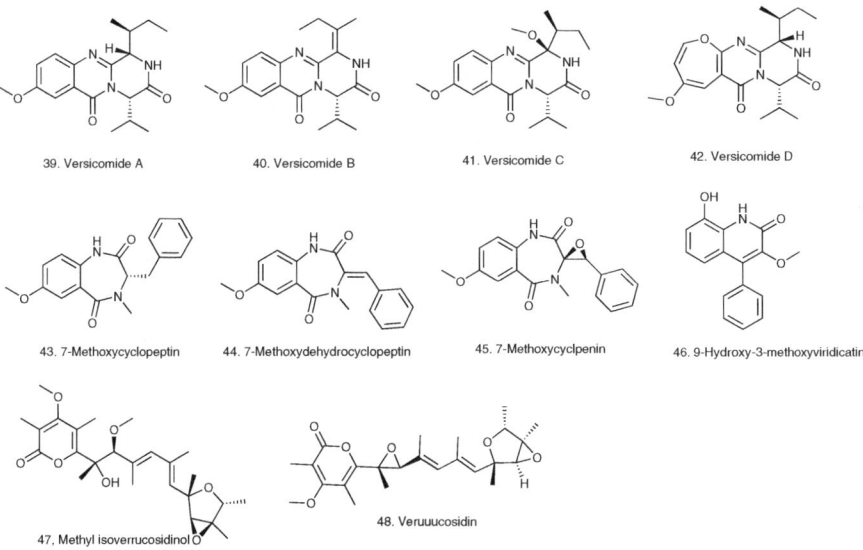

39. Versicomide A   40. Versicomide B   41. Versicomide C   42. Versicomide D

43. 7-Methoxycyclopeptin   44. 7-Methoxydehydrocyclopeptin   45. 7-Methoxycyclpenin   46. 9-Hydroxy-3-methoxyviridicatin

47. Methyl isoverrucosidinol   48. Veruuucosidin

Fig. 18.2 Compounds **39–56**

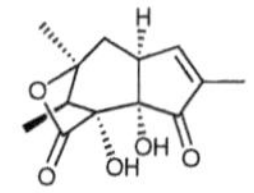
49. Acremostrictin

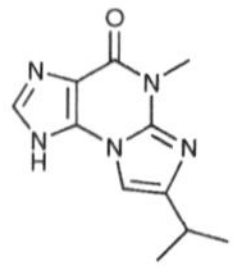
50. Acremolin

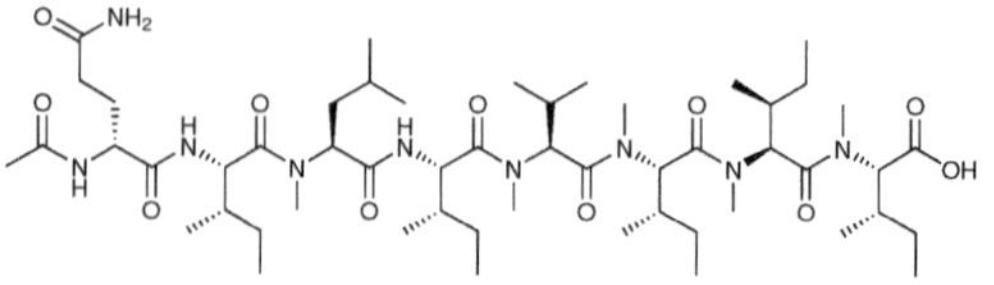
51. RHM1

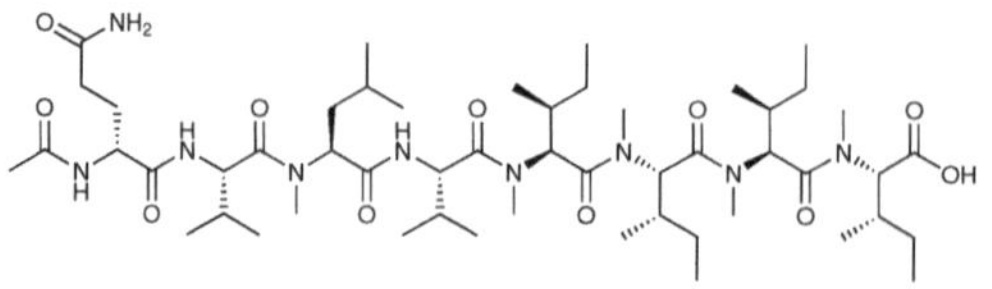
52. RHM2

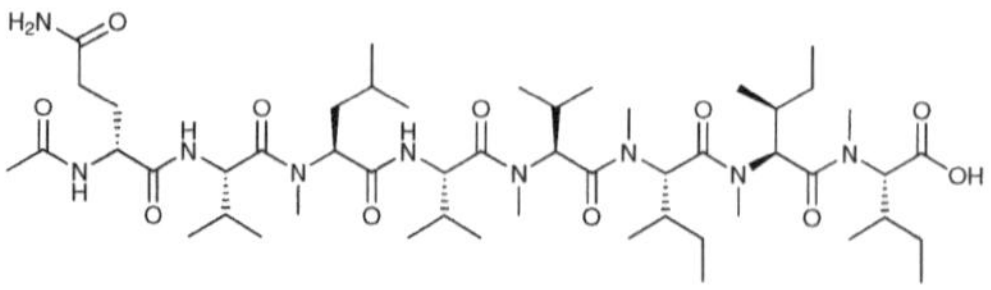
53. RHM3

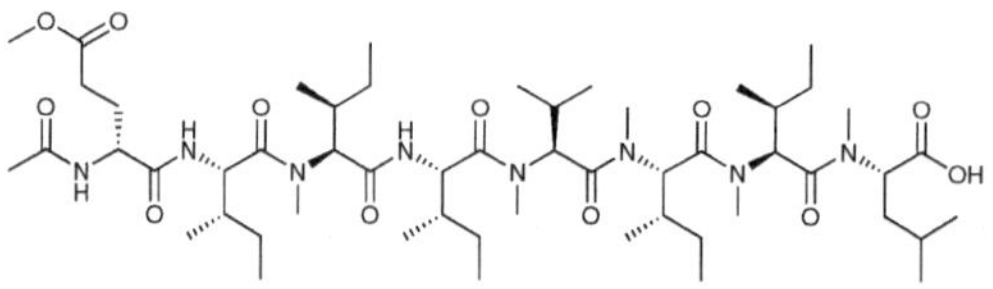
54. RHM4

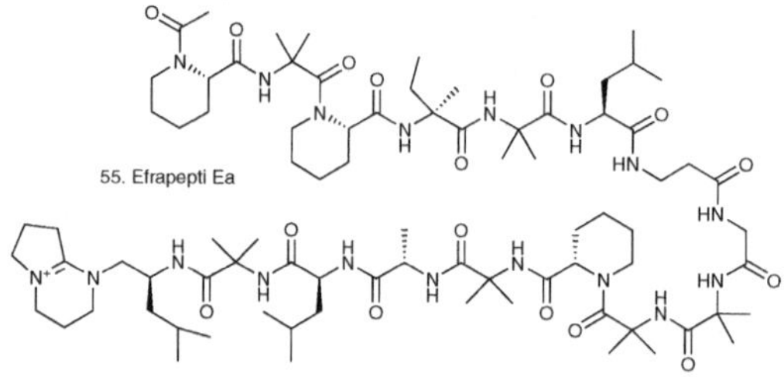
55. Efrapepti Ea

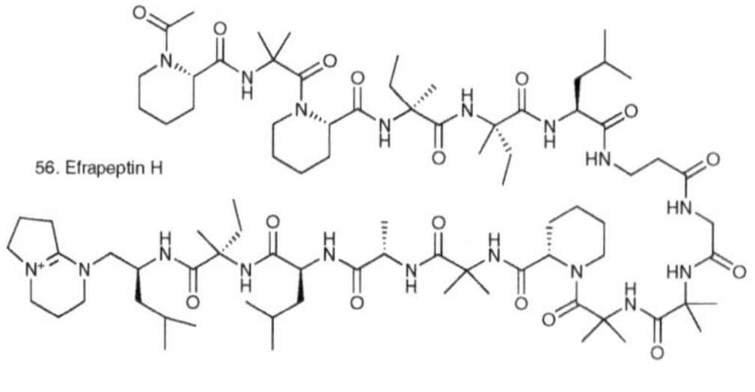
56. Efrapeptin H

Fig. 18.2 (continued)

18.3.4 Bioactive Compounds from Fungi Associated with Marine Invertebrates, Cyanobacteria, and Algae

Several culture-dependent and -independent studies have shown that marine sponges, mollusks, tunicates, algae, and other invertebrates are a rich source of marine fungi (Debbab et al. 2010). These findings are not surprising as fungi often

Table 18.2 Biological activities of compounds **39–48**

Name	Source	#	Bioactivity
Versicomide A	*A. versicolor* XZ-4	**39**	No activity quoted
Versicomide B	*A. versicolor* XZ-4	**40**	No activity quoted
Versicomide C	*A. versicolor* XZ-4	**41**	No activity quoted
Versicomide D	*A. versicolor* XZ-4	**42**	Inhibited *E. coli*
7-Methoxycyclopeptin	*A. versicolor* XZ-4	**43**	Inhibited *E. coli*
7-Methoxydehydrocyclopeptin	*A. versicolor* XZ-4	**44**	Not active against *E. coli*
7-Methoxycyclopenin	*A. versicolor* XZ-4	**45**	Inhibited *E. coli*
9-Hydroxy-3-methoxyviridicatin	*A. versicolor* XZ-4	**46**	Inhibited *E. coli*
Methyl isoverrucosidinol	*Penicillium* sp. Y-50-10	**47**	Inhibited *B. subtilis*
Verrucosidin	*Penicillium verrucosum* var. *cyclopium*	**48**	Neurotoxin

associate with various marine organisms or substrata, assisting with nutrient cycling and the decomposition of organic matter. Marine fungi can be parasitic, commensal, or mutualistic. Ultimately, the interplay between microbe and host produces chemical responses that have profound effects on host microbiomes. These highly evolved responses often result in the production of unique molecular scaffolds that do not bear any resemblance to the chemical space occupied by synthetic compounds and therefore may have clinical potential in human medicine (Boot et al. 2006; Boufridi and Quinn 2018; Debbab et al. 2010; Gonzalez-Medina et al. 2016).

Sponges are the most studied marine habitat for fungi, as they comprise a significant proportion of the benthic community, and 40–60% of their biomass is comprised of associated microbes. Interestingly, abundant amounts of "terrestrial" taxa, such as *Acremonium* and *Penicillium*, are found in these invertebrates, some of which have specifically adapted to the marine environment and some are halogen sensitive, suggesting that some were recently acquired from terrestrial sources (Gal-Hemed et al. 2011). Thus, the highly oxygenated, tricyclic metabolite from a new structural class, acremostrictin (**49**), was produced by a strain of *Acremonium* (now *Sarocladium strictum*) from an unidentified Choristid sponge from Korean waters (Julianti et al. 2011). An unprecedented base, acremolin (**50**), was later isolated from the same strain of *Acremonium*. Acremolin (**50**), a modified guanine base attached to an isoprene unit via a 1H-azirine moiety, exhibited weak cytotoxic activity (Julianti et al. 2012). Its structure was revised later in the year removing the azirine ring, and the revised tri-ring system is shown as structure (**50**) (Banert 2012). *N*-methylated linear octapeptides, RHM1 (**51**) and RHM2 (**52**), were isolated from an atypical strain of *Acremonium* sp. (strain number 021172c) cultured from an *Axinella* sp. collected in Papua New Guinea (Boot et al. 2006). A year later, RHM3 and 4 (**53–54**) as well as the new linear pentadecapeptides efrapeptins Eα and H (**55–56**) were isolated from *Acremonium* sp. (Boot et al. 2007). Efrapeptin Eα (**55**), composed of 15 canonical amino acids, was significantly cytotoxic. However, as shown earlier, efrapeptin H isolated from terrestrial sources was quoted to be bioactive (cf Table 18.1).

Acremonium species associate with other marine organisms in addition to sponges, such as sea cucumbers and algae. Strains of *A. striatisporum* KMM 4401 isolated from the sea cucumber *Eupentacta fraudatrix*, collected in the Sea of Japan, produce derivatives of the virescenosides, though virescenosides A–C (**57–59**) were originally isolated from the terrestrial strain *A. luzulae* (Bellavita et al. 1970; Moussaief et al. 1997). A series of papers from the Elyakov group and his successors in Vladivostok published from 2000 through 2006 describe several of these compounds, including virescenosides M–X (**60–71**), from the same nominal organism, *A. striatisporum* (Afiyatullov et al. 2000, 2004, 2006). Virescenosides V–X (**69–71**) have different unsaturated carbons, either a proton or hydroxyl group at C-6, and either a proton, keto, or hydroxyl group at C-7. There are no reports of their bioactivity.

Structurally unique classes of compounds have been produced by endophytic *Acremonium* strains isolated from marine algae. For example, the novel antioxidant hydroquinone derivatives, 7-isopropenylbicyclo[4.2.0]octa-1,3,5-triene-2,5-diol (also known as acremonium A, **72**) and 7-isopropenylbicyclo[4.2.0]octa-1,3,5-triene-2,5-diol-5-beta-D-glucopyranoside (acremonium A-5-β-D-glucopyranoside, **73**), containing an unusual ring system, were isolated from *Acremonium* sp. isolated from the brown alga *Cladostephus spongius* collected in Spanish waters (Abdel-Lateff et al. 2002). Interestingly, this ring system can only be chemically synthesized from a butane solution of 3-methyl-1,2-dihydronapthalene in the presence of UV light (245 nm), and the mode of cyclization used by *Acremonium* sp. in its biosynthesis is unknown. The same *Acremonium* strain also produced new dihydronapthalenones and acyclic carbocyclic derivatives (structures not shown) with no reported bioactivities. Nevertheless, these are examples of the wide capabilities of this fungus' biosynthetic capabilities.

From the culture broth and mycelia of an *Acremonium* sp. strain cultured from the Caribbean tunicate *Ecteinascidia turbinata* (the source of the EMA and FDA-approved antitumor agent ET743), Belofsky and coworkers isolated the new oxepin-containing pyrimidines, oxepinamides A–C (**74–76**), and the new fumiquinazolines H and I (**77–78**) (Belofsky et al. 2000). The cyclic pentadepsipeptide zygosporamide, composed of α-hydroxyleucic acid and D- and L-amino acids (**79**), was isolated from the culture broth of *Zygosporium masonii* isolated from a marine cyanobacterium collected off the island of Maui, Hawaii (Oh et al. 2006).

Marine fungi have been isolated from "sea hares" (mollusks), and compounds from these have been reported to exhibit (potent) bioactivity. The pericosines A–E (**80–84**) are cyclohexenoids produced by *Periconia byssoides* OUPS-N133 isolated from the gastrointestinal tract of the sea hare *Aplysia kurodai* (Numata et al. 1997; Yamada et al. 2007). Mice inoculated intraperitoneally with P388 followed by *ip* pericosine A (**80**) survived as long as non-treated control mice. Since the initial 1997 report of these unique carbasugar-type metabolites, several syntheses have been published, especially as other bioactive compounds containing a pericosine moiety have been reported (Babu et al. 2014; Boyd et al. 2010; Donohoe et al. 1998; Li and Hou 2014; MuniRaju et al. 2012; Reddy et al. 2012; Tripathi et al. 2011; Usami et al. 2017).

Aside from the FDA-approved cephalosporins, plinabulin (NPI-2358, **85**) is currently the only marine-derived fungal metabolite at a phase III clinical trial (under NCT02504489) against NSCLC cancers. It is a synthetic *tert*-butyl analog of the diketopiperazine halimide (**86**), also known as (−)-phenylahistin, produced by *Aspergillus* sp. CNC-139 isolated from the alga *Halimeda copiosa* collected near the Philippines (Fenical et al. 1999, 2000). This compound was also produced by two terrestrial strains of *A. ustus*, further demonstrating how BGCs may not be unique to an organism and their habitat (Kanoh et al. 1997, 1999a). Kanoh et al. determined that its cytotoxicity originates from its inhibition of tubulin polymerization by interacting with the colchicine-binding site (Kanoh et al. 1999b). To remove chirality and optimize cytotoxic activity, Nereus Pharmaceuticals developed a series of synthetic derivatives, ultimately leading to plinabulin (**85**) (Nicholson et al. 2006) (Fig. 18.3; Table 18.3).

18.3.5 Bioactive Compounds from Driftwood

Marine fungi inhabit wood substrata and play an important role in nutrient regeneration by decomposing dead and decaying organic matter, particularly lignocellulose. Driftwood is rich in carbohydrate polymers and a habitat for grazing organisms tolerating variations in the saline environment. Importantly, driftwood is a stable nutrient source for xylophages, substrata for microbial growth, as well as a point of attachment, facilitating its habitation. Driftwood is relatively unique, as tidal variations change its location over time and vary its exposure to sunlight and the atmosphere.

Fungi that secrete an abundance of lignin-degrading enzymes commonly inhabit driftwood (Kameshwar and Qin 2018a, b). Submerged wood substrata are breeding grounds for members of the orders Halosphaeriales and Lulworthiales, as these environments facilitate the passive release of ascospores and the floatation and attachment of appendaged ascospores (Overy et al. 2014). Additionally, the geographic source of driftwood influences fungal community structure within these types of substrates. Thus, driftwood from the eastern and western regions along the Norwegian coast of the Barents Sea harbors different fungal communities (Rämä et al. 2016). While terrestrial fungal communities that colonize wood have been extensively studied usually due to the economic importance of their "preferred homes," the converse is the case in the corresponding marine environment.

Corollospora, a genus known for rapidly degrading cellulose, has been isolated from driftwood and found to produce bioactive compounds. In 1998, Alvi and coworkers reported the production of a new lactam, pulchellalactam (**87**), from *Corollospora pulchella* (ATCC 62554) isolated from a driftwood sample collected from Peleliu, Palau using a bioactivity screen looking for CD45 phosphatase inhibitors (Alvi et al. 1998). The new phthalide derivative corollosporine (**88**) isolated from *Corollospora maritima* collected from driftwood in the North Sea exhibited antibacterial activity against *S. aureus* SG 511 (Liberra et al. 1998). Since this first report of corollosporine (**88**), other amide derivatives that are more active against

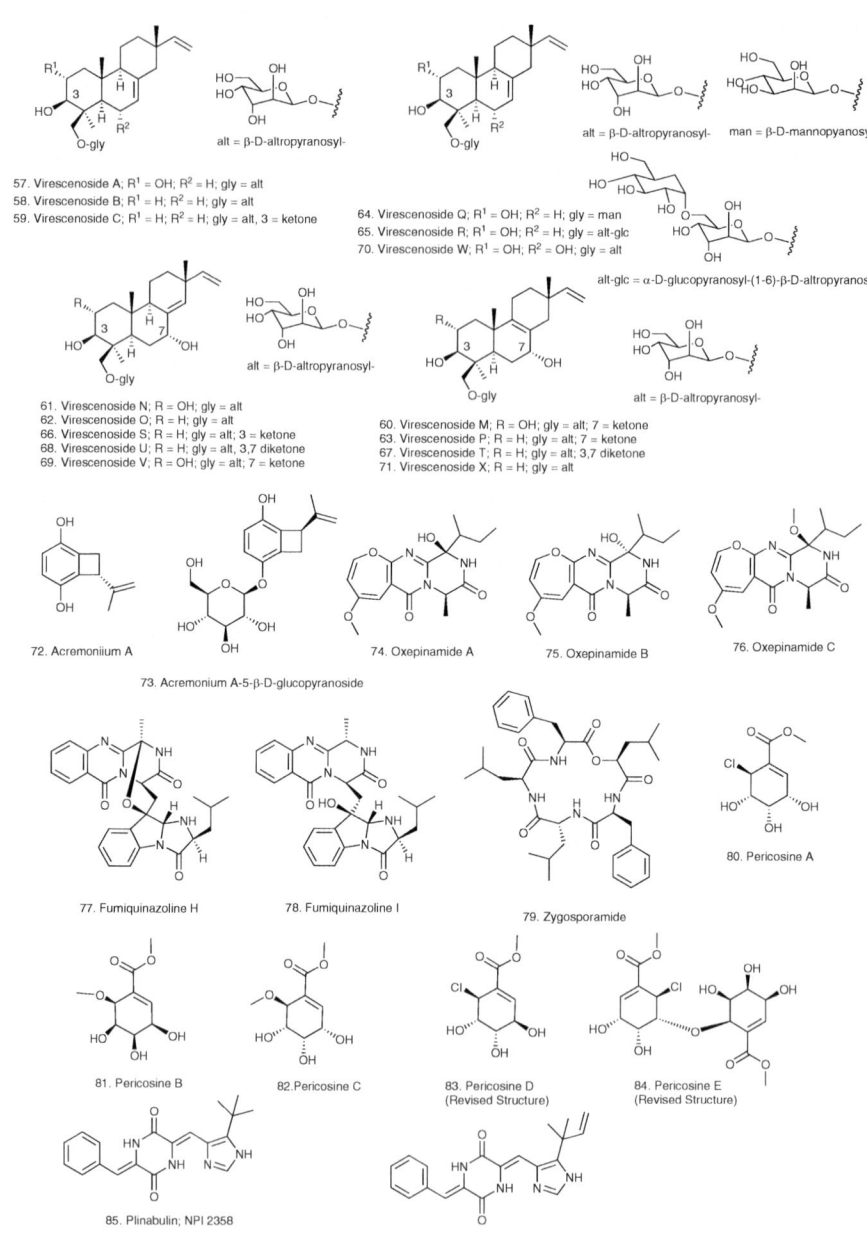

57. Virescenoside A; R¹ = OH; R² = H; gly = alt
58. Virescenoside B; R¹ = H; R² = H; gly = alt
59. Virescenoside C; R¹ = H; R² = H; gly = alt, 3 = ketone

alt = β-D-altropyranosyl-

alt = β-D-altropyranosyl-   man = β-D-mannopyranosyl-

64. Virescenoside Q; R¹ = OH; R² = H; gly = man
65. Virescenoside R; R¹ = OH; R² = H; gly = alt-glc
70. Virescenoside W; R¹ = OH; R² = OH; gly = alt

alt-glc = α-D-glucopyranosyl-(1-6)-β-D-altropyranosy-

alt = β-D-altropyranosyl-

61. Virescenoside N; R = OH; gly = alt
62. Virescenoside O; R = H; gly = alt
66. Virescenoside S; R = H; gly = alt; 3 = ketone
68. Virescenoside U; R = H; gly = alt, 3,7 diketone
69. Virescenoside V; R = OH; gly = alt; 7 = ketone

alt = β-D-altropyranosyl-

60. Virescenoside M; R = OH; gly = alt; 7 = ketone
63. Virescenoside P; R = H; gly = alt; 7 = ketone
67. Virescenoside T; R = H; gly = alt; 3,7 diketone
71. Virescenoside X; R = H; gly = alt

72. Acremoniium A

73. Acremonium A-5-β-D-glucopyranoside

74. Oxepinamide A

75. Oxepinamide B

76. Oxepinamide C

77. Fumiquinazoline H

78. Fumiquinazoline I

79. Zygosporamide

80. Pericosine A

81. Pericosine B

82. Pericosine C

83. Pericosine D
(Revised Structure)

84. Pericosine E
(Revised Structure)

85. Plinabulin; NPI 2358

86. Halimide

Fig. 18.3 Compounds **57–86**

Table 18.3 Biological activities of compounds **49–86**

Name	Source	#	Bioactivity
Acremostrictin	*Acremonium (Sarocladium) strictum*	49	DPPH activity, weak antibacterial activity
Acremolin	*Acremonium (Sarocladium) strictum*	50	Weak inhibition of lung cancer cells
RHM1	*Acremonium* sp.	51	Weak cytotoxicity/antibacterial activity
RHM2	*Acremonium* sp.	52	Weak cytotoxicity
RHM3	*Acremonium* sp.	53	No activity quoted
RHM4	*Acremonium* sp.	54	No activity quoted
Efrapeptin Eα	*Acremonium* sp.	55	Highly cytotoxic against H125 human line
Efrapeptin H	*Acremonium* sp.	56	active from terrestrial sources (Table 18.1)
Virescenoside A	*Acremonium luzulae*	57	No activity quoted
Virescenoside B	*Acremonium luzulae*	58	No activity quoted
Virescenoside C	*Acremonium luzulae*	59	No activity quoted
Virescenoside M	*Acremonium striatisporum*	60	Cytotoxic to sea urchin eggs
Virescenoside N	*Acremonium striatisporum*	61	Cytotoxic to sea urchin eggs; weak
Virescenoside O	*Acremonium striatisporum*	62	Weak activity against Erlich cells
Virescenoside P	*Acremonium striatisporum*	63	Cytotoxic to sea urchin eggs
Virescenoside Q	*Acremonium striatisporum*	64	Weak activity against Erlich cells
Virescenoside R	*Acremonium striatisporum*	65	Cytotoxic to sea urchin eggs
Virescenoside S	*Acremonium striatisporum*	66	Cytotoxic to sea urchin eggs
Virescenoside T	*Acremonium striatisporum*	67	Cytotoxic to sea urchin eggs
Virescenoside U	*Acremonium striatisporum*	68	Cytotoxic to sea urchin eggs
Virescenoside V	*Acremonium striatisporum*	69	No bioactivity reported
Virescenoside W	*Acremonium striatisporum*	70	No bioactivity reported
Virescenoside X	*Acremonium striatisporum*	71	No bioactivity reported
Acremonium A	*Acremonium* sp.	72	Antioxidant activity
Acremonium A-5-β-D-glucopyranoside	*Acremonium* sp.	73	Antioxidant activity
Oxepinamide A	*Acremonium* sp.	74	Excellent anti-inflammatory activity
Oxepinamide B	*Acremonium* sp.	75	No quoted bioactivity
Oxepinamide C	*Acremonium* sp.	76	No quoted bioactivity
Fumiquinazoline H	*Acremonium* sp.	77	Weak anti-*Candida* activity
Fumiquinazoline I	*Acremonium* sp.	78	Weak anti-*Candida* activity
Zygosporamide	*Zygosporium masonii*	79	Active SF-268 and RXF 393 tumor lines
Pericosine A	*Periconia byssoides* OUPS-N133	80	Active against murine P388, in vivo
Pericosine B	*Periconia byssoides* OUPS-N133	81	Cytotoxic; murine P388 and Human lines

(continued)

Table 18.3 (continued)

Name	Source	#	Bioactivity
Pericosine C	*Periconia byssoides* OUPS-N133	**82**	No bioactivity listed
Pericosine D	*Periconia byssoides* OUPS-N133	**83**	Cytotoxic murine P388
Pericosine E	*Periconia byssoides* OUPS-N133	**84**	Alpha-glucosidase inhibition
Plinabulin	*Aspergillus* sp. CNC-139	**85**	Phase III clinical trial against NSCLC
Halimide	*Aspergillus* sp. CNC-139	**86**	Active in vivo against P388

different bacterial strains than their non-amide counterparts have been synthesized using a fungal laccase-based system (Mikolasch and Schauer 2009).

Structurally unique cytotoxic compounds have been reported from wood substrata collected from marine environments. These include the trichothecene sesquiterpene verrol-4-acetate (**89**) isolated from *A. neo-caledoniae* (Roquebert et Dupont n. sp.) in New Caledonia (Laurent et al. 2000). Another new member of the cytotoxic roridin class of trichothecenes, 12,13-deoxyroridin E (**90**), was later isolated from *Myrothecium roridum* TUF 98F42 obtained from submerged wood in Palau (Namikoshi et al. 2001). Further work with the same strain identified the cytotoxic trichothecenes 12′-hydroxyroridin E (**91**) and roridin Q (**99**). In addition, 2′,3′-deoxyroritoxin D (**100**) was also isolated and found to exhibit antimicrobial activity against *S. cerevisiae* (Xu et al. 2006).

Interestingly, the antiviral naphthalenone compounds balticols A–F (**94–99**) were reported from an unspeciated Ascomycete from inshore driftwood in the Baltic Sea (Shushni et al. 2009). These metabolites differ structurally by different degrees of oxygenation and/or methylation. Two years later the same group reported the 12-membered macrolide balticolid (**100**) from the same fungus. Only one paper describing the synthesis of this compound has been published since that initial report (Krishna et al. 2012). Of interest is that balticol A (**94**) and other naphthalene derivatives were later reported from the plant fungal endophyte *Biatriospora marina* cultured from the European white elm *Ulmus laevis* in the Czech Republic (Stodůlková et al. 2015).

Mangrove or tidal swamp fungi are often found on driftwood in intertidal zones. One example is *Fusarium heterosporum* strain CNC-477 from driftwood in the Bahamas. From this, Fenical et al. reported two new related classes of cytotoxic sesterterpenes, the neomangicols A–C (**101–103**) (Renner et al. 1998) and the mangicols A–G (**104–110**) (Renner et al. 2000). Since the identification of these compounds, several syntheses assembling the novel mangicol core have been published (Araki et al. 2004; Chen et al. 2013; Ying and Pu 2014).

The examples shown demonstrate the multiplicity of different structures found produced by "driftwood fungi" but there are many other substrates in the sea that have still yet to be investigated, including volcanic pumice, metal, and even plastics that should be investigated in the search for novel chemistry and accompanying bioactivity (Fig. 18.4; Table 18.4).

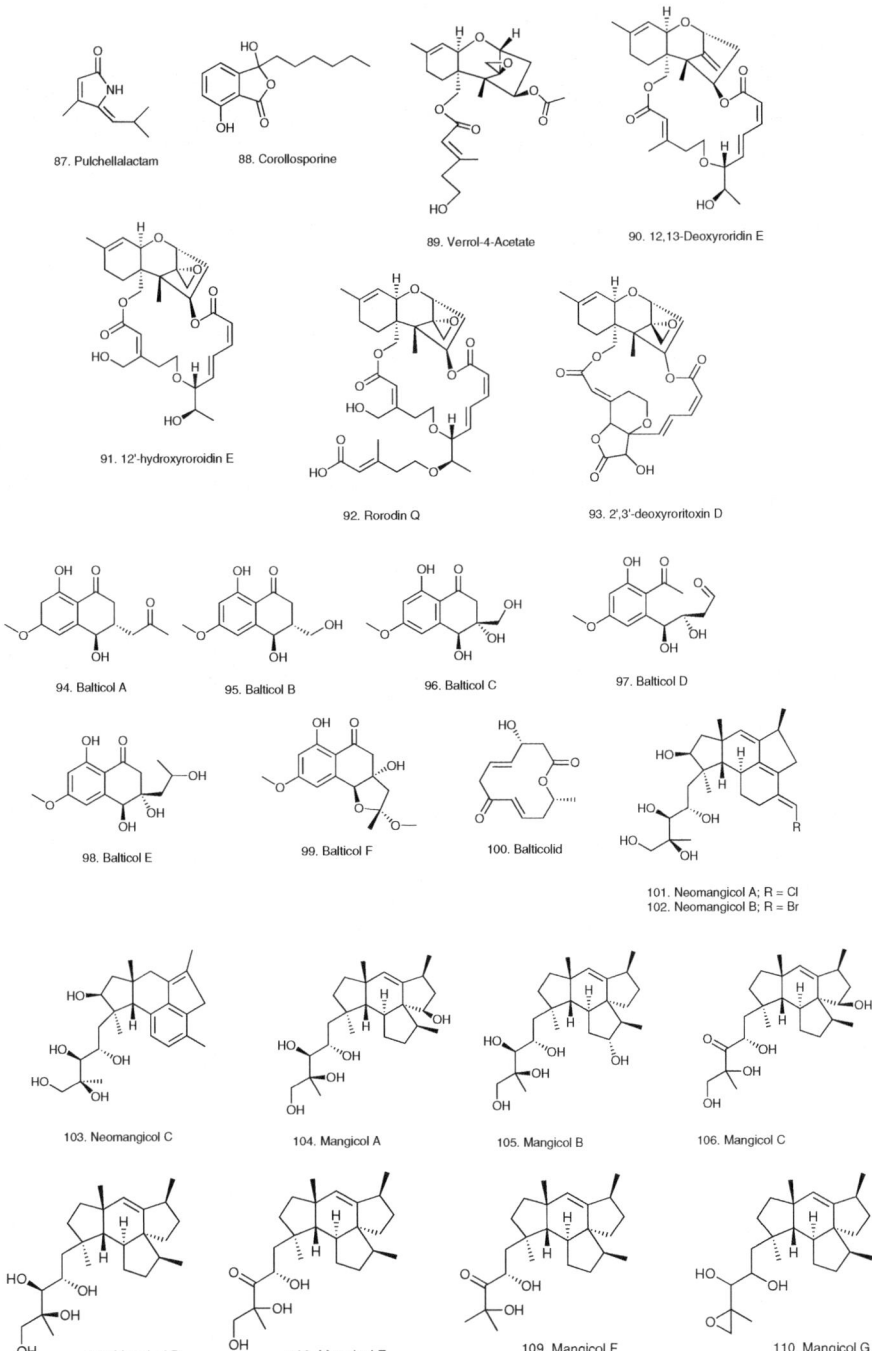

Fig. 18.4 Compounds **87–110**

Table 18.4 Biological activities of compounds **87–110**

Name	Source	#	Bioactivity
Pulchellalactam	*Corollospora pulchella* (ATCC 62554)	**87**	CD45 phosphatase inhibitor
Corollosporine	*Corollospora maritima*	**88**	Antibiotic activity against *S. aureus*
Verrol-4-acetate	*Acremonium neo-caledoniae*	**89**	Cytotoxic against KB cells
12,13-Deoxyroridin E	*Myrothecium roridum* TUF 98F42	**90**	Cytotoxic against L1210 and HL-60 cell lines
12′-Hydroxyroridin E	*Myrothecium roridum* TUF 98F42	**91**	Cytotoxic against L1210 cell line
Roridin Q	*Myrothecium roridum* TUF 98F42	**92**	Cytotoxic against L1210 cell line
2′,3′-Deoxyroritoxin D	*Myrothecium roridum* TUF 98F42	**93**	Antifungal (*S. cerevisiae*)
Balticol A	Ascomycete	**94**	Active against influenza A and HSV-1
Balticol B	Ascomycete	**95**	Active against influenza A and HSV-1
Balticol C	Ascomycete	**96**	Active against influenza A and HSV-1
Balticol D	Ascomycete	**97**	Active against influenza A and HSV-1
Balticol E	Ascomycete	**98**	Active against influenza A and HSV-1
Balticol F	Ascomycete	**99**	Active against influenza A and HSV-1
Balticolid	Ascomycete	**100**	Anti HSV-1
Neomangicol A	*Fusarium heterosporum* strain CNC-477	**101**	Cytotoxic MCF-7, H-60
Neomangicol B	*Fusarium heterosporum* strain CNC-477	**102**	Cytotoxic MCF-7, H-60; *B. subtilis* active
Neomangicol C	*Fusarium heterosporum* strain CNC-477	**103**	Potential decomposition product
Mangicol A	*Fusarium heterosporum* strain CNC-477	**104**	Cytotoxic in human lines; anti-inflammatory
Mangicol B	*Fusarium heterosporum* strain CNC-477	**105**	Cytotoxic in human lines; anti-inflammatory
Mangicol C	*Fusarium heterosporum* strain CNC-477	**106**	Cytotoxic in human tumor cell lines
Mangicol D	*Fusarium heterosporum* strain CNC-477	**107**	Cytotoxic in human tumor cell lines
Mangicol E	*Fusarium heterosporum* strain CNC-477	**108**	Cytotoxic in human tumor cell lines
Mangicol F	*Fusarium heterosporum* strain CNC-477	**108**	Cytotoxic in human tumor cell lines
Mangicol G	*Fusarium heterosporum* strain CNC-477	**110**	Cytotoxic in human tumor cell lines

18.3.6 Bioactive Compounds from Polar Marine Environments

The chemical diversity of polar regions is understudied due to a number of factors, including inaccessibility and the challenges faced when cultivating microbes from psychrophilic sources to obtain enough compound for biological evaluation. Yet, these regions could well be of extreme importance to the industrial and pharmaceutical industries, particularly for novel bioactive chemistry and psychrophilic enzymes (Tiquia and Mormile 2010). These environments have low to ultra-low surface temperatures, strong winds, limited nutrients, frequent freeze-thaw cycles, limited water access, high UV radiation, or combinations of these factors. The Arctic and Antarctic regions are surrounded by the Arctic and the Southern Oceans, respectively. The Arctic Ocean is a basin surrounded by continental land masses, whereas the Southern Ocean is a mixture of all of the world's oceans that surrounds a single isolated land mass. Thus, these polar oceans differ significantly based on their proximity to land and human populations.

Psychrophiles and psychrotolerant microbes have evolved modified gene regulatory systems and metabolic pathways for survival under extreme ecosystems. Fungi in particular play an essential role in this environment, functioning as decomposers, parasites, and mutualists. Fungi are commonly associated with large organisms, such as plants, macroalgae, their woody components, and animals, which are limited in polar environments. These organisms create protected microhabitats that help fungi survive harsh environmental conditions, with an example being the "Lichenosphere" as described by Santiago et al. (2015). Unique natural products have also been found, as organisms chemically adapt to these harsh conditions. For example, an increase in antibiotic activity was observed in the organic extracts of Arctic fungal cultures as the salt concentrations increased, though we should point out that these organisms originate from saline environments, including cold salterns (Sepcic et al. 2011).

The Arctic region is effectively a water basin with an ice covering, with the Arctic Ocean mostly surrounded by the United States (Alaska), Denmark (Greenland), Iceland, Norway, Russia, and Canada. Thus, land and the variable human population have a significant impact on this ocean. Permanent sea ice covers most of this region year-round and sea water temperatures vary from -1.9 °C in winter to 5 °C in summer, a wider temperature range than the Antarctic. These waters are highly stratified into three major water masses with the upper shallow layer being low-salinity layer and the intermediate layer being high-salinity layer with temperatures < 1.5 °C, and the deep salty layer has warmer Atlantic Ocean water (2–4 °C). Thus, the water columns and layers are quite different, and the microbial community is probably quite diverse. The shallow layer is typically frozen for most of the year; the intermediate layer receives water from the ocean inflow from the North Atlantic Current, as well as through the Bering Strait. This environment is strongly influenced by patterns of atmospheric and oceanic circulation that are driven by latitudinal temperature gradients, with seasonal depletion of nutrients.

Given these unique environments, a marine strain (KF970) belonging to the Lindgomycetaceae family, obtained during the 1991 expedition of RV Polarstern to the Greenland coast and the Arctic, was reported to produce the antimicrobial lindgomycin (**111**). Interestingly, another strain (LF327) isolated from a sponge *Halichondria panacea* collected from Kiel Fjord in the Baltic Sea (Germany) also produced the same compound (Wu et al. 2015). The Wu paper states twice that the original microbe (KF970) came from an Antarctic expedition by the Polarstern. From the cruise itinerary, the expedition was from Kiel to the East coast of Greenland and then to Jan Mayen Land, so it was an Arctic microbe. For more information on natural products from the Arctic in general, the publications coming out of the Marbio research group at the Arctic University of Norway are examples of what can be accomplished (Kristoffersen et al. 2018). Also, though they are entitled "from cold water," the two reviews by the Baker group at the University of South Florida, both include small sections on bacteria and fungi, are also indications of what can be found in these "extreme" ecosystems (Lebar et al. 2007; Soldatou and Baker 2017).

The Antarctic marine system is defined by the Southern Ocean, bounded by the Polar Frontal Zone and the frozen continent of Antarctica. Unlike the Arctic, no rivers discharge into the Southern Ocean and it is a major site for the formation of deep-water masses. The Southern Ocean has twice the oceanic surface of the Arctic Ocean, deep, narrow continental shelves, and thinner sea ice, which is expanding, unlike the disappearing ice in the Arctic (Hobbs et al. 2016; McBride et al. 2014). The ocean averages 3000–4000 m depth, to a maximum of 7236 m, with an area of 34.8 million km^2 wide (McBride et al. 2014; Waller et al. 2017). In the summer, a minimum of 7 million km^2 surface ice remains, with greater light exposure than that of the Arctic Ocean. Closer to the seafloor, Antarctic waters are warmer and higher in salinity, but the water column is not as stratified as that of the Arctic Ocean (Downey et al. 2012; Griffiths 2010). Salinity, not temperature, governs its stratification, and there is high nutrient availability. While both the Arctic and Southern Oceans have water temperatures close to freezing (~ -1.9 °C), the Southern Ocean has less variation (0.5–1.5 °C) even in periods of limited ice covering, making this a more thermally stable environment than the Arctic, and in addition, oxygen levels in these waters (> 320 μmol/kg at 50 m) are higher than those of most regions (Orsi and Whitworth 2005).

A number of compounds have been reported from fungal sources isolated from Antarctic marine environments. Penilactones A and B (**112–113**) were reported from a deep sea sediment-derived *P. crustosum* PRB-2 (−526 m in Prydz Bay) (Wu et al. 2012). Their structures, which have opposite stereochemistry, were confirmed in a biomimetic synthesis published a year later (Spence and George 2013). Wu et al. later reported new sesquiterpenes from *Penicillium* sp. PR19N-1 isolated from Prydz Bay sediment (−1000 m) (Wu et al. 2013). These were 1-chloro-3β-acetoxy-7-hydroxy-trinoreremophil-1,6,9-trien-8-one (**114**) and three eremophilane-type sesquiterpenes (**115–117**). This fungal strain was later reported to produce five new eremophilane-type sesquiterpenes (**118–122**) and a new, rare lactam-type eremophilane (**123**) (Lin et al. 2014).

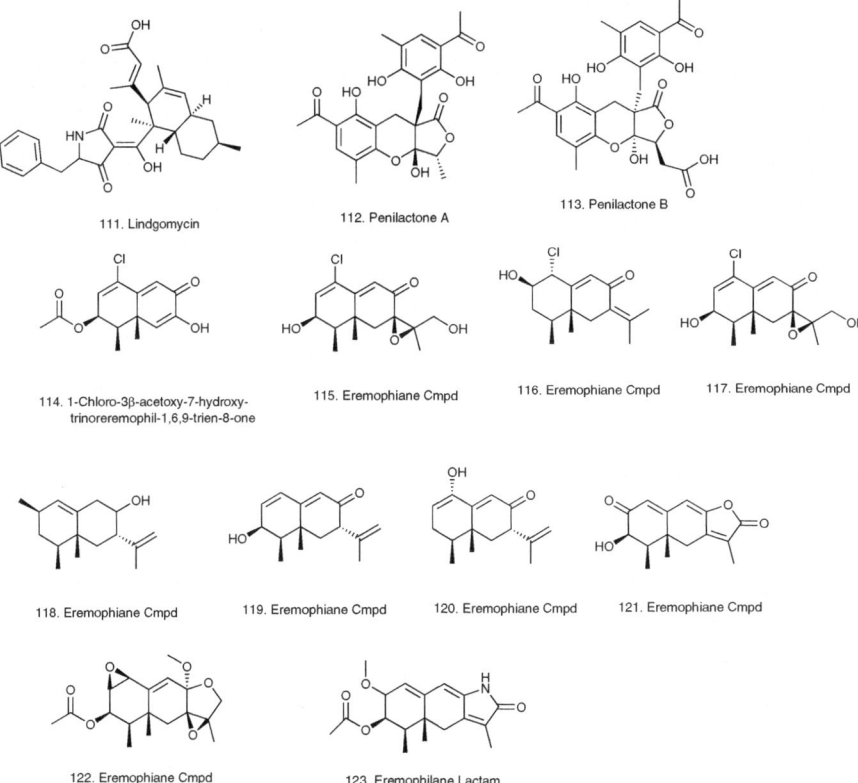

Fig. 18.5 Compounds **111–123**

Before it is too late, we hope to see more fungal products isolated from this area, which is undergoing significant anthropogenic climate change. Antarctica has more endemic species than the Arctic and, with increasing temperatures, will likely be a "hot spot" for species turnover, invasion, and extinction (Griffiths et al. 2017; Smith et al. 2017) (Fig. 18.5; Table 18.5).

18.4 Challenges and Strategies to Discovering Novel Bioactive Molecules from Extremophilic Marine Fungi

There are several reasons why more bioactive secondary metabolites have not been reported from marine fungi (Imhoff 2016). Fungi are morphologically diverse and there are no standardized procedures for their taxonomic identification, which is important for industrial and pharmaceutical applications. Molecular methods have been helpful when there is complete small subunit ribosomal sequence data. Yet, other than sequencing regions of internal transcribed spacers (e.g., ITS1–ITS2)

Table 18.5 Biological activities of compounds **111–123**

Name	Source	#	Bioactivity
Lindgomycin	Lindgomycetaceae family	**111**	Antimicrobial activity
Penilactone A	*P. crustosum* PRB-2	**112**	Weak NF-κB inhibitor
Penilactone B	*P. crustosum* PRB-2	**113**	No bioactivity reported
1-Chloro-3β-acetoxy-7-hydroxy-trinoreremophil-1,6,9-trien-8-one	*Penicillium* sp. PR19N-1	**114**	Cytotoxic; HL-60 and A-549 cell lines
Eremophilane-type sesquiterpene	*Penicillium* sp. PR19N-1	**115**	No bioactivity reported
Eremophilane-type sesquiterpene	*Penicillium* sp. PR19N-1	**116**	No bioactivity reported
Eremophilane-type sesquiterpene	*Penicillium* sp. PR19N-1	**117**	No bioactivity reported
Eremophilane-type sesquiterpene	*Penicillium* sp. PR19N-1	**118**	Cytotoxic; HL-60 and A-549 cell lines
Eremophilane-type sesquiterpene	*Penicillium* sp. PR19N-1	**119**	No bioactivity reported
Eremophilane-type sesquiterpene	*Penicillium* sp. PR19N-1	**120**	No bioactivity reported
Eremophilane-type sesquiterpene	*Penicillium* sp. PR19N-1	**121**	No bioactivity reported
Eremophilane-type sesquiterpene	*Penicillium* sp. PR19N-1	**122**	Cytotoxic; HL-60 and A-549 cell lines
Lactam-type ereophilane	*Penicillium* sp. PR19N-1	**123**	No bioactivity reported

or both ribosomal RNA and protein-coding genes, a species assignment can be unclear when there are no complete ribosomal sequences or verified reference databases. Some researchers use morphology and molecular data to identify a fungal species since the morphological plasticity caused by environmental cues, such as light and temperature, can lead to species exhibiting similar morphotypes. Yet, only a small fraction of marine fungi has been identified on a species level. Morphotypes can be misleading due to hybridization, cryptic speciation, and convergent evolution (Raja et al. 2017). To avoid confusion, post December 31, 2012, a fungus is now only given one name regardless of the stages in its life cycle, and as a result all fungal nomenclature prior to 2013 has to be reconsidered, further confusing taxonomic identification, and leading to problems when the natural products chemistry community attempts to rationalize compound sources. This can lead to the same compound being considered as novel, when in fact it, or close chemical relatives, was reported from terrestrial sources, further compounding the problems involved.

To increase the discovery of more fungal natural products from marine environments, we need high-throughput analytical methods to chemically profile fungal extracts in parallel. To screen numerous isolates, high-throughput metabolomic workflows using liquid chromatography and mass spectrometry have been designed, but they generate large amounts of data. These workflows have even started to incorporate different variables, such as time, genus, and species. While molecular networking sorts through these data to highlight new chemical scaffolds, scientists could also be more targeted in the search for chemical diversity by correlating environment with chemodiversity. More studies are using mass spectrometry to obtain molecular profiles over environmental gradients revealing chemodiverse and phylogenetic "hot spots" (Li et al. 2018; Luzzatto-Knaan et al. 2017). The integration of chemical and spatial data sets have led to the discovery of new metabolites

(Luzzatto-Knaan et al. 2017), and could be extremely useful when looking at fungi in different marine environments. It is all about *location, location, location*. Apparently, environmental factors, such as depth oxygen and nitrate, correlate with the composition of marine fungal communities and explain more variance than geographic distance (Tisthammer et al. 2016). These factors have a 4-dimensional space (season/time, longitudinal and latitudinal dimensions, and depth), and their variation could reveal unique chemical scaffolds. With the explosion of fungal genomes (approximately 8175 organisms sequenced according to the Genomes Online Database in August 2018), we should be able to also connect molecular and geographical information to genomic data (Mukherjee et al. 2017), enabling us to target new fungi instead of continuously isolating the dominant *Penicillium* and *Aspergillus* isolates.

These chemical programs also have to be linked to functional screens for bioactivity rather than a few random screens (those available to the chemist), and bioactivity-driven isolation programs are also required to identify fungal metabolomes of interest. Over 10,000 marine fungal species have been identified, but less than a quarter of those species have been explored for bioactive metabolites.

We also need to target both plentiful and rarer marine fungi, using single-cell genomic sequencing and metagenomic and metatranscriptomic sequencing primers that target the most variable regions of the ribosomal RNA gene. This way, publications are not always on the "easy-to-culture" *Penicillium, Aspergillus, Fusarium,* and *Cladosporium* strains, and we have a more balanced representation of other genera. Importantly, in order to develop systematic methods, the mycologists, ecologists, and natural products research community need to work together. For more information on how natural products chemists and ecologists can share language, resources, and information, one should consult the excellent 2017 review by Reich and Labes (2017).

18.5 Conclusions

Marine natural products are some of the most complex, structurally diverse molecules in chemical space, as they are influenced by their surrounding ecology. Over the last 30 years, there has been an explosion in the number of reports of bioactive compounds from marine fungi due to improved methods for sampling, cultivation, and identification. Examples can be seen in the annual reviews published in Natural Product Reports, initially by Faulkner and then by Blunt et al. (2018).

Due to the limited screening and the sheer time and money required to develop a lead compound, the increase in reports of novel active agents will not become "drug candidates" for a significant number of years. The bottleneck in such programs is the problem of producing enough of a bioactive agent to move through structural characterization and screens to initially develop a viable preclinical candidate. This may well require significant investment in synthetic chemistry as well as large-scale fermentation of precursors. The scientific community, including ecologists, marine mycologists, and (bio)chemists, also needs to come together and decide on how to taxonomically identify fungi. Improved communication will lead to more impactful genomic, metabolomic, bioactivity, and geographic data. Nevertheless, we are hopeful because scientists are now forming fruitful collaborations across disciplines, and there are more research programs centered around marine natural products, from fungi, isolated from both "regular and extreme environments."

References

Abdel-Lateff A, König GM, Fisch KM, Höller U, Jones PG, Wright AD (2002) New antioxidant hydroquinone derivatives from the algicolous marine fungus *Acremonium* sp. J Nat Prod 65:1605–1611

Abraham EP, Newton GGF (1956) Experiments on the degradation of cephalosporin C. Biochem J 62:658–665

Afiyatullov SS, Kuznetsova TA, Isakov VV, Pivkin MV, Prokof'ev NG, Elyakov GB (2000) New diterpenic altrosides of the fungus *Acremonium striatisporum* isolated from a Sea Cucumber. J Nat Prod 63:848–850

Afiyatullov SS, Kalinovsky AI, Kuznetsova TA, Pivkin MV, Prokof'eva NG, Dmitrenok PS, Elyakov GB (2004) New glycosides of the fungus *Acremonium striatisporum* isolated from a Sea Cucumber. J Nat Prod 67:1047–1051

Afiyatullov SS, Kalinovsky AI, Pivkin MV, Dmitrenok PS, Kuznetsova TA (2006) New diterpene glycosides of the fungus *Acremonium striatisporum* isolated from a sea cucumber. Nat Prod Res 20:902–908

Alvi KA, Casey A, Nair BG (1998) Pulchellalactam: a CD45 protein tyrosine phosphatase inhibitor from the marine fungus *Corollospora pulchella*. J Antibiot 51:515–517

Araki K, Saito K, Arimoto H, Uemura D (2004) Enantioselective synthesis of the spirotetracyclic carbon core of mangicols by using a stereoselective transannular Diels–Alder strategy. Angew Chem Int Ed 43:81–84

Babu DC, Rao CB, Venkatesham K, Selvam JJP, Venkateswarlu Y (2014) Toward synthesis of carbasugars (+)-gabosine C, (+)-COTC, (+)-pericosine B, and (+)-pericosine C. Carbohydr Res 388:130–137

Bandani AR, Khambay BPS, Faull JL, Newton R, Deadman M, Butt TM (2000) Production of efrapeptins by *Tolypocladium* species and evaluation of their insecticidal and antimicrobial properties. Mycol Res 104:537–544

Banert K (2012) Acremolin, a stable natural product with an antiaromatic 1H-azirine moiety? A structural reorientation. Tetrahedron Lett 53:6443–6445

Barghoorn ES, Linder DH (1944) Marine fungi: their taxonomy and biology. Farlowia 1:395–467

Bellavita NC, Ceccherelli P, Raffaele M (1970) Structure du virescenoside C, nouveau mettabolite de *Oospora virescens*. Eur J Biochem 15:356–359

Belofsky GN, Anguera M, Jensen PR, Fenical W, Köck M (2000) Oxepinamides A-C and fumiquinazolines H-I: Bioactive metabolites from a marine isolate of a fungus of the genus *Acremonium*. Chem Eur J 6:1355–1360

Biabani MAF, Laatsch H (1998) Advances in chemical studies on low-molecular weight metabolites of marine fungi. J Prakt Chem 340:589–607

Blanchet E, Vansteelandt M, Le Bot R, Egorov M, Guitton Y, Pouchus YF, Grovel O (2014) Synthesis and antiproliferative activity of ligerin and new fumagillin analogs against osteosarcoma. Eur J Med Chem 79:244–250

Blunt JW, Carroll AR, Copp BR, Davis RA, Keyzers RA, Prinsep MR (2018) Marine natural products. Nat Prod Rep 35:8–53

Boot CM, Tenney K, Valeriote FA, Crews P (2006) Highly N-methylated linear peptides produced by an atypical sponge-derived *Acremonium* sp. J Nat Prod 69:83–92

Boot CM, Amagata T, Tenney K, Compton JE, Pietraszkiewicz H, Valeriote FA, Crews P (2007) Four classes of structurally unusual peptides from two marine-derived fungi: structures and bioactivities. Tetrahedron 63:9903–9914

Boufridi A, Quinn RJ (2018) Harnessing the properties of natural products. Annu Rev Pharmacol Toxicol 58:451–470

Boyd DR, Sharma ND, Acaru CA, Malon JF, O'Dowd CR, Allen CCR, Stevenson PJ (2010) Chemoenzymatic synthesis of carbasugars (+)-pericosines A–C from diverse aromatic cis-dihydrodiol precursors. Org Lett 12:2206–2209

Brotzu G (1948) Ricerche su di un nuovo antibiotico. Lav Inst Igiene Cagliari 1948:1–11

Burka LT, Ganguli M, Wilson BJ (1983) Verrucosidin, a tremorgen from *Penicillim verrucosum* var *cyclopium*. J Chem Soc Chem Commun 1983:544–545

Chaichanan J, Wiyakrutta S, Pongtharangkul T, Isarangkul D, Meevootisom V (2014) Optimization of zofimarin production by an endophytic fungus, *Xylaria* sp. Acra L38. Braz J Microbiol 45:287–293

Chakraborty B, Sejpal NV, Payghan PV, Ghoshal N, Sengupta J (2016) Structure-based design-ing of sordarin derivative as potential fungicide with pan-fungal activity. J Mol Graph Model 66:133–142

Chen Z, Song Y, Chen Y, Huang H, Zhang W, Ju J (2012) Cyclic heptapeptides, Cordyheptapeptides C–E, from the marine-derived fungus *Acremonium persicinum* SCSIO 115 and their cytotoxic activities. J Nat Prod 75:1215–1219

Chen W, Tay J-H, Ying J, Sabat M, Yu X-Q, Pu L (2013) A quick access to the spirotricyclic core analogue of mangicol A by a Rh(i)-catalyzed tandem Pauson–Khand/[4+2] cycloaddition. Chem Commun 49:170–172

Cragg GM, Newman DJ (2001) Natural product drug discovery in the next millennium. Pharm Biol 39:8–17

Cui C-B, Kakeya H, Okada G, Onose R, Ubukata M, Takahashi I, Isono K, Osada H (1995) Tryprostatins A and B, novel mammalian cell cycle inhibitors produced by *Aspergillus fumiga-tus*. II. Physicochemical properties and structures. J Antibiot 48:1382–1384

Cui C-B, Kakeya H, Okada G, Osada H (1996a) Novel mammalian cell cycle inhibitors, try-prostratins A, B and other diketopiperazines produced by *Aspergillus fumigatus*. I. Taxonomy, fermentation, isolation and biological properties. J Antibiot 49:527–533

Cui C-B, Kakeya H, Osada H (1996b) Novel mammalian cell cycle inhibitors, spirotryprostatins A and B, produced by *Aspergillus fumigatus*, which inhibit mammalian cell cycle at G2/M phase. Tetrahedron 52:12651–12666

Cui C-B, Kakeya H, Osada H (1996c) Novel mammalian cell cycle inhibitors, tryprostins A, B and other diketopiperazines produced by *Aspergillus fumigatus*. J Antibiot 49:534–540

Cui C-B, Kakeya H, Osada H (1996d) Spirotryprostatin B, a novel mammalian cell cycle inhibitor produced by *Aspergillus fumigatus*. J Antibiot 49:832–835

Cui C-B, Kakeya H, Osada H (1997) Novel mammalian cell cycle inhibitors, cyclotroprostatins A–D, produced by *Aspergillus fumigatus*, which inhibit mammalian cell cycle at G2/M phase. Tetrahedron 53:59–72

Debbab A, Aly AH, Lin WH, Proksch P (2010) Bioactive compounds from marine bacteria and fungi. Microb Biotechnol 3:544–563

Domínguez JM, Martín JJ (1998) Identification of elongation factor 2 as the essential protein tar-geted by sordarins in *Candida albicans*. Antimicrob Agents Chemother 42:2279–2283

Donia M, Hamann MT (2003) Marine natural products and their potential applications as anti-infective agents. Lancet Infect Dis 3:338–348

Donohoe TJ, Blades K, Helliwell M, Waring MJ, Newcombe NJ (1998) The synthesis of (+)-peri-cosine B. Tetrahedron Lett 39:8755–8758

Downey RV, Griffiths HJ, Linse K, Janussen D (2012) Diversity and distribution patterns in high Southern latitude sponges. PLoS One 7:e41672

Fani N, Bordbar AK, Ghayeb Y, Sepehri S (2015) Computational design of tryprostatin-A deriva-tives as novel αβ-tubulin inhibitors. J Biomol Struct Dyn 33:471–486

Fenical W, Jensen PR, Cheng XC (1999) Halimide, a cytotoxic marine natural product, derivatives thereof. Patent, PCT Int. Appl. WO 1999048889 A1, 30 Sept 1999

Fenical W, Jensen PR, Cheng XC (2000) Halimide, a cytotoxic marine natural product, and derivatives thereof. Patent US6069146A, 30 May 2000

Fenical W, LeClair JJ, Hughes CC, Jensen PR, Gaudencio SP, MacMillan JB (2013) The deep oceans as a source for new treatments of cancer. In: Shibaski M, Masamitsu J, Osada H (eds) Chemobiomolecular science. Springer, Berlin, pp 83–91

Fernandez PM, Tabbara SO (2000) Overexpression of the glucose-regulated stress gene GRP78 in malignant but not benign human breast lesions. Breast Cancer Res Treat 59:15–26

Fisher CR, Takai K, Le Bris N (2007) Hydrothermal vent systems. Oceanography 20:14–23

Fröhlich-Nowoisky J, Burrows SM, Xie Z, Engling G, Solomon PA, Fraser MP, Mayol-Bracero OL, Artaxo P, Begerow D, Conrad R, Andreae MO, Després VR, Pöschl U (2012) Biogeography in the air: fungal diversity over land and oceans. Biogeosciences 9:1125–1136

Gal-Hemed I, Atanasova L, Komon-Zelazowska M, Druzhinina IS, Viterbo A, Yarden O (2011) Marine isolates of *Trichoderma* spp. as potential halotolerant agents of biological control for arid-zone agriculture. Appl Environ Microbiol 77:5100–5109

Gonzalez-Medina M, Preto-Martinez FD, Naveja JJ, Mendez-Lucio O, El-Elimat T, Pearce CJ, Oberlies NH, Figueroa M, Medina-Franco JL (2016) Chemoinformatic expedition of the chemical space of natural products. Future Med Chem 8:1399–1412

Griffiths HJ (2010) Antarctic marine biodiversity—what do we know about the distribution of life in the Southern Ocean? PLoS One 5:e11683

Griffiths HJ, Meijers AJS, Bracegirdle TJ (2017) More losers than winners in a century of future Southern Ocean seafloor warming. Nat Clim Change 7:749–754

Gupta S, Krasnoff SB, Roberts DW, Renwick JAA, Brinen LS, Clardy J (1992) Structure of efra-peptins from the fungus *Tolypocladium niveum*: peptide inhibitors of mitochondrial ATPase. J Org Chem 57:2306–2313

Hayakawa Y, Hattori Y, Kawasaki T, Kanoh K, Adachi K, Shizuri Y, Shin-ya K (2008) Efrapeptin J, a new down-regulator of the molecular chaperone GRP78 from a marine *Tolypocladium* sp. J Antibiot 61:365–371

He F, Sun Y-L, Liu K-S, Zhang X-Y, Quian P-Y, Wang Y-F, Qi SH (2012) Indole alkaloids from marine-derived fungus *Aspergillus sydowii* SCSIO 00305. J Antibiot 65:109–111

Hobbs WR, Massom R, Stammerjohn S, Reid P, Williams G, Meier W (2016) A review of recent changes in Southern Ocean sea ice, their drivers and forcings. Glob Planet Change 143:228–250

Imhoff JF (2016) Natural products from marine fungi—still an underrepresented resource. Mar Drugs 14:19

Jain HD, Zhang C, Zhou S, Zhou H, Ma J, Liu X, Liao X, Deveau AM, Dieckhaus CM, Johnson MA, Smith KS, Macdonald TL, Kakeya H, Osada H, Cook JM (2008) Synthesis and structure–activity relationship studies on tryprostatin A, an inhibitor of breast cancer resistance protein. Bioorg Med Chem 16:4626–4651

Jang J-H, Kanoh K, Adachi K, Shizuri Y (2006a) Awajanomycin, a cytotoxic γ-lactone-δ-lactam metabolite from marine-derived *Acremonium* sp. AWA16-1. J Nat Prod 69:1358–1360

Jang J-H, Kanoh K, Adachi K, Shizuri Y (2006b) New dihydrobenzofuran derivative, Awajanoran, from marine-derived *Acremonium* sp. AWA16-1. J Antibiot 59:428–431

Johnson TW Jr, Sparrow FK (1961) Fungi in oceans and estuaries. Publisher J. Cremer, Weinheim, pp 1–668

Jones EBG, Suetrong S, Sakayaroj J, Bahkali AH, Abdel-Wahab MA, Boekhout T, Pang K-L (2015) Classification of marine Ascomycota, Basidiomycota, Blastocladiomycota and Chytridiomycota. Fungal Divers 73:1–72

Julianti E, Oh H, Jang KH, Lee JK, Lee SK, Oh D-C, Oh K-B, Shin J (2011) Acremostrictin, a highly oxygenated metabolite from the marine fungus *Acremonium strictum*. J Nat Prod 74:2592–2594

Julianti E, Oh H, Lee H-S, Oh D-C, Oh K-B, Shin J (2012) Acremolin, a new 1H-azirine metabolite from the marine-derived fungus *Acremonium strictum*. Tetrahedron Lett 53:2885–2886

Kamai Y, Kakuta M, Shibayama T, Fukuoka T, Kuwahara S (2005) Antifungal activities of R-135853, a sordarin derivative, in experimental candiiasis in mice. Antimicrob Agents Chemother 49:52–56

Kameshwar SAK, Qin W (2018a) Comparative study of genome-wide plant biomass-degrading CAZymes in white rot, brown rot and soft rot fungi. Mycology 9:93–105

Kameshwar SAK, Qin W (2018b) Structural and functional properties of pectin and lignin–carbohydrate complexes de-esterases: a review. Bioresour Bioprocess 5:43

Kaneko S, Arai M, Uchida T, Harasaki T, Fukuoka T, Konosu T (2002a) Synthesis and evaluation of N-substituted 1,4-oxazepanyl sordaricins as selective fungal EF-2 inhibitors. Bioorg Med Chem Lett 12:1705–1708

Kaneko S, Uchida T, Shibuya S, Honda T, Kawamoto I, Harasaki T, Fukuoka T, Konosu T (2002b) Synthesis of sordaricin analogues as potent antifungal agents against *Candida albicans*. Bioorg Med Chem Lett 12:803–806

Kanoh K, Kohno S, Asari T, Harada T, Katada J, Muramatsu M, Kawashima H, Sekiya H, Uno I (1997) (−)-Phenylahistin: a new mammalian cell cycle inhibitor produced by *Aspergillus ustus*. Bioorg Med Chem Lett 7:2847–2852

Kanoh K, Kohno S, Katada J, Hayashi Y, Muramatsu M, Un I (1999a) Antitumor activity of phenylahistin in vitro and in vivo. Biosci Biotechnol Biochem 63:1130–1133

Kanoh K, Kohno S, Katada J, Takahashi J, Uno I (1999b) (−)-Phenylahistin arrests cells in mitosis by inhibiting tubulin polymerization. J Antibiot 52:134–141

Kennedy TC, Webb G, Cannell RJ, Kinsman OS, Middleton RF, Sidebottom PJ, Taylor NL, Dawson M, Buss AD (1998) Novel inhibitors of fungal protein synthesis produced by a strain of *Graphium putredinis*. Isolation, characterisation and biological properties. J Antibiot 51:1012–1018

Kohlmeyer J, Kohlmeyer E (1979) Marine mycology: the higher fungi. Academic Press, New York, pp 1–690

Krasnoff SB, Gupta S (1991) Identification and directed biosynthesis of efrapeptins in the fungus *Tolypocladium geodes* gams (Deuteromycotina: Hyphomycetes). J Chem Ecol 17:1953–1962

Krasnoff SB, Gupta S (1992) Efrapeptin production by *Tolypocladium* fungi (Deuteromycotina: Hyphomycetes): intra- and interspecific variation. J Chem Ecol 18:1727–1741

Krishna PR, Prabhakar S, Ramana DV (2012) The first total synthesis of a 12-membered macrolide balticolid. Tetrahedron Lett 53:6843–6845

Kristoffersen V, Rämä T, Isaksson J, Andersen JH, Gerwick WH, Hansen E (2018) Characterization of rhamnolipids Produced by an Arctic marine bacterium from the *Pseudomonas fluorescence* group. Mar Drugs 16:163

Laurent D, Guella G, Roquebert M-F, Farinole F, Mancini I, Pietra F (2000) Cytotoxins, mycotoxins and drugs from a new Deuteromycete, *Acremonium neo-caledoniae*, from the Southwestern lagoon of New Caledonia. Planta Med 66:63–66

Lebar MD, Heimbegner JL, Baker BJ (2007) Cold-water marine natural products. Nat Prod Rep 24:774–797

Li L-S, Hou D-R (2014) Diastereoselective vinylalumination for the synthesis of pericosine A, B and C. RSC Adv 4:91–97

Li X-M, Sun G-X, Chen S-C, Fang Z, Yuan H-Y, Shi Q, Zhu Y-G (2018) Molecular chemodiversity of dissolved organic matter in paddy soils. Environ Sci Technol 52:963–971

Liberra K, Jansen R, Lindequist U (1998) Corollosporine, a new phthalide derivative from the marine fungus *Corollospora maritima* Werderm. 1069. Pharmazie 53:578–581

Lin A, Wu G, Gu Q, Zhu T, Li D (2014) New eremophilane-type sesquiterpenes from an Antarctic deep-sea derived fungus, *Penicillium* sp. PR19 N-1. Arch Pharm Res 37:839–844

López-García P, Philippe H, Gail F, Moreira D (2003) Autochthonous eukaryotic diversity in hydrothermal sediment and experimental microcolonizers at the Mid-Atlantic Ridge. Proc Natl Acad Sci U S A 100:697–702

Luzzatto-Knaan T, Garg N, Wang M, Glukhov E, Peng Y, Ackermann G, Amir A, Duggan BM, Ryazanov S, Gerwick L, Knight R, Alexandrov T, Bandeira N, Gerwick WH, Dorrestein PC (2017) Digitizing mass spectrometry data to explore the chemical diversity and distribution of marine cyanobacteria and algae. eLife 6:e24214

McBride MM, Dalpadado P, Drinkwater KF, Godø OR, Hobday AJ, Hollowed AB, Kristiansen T, Murphy EJ, Ressler PH, Subbey S, Hofmann EE, Loeng H (2014) Krill, climate, and contrasting future scenarios for Arctic and Antarctic fisheries. ICES J Mar Sci 71:1934–1955

McKindles K, Tiquia-Arashiro SM (2012) Functional gene arrays for analysis of microbial communities on ocean platform. In: Tiquia-Arahiro SM (ed) Molecular biological technologies for ocean sensing. Humana Press, Totowa, NJ, Chap. 9, 169–201 p

Mikolasch A, Schauer F (2009) Fungal laccases as tools for the synthesis of new hybrid molecules and biomaterials. Appl Microbiol Biotechnol 82:605–624

Moussaief M, Jacques P, Schaarwachter P, Budzikiewicz H, Thonart P (1997) Cyclosporin C is the main antifungal compound produced by *Acremonium luzulae*. Appl Environ Microbiol 63:1739–1743

Mukherjee S, Stamatis D, Bertsch J, Ovchinnikova G, Verezemsk O, Isbandi M, Thomas AD, Ali R, Sharma K, Kyrpides NC, Reddy TBK (2017) Genomes onLine database (GOLD) v.6: data updates and feature enhancements. Nucleic Acids Res 45:D446–D456

MuniRaju C, Rao JP, Rao BV (2012) Stereoselective synthesis of (+)-pericosine B and (+)-pericosine C using ring closing metathesis approach. Tetrahedron Asymmetry 23:86–93

Nagano Y, Nagahama T, Hatada Y, Nunoura T, Takami H, Miyazaki J, Takai K, Horikoshi K (2010) Fungal diversity in deep-sea sediments—the presence of novel fungal groups. Fungal Ecol 3:316–325

Namikoshi M, Akano K, Meguro S, Kasuga I, Mine Y, Takahashi T, Kobayashi H (2001) A new macrocyclic trichothecene, 12,13-deoxyroridin E, produced by the marine-derived fungus *Myrothecium roridum* collected in Palau. J Nat Prod 64:396–398

Newman DJ (2016) Predominately uncultured microbes as sources of bioactive agents. Front Microbiol 7:1832

Newton GGF, Abraham EP (1955) Cephalosporin C, a new antibiotic containing sulphur and D-α-aminoadipic acid. Nature 175:548

Newton GGF, Abraham EP (1956) Isolation of cephalosporin C, a penicillin-like antibiotic containing d-α-aminoadipic acid. Biochem J 62:651–658

Nicholson B, Lloyd GK, Miller BR, Palladino MA, Kiso Y, Hayash Y, Neuteboom STC (2006) NPI-2358 is a tubulin-depolymerizing agent: in-vitro evidence for activity as a tumor vascular-disrupting agent. Anti-Cancer Drugs 17:25–31

Numata A, Iritani M, Yamada T, Minoura K, Matsumura E, Yamori T, Tsuruo T (1997) Novel anti-tumour metabolites produced by a fungal strain from a sea hare. Tetrahedron Lett 38:8215–8218

Ogita T, Hayashi A, Sato S, Furay K (1987) Antibiotic Zofimarin, manufacture by *Zofiela marina* SNAK21274. JP 624092, 21 Feb 1987

Oh D-C, Jensen PR, Fenical W (2006) Zygosporamide, a cytotoxic cyclic depsipeptide from the marine-derived fungus *Zygosporium masonii*. Tetrahedron Lett 47:8625–8628

Orsi AH, Whitworth T (2005) Hydrographic atlas of the world ocean circulation experiment (WOCE), Southern Ocean, vol 1. Scripps Institution Oceanography, La Jolla, CA

Overy DP, Bayman P, Kerr RG, Bills GF (2014) An assessment of natural product discovery from marine (sensu strictu) and marine-derived fungi. Mycology 5:145–167

Pan C, Shi Y, Auckloo NB, Chen X, Chen AC-T, Tao X, Wu B (2016) An unusual conformational isomer of verrucosidin backbone from a hydrothermal vent fungus, *Penicillium* sp. Y-50-10. Mar Drugs 14:156

Pan C, Shi Y, Chen X, Chen AC-T, Tao X, Wu B (2017) New compounds from a hydrothermal vent crab-associated fungus *Aspergillus versicolor XZ-4*. Org Biomol Chem 15:1155–1163

Raja HA, Miller AN, Pearce CJ, Oberlies NH (2017) Fungal identification using molecular tools: A primer for the natural products research community. J Nat Prod 80:756–770

Rämä T, Davey ML, Nordén J, Halvorsen R, Blaalid R, Mathiassen GH, Alsos IG, Kauserud H (2016) Fungi sailing the Arctic Ocean: Speciose communities in North Atlantic driftwood as revealed by high-throughput amplicon sequencing. Microb Ecol 72:295–304

Ramirez-Llodra E, Brandt A, Danovaro R, De Mol B, Escobar E, German CR, Levin LA, Martinez Arbizu P, Menot L, Buhl-Mortensen P, Narayanaswamy BE, Smith CR, Tittensor DP, Tyler PA, Vanreusel A, Vecchione M (2010) Deep, diverse and definitely different: unique attributes of the world's largest ecosystem. Biogeosciences 7:2851–2899

Reddy YS, Kadigachalam P, Basak RK, John Pal AP, Vankar YD (2012) Total synthesis of (+)-pericosine B and (+)-pericosine C and their enantiomers by using the Baylis–Hillman reaction and ring-closing metathesis as key steps. Tetrahedron Lett 53:132–136

Rédou V, Navarri M, Meslet-Cladière L, Barbier G, Burgaud G (2015) Species richness and adaptation of marine fungi from deep-subseafloor sediments. Appl Environ Microbiol 81:3571–3583

Reich M, Labes A (2017) How to boost marine fungal research: A first step towards a multidisciplinary approach by combining molecular fungal ecology and natural products chemistry. Mar Genomics 36:57–75

Renner MK, Jensen PR, Fenical W (1998) Neomangicols: Structures and absolute stereochemistries of unprecedented halogenated sesterterpenes from a marine fungus of the genus *Fusarium*. J Org Chem 63:8346–8354

Renner MK, Jensen PR, Fenical W (2000) Mangicols: Structures and biosynthesis of a new cass of sesterterpene polyols from a marine fungus of the genus *Fusarium*. J Org Chem 65:4843–4852

Roth FJ Jr, Orpurt PA, Ahearn DG (1964) Occurrence and distribution of fungi in a subtropical marine environment. Can J Bot 42:375–383

Santiago IF, Soares MA, Rosa CA, Rosa LH (2015) Lichensphere: a protected natural microhabitat of the non-lichenised fungal communities living in extreme environments of Antarctica. Extremophiles 19:1087–1097

Sato A, Takahashi, K, Ogita T, Sugano M, Kodama K (1985) Marine Natural Products. Annu. Rpt. Sankyo Res. Lab. pp 1–58

Sepcic K, Zalar P, Gunde-Cimerman N (2011) Low water activity Induces the production of bioactive metabolites in halophilic and halotolerant fungi. Mar Drugs 9:43–58

Shushni MAM, Mentel R, Lindequist U, Jansen R (2009) Balticols A–F, new naphthalenone derivatives with antiviral activity, from an ascomycetous fungus. Chem Biodivers 6:127–137

Singh P (2012) Fungal diversity in deep-sea sediments revealed by culture-dependent and culture-independent approaches. Fungal Ecol 5:543–553

Skropeta D, Wei L (2014) Recent advances in deep-sea natural products. Nat Prod Rep 31:999–1025

Smith KE, Aronson RB, Steffel BV, Amsler MO, Thatje S, Singh H, Anderson J, Brothers CJ, Brown A, Ellis DS, Havenhand JN, James WR, Moksnes P-O, Randolph AW, Sayre-McCord T, McClintock JB (2017) Climate change and the threat of novel marine predators in Antarctica. Ecosphere 8:e02017

Soldatou S, Baker BJ (2017) Cold-water marine natural products, 2006 to 2016. Nat Prod Rep 34:585–626

Spence JTJ, George JH (2013) Biomimetic total synthesis of ent-Penilactone A and Penilactone B. Org Lett 15:3891–3893

Stodůlková E, Man F, Kuzma M, Černý J, Císařová I, Kubátová A, Chudíčková M, Kolařík M, Flieger M (2015) A highly diverse spectrum of naphthoquinone derivatives produced by the endophytic fungus *Biatriospora* sp. CCF 4378. Folia Microbiol (Praha) 60:259–267

Takami H, Inoue A, Fuji F, Horikoshi K (1997) Microbial flora in the deepest sea mud of the Mariana Trench. FEMS Microbiol Lett 152:279–285

Tiquia SM, Mormile M (2010) Extremophiles—a source of innovation for industrial and environmental applications. Environ Technol 31(8–9):823

Tiquia-Arashiro SM (2012) Molecular biological technologies for ocean sensing. Humana Press, Totowa, NJ, 295 p

Tisthammer KH, Cobian GM, Amend AS (2016) Global biogeography of marine fungi is shaped by the environment. Fungal Ecol 19:39–46

Tripathi S, Shaikha AC, Chen C (2011) Facile carbohydrate-based stereocontrolled divergent synthesis of (+)-pericosines A and B. Org Biomol Chem 9:7306–7308

Usami Y, Mizuki K, Kawahata R, Shibano M, Sekine A, Yoneyama H, Harusawa S (2017) Synthesis of natural O-linked carba-disaccharides, (+)- and (−)-pericosine E, and their analogues as α-glucosidase inhibitors. Mar Drugs 15:22

Usui T, Kondoh M, Cui C-B, Mayumi T, Osada H (1998) Tryprostatin A, a specific and novel inhibitor of microtubule assembly. Biochem J 333:543–548

van Dover CL (2000) The ecology of deep-sea hydrothermal vents. Princeton University Press, Princeton, NJ, pp 1–425

van Dover CL, German CR, Speer KG, Parson LM, Vrijenhoek RC (2002) Evolution and biogeography of deep-sea vent and seep invertebrates. Science 295:1253–2157

Vansteelandt M, Blanchet E, Egorov M, Petit F, Toupet L, Bondon A, Monteau F, Bizec B, Thomas OP, Pouchus YF, Bot RL, Grovel O (2013) Ligerin, an antiproliferative chlorinated sesquiterpenoid from a marine-derived *Penicillium* strain. J Nat Prod 76:297–301

Vicente F, Basilio A, Platas G, Collado J, Bills GF, González Del Val A, Martín JJ, Tormo JR, Harris GH, Zink DL, Justice M, Nielsen Kahn J, Peláez F (2009) Distribution of the antifungal agents sordarins across filamentous fungi. Mycol Res 113:754–770

Waller CL, Griffiths HJ, Waluda CM, Thorpe SE, Loaiza I, Moreno B, Pacherres CO, Hughes KA (2017) Microplastics in the Antarctic marine system: An emerging area of research. Sci Total Environ 598:220–227

Wang Y-F, Li Z-L, Zhang L-M, Wu X, Zhang L, Pei Y-H, Jing Y-K, Hua H-M (2012) 2,5-Diketopiperazines from the marine-derived fungus *Aspergillus fumigatus* YK-7. Chem Biodivers 9:385–393

Weiss RF, Lonsdale P, Lupton JE, Bainbridge AE, Craig H (1977) Hydrothermal plumes in the Galapagos Rift. Nature 267:600–603

Wetwitaklung P, Thavanaspong N, Charoenteeraboon J (2009) Chemical constituents and antimicrobial activity of essential oil and extracts of heartwood of *Aquilaria crassna* obtained form water distillation and supercritical fluid carbon dioxide extraction. Silpakorn Univ Sci Technol J 3:25–33

Whang K, Cooke RJ, Okay G, Cha JK (1990) Total syntheis of (+)-verrucosidin. J Am Chem Soc 112:8985–8987

Woehlecke H, Osada H, Herrmann A, Lage H (2003) Reversal of breast cancer resistance protein–mediated drug resistance by tryprostatin A. Int J Cancer 107:721–728

Wu Y, Dockendorff C (2018) Synthesis of a novel bicyclic scaffold inspired by the antifungal natural product sordarin. Tetrahedron Lett 59:3373–3376

Wu G, Ma H, Zhu T, Li J, Gu Q, Li D (2012) Penilactones A and B, two novel polyketides from Antarctic deep-sea derived fungus *Penicillium crustosum* PRB-2. Tetrahedron 68:9745–9749

Wu G, Lin A, Gu Q, Zhu T, Li D (2013) Four new chloro-eremophilane sesquiterpenes from an Antarctic deep-sea derived fungus, *Penicillium* sp. PR19N-1. Mar Drugs 11:1399–1408

Wu B, Wiese J, Labes A, Kramer A, Schmaljohann R, Imhoff JF (2015) Lindgomycin, an unusual antibiotic polyketide from a marine fungus of the Lindgomycetaceae. Mar Drugs 13:4617–4632

Xu J, Takasaki A, Kobayashi H, Oda T, Yamada J, Mangindaan REP, Ukai K, Nagai H, Namikoshi M (2006) Four new macrocyclic trichothecenes from two strains of marine-derived fungi of the genus *Myrothecium*. J Antibiot 59:451–455

Yamada T, Iritani M, Ohishi H, Tanaka K, Minoura K, Doi M, Numata A (2007) Pericosines, antitumour metabolites from the sea hare-derived fungus *Periconia byssoides*. Structures and biological activities. Org Biomol Chem 5:3979–3986

Yamakawa T, Ideue E, Iwaki Y, Sato A, Tokuyama H, Shimokawa J, Fukuyama T (2011) Total synthesis of tryprostatins A and B. Tetrahedron 67:6547–6560

Ying J, Pu L (2014) A facile asymmetric approach to the multicyclic core structure of mangicol A. Chem Eur J 20:16301–16307

Zhang X-Y, Tang G-L, Xu X-Y, Nong X-H, Qi S-H (2014) Insights into deep-sea sediment fungal communities from the East Indian Ocean using targeted environmental sequencing combined with traditional cultivation. PLoS One 9:e109118

Zhao S, Smith KS, Deveau AM, Dieckhaus CM, Johnson MA, Macdonald TL, Cook JM (2002) Biological activity of the tryprostatins and their diastereomers on human carcinoma cell lines. J Med Chem 45:1559–1562

ZoBell CE, Johnson FH (1949) The influence of hydrostatic pressure on the growth and viability of terrestrial and marine bacteria. J Bacteriol 57:179–189

Chapter 19
Synthesis of Metallic Nanoparticles by Halotolerant Fungi

Sonia M. Tiquia-Arashiro (ID)

19.1 Introduction

Halotolerance is the adaptation of living organisms to conditions of high salinity. Halotolerant organisms tend to live in hypersaline environments, along with halophilic organisms. However, they are different from halophilic organisms in that they do not require elevated concentrations of salt to grow. Microbial diversity studies of hypersaline environments including saltern ponds worldwide (Gunde-Cimerman et al. 2000; Butinar et al. 2005; Diaz-Munoz and Montalvo-Rodriguez 2005; Cantrell et al. 2006; Nayak et al. 2012), the Great Salt Lake (Baxter et al. 2005), the Dead Sea (Buchalo et al. 1998; Kis-Papo et al. 2003a, b; Wasser et al. 2003; Bodaker et al. 2010; Nazareth et al. 2012), saline lakes in Inner Mongolia (Pagaling et al. 2009), African soda lakes (Jones and Grant 1999), deep-sea brines (van der Wielen et al. 2005), salt pans or salt marshes (Setati 2010), Mono Lake in California (Steiman et al. 2004), coastal environments of Arctics (Gunde-Cimerman et al. 2005), saline soils of Soos in the Czech Republic (Hujslova et al. 2010), and many others have led to the isolation of halophiles and halotolerant microorganisms.

Fungi hypersaline environments are mostly halotolerant (Gunde-Cimerman et al. 2009; Zajc et al. 2012) rather than halophilic. Thus, they constitute a relatively large and constant part of hypersaline environment communities. Well-studied examples include the yeast *Debaryomyces hansenii* and black yeasts *Aureobasidium pullulans* (Gunde-Cimerman et al. 2009). Surprisingly, more and more cases are being reported of the isolation of halophilic and halotolerant microorganisms from low-salinity environments (Abdel-Hafez et al. 1978; Tiquia et al. 2007; Gunde-Cimerman et al. 2009; Tiquia 2010; Gonsalves et al. 2012). Halophilic and halotolerant fungi

S. M. Tiquia-Arashiro (✉)
Department of Natural Sciences, University of Michigan-Dearborn, Dearborn, MI, USA
e-mail: smtiquia@umich.edu

© Springer Nature Switzerland AG 2019
S. M. Tiquia-Arashiro, M. Grube (eds.), *Fungi in Extreme Environments: Ecological Role and Biotechnological Significance*,
https://doi.org/10.1007/978-3-030-19030-9_19

are often found in unexpected environments such as domestic dishwashers and polar ice, and even on spider webs in desert caves (Gunde-Cimerman et al. 2009).

Halotolerant fungi represent a versatile reservoir of bioactive metabolites and potential source of halotolerant genes that could be used in biotechnology (Tiquia 2010; Tiquia and Mormile 2010). They are a valuable resource of enzymes and other metabolites with stability in harsh conditions of pH, temperature, and/or ionic strength (Margesin and Schinner 2001; Oren 2010). Hence, their use as biocatalysts in the presence of novel nanomaterials is attractive. Combining these bioactive molecules with various nanomaterials like thin layers, nanotubes, and nanospheres results in novel compounds harboring both biological properties of biomolecules and physicochemical characteristics of nanomaterials. While the biosynthesis of nanoparticles in bacteria is well understood (Rai et al. 2012), very few fungal (halotolerant) species have been investigated so far for nanoparticle biosynthesis. In general, the microbiology of fungi is much less investigated, mainly because fungi are difficult to characterize, as their structure complicates the microscopic and mechanistic studies that are required for nanoparticle characterization in it. However, fungi may have some advantages over bacteria for bioprocess, including nanoparticle biosynthesis. Fungi also harbor untapped biological diversity and may provide novel metal reductases for metal detoxification and bioreduction. This review focuses on halotolerant fungi that have been exploited for nanomaterial synthesis; the mechanisms in the nanomaterial fabrication; and possible applications.

19.2   Synthesis of Nanoparticles by Halotolerant Fungi

Reports on nanoparticle synthesis by halotolerant fungi are mostly confined to metallic nanoparticles. These halotolerant fungi capable of synthesizing metallic nanoparticles include *Penicillium fellutatum* (Kathiresan et al. 2009); *Aspergillus niger* (Kathiresan et al. 2009); *Pichia capsulata* (Manivannan et al. 2010); *Yarrowia lipolytica* (Bankar et al. 2009; Pawar et al. 2012); *Rhodosporidium diobovatum* (Seshadri et al. 2011); *Schizosaccharomyces pombe* (Kowshik et al. 2002a, b); *Thraustochytrium* sp. (Asmathunisha and Kathiresan 2013); *Schwanniomyces occidentalis* (Mohite et al. 2016); and *Williopsis saturnus* (Mohite et al. 2017).

Metallic nanoparticles have fascinated scientists for over a century and are now heavily utilized in biomedical sciences and engineering. They are a focus of interest because of their huge potential in nanotechnology (Tiquia-Arashiro and Rodrigues 2016a). Metallic nanoparticles have possible applications in diverse areas such as electronics, cosmetics, coatings, packaging, and biotechnology (Tiquia-Arashiro and Rodrigues 2016b, c). For example, nanoparticles can be induced to merge into a solid at relatively lower temperatures, often without melting, leading to improved and easy-to-create coatings for electronics applications. Typically, nanoparticles possess a wavelength below the critical wavelength of light. This renders them transparent, a property that makes them very useful for applications in cosmetics, coatings, and packaging. Metallic nanoparticles can also be attached to single strands of DNA nondestructively, which opens avenues for medical diagnostic applications (Tiquia-Arashiro and Rodrigues 2016c).

19.2.1 Synthesis of Gold (Au) Nanoparticles

Gold nanoparticles (AuNPs) have attracted attention in biotechnology due to their unique optical and electrical properties, high chemical and thermal ability, and good biocompatibility and potential applications in various life sciences-related applications including biosensing, bioimaging, and drug delivery for cancer diagnosis and therapy (Jiang et al. 2013). Covalently modified gold nanoparticles have attracted a great deal of interest as a drug delivery vehicle. Their predictable and reliable surface modification chemistry, usually through gold-thiol binding, makes the desired functionalization of nanoparticles quite possible and accurate. Recently, many advancements have been made in biomedical applications of AuNPs with better biocompatibility in disease diagnosis and therapeutics (Tiquia-Arashiro and Rodrigues 2016b). AuNPs can be prepared and conjugated with many functionalizing agents, such as polymers, surfactants, ligands, dendrimers, drugs, DNA, RNA, proteins, peptides, or oligonucleotides. Overall, AuNPs is a promising vehicle for drug delivery and therapies.

Yarrowia lipolytica, a halotolerant yeast (Gunde-Cimerman and Zalar 2014), is known to synthesize AuNPs (Apte et al. 2013a; Nair et al. 2013; Tiquia-Arashiro and Rodrigues 2016d). Melanin (a dark-colored pigment from this yeast) plays an important role in its ability to synthesize nanoparticles. Since the inherent content of melanin in *Y. lipolytica* is low, the yeast is induced to overproduce melanin by incubation with a precursor, L-3,4-dihydroxyphenylalanine (L-DOPA). This process mediates the rapid formation of AuNPs. The AuNPs display antibiofilm activity against pathogenic bacteria (Apte et al. 2013a). They also display effective antifungal properties (Apte et al. 2013b). In addition to AuNPs, some strains of *Y. lipolytica* (e.g., *Y. lipolytica* NCIM 3589) can also synthesize CdO and CdS nanostructures in a cell-associated and extracellular manner (Pawar et al. 2012).

19.2.2 Synthesis of Silver (Ag) Nanoparticles

Silver nanoparticles (AgNPs) are already being commercially used as antimicrobial agents. For example, silver NPs are currently found in surgically implanted catheters to reduce the infections caused during surgery, in toys, personal care products, and silverware. The reason for using silver for antimicrobial applications is because silver possesses antifungal, antibacterial, anti-inflammatory, and anticancer effects (Li et al. 2010, 2014; Chernousova and Epple 2013; Tiquia-Arashiro and Rodrigues 2016c; Zhang et al. 2016).

The halotolerant fungus *Penicillium fellutanum* (De Hoog et al. 2005) can produce AgNPs at a faster rate by extracellular means (Kathiresan et al. 2009). Silver nanoparticles are synthesized within 10 min of silver ions being exposed to the *P. fellutanum* culture filtrate. The increase in color intensity of culture filtrate corresponds to the increase in number of nanoparticles formed by reduction of silver ions (Wang et al. 2009). The AgNPs obtained from this assay has a good monodispersity, with maximum synthesis occurring at pH 6.0, temperature of 5 °C, 24 h of incubation time, and silver nitrate concentration of 1 mM and 0.3% NaCl. Most of the AgNPs generated by *P. fellutanum* are spherical in shape with size ranging from 5 to 25 nm. In this study, the enzyme nitrate reductase is secreted by the *P. fellutanum* biomass and is involved in the reduction of the silver ions (Wang et al. 2009).

Other halotolerant fungi capable of synthesizing AgNPs include *Pichia capsulata* (Manivannan et al. 2010), *Aspergillus niger* (Zomorodian et al. 2016), *Yarowinia lipotyca* (Bankar et al. 2009), *Thraustochytrium* sp. (Asmathunisha and Kathiresan 2013), and *Schwanniomyces occidentalis* (Mohite et al. 2016).

19.2.3 Synthesis of Cadmium Sulfide (CdS) Nanoparticles

Cadmium sulfide (CdS) is a II–VI semiconductor which has been synthesized by microorganisms (Prasad and Jha 2010; Kowshik et al. 2002a). It is insoluble in water but soluble in dilute mineral acids. CdS has a bandgap energy of 2.42 eV (Zhang et al. 2007), at room temperature, and it shows great potential for uses in photochemical catalysis, solar cells, nonlinear optical materials, and various luminescence devices (Ma et al. 2007; Tiquia-Arashiro and Rodrigues 2016c). CdS nanocrystals have generated great interest due to their unique size-dependent chemical and physical properties (Zhao et al. 2006). Thus, extensive research has focused on the synthesis of various CdS nanostructures. In the classical studies by Dameron et al. (1989), it was shown that the halotolerant yeasts *Schizosaccharomyces pombe* can produce of CdS nanocrystallites when challenged with cadmium in solution. Short chelating peptides of general structure (γ-Glu-Cys)n-Gly control the nucleation and growth of CdS crystallites to peptide-capped intracellular particles of diameter 20 Å. These quantum CdS crystallites are more monodisperse than CdS particles synthesized chemically. X-ray data indicate that, at this small size, the CdS structure differs from that of bulk CdS and tends towards a six-coordinate rock-salt structure (Dameron et al. 1989).

19.2.4 Synthesis of Lead Sulfide (PbS) Nanoparticles

Semiconductor PbS nanoparticles have attracted great attention in recent decades because of their interesting optical and electronic properties (Bai and Zhang 2009). *Rhodospiridium diobovatum* synthesizes PbS nanoparticles intracellularly with the help of nonprotein thiols (Seshadri et al. 2011). The nanoparticles are in the range of 2–5 nm. Elemental analysis by energy dispersive X-ray (EDAX) reveals that the particles are composed of lead and sulfur in a 1:2 ratio, and that they are capped by a sulfur-rich peptide. Quantitative study of lead uptake through atomic absorption spectrometry reveals that 55% of lead in the medium accumulated in the exponential phase, whereas a further 35% accumulated in the stationary phase; thus, the overall recovery of PbS nanoparticles is 90%. The lead-exposed *R. diobovatum* displayed a marked increase (280% over the control) in nonprotein thiols in the stationary phase. A sulfur-rich peptide is suggested to be the capping agent. In the presence of lead, *R. diobovatum* produces increasing amount of nonprotein thiols during the stationary phase, which are possibly involved in forming the nanoparticles (Seshadri et al. 2011). *Schizosaccharomyces pombe* is another halotolerant fungus capable of synthesizing PbS nanoparticles (Kowshik et al. 2002b).

19.2.5 Synthesis of Zinc Oxide (ZnO) Nanoparticles

Zinc oxide (ZnO) NPs have unique optical and electrical properties. As a wide bandgap semiconductor, they have found more uses in biosensors, nanoelectronics, and solar cells. These NPs are being used in the cosmetic and sunscreen industry due to their transparency and ability to reflect, scatter, and absorb UV radiation and as food additives. Furthermore, zinc oxide NPs are also being considered for use in next-generation biological applications including antimicrobial agents, drug delivery, and bioimaging probes (Jayaseelan et al. 2012). The potential ability of the halotolerant fungi *Pichia kudriavzevii* (Cai et al. 2014) in the synthesis of zinc oxide nanoparticles (ZnO-NPs) was explored recently (Boroumand Moghaddam et al. 2017). The ZnO nanoparticles synthesized by *P. kudriavzevii* possess hexagonal wurtzite structure with an average crystallite size of ~10–61 nm. They are less toxic and displayed antioxidant and antibacterial activities and show strong 1,1-diphenyl-2-picrylhydrazyl (DPPH) radical effect with a dose-dependent activity. The synthesized ZnO nanoparticles also displayed antibacterial activity against both Gram-positive (*Staphylococcus aureus*, *Bacillus subtilis*, and *Staphylococcus epidermidis*) and Gram-negative (*Escherichia coli* and *Serratia marcescens*) bacteria.

19.3 Nanoparticle Formation Mechanisms in Fungi

The nanoparticles are produced either intracellularly or extracellularly. In case of intracellular synthesis, the nanoparticles are produced inside the cells by the reductive pathways. The synthesis occurs when the metal (e.g., Au^{3+}) concentration available is high and when the membrane integrity is compromised to allow for metal ion diffusion within the cell. In extracellular synthesis, the nanoparticles are produced extracellularly when the cell wall reductive enzymes or soluble secreted enzymes are extracted outside the cell and are involved in the reductive process of metal ions. The extracellular biosynthesis of metallic nanoparticles is similar to the enzymatic machinery required for metal detoxification. Overall, nanoparticle biosynthesis is essentially a reduction process followed by a stabilization step (capping). There is no evidence available yet showing that fungi use biosynthetic nanoparticles for their metabolism. The biosynthesis of metal NPs by fungi is a function of heavy metal toxicity resistance mechanisms, whereby toxic metals are converted to nontoxic species and precipitated as metal clusters of nanoscale dimension and defined shape (Narayanan and Sakthivel 2010). Resistance mechanisms include redox enzymes that convert toxic metal ions to inert forms, structural proteins that bind protein. It is proposed that such mechanism works to coordinate synthesis.

One of the enzymes involved in the biosynthesis of metal nanoparticles is the nitrate reductase which reduces the metal ions (Me^{1+}) to the metallic form (Me^0). This enzyme is a NADH- and NADPH-dependent enzyme. Some fungi are known to secrete cofactor NADH- and NADH-dependent enzymes that can be responsible for the biological reduction of Me^{1+} to Me^0 and the subsequent formation of nanoparticles. This reduction is initiated by electron transfer from the NADH by NADH-dependent reductase as electron carrier during which the metal ions gain electrons and are therefore reduced to Me^0. Synthesis of silver nanoparticles using α-NADPH-dependent nitrate reductase and phytochelatin in vitro has been demonstrated by Anil Kumar et al. (2007). The silver ions are reduced in the presence of nitrate reductase, leading to the formation of a stable silver hydrosol 10–25 nm in diameter and stabilized by the capping peptide. Nitrate reductase is suggested to initiate nanoparticle formation by many fungi including *Penicillium* species, while several enzymes, α-NADPH-dependent reductases, nitrate-dependent reductases, and an extracellular shuttle quinone, are implicated in silver NP synthesis for *Fusarium oxysporum*. Jain et al. (2011) indicated that silver NP synthesis for *Aspergillus flavus* occurs initially by a 33 kDa protein followed by a protein (cystein and free amine groups) electrostatic attraction which stabilizes the nanoparticle by forming a capping agent (Soni and Prakash 2011). Several researchers supported nitrate reductase for extracellular synthesis of metallic NPs (Vigneshwaran et al. 2006; Wang et al. 2009; Deepa and Panda 2014; Siddiqi and Husen 2016; Boroumand Moghaddam et al. 2017).

Cadmium sulfide nanoparticle synthesis by yeast involves sequestration of Cd^{2+} by glutathione-related peptides followed by reduction within the cell. Ahmad et al. (2002) reported that cadmium sulfide nanoparticle synthesis by *Fusarium oxysporum* was based on a sulfate reductase (enzyme) process. The nanoparticle formation proceeds by release of sulfate reductase enzymes by *F. oxysporum*, conversion of sulfate ions to sulfide ions that subsequently react with aqueous Cd^{2+} ions to yield highly stable CdS nanoparticles. While the reduction of sulfate to sulfite is known in sulfate-reducing bacteria (which are strictly anaerobic), this is the first report on the secretion of sulfate-reducing enzymes by a fungus. The extracellular synthesis of AgNP by *P. chrysosporium* is attributed to laccase, while intracellular gold nanoparticle synthesis was attributed to ligninase (Sanghi et al. 2011).

Fungi have several advantages over bacteria for NP biosynthesis. They secrete larger amounts of extracellular proteins with diverse functions. The so-called secretome in fungi include all the secreted proteins into the extracellular space (Girard et al. 2013). The high concentration of the fungal secretome has been used for industrial production of homologous and heterologous proteins. For instance, the expression of a functionally active class I fungal hydrophobin from the entomopathogenic fungus *Beauveria bassiana* has been reported (Kirkland and Keyhani 2011). The tripeptide glutathione is a well-known reducing agent involved in metal reduction and is known to participate in cadmium sulfide (CdS) biosynthesis in yeasts and fungi (Chen et al. 2009). However, the knowledge of the fungal secretome is still largely underexplored especially for halophilic and halotolerant fungi. The large and relatively unexplored fungal secretome is an advantage because of the role of extracellular proteins and enzymes it generates have in metal reduction and nanoparticle synthesis.

19.4 Potential Applications

Bionanoparticles have found uses in biomedical and environmental fields. In the biomedical field, these nanoparticles have been investigated for antimicrobial applications (Sondi and Salopek-Sondi 2004), biosensing (Yu et al. 2003; Mckindles and Tiquia-Arashiro 2012; Tiquia-Arashiro 2012), imaging (Boisselier and Astruc 2009), and drug delivery (Muller et al. 2013). In the environmental field, nanoparticles have been investigated for applications in bioremediation of diverse contaminants (Srivastava et al. 2012; Tiquia-Arashiro and Rodrigues 2016c), water treatment (Ma et al. 2012; Yu et al. 2013), improving plant resistance (McKnight et al. 2003; Rai et al. 2012), and production of clean energy (Wan et al. 2015; Tiquia-Arashiro and Rodrigues 2016c). Overall, bionanoparticles have attracted the attention of diverse researchers because their synthesis is more environmentally friendly and produces more homogeneously distributed nanoparticles, and some of them can be easily biodegradable. Although there are several studies investigating the application of fungal-based nanoparticles, they are still way less studied than bacterial-based nanoparticles. Researchers are still identifying the microbiological synthetic pathways of these bionanoparticles. It is expected that with the advancement of the understanding of bionanoparticle synthesis pathways, the application of bionanoparticles will expand to many more fields than biomedical and environmental and they will be potentially applied in diverse nanotechnological industries.

19.5 Conclusions and Future Perspectives

The interest for bionanoparticles has increased in the past years because they present very different properties and functions than synthetic nanoparticles and they tend to be more biocompatible than their inorganic nonbiological counterparts. The "green" method for nanoparticle synthesis, which is rapidly replacing traditional chemical syntheses, is of great interest because of eco-friendliness, economic views, feasibility, and wide range of applications in several areas such as nanomedicine and catalysis medicine. The most obvious disadvantages of biological nanoparticles (BNP) are that they frequently do not withstand high or low temperatures, extreme pH values, high salt concentrations, presence of harsh chemicals and potential environmental conditions that could lead to their hydrolysis (Tiquia-Arashiro and Rodrigues 2016e, f). It is possible, however, that BNPs from extremophiles or extremotolerant microorganisms might overcome these issues. While a variety of prokaryotic and eukaryotic microorganisms have been investigated with respect to their nanoparticle synthetic abilities, the vast biodiversity encountered in the fungal halophilic/halotolerant world has been relatively less explored. Furthermore, most of the studies are related to the synthesis of silver nanoparticles, followed by those of gold and less on cadmium, lead, and zinc. One reason for this could be the relative ease with which the noble metal ions of gold and silver are reduced. Currently, there are no reports on synthesis of nanoparticles of platinum, bismuth, antimony sulfide, and titanium oxide by halotolerant fungi. As summarized in this review, biologically active products from halotolerant fungi represent excellent scaffolds for this purpose. However, there is a need to understand the mechanisms involved in the synthetic process. Another limitation of the studies is that the experiments have been conducted at laboratory scale and there are hardly any efforts for the scale-up of these processes. In the future, these shortcomings need to be addressed in an effective manner to harness the actual nanoparticle synthetic potential of the halotolerant to their full extent.

References

Abdel-Hafez SII, Maubasher AH, Abdel-Fattah HM (1978) Cellulose decomposing fungi of salt marshes in Egypt. Folia Microbiol 23:3744

Ahmad A, Mukherjee P, Mandal D, Senapati S, Khan MI, Kumar R, Sastry M (2002) Enzyme mediated extracellular synthesis of CdS nanoparticles by the fungus, *Fusarium oxysporum*. J Am Chem Soc 124:12108–12109

Anil Kumar S, Abyaneh MK, Gosavi SW, Kulkarni SK, Pasricha R, Ahmad A, Khan MI (2007) Nitrate reductase-mediated synthesis of silver nanoparticles from $AgNO_3$. Biotechnol Lett 29:439–445

Apte M, Girme G, Nair R, Bankar A, Kumar AR, Zinjarde S (2013a) Melanin mediated synthesis of gold nanoparticles by *Yarrowia lipolytica*. Mater Lett 95:149–152

Apte M, Sambre D, Gaikawad S, Joshi S, Bankar A, Kumar AR, Zinjarde S (2013b) Psychrotrophic yeast *Yarrowia lipolytica* NCYC 789 mediates the synthesis of antimicrobial silver nanoparticles via cell-associated melanin. AMB Express 3:32. https://doi.org/10.1186/2191-0855-3-32

Asmathunisha N, Kathiresan K (2013) A review on biosynthesis of nanoparticles by marine organisms. Colloids Surf B Biointerfaces 103:283–287

Bai HJ, Zhang ZM (2009) Microbial synthesis of semiconductor lead sulfide nanoparticles usingimmobilized Rhodobacter sphaeroides. Mater Lett 63:764–766

Bankar AV, Kumar AR, Zinjarde SS (2009) Environmental and industrial applications of *Yarrowia lipolytica*. Appl Microbiol Biotechnol 84:847–865

Baxter BK, Litchfield CD, Sowers K, Griffith JD, Dassarma PA, Dassarma S (2005) Microbial diversity of Great Salt Lake. In: Gunde-Cimerman N, Oren A, Plemenitaš A (eds) Adaptation to life at high salt concentrations in Archaea, Bacteria, and Eukarya, Cellular origin, life in extreme habitats and astrobiology, vol 9. Springer, Dordrecht, pp 9–25

Bodaker I, Sharon I, Suzuki MT, Reingersch R, Shmoish M, Andreishcheva E, Sogin ML, Rosenberg M, Belkin S, Oren A, Béjà O (2010) The dying Dead Sea: comparative community genomics in an increasingly extreme environment. ISME J 4:399–407

Boisselier E, Astruc D (2009) Gold nanoparticles in nanomedicine: preparations, imaging, diagnostics, therapies and toxicity. Chem Soc Rev 38:1759–1782

Boroumand Moghaddam AB, Moniri M, Azizi S, Rahim RA, Ariff AB, Saad WZ, Namvar F, Navaderi M, Mohamad R (2017) Biosynthesis of ZnO nanoparticles by a new *Pichia kudriavzevii* yeast strain and evaluation of their antimicrobial and antioxidant activities. Molecules 22(6):872. https://doi.org/10.3390/molecules22060872

Buchalo AS, Nevo E, Wasser SP, Oren A, Molitoris HP (1998) Fungal life in the extremely hypersaline water of the Dead Sea: first records. Proc R Soc Lond 265:1461–1465

Butinar L, Sonjak S, Zalar P, Plemenitas A, Gunde-Cimerman N (2005) Melanized halophilic fungi are eukaryotic members of microbial communities in hypersaline waters of solar salterns. Bot Mar 48:73–79

Cai F, Chen FY, Tang YB (2014) Isolation, identification of a halotolerant acid red B degrading strain and its decolorization performance. APCBEE Proc 9:131–139

Cantrell SA, Casillas-Martinez L, Molina M (2006) Characterization of fungi from hypersaline environments of solar salterns using morphological and molecular techniques. Mycol Res 110:962–970

Chen YL, Tuan HY, Tien CW, Lo WH, Liang HC, Hu YC (2009) Augmented biosynthesis of cadmium sulfide nanoparticles by genetically engineered *Escherichia coli*. Biotechnol Prog 25:1260–1266

Chernousova S, Epple M (2013) Silver as antibacterial agent: ion, nanoparticle, and metal. Angew Chem Int Ed 52:1636–1653

Dameron CT, Reese RN, Mehra RK, Kortan AR, Carroll PJ, Steigerwald ML, Brus LE, Wingeet DR (1989) Biosynthesis of cadmium-sulfide quantum semiconductor crystallites. Nature 338:596–597

De Hoog GS, Zalar P, van den Ende BG, Gunde-Cimerman N (2005) Relation of human pathogenicity in the fungal tree of life: an overview of ecology and evolution under stress. In: Gunde-Cimerman N, Oren A, Plemenitaš A (eds) Adaptation to life at high salt concentrations in *Archaea*, *Bacteria* and *Eukarya*. Springer, Dordrecht, pp 397–423

Deepa K, Panda T (2014) Synthesis of gold nanoparticles from different cellular fractions of *Fusarium oxysporum*. J Nanosci Nanotechnol 14:3455–3463

Diaz-Munoz G, Montalvo-Rodriguez R (2005) Halophilic black yeasts *Hortaea wernekii* in the Cabo Rojo Solar Salterns: its first record for this extreme environment in Puerto Rico. Caribb J Sci 41:360–365

Girard V, Dieryckx C, Job C, Job D (2013) Secretomes: the fungal strike force. Proteomics 13:597–608

Gonsalves V, Nayak S, Nazareth S (2012) Halophilic fungi in a polyhaline estuarine habitat. J Yeast Fungal Res 3:30–36

Gunde-Cimerman N, Zalar P, de Hoog GS, Plemenitas‌ A (2000) Hypersaline waters in salterns: natural ecological niches for halophilic black yeasts. FEMS Microbiol Ecol 32:235–240

Gunde-Cimerman N, Butinar L, Sonjak S, Turk M, Uršic V, Zalar P, Plemenitaš A (2005) Halotolerant and halophilic fungi from coastal environments in the Arctics. In: Gunde-Cimerman N, Oren A, Plemenitaš A (eds) Adaptation to life at high salt concentrations in *Archaea*, *Bacteria* and *Eukarya*. Springer, Dordrecht, pp 397–423

Gunde-Cimerman N, Ramos J, Plemenitas A (2009) Halotolerant and halophilic fungi. Mycol Res 113:1231–1241

Hujslova M, Kubatova A, Chudickova M, Kolarik M (2010) Diversity of fungal communities in saline and acidic soils in the Soos National Natural Reserve, Czech Republic. Mycol Prog 9:1–15

Jain N, Bhargava A, Majumdar S, Tarafdar JC, Panwar J (2011) Extracellular biosynthesis and characterization of silver nanoparticles using *Aspergillus flavus* NJP08: a mechanism perspective. Nanoscale 3:635–641

Jayaseelan C, Rahuman AA, Kirthi AV, Marimuthu S, Santhoshkumar T, Bagavan A, Gaurav K, Karthik L, Rao KV (2012) Novel microbial route to synthesize ZnO nanoparticles using *Aeromonas hydrophila* and their activity against pathogenic bacteria and fungi. Spectrochim Acta A Mol Biomol Spectrosc 90:78–84

Jiang L, Li X, Liu L, Zhang Q (2013) Thiolated chitosan-modified PLA-PCLTPGS nanoparticles for oral chemotherapy of lung cancer. Nanoscale Res Lett 8:1–11

Jones BE, Grant WD (1999) Microbial diversity and ecology of the Soda Lakes of East Africa. In: Bell CR, Brylinsky M, Johnson-Green P (eds) Proceedings of the 8th international symposium on microbial ecology. Atlantic Canada Society for Microbial Ecology, Halifax, 7 p

Kathiresan K, Manivannan S, Nabeel MA, Dhivya B (2009) Studies on silver nanoparticles synthesized by a marine fungus, *Penicillium fellutanum* isolated from coastal mangrove sediment. Colloids Surf B Biointerfaces 71:133–137

Kirkland BH, Keyhani NO (2011) Expression and purification of a functionally active class I fungal hydrophobin from the entomopathogenic fungus *Beauveria bassiana* in *Escherichia coli*. J Ind Microbiol Biotechnol 38:327–335

Kis-Papo T, Grishkan I, Gunde-Cimerman N, Oren A, Wasser SP, Nevo E (2003a) Spatiotemporal patterns of filamentous fungi in the hypersaline Dead Sea. In: Nevo E, Oren A, Wasser SP (eds) Fungal life in the Dead Sea. Gantner Verlag, Ruggel, pp 271–292

Kis-Papo T, Oren A, Wasser SP, Nevo E (2003b) Survival of filamentous fungi in hypersaline Dead Sea water. Microb Ecol 45:183–190

Kowshik M, Deshmukh N, Vogel W, Urban J, Kulkarni SK, Paknikar KM (2002a) Microbial synthesis of semiconductor CdS nanoparticles, their characterization, and their use in the fabrication of an ideal diode. Biotechnol Bioeng 78:583–588

Kowshik M, Vogel W, Urban J, Kulkarni SK, Paknikar KM (2002b) Microbial synthesis of semiconductor PbS nanocrystallites. Adv Mater 14:815–818

Li WR, Xie XB, Shi QS, Zeng HY, Ou-Yang YS, Chen YB (2010) Antibacterial activity and mechanism of silver nanoparticles on *Escherichia coli*. Appl Microbiol Biotechnol 8:1115–1122

Li CY, Zhang YJ, Wang M, Zhang Y, Chen G, Li L, Wu D, Wang Q (2014) In vivo real-time visualization of tissue blood flow and angiogenesis using Ag_2S quantum dots in the NIR-II window. Biomaterials 35:393–400

Ma RM, Wei XL, Dai L, Huo HB, Qin GG (2007) Synthesis of CdS nanowire networks and their optical and electrical properties. Nanotechnology 18:1–5

Ma H, Hsiao BS, Chu B (2012) Ultrafine cellulose nanofibers as efficient adsorbents for ther-emoval of UO_2^{2+} in water. ACS Macro Lett 1:213–316

Manivannan S, Alikunhi NM, Kandasamy K (2010) In vitro synthesis of silver nanoparticles by marine yeasts from coastal mangrove sediment. Adv Sci Lett 3:1–6

Margesin R, Schinner F (2001) Potential of halotolerant and halophilic microorganisms for bio-technology. Extremophiles 5:73–83

McKindles K, Tiquia-Arashiro SM (2012) Functional gene arrays for analysis of microbial communities on ocean platform. In: Tiquia-Arahiro SM (ed) Molecular biological technologies for ocean sensing. Humana Press, Totowa, NJ, Chap. 9, 169–701 p

McKnight TE, Melechko AV, Griffin GD, Guillorn MA, Merkulov VI, Serna F, Hensley DK, Doktycz MJ, Lowndes DH, Simpson ML (2003) Intracellular integration of synthetic nanostructures with viable cells for controlled biochemical manipulation. Nanotechnology 14:551–556

Mohite P, Apte M, Kumar AR, Zinjarde S (2016) Biogenic nanoparticles from *Schwanniomyces occidentalis* NCIM 3459: mechanistic aspects and catalytic applications. App Biochem Biotechnol 179:583–596

Mohite P, Kumar AR, Zinjarde S (2017) Relationship between salt tolerance and nanoparticle synthesis by Williopsis saturnus NCIM 3298. World J Microbiol Biotechnol 33:163

Muller A, Ni Z, Hessler N, Wesarg F, Muller FA, Kralisch D, Fischer D (2013) The biopolymer bacterial nanocellulose as drug delivery system: investigation of drug loading and release using the model protein albumin. J Pharm Sci 102:579–592

Nair V, Sambre D, Joshi S, Bankar A, Kumar AR, Zinjarde S (2013) Yeast-derived melanin mediated synthesis of gold nanoparticles. J Bionanosci 7:159–168

Narayanan KB, Sakthivel N (2010) Biological synthesis of metal nanoparticles by microbes. Adv Colloid Interface Sci 156:1–13

Nayak S, Gonsalves V, Nazareth S (2012) Isolation and salt tolerance of halophilic fungi from mangroves and solar salterns in Goa–India. Indian J Mar Sci 41:164–172

Nazareth S, Gonsalves V, Nayak S (2012) A first record of obligate halophilic aspergilli from the Dead Sea. Indian J Microbiol 52:22–27

Nina Gunde-Cimerman N, Zalar P (2014) Extremely halotolerant and halophilic fungi inhabit brine in solar salterns around the globe. Food Technol Biotechnol 52(2):170–179

Oren A (2010) Industrial and environmental applications of halophilic microorganisms. Environ Technol 31:825–834

Pagaling E, Wang HZ, Venables M, Wallace A, Grant WD, Cowan DA, Jones BE, Ma Y, Ventosa A, Heaphy S (2009) Microbial biogeography of six salt lakes in Inner Mongolia, China, and a salt lake in Argentina. Appl Environ Microbiol 75:5750–5760

Pawar V, Shinde A, Kumar AR, Zinjarde S, Gosavi S (2012) Tropical marine microbe mediated synthesis of cadmium nanostructures. Sci Adv Mater 4:135–142

Prasad K, Jha AK (2010) Biosynthesis of CdS nanoparticles: an improved green and rapid procedure. J Colloid Interface Sci 342(1):68–72

Rai M, Deshmukh S, Gade A, Elsalam KA (2012) Strategic nanoparticles-mediated gene transfer in plants and animals—a novel approach. Curr Nanosci 8:170–179

Sanghi R, Verma P, Puri S (2011) Enzymatic formation of gold nanoparticles using *Phanerochaete chrysosporium*. Adv Chem Eng Sci 1:154–162

Seshadri S, Saranya K, Kowshik M (2011) Green synthesis of lead sulfide nanoparticles by the lead resistant marine yeast, *Rhodosporidium diobovatum*. Biotechnol Prog 7:1464–1469

Setati ME (2010) Diversity and industrial potential of hydrolase-producing halophilic/halotolerant eubacteria. Afr J Biotechnol 9:1555–1560

Siddiqi KS, Husen A (2016) Fabrication of metal nanoparticles from fungi and metal salts: scope and application. Nanoscale Res Lett 11:98

Sondi I, Salopek-Sondi B (2004) Silver nanoparticles as antimicrobial agent: a case study on *E. coli* as a model for Gram-negative bacteria. J Colloid Interface Sci 275:177–182

Soni N, Prakash S (2011) Factors affecting the geometry of silver nanoparticles synthesis in *Chrysosporium tropicum* and *Fusarium oxysporum*. Am J Nanotechnol 2:112–121

Srivastava S, Kardam A, Raj KR (2012) Nanotech reinforcement onto cellulose fibers: green remediation of toxic metals. Int J Green Nanotechnol 4:46–53

Steiman R, Ford L, Ducros V, Lafond JL, Guiraud P (2004) First survey of fungi in hypersaline soil and water of Mono Lake area (California). Antonie Van Leeuwenhoek 85:69–83

Tiquia SM (2010) Salt-adapted bacteria isolated from the Rouge River and potential for degradation of contaminants and biotechnological applications. Environ Technol 31:967–978

Tiquia SM, Mormile M (2010) Extremophiles—a source of innovation for industrial and environmental applications. Environ Technol 31(8–9):823

Tiquia SM, Davis D, Hadid H, Kasparian S, Ismail M, Sahly R, Shim J, Singh S, Murray KS (2007) Halophilic and halotolerant bacteria from river waters and shallow groundwater along the Rouge River of southeastern Michigan. Environ Technol 28:297–307

Tiquia-Arashiro SM (2012) Molecular biological technologies for ocean sensing. Humana Press, Totowa, NJ, 295 p

Tiquia-Arashiro SM, Rodrigues D (2016a) Extremophiles: applications in nanotechnology. Springer International Publishing, New York, 193 p

Tiquia-Arashiro SM, Rodrigues D (2016b) Nanoparticles synthesized by microorganisms. In: Extremophiles: applications in nanotechnology. Springer International Publishing, New York, pp 1–51

Tiquia-Arashiro SM, Rodrigues D (2016c) Applications of nanoparticles. In: Extremophiles: applications in nanotechnology. Springer International Publishing, New York, pp 163–193

Tiquia-Arashiro SM, Rodrigues D (2016d) Halophiles in nanotechnology. In: Extremophiles: applications in nanotechnology. Springer International Publishing, New York, pp 53–58

Tiquia-Arashiro SM, Rodrigues D (2016e) Thermophiles and psychrophiles in nanotechnology. In: Extremophiles: applications in nanotechnology. Springer International Publishing, New York, pp 89–127

Tiquia-Arashiro SM, Rodrigues D (2016f) Alkaliphiles and acidophiles in nanotechnology. In: Extremophiles: applications in nanotechnology. Springer International Publishing, New York, pp 129–162

van der Wielen PW, Bolhuis H, Borin S, Daffonchio D, Corselli C, Giuliano L, D'Auria G, de Lange GJ, Huebner A, Varnavas SP, Thomson J, Tamburini C, Marty D, McGenity TJ, Timmis KN (2005) The enigma of prokaryotic life in deep hypersaline anoxic basins. Nature 307:121–123

Vigneshwaran N, Ashtaputre NM, Varadarajan PV, Nachane RP, Paralikar KM, Balasubramanya RH (2006) Biological synthesis of silver nanoparticles using the fungus *Aspergillus flavus*. Mater Lett 61:1413–1418

Wan Y, Yang Z, Xiong G, Ruisong G, Liu Z, Luo H (2015) Anchoring Fe_3O_4 nanoparticles on three-dimensional carbon nanofibers toward flexible high-performance anodes for lithium-ion batteries. J Power Sources 294:414–419

Wang CC, Luconi MO, Masi AN, Fernández LP (2009) Derivatized silver nanoparticles as sensor for ultra-trace nitrate determination based on light scattering phenomenon. Talanta 77:1238–1243

Wasser SP, Grishkan I, Kis-Papo T, Buchalo AS, Paul AV, Gunde-Cimerman N, Zalar P, Nevo E (2003) Species diversity of the Dead Sea fungi. In: Nevo E, Oren A, Wasser SP (eds) Fungal life in the Dead Sea. Gantner Verlag, Ruggel, pp 203–270

Yu A, Liang Z, Cho J, Caruso F (2003) Nanostructured electrochemical sensor based on dense gold nanoparticle films. Nano Lett 3:1203–1207

Yu X, Tong S, Ge M, Wu L, Zuo J, Cao C, Song W (2013) Adsorption of heavy metal ions from aqueous solution by carboxylated cellulose nanocrystals. J Environ Sci (China) 25:933–943

Zajc J, Zalar P, Plemenitaš A, Gunde-Cimerman N (2012) The mycobiota of the salterns. In: Raghukumar C (ed) Biology of marine fungi. Progress in molecular and subcellular biology, vol 53. Springer, Berlin, pp 133–158

Zhang YC, Wang GY, Hu XY (2007) Solvothermal synthesis of hexagonal CdS nanostructures from a single-source molecular precursor. J Alloys Compd 437:47–52

Zhang XF, Liu ZG, Shen W, Gurunathan S (2016) Silver nanoparticles: Synthesis, characterization, properties, applications, and therapeutic approaches. Int J Mol Sci 17(9):pii: E1534. https://doi.org/10.3390/ijms17091534

Zhao PQ, Wu XL, Fan JY, Chu PK, Siu GG (2006) Enhanced and tunable blue luminescence from CdS nanocrystal–polymer composites. Scr Mater 55:1123–1126

Zomorodian K, Pourshahid P, Sadatsharifi A, Mehryar P, Pakshir K, Rahimi MJ, Monfared AA (2016) Biosynthesis and characterization of silver nanoparticles by *Aspergillus Species*. Biomed Res Int 2016:5435397, 6 p. https://doi.org/10.1155/2016/5435397

Chapter 20
Cellulases from Thermophilic Fungi: Recent Insights and Biotechnological Potential

Duo-Chuan Li and Anastassios C. Papageorgiou (iD)

20.1 Introduction

Cellulose accounts for 20–50% in dry weight of the plant cell wall material and is the most abundant and renewable nonfossil carbon source on Earth. Enzymatic hydrolysis of cellulose to its constituent monosaccharides has attracted considerable attention in recent years for the production of food and biofuels. Compared to current industrial processes such as heat, mechanical, and acid treatment of cellulose, cellulose degradation by enzymes is considered a more environment-friendly process (Wilson 2009; Plecha et al. 2013). However, cellulose is the most recalcitrant carbohydrate polymer to enzymatic degradation amongst all polysaccharides of the plant cell wall. The enzymatic degradation of cellulose to glucose requires the synergistic action of endocellulases (E.C.3.1.1.4), exocellulases (cellobiohydrolases, CBH, E.C.3.2.1.91), and β-glucosidases (E.C.3.2.1.21). Endoglucanases initiate hydrolysis by cutting internal glycosidic linkages in a random fashion, which results in a rapid decrease of polymer length and a gradual increase in the reducing sugar concentration. Exocellulases act upon either the reducing or the nonreducing ends to release cello-oligosaccharides and cellobiose units. At the end, β-glucosidases cleave cellobiose to release glucose molecules (Vlasenko et al. 2010).

Thermophilic cellulases have become key enzymes for efficient biomass degradation. Their importance stems from the fact that at higher temperatures, cellulose swells and becomes more susceptible to breaking. Thermophilic fungi have received significant attention in the past years as a reservoir of new thermostable enzymes for

D.-C. Li
Department of Mycology, Shandong Agricultural University, Taian, Shandong, China

A. C. Papageorgiou (✉)
Turku Centre for Biotechnology, University of Turku and Åbo Akademi University,
Turku, Finland
e-mail: tassos.papageorgiou@btk.fi

© Springer Nature Switzerland AG 2019
S. M. Tiquia-Arashiro, M. Grube (eds.), *Fungi in Extreme Environments:*
Ecological Role and Biotechnological Significance,
https://doi.org/10.1007/978-3-030-19030-9_20

use in many biotechnological applications (de Cassia Pereira et al. 2015). Thermophilic fungi grow at a temperature of 50 °C or above, and at a minimum of 20 °C or above (Maheshwari et al. 2000). A number of thermophilic fungi have been isolated in recent years, and the cellulases they produce have been characterized at structural and functional level. Recently, the genomes of *Myceliophthora thermophila* and *Thielavia terrestris* were sequenced and revealed high efficiency of their enzymes to hydrolyze all major polysaccharides found in biomass (Berka et al. 2011). In this review, up-to-date information on molecular, genetic, engineering, and structure-function aspects of thermophilic fungal cellulases is presented for the first time and current research efforts to improve their properties for better use in biotechnological applications are highlighted.

20.2 Cloning, Expression, and Regulation

20.2.1 Regulation of Expression

Production of fungal cellulases is commonly induced only in the presence of cellulose as a substrate (Suto and Tomita 2001). However, owing to its insolubility, cellulose is unable to trigger cellulase induction directly. Thus, constitutive cellulase activity at a basal level is required for cellulase induction. Various disaccharides, such as cellobiose, lactose, and sophorose, have been found to strongly induce cellulase expression (Vaheri et al. 1979). A mechanism of self-synthesized cellobiose from uridine diphosphate glucose cellobiose synthetase was recently reported in the cellulolytic fungus *Rhizopus stolonifer* (Zhang et al. 2017). The synthesized cellobiose was able to promote the transcription of cellulase genes. Recently, xylose was also reported to induce cellulase expression in the thermophilic fungus *Thermoascus aurantiacus* (Schuerg et al. 2017).

The cellulase gene expression in fungi has been proposed to be controlled by a repressor/inducer system (Suto and Tomita 2001). In this system, cellulose or other products of degradation act as inducers, whereas glucose or other easily metabolized carbon sources act as repressors. The repressor/inducer system has also been described in thermophilic fungal cellulases (Maheshwari et al. 2000). In the thermophilic fungus *Talaromyces emersonii*, expression of a cellobiohydrolase gene (*cbh2*) was found to be induced by cellulose and repressed by glucose (Murray et al. 2003). Similar regulation patterns are also exhibited by other cellulase genes from thermophilic fungi, such as *T. emersonii cel7* (Grassick et al. 2004), *Humicola grisea* var. *thermoidea cbh* (Pocas-Fonseca et al. 2000), *T. emersonii bg1* (Collins et al. 2007), and *T. aurantiacus cbh1/cel7a* (Benko et al. 2007).

Glucose repression is a common phenomenon in the regulation of fungal cellulase genes and involves the upstream regulatory sequence (URS) (Furukawa et al. 2009; Ilmen et al. 1997). The protein product of the regulatory gene *cre1* in *Trichoderma reesei* is a negatively acting transcription factor that binds to DNA consensus sequence SYGGRG (where S = C or G, Y = C or T, R = A or G) in the URS. In the presence of glucose, the activated CRE1 (a Cys_2His_2 zinc finger protein) binds to the consensus motif and represses cellulase gene transcription (Ilmen et al. 1997; Nakari-Setälä et al. 2009). Modification of *cre1* gene expression by RNA interference has been suggested for the improvement of cellulase expression (Yang et al. 2015). Three new transcription factors (ACEI, ACEII, and XYR1) involved in cellulase gene regulation have been identified in the mesophilic fungus *T. reesei* (Silva-Rocha et al. 2014). ACEI represses expression of all major cellulase genes in the presence of cellulose. ACEII binds to GGSTAAA sequences in the promoter regions of *cbh1* and *cbh2* and induces expression of *cbh1* and *cbh2*. XYR1 is a zinc binuclear cluster protein that binds to a GGCTAA motif in the 5′-upstream region of XYR1-regulated genes and is involved in the induction of all major cellulase genes. Unlike the transcription factors of *T. reesei* cellulase genes, transcription factors of cellulase genes in thermophilic fungi have not been identified so far. However, there are potential binding sites in the 5′-upstream region of these genes (Collins et al. 2007; Hong et al. 2003a; Murray et al. 2003, 2004; Takashima et al. 1996), and CREI genes from two thermophilic fungi (*Talaromyces emersonii* and *T. aurantiacus*) have been cloned (GenBank AF440004 and AY604200, respectively). It is, therefore, likely that cellulase regulation in thermophilic fungi is similar to that in *T. reesei*.

20.2.2   Gene Cloning

About 50 genes encoding thermophilic fungal cellulases have been isolated, analyzed, and expressed. A brief summary is given in Table 20.1. Cellulases are classified into glycoside hydrolase (GH) families 1, 3, 5, 6, 7, 8, 9, 12, 26, 44, 45, 48, and 61 (http://www.cazy.org). Thermophilic fungal cellulases are found in families 1, 3, 5, 6, 7, 12, 45, and 61. GH61 family members are now recognized as Cu(II) ion-dependent lytic polysaccharide monooxygenases (LPMOs) and are included in auxiliary activity families of the CAZy database (Busk and Lange 2015).

The entire sequence of genes encoding thermophilic fungal cellulase contains an open reading frame (ORF) usually interrupted by introns with consensus 5′ and 3′ intron splice sites, a 3′ untranslated sequence (UTS), a 5′ UTS, and a URS or upstream regulatory region (URR). In the URS of thermophilic fungal cellulases, CAAT and TATAA boxes are present. In particular, some sequences involved in the regulation of cellulase expression are found in the URS of thermophilic fungal cellulases, including the consensus motifs (SYGGRG) for CREI/CREA binding, a consensus sequence (GCCARG) for the putative pH-response transcription factor PacC (Murray et al. 2003), and one or more copies of cellulase expression regulator (ACEI, ACEII, and XYR1) binding sites AGGCAAA, GGSTAAA, and GGCATT (Collins et al. 2007; Hong et al. 2003a; Takashima et al. 1996). In addition, putative glycosylation sites (Asn-X-Thr/Ser) in most of the deduced amino acid sequences of thermophilic fungal cellulases have been identified (Collins et al. 2007; Li et al. 2009).

Table 20.1 Some properties of recombinant thermophilic fungal cellulases expressed in heterologous hosts

Fungus	Gene	Family	Host	Optimal pH	pI	Optimal temp (°C)	Thermal stability	Molecular mass (kDa)	References
Acremonium thermophilum	*cel7a*	7	*Trichoderma reesei*	5.5	4.67	60	NR	53.7	Voutilainen et al. (2008)
Chaetomium thermophilum	*cel7a*	7	*Trichoderma reesei*	4	5.05	65	NR	54.6	Voutilainen et al. (2008)
Chaetomium thermophilum	*cbh 3*	7	*Pichia pastoris*	4	5.15	60	$T_{1/2}$: 45 min at 70 °C	50.0	Li et al. (2009)
Humicola grisea	*egl2*	5	*Aspergillus oryzae*	5	6.92	75	80% residual activity at 75 °C for 10 min	42.6	Takashima et al. (1999b)
Humicola grisea	*egl3*	45	*Aspergillus oryzae*	5	5.78	60	75% residual activity after 10 min at 80 °C	32.2	Takashima et al. (1999b)
Humicola grisea	*egl4*	45	*Aspergillus oryzae*	6	6.44	75	75% residual activity after 10 min at 80 °C	24.2	Takashima et al. (1999a)
Humicola grisea var thermoidea	*eg1*	7	*Aspergillus oryzae*	5	6.43	55–60	Stable for 10 min at 60 °C	47.9	Takashima et al. (1996)
Humicola grisea var thermoidea	*cbh1*	7	*Aspergillus oryzae*	5	4.73	60	Stable for 10 min at 55 °C	55.7	Takashima et al. (1996)
Humicola insolens	*avi2*	6	*Humicola insolens*	NR	5.65	NR	NR	51.3	Moriya et al. (2003)
Humicola insolens	*cbhII*	6	*Saccharomyces cerevisiae*	9	NR	57	$T_{1/2}$: 95 min at 63 °C	NR	Heinzelman et al. (2009)
Melanocarpus albomyces	*cel7b*	7	*Trichoderma reesei*	6–8	4.23	NR	NR	50.0	Haakana et al. (2004)
Melanocarpus albomyces	*cel7a*	7	*Trichoderma reesei*	6–8	4.15	NR	NR	44.8	Haakana et al. (2004)
Melanocarpus albomyces	*cel45a*	45	*Trichoderma reesei*	6–8	5.22	NR	NR	25.0	Haakana et al. (2004)

Chaetomium thermophilum	endo45	45	Pichia pastoris	4		70	60 °C for 1 h; residual activity of 65.6% after 1 h at 80 °C	32	Zhou et al. (2017)
Talaromyces emersonii	cel3a	3	Trichoderma reesei	4	3.6	71.5	$T_{1/2}$: 62 min at 65 °C	90.6	Murray et al. (2004)
Talaromyces emersonii	cel7	7	E. coli	5	4.0	68	$T_{1/2}$: 68 min at 80 °C	48.7	Grassick et al. (2004)
Talaromyces emersonii	cel7A	7	Saccharomyces cerevisiae	4–5		65	$T_{1/2}$: 30 min at 70 °C	46.8	Voutilainen et al. (2010)
Thermoascus aurantiacus	cbh1	7	Saccharomyces cerevisiae	6	4.37	65	80% Residual activity for 60 min at 65 °C	48.7	Hong et al. (2003b)
Thermoascus aurantiacus	eg1	5	Saccharomyces cerevisiae	6	4.36	70	Stable for 60 min at 70 °C	37.0	Hong et al. (2003a)
Thermoascus aurantiacus	bgl1	3	Pichia pastoris	5	4.61	70	70% Residual activity for 60 min at 60 °C	93.5	Hong et al. (2007)
Thermoascus aurantiacus	cel7a	7	Trichoderma reesei	5	4.44	65	NR	46.9	Voutilainen et al. (2008)
Myceliophthora thermophila	eg7a	7	Pichia pastoris						Karnaouri et al. (2014)
Myceliophthora thermophila	bgl3a	3	Pichia pastoris	5–6		70	143 min at 60 °C		Karnaouri et al. (2013)
Myceliophthora thermophila	mteg5	5	Pichia pastoris	5–6		70	6.02 h at 60 °C	75	Karnaouri et al. (2017)
Humicola insolens	hicel6C	6	Pichia pastoris	8		60	>90% of initial activity retained after 1 h at 60 °C	41.7	Xu et al. (2015)
Humicola insolens	HiBgl3A HiBgl3B HiBgl3C	3	Pichia pastoris	5.5 6.0 5.5		60 50–55 60	All enzymes highly stable (>80%) for 1 h at 50 °C	95.1 94.6 78.4	Xia et al. (2016)

NR=not reported

20.2.3 Heterologous Expression

Most cloned cellulase genes of thermophilic fungi are well expressed in heterologous host organisms, such as *E. coli*, yeast, and filamentous fungi (Table 20.1). The majority of the recombinant cellulases expressed in yeast and filamentous fungi are glycosylated (Li et al. 2009; Takashima et al. 1999a). Notably, when a gene encoding a β-glucosidase of *Talaromyces emersonii* was cloned into *T. reesei*, the secreted recombinant enzyme contained 17 potential N-glycosylation sites in its functionally active form (Murray et al. 2004). Glycosylation could further contribute to the thermostability improvement of cellulases as previously suggested (Meldgaard and Svendsen 1994). The mechanism, however, is still unknown. It has been reported that N-glycosylation could increase solubility and reduce aggregation (Ioannou et al. 1998; Kayser et al. 2011). Analysis of protein structures deposited in the Protein Data Bank has also suggested a decrease in protein dynamics upon N-glycosylation without significant global or local structural changes (Lee et al. 2015).

20.3 Purification and Characterization

Purified thermophilic fungal cellulases have been characterized in terms of molecular weight, optimal pH, optimal temperature, thermostability, and glycosylation. Usually, thermophilic fungal cellulases are single polypeptides, although some β-glucosidases are dimeric (Gudmundsson et al. 2016; Mamma et al. 2004). The molecular weight of thermophilic fungal cellulases has a wide range (30–250 kDa) with different carbohydrate contents (2–50%). Optimal pH and temperature are similar for the majority of the purified cellulases from thermophilic fungi. Thermophilic fungal cellulases are active in the pH range 4.0–7.0 and have a high temperature maximum at 50–80 °C for activity (Table 20.1). In addition, they exhibit remarkable thermal stability and are stable at 60 °C with longer half-life at 70 °C, 80 °C, and 90 °C than those from other fungi.

The mechanism of protein thermostability has been studied more extensively in thermophilic bacteria and hyperthermophilic archaea (Pack and Yoo 2004; Trivedi et al. 2006; Tiquia and Mormile 2010). However, a common mechanism has not been established so far and several contributors to protein thermostability have been proposed. A recent analysis suggested that an increase in ion pairs on the protein surface and a stronger hydrophobic interior are the major factors of increased protein thermostability (Taylor and Vaisman 2010). Compared with thermophilic proteins from thermophilic bacteria and hyperthermophilic archaea, the understanding of the nature and mechanism of thermostability of proteins from thermophilic fungi is relatively poor. Hence, further studies are necessary for comprehensive understanding of thermostability in cellulases from thermophilic fungi.

20.4   Structure

20.4.1   Primary Structure

A common characteristic of cellulases is their modular structure. Typically, endo-cellulases and cellobiohydrolases are composed of four modules (Fig. 20.1): a signal peptide that mediates secretion, a cellulose-binding domain (CBD) for attachment to the substrate, a catalytic domain (CD) responsible for the hydrolysis of the substrate, and a hinge region (linker) rich in Ser, Thr, and Pro residues. The hinge region is usually post-translationally glycosylated. An example of primary structure variations includes *Talaromyces emersonii* CBHII, which is characterized by a modular structure (Murray et al. 2003), whereas *T. emersonii* CBH1 consists solely of a catalytic domain (Grassick et al. 2004). Similarly, in *Chaetomium thermophilum* CBHs (CBH1, CBH2, and CBH3), CBH1 and CBH2 exhibit a typical CBD, a linker, and a CD. In contrast, CBH3 only comprises a catalytic domain and lacks a CBD and a hinge region (Li et al. 2009). Interestingly, *T. aurantiacus* cellulases compared to other fungal cellulases (e.g., *T. reesei*) lack CBDs (Le Costaouëc et al. 2013). However, cellulases without CBDs can still be efficiently used (Pakarinen et al. 2014).

CBDs are composed of less than 40 amino acids and they interact with cellulose through a flat platform-like hydrophobic binding site that is thought to be complementary to the flat surfaces presented by cellulose crystals (Shoseyov et al. 2006). It has been shown that deletion of the CBDs from *T. reesei* Cel7A and Cel6A and *Humicola grisea* CBH1 greatly reduces enzymatic activity toward crystalline cellulose (Takashima et al. 1998), suggesting that the tight binding to cellulose through the CBDs is necessary for the efficient hydrolysis of crystalline cellulose. Further studies have shown that three aromatic residues in CBD affect its cellulose-binding ability and enzymatic activities. Indeed, substitution of the three aromatic residues of *Humicola grisea* CBH1 CBD with other amino acids has demonstrated their importance in the interplay of high activity of *Humicola grisea* CBH1 on crystalline

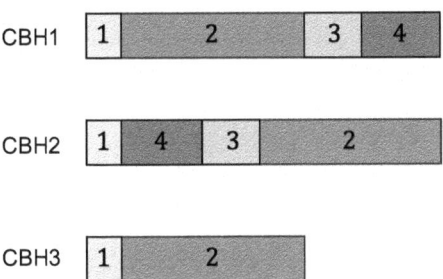

Fig. 20.1 Domain organization of cellobiohydrolases CBH1 (AY861347), CBH2 (AY861348), and CBH3 of *C. thermophilum* (Li et al. 2009). (1) Signal peptide, (2) catalytic domain (CD), (3) ST-rich region, (4) cellulose-binding domain (CBD)

cellulose and high cellulose-binding ability (Takashima et al. 2007). It has been further shown that CBDs promote the enzyme's action on insoluble substrates but are not needed when the enzyme attacks soluble substrates. It has therefore been suggested that CBHs without CBD might play an important role in the hydrolysis of soluble substrates (Li et al. 2009).

20.4.2 Three-Dimensional (3-D) Structure

The 3-D structures of thermophilic fungal cellulases from families 1, 5, 6, 7, 12, and 45 have been reported (Table 20.2) and are described below along with structural information of LPMOs (formerly GH61).

20.4.2.1 Family 1

The crystal structure of a *Humicola insolens* GH1 β-glucosidase has been reported at 2.6 Å resolution (de Giuseppe et al. 2014). In contrast to most β-glucosidases that are inhibited by glucose, this particular GH1 β-glucosidase exhibits tolerance toward glucose. The active site is located in a deep and narrow cavity (Fig. 20.2).

Table 20.2 Thermophilic fungal cellulases with known 3-D structures

Source	Name	Family	Fold	References
Humicola insolens	BG	1	β/α-Barrel	de Giuseppe et al. (2014)
Rasamsonia emersonii	Cel3A	3	β/α-Barrel	Gudmundsson et al. (2016)
Humicola insolens	Cel6A (CBH)	6	β/α-Barrel	Varrot et al. (2003)
Humicola insolens	Cel6B (EG)	6	β/α-Barrel	Davies et al. (2000)
Humicola insolens	EGI	7	β-Sandwich	Davies et al. (1997)
Humicola insolens	Cel7B	7	β-Sandwich	MacKenzie et al. (1998)
Humicola grisea	Cel7B	7	β-Sandwich	Momeni et al. (2014)
Humicola insolens	EGV	45	β-Barrel	Davies et al. (1993)
Humicola grisea	Cel12A	12	β-Sandwich	Sandgren et al. (2004)
Talaromyces emersonii	CBHIB	7	β-Sandwich	Grassick et al. (2004)
Thermoascus aurantiacus	Cel5A	5	β/α-Barrel	Lo Leggio and Larsen (2002)
Melanocarpus albomyces	maEG	45	β-Barrel	Hirvonen and Papageorgiou (2003)
Thielavia terrestris	ttCel45A	45	β-Barrel	Gao et al. (2017)
Melanocarpus albomyces	Cel7B	7	β-Sandwich	Parkkinen et al. (2008)
Thielavia terrestris	GH61E	61	β-Sandwich	Harris et al. (2010)
Thermoascus aurantiacus	TaGH61	61	β-Sandwich	Quinlan et al. (2011)

Fig. 20.2 Ribbon diagram of GH1 β-glucosidase from *Humicola insolens* (PDB id 4mdo; TRS, Tris buffer). Catalytic residues, bound ligands, and disulfide bridges are depicted in stick representation. All figures of the structures were created with CHIMERA (Pettersen et al. 2004)

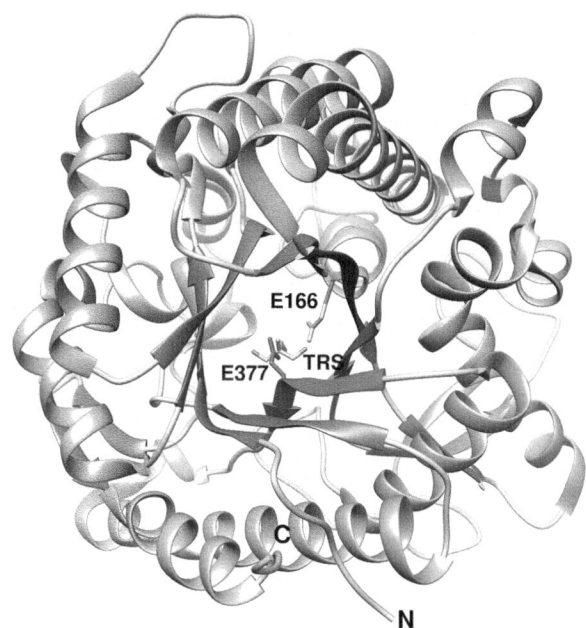

The glucose tolerance was attributed to residues Trp168 and Leu173, which are conserved in glucose-tolerant GH1 enzymes and limit the access of glucose to the −1 subsite by imposing constraints at the +2 subsite.

20.4.2.2 Family 3

The crystal structure of a GH3 β-glucosidase (Cel3A) from *R. emersonii* (*Re*Cel3A) has been determined at 2.2 Å resolution (Gudmundsson et al. 2016). *Re*Cel3A was heterologously expressed in a *H. jecorina* strain and found highly glycosylated with a total of 181 glycosylation residues distributed in 16 glycosylation sites. *Re*Cel3A structure (Fig. 20.3) consists of three domains similar to other GH3 β-glucosidases: an N-terminal triose phosphate isomerase (TIM)-barrel-like domain, a middle $(\alpha/\beta)_6$-sandwich domain, and a C-terminal domain with a fibronectin type III-like fold.

20.4.2.3 Family 5

Family 5 cellulases belong to the endoglucanase type. The overall fold is a common $(\beta/\alpha)_8$-barrel. The structure of *Thermoascus aurantiacus* Cel5A is known (Lo Leggio and Larsen 2002; Van Petegem et al. 2002) and consists solely of a catalytic domain (Fig. 20.4). A substrate-binding cleft rich in Trp residues is visible at the C-terminal end of the barrel. The size and shape of the cleft suggest the binding of

Fig. 20.3 Ribbon diagram
of *Re*Cel3A (PDB id 5ju6)

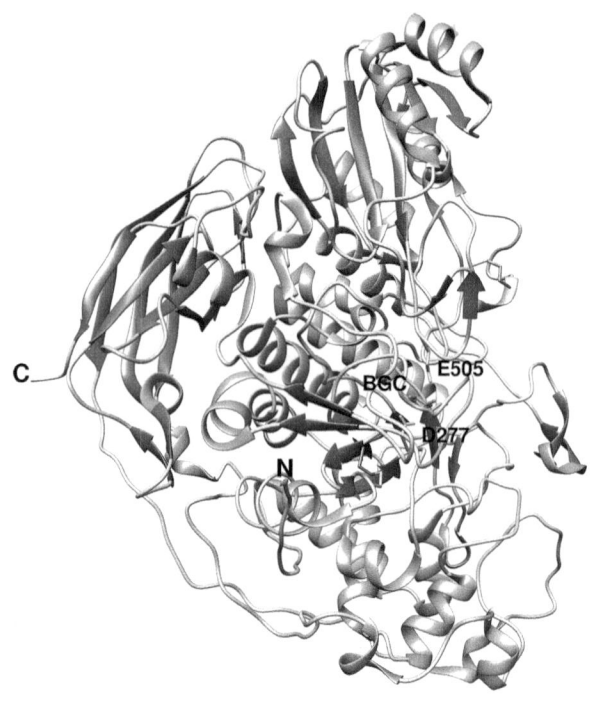

Fig. 20.4 *Thermoascus
aurantiacus* GH5
endoglucanase (PDB id
1gzj)

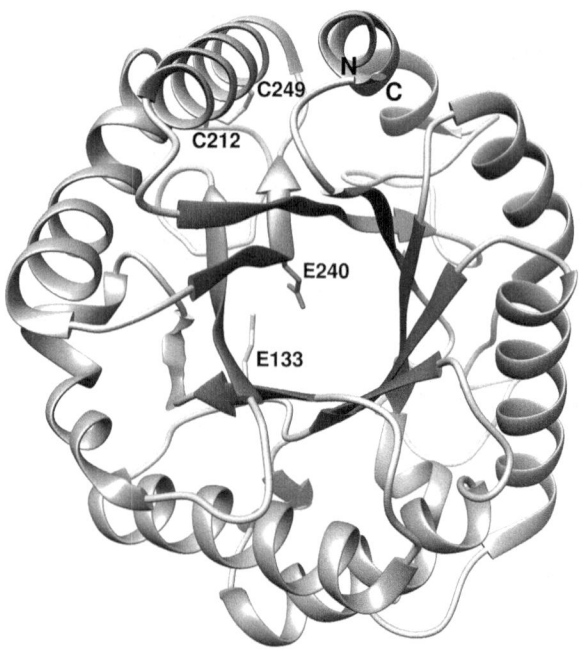

at least seven glucose residues (−4 to +3). Contrary to other GH5 cellulase structures, Cel5A has only a few extra-barrel features, including a short two-stranded β-sheet in β/α-loop 3 and three one-turn helices. The enzyme contains two conserved cysteine residues (Cys212 and Cys249) that form a disulfide bridge.

20.4.2.4   Family 6

Family 6 comprises both endoglucanases and cellobiohydrolases. 3-D structures have been reported for the endoglucanase Cel6B and the cellobiohydrolase Cel6A from *Humicola insolens* (Davies et al. 2000; Varrot et al. 2003). Their structure exhibits a distorted β/α-barrel with the central β-barrel made up of seven instead of eight parallel β-strands (Fig. 20.5). A substrate-binding crevice is formed between strands I and VII. The crevice of the Cel6A contains at least four substrate-binding sites, −2 to +2, whereas that of the Cel6B has six substrate-binding sites, −2 to +4. A significant difference between the endoglucanase Cel6B and the cellobiohydrolase Cel6A is the presence of two extended surface loops that enclose the active site of Cel6A. The absence of these loops in Cel6B results in an open substrate cleft in Cel6B. Owing to this structural difference, endoglucanase can hydrolyze bonds internally in cellulose chains while cellobiohydrolase acts on chain ends.

Fig. 20.5 Ribbon diagram of *Humicola insolens* Cel6B endoglucanase (PDB id 1dys)

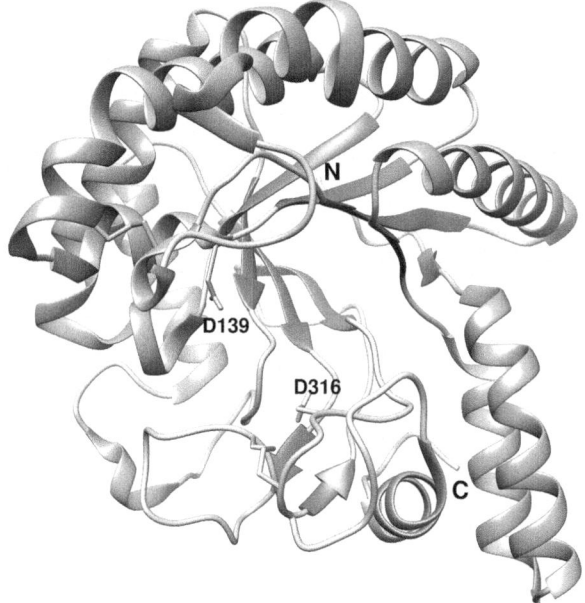

20.4.2.5 Family 7

Similar to GH6, GH7 contains endoglucanases and cellobiohydrolases. 3-D structures of various family members have been reported. However, only a few structures from thermophilic fungi are currently known, including *Talaromyces emersonii* CBHIB (Grassick et al. 2004), *Humicola insolens* EGI (Davies et al. 1997; MacKenzie et al. 1998), *Melanocarpus albomyces* Cel7B (Parkkinen et al. 2008), and *Humicola grisea* Cel7A (Momeni et al. 2014). The overall 3-D architecture of this family exhibits a double β-sandwich. The structure of *M. albomyces* Cel7B (Fig. 20.6), similarly to *Talaromyces emersonii* CBHIB, is a representative of GH7 cellobiohydrolases (Grassick et al. 2004). It consists of two antiparallel β-sheets packed face to face to form a β-sandwich. Owing to their strong curvature, these two β-sheets form the concave and convex surfaces of the sandwich. A main characteristic of Cel7B is an enclosed substrate-binding tunnel formed by extended loops stabilized by nine disulfide bonds. The tunnel is about 50 Å long and contains the substrate-binding sites, −7 to +2 (Parkkinen et al. 2008). *Humicola grisea* Cel7A (PDB id 4csi) has 56% sequence identity with *M. albomyces* Cel7B and exhibits various loop deviations (Momeni et al. 2014) that may contribute to its increased thermostability.

   Humicola insolens EGI has a β-sandwich structure with two large antiparallel β-sheets (Davies et al. 1997; MacKenzie et al. 1998) similar to *M. albomyces* Cel7B

Fig. 20.6 *Melanocarpus albomyces* GH7 cellobiohydrolase in complex with cellotetraose (PDB id 2rg0; CTT=cellotetraose)

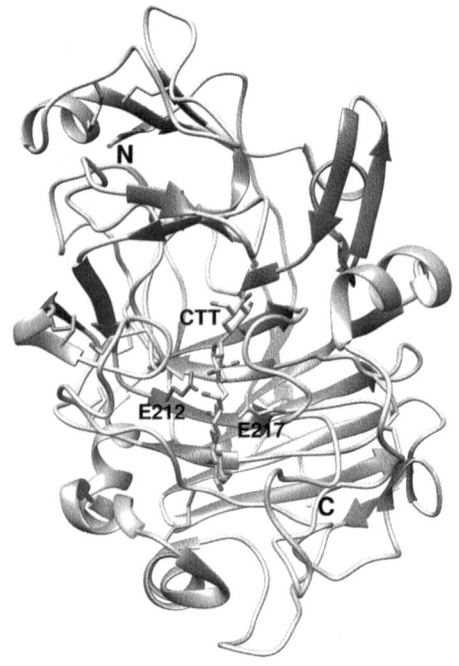

Fig. 20.7 *Humicola insolens* GH7 endoglucanase EGI (PDB id 2a39)

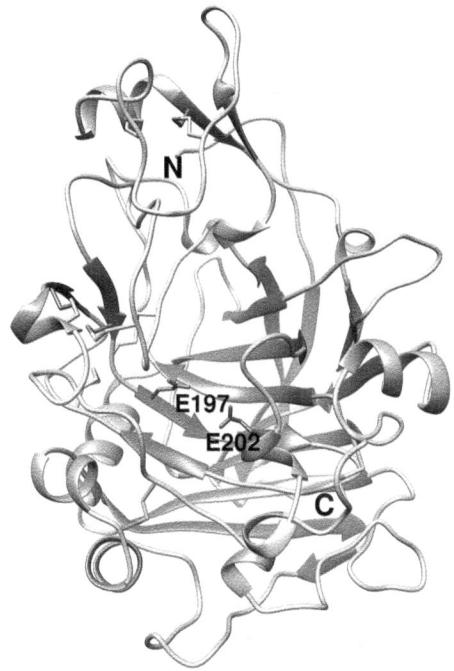

(Fig. 20.7). However, EGI has an open long active site cleft in the center of a canyon formed by the curvature of the β-strands of the β-sandwich while Cel7B has an enclosed substrate-binding tunnel (Davies et al. 1997; Parkkinen et al. 2008), which is similar to GH6 endoglucanases and cellobiohydrolases. *C. thermophilum* CBH3 is a thermostable and single-module cellobiohydrolase with no 3-D structure available (Li et al. 2009). This cellobiohydrolase shares high sequence identity (80%) with *M. albomyces* Cel7B (Parkkinen et al. 2008). Homology modelling has shown that all the key residues in the catalytic site and substrate-binding site as well as the disulfide bonds of *M. albomyces* Cel7B are also present in *C. thermophilum* CBH3.

20.4.2.6 Family 12

The crystal structure of a GH12 fungal cellulase from *Humicola grisea* has been reported (Sandgren et al. 2003). It comprises two antiparallel β-sheets, which pack on top of each other to form a compact curved β-sandwich (Fig. 20.8). The concave face creates an approximately 35 Å-long and unblocked substrate-binding site able to bind at least six glycan residues. Details of the non-covalent interactions between the enzyme and the glucosyl chain in subsites −4 to +2 have been revealed by crystal structures of four enzyme-ligand complexes (Sandgren et al. 2004).

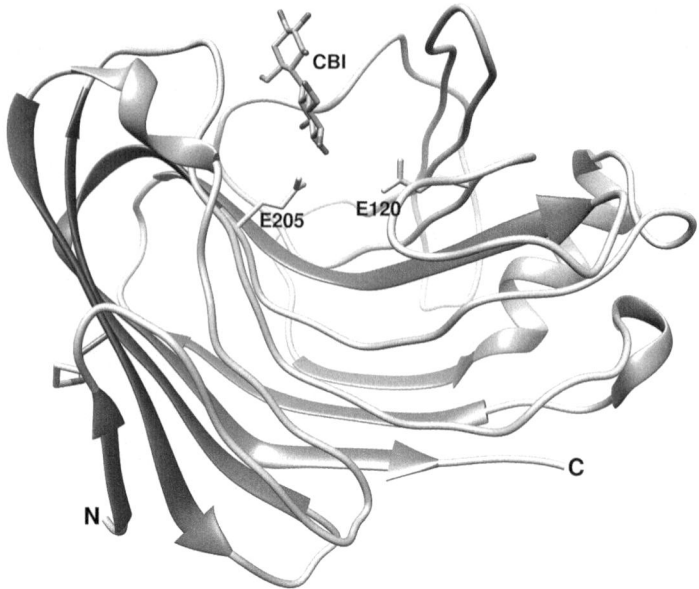

Fig. 20.8 Catalytic domain of *Humicola grisea* GH12 Cel12A in complex with cellobiose (PDB id 1uu4; CBI=cellobiose)

20.4.2.7 Family 45

The structures of three GH45 endoglucanases have been solved: *Humicola insolens* Cel45A (EGV) (Davies et al. 1993), *Melanocarpus albomyces* 20 kDa endogluca-nase (Hirvonen and Papageorgiou 2003; Valjakka and Rouvinen 2003), and *Theilavia terrestris* Cel45 (Gao et al. 2017). These three endoglucanases have simi-lar overall fold, which consists of a six-stranded β-barrel with interconnecting loops (Fig. 20.9). The β-strands are connected with long disulfide-bonded loop structures (seven disulfide bridges) while the rest of the structure is completed by three α-helices. A substrate-binding groove is formed between the β-barrel and the loop structures. This groove, approximately 40 Å long, 10 Å deep, and 12 Å wide, is subdivided into six substrate-binding sites, from −4 to +2 (Hirvonen and Papageorgiou 2003).

20.4.2.8 Lytic Polysaccharide Monooxygenases

LPMOs use a cellulose degradation mechanism different from that of cellulases. The reaction proceeds through an oxidative step that involves hydroxylation of crys-talline cellulose at the C1 or C4 carbon, leading to subsequent cleavage of the gly-cosidic bond. The 3-D structure of LPMOs exhibits a predominantly β-sandwich fold with two twisted antiparallel β-sheets connected through loops of various

Fig. 20.9 *Melanocarpus albomyces* GH45 endoglucanase in complex with cellobiose (PDB id 1oa7)

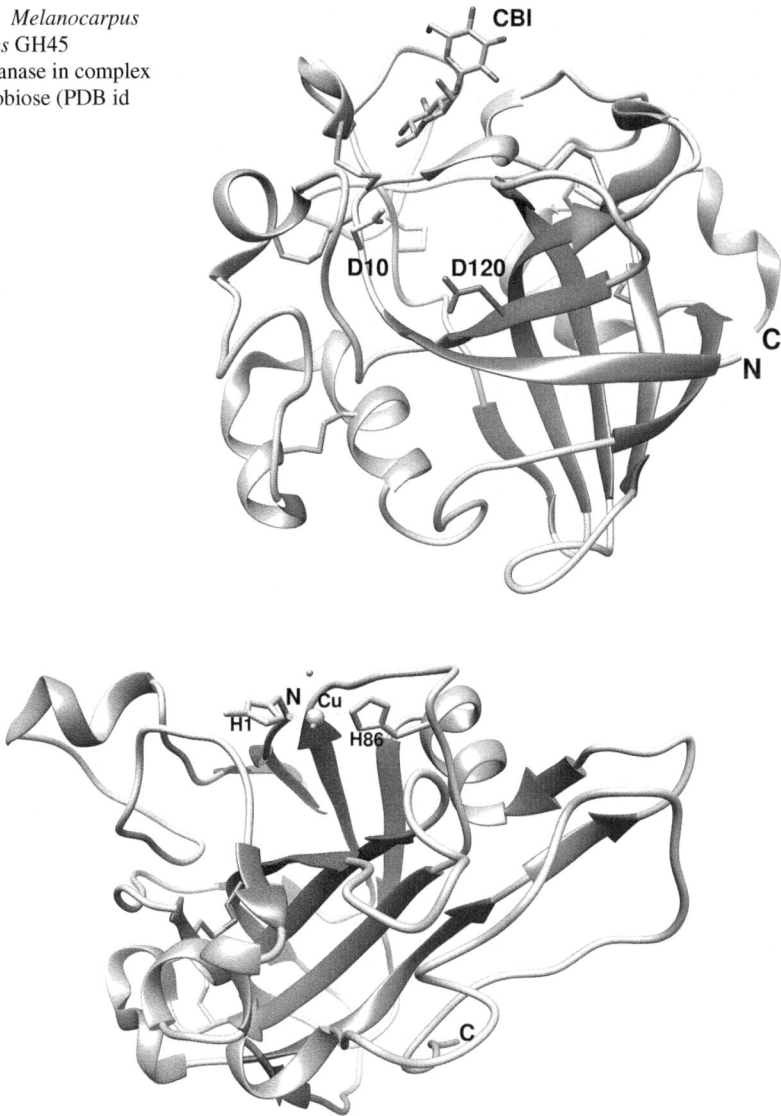

Fig. 20.10 *T. aurantiacus* GH61 (PDB id 2yet). Cu(II) ion and water molecules are shown as spheres

lengths and conformation (Fig. 20.10). The active site is located on a flat solvent-exposed region of the molecule in contrast to traditional cellulases that possess a substrate-binding cleft or tunnel. A tightly bound Cu(II) ion involved in the reaction has been found in the active site. His1, one of the Cu(II) coordinating residues, is methylated in some LPMO structures, including that of *T. aurantiacus* GH61, but not in all (Quinlan et al. 2011). The reason for the methylation is still unknown.

20.5   Improvement of Thermophilic Fungal Cellulases

Cellulose degradation in an efficient and cheap way is a major challenge in biomass conversion by cellulases. Consequently, cellulases need to acquire higher catalytic efficiency on cellulose, higher stability at elevated temperature and at non-physiological pH, and higher tolerance to end-product inhibition (Percival Zhang et al. 2006). Currently, there are two main research approaches of cellulase improvement: structure-based rational site-directed mutagenesis and random mutagenesis through directed evolution. The site-directed mutagenesis requires detailed knowledge of the 3-D protein structure. Conversely, the directed evolution approach is not limited by the lack of the protein 3-D structure but requires an efficient method for high-throughput screening (Labrou 2010).

20.5.1   Thermostability Improvement

Although cellulases from thermophilic fungi are thermostable, further increase of their thermostability is desirable for industrial applications. Improvement of *Melanocarpus albomyces* Cel7B has been pursued by error-prone PCR and 49 positive mutant clones were screened from 14,600 random clones by a robotic high-throughput thermostability screening method (Voutilainen et al. 2007). Two positive thermostable mutants, Ala30Thr and Ser290Thr, showed improved unfolding temperature (T_m) by 1.5 and 3.5 °C, respectively. In addition, the optimum temperature on a soluble substrate for the Ala30Thr mutant was improved by 5 °C. The amino acid alterations are located in the β-strands away from the Cel7B active-site tunnel, which could improve protein packing. In another approach, introduction of additional disulfide bridges to the catalytic module of *Talaromyces emersonii* Cel7A resulted in three mutants with improved thermostability, as shown by Avicel hydrolysis efficiency at 75 °C (Voutilainen et al. 2010).

Structural analysis of thermostable *Humicola grisea* Cel12A has revealed three unusual free cysteines in the enzyme: Cys175, Cys206, and Cys216. Consequently, three Cel12A mutants were constructed by site-directed mutagenesis: Cys175Gly, Cys206Pro, and Cys216Val. It was found that the three free cysteines play a significant role in modulating the stability of the enzyme (Sandgren et al. 2005). More specifically, Cys206Pro and Cys216Val mutations caused a reduction in the T_m of 9.1 °C and 5.5 °C, respectively, compared to the wild-type enzyme. Moreover, when Cys175 was mutated to Gly, the T_m of the enzyme was increased by 1.3 °C. Recent reports of fold-specific thermostability through variations in amino acid compositions of endoglucanases (Yennamalli et al. 2011) have provided additional strategies for thermostability improvement.

Random mutagenesis and recombination of beneficial mutations were employed for the construction of a chimeric Cel6A cellobiohydrolase (Wu and Arnold 2013). Increased hydrophobic interactions and reduced loop flexibility by introduction of Pro residues were found to improve thermostability.

A computational approach, SCHEMA, which uses protein structure data to generate new sequences for special purposes that minimize structure disruption when they are introduced in chimeric proteins, has been employed to create thermostable fungal cellulases (Heinzelman et al. 2009). The high-resolution structure of *Humicola insolens* CBHII (Varrot et al. 2003) used as a template for SCHEMA yielded a collection of highly thermostable CBHII chimeras. Using the computer-generated sequences, a total of 31 new cellulase genes were synthesized and expressed in *S. cerevisiae*. Each purified enzyme was found to be more stable than the most stable parent cellulase from *Humicola insolens*, as measured either by half-life inactivation at 63 °C or by $T_{1/2}$. These findings demonstrated the value of structure-guided recombination for efficient generation of highly stable cellulases.

Improvement of cellulase stability in detergent solutions has also been pursued (Otzen et al. 1999). *H. insolens* Cel45 endoglucanase is inactivated by the anionic surfactant C12-LAS, owing to the surface positive charges of the enzyme (Otzen et al. 1999). Based on the Cel45 crystal structure, surface residues were mutated by site-directed mutagenesis. Introduction of positive charges or removal of negative charges was found to greatly increase detergent sensitivity. The R158E/R196E double mutant, in particular, exhibited synergistic stabilization, presumably by preventing C12-LAS from binding to the protein.

20.5.2   Activity Improvement

Improvement of cellulase activity using site-directed mutagenesis and directed evolution has been pursued in recent years. However, owing to the lack of general rules for site-directed mutagenesis and the limitation of screening methods, there are only a few successful examples of cellulase mutants with significantly higher activity (Percival Zhang et al. 2006). As mentioned earlier, the S290 T mutant of *Melanocarpus albomyces* Cel7B exhibits not only improved thermostability but also a two-fold increase in the rate of Avicel hydrolysis at 70 °C (Voutilainen et al. 2009). Similar results have been obtained for *T. emersonii* Cel7A with site-directed mutagenesis (Voutilainen et al. 2010). Directed evolution of *Chaetomium thermophilum* CBHII produced mutants able to retain >50% of their activity at 80 °C for 1 h while the wild type lost all its activity under the same conditions (Wang et al. 2012).

Recent studies have shown the potential of adding or replacing a CBD to alter the enzyme characteristics and to improve hydrolytic activity (Limon et al. 2001; Shoseyov et al. 2006). *Humicola grisea*, for example, contains two endoglucanases: EGL3 with a CBD and EGL4 without a CBD. The fusion protein, EGL4CBD, which consists of the EGL4 catalytic domain and the EGL3 CBD, shows relatively high activity to carboxymethyl cellulose (Takashima et al. 1999a). *Melanocarpus albomyces* Cel7A, Cel7B, and Cel45A lack a consensus CBD and the associated linker (Haakana et al. 2004). The efficiency of these three cellulases to hydrolyze crystalline cellulose was improved after they were genetically modified to carry the CBD of *T. reesei* CBHI (Szijarto et al. 2008).

20.5.3 Conversion to Glycosynthases

An important development in cellulase engineering is the conversion of cellulases to glycosynthases by site-directed mutagenesis (Shaikh and Withers 2008). Glycosynthases are retaining GH mutants in which the catalytic nucleophile has been replaced by a non-nucleophilic residue. The first glycosynthase from thermophilic fungus was derived from *Humicola insolens* Cel7B following mutation of E197 to Ala. The resultant Cel7B E197A glycosynthase was able to catalyze regio- and stereoselective glycosylation in high yield (Fort et al. 2000). Three mutants of the *H. insolens* Cel7B E197A glycosynthase, E197A/H209A, and E197A/H209G double mutants, and Cel7B E197A/H209A/A211T triple mutant, were subsequently prepared and characterized (Blanchard et al. 2007). These second-generation glycosynthase mutants were rationally redesigned at the +1 subsite to broad the substrate specificity of the glycosynthase. The results showed that E197A/H209A and E197A/H209G preferentially catalyze the formation of a β-(1,4) linkage between two disaccharides. In contrast, the E197A and E197A/H209A/A211T mutants produce predominantly a β-(1,3)-linked tetrasaccharide. This work indicated that the regioselectivity of the glycosylation reaction catalyzed by *Humicola insolens* Cel7B E197A glycosynthase could be modulated by appropriate active site mutations. Use of β-glucosidases for the synthesis of a variety of glycoconjugates, such as alkyl glucosides and aminoglycosides, has gathered momentum in recent years. A GH3 β-glucosidase from the thermophilic fungus *Myceliophthora thermophila* has been studied and found to act as an efficient biocatalyst in alkyl glycoside synthesis (Karnaouri et al. 2013).

20.6 Conclusions and Future Perspectives

Thermophilic fungal cellulases are promising alternatives in biotechnological applications. However, from about 40–50 species of known thermophilic fungi, only a minority of thermophilic fungal cellulases has been characterized so far in detail. A systematic characterization of cellulases is necessary to understand better their thermostability, enzyme mechanism, synergism of the cellulase system, and evolutionary relationships. Site-directed mutagenesis and directed evolution are currently the most preferable approaches for obtaining novel cellulase variants. In addition, the emergence of LPMOs as a complementary cellulose-degrading system offers new opportunities for the future. Further improvement of thermophilic fungal cellulases will assist in developing better and more versatile enzymes for their use alone or in mixtures for biotechnological applications.

Acknowledgements Support by the Chinese National Program for High Technology, Research and Development, the Chinese Project of Transgenic Organisms, and the National Department Public Benefit Research Foundation is acknowledged. ACP thanks Biocenter Finland for infrastructure support.

References

Benko Z, Drahos E, Szengyel Z, Puranen T, Vehmaanpera J, Reczey K (2007) *Thermoascus auran-tiacus* CBHI/Cel7A production in *Trichoderma reesei* on alternative carbon sources. Appl Biochem Biotechnol 137(140):195–204

Berka RM, Grigoriev IV, Otillar R, Salamov A, Grimwood J, Reid I, Ishmael N, John T, Darmond C, Moisan MC, Henrissat B, Coutinho PM, Lombard V, Natvig DO, Lindquist E, Schmutz J, Lucas S, Harris P, Powlowski J, Bellemare A, Taylor D, Butler G, de Vries RP, Allijn IE, van den Brink J, Ushinsky S, Storms R, Powell AJ, Paulsen IT, Elbourne LD, Baker SE, Magnuson J, Laboissiere S, Clutterbuck AJ, Martinez D, Wogulis M, de Leon AL, Rey MW, Tsang A (2011) Comparative genomic analysis of the thermophilic biomass-degrading fungi *Myceliophthora thermophila* and *Thielavia terrestris*. Nat Biotechnol 29:922–927

Blanchard S, Armand S, Couthino P, Patkar S, Vind J, Samain E, Driguez H, Cottaz S (2007) Unexpected regioselectivity of *Humicola insolens* Cel7B glycosynthase mutants. Carbohydr Res 342:710–716

Busk PK, Lange L (2015) Classification of fungal and bacterial lytic polysaccharide monooxygen-ases. BMC Genomics 16:368

Collins CM, Murray PG, Denman S, Morrissey JP, Byrnes L, Teeri TT, Tuohy MG (2007) Molecular cloning and expression analysis of two distinct beta-glucosidase genes, bg1 and aven1, with very different biological roles from the thermophilic, saprophytic fungus *Talaromyces emerso-nii*. Mycol Res 111:840–849

Davies GJ, Dodson GG, Hubbard RE, Tolley SP, Dauter Z, Wilson KS, Hjort C, Mikkelsen JM, Rasmussen G, Schulein M (1993) Structure and function of endoglucanase V. Nature 365:362–364

Davies GJ, Ducros V, Lewis RJ, Borchert TV, Schulein M (1997) Oligosaccharide specificity of a family 7 endoglucanase: insertion of potential sugar-binding subsites. J Biotechnol 57:91–100

Davies GJ, Brzozowski AM, Dauter M, Varrot A, Schulein M (2000) Structure and function of *Humicola insolens* family 6 cellulases: structure of the endoglucanase, Cel6B, at 1.6 A resolu-tion. Biochem J 348(Pt 1):201–207

de Cassia Pereira J, Paganini Marques N, Rodrigues A, Brito de Oliveira T, Boscolo M, da Silva R, Gomes E, Bocchini Martins DA (2015) Thermophilic fungi as new sources for production of cellulases and xylanases with potential use in sugarcane bagasse saccharification. J Appl Microbiol 118:928–939

de Giuseppe PO, Souza TA, Souza FH, Zanphorlin LM, Machado CB, Ward RJ, Jorge JA, Furriel RP, Murakami MT (2014) Structural basis for glucose tolerance in GH1 β-glucosidases. Acta Crystallogr D Biol Crystallogr 70:1631–1639

Fort S, Boyer V, Greffe L, Davies G, Moroz O, Christiansen L, Schulein M, Cottaz S, Driguez H (2000) Highly efficient synthesis of beta(1->4)-oligo- and -polysaccharides using a mutant cel-lulase. J Am Chem Soc 122:5429–5437

Furukawa T, Shida Y, Kitagami N, Mori K, Kato M, Kobayashi T, Okada H, Ogasawara W, Morikawa Y (2009) Identification of specific binding sites for XYR1, a transcriptional activa-tor of cellulolytic and xylanolytic genes in *Trichoderma reesei*. Fungal Genet Biol 46:564–574

Gao J, Huang JW, Li Q, Liu W, Ko TP, Zheng Y, Xiao X, Kuo CJ, Chen CC, Guo RT (2017) Characterization and crystal structure of a thermostable glycoside hydrolase family 45 1,4-β-endoglucanase from *Thielavia terrestris*. Enzym Microb Technol 99:32–37

Grassick A, Murray PG, Thompson R, Collins CM, Byrnes L, Birrane G, Higgins TM, Tuohy MG (2004) Three-dimensional structure of a thermostable native cellobiohydrolase, CBH IB, and molecular characterization of the cel7 gene from the filamentous fungus, *Talaromyces emerso-nii*. Eur J Biochem 271:4495–4506

Gudmundsson M, Hansson H, Karkehabadi S, Larsson A, Stals I, Kim S, Sunux S, Fujdala M, Larenas E, Kaper T, Sandgren M (2016) Structural and functional studies of the glycoside hydrolase family 3 β-glucosidase Cel3A from the moderately thermophilic fungus *Rasamsonia emersonii*. Acta Crystallogr D Struct Biol 72:860–870

Haakana H, Miettinen-Oinonen A, Joutsjoki V, Mantyla A, Suominen P, Vehmaanpera J (2004) Cloning of cellulase genes from Melanocarpus albomyces and their efficient expression in *Trichoderma reesei*. Enzym Microb Technol 34:159–167

Harris PV, Welner D, McFarland KC, Re E, Navarro Poulsen JC, Brown K, Salbo R, Ding H, Vlasenko E, Merino S, Xu F, Cherry J, Larsen S, Lo Leggio L (2010) Stimulation of lignocellulosic biomass hydrolysis by proteins of glycoside hydrolase family 61: structure and function of a large, enigmatic family. Biochemistry 49:3305–3316

Heinzelman P, Snow CD, Smith MA, Yu X, Kannan A, Boulware K, Villalobos A, Govindarajan S, Minshull J, Arnold FH (2009) SCHEMA recombination of a fungal cellulase uncovers a single mutation that contributes markedly to stability. J Biol Chem 284:26229–26233

Hirvonen M, Papageorgiou AC (2003) Crystal structure of a family 45 endoglucanase from *Melanocarpus albomyces*: mechanistic implications based on the free and cellobiose-bound forms. J Mol Biol 329:403–410

Hong J, Tamaki H, Yamamoto K, Kumagai H (2003a) Cloning of a gene encoding a thermostable endo-beta-1,4-glucanase from *Thermoascus aurantiacus* and its expression in yeast. Biotechnol Lett 25:657–661

Hong J, Tamaki H, Yamamoto K, Kumagai H (2003b) Cloning of a gene encoding thermostable cellobiohydrolase from Thermoascus aurantiacus and its expression in yeast. Appl Microbiol Biotechnol 63:42–50

Hong J, Tamaki H, Kumagai H (2007) Cloning and functional expression of thermostable beta-glucosidase gene from Thermoascus aurantiacus. Appl Microbiol Biotechnol 73:1331–1339

Ilmen M, Saloheimo A, Onnela ML, Penttila ME (1997) Regulation of cellulase gene expression in the filamentous fungus *Trichoderma reesei*. Appl Environ Microbiol 63:1298–1306

Ioannou YA, Zeidner KM, Grace ME, Desnick RJ (1998) Human alpha-galactosidase A: glycosylation site 3 is essential for enzyme solubility. Biochem J 332:789–797

Karnaouri A, Topakas E, Paschos T, Taouki I, Christakopoulos P (2013) Cloning, expression and characterization of an ethanol tolerant GH3 beta-glucosidase from *Myceliophthora thermophila*. Peerj 1:e46

Karnaouri AC, Topakas E, Christakopoulos P (2014) Cloning, expression, and characterization of a thermostable GH7 endoglucanase from *Myceliophthora thermophila* capable of high-consistency enzymatic liquefaction. Appl Microbiol Biotechnol 98:231–242

Karnaouri A, Muraleedharan MN, Dimarogona M, Topakas E, Rova U, Sandgren M, Christakopoulos P (2017) Recombinant expression of thermostable processive MtEG5 endoglucanase and its synergism with MtLPMO from *Myceliophthora thermophila* during the hydrolysis of lignocellulosic substrates. Biotechnol Biofuels 10:126

Kayser V, Chennamsetty N, Voynov V, Forrer K, Helk B, Trout BL (2011) Glycosylation influences on the aggregation propensity of therapeutic monoclonal antibodies. Biotechnol J 6:38–44

Labrou NE (2010) Random mutagenesis methods for in vitro directed enzyme evolution. Curr Protein Pept Sci 11:91–100

Le Costaouëc T, Pakarinen A, Várnai A, Puranen T, Viikari L (2013) The role of carbohydrate binding module (CBM) at high substrate consistency: comparison of *Trichoderma reesei* and *Thermoascus aurantiacus* Cel7A (CBHI) and Cel5A (EGII). Bioresour Technol 143:196–203

Lee HS, Qi Y, Im W (2015) Effects of N-glycosylation on protein conformation and dynamics: Protein Data Bank analysis and molecular dynamics simulation study. Sci Rep 5:8926

Li YL, Li H, Li AN, Li DC (2009) Cloning of a gene encoding thermostable cellobiohydrolase from the thermophilic fungus *Chaetomium thermophilum* and its expression in *Pichia pastoris*. J Appl Microbiol 106:1867–1875

Limon MC, Margolles-Clark E, Benitez T, Penttila M (2001) Addition of substrate-binding domains increases substrate-binding capacity and specific activity of a chitinase from *Trichoderma harzianum*. FEMS Microbiol Lett 198:57–63

Lo Leggio L, Larsen S (2002) The 1.62 Å structure of *Thermoascus aurantiacus* endoglucanase: completing the structural picture of subfamilies in glycoside hydrolase family 5. FEBS Lett 523:103–108

MacKenzie LF, Sulzenbacher G, Divne C, Jones TA, Woldike HF, Schulein M, Withers SG, Davies GJ (1998) Crystal structure of the family 7 endoglucanase I (Cel7B) from *Humicola insolens* at 2.2 Å resolution and identification of the catalytic nucleophile by trapping of the covalent glycosyl-enzyme intermediate. Biochem J 335(Pt 2):409–416

Maheshwari R, Bharadwaj G, Bhat MK (2000) Thermophilic fungi: their physiology and enzymes. Microbiol Mol Biol Rev 64:461–488

Mamma D, Hatzinikolaou DG, Christakopoulos P (2004) Biochemical and catalytic properties of two intracellular [beta]-glucosidases from the fungus *Penicillium decumbens* active on flavonoid glucosides. J Mol Catal B Enzym 27:183–190

Meldgaard M, Svendsen I (1994) Different effects of N-glycosylation on the thermostability of highly homologous bacterial (1,3-1,4)-beta-glucanases secreted from yeast. Microbiology 140(Pt 1):159–166

Momeni MH, Goedegebuur F, Hansson H, Karkehabadi S, Askarieh G, Mitchinson C, Larenas EA, Ståhlberg J, Sandgren M (2014) Expression, crystal structure and cellulase activity of the thermostable cellobiohydrolase Cel7A from the fungus *Humicola grisea* var. thermoidea. Acta Crystallogr D Biol Crystallogr 70:2356–2366

Moriya T, Watanabe M, Sumida N, Okakura K, Murakami T (2003) Cloning and overexpression of the *avi2* gene encoding a major cellulase produced by *Humicola insolens* FERM BP-5977. Biosci Biotechnol Biochem 67:1434–1437

Murray PG, Collins CM, Grassick A, Tuohy MG (2003) Molecular cloning, transcriptional, and expression analysis of the first cellulase gene (*cbh2*), encoding cellobiohydrolase II, from the moderately thermophilic fungus *Talaromyces emersonii* and structure prediction of the gene product. Biochem Biophys Res Commun 301:280–286

Murray P, Aro N, Collins C, Grassick A, Penttila M, Saloheimo M, Tuohy M (2004) Expression in *Trichoderma reesei* and characterisation of a thermostable family 3 beta-glucosidase from the moderately thermophilic fungus *Talaromyces emersonii*. Protein Expr Purif 38:248–257

Nakari-Setälä T, Paloheimo M, Kallio J, Vehmaanperä J, Penttilä M, Saloheimo M (2009) Genetic modification of carbon catabolite repression in *Trichoderma reesei* for improved protein production. Appl Environ Microbiol 75:4853–4860

Otzen DE, Christiansen L, Schülein M (1999) A comparative study of the unfolding of the endoglucanase Cel45 from *Humicola insolens* in denaturant and surfactant. Protein Sci 8:1878–1887

Pack SP, Yoo YJ (2004) Protein thermostability: structure-based difference of amino acid between thermophilic and mesophilic proteins. J Biotechnol 111:269–277

Pakarinen A, Haven MO, Djajadi DT, Várnai A, Puranen T, Viikari L (2014) Cellulases without carbohydrate-binding modules in high consistency ethanol production process. Biotechnol Biofuels 7:27

Parkkinen T, Koivula A, Vehmaanpera J, Rouvinen J (2008) Crystal structures of *Melanocarpus albomyces* cellobiohydrolase Cel7B in complex with cello-oligomers show high flexibility in the substrate binding. Protein Sci 17:1383–1394

Percival Zhang YH, Himmel ME, Mielenz JR (2006) Outlook for cellulase improvement: screening and selection strategies. Biotechnol Adv 24:452–481

Pettersen EF, Goddard TD, Huang CC, Couch GS, Greenblatt DM, Meng EC, Ferrin TE (2004) UCSF Chimera—a visualization system for exploratory research and analysis. J Comput Chem 25:1605–1612

Plecha S, Hall D, Tiquia-Arashiro SM (2013) Screening and characterization of soil microbes capable of degrading cellulose from switchgrass (*Panicum virgatum* L.). Environ Technol 34:1895–1904

Pocas-Fonseca MJ, Silva-Pereira I, Rocha BB, Azevedo MO (2000) Substrate-dependent differential expression of *Humicola grisea* var. thermoidea cellobiohydrolase genes. Can J Microbiol 46:749–752

Quinlan RJ, Sweeney MD, Lo Leggio L, Otten H, Poulsen JC, Johansen KS, Krogh KB, Jørgensen CI, Tovborg M, Anthonsen A, Tryfona T, Walter CP, Dupree P, Xu F, Davies GJ, Walton PH (2011) Insights into the oxidative degradation of cellulose by a copper metalloenzyme that exploits biomass components. Proc Natl Acad Sci U S A 108:15079–15084

Sandgren M, Gualfetti PJ, Paech C, Paech S, Shaw A, Gross LS, Saldajeno M, Berglund GI, Jones TA, Mitchinson C (2003) The *Humicola grisea* Cel12A enzyme structure at 1.2 Å resolution and the impact of its free cysteine residues on thermal stability. Protein Sci 12:2782–2793

Sandgren M, Berglund GI, Shaw A, Stahlberg J, Kenne L, Desmet T, Mitchinson C (2004) Crystal complex structures reveal how substrate is bound in the −4 to the +2 binding sites of *Humicola grisea* Cel12A. J Mol Biol 342:1505–1517

Sandgren M, Stahlberg J, Mitchinson C (2005) Structural and biochemical studies of GH family 12 cellulases: improved thermal stability, and ligand complexes. Prog Biophys Mol Biol 89:246–291

Schuerg T, Prahl JP, Gabriel R, Harth S, Tachea F, Chen CS, Miller M, Masson F, He Q, Brown S, Mirshiaghi M, Liang L, Tom LM, Tanjore D, Sun N, Pray TR, Singer SW (2017) Xylose induces cellulase production in *Thermoascus aurantiacus*. Biotechnol Biofuels 10:271

Shaikh FA, Withers SG (2008) Teaching old enzymes new tricks: engineering and evolution of glycosidases and glycosyl transferases for improved glycoside synthesis. Biochem Cell Biol 86:169–177

Shoseyov O, Shani Z, Levy I (2006) Carbohydrate binding modules: biochemical properties and novel applications. Microbiol Mol Biol Rev 70:283–295

Silva-Rocha R, Castro LS, Antoniêto AC, Guazzaroni ME, Persinoti GF, Silva RN (2014) Deciphering the cis-regulatory elements for XYR1 and CRE1 regulators in *Trichoderma reesei*. PLoS One 9:e99366

Suto M, Tomita F (2001) Induction and catabolite repression mechanisms of cellulase in fungi. J Biosci Bioeng 92:305–311

Szijarto N, Siika-Aho M, Tenkanen M, Alapuranen M, Vehmaanpera J, Reczey K, Viikari L (2008) Hydrolysis of amorphous and crystalline cellulose by heterologously produced cellulases of *Melanocarpus albomyces*. J Biotechnol 136:140–147

Takashima S, Nakamura A, Hidaka M, Masaki H, Uozumi T (1996) Cloning, sequencing, and expression of the cellulase genes of *Humicola grisea* var. thermoidea. J Biotechnol 50:137–147

Takashima S, Iikura H, Nakamura A, Hidaka M, Masaki H, Uozumi T (1998) Isolation of the gene and characterization of the enzymatic properties of a major exoglucanase of *Humicola grisea* without a cellulose-binding domain. J Biochem 124:717–725

Takashima S, Iikura H, Nakamura A, Hidaka M, Masaki H, Uozumi T (1999a) Comparison of gene structures and enzymatic properties between two endoglucanases from *Humicola grisea*. J Biotechnol 67:85–97

Takashima S, Iikura H, Nakamura A, Hidaka M, Masaki H, Uozumi T (1999b) Molecular cloning and expression of the novel fungal beta-glucosidase genes from *Humicola grisea* and *Trichoderma reesei*. J Biochem 125:728–736

Takashima S, Ohno M, Hidaka M, Nakamura A, Masaki H, Uozumi T (2007) Correlation between cellulose binding and activity of cellulose-binding domain mutants of *Humicola grisea* cellobiohydrolase 1. FEBS Lett 581:5891–5896

Taylor TJ, Vaisman II (2010) Discrimination of thermophilic and mesophilic proteins. BMC Struct Biol 10(Suppl 1):S5

Tiquia SM, Mormile M (2010) Extremophiles—a source of innovation for industrial and environmental applications. Environ Technol 31(8–9):823

Trivedi S, Gehlot HS, Rao SR (2006) Protein thermostability in Archaea and Eubacteria. Genet Mol Res 5:816–827

Vaheri M, Leisola M, Kauppinen V (1979) Transglycosylation products of cellulase system of *Trichoderma reesei*. Biotechnol Lett 1:41–46

Valjakka J, Rouvinen J (2003) Structure of 20K endoglucanase from *Melanocarpus albomyces* at 1.8 Å resolution. Acta Crystallogr D Biol Crystallogr 59:765–768

Van Petegem F, Vandenberghe I, Bhat MK, Van Beeumen J (2002) Atomic resolution structure of the major endoglucanase from *Thermoascus aurantiacus*. Biochem Biophys Res Commun 296:161–166

Varrot A, Frandsen TP, von Ossowski I, Boyer V, Cottaz S, Driguez H, Schulein M, Davies GJ (2003) Structural basis for ligand binding and processivity in cellobiohydrolase Cel6A from *Humicola insolens*. Structure 11:855–864

Vlasenko E, Schülein M, Cherry J, Xu F (2010) Substrate specificity of family 5, 6, 7, 9, 12, and 45 endoglucanases. Bioresour Technol 101:2405–2411

Voutilainen SP, Boer H, Linder MB, Puranen T, Rouvinen J, Vehmaanperä J, Koivula A (2007) Heterologous expression of *Melanocarpus albomyces* cellobiohydrolase Cel7B, and random mutagenesis to improve its thermostability. Enzyme Microbiol Technol 41:234–243

Voutilainen SP, Puranen T, Siika-Aho M, Lappalainen A, Alapuranen M, Kallio J, Hooman S, Viikari L, Vehmaanpera J, Koivula A (2008) Cloning, expression, and characterization of novel thermostable family 7 cellobiohydrolases. Biotechnol Bioeng 101:515–528

Voutilainen SP, Boer H, Alapuranen M, Janis J, Vehmaanpera J, Koivula A (2009) Improving the thermostability and activity of *Melanocarpus albomyces* cellobiohydrolase Cel7B. Appl Microbiol Biotechnol 83:261–272

Voutilainen SP, Murray PG, Tuohy MG, Koivula A (2010) Expression of *Talaromyces emersonii* cellobiohydrolase Cel7A in *Saccharomyces cerevisiae* and rational mutagenesis to improve its thermostability and activity. Protein Eng Des Sel 23:69–79

Wang XJ, Peng YJ, Zhang LQ, Li AN, Li DC (2012) Directed evolution and structural prediction of cellobiohydrolase II from the thermophilic fungus *Chaetomium thermophilum*. Appl Microbiol Biotechnol 95:1469–1478

Wilson DB (2009) Cellulases and biofuels. Curr Opin Biotechnol 20:295–299

Wu I, Arnold FH (2013) Engineered thermostable fungal Cel6A and Cel7A cellobiohydrolases hydrolyze cellulose efficiently at elevated temperatures. Biotechnol Bioeng 110:1874–1883

Xia W, Bai Y, Cui Y, Xu X, Qian L, Shi P, Zhang W, Luo H, Zhan X, Yao B (2016) Functional diversity of family 3 β-glucosidases from thermophilic cellulolytic fungus *Humicola insolens* Y1. Sci Rep 6:27062

Xu X, Li J, Zhang W, Huang H, Shi P, Luo H, Liu B, Zhang Y, Zhang Z, Fan Y, Yao B (2015) A neutral thermostable β-1,4-glucanase from *Humicola insolens* Y1 with potential for applications in various industries. PLoS One 10:e0124925

Yang F, Gong Y, Liu G, Zhao S, Wang J (2015) Enhancing cellulase production in thermophilic fungus *Myceliophthora thermophila* ATCC42464 by RNA interference of *cre1* gene expression. J Microbiol Biotechnol 25:1101–1107

Yennamalli RM, Rader AJ, Wolt JD, Sen TZ (2011) Thermostability in endoglucanases is fold-specific. BMC Struct Biol 11:10

Zhang Y, Tang B, Du G (2017) Self-induction system for cellulase production by cellobiose produced from glucose in *Rhizopus stolonifer*. Sci Rep 7:10161

Zhou QZ, Ji P, Zhang JY, Li X, Han C (2017) Characterization of a novel thermostable GH45 endoglucanase from *Chaetomium thermophilum* and its biodegradation of pectin. J Biosci Bioeng 124:271–276

Chapter 21
β-Galactosidases from an Acidophilic Fungus, *Teratosphaeria acidotherma* AIU BGA-1

Kimiyasu Isobe ⓘ **and Miwa Yamada**

21.1 Introduction

Compared with bacteria, fungi generally grow well at a slightly lower pH range with a lower optimal growth temperature of 25–30 °C. Because of these biological characteristics, fungi have been used to discover some acidophilic enzymes, but the isolated enzymes are not always stable at an extremely low pH range. In our previous investigation of screening for enzymes with high activity and stability at an extremely low pH range, we isolated >100 fungal strains by culturing at pH 1.0 or 2.5 and 42–45 °C, and we obtained the following results (Isobe et al. 2006). Of these isolated strains, four strains that were identified as *Aspergillus niger* produced a catalase (EC 1.11.1.6) with high activity at pH 3.0 and stability at pH 2.0. This result indicated that enzymes with high activity and stability at an extremely low pH range can be obtained from fungal strains capable of growing at extremely acidic and rather high-temperature environments. We next applied the cultivation conditions we used to the screening of extremely acidophilic β-D-galactosidases (β-D-galactoside galactohydrolase, EC 3.2.1.23).

β-D-Galactosidase, which is commonly known as lactase, catalyzes the hydrolysis of lactose into glucose and galactose. This enzyme is naturally present in the human intestine, and it contributes to the hydrolysis of lactose in milk or dairy products. Lactose-intolerant individuals, who have only low levels of this enzyme, cannot efficiently digest lactose. Microbial β-D-galactosidases are thus used to produce milk and other dairy products with low-lactose contents. Enzymes are also applied to increase the lactose solubility or sweetness of foods, because lactose is the main carbohydrate with relatively low sweetness in milk and whey.

K. Isobe (✉) · M. Yamada
Department of Biological Chemistry and Food Science, Iwate University, Morioka, Japan
e-mail: kiso@iwate-u.ac.jp

© Springer Nature Switzerland AG 2019
S. M. Tiquia-Arashiro, M. Grube (eds.), *Fungi in Extreme Environments: Ecological Role and Biotechnological Significance*,
https://doi.org/10.1007/978-3-030-19030-9_21

Many microbial β-D-galactosidases with different properties have been demonstrated to use lactose in milk or whey. For example, enzymes with a neutral optimum pH have been used to hydrolyze lactose in milk (Biermann and Glantz 1968; Cavaille and Combes 1955; Kim et al. 2003; Laderoa et al. 2002). Enzymes with an optimum pH in the acidic range have been developed for use in the processing of acidic whey and its permeate (El-Gindy 2003; Gonzalez and Monsan 1991; Hatzinikolaou et al. 2005; Nagy et al. 2001; Nakkharat and Haltrich 2006; Shaikh et al. 1999; Takenishi et al. 1983). Some acidophilic enzymes have also been used as a digestive supplement for lactose-intolerant individuals (O'Connell and Walsh 2008; Wang et al. 2009). Several β-D-galactosidases with transgalactosylation activity have been studied for the production of prebiotic galacto-oligosaccharides or structural and functional modifications of food materials (Chakraborti et al. 2000; Gupta et al. 2012; Lu et al. 2009; Wang et al. 2014). Thus, β-D-galactosidase is one of the most important enzymes in the food, dairy, baking, and pharmaceutical industries, and microorganisms have been assessed as potential sources of different types of β-D-galactosidase.

We demonstrated that the acidophilic fungus *Teratosphaeria acidotherma* AIU BGA-1, which was isolated from an acidic and high-temperature hot spring in Japan, produced not only one acidophilic β-D-galactosidase, but also two extremely acidophilic enzymes and one alkalophilic enzyme in the mycelia (Isobe et al. 2013a). This chapter describes our screening for a new fungal producer of β-D-galactosidases and its optimal cultivation conditions for the production of β-D-galactosidases. The specific characteristics of two extremely acidophilic enzymes and one alkalophilic enzyme are also summarized and compared with those of other microbial β-D-galactosidases.

21.2   Materials and Methods

21.2.1   Isolation of a Fungal Producer of Extremely Acidophilic β-D-Galactosidase

Fungal strains growing at an extremely low pH range were isolated from different acidic sites (Isobe et al. 2013a). Each suspension of soil collected from the acidic and/or high-temperature sites was spread on an agar plate of glucose medium consisting of 2.0% glucose, 0.2% corn steep liquor, 0.2% NH_4NO_3, 0.1% KH_2PO_4, 0.1% Na_2HPO_4, 0.05% $MgSO_4 \cdot 7H_2O$, 0.15% $Ca(NO_3)_2 \cdot 4H_2O$, and 0.1% $FeCl_3 \cdot 6H_2O$, pH 1.0 or pH 2.5, and incubated at 35–45 °C for 7 days. The fungal strains that grew on the agar plates were isolated and inoculated into 10 mL of a lactose medium that had the same components as the glucose medium except that glucose was replaced by lactose and the pH was 4.0. The cultivation was performed at 30 °C for 2 days with shaking.

Aspergillus niger strains from the Japan Collection of Microorganisms (JCM) were also incubated with the lactose medium under the same conditions. The selected strains were then inoculated into 150 mL of the lactose medium in a 500-mL culture flask and incubated at 30 °C for 2 days with shaking.

The mycelia of each strain were collected by filtration and suspended with 10 mM potassium phosphate buffer (KPB), pH 7.0, and disrupted using a Multi-Beads Shocker (Yasui Kikai, Osaka, Japan) below 5 °C for 8 min. The supernatant solution was collected by centrifugation at $10,000 \times g$ for 10 min and then used for the assay of β-D-galactosidase activity at pH 1.5, 4.5, and 7.0. The strain that exhibited the highest relative activity of β-D-galactosidase at pH 1.5 to that at pH 4.5 was selected as a producer of extremely acidophilic β-D-galactosidases (Isobe et al. 2013a).

21.2.2 Taxonomic Studies of the Selected Strain

The identification of the selected strain was performed at TechnoSuruga Laboratory (Shizuoka, Japan) (Isobe et al. 2013a). In brief, the strain was incubated on a potato-dextrose agar plate (Nihon Seiyaku, Tokyo, Japan) and an oatmeal agar plate (Becton Dickinson, Lincoln Park, NJ) at 25 °C in the dark, and the morphological characteristics were observed with both a compound microscope and a stereomicroscope. The sequences of 28S rDNA-D1/D2 and ITS-5.8S rDNA were analyzed using a PrimeSTAR HS DNA polymerase (Takara Bio, Otsu, Japan), an ABI BigDye Terminator v3.1 Kit (Applied Biosystems, Foster City, CA), and an ABI PRISM 3130x1 Genetic Analyzer System (Applied Biosystems). The sequence alignment and calculation of the homology levels were carried out using the GenBank, DDBJ, and EMBL databases.

21.2.3 Cultivation of the Selected Strain

The selected strain was incubated with the lactose medium, pH 4.0, at 30 °C for 4 days to obtain the acidophilic enzymes or with the lactose medium, pH 5.0, for 2 days to obtain the alkalophilic enzyme (Chiba et al. 2015; Isobe et al. 2013b; Yamada et al. 2017).

21.2.4 Enzyme Activity Assay

The β-D-galactosidase activity was assayed using 2-nitrophenyl-β-D-galactopyranoside (2-NPGA) dissolved with 0.1 M CH_3COONa-HCl, pH 1.5, 0.1 M CH_3COOH-NaOH, pH 4.5, 0.1 M KPB, pH 7.0, or 0.1 M NH_4Cl-NH_3, pH

8.5. Ten millimolar of 2-NPGA was incubated with the enzyme at 37 °C for 15 or 60 min at pH 1.5, 4.5, 7.0, or 8.5. The reaction was terminated by adding 10% Na_2CO_3 (final concentration, 5%), and the 2-nitrophenol released was quantified by measuring the absorbance at 420 nm. One unit of enzyme activity was defined as the amount of enzyme catalyzing the release of one micromole of 2-nitrophenol per minute under the above conditions. The molar absorptivity for the dye formed under the above conditions was 4.6×10^3 M^{-1} cm^{-1} (Isobe et al. 2013b; Chiba et al. 2015; Yamada et al. 2017).

The lactose hydrolyzing activity was spectrophotometrically assayed by measuring the glucose formation velocity from lactose at 555 nm by coupling with glucose oxidase and peroxidase (Isobe et al. 2013b).

21.2.5 Isolation and Purification of β-D-Galactosidases

As described further in the Results section below, the β-D-galactosidase activity was separated into three peaks by ion-exchange column chromatography (Fig. 21.2a), and one of these peaks was further separated into two peaks by hydrophobic column chromatography (Fig. 21.2b) (Isobe et al. 2013a). Of these four enzymes, two enzymes were extremely acidophilic and one enzyme was alkalophilic. Therefore, these three enzymes were purified to a homogeneous state.

21.2.5.1 Enzyme 1: Alkalophilic Enzyme with Optimal pH 8.0

The mycelia harvested at 2 days of cultivation with the lactose medium at pH 5.0 were disrupted with glass beads in 10 mM KPB, pH 7.0, by a Multi-Beads Shocker, and the supernatant collected by centrifugation was used as a crude enzyme solution. The enzyme solution was then applied onto a DEAE-Toyopearl column equilibrated with 20 mM KPB, pH 6.7. The adsorbed enzyme was eluted by a linear gradient of 20 mM KPB, pH 6.7, and the same buffer containing 0.1 M NaCl, and the active fractions with high activity at pH 7.0 were collected (the first peak in Fig. 21.2a). The eluate was applied to a Phenyl-Toyopearl column, and the adsorbed enzyme was eluted by a linear gradient of 10 mM KPB, pH 7.5, containing 1.2 M and 0.4 M ammonium sulfate. The eluate was applied to a hydroxyapatite column equilibrated with 20 mM KPB, pH 7.5, and the adsorbed enzyme was eluted by a linear gradient of 20 and 90 mM KPB, pH 7.5 (Yamada et al. 2017).

21.2.5.2 Enzyme 2a: Acidophilic Enzyme with Optimal pH 1.0

The buffer used was KPB, pH 6.5, containing 0.5 mM phenylmethanesulfonyl fluoride (PMSF). Mycelia harvested at 4 days of cultivation with the lactate medium, pH 4.0, were disrupted with glass beads in 10 mM buffer, and the supernatant collected by centrifugation was used as a crude enzyme solution. The enzyme proteins precipitated between 45% and 80% saturated ammonium sulfate were then collected from the crude enzyme solution. The precipitate was dissolved with 40 mM buffer and dialyzed against the same buffer solution. The dialyzed enzyme solution was applied to a DEAE-Toyopearl column, and the adsorbed enzymes were eluted by a linear gradient of 40 mM buffer and the same buffer containing 0.1 M NaCl (active fractions with high activity at pH 1.5 were collected; the second peak in Fig. 21.2a). This partial purification from mycelia was carried out two times, and both eluates were mixed.

To the mixed enzyme solution, 1.2 M solid ammonium sulfate was added, and the enzyme solution was applied to a Phenyl-Toyopearl column equilibrated with 20 mM buffer containing 1.2 M ammonium sulfate. The adsorbed enzyme was eluted by a linear gradient of the same equilibrium buffer and 20 mM buffer. The active fractions with high activity at pH 1.5 were collected (the first peak in Fig. 21.2b) and dialyzed against 10 mM buffer. The dialyzed enzyme solution was applied to a GigaCapQ-Toyopearl column, and the adsorbed enzyme was eluted by a linear gradient with 70 mM buffer and the same buffer containing 0.2 NaCl.

The active fractions were applied again to a Phenyl-Toyopearl column, and the adsorbed enzyme was eluted by a linear gradient of 20 mM buffer containing 1.2 and 0.5 M ammonium sulfate. The eluate was dialyzed and applied again to a GigaCapQ-Toyopearl column equilibrated with 40 mM buffer. The enzyme was eluted by a linear gradient of 40 mM buffer, pH 6.5 and 3.0, and active fractions were concentrated by ultrafiltration. The concentrated enzyme solution was applied to a Toyopearl HW-55 column equilibrated with 50 mM buffer without PMSF (Chiba et al. 2015).

21.2.5.3   Enzyme 3: Acidophilic Enzyme with Optimal pH 3.5

The purification steps from ammonium sulfate fractionation to Phenyl-Toyopearl column chromatography were performed under the same conditions as those used for enzyme 2a, and the third peak of the DEAE-Toyopearl column chromatography was collected (Fig. 21.2a). The eluate from the Phenyl-Toyopearl column was applied again to a DEAE-Toyopearl column, and the enzyme was eluted under the same conditions as those used for the first DEAE-Toyopearl column chromatography. Then, the eluate was applied to a 4-aminophenyl-β-D-galactopyranoside-Toyopearl column, and the adsorbed enzyme was eluted by a linear gradient of 10 mM KPB, pH 6.5, and the same buffer containing 0.15 NaCl. The eluate was concentrated and applied to a Toyopearl HW-55 column under the same conditions as those used for enzyme 2a (Isobe et al. 2013b).

21.2.6 Thin-Layer Chromatography (TLC) Analysis of the Reaction Products from Lactose

TLC analysis of the products by lactose hydrolysis was performed using a silica gel plate (Silica Gel 60F 254; Merck, Darmstadt, Germany) and a solvent composed of *n*-butanol:*n*-propanol:ethanol:water (2:3:3:2). The spots were detected by heating for a few minutes on a hot plate after the plate was sprayed with a solution containing 20 mg of diphenylamine, 20 μL of aniline, and 0.1 mL of phosphoric acid in 1.0 mL of acetone (Isobe et al. 2013b).

21.2.7 Buffer for Optimal pH and pH Stability

The effects of enzyme activity and stability were analyzed with CH_3COONa-HCl, pH 1.0–4.5; CH_3COONa-CH_3COOH, pH 4.0–5.5; KPB, pH 5.5–8.0; and NH_4Cl-NH_4OH, pH 8.0–10.0 (Isobe et al. 2013b).

21.3 Results

21.3.1 Isolation of a Fungal Producer of Extremely Acidophilic β-D-Galactosidase

We incubated each of the fungal strains isolated on the glucose medium and the *A. niger* strains from the JCM with the lactose medium, pH 4.0, for 2 days, and assayed the β-D-galactosidase activity at pH 1.5, 4.5, and 7.0 using the crude enzyme solution. Of the strains tested, 14 strains exhibited >5 mU per mL at each of the above three pH values, and we divided the 14 strains into three groups based on the ratios of the enzyme activities at pH 1.5, 4.5, and 7.0 (Table 21.1). One strain exhibited high activity at both pH 1.5 and pH 7.0 compared to that at pH 4.5 (group A). Three strains exhibited similar enzyme activities at both pH 1.5 and 4.5, and the values were much higher than those at pH 7.0 (group B). Ten strains exhibited higher activity at pH 4.5 than that at pH 1.5 and pH 7.0 (group C). The five *A. niger* strains (JCM nos. 1922, 5546, 5548, 5634, and 5697) were classified into group C. These results indicated that strains of groups A and B might produce acidophilic β-D-galactosidase in the mycelia.

We then incubated each crude enzyme solution from the four strains in groups A and B at pH 1.5 and 5.0 for 3 h at 37 °C, and the pH stability of β-D-galactosidase was analyzed. The enzyme from the single group A strain remained >90% and 95% of the original activity at pH 1.5 and 5.0, respectively, whereas the enzymes from the three group B strains were stable at pH 5.0, but not at pH 1.5 (data not shown). These results indicated that the group A strain might produce a β-D-galactosidase

Table 21.1 β-ᴅ-Galactosidase activity of isolated fungi and *A. niger* strains (Isobe et al. 2013a)

Group	Strain	β-ᴅ-Galactosidase activity (mU/mL)			Relative activity (%)	
		pH 1.5	pH 4.5	pH 7.0	pH 1.5/4.5	pH 7.0/4.5
A	1	12.1	5.50	7.43	220	135
	1	107	103	21.3	104	21
B	2	71.5	78.8	14.2	91	18
	3	59.1	55.8	11.7	106	21
	1	10.3	40.5	9.78	25	24
	2	10.2	42.0	8.64	24	21
	3	6.76	38.9	7.07	17	18
	4	6.52	30.5	7.43	21	24
	5	6.28	28.1	7.07	22	25
C	JCM 5697	8.09	28.0	6.16	29	22
	JCM 5634	23.5	88.1	22.5	27	26
	JCM 5548	9.84	36.8	8.33	27	23
	JCM 5546	6.40	33.6	7.00	19	21
	JCM 1922	15.9	68.8	12.9	23	19

Each strain was incubated with the lactose medium, pH 4.0, at 30 °C for 2 days. The enzyme activities of crude enzyme solution were assayed under standard assay conditions. The enzyme activities are expressed as mU/mL of the crude enzyme solution.

that is stable at an extremely low pH range. We therefore selected the group A strain and used it in our subsequent investigations (Isobe et al. 2013a).

21.3.2 Identification of the Selected Strain

The sequence of 28S rDNA-D1/D2 of the selected strain was 100% identical to that of *T. acidotherma* NBRC106057T, NBRC106058, NBRC106059, and NBRC106060 (AB537898, AB537899, AB537900, and AB537901, respectively) (Yamazaki et al. 2010), whereas the similarity to the other *Teratosphaeria* strains was <95%. The sequence of ITS-5.8S rDNA was also 100% identical to that of the above four strains, but not to that of the other *Teratosphaeria* strains.

When we incubated the selected strain at 25 °C for 9 days, it grew well at pH 2.0 on the agar plate, and the colonies were velvet with a grayish-yellow color on the potato dextrose agar plate and an olive color on the oatmeal agar plate. The vegetative hyphae had septa, and a part of the hyphae formed thick-walled cells and meristematic cells. These biological traits and morphological characteristics supported the genetic results, although the formation of an ascoma was not recognized by cultivation on an agar plate at 25 °C for 6 weeks. We therefore concluded that the selected strain belongs to *T. acidotherma*, and we named it *T. acidotherma* AIU BGA-1 (Isobe et al. 2013a).

21.3.3 Optimal Cultivation Conditions of the Selected Strain for β-D-Galactosidase Production

When we cultivated the BGA-1 strain with the medium containing glucose or lactose, the β-D-galactosidase production in the lactose medium was much higher than that in the glucose medium. Thus, β-D-galactosidase was induced by lactose (Isobe et al. 2013a).

The effect of the pH on the enzyme production was then analyzed by culturing with the lactose medium at an initial pH of 2.0–6.0. The strain grew well at pH 2.0–4.0, and the growth speed slowed down remarkably at pH 5.0 and 6.0. The enzyme production at pH 2.0–4.0 was also higher than that at pH 5.0 and 6.0. The relative values of enzyme activity at pH 1.5 to that at pH 4.5 in the cultures at pH 2.0–4.0 were also higher than those in the cultures at pH 5.0 and pH 6.0. In contrast, the relative enzyme activity at pH 7.0 to pH 4.5 became higher by shifting the initial pH from 3.0 to 6.0 (Table 21.2). These results indicated that more than one β-D-galactosidase including an extremely acidophilic enzyme was produced by the BGA-1 strain, and the product amounts of enzymes were affected by the initial pH of the lactose medium (Isobe et al. 2013a).

We also analyzed the effect of cultivation time on the enzyme production by cultivation with the lactose medium, pH 4.0, for 4 days. The enzyme activity at pH 1.5 was higher than that at pH 4.5 and 7.0 at 2 days of cultivation, and the enzyme activities at pH 1.5 and 4.5 were increased by prolonging the cultivation time, but the activity at pH 7.0 was not increased (Fig. 21.1). These results also indicated that the BGA-1 strain produced multiple β-D-galactosidases including an extremely acidophilic enzyme. We thus cultivated the strain with the lactose medium, pH 4.0, for 2 days and identified the β-D-galactosidases in the mycelia (Isobe et al. 2013a).

Table 21.2 The effect of initial pH on the β-D-galactosidase production by *T. acidotherma* AIU BGA-1 (Isobe et al. 2013a)

Initial pH	Growth (g/100 mL broth)	β-D-Galactosidase activity (U/100 mL broth)			Relative activity (%)	
		pH 1.5	pH 4.5	pH 7.0	pH 1.5/4.5	pH 7.0/4.5
2.0	0.92	1.01	0.87	0.30	116	34
3.0	2.13	2.56	2.27	0.91	113	40
4.0	1.70	2.11	1.64	0.88	135	54
5.0	0.36	0.23	0.29	0.17	77	56
6.0	0.22	0.092	0.094	0.10	98	106

T. acidotherma AIU BGA-1 was incubated with the lactose medium, pH 2.0–6.0, at 30 °C for 4 days, and the β-D-galactosidase activity was assayed at pH 1.5, 4.5, and 7.0. Data are the mean values of three independent cultivations, std. dev. <15%.

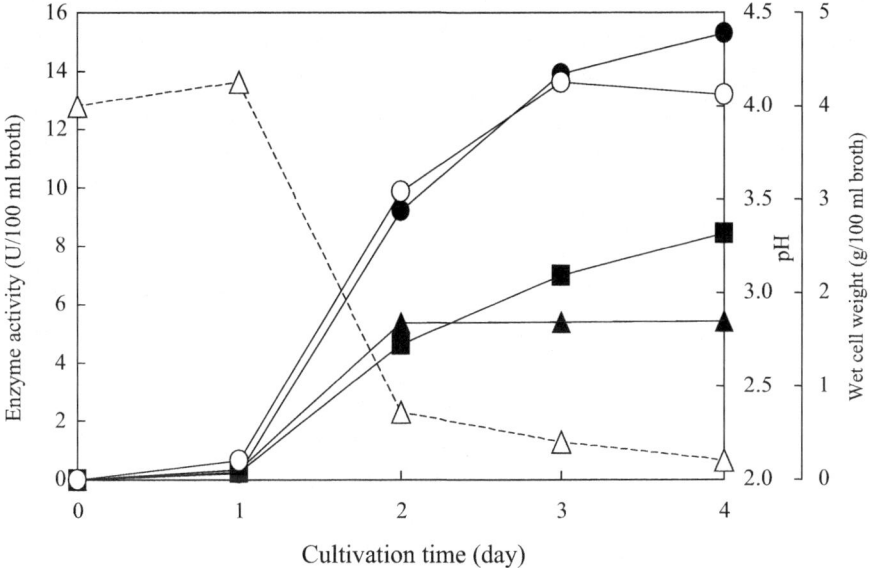

Fig. 21.1 The production of β-D-galactosidase by *T. acidotherma* AIU BGA-1 (Isobe et al. 2013a). The strain was incubated in a 3 L culture flask containing 1 L of the lactose medium at 30 °C for 4 days, and β-D-galactosidase activity was assayed each day. Enzyme activity at pH 1.5, *closed circles*; pH 4.5, *closed squares*; pH 7.0, *closed triangles*. Cell growth, *open circles*; The pH of the culture broth, *dotted line with open triangles*

21.3.4   Identification of Multiple Forms of β-D-Galactosidase

The crude enzyme solution from mycelia harvested after 2 days of cultivation was applied to DEAE-Toyopearl column chromatography, and the β-D-galactosidase activity of each fraction was assayed at pH 1.5, 4.5, and 7.0. The enzyme activity was separated into three peaks (Fig. 21.2a). The first eluate exhibited high activity at pH 7.0 (enzyme 1). The second eluate contained a discernible shoulder, and the main part exhibited high activity at pH 1.5 and low activity at pH 4.5 and 7.0, whereas the shoulder part exhibited similar activity at pH 1.5 and 4.5 (enzymes 2a plus 2b). The third eluate exhibited high activity at pH 1.5 and 4.5 (enzyme 3). The second eluate with a shoulder was further separated into two peaks by Phenyl-Toyopearl column chromatography (Fig. 21.2b), in which the former eluate exhibited high activity at pH 1.5 (enzyme 2a), and the latter eluate exhibited high activity at pH 4.5 (enzyme 2b). It was thus revealed that the BGA-1 strain produced four β-D-galactosidases.

We then analyzed the pH activity profiles of the four β-D-galactosidases between 1.0 and 8.0 by incubating at 30 °C for 60 min using the above partially purified enzymes. Enzymes 1, 2a, 2b, and 3 exhibited the highest activity at pH 8.0, 1.0–1.5, 4.0–5.5, and 2.0–4.0, respectively. Thus, enzymes 2a and 3 are acidophilic β-D-galactosidases with high activity at an extremely low pH range, and enzyme 1 is an alkalophilic β-D-galactosidase (Isobe et al. 2013a). We therefore focused on these three enzymes (1, 2a, and 3) and purified them.

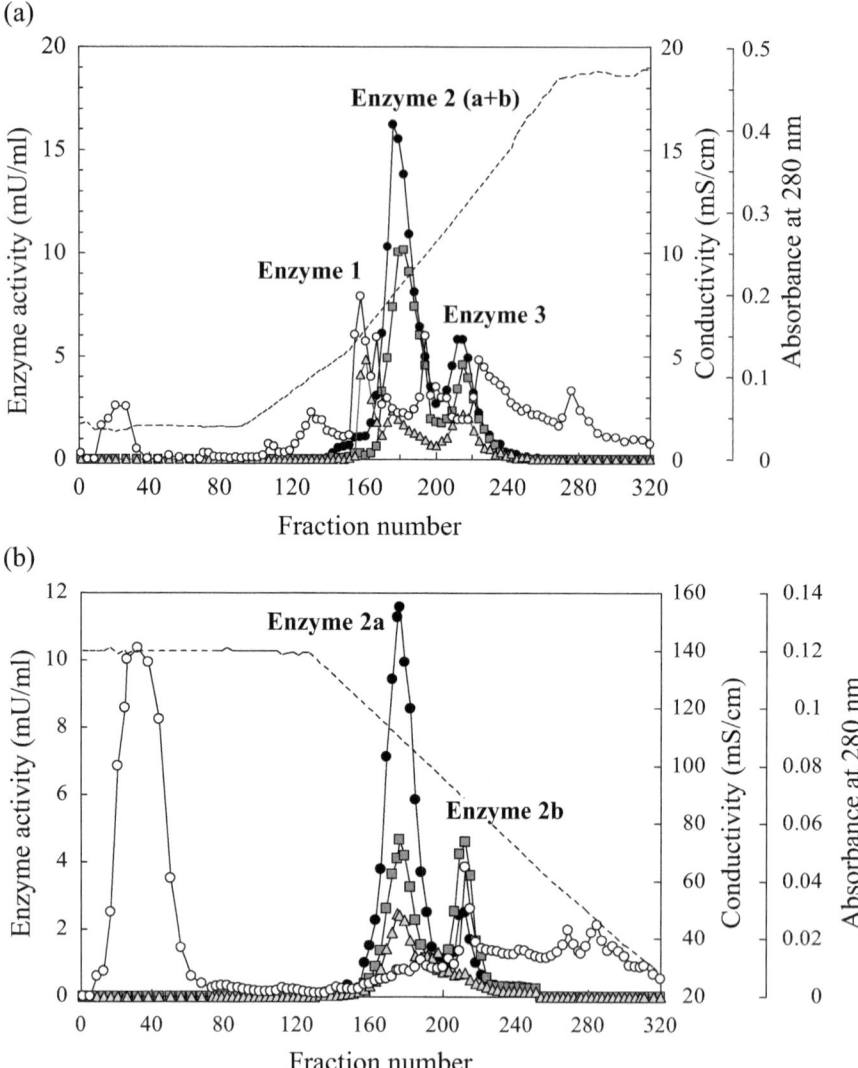

Fig. 21.2 Elution profiles of the β-D-galactosidases by column chromatography with DEAE-Toyopearl (**a**) and Phenyl-Toyopearl (**b**) columns (Isobe et al. 2013a). (**a**) The crude enzyme solution was applied to a DEAE-Toyopearl column equilibrated with 10 mM KPB, pH 6.5, and adsorbed enzymes were eluted by a linear gradient of 10 mM KPB pH 6.5 and 0.16 M NaCl. (**b**) To the second eluate from a DEAE-Toyopearl column, 1.2 M solid ammonium sulfate was added, and the enzyme solution was applied to a Phenyl-Toyopearl column. The adsorbed enzymes were eluted by a linear gradient of 20 mM KPB, pH 6.5, containing 1.2 M ammonium sulfate and 20 mM buffer. Enzyme activity at pH 1.5, *closed circles*; pH 4.5, *gray squares*; pH 7.0, gray *triangles*. The absorbance value at 280 nm, *open circles*; conductivity, *dotted lines*

21.3.5 Purification of the Three Enzymes

Since the pH stability of enzymes is important for enzyme purification, we analyzed the pH stability of the above three enzymes by incubation at pH 1.5, 5.0, and 7.0 for 3 h at 37 °C. Enzymes 2a and 3 retained >95% of the original activity at these three pH values. In contrast, enzyme 1 was stable at pH 7.0, but unstable at pH 1.5 and 5.0 (Isobe et al. 2013a). We thus purified enzymes 1, 2a, and 3 to the homogeneous state using different pH values and methods as described in the Materials and Methods section above. The specific activity of purified enzymes 1, 2a, and 3 was 4.05, 8.03, and 5.48 U/mg of protein at pH 8.5, 1.5, and 4.5, respectively (Yamada et al. 2017; Chiba et al. 2015; Isobe et al. 2013b). These purified enzymes were used for subsequent studies.

21.3.6 Molecular Mass and N-Terminal Amino Acid Sequence

The molecular mass of the native and denatured enzyme 1 was estimated to be 180 kDa by gel filtration and 120 kDa by sodium dodecyl sulfate-polyacrylamide gel electrophoresis (SDS-PAGE), indicating that enzyme 1 has a monomeric structure (Yamada et al. 2017). The molecular mass of enzyme 2a was also estimated to be 180 kDa by gel filtration. The denatured enzyme 2a was separated into two protein bands with the molecular masses of 120 and 66 kDa on SDS-PAGE (Chiba et al. 2015). The native enzyme 3 showed a molecular mass of 140 kDa by gel filtration, and the denatured state was separated into two protein bands with the molecular masses of 86 and 50 kDa on SDS-PAGE (Isobe et al. 2013b). Thus, enzymes 2a and 3 were composed of a heterodimer.

The N-terminal amino acid sequences of the large subunit of enzyme 2a and the small subunit of enzyme 3 were found to be SPNLQDIVTVDGESY and NTRMIIFNDK, respectively; those of the small subunit of enzyme 2a and the large subunit of enzyme 3 were not identified (Chiba et al. 2015; Isobe et al. 2013b). The N-terminal amino acid sequence of enzyme 1 was also not identified (Yamada et al. 2017).

21.3.7 Effect of pH on Enzyme Activity and Stability

The pH activity profiles of three enzymes were analyzed in the pH range of 1.0–10.0 by incubation at 37 °C for 15 min. Enzyme 1 exhibited the highest activity at pH 8.0, and the enzyme activity at pH 6.5 and pH 10.0 was 50% of that at pH 8.0 (Yamada et al. 2017). Enzyme 2a exhibited the highest activity at pH 1.0, and the enzyme activity at pH 3.5–4.0 was 50% of that at pH 1.0 (Chiba et al. 2015). Enzyme 3 exhibited the highest activity at pH 3.0–3.5, and the enzyme activity at pH 1.0, 1.5, and 6.0 was 50%, 85% and 80% of that at pH 3.5, respectively (Isobe et al. 2013b) (Fig. 21.3a).

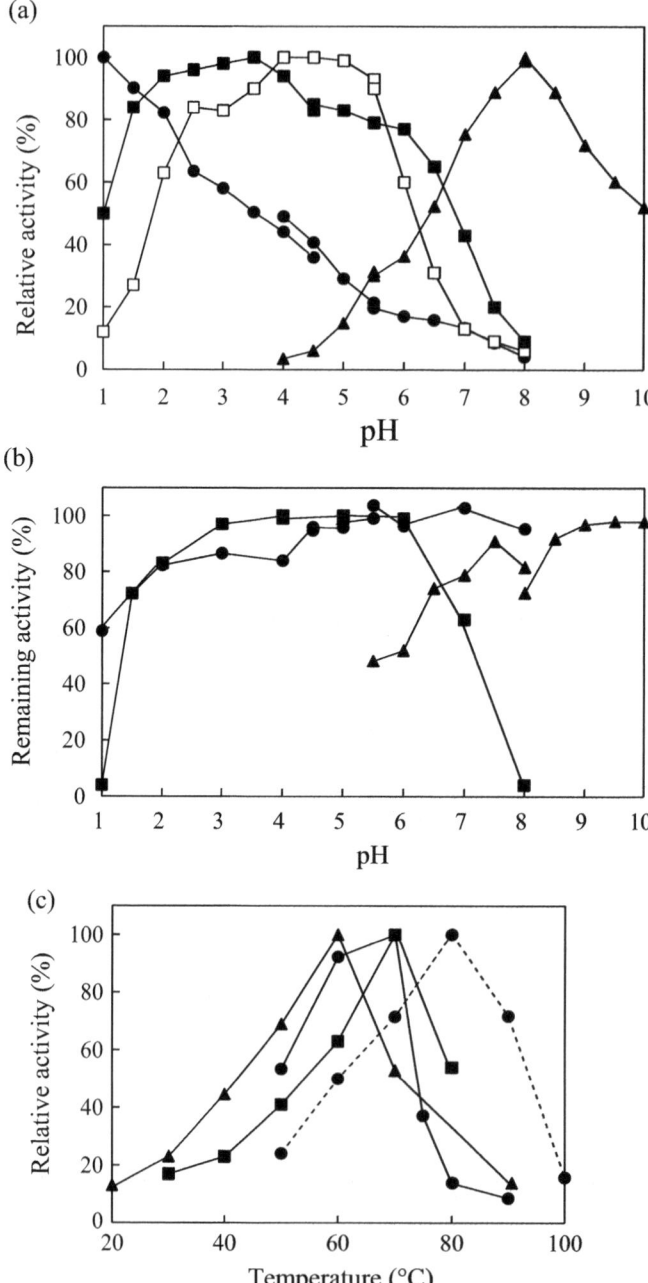

Fig. 21.3 Effects of pH and temperature on the activity and stability of the β-D-galactosidases from *T. acidotherma* AIU BGA-1 (Isobe et al. 2013b; Chiba et al. 2015; Yamada et al. 2017). (**a**) Effects of pH on enzyme activity. The β-D-galactosidase activities of enzymes 1, 2a, and 3 were

These three enzymes also exhibited different pH stability. When alkalophilic enzyme 1 was incubated at 40 °C for 30 min, >80% of the original activity remained in the pH range of 7.5–10.0, but <50% of the original activity remained below pH 6.0 (Yamada et al. 2017). When enzyme 2a was incubated at 50 °C for 30 min, >80% of the original activity remained in the pH range of 2.0–8.0, and 60% of the original activity remained at pH 1.0 (Chiba et al. 2015). Enzyme 3 retained >80% of the original activity in the pH range 2.0–6.0 by incubation at 50 °C for 60 min, and 70% and 60% of the original activity at pH 1.5 and 7.0, respectively (Isobe et al. 2013b) (Fig. 21.3b).

21.3.8 Effect of Temperature on Enzyme Activity

We analyzed the optimal temperature of the three enzymes at the optimal pH of each enzyme. Enzymes 1, 2a, and 3 exhibited the highest activity at 60, 70, and 70 °C at pH 8.5, 1.0, and 4.5, respectively. Since the stability of enzyme 2a at above pH 2.0 was higher than that at pH 1.0, we also analyzed the optimal temperature at pH 4.5, and its highest activity was obtained at 80 °C (Yamada et al. 2017; Chiba et al. 2015; Isobe et al. 2013b) (Fig. 21.3c).

21.3.9 Substrate Specificity and Kinetic Properties

We analyzed the substrate specificity of enzymes 1, 2a, and 3 at the optimal pH of each enzyme using lactose, 2-NPGA, 4-NPGA, 4-nitrophenyl β-D-glucopyranoside (4-NPGL), 4-nitrophenyl β-D-xylopyranoside (4-NPXY), 4-nitrophenyl β-D-fucopyranoside (4-NPFU), 4-nitrophenyl β-D-mannopyranoside (4-NPMA), and 4-nitrophenyl β-L-arabinofuranoside (4-NPAR). Enzyme 1 exhibited the highest activity toward 4-NPGA, and the apparent K_m value was estimated to be 0.49 mM at pH 8.5. Enzyme 1 also exhibited activity toward 2-NPGA, whose K_m value was

---

Fig. 21.3 (continued) assayed using the purified enzymes under standard conditions using the indicated buffer and pH. The β-D-galactosidase activity of enzyme 2b was assayed using partially purified enzyme 2b (the eluate from Phenyl-Toyopearl column chromatography). The relative activities are shown as a percentage of the highest activity for each enzyme. (**b**) The pH stability. Enzymes 1, 2a, and 3 were incubated at 40 °C for 30 min, 50 °C for 30 min, and 50 °C for 60 min, respectively, without substrate. The enzyme activities before and after heating were assayed under the standard assay conditions of each enzyme. The remaining activities are shown by the percentage to the enzyme activity without heating. (**c**) Effects of temperature on enzyme activity. The β-D-galactosidase activities of enzymes 1, 2a, and 3 were assayed at the optimal pH of each enzyme at the indicated temperatures. The activity of enzyme 2a was also assayed at pH 4.5 (*dotted line with closed circles*). The β-D-galactosidase activity of enzyme 1, *closed triangles*; enzyme 2a, *closed circles*; enzyme 2b, *open squares*; and enzyme 3, *closed squares*

Table 21.3 Substrate specificity and kinetic values of enzymes 1, 2a, and 3 (Yamada et al. 2017; Chiba et al. 2015; Isobe et al. 2013b)

	Enzyme 1		Enzyme 2a				Enzyme 3	
	Relative activity (%) (10 mM)	K_m (mM)	Relative activity (%) (10 mM)		K_m (mM)		Relative activity (%) (20 mM)	K_m (mM)
Substrate	pH 8.5	pH 8.5	pH 1.5	pH 4.5	pH 1.5	pH 4.5	pH 4.5	pH 4.5
2-NPGA	37.8	0.35	59.2	31.8	0.82	0.43	62.5	0.19
4-NPGA	100	0.49	23.8	86.7	0.46	0.41	100	1.2
4-NPFU	<5	n.d.	100	100	n.d.	n.d.	10	n.d.
Lactose		701			618			170

Enzyme activities for 2-NPGA, 4-NPGA, and 4-NPFU were assayed under standard assay conditions at the optimal pH or stable pH of each enzyme. The relative activities are expressed as a percentage of the highest enzyme activity. The lactose-hydrolyzing activity was assayed by the glucose oxidase method. The substrate concentrations for estimating the K_m value for 2-NPGA, 4-NPGA, and lactose are described in references (Yamada et al. 2017; Chiba et al. 2015; Isobe et al. 2013b). n.d., not detectable. *2-NPGA* 2-nitrophenyl-β-D-galactopyranoside, *4-NPGA* 4-nitrophenyl-β-D-galactopyranoside, *4-NPFU* 4-nitrophenyl β-D-fucopyranoside

estimated to be 0.35 mM, but it did not exhibit activity toward the other nitrophenyl pyranosides tested. The K_m value for lactose was estimated as 701 mM by the glucose oxidase method (Yamada et al. 2017).

Enzyme 2a exhibited high activity toward 4-NPFU, 2-NPGA, and 4-NPGA at pH 1.5 and 4.5. Among these substrates, 4-NPFU was the best substrate, but its K_m value was not estimated because the reaction rate for 4-NPFU increased almost linearly up to 40 mM 4-NPFU. The apparent K_m value for 2-NPGA was estimated as 0.82 mM at pH 1.5, and 0.43 mM at pH 4.5, respectively. That for 4-NPGA was 0.46 mM at pH 1.5, and 0.41 mM at pH 4.5. The value for lactose was 618 mM at pH 1.5 (Chiba et al. 2015).

Enzyme 3 exhibited the highest activity for 4-NPGA, and 2-NPGA was also a good substrate at pH 4.5. The apparent K_m values for 2-NPGA, 4-NPGA, and lactose were estimated as 0.19, 1.2, and 170 mM, respectively (Isobe et al. 2013b) (Table 21.3).

21.3.10   Inhibitor Specificity

Enzyme 1 was activated and stabilized by Mn^{2+} but inhibited by chelating reagents (EDTA and *o*-phenanthroline) and metals (Cu^{2+} and Ni^{2+}) (Yamada et al. 2017). Enzymes 2a and 3 were not remarkably affected by carbonyl or the chelating and sulfhydryl reagents or metals tested (Chiba et al. 2015; Isobe et al. 2013b).

21.3.11 Hydrolysis of Lactose, Milk, and Whey

When enzyme 1 was incubated with lactose at 37 °C for 20 h at pH 8.5, galactose and glucose were identified as the products from lactose, and several other spots were detected at a different position from those of the galactose and glucose by TLC analysis (Fig. 21.4a). This result indicated that enzyme 1 would have galacto-oligosaccharide formation activity at the alkaline pH range (Yamada et al. 2017). In the

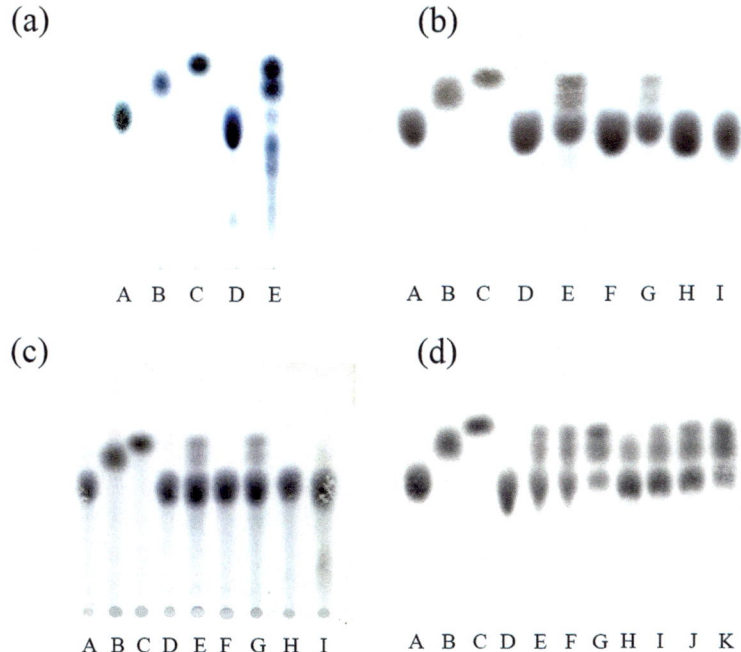

Fig. 21.4 TLC analysis of the reaction products from lactose by hydrolysis with β-ᴅ-galactosidases from *T. acidotherma* AIU BGA-1 (Yamada et al. 2017; Chiba et al. 2015; Isobe et al. 2013b). Enzymes 1, 2b, and 3 were each incubated with lactose, and the reaction was terminated by boiling. The reaction products were detected as described in the Materials and Methods section. (**a**) Enzyme 1 (16 mU) was incubated with 0.3 M lactose and 2 mM MnCl₂ at 37 °C and pH 8.5 for 20 h. A, lactose; B, galactose; C, glucose; D, lactose without enzyme 1; E, lactose with enzyme 1. (**b**) Enzyme 2a (4.2 mU) was incubated with 0.1 M lactose at 37 °C for 20 h at pH 1.5, 4.5, and 7.0. A, lactose; B, galactose; C, glucose; D, lactose incubated at pH 1.5 without enzyme; E, lactose incubated at pH 1.5 with enzyme 2a; F, lactose incubated at pH 4.5 without enzyme; G, lactose incubated at pH 4.5 with enzyme 2a; H, lactose incubated at pH 7.0 without enzyme; I, lactose incubated at pH 7.0 with enzyme 2a. (**c**) Enzyme 3 (4.2 mU) was incubated with 0.1 M lactose under the same conditions as those used for enzyme 2a. The lines of TLC are also the same as that in panel (**b**). (**d**) Milk or whey from yogurt adjusted to pH 1.5 was incubated with enzyme 2a (50–300 mU) at 50 °C for 20 h. A, lactose; B, galactose; C, glucose; D, milk without enzyme; E, milk with 50 mU of enzyme 2a; F, milk with 100 mU of enzyme 2a; G, milk with 300 mU of enzyme 2a; H, whey without enzyme; I, whey with 50 mU of enzyme 2a; J, whey with 100 mU of enzyme 2a; K, whey with 300 mU of enzyme 2a

cases of enzymes 2a and 3, the formation of galactose and glucose was also identified from lactose by incubation at pH 1.5 and 4.5, but other new spots were not detected. Neither enzyme hydrolyzed lactose by incubation at pH 7.0 (Chiba et al. 2015; Isobe et al. 2013b) (Fig. 21.4b, c). These results agreed with the pH activity profiles of both enzymes. We then incubated enzyme 2a with milk or whey at pH 1.5, and we observed that the enzyme efficiently catalyzed the hydrolysis of lactose in milk and whey into galactose and glucose at an extremely low pH range (Chiba et al. 2015) (Fig. 21.4d).

21.4 Discussion

There is increasing demand for enzymes with high activity and stability at a low pH range in order to expand the application of enzymes into extremely low pH ranges. Since fungi generally grow well in the low pH range, they have attracted attention as producers of such enzymes. However, there have been only a few reports on the isolation of fungal strains that produce extremely acidophilic enzymes. We isolated *T. acidotherma* AIU BGA-1 from an acidic hot spring as a new producer of extremely acidophilic β-D-galactosidases. This strain grew well in the pH range 2.0–4.0, and its growth rate slowed down at pH 5.0 and 6.0. The production of β-D-galactosidases was also highest by incubation at pH 3.0 and decreased by shifting the initial pH to 5.0–6.0. These results are in agreement with our previous finding that when fungal strains capable of growing well under extremely acidic conditions at high temperature were cultivated with a liquid medium with the initial pH of 3.0, the final pH of the culture broth varied widely between 2.5 and 9.0, and enzymes with high activity and stability at the low pH range were obtained in the strains maintained at pH below 6.0 (Isobe et al. 2006). Thus, the isolation of fungal strains that grow well under extremely low pH and high temperature conditions was effective to obtain extremely acidophilic enzymes.

   T. acidotherma AIU BGA-1 showed interesting enzyme production characteristics. This strain produced four β-D-galactosidases with different pH activity profiles in the mycelia, two extremely acidophilic enzymes, one acidophilic enzyme, and one alkalophilic enzyme, although the strain grew well in the acidic pH range. Regarding the production of multiple forms of β-D-galactosidase, it was reported that *A. niger* produced three extracellular β-D-galactosidases with different molecular masses (Widmer and Leuba 1979). All of these enzymes exhibited high activity at the pH range 2.0–5.0, and their pI values and amino acid compositions were similar, whereas their carbohydrate contents differed. It was thus presumed that the multiple forms of the *A. niger* β-D-galactosidase are due to the differences of the carbohydrate content.

   It was also reported that *Aspergillus carbonalius* produced two extracellular β-D-galactosidases with different pH activity profiles (one is optimal at pH 2.5–3.5 and the other is optimal at pH 3.0–3.5; both enzymes have no activity at pH 1.5) and different molecular masses (O'Connell and Walsh 2008). Although both enzymes are glycoproteins, their amino acid compositions differ. Thus, multiple forms of the *A. carbonalius* β-D-galactosidase are related to the differences in amino acid and carbohydrate compositions. Multiple forms of β-D-galactosidase were also demonstrated from a commercial enzyme powder produced by *Saccharomyces lactis* (Mbuyi-Kalala et al. 1988). The enzyme powder contained four β-D-galactosidases with different molecular mass and carbohydrate content values. The β-D-galactosidases exhibited high activity at the neutral pH range, and the multiplicity of the *Saccharomyces* β-D-galactosidase was affected by the differences in carbohydrate content in a manner similar to that of the *A. niger* β-D-galactosidases. In contrast to these strains, *T. acidotherma* AIU BGA-1 produced four intracellular β-D-galactosidases with widely different pH activity profiles: two were extremely acidophilic, one was acidophilic, and one was alkalophilic. Thus, *T. acidotherma* AIU BGA-1 is the first fungal producer of multiple forms of intracellular β-D-galactosidase with widely different pH activity profiles.

Among the three acidophilic β-D-galactosidases from *T. acidotherma* AIU BGA-1, enzyme 2a exhibited the highest activity at pH 1.0. Enzyme 3 exhibited the highest activity at pH 3.0–3.5, and the enzyme activity at pH 1.0 was 50% of that at pH 3.0. Enzyme 2b exhibited the highest activity at pH 4.0–5.0, and the enzyme activity at pH 2.0 and pH 1.0 was 63% and 12% of that at pH 4.0, respectively (Fig. 21.3a). Thus, enzymes 2a and 3 are extremely acidophilic.

Many acidophilic β-D-galactosidases with high activity at a low pH range have been identified from different fungal strains (Table 21.4). Most of those acidophilic enzymes exhibit maximal activity in the pH range 2.5–5.5 (Gonzalez and Monsan 1991; Hatzinikolaou et al. 2005; Isobe et al. 2013b; Manzanares et al. 1998; Nagy et al. 2001; O'Connell and Walsh 2008; Shaikh et al. 1999; Takenishi et al. 1983; Widmer and Leuba 1979). An acidophilic β-galactosidase with maximum activity at pH 1.5 was isolated from *Bispora* sp., and its enzyme activity at pH 1.0 was 60–70% of that at pH 1.5 (Wang et al. 2009). Thus, enzyme 2a from *T. acidotherma* AIU BGA-1 exhibited the lowest optimal pH value among the microbial β-D-galactosidases reported to date. Enzyme 2a also exhibited its highest stability and highest optimal temperature at extremely acidic pH values among the known microbial β-D-galactosidases.

With respect to the physicochemical properties of microbial β-D-galactosidases, enzymes 2a and 3 differ from the acidophilic enzymes from *Penicillium chrysogenum* (Nagy et al. 2001) and *Rhizopus* sp. (Shaikh et al. 1999) in terms of molecular mass and subunit compositions. Enzyme 2a has a 180 kDa molecular mass with two heterosubunits of 120 and 66 kDa. Enzyme 3 has a 140 kDa molecular mass composed of two heterosubunits of 86 and 50 kDa (Chiba et al. 2015; Isobe et al. 2013b). In contrast, the enzyme from *P. chrysogenum* and the enzyme from *Rhizopus* sp. have a 270 kDa molecular mass with four identical subunits and a 250 kDa molecular mass with two identical subunits, respectively. Enzymes 2a and 3 from

Table 21.4 Comparison of characteristics of acidophilic β-D-galactosidases from fungal strains

Origin	Mol. mass (kDa)	Subunit (kDa)	K_m value (mM)		Active range (pH)	Opt-pH	pH-Stab	Opt-temp (°C)	Inhibitor	pI	Refs.
			2-NPG	Lactose							
Enzyme 2a	180	66, 120	0.43		1.0–4.5	<1.0	1.0–8.0	80	Non	4.6	Chiba et al. (2015)
Enzyme 3	140	50, 86	0.19		1.0–7.0	3.0–4.0	1.5–7.0	70	Non	4.1	Isobe et al. (2013b)
A. niger				32–89	2.5–6.0	3.0–3.5	3.0–6.0	65			Hatzinikolaou et al. (2005)
A. niger	124–173		2.4	85–125	1.7–5.2	3.5		70		4.6	Widmer and Leuba (1979)
A. niger	93	93	1.3			4.0	3.0–5.0	60–65		4.6	Manzanares et al. (1998)
A. fonsecaeus	126		1.8	61	2.0–6.0	2.6–4.5				4.2	Gonzalez and Monsan (1991)
P. chrysogenum	270	66	1.8		3.5–6.5	4				4.6	Nagy et al. (2001)
P. multicolor	126	130	0.6	8.9	2.0–6.0	4.0	3.5–7.5	60	Hg^{2+}, Cu^{2+}		Takenishi et al. (1983)
Rhizopus sp.	250	120	1.3	50		4.5	3.5–7.5	60	Hg^{2+}, Cu^{2+}	4.2	Shaikh et al. (1999)
T. thermophilus	49	50	24	18	4.5–7.0	5.5–6.0	5.5–7.5	50	Cu^{2+}	4.5	Nakkharat and Haltrich (2006)
A. carbonarius	139	110	2.6	83	3.0–6.0	3.0–3.5		55			O'Connell and Walsh (2008)
A. carbonarius	152	120	0.56	309	2.0–5.5	2.5–3.0		65			O'Connell and Walsh (2008)
Bisopra sp.	130	130	5.2	0.31	1.0–3.0	1.5	1.5–6.0	60	SDS, ME		Wang et al. (2009)

Talaromyces acidotherma AIU BGA-1 are also clearly different from acidophilic enzymes from *A. niger* (Manzanares et al. 1998; Widmer and Leuba 1979), *T. thermophilus* (Nakkharat and Haltrich 2006), *Penicillium multicolor* (Takenishi et al. 1983), *Aspergillus fonsecaeus* (Gonzalez and Monsan 1991), *Aspergillus carbonarius* (O'Connell and Walsh 2008), and *Bispora* sp. (Wang et al. 2009), because these seven enzymes each has a monomeric structure.

Although *T. acidotherma* AIU BGA-1 grew well at pH 2.0–4.0, the strain also produced alkalophilic β-D-galactosidase (enzyme 1) with optimal activity at pH 8.0 and high stability at pH 7.0–10.0 (Yamada et al. 2017). To date, many acidophilic and neutrophilic β-D-galactosidases have been isolated from fungal strains, but alkalophilic β-galactosidases with maximal activity at pH 7.5–10.5 have been derived only from bacterial strains such as *Meiothermus ruber* DSM 1279 (Gupta et al. 2012), *Bacillus* sp. MTCC 3088 (Chakraborti et al. 2000), *Enterobacter cloacae* B5 (Lu et al. 2009), *Pseudoalteromonas* sp. (Fernandes et al. 2002), *Arthrobacter* sp. ON14 (Xu et al. 2011), and *Arthrobacter psychrolactophilus* F2 (Nakagawa et al. 2007) (Table 21.5). Thus, *T. acidotherma* AIU BGA-1 is the first fungal producer of alkalophilic β-D-galactosidase.

Among these alkalophilic β-D-galactosidases, the enzymes from *M. ruber* DSM 1279, *Bacillus* sp. MTCC 3088, and *E. cloacae* B5 exhibited not only lactose hydrolysis activity but also transgalactosylation activity to catalyze the formation of galacto-oligosaccharide (Chakraborti et al. 2000; Gupta et al. 2012; Lu et al. 2009; Wang et al. 2014). Since *T. acidotherma* enzyme 1 also produced galacto-oligosaccharides by incubation with lactose, the enzyme would catalyze the same reactions as those by bacterial alkalophilic β-D-galactosidases, whereas the K_m value of enzyme 1 for lactose was higher than that of bacterial enzymes.

Regarding physicochemical properties, *T. acidotherma* enzyme 1 has a monomeric structure with a 180-kDa molecular mass. On the other hand, enzymes from *M. ruber* DSM 1279, *Bacillus* sp. MTCC 3088, and *E. cloacae* B5 have the molecular mass of 190, 484, and 442 kDa, respectively, and are composed of hetero- or identical subunits (Chakraborti et al. 2000; Gupta et al. 2012; Lu et al. 2009). Thus, alkalophilic β-D-galactosidase from *T. acidotherma* AIU BGA-1 catalyzes the same reaction, but the structure is different from the structures of these three bacterial enzymes. The optimal temperature of *T. acidotherma* enzyme 1 is also significantly different from those of the alkalophilic β-D-galactosidases from *E. cloacae* (Lu et al. 2009), *Pseudoalteromonas* sp. (Fernandes et al. 2002), *Arthrobacter* sp. (Xu et al. 2011), and *A. psychrolactophilus* (Nakagawa et al. 2007) (Table 21.5).

Table 21.5 Comparison of characteristics of alkalophilic β-d-galactosidases from microorganisms

Origin	Mol. mass (kDa)	Subunit (kDa)	K_m value (mM) 2-NPGA	K_m value (mM) Lactose	Opt-pH	Opt-temp (°C)	Inhibitor (metals)	Activator (metals)	Refs.
Enzyme 1	180	120	0.35	701	8.0	60	Cu^{2+}, Ni^{2+}	Mn^{2+}	Yamada et al. (2017)
M. ruber	190	46	5.0	333	8.0	65	Cu^{2+}, Cd^{2+}, Fe^{2+}	Mn^{2+}, Zn^{2+}	Gupta et al. (2012)
Bacillus sp.	484	115, 87, 72, 46, 41			8.0	60	Cu^{2+}, Ni^{2+}, Hg^{2+}, Ag^{2+}	Mg^{2+}	Chakraborti et al. (2000)
E. cloacae	442	119	0.01	0.3	7.5–10.5	35			Lu et al. (2009)
Pseudoalteromonas sp.	513	110	0.16		9.0	26	Cu^{2+}, Zn^{2+}, Hg^{2+}	Mg^{2+}, Mn^{2+}	Fernandes et al. (2002)
Arthrobacter sp.		116			8.0	15	Cu^{2+}, Zn^{2+}	Mg^{2+}, Mn^{2+}	Xu et al. (2011)
A. psychrolactophilus		130	2.7	42.1	8.0	10			Nakagawa et al. (2007)

21.5   Conclusions

The strain *T. acidotherma* isolated from an extremely low-pH and high-temperature environment grew well at a low pH range and produced four intracellular β-D-galactosidases by incubation with lactose: two extremely acidophilic enzymes, one acidophilic enzyme, and one alkalophilic enzyme. This strain will be useful to obtain β-D-galactosidases with different pH activity profiles by one cultivation. Among these enzymes, the two extremely acidophilic enzymes and the single alkalophilic enzyme are novel β-D-galactosidases. The extremely acidophilic enzymes are desirable for the efficient hydrolysis of lactose in milk or whey at an extremely low pH range. The alkalophilic enzyme would be applicable to the production of galacto-oligosaccharides. Thus, the β-D-galactosidases from *T. acidotherma* AIU BGA-1 have great advantages for industrial applications.

References

Biermann L, Glantz MD (1968) Isolation and characterization of β-galactosidase from *Saccharomyces lactis*. Biochim Biophys Acta 167:373–377

Cavaille D, Combes D (1955) Characterization of β-galactosidase from *Kluyveromyces lactis*. Biotechnol Appl Biochem 22:55–64

Chakraborti S, Sani RK, Banerjee UC, Sobti RC (2000) Purification and characterization of a novel β-galactosidase from *Bacillus* sp. MTCC 3088. J Ind Microbiol Biotechnol 24:58–63

Chiba S, Yamada M, Isobe K (2015) Novel acidophilic β-galactosidase with high activity at extremely acidic pH region from *Teratosphaeria acidotherma* AIU BGA-1. J Biosci Bioeng 120:263–267

El-Gindy A (2003) Production, partial purification and some properties of β-galactosidase from *Aspergillus carbonarius*. Folia Microbiol 48:581–584

Fernandes S, Geueke B, Delgado O, Coleman J, Hatti-Kaul R (2002) β-Galactosidase from a cold-adapted bacterium: purification, characterization and application for lactose hydrolysis. Appl Microbiol Biotechnol 58:313–321

Gonzalez R, Monsan P (1991) Purification and some properties of β-galactosidase from *Aspergillus fonsecaeus*. Enzyme Microb Technol 13:349–352

Gupta R, Govil T, Capalash N, Sharma P (2012) Characterization of a glycoside hydrolase family 1 β-galactosidase from hot spring metagenome with transglycosylation activity. Appl Biochem Biotechnol 168:1681–1693

Hatzinikolaou D, Katsifas E, Mamma D, Karagouni A, Christakopoulos P, Kekos D (2005) Modeling of the simultaneous hydrolysis-ultrafiltration of whey permeate by a thermostable β-galactosidase from *Aspergillus niger*. Biochem Eng J 24:161–172

Isobe K, Inoue N, Takamatsu Y, Kamada K, Wakao N (2006) Production of catalase by fungi growing at low pH and high temperature. J Biosci Bioeng 101:73–76

Isobe K, Takahashi N, Chiba S, Yamashita M, Koyama T (2013a) Acidophilic fungus, *Teratosphaeria acidotherma* AIU BGA-1, produces multiple forms of intracellular β-galactosidase. J Biosci Bioeng 116:171–174

Isobe K, Yamashita M, Chiba S, Takahashi N, Koyama T (2013b) Characterization of new β-galactosidase from acidophilic fungus, *Teratosphaeria acidotherma* AIU BGA-1. J Biosci Bioeng 116:293–297

Kim CS, Ji ES, Oh DK (2003) Expression and characterization of *Kluyveromyces lactis* β-galactosidase in *Escherichia coli*. Biotechnol Lett 25:1769–1774

Laderoa M, Santosa A, Garcíab JL, Carrascosac AV, Pesselac BCC, García-Ochoaa F (2002) Studies on the activity and the stability of β-galactosidases from *Thermus* sp strain T2 and from *Kluyveromyces fragilis*. Enzyme Microb Technol 30:392–405

Lu LL, Xiao M, Li ZY, Li YM, Wang FS (2009) A novel transglycosylating β-galactosidase from *Enterobacter cloacae* B5. Process Biochem 44:232–236

Manzanares P, de Graaff LH, Visser J (1998) Characterization of galactosidases from *Aspergillus niger*: purification of a novel α-galactosidase activity. Enzyme Microb Technol 22:383–390

Mbuyi-Kalala A, Schnek AG, Leonis J (1988) Separation and characterization of four enzyme forms of β-galactosidase from *Saccharomyces lactis*. Eur J Biochem 178:437–443

Nagy Z, Kiss T, Szentirmai A, Biro S (2001) β-Galactosidase of *Penicillium chrysogenum*: production, purification, and characterization of the enzyme. Protein Expr Purif 21:24–29

Nakagawa T, Ikehata R, Myoda T, Miyaji T, Tomizuka N (2007) Overexpression and functional analysis of cold-active β-galactosidase from *Arthrobacter psychrolactophilus* strain F2. Protein Expr Purif 54:295–299

Nakkharat P, Haltrich D (2006) Purification and characterization of an intracellular enzyme with β-glucosidase and β-galactosidase activity from the thermophilic fungus *Talaromyces thermophilus* CBS 236.58. J Biotechnol 123:304–313

O'Connell S, Walsh G (2008) Application relevant studies of fungal β-galactosidases with potential application in the alleviation of lactose intolerance. Appl Biochem Biotechnol 149:129–138

Shaikh SA, Khire JM, Khan MI (1999) Characterization of a thermostable extracellular β-galactosidase from a thermophilic fungus *Rhizomucor* sp. Biochim Biophys Acta 1472:314–322

Takenishi S, Watanabe Y, Miwa T, Kobayashi R (1983) Purification and some properties of β-galactosidase from *Penicillium multicolor*. Agric Biol Chem 47:2533–2540

Wang H, Luo H, Bai Y, Wang Y, Yang P, Shi P, Zhang W, Fan Y, Yao B (2009) An acidophilic β-galactosidase from *Bispora* sp. MEY-1 with high lactose hydrolytic activity under simulated gastric conditions. J Agric Food Chem 57:5535–5541

Wang SD, Guo GS, Li L, Cao LC, Tong L, Ren GH, Liu YH (2014) Identification and characterization of an unusual glycosyltransferase-like enzyme with β-galactosidase activity from a soil metagenomic library. Enzyme Microb Technol 57:26–35

Widmer F, Leuba JL (1979) β-Galactosidase from *Aspergillus niger*. Separation and characterization of three multiple forms. Eur J Biochem 100:559–567

Xu K, Tang X, Gai Y, Mehmood M, Xiao X, Wang F (2011) Molecular characterization of cold-inducible β-galactosidase from *Arthrobacter* sp. ON14 isolated from Antarctica. J Microbiol Biotechnol 21:236–242

Yamada M, Chiba S, Endo Y, Isobe K (2017) New alkalophilic β-galactosidase with high activity in alkaline pH region from *Teratosphaeria acidotherma* AIU BGA-1. J Biosci Bioeng 123:15–19

Yamazaki A, Toyama K, Nakagiri A (2010) A new acidophilic fungus, *Teratosphaeria acidotherma* (Capnodiales, Ascomycota) from a hot spring. Mycoscience 51:443–455

Chapter 22
Fungi from Extreme Environments: A Potential Source of Laccases Group of Extremozymes

Om Prakash, Kapil Mahabare, Krishna Kumar Yadav, and Rohit Sharma (iD)

22.1 Introduction

Laccases are a mixture of synergistic enzymes and include laccase, phenol oxidase, phenol peroxidase, lignin peroxidase, manganese peroxidase, tyrosinase, etc. They are copper-containing blue phenol oxidase and are common among various groups of organisms including bacteria, fungi and plants (Mayer 2006). Various groups of plants, animals and microbes produce phenol oxidases, both intracellular and extracellular, for a variety of purposes. Fungi, the second largest members of eukaryotic kingdom produce variety of phenol oxidases. Among all the fungi, *Ascomycota* and *Basidiomycota* particularly produce intracellular phenol oxidases and use them to synthesize protective compounds like melanin (pigmentation), in spore formation and detoxification of toxic compounds from their environment. Phenol oxidase enzymes are also responsible for fungal pathogenicity due to their plant cell wall lignin degradation potential. These enzymes hydrolyze lignocelluloses present in agro-waste, especially facilitating the degradation of lignin component which is the most complex constituent of plant cell wall. These are non-specific enzymes which can act on variety of phenolic substances. Hence, they are flexible and can be used in a range of industrial processes (Nigam 2013). These enzymes are well known in bioremediation, industrial effluent treatment containing hazardous chemicals like dyes, phenols and other xenobiotic compounds (Robinson et al. 2001a, b; Robinson and Nigam 2008; Dahiya et al. 2001). It is quite popular that the leather industry has adopted eco-friendly methods for tanning process using keratinases instead of chemicals. Similarly pulp and paper industries have adopted treatment of wood pulp by ligninolytic enzymes for lignin degradation. These ligninolytic enzymes are also

O. Prakash · K. Mahabare · K. K. Yadav · R. Sharma (✉)
National Centre for Microbial Resource (NCMR), National Centre for Cell Science,
S.P. Pune University, Ganeshkhind, Pune, Maharashtra, India
e-mail: rohit@nccs.res.in

© Springer Nature Switzerland AG 2019   441
S. M. Tiquia-Arashiro, M. Grube (eds.), *Fungi in Extreme Environments:
Ecological Role and Biotechnological Significance*,
https://doi.org/10.1007/978-3-030-19030-9_22

being used in the wine, fruit juice and denim industries (Dahiya et al. 1998). Recently, concerns for decontamination of hazardous chemicals and environmental contaminants have been increased, especially in developing countries. India is also taking immense efforts in this direction to treat these hazardous compounds before releasing them in the environment. There are many compounds which are causing toxic effects on health and also damaging environment due to their presence in water bodies. This adversely affects soil and water microbiota, aquatic life, plants and human health. In humans, it is reported to cause cardiac toxicity, liver and kidney damage, neurotoxicity, reproductive and developmental toxicity and reduced blood pressure. Such compounds have long persistence in the environment and also show bioaccumulation and biomagnification in plants and animals. As discussed earlier, these are naturally present in fruits and part of plant components. Many phenolic compounds are also produced by humans through manufacturing of various day-to-day materials. At present, aquatic environments including rivers, pond, lakes, etc. are the most affected habitats in terms of phenolic contamination. These compounds enter the water bodies through natural, industrial, domestic and agricultural practices (Wallace 1996). Decomposition is not a problem for naturally produced phenols of plants and fruits origin because nature has developed the mechanism for their degradation. Wastewater from industries is really a matter of concern as it accumulates phenolic compounds in large quantity, which is difficult to manage. These compounds are also present in effluent of chemical and pharmaceutical industries, including petroleum refineries, petrochemical industries, coal gasification, resin manufacturing industries, dye synthesis units, pharmaceutical industries, pulp and paper mills, etc. In several countries, due to lack of standard norms, these compounds are gradually increasing in water bodies and creating health hazards in rivers. In addition, phenolic compounds also undergo transformation due to presence of other compounds in the aquatic body and microbial activities (Kulkarni and Kaware 2013). Due to their harmful effects, there is an urgent need to remove them from the environment (Huang et al. 2015; Nuhoglu and Yalcin 2005). Phenolic compounds are diverse in nature but in this chapter we will give the emphasis on three main types of compounds, namely phenol, cresol and alkyl phenol.

Many industries like pulp and paper are nowadays focusing on research on fungal and bacterial based bioremediation strategies due to their ability to synthesize polyphenol oxidase. It can be used for degradation of lignocelluloses and residual polyaromatic hydrocarbons for production of pulp, transformation of fuels and bioremediation of soils contaminated with toxic products (Duran et al. 2002; Claus 2003, 2004; Rabinovich et al. 2004; Masai et al. 2007) (Table 22.1). Fungi are the most potent producers of enzymes involved in lignin degradation. They require these enzymes for penetration into the plants cell wall, degradation of wood and litter biomass, etc. Pathogenic fungi and wood-rotting fungi have been studied in detail for the production of laccases (Robinson et al. 2001a, b; Robinson and Nigam 2008). These enzymes have been produced economically using several agricultural wastes as substrates on a large scale (Nigam and Pandey 2009).

Table 22.1 Use of laccases in bioremediation process of various environmental pollutants

Serial No.	Substrates	References
1	Xenobiotic compounds	Ullah et al. (2000), Schultz et al. (2001), and Bollag et al. (2003)
2	Synthetic dyes	Abadulla et al. (2000), Nagai et al. (2002), Claus et al. (2002), Soares et al. (2002), Peralta-Zamora et al. (2003), Wesenberg et al. (2003), and Zille et al. (2003)
3	Pesticides	Jolivalt et al. (2000), Torres et al. (2003)
4	Polycyclic aromatic hydrocarbons	Majcherczyk and Johannes (2000), Cho et al. (2002), and Pozdnyakova et al. (2004)
5	Bleaching of kraft pulp	Balakshin et al. (2001), Lund et al. (2003), and Sigoillot et al. (2004)
6	Detoxify agricultural soil	D'Annibale et al. (2000), Tsioulpas et al. (2002), and Velazquez-Cedeno et al. (2002)

Among fungi, they are mainly produced by members of *Basidiomycota* and *Ascomycota* while *Zygomycota*, *Chytridiomycota* and *Glomeromycota* are not reported to produce them. Several members of *Ascomycota* and *Basidiomycota* produce these enzymes, *viz.*, *Podospora anserine*, *Sclerotinia sclerotium*, *Pleurotus ostreatus*, *P. sapidus*, *Agaricus bisporus*, *Lentinus edodes*, *Schizophyllum commune*, *Trichoderma versicolor* and many other wood-rotting fungi (Bodke et al. 2012). There is only one report of slime moulds, *Physarum polycephalum* (Daniel et al. 1963), for this activity.

Fungi from extreme environment have also been studied to obtain suitable enzymes which can work in the harsh conditions of fermentation and in the conditions of paper-pulp delignification and waste treatment. Pulp and paper industries usually use thermophilic enzymes. Five thermophilic laccase enzyme isoforms were isolated, purified and characterized from xerophytic plants *Cereus pterogonus* and *Opuntia vulgaris* (Gali and Kotteazeth 2012, 2013; Kumar and Srikumar 2011, 2012). Different forms of laccases with extraordinary properties have been obtained from fungi like *Steccherinum ochraceum* and *Polyporus versicolor* (Chernykh et al. 2008; Nigam 2013). Several fungi, *viz.*, *Curvularia lonarensis, Penicillium* sp., and *Trametes* sp., have been reported from various extreme environments (thermophiles, alkaliphiles, psychrophiles, marine fungi, etc.) which have been studied for the laccase enzyme production potentials (Sharma et al. 2016; Dhakar et al. 2014; Dhakar and Pandey 2013) which we discuss below in detail.

Fungal based biotechnology is still in the developmental stage since past few decades and has improved significantly. Fungi from terrestrial origin have diverse properties and are used in the production of antibiotics, extracellular enzymes, organic acids, etc. (Pointing and Hyde 2001) (Fig. 22.1a). In the past couple of decades, fungi have also been used as 'cell factories' due to the advancement in molecular and genetic tools (Punt et al. 2002). Only few studies are available on fungal laccases from extremophilic fungi as compared to terrestrial and mesophilic

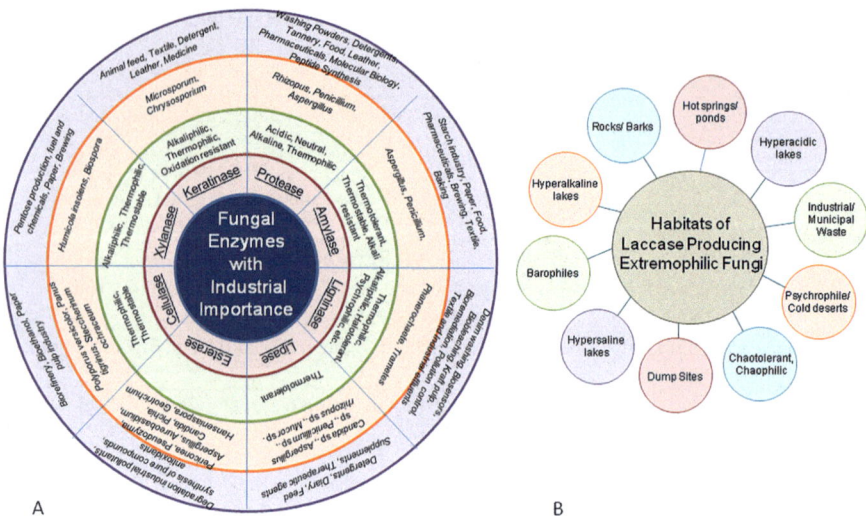

Fig. 22.1 (**a**) Scope of industrial applications of various enzymes produced by fungi in industries, viz., textile, paper pulp, food and feed, and pharmaceutical industries. (**b**) Habitats of laccase-producing fungi from extreme environments

counterparts. What prospects do the extremophilic fungi have in such situation? Why and how these fungi have become more important target group of organisms for pharmaceutical and environmental perspectives? This chapter discussed importance of extremophilic fungi in production of laccases in association with description of fungal diversity from extremophilic environments as it helps to understand the distribution of laccases producing fungi. The emphasis is on (a) molecular biology and genetics of fungal laccases, (b) factors affecting production of laccases from extremophilic fungi and (c) recent advances on fungal laccases and potential of fungal laccases from extreme habitats.

22.2 Laccases Producing Fungi from Extreme Habitats

Fungi living in extreme environment such as high or low temperatures (thermophiles and psychrophiles), high salinity (halophiles), acidic or alkaline pH values (acidophiles and alkaliphiles, respectively), anoxygenic conditions (anaerobic fungi), high pressures (barophiles), etc. (Poli et al. 2017; Dalmaso et al. 2015; Magan 2007) are known as extremophilic fungi (Fig. 22.1b). Many reports are available on extracellular phenol oxidase by fungi from different habitats (Crognale et al. 2012) but only a few studies on laccase-producing fungi from extreme environment like marine, hot springs and soda lakes are available.

22.2.1   Laccase from Alkaliphilic Fungi

The hyperalkaline habitat has both ecological and industrial significance as in high alkaline condition very few fungi can grow. It has been reported that most of the municipal wastewater treatment plants and effluents from industries have high alkalinity, and high concentration of metal ions. Hence fungi surviving in such conditions and with laccase-producing capacity can work as excellent bioinoculant for bioaugmentation-based bioremediation. Functional metagenomic studies of Soda Lake have showed that many uncultured fungi have laccases-like Cu-oxidase encoded with potential in degradation of phenolic compounds (Vavourakis et al. 2016; Ausec et al. 2011). Crognale et al. (2012) had also reported phenol oxidase-producing halotolerant fungi from olive brine wastewater. Sharma et al. (2016) isolated 104 fungal strains from Lonar lake (Fig. 22.2), a hyperalkaline habitat, and 14 were positive for enzyme production in primary screening using 2,2'-azino-bis-(3-ethylbenzthiazoline-6-sulfonic acid (ABTS) as substrate. It included *Fusarium equiseti*, *Curvularia lonarensis*, *Cladosporium funiculosum*, *Cladosporium oxysporum*, *Cladosporium halotolerans*, *Aspergillus niger*, a probable novel *Cladorrhinum* species and an unidentified fungus. Among these *Fusarium* sp. MEF008, *Curvularia lonarensis* MEF018 (Fig. 22.2), *Cladorrhinum* sp. MEF109 and *Cladosporium* sp. MEF135

Fig. 22.2 View of Lonar Lake, an alkaliphilic lake located at Buldhana district of Maharashtra state, India; (inset) plate showing culture (MEF018) positive for phenol oxidase production containing ABTS as substrate

Table 22.2 List of phenol oxidase-producing fungi isolated from Lonar Lake

Serial No.	Strain Id's	Identification	Isolated from	Phenol oxidase
1	MEF008	*Fusarium equiseti*	Lonar Lake	+++
2	MEF018	*Curvularia lonarensis*	Lonar Lake	++++
3	MEF040	*Cladosporium funiculosum*	Lonar Lake	+
4	MEF041	*Cladosporium oxysporum*	Lonar Lake	+
5	MEF135	*Cladosporium oxysporum*	Lonar Lake	+++
6	MEF062	Unidentified	Lonar Lake	+
7	MEF073	*Cladosporium halotolerans*	Lonar Lake	+
8	MEF095	*Aspergillus niger*	Lonar Lake	+
9	MEF104	*Cladorrhinum* sp. nov.	Lonar Lake	+
10	MEF121	*Cladorrhinum* sp. nov.	Lonar Lake	+
11	MEF127	*Cladorrhinum* sp. nov.	Lonar Lake	+
12	MEF109	*Cladorrhinum* sp.	Lonar Lake	+++
13	MEF133	*Cladorrhinum* sp.	Lonar Lake	+
14	MEF134	*Cladorrhinum* sp.	Lonar Lake	+

were higher phenol oxidase producer (Table 22.2). *Curvularia lonarensis* MEF018, an alkaliphilic fungus with potential to be exploited industrially, produced laccases at 40 °C, pH 12–14, and at salinity of 3%. While working on Lonar Lake, we observed that the fungi which were collected from the banks of the lake with wooden debris had showed phenol oxidase activity. The lake which is reported to be formed by meteor impact is surrounded by trees of *Acacia* sp. The wooden debris of trees falls on the lake water, leaching phenol in the lake water with time. In due course of time, fungi colonizing the wooden debris have developed the ability to produce laccase enzymes, thus playing an important role in the lake ecosystem. Such ecological adaptations of fungal strains have helped them to develop capacity to produce metabolites and enzymes which act at high temperature, pH, or salt concentrations.

22.2.2 Laccase from Marine Fungi

The applications of laccases in degradation of xenobiotics by aquatic, obligate marine (and marine-derived) fungi have been observed (Martin et al. 2009; Junghanns et al. 2009; Pointing and Hyde 2000; Li et al. 2002). These marine fungi produce unique secondary metabolites and enzymes not reported from fungi residing in terrestrial habitats (Jensen and Fenical 2002). D'Souza-Ticlo et al. (2009a, b) reported that a marine isolate of *Cerrena unicolor* MTCC 5159 produces halotolerant laccase and also degrades raw textile mill effluents (Verma et al. 2010). Generally, marine fungi are able to grow on decaying lignocelluloses substrates like branches, leaves, and woods of mangroves which include mostly the members of *Ascomycota* and with few exceptions of species of *Basidiomycota* (Hyde and Jones 1988). Marine fungi play an important role in the degradation of mangrove leaves, wood pieces and other wooden debris on the shores, thus forming detritus. These fungi play a significant role in the mineralization in the tropical marine ecosystem. However, the information related to marine laccase is still sparse and needs more work on the characterization of the type of lignin-modifying enzymes present in marine ecosystems. Raghukumar et al. (1994) isolated 17 fungi from marine habitats, out of which 12 were laccase positive which included *Gliocladium* sp., *Sordaria flmicola*, *Gongronella* sp., *Aigialus grandis*, *Halosarpheia ratnagiriensis*, *Verruculina enalia*, *Cirrenalia pygmea*, *Zalerion varium* and *Hypoxylon oceanicum*. Jaouani et al. (2014) have explored the fungal diversity of Sebkha El Melah, a Saharan salt flat located in southern Tunisia and isolated 21 moderately halotolerant fungi. It included 15 taxa belonging to 6 genera of *Ascomycota*, viz., *Cladosporium* spp., *Alternaria* spp., *Aspergillus* spp., *Penicillium* spp., *Ulocladium* sp. and *Engyodontium* sp. Three species out of 15 showed laccase activities at 10% NaCl, viz., *Cladosporium halotolerans*, *Cladosporium sphaerospermum* and *Penicillium canescens*. Laccase production at 10% salt by these strains is of biotechnological interest, especially in bioremediation of organic pollutants in high salt-contaminated environments.

22.2.3 Laccase from Thermophilic Fungi

Now we know that life can exist in extreme environments and molecular studies related to their survival mechanisms in extreme condition shed new insight about their survival strategies in extreme habitats. The stabilization of processes due to thermal stress is because of multiple reasons and involves DNA, RNA, proteins, ribosomes and enzymes (Poli et al. 2017). Thermophilic fungi received immense attention due to their ability to produce enzymes suitable for industrial purposes. Species belonging to genus *Corynascus* (*Myceliopthora*) have been of interest to mycologist as it produces thermostable enzymes. For example, *Corynascus thermophilus* (basionym: *Thielavia thermophila*) produced thermostable laccases with high activity and ability to express in various hosts (Berka et al. 1997; Bulter et al. 2003; Babot et al. 2011). Laccases produced by *C. thermophilus* ATCC 42464 are completely characterized, patented and genome sequenced (Bhat and Maheshwari 1987; Roy et al. 1990; Sadhukhan et al. 1992; Badhan et al. 2007; Beeson et al. 2011). However, there is no other report of any thermophilic fungi which is so extensively studied for laccase production. It shows the scarcity of thermophilic laccase-producing strains available so far.

22.2.4 Laccase from Deep-Sea Sponge Fungi

Studies on fungal diversity of marine sponges have been reported (Baker et al. 2009; Wang et al. 2008; Richards et al. 2012; He et al. 2014). Members of *Eurotiales*, *Capnoidales*, *Pleosporales* and *Hypocreales* have been identified and found to be associated with various sponges (Suryanarayanan 2012). Aspergillus and Penicillium genera are ubiquitous with marine sponges whereas other genera which are associated with sponges but not that frequent include Alternaria, Acremonium, Beauveria, Cladosporium, Curvularia, Eurotium, Fusarium, Gymnascella, Paecilomyces, Petriella, Pichia, Spicellum and Trichoderma (Suryanarayanan 2012). Many marine fungi isolated from sponges have been screened for their lignocelluloytic activities (Bucher et al. 2004; Bianchi 2011; Richards et al. 2012). Fungi with lignocellulolytic activity from sea and other marine habitats like mangrove forests and sponges have been reported (Baker et al. 2009; Bonugli-Santos et al. 2010a, b). Batista-García et al. (2017) isolated fungi from sponges, *Stelletta normani* (*Demospongiae*, *Astrophorida*, *Ancorinidae*), from a depth of 751 m from Irish waters in the North Atlantic Ocean. Three halotolerant strains were isolated and identified which displayed laccase production along with other enzymes (CMCase and xylanase), viz., *Cadophora* sp. TS2, *Emericellopsis* sp. TS11 and *Pseudogymnoascus* sp. TS 12. These strains also showed psychrotolerance with optimal growth at 20 °C. Such strains are of immense importance both ecologically and industrially as they play significant role in maintenance of ecosystem and in development of industrial biotechnology.

22.2.5   Laccase from Lichen

Lichen is a symbiotic association between a fungus and cyanobacterium. These are the microbes which preliminarily colonize on rocky substrates which is also considered an extreme environment due to lack of nutritional substrate. They are involved in the weathering of rocks and conversion and accumulation of organic matter forming the primitive soil (Nash 1996; Chen et al. 2000). Zavarzina and Zavarzin (2006) while studying the formation of primitive soil under vegetation observed that many lichens have the ability to produce and release phenol oxidases in environment.

Little information is available on the laccase isoforms from lichens. Extracellular laccase activity is considered to be due to combination of multi-copper oxidases, phenol peroxides and tyrosinases (Laufer et al. 2006a). Studies on lichen *Peltigera malacea* (a member of order *Peltigerales*) showed that the active form is a tetramer with a high molecular mass of 340 kD (Laufer et al. 2009). Work on lichens *Solorina crocea* and *Peltigera apthosa* shows that both contained a dimeric laccase (ca. 170 kD) and a monomeric form (ca.85 kD) (Lisov et al. 2007). Recently, Laufer et al. (2006b) and Zavarzina and Zavarzin (2006) reported that many lichens within the suborder Peltigerineae show high rates of extracellular laccase activity. In other groups of fungi laccases have been reported from a mass range of 60–70 kD, but laccases from lichen are heavier in the range of 200–350 kD (Baldrian 2006; Laufer et al. 2009). It has been found that almost all members of *Peltigerineae* family of lichens show some degree of laccase activity. Beckett et al. (2012, 2013) also reported strong peroxidase activity in various genera of *Peltigerales* order like

Lobaria, *Pseudocyphellaria* and *Sticta* and non-Peltigeralean lichens. They also showed that high laccase activity was present in the cell walls of thalli. However, their role in biology of lichens still needs more work as most of them grow on oligotrophic conditions, viz., rocks, bark, etc.

22.2.6 Laccase from Psychrophilic Fungi

Several fungi are capable of producing extremozymes at varying temperature, pH and salt range. It is known that they play an important role in biodegradation in low-temperature habitats. Dhakar and Pandey (2013) and Dhakar et al. (2014) studied the production of laccases by thermotolerant *Trametes hirsuta* (MTCC 11397) and *Penicillium pinophilum* (MCC 1049) isolated from a glacial site in Indian Himalayan Region (IHR). Such features make the strains efficient for the degradation in extreme conditions. However, as per literature survey very few studies have been done on psychrophilic fungal laccases.

22.2.7 Laccase from Fungi Inhabiting Dumping Sites

Ndahebwa Muhonja et al. (2018) studied the molecular and biochemical aspect of characterization of low-density polyethene (LDPE)-degrading fungi from Dandora dumpsite, Nairobi. They isolated ten fungal isolates and screened for their ability to produce extracellular laccase. *Aspergillus fumigatus* B2, 2 exhibited the highest presence of laccase which is reported to play a role in degradation of polyethene. In another study, Sumathi et al. (2016) studied the degradation of polyvinyl chloride (PVC) by laccase using a fungus *Cochliobolus* sp., isolated from plastic dumped soils near plastic industry in Renigunta near to Tirupati, Chittoor district of Andhra Pradesh, India. Plastic waste has become one of the worst man-made problems accounting for 20–30% of municipal solid waste in landfill sites. These are extreme man-made habitats wherein there is large concentration of metal toxicity, gases, etc. These studies demonstrate that fungi isolated from such habitats have potential application for bioremediation as there was significant difference in the Fourier-transform infrared spectroscopy (FTIR), Gas chromatography–mass spectrometry (GC-MS), Scanning Electron Microscope (SEM) results of control and *Cochliobolus* species-treated low-molecular-weight PVC. There is a need to conduct more studies in such extreme environments to isolate potential strains with desirable properties of bioremediations like PVC degradation and develop environment-friendly technology.

There is an immense hidden potential present in diverse group of fungi found in extreme environments (Tiquia and Mormile 2010). They are still not exploited for laccase to their complete potential due to difficulty in the culturing of such fungi in laboratory. In recent times, metagenomics has been regarded as a powerful omics tool, by which we can conduct diversity analysis of any microbe including fungi by direct DNA extraction and sequencing from different matrix (Barone et al. 2014; Handelsman 2004). Studies on functional aspect mainly focused on enzyme encoding gene(s) and discovery of novel biocatalysts, secondary metabolites and bioactive compounds (Wong 2010). Metagenomic approach can be used to study the gene encoding novel laccase enzymes for industrial production. At present, such studies are mainly focussed on extreme habitats from which it seems too difficult to cultivate the fungal population. It has been extremely effective in the discovery of novel extremophilic enzymes discovered from marine habitats, cold-adapted enzymes, thermophilic homologs, etc. Metagenomic tools are more commonly used with bacteria as compared with fungi. Fang et al. (2012) discovered a novel laccase with alkaline activity from bacterium. Miyazaki (2005) detected copper-inducible laccase activity in *Thermus thermophilus* HB27. It became possible by searching the genome databases of aerobic thermophilic bacteria for laccases and an open reading frame (OPR) annotated TTC1370 in *T. thermophilus* HB27 (Henne et al. 2004). Suryanarayanan et al. (2012) and Kunamneni et al. (2008) have also reported laccases from fungal endophytes of plants, which may not be an extreme habitat but surely a unique one. However most of the fungal diversity of extreme environments is still unexplored because of the imitating of the extreme conditions. Moreover, it is estimated that very less fungi of extreme environments are known; hence, a lot of fungal diversity still remains to be explored and exploited for laccase production. Use of metagenomics in fungi will provide an opportunity of culture-independent study of fungal diversity of extreme environments and its biotechnological application enzymes such as laccase (Fig. 22.3a).

22.3 Biological and Ecological Role of Laccase in Extreme Habitats

Extreme environments pose severe physicochemical conditions to the microbes accompanied by low molecular diffusion, macromolecular interactions and low metabolic rate. Hence, to survive in extreme climatic and environmental conditions, microbes including fungi need physiological adaptations. These adaptations facilitate in cellular functions and metabolic reactions. In such habitats fungi and other microbes possess proteins and enzymes which are robust in nature and help them to survive in harsh conditions. Fungi follow absorptive mode of nutrition and enzymes play an important role in breaking down the complex food material into simpler ones for absorption purpose. In addition, it also helps in invasion of plants during pathogenesis (fungal virulence factors) by plant-wounding response and pathogen defence (Beckett et al. 2005). The role of laccase has been studied in detail in white-rot fungi, where they participate in reactions that produce reactive oxygen species (ROS) such as the superoxide (O_2^-) and hydroxyl radicals ($-OH$). These molecules are involved in lignin degradation (Hammel et al. 2002; Leonowicz et al. 2001).

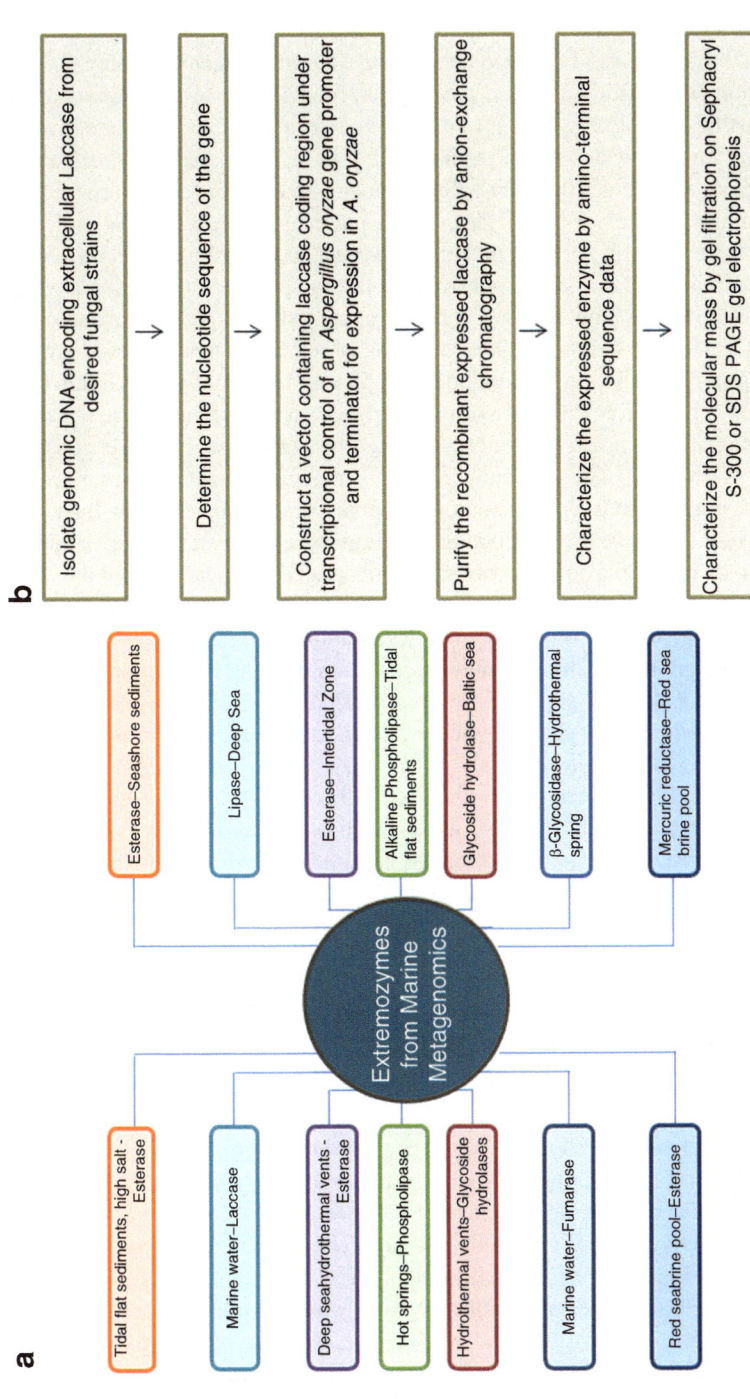

Fig. 22.3 (a) Metagenomics have helped in the discovery of several extremozymes from obligate marine or associated environment; (b) flowchart of methodology used for characterization of gene encoding laccase enzyme

Fungal laccases are reported to be involved in development stages of fungi like morphogenesis and melanin production which is also related to the pathogen factor (Baldrian 2006). Laccases have also been reported to have other physiological roles like development of fruiting bodies, sporulation, fungal spore pigmentation and cell wall reconstitution (Alcalde 2007). In extreme environments, laccases are involved in polymerization and depolymerization of humic acids in sediments of marine habitats (Zavarzina et al. 2004). The dothideaceous black yeasts synthesize DHN-melanin. In the last step of the DHN-melanin pathway where 1,8-DHN molecules conjoin to form melanin polymer and enzymes such as phenol oxidases (tyrosinase and laccases), peroxidases are responsible for the same (Yurlova et al. 2008).

Lignin is a complex plant cell wall component consisting of phenylpropanoid units linked by C–C and C–O bonds. Since very less organisms secrete laccases extracellularly fungi with laccase production potential play an important role in lignin depolymerization. Although enzymes are substrate specific, an important aspect of laccases enzyme is that it has a broad range of substrates. This is because it consists of a group of enzymes which includes lignin peroxidase, manganese peroxidases, laccases, tyrosinases, phenol oxidase, etc. Authors observed that fungi of Lonar Lake were playing an important role by contributing in the carbon cycle by degrading the phenolic compounds released from the woods, thus playing a significant role in the alkaliphilic ecosystem. In man-made extreme environments such as plastic-dumping sites and landfills fungi are involved in polymer degradation through depolymerization and polymer is broken into smaller subunits. Some fungi growing in such habitats utilize the plastic materials as a source of carbon. *Pestalotiopsis microspora*, *Fusarium solani*, *Alternaria solani*, *Spicaria* spp., *Penicillium oxalicum*, *P. chrysogenum*, *Aspergillus fumigatus*, *A. terreus* and *A. flavus* isolates were found to grow on polyester polyurethane (PUR) as the sole carbon (Russell et al. 2011; Kale et al. 2015; Ojha et al. 2017; Ibrahim et al. 2011). The discovery of new laccases and Thurston's (1994) study have extensively reviewed the role of laccase in the biology of the fungi, but further studies on laccases using advance method will further extend our knowledge about their role in fungal biology.

22.4   Potential of Laccase Enzymes

The pulp consists of cellulosic fibres of wood, crops and waste paper. It is made from mechanical and chemical processes that separate cellulose fibres from rest of the wood. It includes application of harsh conditions like high temperatures (~80 °C), alkaline pH and use of chemicals. Developed countries are pushing for use of eco-friendly methods as an alternative of chemical methods, and enzymatic bio-pulping is being considered a viable option. It is eco-friendly, safer and a profitable solution for the paper and pulp industry using stable hyperthermophilic/alkaline enzymes. Sarmiento et al. (2015) have listed selected examples of in-development and commercially available hot and cold-adapted extremozymes (Table 22.3). The paper and pulp industries are using laccases for bio-bleaching and degradation

Table 22.3 Examples of commercially available cold-active and thermostable enzymes (adopted and modified from Sarmiento et al. 2015)

Market	Enzyme	Commercially available	Uses
Cold-active enzymes			
Molecular biology	Alkaline phosphatases	Antarctic phosphatase (New England Biolabs Inc.)	Dephosphorylation of 5′ end of a linearized fragment of DNA
	Uracil-DNA N-glycosylases (UNGs)	Uracil-DNA N-glycosylase (UNG) (ArcticZymes), Antarctic Thermolabile UDG (New England Biolabs Inc.) and	Release of free uracil from uracil-containing DNA
	Nucleases	Cryonase (Takara-Clontech)	Digestion of all types of DNA and RNA
Detergent	Lipases	Lipoclean®, Lipex®, Lipolase® Ultra, Kannase, Liquanase®, Polarzyme®, (Novozymes)	Breaking down of lipid stains
	Proteases	Purafect® Prima, Properase®, Excellase (Genencor)	Breaking down of protein stains
	Amylases	Stainzyme® Plus (Novozymes), Preferenz™ S100 (DuPont), Purafect® OxAm (Genencor)	Breakdown starch-based stains
	Cellulases	Rocksoft™ Antarctic, Antarctic LTC (Dyadic), UTA-88 and UTA-90 (Hunan Youtell Biochemical), Retrocell Recop and Retrocell ZircoN (EpyGen Biotech), Celluzyme®, Celluclean® (Novozymes)	Wash of cotton fabrics
	Mannanases	Mannaway® (Novozymes), Effectenz™ (DuPont)	Degradation of mannan or gum
	Pectate lyases	XPect® (Novozymes)	Pectin-stain removal activity
Textile	Amylases	Optisize® COOL and Optisize NEXT (Genencor/DuPont)	Desizing of woven fabrics
	Cellulases	Primafast® GOLD HSL IndiAge® NeutraFlex, PrimaGreen® EcoLight 1 and PrimaGreen® EcoFade LT100 (Genencor/DuPont)	Bio-finishing combined with dyeing of cellulosic fabrics
Food and beverages	Pectinases	Novoshape® (Novozymes), Pectinase 62L (Biocatalysts), Lallzyme® (Lallemand)	Fermentation of beer and wine, breadmaking, and fruit juice processing

(continued)

Table 22.3 (continued)

Market	Enzyme	Commercially available	Uses
Other	Catalase	Catalase (CAT), (Swissaustral)	Textile, cosmetic applications

Thermostable enzymes

Market	Enzyme	Commercially available	Uses
Food and beverages	Amylases	Avantec®, Termamyl® SC, Liquozyme®, Novamyl®, Fungamyl® (Novozymes), Fuelzyme®	Enzymatic starch hydrolysis to form syrups. Applied in processes, such as baking, brewing, preparation of digestive aids, production of cakes and fruit juices
	Glucoamylases	Spirizyme® (Novozymes)	Used on liquefied starch-containing substrates
	Glucose (xylose) isomerases	Sweetzyme® (Novozymes)	Isomerization equilibrium of glucose into fructose
	Proteases	Protease PLUS	Applied in brewing to hydrolyze most proteins
	Amyloglucosidases	GlucoStar PLUS (Dyadic)	Used in processing aids
	Xylanases, cellulases, pectinases, mannanases, β-xylosidases, α-l-arabinofuranosidases, amylases, protease, other	CeluStar XL, BrewZyme LP, Dyadic Beta Glucanase BP CONC, Dyadic xylanase PLUS, Xylanase 2XP CONC, AlphaStar CONC and Protease AP CONC. (Dyadic), Panzea BG, Panzea 10× BG, Panzea Dual (Novozymes), Cellulase 13P (Biocatalysts)	Hydrolysis of hemicellulose and cellulose to lower molecular weight polymers in brewing
	Lipases and xylanases	Lipopan® and Pentopan® (Novozymes)	Stronger dough in bakery
	Glucose oxidases	Gluzym® (Novozymes)	Stronger gluten in bakery
Pulp and paper	Xylanases	Luminase® PB-100 and PB-200 (Verenium), Xylacid® (Varuna Biocell), Xyn 10A (Megazyme)	Bio-bleaching
	Laccases	Laccase (Novozyme)	Bio-bleaching
	Lipases and esterases	Lipase B Lipozyme® CALB L, Lipase A NovoCor®AD L, Resinase™ HT and Resinase A2X (Novozymes), Optimyze® (Buckman Laboratories)	Pitch control
	Cellulases/hemicellulases preparations	FibreZyme® G5000, FibreZyme® LBL CONC, FibreZyme™ LDI and FibreZyme™ G4 (Dyadic)	Modify cellulose and hemicellulose components of virgin and recycled pulps
	Amylase	Dexamyl-HTP (Varuna Biocell)	Modified starch of coated paper

of lignin (which renders dark colour to pulp), thus improving the colour of the product. According to Virk et al. (2012), laccases also help in removal of tacky materials containing resins from the wood. Fungal laccases have been more popular for use in bio-bleaching as compared to laccases of bacterial origin (Baldrian 2004, 2006). Fungi from extreme environment are considered as an important source of commercial enzymes such as amylase, lipases, protease and cellulase. Novozymes are such thermostable laccases produced from thermophilic fungus *Corynascus thermophilus*. AB Vista (Wiltshire, UK) has the patent for thermostable laccase enzyme effective between 30 °C and 80 °C (Paloheimo et al. 2006; Sarmiento et al. 2015). For commercial production, the laccase gene from *M. thermophila* is cloned in *Aspergillus oryzae* and the enzyme is active up to 70 °C (Xu et al. 1996; Berka et al. 1997) (Fig. 22.3b). Novozyme commercially produces laccase enzyme (EC number 1.10.32) Novozym® 51003, from fungus *Aspergillus oryzae*. It is produced in liquid form and oxidizes various phenols, anilines, benzenethiols, metal ion complexes and other compounds into quinones or other oxidized compounds, with concomitant reduction of dioxygen to water. Another product Novozymes DeniLite® is being used for enzymatic bleaching solutions altering the indigo colour through oxidation. It has made denim bleaching safer, eco-friendly and more sustainable. The rapid action of enzymes coupled with low working temperature of peroxidase enzyme helps in the production of more durable denims.

22.5   Current Challenges and Conclusions

Today, with increasing awareness of climate change and sustainable development, there is an inadvertent pressure on biotechnology to deliver eco-friendly solutions to processes presently employing chemical methods. Biotechnological industries are utilizing a variety of enzymes as solutions to various industrial processes. Many of these are synthesized commercially using fungal strains selected after large-scale screening. Research on extreme environments has helped in the selection of particular fungal strain with desired property. These have also been optimized for the production of high-quality laccase on a large scale for industrial applications. There is a huge requirement of laccases for the industries working in the field of waste management, whether it is biomedical, agriculture, or municipal waste and paper pulp industries (Fig. 22.4). However, there is a need to study various extreme environments with the aim of isolating fungi having potential of laccase production and capable of acting in conditions of high pH, temperature and still maintain high activity. Recent studies in molecular biology and genetics have helped in inserting and expressing the active factor or gene of the desired fungi into bacteria or yeast for rapid and easy production. This has helped in production of laccase of desired property like thermostability, tolerance to acidic or alkaline environments and metal toxicity.

As discussed above, at present most of the studies have been focussed on the laccase produced by white-rot fungi-like species of *Fomes*, *Panus* and *Phanerochaete* (Papinutti et al. 2008; Quaratino et al. 2007; Dahiya et al. 2001) which are mostly

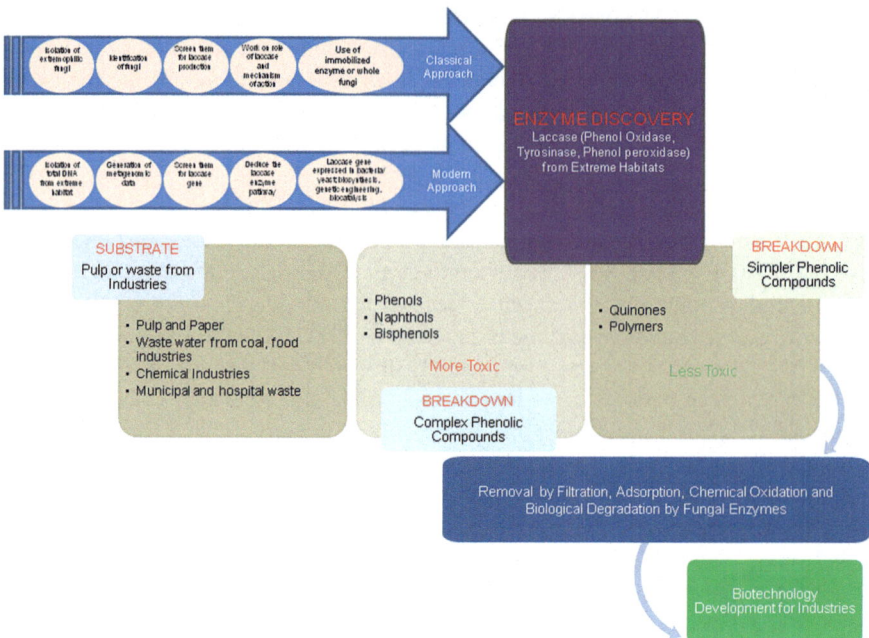

Fig. 22.4 Extremophilic fungi have immense potentials for development of sustainable environmental technologies including biodegradation of phenolic pollutants and provide waste management solutions to various industries. This is well supported by advances in microbial techniques enabling the cloning of gene or gene clusters involved in the biosynthesis of laccases group of enzymes

mesophilic. Hence, there is a need to focus studies on fungi from extreme environments like soda lakes, hot springs, marine habitats, etc. for their isolation and screening for active laccase production. There are more studies on extremophilic bacteria as compared to extremophilic fungi. It may be due to optimized isolation strategies being used and incubation methods being practised and refined in case of bacteria but less practised in case of fungi. Recent techniques of metagenomics can also be employed to know the fungal population of such habitats and then design isolation strategies accordingly (Fig. 22.4). Many researchers have employed strategies to directly understand the functional aspect of an ecosystem using BIOLOG and API strips (Oest et al. 2018; Patel et al. 2019; Tiquia 2010, 2011). Similar strategy can be used in extreme environments that will help to know the laccase activity of the habitat, thus giving an idea of the laccase-producing abilities of the fungal strains inhabiting such habitats. Amplification of laccase gene using specific primers is another strategy for evaluating the capacity of fungal strains for laccase production. Such strategies may yield fungal strains producing laccases with unusual properties useful in industrial applications. There is a growing demand for novel and robust laccases for biotechnology in industrial applications, i.e. biofuel production, pulp and paper industries and eco-friendly municipal waste treatment. Hence, researchers should give attention and make efforts to discover the novel fungi with the capacity to produce laccases from extreme environments.

Acknowledgement   The authors thank Department of Biotechnology, New Delhi, for financial support for the establishment of National Centre for Microbial Resource (NCMR), Pune, wide grant letter no. BT/Coord.II/01/03/2016 dated 6th April 2017.

References

Abadulla E, Tzanov T, Costa S, Robra KH, Cavaco-Paulo A, Gubitz GM (2000) Decolorization and detoxification of textile dyes with a laccase from *Trametes hirsuta*. Appl Environ Microbiol 66:3357–3362

Alcalde M (2007) Laccases: biological functions, molecular structure and industrial applications. In: Industrial enzymes. Springer, Dordrecht, pp 461–476

Ausec L, Zakrzewski M, Goesmann A, Schlüter A, Mandic-Mulec I (2011) Bioinformatic analysis reveals high diversity of bacterial genes for laccase-like enzymes. PLoS One 6:e25724

Babot ED, Rico A, Rencoret J, Kalum L, Lund H, Romero J, José C, Martínez ÁT, Gutiérrez A (2011) Towards industrially-feasible delignification and pitch removal by treating paper pulp with *Myceliophthora thermophila* laccase and a phenolic mediator. Bioresour Technol 102(12):6717–6722

Badhan AK, Chadha BS, Kaur J, Saini HS, Bhat MK (2007) Production of multiple xylanolytic and cellulolytic enzymes by thermophilic fungus *Myceliophthora* sp. IMI 387099. Bioresour Technol 98(3):504–510

Baker PW, Kennedy J, Dobson ADW, Marchesi JR (2009) Phylogenetic diversity and antimicrobial activities of fungi associated with *Haliclona simulans* isolated from Irish coastal waters. Marine Biotechnol 11(4):540–547

Balakshin M, Capanema E, Chen CL, Gratzl J, Kirkman A, Gracz H (2001) Biobleaching of pulp with dioxygen in the laccase-mediator system–reaction mechanisms for degradation of residual lignin. J Mol Catal B 13:1–16

Baldrian P (2004) Increase of laccase activity during interspecific interactions of white-rot fungi. FEMS Microbiol Ecol 50:245–253

Baldrian P (2006) Fungal laccases—occurrence and properties. FEMS Microbiol Rev 30(2):215–242

Barone R, De Santi C, Palma Esposito F, Tedesco P, Galati F, Visone M, Di Scala A, De Pascale D (2014) Marine metagenomics, a valuable tool for enzymes and bioactive compounds discovery. Front Mar Sci 1:38

Batista-García RA, Sutton T, Jackson SA, Tovar-Herrera OE, Balcázar-López E, del Rayo Sánchez-Carbente M, Sánchez-Reyes A, Dobson AD, Folch-Mallol JL (2017) Characterization of lignocellulolytic activities from fungi isolated from the deep-sea sponge Stelletta normani. PLoS one. 12(3): e0173750

Beckett RP, Minibayeva FV, Laufer Z (2005) Extracellular reactive oxygen species production by lichens. Lichenologist 37(5):397–407

Beckett RP, Minibayeva FV, Liers C (2012) Occurrence of high tyrosinase activity in the non-Peltigeralean lichen *Dermatocarpon miniatum* (L.) W. Mann. Lichenologist 44:827–832

Beckett RP, Minibayeva FV, Liers C (2013) On the occurrence of peroxidase and laccase activity in lichens. Lichenologist 45(2):277–283

Beeson WT, Iavarone AT, Hausmann CD, Cate JH, Marletta MA (2011) Extracellular aldonolactonase from *Myceliophthora thermophila*. Appl Environ Microbiol 77(2):650–656

Berka RM, Schneider P, Golightly EJ, Brown SH, Madden M, Brown KM, Halkier T, Mondorf K, Xu F (1997) Characterization of the gene encoding an extracellular laccase of *Myceliophthora thermophila* and analysis of the recombinant enzyme expressed in *Aspergillus oryzae*. Appl Environ Microbiol 63(8):3151–3157

Bhat KM, Maheshwari R (1987) *Sporotrichum* thermophile growth, cellulose degradation, and cellulase activity. Appl Environ Microbiol 53(9):2175–2182

Bianchi T (2011) The role of terrestrially derived organic carbon in the coastal ocean: a changing paradigm and the priming effect. PNAS 108(49):19,473–19,481

Bodke PM, Senthilarasu G, Raghukumar S (2012) Screening diverse fungi for laccases of varying properties. Indian J Microbiol 52(2):247–250

Bollag JM, Chu HL, Rao MA, Gianfreda L (2003) Enzymatic oxidative transformation of chlorophenol mixtures. J Environ Qual 32:63–69

Bonugli-Santos RC, Durrant LR, da Silva M, Sette LD (2010a) Production of laccase, manganese peroxidase and lignin peroxidase by Brazilian marine-derived fungi. Enzyme Microb Technol 46(1):32–37

Bonugli-santos RC, Durrant LR, Sette LD (2010b) Laccase activity and putative laccase genes in marine derived basidiomycetes. Fungal Biol 114(10):863–872

Bucher VVC, Hyde KD, Pointing SB, Reddy CA (2004) Production of wood decay enzymes, mass loss and lignin solubilization in wood by marine ascomycetes and their anamorphs. Fungal Diversity 15:1–14

Bulter T, Alcalde M, Sieber V, Meinhold P, Schlachtbauer C, Arnold FH (2003) Functional expression of a fungal laccase in *Saccharomyces cerevisiae* by directed evolution. Appl Environ Microbiol 69(2):987–995

Chen J, Blume H-P, Beyer L (2000) Weathering of rocks induced by lichen colonization—a review. Catena 39:121–146

Chernykh A, Myasoedova N, Kolomytseva M, Ferraroni M, Briganti F, Scozzafava A, Golovleva L (2008) Laccase isoforms with unusual properties from the basidiomycete *Steccherinum ochraceum* strain 1833. J Appl Microbiol 105:2065–2075

Cho SJ, Park SJ, Lim JS, Rhee YH, Shin KS (2002) Oxidation of polycyclic aromatic hydrocarbons by laccase of *Coriolus hirsutus*. Biotechnol Lett 24:1337–1340

Claus H (2003) Laccases and their occurrence in prokaryotes. Arch Microbiol 179:145–150

Claus H (2004) Laccases: structure, reactions, distribution. Micron 35:93–96

Claus H, Faber G, Konig H (2002) Redox-mediated decolorization of synthetic dyes by fungal laccases. Appl Microbiol Biotechnol 59:672–678

Crognale S, Pesciaroli L, Petruccioli M, D'Annibale A (2012) Phenol oxidase-producing halotolerant fungi from olive brine wastewater. Process Biochem 47(9):1433–1437

D'Annibale A, Stazi SR, Vinciguerra V, Sermanni GG (2000) Oxirane-immobilized *Lentinula edodes* laccase: stability and phenolics removal efficiency in olive mill wastewater. J Biotechnol 77:265–273

D'Souza-Ticlo D, Garg S, Raghukumar C (2009a) Effects and interactions of medium components on laccase from a marine-derived fungus using response surface methodology. Mar Drugs 7(4):672–688

D'Souza-Ticlo D, Sharma D, Raghukumar C (2009b) A thermostable metal-tolerant laccase with bioremediation potential from a marine-derived fungus. Marine Biotechnol 11(6):725–737

Dahiya JS, Singh D, Nigam P (1998) Characterisation of laccase produced by *Coniotherium minitans*. J Basic Microbiol 38:349–359

Dahiya J, Singh D, Nigam P (2001) Decolourisation of synthetic and spentwash-melanoidins using the white-rot fungus *Phanerochaete chrysosporium* JAG-40. Bioresour Technol 78:95–98

Dalmaso G, Ferreira D, Vermelho A (2015) Marine extremophiles: a source of hydrolases for biotechnological applications. Mar Drugs 13:1925–1965

Daniel JW, Babcock KL, Sievert AH, Rusch HP (1963) Organic requirements and synthetic media for growth of the myxomycete *Physarum polycephalum*. J Bacteriol 86(2):324–331

Dhakar K, Pandey A (2013) Laccase production from a temperature and pH tolerant fungal strain of *Trametes hirsuta* (MTCC 11397). Enzyme Res 2013:869062

Dhakar K, Jain R, Tamta S, Pandey A (2014) Prolonged laccase production by a cold and pH tolerant strain of *Penicillium pinophilum* (MCC 1049) isolated from a low temperature environment. Enzyme Res 2014:120708

Duran N, Rosa MA, Dannibale A, Gianfreda L (2002) Applications of laccases and tyrosinases (phenol oxidases) immobilized on different supports: a review. Enzyme Microb Technol 31:907–931

Fang ZM, Li TL, Chang F, Zhou P, Fang W, Hong YZ, Zhang XC, Peng H, Xiao YZ (2012) A new marine bacterial laccase with chloride-enhancing, alkaline-dependent activity and dye decolorization ability. Bioresour Technol 111:36–41

Gali NK, Kotteazeth S (2012) Isolation, purification and characterization of thermophilic laccase from xerophyte *Cereus pterogonus*. Chem Nat Compd 48:451–456

Gali NK, Kotteazeth S (2013) Biophysical characterization of thermophilic laccase from the xerophytes: *Cereus pterogonus* and *Opuntia vulgaris*. Cellulose 20:115–125

Hammel KE, Kapich AN, Jensen KA Jr, Ryan ZC (2002) Reactive oxygen species as agents of wood decay by fungi. Enzyme Microb Technol 30(4):445–453

Handelsman J (2004) Metagenomics: application of genomics to uncultured microorganisms. Microbiol Mol Biol Rev 68:669–685

He L, Liu F, Karuppiah V, Ren Y, Li Z (2014) Comparisons of the fungal and protistan communities among different marine sponge holobionts by pyrosequencing. Microb Ecol 67:951–961

Henne A, Bruggemann H, Raasch C, Wiezer A, Hartsch T, Liesegang H, Johann A, Lienard T, Gohl O, Martinez-Arias R, Jacobi C, Starkuviene V, Schlenczeck S, Dencker S, Huber R, Klenk HP, Kramer W, Merkl R, Gottschalk G, Fritz HJ (2004) The genome sequence of the extreme thermophile *Thermus thermophilus*. Nat Biotechnol 22:547–553

Huang DL, Wang C, Xu P, Zeng GM, Lu BA, Li NJ, Huang C, Lai C, Zhao MH, Xu JJ, Luo XY (2015) A coupled photocatalytic–biological process for phenol degradation in the *Phanerochaete chrysosporium*–oxalate–Fe3O4 system. Int Biodeter Biodegr 97:115–123

Hyde KD, Jones EB (1988) Marine mangrove fungi. Marine Ecology 9(1):15–33

Ibrahim IN, Maraqa A, Hameed KM, Saadoun IM, Maswadeh HM (2011) Assessment of potential plastic-degrading fungi in Jordanian habitats. Turkish J Biol 35(5):551–557

Jaouani A, Neifar M, Prigione V, Ayari A, Sbissi I, Ben Amor S, Ben Tekaya S, Varese GC, Cherif A, Gtari M (2014) Diversity and enzymatic profiling of halotolerant micromycetes from Sebkha El Melah, a Saharan salt flat in southern Tunisia. Biomed Res Int 2014:439,197

Jensen PR, Fenical W (2002) Secondary metabolites from marine fungi. In: Hyde KD (ed) Fungi in marine environments. Fungal Diversity Press, Hong Kong, China, pp 293–315

Jolivalt C, Brenon S, Caminade E, Mougin C, Pontie M (2000) Immobilization of laccase from *Trametes versicolor* on a modified PVDF microfiltration membrane: characterization of the grafted support and application in removing a phenylurea pesticide in wastewater. J Membr Sci 180:103–113

Junghanns C, Pecyna MJ, Bohm D, Jehmlich N, Martin C, von Bergen M, Schauer F, Hofrichter M, Schlosser D (2009) Biochemical and molecular genetic characterisation of a novel laccase produced by the aquatic ascomycete *Phoma* sp. UHH 5-1-03. Appl Microbiol Biotechnol 84:1095–1105

Kale SK, Deshmukh AG, Dudhare MS, Patil VB (2015) Microbial degradation of plastic: a review. J Biochem Technol 6(2):952–961

Kulkarni SJ, Kaware JP (2013) Review on research for removal of phenol from wastewater. Int J Sci Res Publ 3(4):1–5

Kumar GN, Srikumar K (2011) Thermophilic laccase from xerophyte species *Opuntia vulgaris*. Biomed Chromatogr 25:707–711

Kumar GN, Srikumar K (2012) Characterization of xerophytic thermophilic laccase exhibiting metal ion-dependent dye decolorization potential. Appl Biochem Biotechnol 167:662–676

Kunamneni A, Camarero S, García-Burgos C, Plou FJ, Ballesteros A, Alcalde M (2008) Engineering and applications of fungal laccases for organic synthesis. Microb Cell Fact 7(1):32

Laufer Z, Beckett RP, Minibayeva FV (2006a) Co-occurrence of the multicopper oxidases tyrosinase and laccase in lichens in sub-order Peltigerineae. Ann Bot 98(5):1035–1042

Laufer Z, Beckett RP, Minibayeva FV, Lüthje S, Böttger M (2006b) Occurrence of laccases in lichenized ascomycetes of the *Peltigerineae*. Mycol Res 110(7):846–853

Laufer Z, Beckett RP, Minibayeva FV, Lüthje S, Böttger M (2009) Diversity of laccases from lichens in suborder *Peltigerineae*. Bryologist 112(2):418–426

Leonowicz A, Cho N-S, Luterek J, Wilkolazka A, Wojtas-Wasilewska M, Matuszewska A, Hofrichter M, Wesenberg D, Rogalski J (2001) Fungal laccase: properties and activity on lignin. J Basic Microbiol 41:185–227

Li X, Kondo R, Sakai K (2002) Studies on hypersaline-tolerant white-rot fungi. II. Biodegradation of sugarcane bagasse with marine fungus *Phlebia* sp. MG-60. J Wood Sci 48:159–162

Lisov AV, Zavarzina AG, Zavarzin AA, Leontievsky AA (2007) Laccases produced by lichens of the order *Peltigerales*. FEMS Microbiol Lett 275:46–52

Lund M, Eriksson M, Felby C (2003) Reactivity of a fungal laccase towards lignin in softwood kraft pulp. Holzforschung 57:21–26

Magan N (2007) Fungi in extreme environments. Mycota 4:85–103

Majcherczyk A, Johannes C (2000) Radical mediated indirect oxidation of a PEG-coupled polycyclic aromatic hydrocarbon (PAH) model compound by fungal laccase. Biochim Biophys Acta 1474:157–162

Martin C, Corvini PF, Vinken R, Junghanns C, Krauss G, Schlosser D (2009) Quantification of the influence of extracellular laccase and intracellular reactions on the isomer-specific biotransformation of the xenoestrogen technical nonylphenol by the aquatic hyphomycete *Clavariopsis aquatica*. Appl Environ Microbiol 75:4398–4409

Masai E, Katayama Y, Fukuda M (2007) Genetic and biochemical investigations on bacterial catabolic pathways for lignin–derived aromatic compounds. Biosci Biotech Bioch 71:1–15

Mayer AM (2006) Polyphenol oxidases in plants and fungi: going places? A review. Phytochemistry 67(21):2318–2331

Miyazaki K (2005) A hyperthermophilic laccase from *Thermus thermophilus* HB27. Extremophiles 9(6):415–425

Nagai M, Sato T, Watanabe H, Saito K, Kawata M, Enei H (2002) Purification and characterization of an extracellular laccase from the edible mushroom *Lentinula edodes*, and decolorization of chemically different dyes. Appl Microbiol Biotechnol 60:327–335

Nash T (ed) (1996) Lichen Biology. Cambridge Univ. Press, Cambridge

Ndahebwa Muhonja C, Magoma G, Imbuga M, Makonde HM (2018) Molecular characterization of Low-Density Polyethene (LDPE) degrading bacteria and fungi from Dandora Dumpsite, Nairobi, Kenya. Int J Microbiol 2018:4167845

Nigam PS (2013) Microbial enzymes with special characteristics for biotechnological applications. Biomolecules 3(3):597–611

Nigam P, Pandey A (2009) Biotechnology for agro-industrial residues utilisation. Springer Science Business Media B.V., pp 1–466

Nuhoglu A, Yalcin B (2005) Modelling of phenol removal in a batch reactor. Process Biochem 40(3–4):1233–1239

Oest A, Alsaffar A, Fenner M, Azzopardi D, Tiquia-Arashiro SM (2018) Patterns of change in metabolic capabilities of sediment microbial communities along river and lake ecosystems. J Inter Microbiol 2018:6234931. https://doi.org/10.1155/2018/6234931

Ojha N, Pradhan N, Singh S, Barla A, Shrivastava A, Khatua P, Rai V, Bose S (2017) Evaluation of HDPE and LDPE degradation by fungus, implemented by statistical optimization. Sci Rep 7:39,515

Paloheimo M, Puranen T, Valtakari L, Kruus K, Kallio J, Maentylae A, Fagerstrom R, Ojapalo P, Vehmaanpera J (2006) Novel laccase enzymes and their uses. US Patent No. 77321784 B2. United States Paten and Trademark Office

Papinutti L, Dimitriu P, Forchiassin F (2008) Stabilization studies of *Fomes sclerodermeus* laccases. Bioresour Technol 99(2):419–424

Patel D, Gismondi R, Alsaffar A, Tiquia-Arashiro SM (2019) Applicability of API ZYM to capture seasonal and spatial variabilities in lake and river sediments. Environ Technol. https://doi.org/1 0.1080/09593330.2018.1468492

Peralta-Zamora P, Pereira CM, Tiburtius ERL, Moraes SG, Rosa MA, Minussi RC, Durán N (2003) Decolorization of reactive dyes by immobilized laccase. Appl Catal B 42:131–144

Pointing SB, Hyde KD (2000) Lignocellulose-degrading marine fungi. Biofouling 15:221–229

Pointing SB, Hyde KD (2001) Bio-exploitation of filamentous fungi. Fungal Diversity Res Ser 6:1–467

Poli A, Finore I, Romano I, Gioiello A, Lama L, Nicolaus B (2017) Microbial diversity in extreme marine habitats and their biomolecules. Microorganisms 5(2):25

Pozdnyakova NN, Rodakiewicz-Nowak J, Turkovskaya OV (2004) Catalytic properties of yellow laccase from *Pleurotus ostreatus* D1. J Mol Catal B 30:19–24

Punt PJ, Biezen NV, Conesa A, Albers A, Mangnus J, Hondel C (2002) Filamentous fungi as cell factories for heterologous protein production. Trends Biotechnol 20:200–206

Quaratino D, Federici F, Petruccioli M, Fenice M, D'Annibale A (2007) Production, purification and partial characterisation of a novel laccase from the white-rot fungus *Panus tigrinus* CBS 577.79. Antonie van Leeuwenhoek 91(1):57–69

Rabinovich ML, Bolobova AV, Vasilchenko LG (2004) Fungal decomposition of natural aromatic structures and xenobiotics: a review. Appl BiochemMicrobiol 40:1–17

Raghukumar C, Raghukumar S, Chinnaraj A, Chandramohan D, D'souza TM, Reddy CA (1994) Laccase and other lignocellulose modifying enzymes of marine fungi isolated from the coast of India. Botanica Marina 37(6):515–524

Richards TA, Jones MDM, Leonard G, Bass D (2012) Marine fungi: their ecology and molecular diversity. Ann Rev Mar Sci 4:495–522

Robinson T, Nigam P (2008) Remediation of textile dye-waste water using a white rot fungus *Bjerkandera adusta* through solid state fermentation (SSF). Appl Biochem Biotechnol 151:618–628

Robinson T, Chandran B, Nigam P (2001a) Studies on the production of enzymes by white-rot fungi for the decolourisation of textile dyes. Enzyme Microb Technol 29:575–579

Robinson T, Chandran B, Nigam P (2001b) Studies on the decolourisation of an artificial effluent through lignolytic enzyme production by white-rot fungi in N-rich and N-limited media. Appl Microbiol Biotechnol 57:810–813

Roy SK, Dey SK, Raha SK, Chakrabarty SL (1990) Purification and properties of an extracellular endoglucanase from *Myceliophthora thermophila* D-14 (ATCC 48104). Microbiology 136(10):1967–1971

Russell JR, Huang J, Anand P, Kucera K, Sandoval AG, Dantzler KW, Hickman DS, Jee J, Kimovec FM, Koppstein D, Marks DH, Mittermiller PA, Nunez SJ, Santiago M, Townes MA, Vishnevetsky M, Williams NE, Nunez Vargas MP, Boulanger LA, Slack CB, Strobell SA (2011) Biodegradation of polyester polyurethane by endophytic fungi. Appl Environ Microbiol 77(17):6076–6084

Sadhukhan RA, Roy SK, Raha SK, Manna SU, Chakrabarty S (1992) Induction and regulation of alpha-amylase synthesis in a cellulolytic thermophilic fungus *Myceliophthora thermophila* D14 (ATCC 48104). Indian J Exp Biol 30(6):482–486

Sarmiento F, Peralta R, Blamey JM (2015) Cold and hot extremozymes: industrial relevance and current trends. Front Bioeng Biotechnol 3:148

Schultz A, Jonas U, Hammer E, Schauer F (2001) Dehalogenation of chlorinated hydroxybiphenyls by fungal laccase. Appl Environ Microbiol 67:4377–4381

Sharma R, Prakash O, Sonawane MS, Nimonkar Y, Golellu PB, Sharma R (2016) Diversity and distribution of phenol oxidase producing fungi from soda lake and description of *Curvularia lonarensis* sp. nov. Front Microbiol 7:1847

Sigoillot C, Record E, Belle V, Robert JL, Levasseur A, Punt PJ, van den Hondel CAMJ, Fournel A, Sigoillot JC, Asther M (2004) Natural and recombinant fungal laccases for paper pulp bleaching. Appl Microbiol Biotechnol 64:346–352

Soares GMB, Amorim MTP, Hrdina R, Costa-Ferreira M (2002) Studies on the biotransformation of novel disazo dyes by laccase. Process Biochem 37:581–587

Sumathi T, Viswanath B, Sri Lakshmi A, SaiGopal DV (2016) Production of laccase by *Cochliobolus* sp. isolated from plastic dumped soils and their ability to degrade low molecular weight PVC. Biochem Res Int 2016:9519527

Suryanarayanan TS (2012) The diversity and importance of fungi associated with marine sponges. Bot Mar 55(6):553–564

Suryanarayanan TS, Thirunavukkarasu N, Govindarajulu MB, Gopalan V (2012) Fungal endophytes: an untapped source of biocatalysts. Fungal Diversity 54(1):19–30

Thurston CF (1994) The structure and function of fungal laccases. Microbiology 140(1):19–26

Tiquia SM (2010) Metabolic diversity of the heterotrophic microorganisms and potential link to pollution of the Rouge River. Environ Pollut 158:1435–1443

Tiquia SM (2011) Extracellular hydrolytic enzyme activities of the heterotrophic microbial communities of the Rouge River: an approach to evaluate ecosystem response to urbanization. Microb Ecol 62(3):679–689

Tiquia SM, Mormile M (2010) Extremophiles—A source of innovation for industrial and environmental applications. Environ Technol 31(8-9):823

Torres E, Bustos-Jaimes I, Le Borgne S (2003) Potential use of oxidative enzymes for the detoxification of organic pollutants. Appl Catal B 46:1–15

Tsioulpas A, Dimou D, Iconomou D, Aggelis G (2002) Phenolic removal in olive oil mill wastewater by strains of Pleurotus spp. in respect to their phenol oxidase (laccase) activity. Bioresour Technol 84:251–257

Ullah MA, Bedford CT, Evans CS (2000) Reactions of pentachlorophenol with laccase from *Coriolus versicolor*. Appl Microbiol Biotechnol 53:230–234

Vavourakis CD, Ghai R, Rodriguez-Valera F, Sorokin DY, Tringe SG, Hugenholtz P et al (2016) Metagenomic insights into the uncultured diversity and physiology of microbes in four hypersaline Soda Lake brines. Front Microbiol 7:211

Velazquez-Cedeno MA, Mata G, Savoie JM (2002) Waster educing cultivation of *Pleurotus ostreatus* and *Pleurotus pulmonarius* on coffee pulp: changes in the production of some lignocellulolytic enzymes. World J Microbiol Biotechnol 18:201–207

Verma AK, Raghukumar C, Verma P, Shouche YS, Naik CG (2010) Four marine-derived fungi for bioremediation of raw textile mill effluents. Biodegradation 21(2):217–233

Virk AP, Sharma P, Capalash N (2012) Use of laccase in pulp and paper industry. Biotechnol Prog 28(1):21–32

Wallace J (1996) Phenol. In: Kroschwitz JI, Howe-Grant M (ed) Kirk-Othmer encyclopedia of chemical technology, 4th edn. Wiley, New York, pp 592–602

Wang G, Li Q, Zhu P (2008) Phylogenetic diversity of culturable fungi associated with the Hawaiian sponges *Suberites zeteki* and *Gelliodes fibrosa*. Antonie Van Leeuwenhoek 93(1–2):163–174

Wesenberg D, Kyriakides I, Agathos SN (2003) White-rot fungi and their enzymes for the treatment of industrial dye effluents. Biotechnol Adv 22:161–187

Wong D (2010) Applications of metagenomics for industrial bioproducts. In: Marco D (ed) Metagenomics: theory, methods and applications, 1st edn. Horizon Scientific Press, Norwich, UK, pp 141–158

Xu F, Shin W, Brown SH, Wahleithner JA, Sundaram UM, Solomon EI (1996) A study of a series of recombinant fungal laccases and bilirubin oxidase that exhibit significant differences in redox potential, substrate specificity, and stability. Biochim Biophys Acta 1292(2): 303–311

Yurlova NA, De Hoog GS, Fedorova LG (2008) The influence of ortho-and para-diphenoloxidase substrates on pigment formation in black yeast-like fungi. Stud Mycol 61:39–49

Zavarzina AG, Zavarzin AA (2006) Laccase and tyrosinase activities in lichens. Microbiology 75(5):546–556

Zavarzina AG, Leontievsky AA, Golovleva LA, Trofimov SYA (2004) Transformation of soil humic acids by blue laccase of *Panus tigrinus* 8/18: an *in vitro* study. Soil Biol Biochem 36:359–369

Zille A, Tzanov T, Gubitz GM, Cavaco-Paulo A (2003) Immobilized laccase for decolorization of Reactive Black 5 dyeing effluent. Biotechnol Lett 25:1473–1477

Part IV
Bioenergy and Biofuel Synthesis

Chapter 23
Lignocellulose-Degrading Thermophilic Fungi and Their Prospects in Natural Rubber Extraction from Plants

Shomaila Sikandar (iD)**, Imran Afzal, Naeem Ali, and Katrina Cornish**

23.1 Introduction

Fungi are pervasive in nature and are established across a wide range of biological/ environmental habitats. To establish they must produce the essential extracellular enzymes in their immediate environment to enable the primary resource capture required for their growth and development. Temperature, salt concentration, pH, gas balance, and water availability are the fundamental abiotic aspects generally affecting their biological role and success (Magan and Aldred 2008). The prospects of isolating new extremophilic fungal species increase when samples are collected from extreme temperatures or pressures, high-radiation environments, or environments with limited nutrients or high salinity. In extreme environments, they develop different survival strategies for reproduction and growth (Timling and Taylor 2012; Gunde-Cimerman and Zalar 2014). Moreover, because fungi have the capability to grow

S. Sikandar (✉)
Faculty of Biology, Department of Biology, Lahore Garrison University, Lahore, Pakistan

Faculty of Biological Sciences, Department of Microbiology, Quaid-i-Azam University, Islamabad, Pakistan

I. Afzal
Faculty of Biology, Department of Biology, Lahore Garrison University, Lahore, Pakistan

N. Ali
Faculty of Biological Sciences, Department of Microbiology, Quaid-i-Azam University, Islamabad, Pakistan

K. Cornish
Food, Agricultural and Biological Engineering, Ohio Agricultural Research and Development Center, The Ohio State University, Wooster, OH, USA

Horticulture and Crop Science, Ohio Agricultural Research and Development Center, The Ohio State University, Wooster, OH, USA

© Springer Nature Switzerland AG 2019
S. M. Tiquia-Arashiro, M. Grube (eds.), *Fungi in Extreme Environments: Ecological Role and Biotechnological Significance*,
https://doi.org/10.1007/978-3-030-19030-9_23

465

over a wide range of temperature conditions, they are categorized as psychrophiles (−15–10 °C), mesophiles (20–45 °C), and thermophiles (above 45 °C). A heterogeneous group of genera, including Ascomycetes, Phycomycetes, Mycelia, Sterilia, and Fungi Imperfecti, contain numerous thermotolerant and thermophilic forms (Cooney and Emerson 1964; Mouchacca 1997).

In addition, thermophilic fungi are often found as members of microbial communities colonizing a wide range of diverse substrates, particularly damp organic substrates such as straw-based composts, hay, and tropical soils. Therefore, they form important components during succession on, and colonization of, a wide variety of substrata (Allen and Emerson 1949; Miehe 1907). *Mucor pusillus,* the first identified thermophilic fungi, was isolated from bread (Lindt 1886). *Thermomyces lanuginosus,* previously known as *Humicola lanuginose,* first isolated in 1899 is broadly disseminated and normally isolated from self-heating masses of organic debris (Cooney and Emerson 1964; Emerson 1968; Krause et al. 2003; Tiquia 2005; Sikandar et al. 2017). Different species of thermophilic fungi were isolated from self-heating hay (Miehe 1907), namely, *Thermoascus aurantiacus, Thermomyces lanuginosus, Thermoidium sulfureum,* and *Mucor pusillus,* and these have also been isolated from several other natural substrates (Noack 1920; Miehe 1930).

In many biotechnological applications, thermophilic fungi have gained substantial attention for new sources of thermostable enzymes (thermozymes) particularly for lignocellulosic biomass degradation (Xie et al. 2014; Tiquia-Arashiro and Mormile 2013). Plant cell walls consist of up to 70% of cellulose, hemicellulose, and pectin (Jørgensen et al. 2007; Plecha et al. 2013). Lignocellulosic biomass consists of agro-industrial, agricultural, forestry, and food wastes that are plentiful, renewable, and low-cost energy sources. These lignocellulosic wastes accumulate in large quantities and can cause environmental problems (Chaudhary et al. 2012). Lignocellulose-degrading fungi are ubiquitous and gain nutrition by the degradation of plant biomass by producing degrading enzymes such as xylanases, cellulases, amylases, mannanase, pectinases, chitinases, proteases, lipases, ligninase, esterases, and phytases (van den Brink and de Vries 2011). In addition, thermophilic fungi produce enzymes which are more thermostable than those of mesophilic fungi, because they are more resistant to proteolysis and chemical denaturation (Table 23.1). Fungal thermozymes have been characterized by their ability to resist proteolysis and tolerate extreme conditions in the presence of denaturing agents, organic solvents, and high salinity (Sunna and Bergquist 2003; Raddadi et al. 2015). Proposed mechanisms of their protein thermostability include tighter packing or compactness (Russell et al. 1994), greater rigidity (Bogin et al. 1998), increased hydrogen bonding (Vogt et al. 1997), and deleted or shortened peptide loops (Russell and Sternberg 1997). A complete analysis of physiology, epigenetics, and metagenomics of fungal consortia showed the underlying production mechanisms of cell wall-degrading enzymes (Guerriero et al. 2015). Genomic taxonomic studies of thermophilic fungi led to their improved classification and deposition in culture collections and their sequences in online nucleotide public databases (de Oliveira et al. 2015).

Table 23.1 Comparison of the growth parameters of some mesophilic and thermophilic fungi

Fungus	Type of fungus	Growth temperature (°C)	References
Neurospora crassa	Mesophile	30	Alberghina (1973)
Trichoderma reesei	Mesophile	30	Mandels and Sternberg (1976)
Thermomyces lanuginosus	Thermophile	50	Rajasekaran and Maheshwari (1993)
Sporotrichum thermophile	Thermophile	50	Rajasekaran and Maheshwari (1993)
Trichoderma viride	Thermophile	50	Rajasekaran and Maheshwari (1993)
Aspergillus niger	Mesophile	30	Rajasekaran and Maheshwari (1993)
Trichoderma reesei	Mesophile	30	Mandels and Sternberg (1976)
Talaromyces thermophilus	Mesophile	33	Wright et al. (1983)
Scytalidium thermophilum	Thermophile	45	Sundaram (1986)
Mucor pusillus	Thermophile	55	Arima et al. 1967
*Malbranchea pulchella*var. *sulfurea*	Thermophile	45	Ong and Gaucher (1976)
Humicola lanuginosa strain Y-38	Thermophile	60	Liu et al. (1973)
Penicillium duponti	Mesophile	25	Malcolm and Shepherd (1972)
Pencillium chrysogenum	Mesophile	30	Renosto et al. (1985)

Lignocellulose-degrading thermophilic fungi and their thermozymes have been widely studied for the bioconversion of the lignocellulosic material into value-added products (Xie et al. 2014; Znameroski and Glass 2013). Moreover, they have proven to be an important and efficient source of thermozymes from lignocellulosic biomass that can be useful in many processes, including extraction of natural rubber from rubber-bearing plants (Sikandar et al. 2017). Natural rubber is a biopolymer consisting of 320 to >35,000 isoprene molecules. *Taraxacum kok-saghyz* (TK) also recognized as the rubber dandelion or "Buckeye Gold" is used as an alternative source of rubber plant by multiple commercial groups (van Beilen and Poirier 2007). TK roots comprise 2 to >36% of rubber, along with 25–40% inulin (polysaccharide of fructose), on a dry-weight basis (Buranov and Elmuradov 2009; van Beilen and Poirier 2007). Thus, the potential of lignocellulose-degrading thermophilic fungi to utilize plant biomass to produce cellulases, pectinases, and hemicellulases towards natural rubber extraction needs further investigation. The reduction in cost of enzyme production may lead to an innovation in the commercialization of natural rubber extraction from rubber-bearing plants. For the first time, this review assembles information on the efficiency of lignocellulose-degrading thermophilic fungi and their thermozymes towards an intractable problem associated with previous methods for rubber extraction from rubber-producing dandelions.

23.2 Lignocellulose-Degrading Fungal Thermozymes

Lignocellulose is a tightly, covalently, and non-covalently interconnected material mainly composed of cellulose, pectin, hemicelluloses, and lignin. Wood, grass, municipal solid wastes, agricultural residues, and paper industry wastes are primary sources (Howard et al. 2003; Mussatto and Teixeira 2010). Approximately 70% of lignocellulosic biomass is used in pulp and paper, food, feed, and biofuel, the four largest industrial sector biomass applications (Jørgensen et al. 2007). Accumulation of industrial waste and agricultural residues causes serious pollution but they could be (1) used as substrates for enzyme production or (2) fermented into alcohol or glucose for chemical syntheses or fuels (Kumakura et al. 1988; Nguyen et al. 2013; Tiquia-Arashiro 2014; Pomaranski and Tiquia-Arashiro 2016). Enzymes from lignocellulose-degrading thermophilic fungi frequently have been considered to evaluate their utility in industrial bioprocesses and to characterize and compare physicochemical properties of enzymes from mesophilic fungi. Enzymes secreted into growth media and prepared as culture filtrates have been more often studied than intracellular enzymes. The ability of thermophilic fungi to degrade lignocellulosic biomass at higher temperatures than mesophiles has several advantages: high temperature fermentation using thermophiles decreases substrate viscosity in fermenters, shortens the reaction time, and limits contamination (Blumer-Schuette et al. 2014). Thermostable enzymes require shorter reaction times for maximum plant biomass degradation compared to mesophillic conditions as in *Aspergillus* and *Trichoderma*. Extracellular cell wall-degrading enzymes from thermophilic fungi are comprised of a two-enzyme system (Maheshwari et al. 2000), hydrolases which degrade polysaccharides (Fig. 23.1) and oxidative and extracellular ligninolytic enzymes which open phenyl rings and degrade lignin.

23.2.1 Cellulases

In fungi, three hydrolytic enzymes are responsible for degrading cellulose: (1) endo-(1,4)-β-D-glucanase (carboxymethyl cellulase, endocellulase, endoglucanase, [EC 3.2.1.4]) which randomly cuts β-linkages, normally in cellulose amorphous parts; (2) β-glucosidase (cellobiase [EC 3.2.1.21]) which releases glucose from cellobiose and short-chain cello-oligosaccharides; and (3) exo-(1,4)-β-D-glucanase (exocellulase, avicelase, cellobiohydrolase, microcrystalline cellulase, [EC 3.2.1.91]) which releases cellobiose either from the reducing or the nonreducing end of crystalline cellulose (Bhat and Bhat 1997).

The levels of extracellular cellulases determine the extent of cellulose solubilization, and this led to attempts to develop practical processes to convert cellulose to glucose by isolating and screening cellulase-producing fungi (Mandels 1975; Mandels and Sternberg 1976). Although cellulose was rapidly degraded by some species of thermophilic fungi, the cellulase activity of their culture filtrates was generally low (Mandels 1975). In contrast, the activity of cellulases produced by the thermophilic

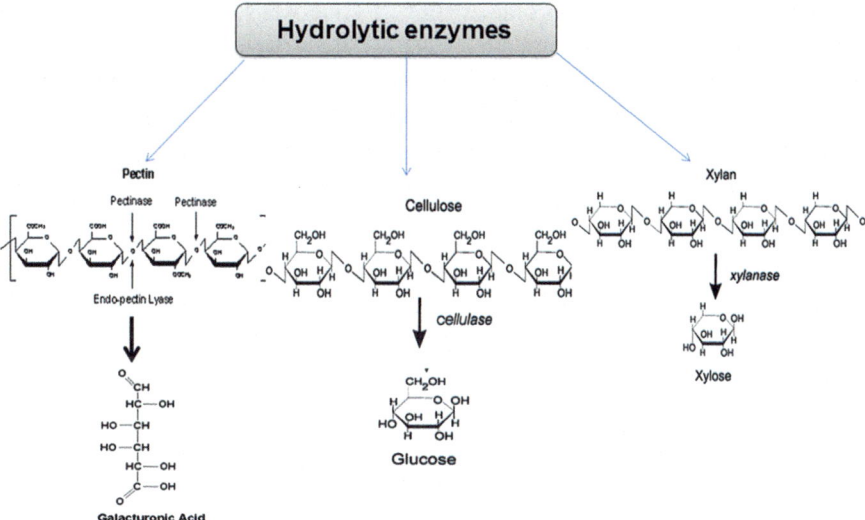

Fig. 23.1 Generalized view of polymeric substances and their products derived from hydrolytic enzymes

fungi *Talaromyces emersonii* (Folan and Coughlan 1978) and *Sporotrichum thermophile* (Coutts and Smith 1976) was closely similar to that produced by *Trichoderma reesei,* the maximum cellulase-producing fungal strain. However, apparent cellulase production is clearly affected by assay method (Oberson et al. 1992). Bhat and Maheshwari (1987) explained that the culture filtrates of endoglucanase and exoglucanase activities of *S. thermophile* strain were lower than that of *T. reesei.* However, in spite of these lower activities, *S. thermophile* grew at five times faster and degraded cellulose more rapidly than *T. reesei.* The vigorous growth of *S. thermophile* was similar on glucose and insoluble cellulose. Based on these observations, the rate or extent of cellulolysis depends on the secreted levels of cellulase activities. As is the case with mesophilic fungi, multiple cellulase enzymes are produced by thermophilic fungi and, in most, crystalline cellulose is the preferred substrate over impure or amorphous forms (Romanelli et al. 1975; McHale and Coughlan 1981; Ganju et al. 1989). However, high cellulase and xylanase activities were secreted by *Thermoascus aurantiacus* (Khandke et al. 1989b; Kawamori et al. 1987), *Humicola insolens* (Hayashida and Yoshioka 1980), and *Thermomyces lanuginosus* (Sikandar et al. 2017), even on lignocellulosic and hemicellulosic substrates.

23.2.2   Xylanases

Xylan, a major component of hemicellulose, is the most plentiful polysaccharide in nature, and a variety of hydrolytic enzymes are mandatory for its complete degradation. These include endoxylanases (EC 3.2.1.8), which hydrolyze β-1,4-linked xylose (the xylan backbone) and β-xylosidases (EC 3.2.1.37), which cleave xylo-oligomers (Biely 1985). Thermophilic fungi, producing xylanases, have attracted considerable attention in numerous biotechnological applications, particularly lignocellulosic biomass degradation by enzymatic degradation of xylan from lignin carbohydrates and for bio-bleaching of pulp in the paper industry. Xylanase production can be achieved on a wide range of carbon sources including pure xylan (Gomes et al. 1993b), and natural substrates rich in xylan such as wheat bran (Yoshioka et al. 1981; Sikandar et al. 2017), corn cobs (Bennett et al. 1998), and sugarcane bagasse (Prabhu and Maheshwari 1999). Paper waste was a unique carbon source which induced high levels of xylanases in the thermophilic fungi *Humicola lanuginosa* (Anand et al. 1990) and *Thermoascus aurantiacus* (Khandke et al. 1989a). *T. lanuginosus* also is a good producer of xylanases and pectinases (Puchart et al. 1999; Singh et al. 2003; Sikandar et al. 2017). In addition, some xylanase-producing *Thermoascus aurantiacus* and *Thermomyces lanuginosus* fungal strains are active above at 70 °C. The majority of fungal xylanases are endoxylanases, and optimal temperature for most thermophilic xylanases ranges from 55 °C to 65 °C (Table 23.2). The thermostability of xylanases from *T. lanuginosus* was known due to the presence of an extra disulfide bridge which was not present in mesophilic xylanases (Eswaramoorthy et al. 1994).

Table 23.2 Optimum temperatures of hydrolytic enzymes of different thermophilic fungi

Thermophilic	Optimum temperature (°C)			References
	Cellulases	Xylanases	Pectinases	
Mucor pusillus	–	–	55	Arima et al. (1967)
M. miehei	–	–	–	Ottese and Rickert (1970)
Penicillium duponti		–	40	Martin et al. (2004)
*Malbranchea pulchella*var. *sulfurea*	–	–	–	Ong and Gaucher (1973, 1976)
Humicola lanuginosa	–	65	–	Ong and Gaucher (1973, 1976)
Bacillus thermoproteolyticus	–	–	–	Voordouw et al. (1974)
Sporotrichum thermophile	63	–	–	Coutts and Smith (1976)
Talaromyces emersonii	75–80	60	–	Folan and Coughlan (1978)
Thermoascus aurantiacus	65	70–80	65	Khandke et al. (1989b)
Humicola insolens	50	50–65	–	Hayashida and Yoshioka (1980)
Thermoascus lanuginosus	55	70–80	40–50	Puchart et al. (1999), Sikandar et al. 2017
Melanocarpus albomyces	–	65	–	Prabhu and Maheshwari (1999)
Paecilomyces varioti	–	65	–	Krishnamurthy (1989)
Thermomucor indicaeseudaticae	–	–	45	Martin et al. (2010)
Aspergillus niger	–	–	40	Akhter et al. (2011)

23.2.3 Pectinases

Pectin is an essential component of plant cell wall and can be hydrolyzed by pectinases by the cleavage of α-1,4 linkages of polygalacturonic acid. Thermophilic fungi producing pectinases are found in decaying fruits and vegetable matter, and very few of them have been isolated and characterized (Adams and Deploey 1978; Sikandar et al. 2017). However, when 40 thermophilic fungal species (previously reported to produce pectinases) belonging to seven genera were cultured on pectin as a carbon source in a medium, no activity was detected in most of the cultures (Inamdar 1987). *T. aurantiacus* produces most stable polygalacturonases, active above 65 °C. Temperature stability remains a significant limitation to commercial utility.

23.3 Natural Rubber

Natural rubber is the important biological materials used in different nonfood applications. It is a polymer containing isoprene units which are linked together in a 1,4-*cis* configuration. Although rubber is produced in over 2500 plant species, commercial production of rubber is almost still exclusively from *Hevea brasiliensis*, the Para rubber tree (Cataldo 2000). The rubber-bearing dandelion *Taraxacum kok-saghyz* holds significant quantities of rubber in its roots (up to 36% on a dry-weight basis (Buranov et al. 2005) although most genotypes contain much less rubber than this (Cornish et al. 2016; Martinez et al. 2015; Ramirez-Cadavid et al. 2017).

In rubber-bearing plants, rubber is formed in rubber particles in the cytosol of ordinary parenchyma cells, or more commonly as free-flowing latex formed in latex vessels (laticifers) in the same anatomical zone as the phloem (Martinez et al. 2015). In TK roots, carbohydrate components of plant biomass such as cellulose, hemicellulose, and pectin are present (Ramirez-Cadavid et al. 2017). Rubber dandelion roots also contain a major amount of inulin and sugars (up to 40%) that can be converted into ethanol or other chemicals (Ramirez-Cadavid et al. 2017; Ramirez-Cadavid et al. 2018; Ujor et al. 2015). The utilization of biopolymers (celluloses, hemicelluloses, and lignin) and bio-products (phenolic compounds) from the remaining biomass signifies an opportunity for increased commercialization activities. Solid rubber can be extracted from dried tissue by several methods including simultaneous or sequential organic solvent extraction (Eagle 1981; Black et al. 1983; Hamerstrand and Montgomery 1984; Schloman et al. 1988), water extraction without (Eskew and Edwards 1946) or with hydrolytic enzymes (Sikandar et al. 2017), or dry milling (Buranov 2009). A new aqueous method was recently reported (Sikandar et al. 2017) using lignocellulose-degrading hydrolytic enzymes from *T. lanuginosus* STm, which were used to enhance the extracted yield and purity of rubber from rubber dandelion (Fig. 23.2).

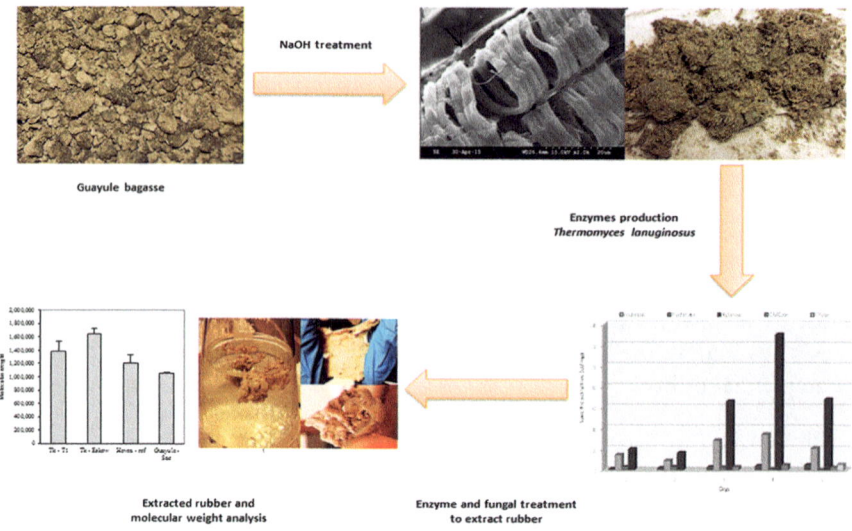

Fig. 23.2 Application of alkali-treated plant biomass for extraction of natural rubber using fungal hydrolytic enzymes

23.4   Natural Rubber Extraction by Fungal Thermozymes

Enzymes from lignocellulose-degrading thermophilic fungi provide diverse biotechnological opportunities for biotransformations and biocatalysis due to their stability across a broad range of temperature, ionic strength, pH, and salinity, and their capability to function in organic solvents, which denature most enzymes from other sources (Adrio and Demain 2014; Karmakar and Ray 2011). These enzymes are used in different industrial and commercial products (Adrio and Demain 2014) due to their reproducibility, high performance, and economic viability (Gurung et al. 2013). Enzyme production using lignocellulosic biomass provides both economic and environmental advantages. According to many previous studies, *T. lanuginosus* produced high levels of xylanases (Gomes et al. 1993a; Jensen et al. 1987; Li et al. 1997). Puchart et al. (1999) also reported maximum xylanase activity (1.0 U/mL) by *T. lanuginosus* using corncob as a carbon source. Previous studies reported that lignocellulosic biomass is a superior substrate for producing lignocellulose-degrading enzymes (Damaso et al. 2000). Newly isolated *T. lanuginosus* STm using lignocellulosic biomass produced high levels of xylanases (67.4 U/mg) and cellulases (16.7 U/mg) (Sikandar et al. 2017). *T. lanuginosus* STm efficiently grows and degrades different lignocellulosic biomass at high temperature, which suggests its efficacy to degrade other types of plant biomass containing rubber.

Conventional methods previously used for rubber extraction included both solvent and water-based methods. In sequential and simultaneous organic solvent extraction methods, acetone is used to extract low-molecular-weight organic compounds, such as terpene resins, whereas rubber extraction requires nonpolar solvents such as cyclohexane, hexane, pentan, or chloroform (Black et al. 1983; Schloman et al. 1988; Eagle 1981; Hamerstrand and Montgomery 1984). However, green dry-milling (Buranov 2009) and benign water-milling (Eskew and Edwards 1946) processes can also be used to extract solid natural rubber. To reduce cost and environment-associated problems with chemical methods of rubber extraction, use of lignocellulose-degrading fungal thermozymes is encouraged. In-house production of fungal thermozymes makes the overall process of natural rubber extraction cost effective. Enzymatic hydrolysis of TK roots is an effective biotechnological process, also used for the production of value-added products. The effectiveness of *T. lanuginosus* STm and its enzymes towards rubber extraction from TK roots is due to physical and biochemical combination (Sikandar et al. 2017), and fungal unique enzymatic fingerprint, wherein the combination of excreted enzymes is capable of disrupting the tight linkages (likely covalent) between the biomass and the rubber.

23.5   Conclusions and Future Perspectives

Thermophilic fungi have been previously isolated and used as sources of novel industrial enzymes owing to their ability to thrive in extreme conditions. In particular, lignocellulose-degrading thermophilic fungi have found applications in the degradation of complex lignocellulosic biomass due to their high resistance under extreme temperature, chemicals, organic solvents, and pH. They have the ability to degrade recalcitrant plant biomass by producing cellulases, pectinases, and xylanases, respectively. The expansion of novel industrial processes built on fungal thermozymes, and the growing demand by industries for novel enzymes, incentivizes new discovery of useful thermophilic fungi. *Thermomyces lanuginosus* STm is a sustainable hydrolytic enzyme source that was used to the extraction of natural rubber from plants containing rubber by increasing its overall yield compared to other conventional processes previously used. Moreover, this bioprocess is more prospective towards cost-effective and environmentally friendly approach. In future, a multifaceted approach including characterization of lignocellulosic biomass for enzyme production, environmental metagenomics, fungal proteomics, and synergistic analysis of natural and genetically optimized thermozymes is required to efficiently and cost-effectively exploit lignocellulose-degrading thermophilic fungi.

References

Adams PR, Deploey JJ (1978) Enzymes produced by thermophilic fungi. Mycologia 70:906–910
Adrio JL, Demain AL (2014) Microbial enzymes: tools for biotechnological processes. Biomolecules 4:117–139
Alberghina FA (1973) Growth regulation in *Neurospora crassa*. Effects of nutrients and of temperature. Arch Microbiol 89:83–94

Allen PJ, Emerson R (1949) Guayule rubber microbiological improvement by shrub retting. Ind Eng Chem 41:346–365

Akhter N, Morshed MA, Uddin A, Begum F, Sultan T, Azad AK (2011) Production of pectinase by *Aspergillus niger* cultured in solid state media. Int J Biosci 1(1):33–42

Anand L, Krishnamurthy S, Vithayathil PJ (1990) Purification and properties of xylanase from the thermophilic fungus, *Humicola lanuginosa* (Griffon and Maublanc) Bunce. Arch Biochem Biophys 276:546–553

Arima K, Iwasaki S, Tamura G (1967) Milk clotting enzyme from microorganisms. I. Screening test and the identification of the potent fungus. Agric Biol Chem 31:540–545

Bennett NA, Ryan J, Biely P, Vrsanska M, Kremnicky L, Macris BJ, Kekos D, Christakopoulos P, Katapodis P, Claeyssens M, Nerinckx W, Ntauma P, Bhat MK (1998) Biochemical and catalytic properties of an endoxylanase purified from the culture filtrate of *Thermomyces lanuginosus* ATCC 46882. Carbohydr Res 306:445–455

Bhat MK, Bhat S (1997) Cellulose degrading enzymes and their potential industrial applications. Biotechnol Adv 15:583–620

Bhat KM, Maheshwari R (1987) Sporotrichum thermophile: growth, cellulose degradation, and cellulase activity. Appl Environ Microbiol 53:2175–2182

Biely P (1985) Microbial xylanolytic systems. Trends Biotechnol 3:286–290

Black L, Hamerstrand G, Nakayama F, Rasnik B (1983) Gravimetric analysis for determining the resin and rubber content of guayule. Rubber Chem Technol 56:367–371

Blumer-Schuette SE, Brown SD, Sander KB, Bayer EA, Kataeva I, Zurawski JV, Conway JM, Adams MW, Kelly RM (2014) Thermophilic lignocellulose deconstruction. FEMS Microbiol Rev 38:393–448

Bogin O, Peretz M, Hacham Y, Korkhin Y, Frolow F, Kalb AJ, Burstein Y (1998) Enhanced thermal stability of *Clostridium beijerinckii* alcohol dehydrogenase after strategic substitution of amino acid residues with prolines from the homologous thermophilic *Thermoanaerobacter brockii* alcohol dehydrogenase. Protein Sci 7(5):1156–1163

Buranov AU (2009) Process for recovering rubber from rubber-bearing plants with a gristmill. Google Patents

Buranov AU, Elmuradov BJ (2009) Extraction and characterization of latex and natural rubber from rubber-bearing plants. J Agric Food Chem 58:734–743

Buranov A, Elmuradov B, Shakhidoyatov K, Anderson F, Lawrence J (2005) Rubber-bearing plants of Central Asia. In: Industrial crops and rural development. Proceedings of 2005 Annual Meeting of the Association for the Advancement of Industrial Crops: International Conference on Industrial Crops and Rural Development 17–21

Cataldo F (2000) Guayule rubber: a new possible world scenario for the production of natural rubber. Prog Rubber Plast Technol 16:31–59

Chaudhary G, Singh LK, Ghosh S (2012) Alkaline pretreatment methods followed by acid hydrolysis of *Saccharum spontaneum* for bioethanol production. Bioresour Technol 124:111–118

Cooney DG, Emerson R (1964) Thermophilic fungi: an account of their biology, activities and classification. Freeman, San Francisco, CA, pp 80–88

Cornish K, Kopicky SL, McNulty SK, Amstutz N, Chanon AM, Walker S, Kleinhenz MD, Miller AR, Streeter JG (2016) Temporal diversity of *Taraxacum kok-saghyz* plants reveals high rubber yield phenotypes. Biodiversitas 17:847–856

Coutts AD, Smith RE (1976) Factors influencing the production of cellulases by *Sporotrichum thermophile*. Appl Environ Microbiol 31:819–825

de Oliveira TB, Gomes E, Rodrigues A (2015) Thermophilic fungi in the new age of fungal taxonomy. Extremophiles 19(1):31–37

Damaso MC, Andrade CM, Pereira N (2000) Use of corncob for endoxylanase production by thermophilic fungus *Thermomyces lanuginosus* IOC-4145. Appl Biochem Biotechnol 84:821–834

Eagle F (1981) Guayule. Rubber Chem Technol 54:662–684

Emerson R (1968) Thermophiles. In: Answorth GC, Sussman AS (eds) The fungi: an advanced treatise. Academic Press, London, p 105. 128

Eskew RK, Edwards PW 1946. Process for recovering rubber from fleshy plants. Google Patents

Eswaramoorthy S, Vithayathil PJ, Viswamitra MA (1994) Crystallization and preliminary X-ray crystallographic studies of thermostable xylanase crystals isolated from *Paecilomyces varioti*. J Mol Biol 243:806–808

Folan MA, Coughlan MP (1978) The cellulase complex in the culture filtrate of the thermophyllic fungus, *Talaromyces emersonii*. Int J Biochem 9:717–722

Ganju RK, Murthy SK, Vithayathil PJ (1989) Purification and characterization of two cellobiohydrolases from *Chaetomium thermophile* var. coprophile. Biochim Biophys Acta 993:266–274

Gomes J, Gomes I, Kreiner W, Esterbauer H, Sinner M, Steiner W (1993a) Production of high level of cellulase-free and thermostable xylanase by a wild strain of *Thermomyces lanuginosus* using beechwood xylan. J Biotechnol 30:283–297

Gomes J, Purkarthofer H, Hayn M, Kapplmüller J, Sinner M, Steiner W (1993b) Production of a high level of cellulase-free xylanase by the thermophilic fungus *Thermomyces lanuginosus* in laboratory and pilot scales using lignocellulosic materials. Appl Microbiol Biotechnol 39:700–707

Guerriero G, Hausman JF, Strauss J, Ertan H, Siddiqui KS (2015) Destructuring plant biomass: focus on fungal and extremophilic cell wall hydrolases. Plant Sci 234:180–193

Gunde-Cimerman N, Zalar P (2014) Extremely halotolerant and halophilic fungi inhabit brine in solar salterns around the globe. Food Tech Biotechnol 52:170–179

Gurung N, Ray S, Bose S, Rai V (2013) A broader view: microbial enzymes and their relevance in industries, medicine, and beyond. Biomed Res Int 2013:329121

Hamerstrand G, Montgomery R (1984) Pilot-scale guayule processing using countercurrent solvent extraction equipment. Rubber Chem Technol 57:344–350

Hayashida S, Yoshioka H (1980) Production and purification of thermostable cellulases from *Humicola insolens* YH-8. Agric Biol Chem 44:1721–1728

Howard R, Abotsi E, Van Rensburg EJ, Howard S (2003) Lignocellulose biotechnology: issues of bioconversion and enzyme production. Afr J Biotechnol 2:602–619

Inamdar A (1987) Polygalacturonase from *Thermoascus aurantiacus*: isolation and functional characteristics. Ph.D thesis. Indian Institute of Science, Bangalore

Jensen B, Olsen J, Allermann K (1987) Effect of media composition on the production of extracellular amylase from the thermophilic fungus *Thermomyces lanuginosus*. Biotechnol Lett 9:313–316

Jørgensen H, Vibe-Pedersen J, Larsen J, Felby C (2007) Liquefaction of lignocellulose at high-solids concentrations. Biotechnol Bioeng 96:862–870

Karmakar M, Ray RR (2011) Current trends in research and application of microbial cellulases. Res J Microbiol 6:41–53

Kawamori M, Takayama K-I, Takasawa S (1987) Production of cellulases by a thermophilic fungus *Thermoascus aurantiacus* A-131. Agric Biol Chem 51:647–654

Khandke KM, Vithayathil PJ, Murthy SK (1989a) Degradation of larchwood xylan by enzymes of a thermophilic fungus, *Thermoascus aurantiacus*. Arch Biochem Biophys 274:501–510

Khandke KM, Vithayathil PJ, Murthy SK (1989b) Purification of xylanase, β-glucosidase, endocellulase, and exocellulase from a thermophilic fungus, *Thermoascus aurantiacus*. Arch Biochem Biophys 274:491–500

Krause MS, De Ceuster TJJ, Tiquia SM, Michel FC Jr, Madden LV, Hoitink HAJ (2003) Isolation and characterization of Rhizobacteria from composts that suppress the severity of bacterial leaf spot of radish. Phytopathology 93(10):1292–1300

Krishnamurthy S (1989) Purification and properties of xylanases and β-glucosidases elaborated by the thermophilic fungus *Paecilomyces varioti* Bainier. Ph.D. thesis. Indian Institute of Science, Bangalore

Kumakura M, Kasai N, Tamada M. and Kaetsu I (1988). Method of pretreatment in saccharification and fermentation of waste cellulose resource. Google Patents

Li DC, Yang YJ, Shen CY (1997) Protease production by the thermophilic fungus *Thermomyces lanuginosus*. Mycol Res 101:18–22

Lindt W (1886) Mitteilungen über einige neue pathogene Shimmelpilze. Arch Exp Pathol Pharmakol 21:269–298

Liu W-H, Beppu T, Arima K (1973) Physical and chemical properties of the lipase of thermophilic fungus *Humicola lanuginosa* S-38. Agric Biol Chem 37:2493–2499

Magan N, Aldred D (2008) Environmental fluxes and fungal interactions: maintaining a competitive edge. In: British mycological society symposia series. Academic Press, Vol. 27, pp. 19–35

Maheshwari R, Bharadwaj G, Bhat MK (2000) Thermophilic fungi: their physiology and enzymes. Microbiol Mol Biol Rev 64:461–488

Malcolm AA, Shepherd MG (1972) Purification and properties of *Penicillium* glucose-6-phosphate dehydrogenase. Biochem J 128:817–831

Mandels M (1975) Microbial sources of cellulase. Biotechnol Bioeng Symp 5:81–105

Mandels M, Sternberg D (1976) Recent advances in cellulase technology. J Ferment Technol 54:267–286

Martin N, de Souza SR, da Silva R, Gomes E (2004) Pectinase production by fungal strains in solid-state fermentation using agro-industrial bioproduct. Braz Arch Biol Technol 47(5):813–819

Martin N, Guez MAU, Sette LD, Da Silva R, Gomes E (2010) Pectinase production by a Brazilian thermophilic fungus Thermomucor indicae-seudaticae N31 in solid-state and submerged fermentation. Microbiology 79(3):306–313

Martinez M, Poirrier P, Chamy R, Prüfer D, Schulze-Gronover C, Jorquera L, Ruiz G (2015) *Taraxacum officinale* and related species—an ethnopharmacological review and its potential as a commercial medicinal plant. J Ethnopharmacol 169:244–262

McHale A, Coughlan MP (1981) The cellulolytic system of *Talaromyces emersonii*. Identification of the various components produced during growth on cellulosic media. Biochim Biophys Acta 662:145–151

Miehe H (1907) Die Sel bsterhitzung des Heus Eine biologische Studie. Gustav Fischer Verlag, Jena, Germany

Miehe H (1930) Die Wärmebildung von Reinkulturen im Hinblick auf die Ätiologie der Selbsterhitzung pflanzlicher Stoffe. Arch Mikrobiol 1:78–118

Mouchacca J (1997) Thermophilic fungi biodiversity and taxonomic status. Cryptogamie Mycol 18:19–69

Mussatto SI, Teixeira J (2010) Lignocellulose as raw material in fermentation processes. In: Méndez-Vilas A (ed) Current research, technology and education topics in applied microbiology and microbial biotechnology, vol 2. Formatex Research Center, pp 897–907

Nguyen S, Ala F, Cardwell C, Cai D, McKindles KM, Lotvola A, Hodges S, Deng Y, Tiquia-Arashiro SM (2013) Isolation and screening of carboxydotrophs isolated from composts and their potential for butanol synthesis. Environ Technol 34:1995–2007

Noack K (1920) Der Betriebstoffwechsel der thermophilen Pilze. Jahrb Wiss Bot 59:593–648

Oberson J, Binz T, Fracheboud D, Canevascini G (1992) Comparative investigation of cellulose-degrading enzyme systems produced by different strains of *Myceliophthora thermophila* (Apinis) v. Oorschot. Enzyme Microb Technol 14:303–312

Ong PS, Gaucher GM (1973) Protease production by thermophilic fungi. Can J Microbiol 19:129–133

Ong PS, Gaucher GM (1976) Production, purification and characterization of thermomycolase, the extracellular serine protease of the thermophilic fungus *Malbranchea pulchella* var. *sulfurea*. Can J Microbiol 22:165–176

Ottesen M, Rickert W (1970) The isolation and partial characterization of an acid protease produced by *Mucor miehei*. C R Trav Lab Carlsberg 37:301–325

Plecha S, Hall D, Tiquia-Arashiro SM (2013) Screening and characterization of soil microbes capable of degrading cellulose from Switchgrass (*Panicum virgatum* L.). Environ Technol 34:1895–1904

Pomaranski E, Tiquia-Arashiro SM (2016) Butanol tolerance of carboxydotrophic bacteria isolated from manure composts. Environ Technol 37(15):1970–1982

Prabhu KA, Maheshwari R (1999) Biochemical properties of xylanases from a thermophilic fungus, *Melanocarpus albomyces*, and their action on plant cell walls. J Biosci 24:461–470

Puchart VR, Katapodis P, Biely P, Kremnický LR, Christakopoulos P, Vršanská M, Kekos D, Macris BJ, Bhat MK (1999) Production of xylanases, mannanases, and pectinases by the thermophilic fungus *Thermomyces lanuginosus*. Enzyme Microb Technol 24:355–361

Raddadi N, Cherif A, Daffonchio D, Mohamed N, Fava F (2015) Biotechnological applications of extremophiles, extremozymes and extremolytes. Appl Microbiol Biotechnol 99:7907–7913

Rajasekaran AK, Maheshwari R (1993) Thermophilic fungi: an assessment of their potential for growth in soil. J Biosci 18(3):345

Ramirez-Cadavid DA, Cornish K, Michel FC Jr (2017) *Taraxacum kok-saghyz* (TK): Compositional analysis of a feedstock for natural rubber and other bioproducts. Ind Crop Prod 107:624–640

Ramirez-Cadavid DA, Valles-Ramirez S, Cornish K, Michel FC Jr (2018) Simultaneous quantification of rubber, inulin, and resins in *Taraxacum kok-saghyz* (TK) roots by sequential solvent extraction. Ind Crop Prod 122:647–656

Renosto F, Seubert PA, Knudson P, Segel IH (1985) APS kinase from *Penicillium chrysogenum*. Dissociation and reassociation of subunits as the basis of the reversible heat inactivation. J Biol Chem 260(3):1535–1544

Romanelli RA, Houston CW, Barnett SM (1975) Studies on thermophilic cellulolytic fungi. Appl Microbiol 30:276–281

Russell RB, Sternberg MJ (1997) Two new examples of protein structural similarities within the structure-function twilight zone. Protein Eng 10(4):333–338

Russell RJ, Hough DW, Danson MJ, Taylor GL (1994) The crystal structure of citrate synthase from the thermophilic archaeon, *Thermoplasma acidophilum*. Structure 2(12):1157–1167

Schloman WW, Carlson DW, Hilton AS (1988) Guayule extractables: influence of extraction conditions on yield and composition. Biomass 17:239–249

Sikandar S, Ujor VC, Ezeji TC, Rossington JL, Michel FC, McMahan CM, Cornish K (2017) *Thermomyces lanuginosus* STm: A source of thermostable hydrolytic enzymes for novel application in extraction of high-quality natural rubber from Taraxacum kok-saghyz (Rubber dandelion). Ind Crop Prod 103:161–168

Singh S, Madlala AM, Prior BA (2003) *Thermomyces lanuginosus*: properties of strains and their hemicellulases. FEMS Microbiol Rev 27:3–16

Sundaram TK (1986) Physiology and growth of thermophilic bacteria. In: Brock D (ed) Thermophiles: general, molecular, and applied microbiology. John Wiley & Sons, Inc., New York, N.Y, pp 75–106

Sunna A, Bergquist PL (2003) A gene encoding a novel extremely thermostable 1,4-beta-xylanase isolated directly from an environmental DNA sample. Extremophiles 7:63–70

Timling I, Taylor DL (2012) Peeking through a frosty window molecular insights into the ecology of Arctic soil fungi. Fungal Ecol 5:419–429

Tiquia SM (2005) Microbial community dynamics in manure composts based on 16S and 18S rDNA T-RFLP profiles. Environ Technol 26(10):1104–1114

Tiquia-Arashiro SM (2014) Thermophilic carboxydotrophs and their biotechnological applications. Springerbriefs in microbiology: extremophilic microorganisms. Springer International Publishing, p. 131

Tiquia-Arashiro SM, Mormile M (2013) Sustainable technologies: Bioenergy and biofuel from biowaste and biomass. Environ Technol 34(13):1637–1805

Ujor V, Bharathidasan AK, Michel FC Jr, Ezeji TC, Cornish K (2015) Butanol production from inulin-rich chicory and *Taraxacum kok-saghyz* extracts: determination of sugar utilization profile of *Clostridium saccharobutylicum* P262. Ind Crop Prod 76:739–748

van den Brink J, de Vries RP (2011) Fungal enzyme sets for plant polysaccharide degradation. Appl Microbiol Biotechnol 91(6):1477–1492

Van Beilen JB, Poirier Y (2007) Guayule and Russian dandelion as alternative sources of natural rubber. Crit Rev Biotechnol 27:217–231

Vogt G, Woell S, Argos P (1997) Protein thermal stability, hydrogen bonds, and ion pairs. J Mol Biol 269(4):631–643

Voordouw G, Gaucher GM, Roche RS (1974) Physicochemical properties of thermomycolase, the thermostable, extracellular, serine protease of the fungus *Malbranchea pulchella*. Can J Biochem 52:981–990

Wright CH, Kafkewitz DA, Somberg EW (1983) Eucaryote thermophily: role of lipids in the growth of Talaromyces thermophilus. J Bacteriol 156(2):493–497

Xie S, Syrenne R, Sun S, Yuan JS (2014) Exploration of natural biomass utilization systems (NBUS) for advanced biofuel—from systems biology to synthetic design. Curr Opin Biotechnol 27:195–203

Yoshioka H, Nagato N, Chavanich S, Nilubol N, Hayashida S (1981) Purification and properties of thermostable xylanase from *Talaromyces byssochlamydoides* YH-50. Agric Biol Chem 45:2425–2432

Znameroski EA, Glass NL (2013) Using a model filamentous fungus to unravel mechanisms of lignocellulose deconstruction. Biotechnol Biofuels 6:6

Chapter 24
Thermophilic Fungi and Their Enzymes for Biorefineries

Abha Sharma, Anamika Sharma, Surender Singh, Ramesh Chander Kuhad, and Lata Nain (iD)

24.1 Introduction

Increasing prices of crude oil, depletion of nonrenewable energy reserves, as well as environmental concerns related to air pollution and global warming have focused on the development of alternative renewable energy resources. However, the issue of food versus fuel makes first-generation biofuels indefensible, thereby generating an overwhelming research interest towards exploitation of abundant lignocellulosic biomass for the production of biofuel as well as other commodity chemicals in a biorefinery approach (Hasunuma and Kondo 2012; Kumar et al. 2008). Biorefinery can also be defined as a facility that integrates biomass conversion processes and equipment to produce fuels, power, heat, and value-added chemicals from biomass with the aid of microorganisms and their enzymes (Fernando et al. 2006; Kamm and Kamm 2004; Nguyen et al. 2013; Pomaranski and Tiquia-Arashiro 2016).

Lignocellulosic biomass is the most abundant and inexhaustible biomass on earth, which holds enormous potential for sustainable production of chemicals and fuels in an eco-friendly manner (Somerville et al. 2010; Taarning et al. 2011; Tiquia-Arashiro and Mormile 2013). The various types of lignocellulosic raw materials include wheat straw, rice straw, palm, corncobs, corn stems and husk, etc. having varying amount of lignocellulosic components. The lignocellulosic biomass has higher amount of oxygen, and lower fractions of hydrogen and carbon with respect

A. Sharma · A. Sharma · L. Nain (✉)
Division of Microbiology, ICAR-Indian Agricultural Research Institute, New Delhi, India

S. Singh
Division of Microbiology, ICAR-Indian Agricultural Research Institute, New Delhi, India

Department of Microbiology, Central University of Haryana, Mahendergarh, India

R. C. Kuhad
Department of Microbiology, Central University of Haryana, Mahendergarh, India

© Springer Nature Switzerland AG 2019   479
S. M. Tiquia-Arashiro, M. Grube (eds.), *Fungi in Extreme Environments:*
Ecological Role and Biotechnological Significance,
https://doi.org/10.1007/978-3-030-19030-9_24

to petroleum resources and as a result more classes of products can be obtained from degradation of lignocellulosic based biorefineries than petroleum-based ones (Plecha et al. 2013; Isikgor and Becer 2016; Tiquia-Arashiro 2014a).

Among microorganisms, fungi with the help of their extracellular enzymes can carry out efficient degradation of plant biomass as it is also their principal carbon source in natural habitat. Consequently, fungi and their enzymes are very attractive with respect to the concept of biorefinery, wherein they are used for the degradation of lignocellulosic biomass with an aim of generating biofuels and other platform chemicals (Kuhad et al. 2010). A large number of enzymes that carry out lignocellulose degradation, viz. cellulases, xylanases, and ligninases, have been purified and characterized from fungi, mainly ascomycetes and basidiomycetes, and in many cases the corresponding enzymes have been cloned (de Vries and Visser 2001). Cellulases act on cellulosic chains of the lignocellulose-releasing monomeric glucose, while xylanases hydrolyze hemicelluloses of the substrate to xylose. The released monomeric sugars are fermented with microorganisms to produce biofuels and commodity chemicals in a biorefinery concept. However, the cellulose microfibrils in the plant cell wall are interrupted by hemicellulose and are surrounded by a lignin matrix (Penjumras et al. 2014). The resistance of lignin to degradation makes enzymatic access to cellulose a major obstacle in the saccharification, thereby attracting considerable interest in the methods employed for breakdown of lignin (Bugg et al. 2011). Pertaining to this, pretreatment methods that remove lignin and hemicellulose from plant material can be divided into the following categories: physical (milling and grinding), physicochemical (steam pretreatment or autohydrolysis, hydrothermolysis, and wet oxidation), chemical (alkali, dilute acid, oxidizing agents, and organic solvents), biological, or a combination of these (Gupta et al. 2011). Among these, biological pretreatment methods indeed provide an eco-friendly alternative to chemical and physical pretreatments, improving biorefinery economics by reducing pretreatment costs and alleviating inhibitor formation (Deswal et al. 2014). Biological pretreatment of lignocellulosic biomass mainly employs white-rot fungi, which degrade lignin and hemicelluloses selectively but very little of cellulose with the aid of their extracellular hemicellulases and ligninases (Sanchez and Demain 2011). Furthermore, due to the complexity of lignin chemical structure and its enzymatic modifications, a variable number of low molecular weight compounds can be released from its microbial degradation, which would be potentially a rich source of renewable aromatic chemicals to be used in food and flavor industry, and for fine chemicals as well as material synthesis (Fisher and Fong 2014).

On the other hand, despite the urgency for developing biorefineries intended to produce fuel and chemicals from renewable resources, low-cost processing technologies which efficiently convert lignocellulosic biomass energy into liquid fuels do not yet exist (Madadi et al. 2017). The high cost of exogenous enzymes used in the process is one of the major limiting factors that make the process expensive and unattainable. Erstwhile, the use of thermophilic enzymes for deconstruction of lignocelluloses provides a possible solution to the above problem, as these enzymes typically have a higher specific activity as well as higher stability in conditions of elevated temperature prevalent in industrial environments. As a result, they allow extended hydrolysis times and thus decrease the amount of enzyme needed for saccharification (Yeoman et al. 2010), making the process efficient as well as economical. Additionally, these enzymes promote better penetration and cell wall disorganization of the substrate (Turner et al. 2007). Moreover, microbial contamination risks are significantly reduced at elevated temperatures, and finally these enzymes can typically be kept at room temperature without loss of activity, reducing refrigeration cost during transport as well as storage. These advantages attributed to thermostable enzymes are noteworthy as almost one-third of the projected process costs in biomass conversions are associated with enzyme production (Haki and Rakshit 2003; Tiquia-Arashiro 2014b). Therefore, prospecting of novel thermophilic fungi which produce thermostable lignocellulolytic enzymes will be highly advantageous for the successful implementation of biorefineries (Kumar et al. 2008). Thermophilic microbes growing at temperature of 50–80 °C in extreme habitats are source of highly active and thermostable enzymes (Zeldes et al. 2015; Tiquia-Arashiro and Rodrigues 2016). Similarly, engineering existing mesophilic lignocellulolytic microbial enzymes to function efficiently at raised temperatures in order to boost reaction rates and exploit several other advantages of a high-temperature process is another potential alternative (Trudeau et al. 2014).

Keeping all of the above in mind, it can be concluded that thermophilic fungi and their enzymes might be used for the construction of multiproduct modern day biorefineries from lignocellulosics. This chapter discusses the potential and possibilities of thermostable lignocellulolytic enzymes, developed or isolated from thermophilic fungi for bioconversion of lignocellulosic raw materials into biofuel and other commodity chemicals with a biorefining perspective.

24.2 Lignocellulosic Biomass Components and Fungal Enzymes for Their Deconstruction

Lignocellulose is a complex of lignin, cellulose, and hemicellulose present in the plant cell walls (Gupta et al. 2011). Lignin is a highly heterogeneous aromatic polymer that is built from three phenylpropanoid precursors, p-coumaryl alcohol, coniferyl alcohol, and sinapyl alcohol (Higuchi 1993). On the other hand, cellulose is a linear polysaccharide consisting of β-1,4-linked D-glucose residues (Carpita and Gibeaut 1993), while hemicellulose consists of a backbone of β-1,4-linked D-xylose residues with a number of side groups (Scheller and Ulvskov 2010). Covalent and non-covalent linkages between the polysaccharides and lignin create the intricate structure that provides strength to the plant cell wall and also acts as a defense against microbial attack. However, this strong network can be broken down with the help of fungal enzymes into monomeric sugars, which serve as carbon sources for the fungi to produce hydrolytic enzymes (Gupta et al. 2011).

Lignin degradation in plant material involve three enzymes, namely, lignin peroxidase (LiP), manganese peroxidase (MnP), and versatile peroxidase (VP). LiP (E.C. 1.11.1.14) is a H_2O_2-dependent oxidative enzyme with a wide substrate range of both phenolic and non-phenolic aromatic compounds that includes the propyl side chains of lignin and the aromatic rings of lignin model compounds (Falade et al. 2016). On the other hand, MnP (E.C. 1.11.1.13) relies upon the generation of Mn^{3+} as a diffusible charge-transfer mediator and can reduce amines, dyes, and phenolic lignin model compounds (Fisher and Fong 2014). VPs are skilled of both LiP and MnP (manganese-independent and -dependent) catalytic activities, cleaving high-redox-potential non-phenolics, as well as lower potential aromatics and amines (Pérez-Boada et al. 2005). However, despite their ability to mineralize lignin, peroxidases are generally not used for degradation of plant biomass as the enzymes have a preference towards coupling of phenoxy radicals, leading to polymerization rather than depolymerization of lignin samples under in vitro conditions (Conesa et al. 2002). On the other hand, fungal laccases in presence of synthetic mediators like 2,2'-azino-bis(3-ethylbenzothiazoline-6-sulphonic acid) (ABTS) and hydroxybenzotriazole (HBT) mimic their in vivo lignin-delignifying role and catalyze oxidative degradation of lignin in plant biomass also under in vitro conditions (Jeon and Chang 2013). Laccases (EC 1.10.3.2) are multi-copper oxidases produced by fungi, bacteria, plants, and even insects that catalyze the one-electron oxidation of four equivalents of reducing substrate, with the corresponding four-electron reduction of atmospheric oxygen to water (Yang et al. 2017). Fungal laccases have higher redox potentials than laccases from other species (plants, bacteria, and insects) and thus are of great implication for in vitro oxidative depolymerization reactions (Pogni et al. 2015). A large number of laccases have been isolated from many basidiomycetes fungi for delignification of plant biomass leaving behind white cellulose, which is hydrolyzed further by the action of cellulases and hemicellulases for the release of monomeric sugars for bioconversion to fuels or commodity chemicals (Hatakka and Hammel 2011).

Efficient cellulose hydrolysis of plant biomass requires the concerted action of three different classes of enzymes, namely endoglucanases (E.C. 3.2.1.4), exoglucanases (E.C. 3.2.1.91), and β-glucosidases (E.C. 3.2.1.21), collectively called as cellulases. Endoglucanases randomly hydrolyze internal glycosidic linkages, resulting in rapid decrease in polymer length while exoglucanases or cellobiohydrolases (CBHs) hydrolyze cellulose chains by removing mostly cellobiose (repeating structural unit that makes up the cellulose chain comprising two β-1,4-linked glucose molecules) from either reducing or nonreducing ends, resulting in rapid release of reducing ends. The endo- and exoglucanases degrade cellulose to cellobiose, after which β-glucosidases hydrolyze cellobiose to free glucose molecules (Yeoman et al. 2010). The cellulose-decomposing fungi include members of the Ascomycota, Basidiomycota, Deuteromycota, as well as some chytrids that occur in the rumen of some animals.

Hemicellulose is the second most abundant renewable polysaccharide after cellulose, having a linear backbone of β-1, 4-linked xyloses (Walia et al. 2017). It is a heteropolysaccharide containing O-acetyl, arabinosyl, and 4-O-methyl-D-glucuronic acid substituent. The complete and efficient enzymatic hydrolysis of this complex polymer requires an array of enzymes with diverse specificity and modes of action. Endo-1,4-β D-xylanase (E.C. 3.2.1.8) randomly cleaves the xylan backbone; β-D-xylosidases (E.C. 3.2.1.37) cleaves xylobiose, whereas the removal of the side groups is catalyzed by α-L-arabinofuranosidases (E.C. 3.2.1.55), D-glucouronidases (E.C. 3.2.1.139), and acetylxylan esterases (E.C. 3.1.1.72). The existence of such a multifunctional xylano-lytic enzyme system is relatively common in fungi and a large number of xylanases have been purified and characterized from many ascomycetes (Driss et al. 2012).

24.3 Thermophilic Fungi Involved in Lignocellulosic Degradation

Microorganisms, based on their optimal growth temperatures, can be divided into three main groups: psychrophiles (below 20 °C), mesophiles (moderate temperatures), and thermophiles (high temperatures, above 55 °C) (Turner et al. 2007). Thermostable enzymes are derived from thermophilic microbes as cellular components (enzymes, proteins, and nucleic acids) of thermophilic organisms are also thermostable. In biorefining, renewable resources such as lignocellulosic biomass are utilized for extraction of intermediates or for direct bioconversion into chemicals, commodities, and fuels by the aid of lignocellulolytic enzymes (Fernando et al. 2006). Enzymes that are more thermostable use shorter reaction times for the complete saccharification of plant polysaccharides compared to the mesophilic ones (van den Brink et al. 2013). In this regard, three thermophilic fungi (*Aspergillus nidulans*, *Scytalidium thermophilum*, and *Humicola* sp.) isolated from wheat straw, farm yard manure, and soil showed high cellobiase, CMCase, xylanase, and FPase activities (Kumar et al. 2008). Further, these three fungi were used to compost a mixture (1:1) of paddy straw and lignin-rich soybean trash for 3 months and it was observed that the fungal consortium was effective in converting paddy straw into good-quality compost (Kumar et al. 2008). In another study, thermophillic bacterial-fungal communities were developed to deconstruct switchgrass for biofuel production, and it was observed that *Aspergillus* and *Penicillium* are the dominant fungi present in thermophilic consortia (Jain et al. 2017a). In yet another study, compost inoculated with a psychrotrophic-thermophilic complex microbial agent comprising of a psychrotrophic bacterium consortium (PBC) and a thermophilic cellulolytic fungi consortium (TCFC) in cold-climate conditions showed greater decrease in total organic carbon and C:N ratios, as well as significant increase in total nitrogen, degradation of cellulose and lignin than in only PBC-inoculated compost (Gou et al. 2017).

24.4 Thermostable Cellulases

Fungi are a rich source of thermostable cellulases, and several cellulases have been characterized from a number of fungal species (Table 24.1). Interestingly, a collection of highly thermostable cellobiohydrolases chimeras were obtained by structure-guided recombination of three fungal class II cellulases from thermophilic fungus *Humicola insolens* which were expressed heterologously in *Saccharomyces cerevisiae*. It was observed that five of these chimeras had better half-lives of thermal inactivation at 63 °C, which was even greater than the most stable parent, CBH II enzyme from the thermophilic fungus *Humicola insolens*, suggesting that the chimera collection contain hundreds of highly stable cellulases (Heinzelman et al. 2009).

Couturier and co-workers (2011) identified the gene encoding a thermostable typical multi-modular glycoside hydrolase family 45 endoglucanase in *Pichia pastoris* GS115 genome. The study revealed the first characterized endo-β-1,4 glucanase from yeast, whose thermostability is promising for biotechnological applications related to the saccharification of lignocellulosic biomass (Couturier et al. 2011). In yet another interesting report, the cellulolytic potential of 16 thermophilic fungi from three ascomycete orders *Sordariales*, *Eurotiales*, and *Onygenales* and from the zygomycete order *Mucorales* was investigated, thereby covering all fungal orders that include thermophiles (Busk and Lange 2013).

Table 24.1 Thermostable cellulases isolated and characterized from thermophilic fungi

Source	Name	Enzyme	Temperature optima (°C)	pH	References
Compost	*Aspergillus terreus* M11	Endoglucanase and β-glucosidase	70	3	Gao et al. (2008)
Recombinant (expressed in *P. pastoris*)	*Chaetomium thermophilum*	Cellobiohydrolase	60	5.0	Li et al. (2009)
Compost	*Aspergillus fumigatus*	Endoglucanases	50	5	Liu et al. (2011)
Recombinant (expressed in *Pichia pastoris*)	*Hypocrea jecorina*	Endoglucanase	70–80	5	Trudeau et al. (2014)
Rotten wood sample	*Aspergillus fumigates*	β-glucosidase	60	7.5	Srivastava et al. (2016)
Marine sponge samples (*Stelletta normani*) collected in Irish territorial water in North Atlantic Ocean	*Cadophora* sp. *Emericellopsis* sp. *Pseudogymnoascus* sp.	CMCase CMCase CMCase	70 60 60	6.0 6.0 6.0	Batista-García et al. (2017)
ATCC	*M. thermophila* ATCC 42464	Endoglucanase	70	–	Karnaouri et al. (2017)

Trudeau and co-workers (2014) created a stable endoglucanase derived from *Hypocrea jecorina* (anamorph *Trichoderma reesei*) Cel5A by a combination of stabilizing mutations, which were identified using consensus design, chimera studies, and structure-based computational methods. The designed endoglucanase had a 17 °C higher optimal temperature than the wild-type enzyme and hydrolyzed 1.5 times as much cellulose over a period of 60 h at its optimal temperature compared to the wild type at its optimal temperature. Furthermore, the thermostable mixture produced three times as much total sugar as the best mixture of the wild-type enzymes operating at their optimum temperature of 60 °C, clearly demonstrating the advantage of higher temperature cellulose hydrolysis (Trudeau et al. 2014). On the other hand, thermophilic cellulase (0.28 FPU/mL) produced from *Myceliophthora thermophila* under submerged cultivation conditions was used for saccharification of household food waste followed by fermentation with yeast resulting in 19.27 g/L of ethanol (Matsakas and Christakopoulos 2015). Production of thermostable cellulase from *Thermoascus auranticus* RCKK was improved by optimizing process conditions under simultaneous saccharification and fermentation (SSF) using central composite design of response surface methodology, and the crude enzyme was found to very efficiently hydrolyze office waste paper, algal pulp, and biologically treated wheat straw at 60 °C with sugar release of about 830 mg/g, 285 mg/g, and 260 mg/g of the substrate, respectively (Jain et al. 2017b). Similarly, a thermostable β-glucosidase isolated from *Thermotoga napthophila* RUK-10 possess catalytic activity for cellobiose hydrolysis with high potential in biomass conversion (Zhang et al. 2017). However, 19 thermophilic cellulolytic isolates from Algerian soil were found to hold great potential in the recycling of cellulosic biomass for bioenergy production when tested for the degradation of cellulosic biomass (printable paper, filter paper, and cotton) for 14 days of incubation at 60 °C (Khelil and Cheba 2014). Furthermore, a novel thermophilic β-glucosidase was reported from *Thermotoga napthophila* RKU-10 and used for the synthesis of prebiotic galactotrisaccharides at 75 °C, pH 6.5 (Yang et al. 2018). Thermophilic cellulolytic cocktails produced from white-rot fungus, *Inonotus obliquus*, under SSF conditions demonstrated activities of CMCase, Fpase, and β-glucosidase as 27.15 IU/g, 3.16 IU/g, and 2.53 IU/g, respectively, with high catalytic activity at 40–60 °C, releasing 130.24 mg/g of reducing sugars from raw wheat straw (Xu et al. 2018).

24.5 Thermostable Xylanases

Fungi are important sources of hemicellulases as they produce higher titers as compared to yeasts and bacteria. Thermostable β-D-xylosidases have been characterized from many fungal species with optimum temperature ranging from 60 °C to 75 °C (Table 24.2). Interestingly, gene for xylanase (MpXyn10A) was overexpressed from thermophilic fungus *Malbranchea pulchella* in *Aspergillus nidulans* followed by characterization of the expressed protein by Ribeiro and

Table 24.2 Thermostable xylanases isolated and characterized from thermophilic fungi

Source	Name	Enzyme	Temperature optima (°C)	pH	References
Soil sample collected from South Africa	*Thermomyces lanuginosus-SSBP*	β-D-Xylanase	65–70	6.5	Lin et al. (1999)
Forest area	*Thermoascus aurantiacus*	Xylanase	75	5.0–5.5	Silva et al. (2005)
ATCC	*Humicola brevis*	Xylanase	65–75	5.0–6.5	Masui et al. (2012)
Maize silage	*Rhizomucor pusillus*	Xylanase	70	6	Robledo et al. (2016)
Compost	*Mycothermus thermophilus*	Xylanase	65	6.0–6.5	Ma et al. (2017)
Sugarcane bagasse compost	*Thielavia terrestris*	Xylanase	85	5.5	García-Huante et al. (2017)
Compost	*Thermoascus aurantiacus M-2*	Xylanase	75	5	Ping et al. (2018)

co-workers (2014). The authors observed optimum activity of the enzyme at pH 5.8 and 80 °C with promising results in degradation of sugarcane bagasse (Ribeiro et al. 2014). However, the thermostability of a fungal GH11 xylanase was improved via site-directed mutagenesis guided by sequence and structural analysis. The recombinant xylanase, Xyn-MUT, was constructed by substituting three residues (N207S, G208S, A210S) with serine at the C-terminus of XynCDBFV and it was concluded that the single-point mutations gave rise to improved thermostability (Han et al. 2017).

24.6 Thermostable Laccases

Among the fungi, certain species express laccases with exceptionally high thermal stability (Table 24.3). Additionally, due to the presence of numerous laccase isoforms owing to multigene family of the enzyme within the same as well as different fungal species, isozymes with different thermal stabilities have been isolated from the same fungus. For instance, three laccase isozymes of *Pleurotus ostreatus* were found to differ significantly in their thermal stabilities with $T_{1/2}$ at 60 °C for 200 min for POXA1, while 30 min for POXC, and only 10 min for POXA2 laccase (Palmieri et al. 1997). In another study, wood-decaying basidiomycete *Steccherinum ochraceum* isolate 1833 was reported to produce three highly thermostable laccase isoforms (I, II, III) with maximum activities in the range of 75–80 °C (Chernykh et al. 2008). On the other hand, a thermostable laccase isoform Pplcc2 from brown-rot fungus *Postia placenta* Mad-698-R was heterologously expressed in *Pichia pastoris* GS115 followed by its purification and characterization (An et al. 2015). However, a thermophilic lignin-degrading microbiota was

Table 24.3 Thermostable laccases isolated and characterized from thermophilic fungi

Source	Name	Enzyme	Temperature optima (°C)	pH	References
Wood sample	*Trametes versicolor*	Laccase	70	3	Zhu et al. (2011)
Coal sample	*Cladosporium cladosporioides*	Laccase	40–70	3.5	Halaburgi et al. (2011)
Culture collection of the Universidad Autónoma de Nuevo León of Mexico	*Pycnoporus sanguineus*	Laccase	50–60	3.0–5.0	Ramírez-Cavazos et al. (2014)
ATCC	*Coprinopsis cinerea*	Laccase	60	6.5	Pan et al. (2014)
Recombinant (expressed in *Komagataella pastoris*)	*Trametes trogii*	Laccase	60–70	4.0–5.0	Campos et al. (2016)
Universidad Autónoma de Nuevo León, Mexico, culture collection	*Pycnoporus sanguineus CS43*	Laccase	60	–	Orlikowska et al. (2018)

developed and characterized using enrichment technique on Douglas fir at 55 °C. The results identified genera *Talaromyces, Aspergillus*, and *Byssochlamys* as the dominant fungi associated with lignin degradation under thermophilic conditions (Ceballos et al. 2017).

24.7 Application of Thermophilic Lignocellulolytic Enzymes in Biorefineries

Higher thermostability of the above-discussed lignocellulolytic enzymes allows saccharification of biomass polysaccharides at elevated temperatures. Consequently, the reaction times are shortened, mass transfer as well as substrate viscosity are increased, and as a result efficiency and overall economics of lignocellulose-based biorefineries have improved (Berka et al. 2011). The use of thermophilic enzymes for the production of biofuel and platform chemicals in a biorefinery approach is discussed below:

24.7.1 Biofuel

Biofuel production from lignocellulosic biomass (Fig. 24.1) can be done in three ways: (1) separate hydrolysis and fermentation (SHF), (2) simultaneous saccharification and fermentation (SSF), and (3) consolidated bioprocessing (CBP).

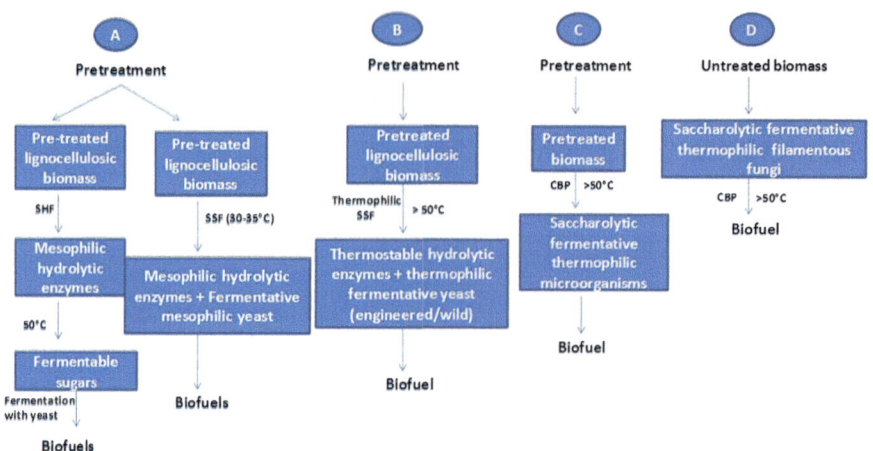

Fig. 24.1 Biofuel production from lignocellulosic biomass (**a**) SHF, wherein hydrolysis of pretreated substrate is carried out by mesophilic enzymes at 50 °C followed by fermentation with yeast at 30 °C and SSF, wherein enzymatic hydrolysis as well as fermentation of pretreated substrate are carried out simultaneously at 30–35 °C; (**b**) thermophilic SSF, wherein the pretreated substrate is simultaneously hydrolyzed as well as fermented at elevated temperatures using thermostable hydrolytic enzymes and thermotolerant yeast strains; (**c**) CBP of pretreated substrate by saccharolytic fermentative thermophilic microbes (bacteria/yeast) at 50 °C; (**d**) CBP of untreated substrate by saccharolytic fermentative thermophilic filamentous fungi at 50 °C

24.7.1.1 Separate Hydrolysis and Fermentation (SHF) Process

Production of biofuel via SHF process involves three main steps: (1) pretreatment of the biomass; (2) enzymatic hydrolysis of pretreated substrate by cellulases and hemicellulases to release monomeric sugars followed by (3) fermentation with yeast to produce alcohol.

The first step of the process, i.e., pretreatment of lignocellulose, can be performed by different methods, viz. physical, chemical, and biological. High temperature and acid have been employed initially in pretreatment of lignocellulosic residues at industrial scales. However, this approach is expensive, slow, and inefficient (Rubin 2008). In addition, the overall yield of the fermentation process will be decreased because the pretreatment releases fermentation inhibitors such as weak acids, furan, and phenolic compounds (Palmqvist and Hahn-Hägerdal 2000). On the other hand, employing biological pretreatment strategy for lignin deconstruction of plant biomass using microorganism, the problem of inhibitors can be overcome with added economic and environmental benefits (Dashtban et al. 2009). In the biological pretreatment process, white-rot fungi are mostly used to degrade lignin and hemicelluloses present in the plant biomass. Recently, three white-rot fungi—*P. florida*, *C. caperata* RCK2011, and *Ganoderma* sp. rckk-02—were used for the degradation of lignin in sugarcane bagasse and it was observed that the biologically pretreated substrate when hydrolyzed with crude cellulase from brown-rot fungus *Fomitopsis* sp.RCK 2011 released 2.4-fold higher sugars than the untreated substrate

(Deswal et al. 2014). Besides, application of thermophilic fungal species in lignocellulose pretreatment at industrial scales will result in further energy savings as the costly cooling after steam pretreatment is avoided. For example, thermophilic fungal species such as *Sporotrichum thermophile* (Bhat and Maheshwari 1987) and *Thermoascus aurantiacus* (Gomes et al. 2000) have been proposed as good candidates for pretreatment as well as bioconversion of lignocellulosic residues to sugars at industrial scales.

In the second step, i.e., hydrolysis step of the process, capabilities of enzymes (cellulases and xylanases) from various filamentous fungi, including members of the genera *Aspergillus, Rhizopus, Monilia, Neurospora, Fusarium, Trichoderma,* and *Mucor,* have been explored for the successful production of ethanol from biomass (Madadi et al. 2017). However, these enzymes are generally obtained from the fermentation of mesophilic fungal sources and thus do not work efficiently at higher temperatures >50 °C (Yennamalli et al. 2013). Therefore, to increase process economics as well as process efficiency, there is an intense interest in exploiting the potential of thermostable bioprocessing enzymes. In this regard, two enzyme mixtures, a mesophilic and a thermostable, were exposed to typical process conditions of ethanol production (temperatures from 55 °C to 65 °C and up to 5% ethanol) and then analyzed by enzyme activity assay as well as SDS-PAGE and it was observed that the thermostable and mesophilic mixture remained active at up to 65 °C and 55 °C, respectively (Skovgaard and Jørgensen 2013). In another study by Long and co-workers (2018), hydrolysis of biomass (corn stover, poplar, and kraft pulp) was studied at high temperatures (85 °C) by thermostable xylanase (Xyn10A) from *Thermotoga thermarum* DSM 5069 followed with saccharification step by commercial cellulase. The authors observed that high-temperature xylanase treatment considerably increased cellulose accessibility/hydrolyzability towards cellulases, with smoothed fiber surface morphology, compared with commercial xylanase treatment at 50 °C. The authors concluded that the increased temperature during thermostable xylanase treatment facilitated viscosity reduction of biomass slurry, which exhibited more benefits during hydrolysis of various steam-pretreated substrates at increased solid content (Long et al. 2018).

In the final step of biofuel production, hydrolytic products including monomeric hexoses (glucose, mannose, and galactose) and pentoses (xylose and arabinose) are fermented to ethanol using yeast strains. Among these hydrolytic products, glucose is normally the most abundant, followed by xylose or mannose and other sugars at lower concentration. *Saccharomyces cerevisiae* is the most frequently and traditionally used microorganism for fermenting ethanol from starch-based residues at industrial scales (Hahn-Hägerdal et al. 2007). However, *S. cerevisiae* is unable to efficiently utilize xylose as the sole carbon source or ferment it to ethanol (Chu and Lee 2007). To make industrial lignocellulosic bioconversion more economically feasible, it is necessary to choose microorganisms capable of fermenting both glucose and xylose. This can be achieved by metabolic engineering of *S. cerevisiae* with genes from other xylose-fermenting yeasts like *Pichia stipitis* (Hahn-Hägerdal et al. 2007) as well with the help of nonrecombinant (e.g., adaptation) techniques (Dashtban et al. 2009).

24.7.1.2 Simultaneous Saccharification and Fermentation (SSF) Process

The last two steps of bioconversion of pretreated lignocellulolytic residues to ethanol (hydrolysis and fermentation) can be performed separately (SHF) or simultaneously (SSF). In the separate hydrolysis and fermentation (SHF), the hydrolysate products will be fermented to ethanol in a separate process. The advantage of this method is that both processes can be optimized individually (45–50 °C for hydrolysis, whereas it is 30 °C for fermentation), while drawbacks include costly addition of β-glucosidase to overcome end product inhibition caused by accumulation of enzyme-inhibiting end products (cellobiose and glucose) during the hydrolysis. Alternatively, in simultaneous saccharification and fermentation (SSF), the end products after the pretreatment step are directly converted into ethanol. Therefore, addition of high amounts of β-glucosidase is not necessary and this reduces the process cost (Stenberg et al. 2000). Nevertheless, the main downside of SSF is the need to compromise processing conditions such as temperature and pH at suboptimal levels for each individual step (saccharification and fermentation). Therefore, finding thermotolerant yeasts which can work efficiently at higher temperatures required for optimum enzymatic hydrolysis is necessary for the development of an efficient SSF process (Choudhary et al. 2016). Alternatively, development of recombinant yeast strains with improved thermotolerance will also enhance the performance of SSF substantially.

Hari Krishna and co-workers (2001) compared thermotolerant yeast *Kluyveromyces fragilis* NCIM 3358 with *Saccharomyces cerevisiae* NRRL-Y-132 and found that *K. fragilis* perform better in the SSF process at 42 °C resulting in high yields of ethanol (2.5–3.5% w/v) compared to *S. cerevisiae* (2.0–2.5% w/v) (Hari Krishna et al. 2001). In another study, steam-pretreated lignocellullosic material (eucalyptus, poplar, bagasse, sweet sorghum, mustard, and wheat straw) was used for ethanol production via SSF process using thermotolerant *K. marxianus* at 42 °C (Ballesteros et al. 2004). Huang and D'Andrea (2006) produced 40 g/L ethanol from 161 g/L of paper sludge organic material containing 66% (w/w) glucan in an SSF process at 42 °C using *S. cerevisiae* TJ14 strain in conjunction with cellulase produced by filamentous fungus *Acremonium cellulolyticus* (Huang and D'Andrea 2006). Interestingly, a respiratory-deficient mutant of *Candida glabrata* produced 17.0 g/L ethanol from 50.0 g/L Avicel microcrystalline cellulose at 42 °C under aerobic conditions (Watanabe et al. 2009). At the same time, a transformation system was constructed to express an *Aspergillus aculeatus* β-glucoside (BGL) gene in a thermotolerant strain of *P. kudriavzevi*, which was also acid and ethanol tolerant. The transformant was found to produce 29 g/L ethanol from 100 g/L Avicel microcrystalline cellulose in simultaneous saccharification and fermentation at 40 °C without addition of BGL (Kitagawa et al. 2010). On the other hand, a newly isolated thermotolerant ethanologenic yeast strain, *Pichia kudriavzevii* IPE100, produced ethanol with a theoretical yield of 85% with glucose at 42 °C (Kwon et al. 2011). In yet another study, mutation screening for thermotolerance was performed in *S. cerevisiae* strains using a proofreading-deficient DNA polymerase or ultraviolet (UV) irradiation, which resulted in *S. cerevisiae* mutants that grow at temperatures up to 40–42 °C (Abe et al. 2009). However, thermotolerance as well as ethanol tolerance of *S. cerevisiae* were improved using genome shuffling approach by a combination of protoplast fusion and UV irradiation (Shi et al. 2009).

SSF of three delignified lignocellulosic biomass, viz. rice straw, wheat straw, and sugarcane bagasse, was performed at 42 °C using a newly isolated themotolerant yeast, *Kluyveromyces* sp., in conjunction with in-house cellulases from *Aspergillus terreus* (Narra et al. 2015). Further, ten thermophilic yeast strains (capable of growth at 40 °C) were isolated from diverse sources, belonging to various genera like *Saccharomyces*, *Candida*, *Pichia*, and *Wickerhamomyces*. The authors observed that *Saccharomyces cerevisiae* JRC6, isolated from distillery waste, produced ethanol with 88.3% and 89.1% theoretical efficiency at 40 °C and 42 °C, respectively, from glucose (Choudhary et al. 2017). Likewise, 5 thermotolerant yeasts, designated *Saccharomyces cerevisiae* KKU-VN8, KKU-VN20, and KKU-VN27, *Pichia kudriavzevii* KKU-TH33, and *P. kudriavzevii* KKU-TH43 out of 234 yeast isolates from Greater Mekong Sub region (GMS) countries, i.e., Thailand, The Lao People's Democratic Republic (Lao PDR), and Vietnam, were selected for ethanol production at 45 °C (Techaparin et al. 2017).

24.7.1.3   Consolidated Bioprocessing (CBP) Process

Consolidated bioprocessing (CBP) aims to minimize all bioconversion steps of biofuel production into one step in a single reactor using one or more microorganisms. CBP operation featuring cellulase production, cellulose/hemicellulose hydrolysis, and fermentation of 5- and 6-carbon sugars in one step has shown the potential to provide the lowest cost for biological conversion of cellulosic biomass to fuels (Singh et al. 2017). In CBP, the whole process is carried out at a single elevated temperature rather than in three steps (pretreatment, hydrolysis, and fermentation) involving different temperatures. This high-temperature-based biomass processing seems challenging but could be the most rewarding approach for bioethanol production in near future and thermophilic microbes and their enzymes can play an important role in its successful implementation (Acharya and Chaudhary 2012). In past, two approaches have been used for CBP: (1) engineering a cellulase producer to be ethanologenic or (2) engineering an ethanologen, such as *S. cerevisiae* or *Zymomonas mobilis*, to be cellulolytic. But there are various difficulties and challenges in the conversion of a candidate microorganism using gene transfer technology due to the adverse effects of the co-expression of multiple heterologous genes on cell performance, the modulation of simultaneous co-expression of multiple genes at the transcription level and improper folding of some secretory proteins (Xu et al. 2009). On the other hand, search for a native CBP microorganism displaying high levels of alcohol/sugar tolerance, thermotolerance, as well as the ability to utilize multiple sugars is still going on.

CBP of lignocellulosic biomass to ethanol using thermophilic bacteria, *Clostridium thermocellum*, providing a promising solution for efficient lignocellulose conversion without addition of exogenous enzymes was performed (Svetlitchnyi et al. 2013). However, despite the potential of a number of bacteria to ferment hexose and pentose sugars to ethanol, it is difficult in practice to maintain anaerobic conditions in large-scale fermentation restricting the use of thermophilic anaerobes. Alternatively, use of filamentous fungi in the CBP process will go a long way as even the pretreatment step (Fig. 24.2) can be eliminated due to their delignification ability. In this regard Ali and co-workers reviewed various fungi with the potential of being used in CBP for biofuel production. These included *Trichoderma reesei, Fusarium oxysporum, Mucor indicus, Monilia sp., Rhizopus oryzae, Paecilomyces* sp., *Aspergillus oryzae, Neocallimastix patriciarum,* and *Neurospora crassa.* All these filamentous fungi possessed the ability to assimilate and metabolize numerous sugars, both hexose and pentose types. Furthermore, these fungi have a greater degree of thermotolerance than many bacteria and can grow at closer to the optimal temperature of enzymatic hydrolysis, i.e., 40–50 °C (Ali et al. 2016).

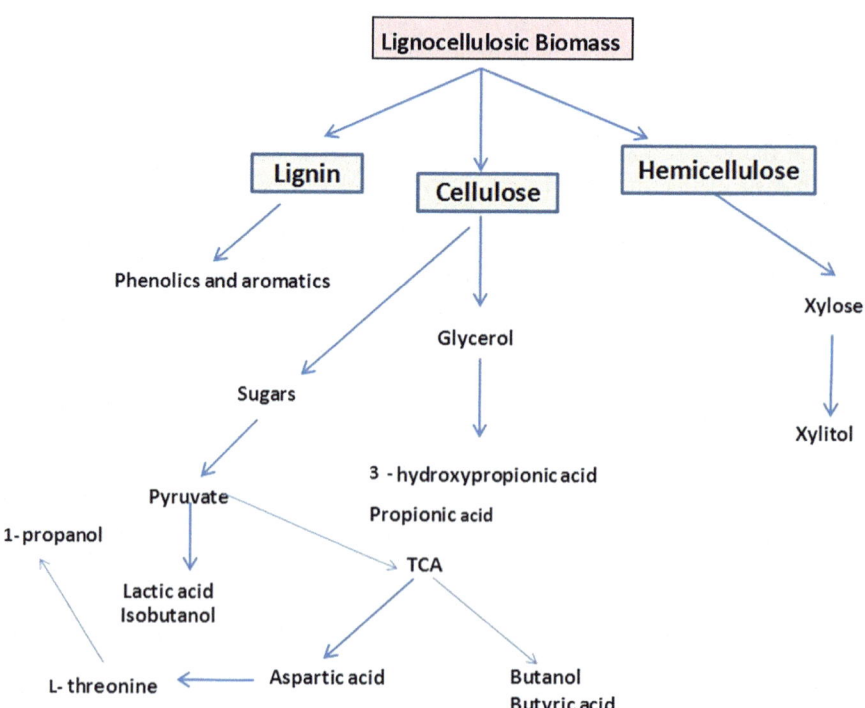

Fig. 24.2 Platform chemical production from lignocellulosic biomass

24.7.2 Platform Chemicals

Increased dependency on the depleting and highly priced petroleum-based products as well as the issue of greenhouse gas emission have raised extensive interest in microbial fermentation processes to produce platform chemicals from renewable resources (Roa Engel et al. 2008). A platform chemical is an intermediate molecule, with a structure able to generate a number of derivatives (Bomtempo et al. 2017) as shown in Fig. 24.2. These biochemicals can be produced from biomass (a) by separate hydrolysis and fermentation of the biomass via intermediate sugars or (b) by simultaneous hydrolysis and fermentation of biomass using thermophilic microbes and their enzymes. Various thermophilic fungi per se have been used for biochemical production from biomass (Table 24.4). A range of platform chemicals being synthesized from biomass are also discussed below:

24.7.2.1 3-Hydroxypropionic Acid

3-Hydroxypropionic acid (3-HP) is an attractive platform chemical, which can be used to produce a variety of commodity chemicals, such as acrylic acid and acrylamide. In a study, 3-HP was isolated from the extracts obtained from submerged cultures of several endophytic fungi, namely *Phomopsis phaseoli*, *Melanconium betulinum*, *Betula pendula*, and *B. pubescens* (Schwarz et al. 2004). However, engineered *Saccharomyces cerevisiae* was also used for the production of 3-HP (Chen et al. 2014; Kildegaard et al. 2015).

24.7.2.2 Itaconic, Fumaric, Ferulic and Malic Acid

Fumaric acid (31%) was produced along with chitin (0.21 g chitin/g biomass) from dairy manure using pelletized filamentous fungus, *Rhizopus oryzae* ATCC 20344 (Liao et al. 2008), while itaconic acid (25 g/L) was produced by fermentation with *Ustilago maydis* on pretreated cellulose and hemicellulosic fractions (Klement et al. 2012). L-Malic acid (titers of 106 g/L and productivity of 0.88 g/L/h) was produced from *Aspergillus* sp. N1-14' grown in medium containing glucose (Zhou et al.

Table 24.4 Biochemical production from biomass using thermophilic fungi

Biochemical	Thermophilic fungi	Yield	References
3-Hydroxy propionic acid	*Saccharomyces cerevisiae*	463 mg/L	Chen et al. (2014)
(5S)-hydroxyhexane-2-one	*Thermus thermophilus*	3.9 g/L/h	Diederichs et al. (2015)
Chitooligosaccharides	*Humicola grisea*	–	Kumar et al. (2017)
Rubber extraction	*Thermomyces lanuginosus*	90 mg/g Dry TK root	Sikandar et al. (2017)

2000). However, ferulic acid was extracted from corn cobs using thermostable xylanases and esterases from *Thermobifida fusca* (Huang et al. 2011). The process was achieved by cloning a gene (*axe*) encoding the thermostable acetyl xylan esterase (AXE) from *Thermobifida fusca* NTU22 into a *Yarrowia lipolytica* host strain and the recombinant expression resulted in extracellular esterase production at levels as high as 70.94 U/mL, approximately 140 times higher than that observed in a *Pichia pastoris* expression system. Further, the incubation of corncob with *T. fusca* xylanase (Tfx) for 12 h and then with the (Acetyl xylan esterase) AXE for an additional 12 h resulted in accumulation of 396 µM ferulic acid in the culture broth (Huang et al. 2011).

Two *Aspergillus* sp. (*A. terreus* and *A. oryzae*) yielded fumaric acid (0.54 mg) and itaconic acid (0.11 mg) by solid-state fermentation and simultaneous saccharification and fermentation (Jiménez-Quero et al. 2016).

24.7.2.3 Xylitol

Saccharomyces cerevisiae and *Candida tropicalis* were used separately and as co-culture for simultaneous saccharification and fermentation of 5–20% corn cobs. The authors obtained 21, 20, and 15 g/L of xylitol from 200 g/L corn cobs from cultures of *C. tropicalis*, co-culture, and *S. cervisiae*, respectively (Latif and Rajoka 2001). Fernandes and co-workers also cloned xylitol and L-arabitol dehydrogenase genes heterologously from thermophilic fungus *Talaromyces emersonii* expressed in yeast. The authors suggested that the genes may be valuable in the production of xylitol, L-arabitol, and ethanol from renewable resources rich in pentose sugars (Fernandes et al. 2010).

24.7.2.4 Glycolic Acid

Glycolic acid has attracted attention as platform chemical, especially as building block for polyglycolate (PGA), a polymer with high gas permeability and mechanical strength, and therefore excellent properties as packaging material (Grayson et al. 2004). De novo synthesis of glycolic acid from renewable resources, apparently more favorable, has been achieved in the yeasts *S. cerevisiae* and *Kluyveromyces lactis* (Koivistoinen et al. 2013).

24.7.2.5 Lactic Acid

Lactic acid is an important platform chemical for producing polylactic acid (PLA) and other value-added products (Grayson et al. 2004). In this regard, lactic acid-producing bacteria (LAB) have received wide attention for production of lactic acid because of their high growth rate and product yield but these bacteria have complex nutrient requirements because of their limited ability to synthesize B-vitamins and amino acids, making supplementation of sufficient nutrients such as yeast extracts to media necessary (Chopin 1993). Moreover, LAB produce both the isomers of lactic acid (D-Lactic acid and L-Lactic acid). On the other hand, fungal *Rhizopus* species have also attracted a great interest for lactic acid production as unlike the LAB, lactic acid-producing *Rhizopus* strains generate L-lactic acid as a sole isomer of lactic acid (Yoshihara et al. 2003). Lactic acid exists naturally in two optical isomers: D(−)-lactic acid and L(+)-lactic acid. Elevated levels of the D-isomer are harmful to humans and thus L(+)-lactic acid is the preferred isomer for food-related and pharmaceutical industries (Zhang et al. 2007). Furthermore, fermentation with filamentous fungi makes separation of fungal biomass from the fermentation broth easier because of their filamentous or pellet forms, leading to a simple and inexpensive downstream processing. In addition, as a by-product from lactic acid production, fungal biomass of *Rhizopus* strains can also be used for fungal chitosan production (Yoshihara et al. 2003) which can be used as an additive in animal feeds to improve the feed quality (Kusumaningtyas et al. 2006). Many studies have explored the possibility of producing lactic acid from lignocellulosic substrates by *Rhizopus* species (Maas et al. 2006; Miura et al. 2004; Ruengruglikit and Hang 2003; Woiciechowski et al. 1999).

24.8 Conclusions

The issue of environmental crisis, sustainability of future generations, as well as increasing prices of existing nonrenewable reserves have led to the development of fuels and chemicals from alternative renewable energy resources. Lignocellulosic biomass is a potential low-cost renewable raw material for the production of chemicals and biofuels through bioconversion of the three major chemical components: cellulose, hemicellulose, and lignin. Deconstruction of these lignocellulosic components with the aid of microbes and their enzymes for the production of biochemicals and biofuels makes the foundation of biorefineries. Use of fungal enzyme cocktails is one of the most promising ways to convert the lignocellulosic biomass into the reducing sugars for industrial utilizations, due to the fungi's capability to secrete high titers of lignocellulolytic enzymes. However, expensive enzyme production processes, poor stability, and low efficiency of currently used enzymes in industrial conditions still present a major obstacle in development of commercial biorefineries. In this regard, thermostable enzymes that hydrolyze lignocellulose to its component sugars at elevated temperatures have many advantages for improving the conversion rate of biomass over their mesophilic counterparts. Their robust thermostabilities make them better suited for the harsh industrial conditions as well as for better and efficient deconstruction of lignocellulose to fermentable products. Therefore, it can be concluded that the establishment of biorefineries using thermophilic fungi and their enzymes is the most sought after and advantageous alternative to the success of modern day biorefineries.

24.9   Future Perspectives

Exploration of highly efficient thermostable enzyme systems which improve the efficiency and economics of bioconversion of lignocellulosic biomass into value-added products is very crucial for the implementation of commercial biorefineries. Bioengineering of fermentative mesophilic organisms to enhance their thermostability for biofuel production in consolidated bioprocessing is a promising approach to enhance bioconversion efficiency and reduction of cost involved. Although a lot of progress has been made on these lines with respect to biofuel production, development of commodity chemicals from biomass using thermophilic fungi and their enzymes is still under progress and a lot more efforts are needed to develop efficient bioprocess for the production of platform chemicals in a biorefinery approach. This will involve the development of "one-step biorefineries" using fermentative thermotolerant fungi, which secrete all the three enzymes required for lignocellulose deconstruction as well as production of biofuel and platform chemicals from biomass in a single pot. Use of modern genetic tools for metabolic engineering may further enhance the productivity, thereby making the process cost effective.

References

Abe H, Fujita Y, Takaoka Y, Kurita E, Yano S, Tanaka N, Nakayama K-i (2009) Ethanol-tolerant *Saccharomyces cerevisiae* strains isolated under selective conditions by over-expression of a proofreading-deficient DNA polymerase delta. J Biosci Bioeng 108:199–204

Acharya S, Chaudhary A (2012) Bioprospecting thermophiles for cellulase production: a review. Braz J Microbiol 43:844–856

Ali SS, Nugent B, Mullins E, Doohan FM (2016) Fungal-mediated consolidated bioprocessing: the potential of *Fusarium oxysporum* for the lignocellulosic ethanol industry. AMB Express 6:13

An H, Xiao T, Fan H, Wei D (2015) Molecular characterization of a novel thermostable laccase PPLCC2 from the brown rot fungus *Postia placenta* MAD-698-R. Electron J Biotechnol 18:451–458

Ballesteros M, Oliva JM, Negro MJ, Manzanares P, Ballesteros I (2004) Ethanol from lignocellulosic materials by a simultaneous saccharification and fermentation process (SFS) with *Kluyveromyces marxianus* CECT 10875. Process Biochem 39:1843–1848

Batista-García RA et al (2017) Characterization of lignocellulolytic activities from fungi isolated from the deep-sea sponge *Stelletta normani*. PLoS One 12:e0173750

Berka RM, Sutton T, Jackson SA, Tovar-Herrera OE, Balcázar-López E, Sánchez-Carbente MD, Sánchez-Reyes A, Dobson AD, Folch-Mallol JL (2011) Comparative genomic analysis of the thermophilic biomass-degrading fungi *Myceliophthora thermophila* and *Thielavia terrestris*. Nat Biotechnol 29:922–927

Bhat KM, Maheshwari R (1987) *Sporotrichum thermophile* growth, cellulose degradation, and cellulase activity. Appl Environ Microbiol 53:2175–2182

Bomtempo J-V, Alves FC, Oroski FA (2017) Developing new platform chemicals: what is required for a new bio-based molecule to become a platform chemical in the bioeconomy? Faraday Discuss 202:213–225

Bugg TDH, Ahmad M, Hardiman EM, Rahmanpour R (2011) Pathways for degradation of lignin in bacteria and fungi. Nat Prod Rep 28:1883–1896

Busk P, Lange L (2013) Cellulolytic potential of thermophilic species from four fungal orders. AMB Exp 3:47

Campos PA, Levin LN, Wirth SA (2016) Heterologous production, characterization and dye decolorization ability of a novel thermostable laccase isoenzyme from *Trametes trogii* BAFC 463. Process Biochem 51:895–903

Carpita NC, Gibeaut DM (1993) Structural models of primary cell walls in flowering plants: consistency of molecular structure with the physical properties of the walls during growth. Plant J 3:1–30

Ceballos SJ, Yu C, Claypool CT, Singer SW, Simmons BA, Thelen MP, Simmons CW, Vandergheynst J (2017) Development and characterization of a thermophilic, lignin degrading microbiota. Process Biochem 63:193–203

Chen Y, Bao J, Kim I-K, Siewers V, Nielsen J (2014) Coupled incremental precursor and co-factor supply improves 3-hydroxypropionic acid production in *Saccharomyces cerevisiae*. Metab Eng 22:104–109

Chernykh A, Myasoedova N, Kolomytseva M, Ferraroni M, Briganti F, Scozzafava A, Golovleva L (2008) Laccase isoforms with unusual properties from the basidiomycete *Steccherinum ochraceum* strain 1833. J Appl Microbiol 105:2065–2075

Chopin A (1993) Organization and regulation of genes for amino acid biosynthesis in lactic acid bacteria. FEMS Microbiol Rev 12:21–37

Choudhary J, Singh S, Nain L (2016) Thermotolerant fermenting yeasts for simultaneous saccharification fermentation of lignocellulosic biomass. Electron J Biotechnol 21:82–92

Choudhary J, Singh S, Nain L (2017) Bioprospecting thermotolerant ethanologenic yeasts for simultaneous saccharification and fermentation from diverse environments. J Biosci Bioeng 123:342–346

Chu BC, Lee H (2007) Genetic improvement of *Saccharomyces cerevisiae* for xylose fermentation. Biotechnol Adv 25:425–441

Conesa A, Punt PJ, van den Hondel CAMJJ (2002) Fungal peroxidases: molecular aspects and applications. J Biotechnol 93:143–158

Couturier M, Feliu J, Haon M, Navarro D, Lesage-Meessen L, Coutinho PM, Berrin J-G (2011) A thermostable GH45 endoglucanase from yeast: impact of its atypical multimodularity on activity. Microb Cell Fact 10:103

Dashtban M, Schraft H, Qin W (2009) Fungal bioconversion of lignocellulosic residues; opportunities and perspectives. Int J Biol Sci 5:578–595

de Vries RP, Visser J (2001) *Aspergillus* enzymes involved in degradation of plant cell wall polysaccharides. Microbiol Mol Biol Rev 65:497–522

Deswal D, Gupta R, Nandal P, Kuhad R (2014) Fungal pretreatment improves amenability of lig-nocellulosic material for its saccharification to sugars. Carbohydr Polym 99:264–269

Diederichs S, Linn K, Lückgen J, Klement T, Grosch JH, Honda K, Ohtake H, Büchs J (2015) High-level production of (5S)-hydroxyhexane-2-one by two thermostable oxidoreductases in a whole-cell catalytic approach. J Mol Catal B: Enzym 121:37–44

Driss D, Bhiri F, ghorbel R, Chaabouni SE (2012) Cloning and constitutive expression of His-tagged xylanase GH 11 from *Penicillium occitanis* Pol6 in *Pichia pastoris* X33: Purification and characterization. Protein Expr Purif 83:8–14

Falade AO, Nwodo UU, Iweriebor BC, Green E, Mabinya LV, Okoh AI (2016) Lignin peroxidase functionalities and prospective applications. MicrobiologyOpen 6:e00394

Fernandes S, Tuohy MG, Murray PG (2010) Cloning, heterologous expression, and characterization of the xylitol and L-arabitol dehydrogenase genes, Texdh and Telad, from the thermophilic fungus *Talaromyces emersonii*. Biochem Genet 48:480–495

Fernando S, Adhikari S, Chandrapal C, Murali N (2006) Biorefineries: current status, challenges, and future direction. Energy Fuel 20:1727–1737

Fisher AB, Fong SS (2014) Lignin biodegradation and industrial implications. Bioengineering 1:92–112

Gao J, Weng H, Zhu D, Yuan M, Guan F, Xi Y (2008) Production and characterization of cellulolytic enzymes from the thermoacidophilic fungal *Aspergillus terreus* M11 under solid-state cultivation of corn stover. Bioresour Technol 99:7623–7629

García-Huante Y et al (2017) The thermophilic biomass-degrading fungus *Thielavia terrestris* Co3Bag1 produces a hyperthermophilic and thermostable β-1,4-xylanase with exo- and endo-activity. Extremophiles 21:175–186

Gomes I, Gomes J, Gomes DJ, Steiner W (2000) Simultaneous production of high activities of thermostable endoglucanase and beta-glucosidase by the wild thermophilic fungus *Thermoascus aurantiacus*. Appl Microbiol Biotechnol 53:461–468

Gou C, Wang Y, Zhang X, Lou Y, Gao Y (2017) Inoculation with a psychrotrophic-thermophilic complex microbial agent accelerates onset and promotes maturity of dairy manure-rice straw composting under cold climate conditions. Bioresour Technol 243:339–346

Grayson ACR, Voskerician G, Lynn A, Anderson JM, Cima MJ, Langer R (2004) Differential degradation rates in vivo and in vitro of biocompatible poly(lactic acid) and poly(glycolic acid) homo- and co-polymers for a polymeric drug-delivery microchip. J Biomater Sci Polym Ed 15:1281–1304

Gupta R, Mehta G, Khasa YP, Kuhad RC (2011) Fungal delignification of lignocellulosic biomass improves the saccharification of cellulosics. Biodegradation 22:797–804

Hahn-Hägerdal B, Karhumaa K, Fonseca C, Spencer-Martins I, Gorwa-Grauslund MF (2007) Towards industrial pentose-fermenting yeast strains. Appl Microbiol Biotechnol 74:937–953

Haki GD, Rakshit SK (2003) Developments in industrially important thermostable enzymes: a review. Bioresour Technol 89:17–34

Halaburgi VM, Sharma S, Sinha M, Singh TP, Karegoudar TB (2011) Purification and characterization of a thermostable laccase from the ascomycetes *Cladosporium cladosporioides* and its applications. Process Biochem 46:1146–1152

Han N, Miao H, Ding J, Li J, Mu Y, Zhou J, Huang Z (2017) Improving the thermostability of a fungal GH11 xylanase via site-directed mutagenesis guided by sequence and structural analysis. Biotechnol Biofuels 10:133

Hari Krishna S, Janardhan Reddy T, Chowdary GV (2001) Simultaneous saccharification and fermentation of lignocellulosic wastes to ethanol using a thermotolerant yeast. Bioresour Technol 77:193–196

Hasunuma T, Kondo A (2012) Consolidated bioprocessing and simultaneous saccharification and fermentation of lignocellulose to ethanol with thermotolerant yeast strains. Process Biochem 47:1287–1294

Hatakka A, Hammel KE (2011) Fungal Biodegradation of Lignocelluloses. In: Industrial Applications. The Mycota. Springer, Berlin, Heidelberg, pp 319–340

Heinzelman P, Snow CD, Wu I, Nguyen C, Villalobos A, Govindarajan S, Minshull J, Arnold FH (2009) A family of thermostable fungal cellulases created by structure-guided recombination. PNAS 106:5610–5615

Higuchi T (1993) Biodegradation mechanism of lignin by white-rot basidiomycetes. J Biotechnol 30:1–8

Huang TT, D'Andrea AD (2006) Regulation of DNA repair by ubiquitylation. Nat Rev Mol Cell Biol 7:323–334

Huang Y-C, Chen Y-F, Chen C-Y, Chen W-L, Ciou Y-P, Liu W-H, Yang C-H (2011) Production of ferulic acid from lignocellulolytic agricultural biomass by *Thermobifida fusca* thermostable esterase produced in *Yarrowia lipolytica* transformant. Bioresour Technol 102:8117–8122

Isikgor F, Becer CR (2016) Lignocellulosic biomass: a sustainable platform for the production of bio-based chemicals and polymers. Polym Chem 6:4497–4559

Jain A, Pelle HS, Baughman WH, Henson JM (2017a) Conversion of ammonia-pretreated switchgrass to biofuel precursors by bacterial–fungal consortia under solid-state and submerged-state cultivation. J Appl Microbiol 122:953–963

Jain KK, Kumar S, Deswal D, Kuhad RC (2017b) Improved production of thermostable cellulase from *Thermoascus aurantiacus* RCKK by fermentation bioprocessing and its application in the hydrolysis of office waste paper, algal pulp, and biologically treated wheat straw. Appl Biochem Biotechnol 181:784–800

Jeon J-R, Chang Y-S (2013) Laccase-mediated oxidation of small organics: bifunctional roles for versatile applications. Trends Biotechnol 31:335–341

Jiménez-Quero A, Pollet E, Zhao M, Marchioni E, Averous L, Phalip V (2016) Fungal fermentation of lignocellulosic biomass for itaconic and fumaric acid production. J Microbiol Biotechnol 28:1–8

Kamm B, Kamm M (2004) Principles of biorefineries. Appl Microbiol Biotechnol 64:137–145

Karnaouri A, Muraleedharan MN, Dimarogona M, Topakas E, Rova U, Sandgren M, Christakopoulos P (2017) Recombinant expression of thermostable processive MtEG5 endoglucanase and its synergism with MtLPMO from *Myceliophthora thermophila* during the hydrolysis of lignocellulosic substrates. Biotechnol Biofuels 10:126

Khelil O, Cheba B (2014) Thermophilic cellulolytic microorganisms from Western Algerian sources: promising isolates for cellulosic biomass recycling. Procedia Technol 12:519–528

Kildegaard KR, Wang Z, Chen Y, Nielsen J, Borodina I (2015) Production of 3-hydroxypropionic acid from glucose and xylose by metabolically engineered *Saccharomyces cerevisiae*. Metab Eng Commun 2:132–136

Kitagawa T et al (2010) Construction of a beta-glucosidase expression system using the multistress-tolerant yeast *Issatchenkia orientalis*. Appl Microbiol Biotechnol 87:1841–1853

Klement T, Milker S, Jäger G, Grande PM, Domínguez de María P, Büchs J (2012) Biomass pretreatment affects *Ustilago maydis* in producing itaconic acid. Microb Cell Fact 11:43

Koivistoinen OM, Kuivanen J, Barth D, Turkia H, Pitkänen J-P, Penttilä M, Richard P (2013) Glycolic acid production in the engineered yeasts *Saccharomyces cerevisiae* and *Kluyveromyces lactis*. Microb Cell Fact 12:82

Kuhad RC, Mehta G, Gupta R, Sharma KK (2010) Fed batch enzymatic saccharification of newspaper cellulosics improves the sugar content in the hydrolysates and eventually the ethanol fermentation by *Saccharomyces cerevisiae*. Biomass Bioenergy 34:1189–1194

Kumar A, Gaind S, Nain L (2008) Evaluation of thermophilic fungal consortium for paddy straw composting. Biodegradation 19:395–402

Kumar M, Brar A, Vivekanand V, Pareek N (2017) Production of chitinase from thermophilic *Humicola grisea* and its application in production of bioactive chitooligosaccharides. Int J Biol Macromol 104:1641–1647

Kusumaningtyas E, Widiastuti R, Maryam R (2006) Reduction of aflatoxin B1 in chicken feed by using *Saccharomyces cerevisiae, Rhizopus oligosporus* and their combination. Mycopathologia 162:307–311

Kwon Y-J, Ma A-Z, Li Q, Wang F, Zhuang G-Q, Liu C-Z (2011) Effect of lignocellulosic inhibitory compounds on growth and ethanol fermentation of newly-isolated thermotolerant *Issatchenkia orientalis*. Bioresour Technol 102:8099–8104

Latif F, Rajoka MI (2001) Production of ethanol and xylitol from corn cobs by yeasts. Bioresour Technol 77:57–63

Li YL, Li H, Li AN, Li DC (2009) Cloning of a gene encoding thermostable cellobiohydrolase from the thermophilic fungus *Chaetomium thermophilum* and its expression in *Pichia pastoris*. J Appl Microbiol 106:1867–1875

Liao W, Liu Y, Frear C, Chen S (2008) Co-production of fumaric acid and chitin from a nitrogen-rich lignocellulosic material—dairy manure—using a pelletized filamentous fungus *Rhizopus oryzae* ATCC 20344. Bioresour Technol 99:5859–5866

Lin J, Ndlovu LM, Singh S, Pillay B (1999) Purification and biochemical characteristics of β-D-xylanase from a thermophilic fungus, *Thermomyces lanuginosus*-SSBP. Biotechnol Appl Biochem 30:73–79

Liu D, Zhang R, Yang X, Xu Y, Tang Z, Tian W, Shen Q (2011) Expression, purification and characterization of two thermostable endoglucanases cloned from a lignocellulosic decomposing fungi *Aspergillus fumigatus* Z5 isolated from compost. Protein Expr Purif 79:176–186

Long L, Tian D, Zhai R, Li X, Zhang Y, Hu J, Wang F, Saddler J (2018) Thermostable xylanase-aided two-stage hydrolysis approach enhances sugar release of pretreated lignocellulosic biomass. Bioresour Technol 257:334–338

Ma R, Bai Y, Huang H, Luo H, Chen S, Fan Y, Cai L, Yao B (2017) Utility of Thermostable xylanases of *Mycothermus thermophilus* in generating prebiotic xylooligosaccharides. J Agric Food Chem 65:1139–1145

Maas RHW, Bakker RR, Eggink G, Weusthuis RA (2006) Lactic acid production from xylose by the fungus *Rhizopus oryzae*. Appl Microbiol Biotechnol 72:861–868

Madadi M et al (2017) Recent status on enzymatic saccharification of lignocellulosic biomass for bioethanol production. Electron J Biol 13:135–143

Masui DC, Zimbardi ALRL, Souza FHM, Guimarães LHS, Furriel RPM, Jorge JA (2012) Production of a xylose-stimulated β-glucosidase and a cellulase-free thermostable xylanase by the thermophilic fungus *Humicola brevis* var. thermoidea under solid state fermentation. World J Microbiol Biotechnol 28:2689–2701

Matsakas L, Christakopoulos P (2015) Ethanol production from enzymatically treated dried food waste using enzymes produced on-site. Sustainability (Switzerland) 7:1446–1458

Miura S, Dwiarti L, Arimura T, Hoshino M, Tiejun L, Okabe M (2004) Enhanced production of L-lactic acid by ammonia-tolerant mutant strain *Rhizopus* sp. MK-96-1196. J Biosci Bioeng 97:19–23

Narra M, James JP, Balasubramanian V (2015) Simultaneous saccharification and fermentation of delignified lignocellulosic biomass at high solid loadings by a newly isolated thermotolerant *Kluyveromyces* sp. for ethanol production. Bioresour Technol 179:331–338

Nguyen S, Ala F, Cardwell C, Cai D, McKindles KM, Lotvola A, Hodges S, Deng Y, Tiquia-Arashiro SM (2013) Isolation and screening of carboxydotrophs isolated from composts and their potential for butanol synthesis. Environ Technol 34:1995–2007

Orlikowska M, Rostro-Alanis MDJ, Bujacz A, Hernández-Luna C, Rubio R, Parra R, Bujacz G (2018) Structural studies of two thermostable laccases from the white-rot fungus *Pycnoporus sanguineus*. Int J Biol Macromol 107:1629–1640

Palmieri G, Giardina P, Bianco C, Scaloni A, Capasso A, Sannia G (1997) A novel white laccase from *Pleurotus ostreatus*. J Biol Chem 272:31301–31307

Palmqvist E, Hahn-Hägerdal B (2000) Fermentation of lignocellulosic hydrolysates. I: Inhibition and detoxification. Bioresour Technol 74:17–24

Pan K, Zhao N, Yin Q, Zhang T, Xu X, Fang W, Hong Y, Fang Z, Xiao Y (2014) Induction of a laccase Lcc9 from *Coprinopsis cinerea* by fungal coculture and its application on indigo dye decolorization. Bioresour Technol 162:45–52

Penjumras P, Rahman RBA, Talib RA, Abdan K (2014) Extraction and characterization of cellulose from durian rind. Agric Agric Sci Procedia 2:237–243

Pérez-Boada M, Ruiz-Dueñas FJ, Pogni R, Basosi R, Choinowski T, Martínez MJ, Piontek K, Martínez AT (2005) Versatile peroxidase oxidation of high redox potential aromatic compounds: site-directed mutagenesis, spectroscopic and crystallographic investigation of three long-range electron transfer pathways. J Mol Biol 354:385–402

Ping L, Wang M, Yuan X, Cui F, Huang D, Sun W, Zou B, Huo S, Wang H (2018) Production and characterization of a novel acidophilic and thermostable xylanase from *Thermoascus aurantiacus*. Int J Biol Macromol 109:1270–1279

Plecha S, Hall D, Tiquia-Arashiro SM (2013) Screening and characterization of soil microbes capable of degrading cellulose from switchgrass (*Panicum virgatum* L.). Environ Technol 34:1895–1904

Pogni R, Baratto MC, Sinicropi A, Basosi R (2015) Spectroscopic and computational characterization of laccases and their substrate radical intermediates. Cell Mol Life Sci 72:885–896

Pomaranski E, Tiquia-Arashiro SM (2016) Butanol tolerance of carboxydotrophic bacteria isolated from manure composts. Environ Technol 37(15):1970–1982

Ramírez-Cavazos LI, Junghanns C, Nair R, Cárdenas-Chávez DL, Hernández-Luna C, Agathos SN, Parra R (2014) Enhanced production of thermostable laccases from a native strain of *Pycnoporus sanguineus* using central composite design. J Zhejiang Univ Sci B 15:343–352

Ribeiro LFC, De Lucas RC, Vitcosque GL, Ribeiro LF, Ward RJ, Rubio MV, Damásio AR, Squina FM, Gregory RC, Walton PH, Jorge JA, Prade RA, Buckeridge MS, Polizeli M de L (2014) A novel thermostable xylanase GH10 from *Malbranchea pulchella* expressed in *Aspergillus nidulans* with potential applications in biotechnology. Biotechnol Biofuels 7:115

Roa Engel CA, Straathof AJJ, Zijlmans TW, van Gulik WM, van der Wielen LAM (2008) Fumaric acid production by fermentation. Appl Microbiol Biotechnol 78:379–389

Robledo A, Aguilar CN, Belmares-Cerda RE, Flores-Gallegos AC, Contreras-Esquivel JC, Montañez JC, Mussatto SI (2016) Production of thermostable xylanase by thermophilic fungal strains isolated from maize silage. CyTA—J Food 14:302–308

Rubin EM (2008) Genomics of cellulosic biofuels. Nature 454:841–845

Ruengruglikit C, Hang DY (2003) L(+)-Lactic acid production from corncobs by *Rhizopus oryzae* NRRL-395. Lwt—Food Sci Technol 36:573–575

Sanchez S, Demain AL (2011) Enzymes and bioconversions of industrial, pharmaceutical, and biotechnological significance. Org Process Res Dev 15:224–230

Scheller HV, Ulvskov P (2010) Hemicelluloses. Annu Rev Plant Biol 61:263–289

Schwarz M, Köpcke B, Weber RWS, Sterner O, Anke H (2004) 3-Hydroxypropionic acid as a nematicidal principle in endophytic fungi. Phytochemistry 65:2239–2245

Shi DJ, Wang CL, Wang KM (2009) Genome shuffling to improve thermotolerance, ethanol tolerance and ethanol productivity of *Saccharomyces cerevisiae*. J Ind Microbiol Biotechnol 36:139–147

Sikandar S, Ujord VC, Ezeji TC, Rossington JL, Michel FC Jr, .McMahan, CM, Ali N, Cornish C (2017) *Thermomyces lanuginosus* STm: a source of thermostable hydrolytic enzymes for novel application in extraction of high-quality natural rubber from *Taraxacum kok-saghyz* (Rubber dandelion). Ind Crop Prod 103:161-168.

Silva RD, Lago ES, Merheb CW, Macchione MM, Park YK, Gomes E (2005) Production of xylanase and CMCase on solid state fermentation in different residues by *Thermoascus aurantiacus* miehe. Braz J Microbiol 36:235–241

Singh N, Mathur AS, Tuli DK, Gupta RP, Barrow CJ, Puri M (2017) Cellulosic ethanol production via consolidated bioprocessing by a novel thermophilic anaerobic bacterium isolated from a Himalayan hot spring. Biotechnol Biofuels 10:73

Skovgaard PA, Jørgensen H (2013) Influence of high temperature and ethanol on thermostable lignocellulolytic enzymes. J Ind Microbiol Biotechnol 40:447–456

Somerville C, Youngs H, Taylor C, Davis SC, Long SP (2010) Feedstocks for lignocellulosic biofuels. Science 329:790–792

Srivastava N, Srivastava M, Mishra PK, Ramteke PW (2016) Application of ZnO nanoparticles for improving the thermal and pH stability of crude cellulase obtained from *Aspergillus fumigatus* AA001. Front Microbiol 7:514

Stenberg K, Bollók M, Réczey K, Galbe M, Zacchi G (2000) Effect of substrate and cellulase concentration on simultaneous saccharification and fermentation of steam-pretreated softwood for ethanol production. Biotechnol Bioeng 68:204–210

Svetlitchnyi VA, Kensch O, Falkenhan DA, Korseska SG, Lippert N, Prinz M, Sassi J, Schickor A, Curvers S (2013) Single-step ethanol production from lignocellulose using novel extremely thermophilic bacteria. Biotechnol Biofuels 6:31

Taarning E, Osmundsen CM, Yang X, Voss B, Andersen SI, Christensen CH (2011) Zeolite-catalyzed biomass conversion to fuels and chemicals. Energ Environ Sci 4:793–804

Techaparin A, Thanonkeo P, Klanrit P (2017) High-temperature ethanol production using thermotolerant yeast newly isolated from Greater Mekong Subregion. Braz J Microbiol 48:461–475

Tiquia-Arashiro SM (2014a) Thermophilic carboxydotrophs and their biotechnological applications. In: Springerbriefs in microbiology: extremophilic microorganisms. Springer International Publishing, p. 131

Tiquia-Arashiro SM (2014b) Biotechnological applications of thermophilic carboxydotrophs. In: Thermophilic carboxydotrophs and their applications in biotechnology. Chapter 4. Springer International Publishing, pp. 29–101

Tiquia-Arashiro SM, Mormile M (2013) Sustainable technologies: bioenergy and biofuel from biowaste and biomass. Environ Technol 34(13):1637–1805

Tiquia-Arashiro SM, Rodrigues D (2016) Thermophiles and psychrophiles in nanotechnology. In: Extremophiles: applications in nanotechnology. Springer International Publishing, pp. 89–127

Trudeau DL, Lee TM, Arnold FH (2014) Engineered thermostable fungal cellulases exhibit efficient synergistic cellulose hydrolysis at elevated temperatures. Biotechnol Bioeng 111(12):2390–2397

Turner P, Mamo G, Karlsson EN (2007) Potential and utilization of thermophiles and thermostable enzymes in biorefining. Microb Cell Fact 6:9

van den Brink J, van Muiswinkel GCJ, Theelen B, Hinz SWA, de Vries RP (2013) Efficient plant biomass degradation by thermophilic fungus *Myceliophthora heterothallica*. Appl Environ Microbiol 79:1316–1324

Walia A, Guleria S, Mehta P, Chauhan A, Parkash J (2017) Microbial xylanases and their industrial application in pulp and paper biobleaching: a review. 3 Biotech 7:11

Watanabe I, Nakamura T, Shima J (2009) Characterization of a spontaneous flocculation mutant derived from *Candida glabrata*: a useful strain for bioethanol production. J Biosci Bioeng 107:379–382

Woiciechowski AL, Soccol CR, Ramos LP, Pandey A (1999) Experimental design to enhance the production of l-(+)-lactic acid from steam-exploded wood hydrolysate using *Rhizopus oryzae* in a mixed-acid fermentation. Process Biochem 34:949–955

Xu Q, Singh A, Himmel ME (2009) Perspectives and new directions for the production of bioethanol using consolidated bioprocessing of lignocellulose. Curr Opin Biotechnol 20:364–371

Xu X, Lin M, Zang Q, Shi S (2018) Solid state bioconversion of lignocellulosic residues by Inonotus obliquus for production of cellulolytic enzymes and saccharification. Bioresour Technol 247:88–95

Yang J, Li W, Ng TB, Deng X, Lin J, Ye X (2017) Laccases: production, expression regulation, and applications in pharmaceutical biodegradation. Front Microbiol 8:832

Yang J, Gao R, Zhou Y, Anankanbil S, Li J, Xie G, Guo Z (2018) β-Glucosidase from *Thermotoga naphthophila* RKU-10 for exclusive synthesis of galactotrisaccharides: Kinetics and thermodynamics insight into reaction mechanism. Food Chem 240:422–429

Yennamalli RM, Rader AJ, Kenny AJ, Wolt JD, Sen TZ (2013) Endoglucanases: insights into thermostability for biofuel applications. Biotechnol Biofuels 6:136

Yeoman CJ, Han Y, Dodd D, Schroeder CM, Mackie RI, Cann IKO (2010) Thermostable enzymes as biocatalysts in the biofuel industry. Adv Appl Microbiol 70:1–55

Yoshihara K, Shinohara Y, Hirotsu T, Izumori K (2003) Chitosan productivity enhancement in *Rhizopus oryzae* YPF-61A by D-psicose. J Biosci Bioeng 95:293–297

Zeldes BM, Keller MW, Loder AJ, Straub CT, Adams MWW, Kelly RM (2015) Extremely thermophilic microorganisms as metabolic engineering platforms for production of fuels and industrial chemicals. Front Microbiol 6:1209

Zhang ZY, Jin B, Kelly JM (2007) Production of lactic acid from renewable materials by *Rhizopus* fungi. Biochem Eng J 35:251–263

Zhang Z, Wang M, Gao R, Yu X, Chen G (2017) Synergistic effect of thermostable β-glucosidase TN0602 and cellulase on cellulose hydrolysis. 3 Biotech 7:54

Zhou X, Wu Q, Cai Z, Zhang J (2000) Studies on the correlation between production of L-malic acid and some cytosolic enzymes in the L-malic acid producing strain *Aspergillus* sp. N1-14. Wei Sheng Wu Xue Bao 40:500–506

Zhu Y, Zhang H, Cao M, Wei Z, Huang F, Gao P (2011) Production of a thermostable metal-tolerant laccase from *Trametes versicolor* and its application in dye decolorization. Biotechnol Bioproc E 16:1027

Part V
Bioremediation and Biosolids Treatment

Chapter 25
Acidomyces acidophilus: Ecology, Biochemical Properties and Application to Bioremediation

Wai Kit Chan, Dirk Wildeboer, Hemda Garelick, and Diane Purchase (iD)

25.1 Introduction

Extremophilic fungi have been successfully isolated from highly polluted or extreme environments such as mining wastewaters of pH 1.1 (Oggerin et al. 2013), a uranium mine (Vázquez-Campos et al. 2014) and a volcanic geothermal system (Chiacchiarini et al. 2010). Hitherto, the data on extremophilic fungi diversity in extremely acidic conditions is limited; however, melanised and meristematic fungi have been found to be the prevailing groups in such habitats (Baker et al. 2004; López-Archilla et al. 2004; Selbmann et al. 2008; Hujslová et al. 2010). For example, *Hortaea acidophila* was isolated from brown coal that contains humic and fulvic acid at pH around 0.6 (Hölker et al. 2004); two acidophilic strains, *Hortaea werneckii* and *Acidomyces acidophilum,* were isolated from Rio Tinto in Spain, where the mean pH value was 2.3, and which contained high concentrations of Fe, Cu, Zn, As, Mn and Cr (Zettler et al. 2002). Another black fungus, *Exophiala sideris*, was isolated from an ancient gold arsenic mine polluted with akylbenzenes in Złoty Stok, Poland, for which the authors claimed that it can be used potentially for the bioremediation of metalloids (Seyedmousavi et al. 2011).

Acidomyces acidophilus is a pigmented ascomycete capable of growing in extremely acidic conditions (Sigler and Carmichael 1974). Its melanin-containing cell walls offer the fungus protection from adverse environmental conditions such as extreme pH, temperature and toxins (Jacobson et al. 1995; Martin et al. 1990; Tetsch et al. 2005; Hujslová et al. 2013). This protection also provides the fungus a certain level of resistance to oxidative stress (Jung et al. 2006). *A. acidophilus* is an anamorphic fungus, hyphomycete, and phylogenetically belongs to the class

W. K. Chan · D. Wildeboer · H. Garelick · D. Purchase (✉)
Faculty of Science and Technology, Department of Natural Sciences, Middlesex University, London, UK
e-mail: d.purchase@mdx.ac.uk

© Springer Nature Switzerland AG 2019
S. M. Tiquia-Arashiro, M. Grube (eds.), *Fungi in Extreme Environments: Ecological Role and Biotechnological Significance,*
https://doi.org/10.1007/978-3-030-19030-9_25

Dothideomycetes, order *Capnodiales*. The colonies are dark and compact, slow growing in neutral medium but faster in acidic medium. Their septate, scarcely branched, brown and thick-walled hyphae eventually convert into a meristematic mycelium. Conidia are produced by arthric disarticulation of hyphae. The morphologies of *Acidomyces* species were not easily described using microscopy because of its tendency to convert to highly melanised meristematic growth, produce scarcely disarticulating clumps of cells, or tend to appear without any conidiation but entirely hyphal (Selbmann et al. 2008).

The intracellular enzymatic regulation mechanisms in these extremophilic fungi, especially *A. acidophilus*, of low pHs, reactive oxygen species and extreme temperatures are now widely considered as valuable resources for their exploitation in novel biotechnological applications (such as cleaning of contaminated soils or water) and in the field of bio-catalysis (Tiquia-Arashiro and Rodrigues 2016). They also generated research interests on biomolecule stability investigation and the design and synthesis of proteins that are relevant for various industrial applications (Polizeli et al. 2005; Hess 2008; Selbmann et al. 2008).

To survive in such adverse conditions, *A. acidophilus* has developed various approaches and evolved different mechanisms to cope with and even thrive in these harsh ecological niches. Although *A. acidophilus* is a remarkable fungus, it is not well studied and there is a dearth of information in its biochemical properties and applications. This chapter aims to provide a better understanding of *A. acidophilus* and highlights its potential in sustainable environmental biotechnology. We explore the ecological niches of *A. acidophilus* as well as its remarkable adaptation to survival in the extreme environment, providing an insight into their evolutionary ability in using their flexible ecological plasticity to exploit and thrive in new environments. With the advancement of 'omics' tools, genetic and proteomics information of this fungus has begun to emerge, and we present up-to-date information on this much-neglected fungus. The chapter provides an in-depth review on *A. acidophilus* literature including a discussion of its potential in bioremediation and explores its application in other research areas. We hope to bring this fungus to a wider scientific attention.

25.2   Ecology of *Acidomyces acidophilus*

'Extreme' conditions are taken to be those that deviate from normal physicochemical limits supporting growth of any organisms. In extremely low pH and toxic conditions, stressing conditions can cause detrimental effects on microorganisms' metabolic processes and enzyme- and protein-mediated biological functions (Selbmann et al. 2013). However, many eukaryotes such as fungi have the ability to thrive under stressing conditions that are hostile to other living organisms. Many fungal species, of an unexpected degree of diversity, were discovered living in extreme habitats such as volcanic soils (Appoloni et al. 2008), extremely cold and ice-free deserts in the Antarctica (Friedmann et al. 1993, Selbmann et al. 2008), in geothermal and humid soils (Hujslová et al. 2013), in acid mine drainage (Baker et al. 2004), in hypersaline environments (Plemenitaš et al. 2014) or in extremely polluted mining soils (Chan et al. 2018). Most of the acidophilic and psychrophilic black meristematic yeasts and fungi species are the best examples of common inhabitants in these stressful environments (Baker et al. 2004; Hujslová et al. 2010). Organisms that have the ability to adapt and grow in an extremely acidic environment, pH <4, are known as acidophiles. This ecological niche can usually be found in heavily polluted sites created by prolonged anthropogenic activities, which release toxic pollutants that contaminate and change the pH conditions of the surrounding soils or water.

This extreme environment is where the black fungus *Acidomyces acidophilus* was first isolated, from a highly acidic, sulphate-containing industrial water by Starkey and Waksman (1943). Subsequently, more *Acidomyces* species were isolated and identified successfully from various extreme environments. These are summarised in Table 25.1. All the *A. acidophilus* strains show the common feature of being acidophilic and thrive in highly acidic ecological niche between pH 1 and 3. Selbmann et al. (2013) showed that acidity plays a pivotal role in maintaining *A. acidophilus* growth, where optimal growth was observed between pH 3 and 5. Another species, *Acidomyces richmondensis*, was isolated from sulphuric ore acid mine drainage in Richmond, USA, at pH 0.5 and 0.9 and thermophilic temperature in the range of 35–57 °C by Baker et al. (2004). However, the optimal growth temperature for most *A. acidophilus* is 25 °C (Selbmann et al. 2008; Hujslová et al. 2013). *A. acidophilus* was not only able to tolerate extremely low pH, but it has also developed tolerance towards high concentration of toxic metals and metalloids stresses such as Al, As, Cu, Fe or U (Starkey and Waksman 1943; Gould et al. 1974; Gimmler et al. 2001; Yamazaki et al. 2010; Vázquez-Campos et al. 2014; Chan et al. 2018).

Fungi have developed immense stress resistance to cope with and tolerate extreme conditions through different adaptation strategies and mechanisms. Generally, fungi have the ability to adapt to new environments, due to their remarkable phenotypic plasticity, by performing a process called 'ecological fitting', as proposed by Agosta and Klemens (2008). As Capnodiales black yeast-like fungi, *A. acidophilus* strains are characteristically polymorphic and display extraordinary ecological, biological and morphological plasticity. In the presence of various stress factors, fungi can perform specific adaptation strategies such as meristematic growth, melanisation, alterations of cell wall structure or isodiametric expansion, ensuring optimisation of the volume-to-surface ratio and allowing them to tolerate and thrive under stress conditions (Hujslová et al. 2013; Selbmann et al. 2013; Gunde-Cimerman et al. 2005; Wollenzien et al. 1995; Sterflinger and Hain 1999).

The high degrees of melanisation, phenolic polymers and thick cell walls (Figueras et al. 1996) in *A. acidophilius* result in oxygen-containing functional groups such as carboxyl, sulphhydryl, phosphate, methoxyl and carbonyl enhancing their survival and their ability to withstand extreme pH and toxic compounds

Table 25.1 Sources of previously isolated *Acidomyces acidophilus* strains

Species	Geography	Strain no.	Source	ITS	References
Acidomyces acidophilus	Czech Republic	MH1206 MH931, MH933, MH1098 MH1085	Sulphur- and humic acid-rich brown coal Sulphur-rich coal layers Exposed clay sediment, sulphur-rich coal layers	JQ172741	Hujslová et al. (2013)
	Iceland	MH1091, MH1092, MH1102, MH1109	Geothermal area pH 1.1–2.5		
		AK72/03 = CCF3679 MH932	Peat bogs, mineral fens, salt marshes (acidic)	FJ430711 JQ172742	
	England	DSM 105253	Tin mining soil pH 1	KT727926	Chan et al. (2018)
Acidomyces acidophilum	Germany	CBS335.97	Acidophilic algae *Dunaliella acidophila* pH 1.0	AJ244237	Gimmler et al. (2001)
Acidomyces acidophilum (deposited as *Scytalidium acidophilum*)	Canada	CBS 270.74	Soil from acidic elemental sulphur close to natural gas purification pH 1.1	–	Sigler and Carmichael (1974)
Acidomyces acidophilum (deposited as *Botryomyces caespitosus*)	Germany	dH13081 = det 106/2023 (supplied by GC Frisvad)	2 N Sulphuric acid pH 1.0	–	Starkey and Waksman (1943)
	Denmark	CBS899.87	Pyrite ore acidic drainage pH 2.0	–	Selbmann et al. (2008)
	Iceland	dH11526 = det 237-1999 (supplied by S Gross, Berlin)	Volcanic soil	–	
Acidomyces sp.	The Netherlands	dH13119	Acidic industrial process water pH 1.5	–	
Acidomyces richmondensis	USA	MB511298	Sulphuric ore acid mine drainage	AY374298-300	Baker et al. (2004)

in their habitual environment (Chan et al. 2018). All the *A. acidophilus* isolated from extreme environment showed to be highly melanised (Selbmann et al. 2008), indicating that melanin is an important protective factor in increasing their resistance to reactive oxygen species (ROS), toxic metals, extreme pH and temperature, and UV radiation. The highly melanised cell wall of *A. acidophilus* is shown in Fig. 25.1b. The hydrophobicity and negative charges of melanin offer the fungus protection from oxidative stress (Jung et al. 2006). Another interesting property of melanin is that it can shield organisms from ionising radiation. Since melanin has a stable free radical population, it is thought that the radio-protective properties are due to the scavenging of free radicals generated by radiation (Eisenman and Casadevall 2012). In addition, the controlled dissipation of high-energy recoil electrons by melanin prevents secondary ionisations and the generation of damaging free radical species (Schweitzer et al. 2009). The dynamic architecture of thick cell wall found in *A. acidophilus* possesses an intracellular pH controlling system and in extreme pH conditions it maintains the cytoplasm to a near-neutral pH (Hesse et al. 2002; Bignell 2012). These features coupled with its ability to change polarity (Yoshida et al. 1996) in *A. acidophilus* morphologies resulted from convergent evolution of its long-term exposure to extreme ecological niches (Selbmann et al. 2008).

(a)   (b)   (c)

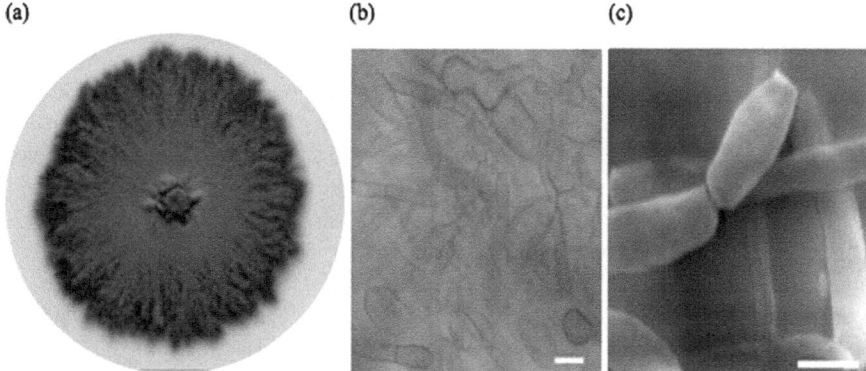

Fig. 25.1 Morphological features of the *Acidomyces acidophilus* WKC-1 (**a**) colony of the isolated fungal strain in CDA medium, (**b**) Filamentous hyphae of the strain with intercalary and unbranched hyphae with melanised and thick-walled cells and (**c**) Hyphae terminal swelling cell using scanning electron microscope (SEM) at a magnification of 1000× (**b**) and 2200× (**c**), scale bar in (**b**) and (**c**) 2 μm (Chan et al. 2018). Image reproduced with permission of the rights holder, Springer

25.3 Biochemical Properties of *A. acidophilus*

The biochemical properties expressed by *A. acidophilus* are some of the key strategies leading to their remarkable survival ability in extreme environment. In a mine in Richmond, USA, an isolated *Acidomyces* sp. (*A. richmondensis*) produced a variety of novel putative fungal glycosyl hydrolases and secreted other enzymes extracellularly to resist the highly acidic (pH <1), elevated metal concentration ($\approx$200 mM Fe) and thermophilic (40–50 °C) conditions of the acid mine drainage (Baker et al. 2004). As for the ability to degrade aromatic compounds, some *A. acidophilus* strains (CBS 270.74, CBS 335.97 and CBS 899.87) tested by Selbmann et al. (2013) have shown the ability to use 4-hydroxybenzene (4HB) and protocatechuic acid (PCA) in acidic condition and phenylacetic acid (PAA) was efficiently metabolised at pH 7. This illustrates the potential of *A. acidophilus* to utilise these compounds as their carbon source. It is also noted that strain CBS 335.97 is the only candidate that showed a slight growth on catechol (CAT) in acidic condition. All the tested *A. acidomyces* strains were able to produce the enzymes lipase and amylase even in acidic condition such as pH 3. Another study by Ervin and Wolfe (2016) suggested that *Acidomyces* sp. isolated from a geothermal spring lake contains viable sets of hemicellulose- and lignin-degrading extracellular enzymes that function in acidic and thermophilic conditions for lignocellulosic degradation.

In a metabolomic study performed on *A. richmondensis,* from acid mine drainage (AMD) biofilm (Mosier et al. 2016), its genome revealed a gene that was involved in the biosynthesis and degradation of taurine metabolites. When exposed to environmental stressors, taurine is involved in several physiological roles such as protecting proteins, nucleic acids and membranes against ROS (Andres and Bertin 2016). Besides taurine, *A. richmondensis* also contained genes that were involved in the biosynthesis and degradation of trehalose, including a betaine-aldehyde dehydrogenase gene. These encoded genes play important roles in the metabolism of compatible solutes, especially in the high-ionic-strength AMD waters (Druschel et al. 2004). Genome-encoded complete tricarboxylic acid (TCA), glycolysis and pentose phosphate pathways were found in the *A. richmondensis* during the metaproteogenomic analysis by Mosier et al. (2016) as well as many genes that were involved in the metabolism of fructose, mannose, galactose, starch and sucrose. In a proteomic functional analysis on *A. richmondensis*, it was predicted that its genes encode for 350 carbon active enzymes (CAZymes), and either involve in degradation, modification or creating glycosidic bonds (Lombard et al. 2014). These CAZymes include glycosyl transferases, glycoside hydrolases, carbohydrate esterases and carbohydrate-binding modules, and are involved in the formation of glycosidic bonds, hydrolysis/rearrangement of glycosidic bonds, hydrolysis of carbohydrate esters and carbohydrate-binding activity, respectively (Mosier et al. 2016). As one of the key components of acid-stable enzyme cocktails, *A. richmondensis* is still able to produce these glycosyl hydrolases compared to other non-extremophilic fungi that are involved in the bioenergy production even under extreme pH and toxic conditions.

25.4 Application to Bioremediation

The rich biochemical properties of *A. acidophilus* enable it to tolerate not only extreme pH but also toxic compounds in highly contaminated soil or water. It shows a great biotechnological potential, especially in bioremediation applications (Selbmann et al. 2013; Chan et al. 2018). Several extremophile fungi *such as Coniochaeta fodinicola, Hortaea acidophilia, Teratosphaeria acidotherma* and *A. acidophilus* were able to produce novel enzymes and other metabolites to survive in harsh environments which can be applied in bioremediation, primarily by removing toxic metals and metalloids from soil and water contamination caused by anthropogenic activities such as mining (Tetsch et al. 2005; Stierle et al. 2006; Wang et al. 2010; Luo et al. 2010; Yang et al. 2011; Selbmann et al. 2013; Boonen et al. 2014; Hujslová 2015).

The study carried out by Mosier et al. (2016) showed that *A. richmondensis* encoded for and expressed a great number of genes that are involved in heavy metal transport and detoxification. The proteomic and transcriptomic analyses on *A. richmondensis* revealed a wide range of metal transporters specific to iron, copper, zinc, magnesium, calcium and nickel. Besides these transporters, it also contained genes that are significant in heavy metal chelation, including a ferrochelatase; a siderophore-dependent iron transporter, sideroflexin; and several transcripts for ferric-chelate reductase. Detoxification of exogenous cyanide and/or cyanide by-products of other cellular metabolic reaction genes such as cyanide hydratase, cyanide nitrilase and cyanate lyases was encoded and expressed by *A. richmondensis*. They also found other genes encoded within *A. richmondensis* that are involved in arsenic detoxification process such as arsenate reductase (ArsH, an NADPH-dependent FMN reductases), membrane-associated arsenite permeases (ArsB or ACR_3) and arsenite-translocating ATPases (ArsA). Other biotransformation genes involved in arsenate reduction such as arsenite methyltransferase, genes encoded for methylarsonite methyltransferase activity as well as genes encoding glutathione S-transferase were also expressed. These proteins found in *A. richmondensis* are heavily involved in the reduction of arsenate to arsenite (Zakharyan et al. 2005; Ventura-Lima et al. 2011; Pantoja Munoz 2014; Andres and Bertin 2016).

The study by Chan et al. (2018) on biosorption of arsenic and antimony by *A. acidophilus* WKC-1 showed that $-OH$, $-NH$, $-CH$, $-SO_3$ and PO_4 functional groups are identified as the key biosorption binding sites for As^{5+} and Sb^{5+}. The isolate WKC-1 showed a high resistance to and high percentage of As^{5+} removal of 70.30%, one of the highest reported in *A. acidophilus* species, when cultivated in 100 mg L^{-1} As^{5+} concentration after 21 days. The tolerance of the isolated *A. acidophilus* WKC-1 strain to low pH and high As concentration together with its capacity to remove approximately 170 mg As^{5+} per gram dry biomass made it a potential candidate to be used in bioremediation of As.

Despite the robustness of this fungus and its ability to survive in extremely adverse environment, *A. acidophilus* remained a poorly researched organism, and further work is required to fully explore its potential in bioremediation and other biotechnological applications.

25.5 Future Perspectives

The remarkable tolerance of *A. acidophilus* to adverse conditions and its significant variability and ability to mediate in extreme acidity ecosystems where most forms of life cannot may open up research opportunities and exploration for planetary and astrobiological studies in the search of life on acidic and hot planets such as our sister planet, Venus.

Although *A. acidophilus* has proven to be effective in removing metalloids such as arsenic in soil in laboratory settings, in order to apply it to remediate contaminated soil effectively it needs to be scaled up accordingly. The slow growth rate of *A. acidophilus* is the bottleneck in scaling up; however it has been shown by Chan et al. (2018) that growing the fungi on a rotator using liquid medium has significantly reduced the growth period (from 28 to 3 days). Recent advances in molecular biology and biotechnology techniques can be applied to genetic modify *A. acidophilus* to improve its remediation capabilities. The genes involved in detoxification of heavy metals in *A. acidophilus* can be cloned and expressed in a fast growing host. Various genetic engineering approaches have been developed to optimise mycoremediation by engineer-improved fungi and enzymes. *A. acidophilus* contains several genes that encode glutathione S-transferase and through genetic splicing or gene regulations these important genes that reduce the toxicity of arsenic from arsenate to arsenite can be overexpressed or used to construct new metabolic pathways in other fungi, allowing the novel enzymes/genes that are produced by *A. acidophilus* to be used to remediate not only metals but also other pollutants.

With the advancement of proteomics and metagenomics studies, researchers are now able to identify stress-associated gene biomarkers and/or enzymes that are regulated and expressed in the *A. acidophilus* genome when exposed to extreme conditions. These novel pharmaceutical extremozymes and biologically active compounds produced by *A. acidophilus* that can tolerate and perform biological activity in low pH could have important applications in other applied scientific areas such as ecotoxicology.

25.6 Conclusions

The extreme ecology *of A. acidophilus* has allowed it to develop different adaptive mechanisms to persist in hostile environments through evolutionary processes. The morphology of *A. acidophilus* such as its melanised thick cell wall, phenotypic plasticity and ability to produce novel enzymes play vital roles to its survivability by providing protection and performing specific adaptation strategies to cope stressing conditions. Heavy metal transporters and various detoxification genes found in *A. acidophilus* drive its great mycoremediation potential in removing toxic compounds in the environment. However, their biotechnological capabilities are not fully utilised. Understanding their ecology and biochemistry in extreme environments could help to provide an insight into and maximise the potential application of this remarkable fungus.

References

Agosta SJ, Klemens JA (2008) Ecological fitting by phenotypically flexible genotypes: implications for species associations, community assembly and evolution. Ecol Lett 11:1123–1134

Andres J, Bertin PN (2016) The microbial genomics of arsenic. FEMS Microbiol Rev 40:299–322

Appoloni S, Lekberg Y, Tercek MT, Zabinski CA, Redecker D (2008) Molecular community analysis of arbuscular mycorrhizal fungi in roots of geothermal soils in Yellowstone National Park (USA). Microb Ecol 56:649

Baker BJ, Lutz MA, Dawson SC, Bond PL, Banfield JF (2004) Metabolically active eukaryotic communities in extremely acidic mine drainage. Appl Environ Microbiol 70:6264–6271

Bignell E (2012) The molecular basis of pH sensing, signalling, and homeostasis in fungi. Adv Appl Microbiol 79:1–18

Boonen F, Vandamme AM, Etoundi E, Pigneur LM, Housen I (2014) Identification and characterization of a novel multicopper oxidase from *Acidomyces acidophilus* with ferroxidase activity. Biochimie 102:37–46

Chan WK, Wildeboer D, Garelick H, Purchase D (2018) Competition of As and other Group 15 elements for surface binding sites of an extremophilic *Acidomyces acidophilus* isolated from a historical tin mining site. Extremophiles 22:795–809

Chiacchiarini P, Lavalle L, Giaveno A, Donati E (2010) First assessment of acidophilic microorganisms from geothermal Copahue–Caviahue system. Hydrometallurgy 104:334–341

Druschel GK, Baker BJ, Gihring TM, Banfield JF (2004) Acid mine drainage biogeochemistry at Iron Mountain, California. Geochem Trans 5:13

Eisenman HC, Casadevall A (2012) Synthesis and assembly of fungal melanin. Appl Microbiol Biotechnol 93:931–940

Ervin B, Wolfe G (2016) Acido-thermotolerant fungi from Boiling Springs Lake, LVNP: potential for lignocellulosic biofuels. Am Mineral 101:2484–2497

Figueras MJ, De Hoog GS, Takeo K, Guarro J (1996) Stationary phase development of *Trimmatostroma abietis*. Antonie Van Leeuwenhoek 69:217–222

Friedmann EI, Kappen L, Meyer MA, Nienow JA (1993) Long-term productivity in the cryptoendolithic microbial community of the Ross Desert, Antarctica. Microb Ecol 25:51–69

Gimmler H, De Jesus J, Greiser A (2001) Heavy metal resistance of the extreme acidotolerant filamentous fungus *Bispora* sp. Microb Ecol 42:87–98

Gould WD, Fujikawa JI, Cook FD (1974) A soil fungus tolerant to extreme acidity and high salt concentrations. Can J Microbiol 20:1023–1027

Gunde-Cimerman N, Oren A, Plemenitaš A (2005) Adaptation to life at high salt concentrations in archaea, bacteria, and eukarya, vol 9. Springer Publisher, Dordrecht, pp 1–6

Hess M (2008) Thermoacidophilic proteins for biofuel production. Trends Microbiol 16:414–419

Hesse SJ, Ruijter GJG, Dijkema C, Visser J (2002) Intracellular pH homeostasis in the filamentous fungus *Aspergillus niger*. Eur J Biochem 269:3485–3494

Hölker U, Bend J, Pracht R, Tetsch L, Müller T, Höfer M, Hoog GS (2004) *Hortaea acidophila*, a new acid-tolerant black yeast from lignite. Antonie Van Leeuwenhoek 86:287–294

Hujslová M (2015) Diversity and taxonomy of fungi inhabiting extremely acidic and saline soils of natural and anthropogenic origin in the Czech Republic. Doctoral dissertation. Retrieved from http://hdl.handle.net/20.500.11956/82619

Hujslová M, Kubátová A, Chudíčková M, Kolařík M (2010) Diversity of fungal communities in saline and acidic soils in the Soos National Natural Reserve, Czech Republic. Mycol Progr 9:1–15

Hujslová M, Kubátová A, Kostovčík M, Kolařík M (2013) *Acidiella bohemica* gen. et sp. nov. and *Acidomyces* spp.(Teratosphaeriaceae), the indigenous inhabitants of extremely acidic soils in Europe. Fungal Diversity 58:33–45

Jacobson ES, Hove E, Emery HS (1995) Antioxidant function of melanin in black fungi. Infect Immun 63:4944–4945

Jung WH, Sham A, White R, Kronstad JW (2006) Iron regulation of the major virulence factors in the AIDS-associated pathogen *Cryptococcus neoformans*. PLoS Biol 4:410

Lombard V, Golaconda-Ramulu H, Drula E, Coutinho PM, Henrissat B (2014) The carbohydrate-active enzymes database (CAZy) in 2013. Nucleic Acids Res 42:490–495

López-Archilla AI, González AE, Terrón MC, Amils R (2004) Ecological study of the fungal populations of the acidic Tinto River in southwestern Spain. Can J Microbiol 50:923–934

Luo H, Yang J, Li J, Shi P, Huang H, Bai Y, Fan Y, Yao B (2010) Molecular cloning and characterization of the novel acidic xylanase XYLD from *Bispora* sp. MEY-1 that is homologous to family 30 glycosyl hydrolases. Appl Microbiol Biotechnol 86:1829–1839

Martin AM, Chintalapati SP, Patel TR (1990) Extraction of bitumens and humic substances from peat and their effects on the growth of an acid-tolerant fungus. Soil Biol Biochem 22:949–954

Mosier AC, Miller CS, Frischkorn KR, Ohm RA, Li Z, LaButti K, Lapidus A, Lipzen A, Chen C, Johnson J, Lindquist EA (2016) Fungi contribute critical but spatially varying roles in nitrogen and carbon cycling in acid mine drainage. Front Microbiol 7:238

Oggerin M, Tornos F, Rodríguez N, Del Moral C, Sánchez-Román M, Amils R (2013) Specific jarosite biomineralization by *Purpureocillium lilacinum*, an acidophilic fungi isolated from Río Tinto. Environ Microbiol 15:2228–2237

Pantoja Munoz L (2014) The mechanisms of arsenic detoxification by the green microalgae chlorella vulgaris. Doctoral dissertation, Middlesex University. Retrieved from http://eprints.mdx.ac.uk/13252/

Plemenitaš A, Lenassi M, Konte T, Kejžar A, Zajc J, Gostinčar C, Gunde-Cimerman N (2014) Adaptation to high salt concentrations in halotolerant/halophilic fungi: a molecular perspective. Front Microbiol 5:199

Polizeli MLTM, Rizzatti ACS, Monti R, Terenzi HF, Jorge JA, Amorim DS (2005) Xylanases from fungi: properties and industrial applications. Appl Microbiol Biotechnol 67:577–591

Schweitzer AD, Howell RC, Jiang Z, Bryan RA, Gerfen G, Chen CC, Mah D, Cahill S, Casadevall A, Dadachova E (2009) Physico-chemical evaluation of rationally designed melanin as novel nature-inspired radioprotectors. PLoS One 4:7229

Selbmann L, De Hoog GS, Zucconi L, Isola D, Ruisi S, van den Ende AGC, Ruibal C, De Leo F, Urzì C, Onofri S (2008) Drought meets acid: three new genera in a dothidealean clade of extremotolerant fungi. Stud Mycol 61:1–20

Selbmann L, Egidi E, Isola D, Onofri S, Zucconi L, de Hoog GS, Chinaglia S, Testa L, Tosi S, Balestrazzi A, Lantieri A (2013) Biodiversity, evolution and adaptation of fungi in extreme environments. Plant Biosyst 147:237–246

Seyedmousavi S, Badali H, Chlebicki A, Zhao J, Prenafeta-Boldu FX, De Hoog GS (2011) *Exophiala sideris*, a novel black yeast isolated from environments polluted with toxic alkyl benzenes and arsenic. Fungal Biol 115:1030–1037

Sigler L, Carmichael JW (1974) A new acidophilic *Scytalidium*. Can J Microbiol 20:267–268

Starkey RL, Waksman SA (1943) Fungi tolerant to extreme acidity and high concentrations of copper sulfate. J Bacteriol 45:509

Sterflinger K, Hain M (1999) In situ hybridization with rRNA targeted probes as a new tool for the detection of black yeasts and meristematic fungi. Stud Mycol 43:23–30

Stierle AA, Stierle DB, Kelly K (2006) Berkelic acid, a novel spiroketal with selective anticancer activity from an acid mine waste fungal extremophile. J Org Chem 71:5357–5360

Tetsch L, Bend J, Janßen M, Hölker U (2005) Evidence for functional laccases in the acidophilic ascomycete *Hortaea acidophila* and isolation of laccase-specific gene fragments. FEMS Microbiol Lett 245:161–168

Tiquia-Arashiro SM, Rodrigues D (2016) Alkaliphiles and acidophiles in nanotechnology. In: Extremophiles: applications in nanotechnology. Springer International Publishing, Cham, Switzerland, pp 129–162

Vázquez-Campos X, Kinsela AS, Waite TD, Collins RN, Neilan BA (2014) *Fodinomyces uranophilus* gen. nov. sp. nov. and *Coniochaeta fodinicola* sp. nov., two uranium mine-inhabiting Ascomycota fungi from northern Australia. Mycologia 106:1073–1089

Ventura-Lima J, Bogo MR, Monserrat JM (2011) Arsenic toxicity in mammals and aquatic animals: a comparative biochemical approach. Ecotoxicol Environ Saf 74:211–218

Wang H, Luo H, Li J, Bai Y, Huang H, Shi P, Fan Y, Yao B (2010) An α-galactosidase from an acidophilic *Bispora* sp. MEY-1 strain acts synergistically with β-mannanase. Bioresour Technol 101:8376–8382

Wollenzien U, De Hoog GS, Krumbein WE, Urzi C (1995) On the isolation of microcolonial fungi occurring on and in marble and other calcareous rocks. Sci Total Environ 167:287–294

Yamazaki A, Toyama K, Nakagiri A (2010) A new acidophilic fungus *Teratosphaeria acidotherma* (Capnodiales, Ascomycota) from a hot spring. Mycoscience 51:44–455

Yang J, Luo H, Li J, Wang K, Cheng H, Bai Y, Yuan T, Fan Y, Yao B (2011) Cloning, expression and characterization of an acidic endo-polygalacturonase from *Bispora* sp. MEY-1 and its potential application in juice clarification. Process Biochem 46:272–277

Yoshida S, Takeo K, De Hoog GS, Nishimura K, Miyaji M (1996) A new type of growth exhibited by *Trimmatostroma abietis*. Antonie Van Leeuwenhoek 69:211–215

Zakharyan RA, Tsaprailis G, Chowdhury UK, Hernandez A, Aposhian HV (2005) Interactions of sodium selenite, glutathione, arsenic species, and omega class human glutathione transferase. Chem Res Toxicol 18:1287–1295

Zettler LA, Gómez F, Zettler E, Keenan BG, Amils R, Sogin ML (2002) Microbiology: eukaryotic diversity in Spain's River of Fire. Nature 417:137

Chapter 26
Bioremediation Abilities of Antarctic Fungi

María Martha Martorell (iD)**, Lucas Adolfo Mauro Ruberto,
Lucía Inés Figueroa de Castellanos, and Walter Patricio Mac Cormack**

26.1 Introduction

Cold environments represent a special challenge for life, even for the ubiquitous and metabolically versatile microorganisms. Therefore, bioremediation processes are hardly limited by low temperature. Most of the reported bioremediation-related information refers to temperate regions. However, during the last 20 years the research regarding such processes in cold environment has been reported for different cold regions all around the world, including the Alps, the Arctic, and Antarctica (de Jesús et al. 2015). Despite their amazing growth potential as well as enormous catabolic capabilities yeasts have been, among microorganisms, poorly studied as bioremediation tools.

This book chapter focuses on one specific extreme and permanently cold environment, Antarctica. General information on bioremediation and its use in cold environments is introduced. Also, the main contaminants in the Antarctic continent are summarized. For bioremediation in this continent, we propose, as a study case,

M. M. Martorell (✉) · W. P. M. Cormack
Instituto Antártico Argentino (IAA), General San Martín, Buenos Aires, Argentina

Universidad de Buenos Aires, Ciudad Autónoma de Buenos Aires, Argentina

L. A. M. Ruberto
Instituto Antártico Argentino (IAA), General San Martín, Buenos Aires, Argentina

Universidad de Buenos Aires, Ciudad Autónoma de Buenos Aires, Argentina

Instituto de Nanobiotecnología (NANOBIOTEC-UBA-CONICET),
Ciudad Autónoma de Buenos Aires, Argentina

L. I. F. de Castellanos
Planta Piloto de Procesos Industriales Microbiológicos (PROIMI-CONICET),
San Miguel de Tucumán, Tucumán, Argentina

Universidad Nacional de Tucumán (UNT), San Miguel de Tucumán, Tucumán, Argentina

© Springer Nature Switzerland AG 2019   517
S. M. Tiquia-Arashiro, M. Grube (eds.), *Fungi in Extreme Environments:*
Ecological Role and Biotechnological Significance,
https://doi.org/10.1007/978-3-030-19030-9_26

the use of autochthonous yeasts with multiple bioremediation abilities, as they can be used to treat several sources of contamination in this region which is under a strong policy (Antarctic Protocol of Environmental Protection) related to management and environmental protection.

26.2   Bioremediation in Cold Environments

Bioremediation is defined as the use of living organisms or its components (such as enzymes) to reduce, eliminate, or transform toxic compounds from the environment (Tyagi et al. 2011). The goal of bioremediation is to speed up the natural biodegradation process that happens in most contaminated environments by optimizing some conditions through bioaugmentation or biostimulation. Bioremediation is considered an eco-friendly and economically effective cleanup system, and for these characteristics it is worldwide considered as a socially accepted process (Margesin 2014).

For organic contaminants, bioremediation is mainly based on the ability of microorganisms to use these organic compounds as their carbon and energy source (biodegradation) and therefore transform these contaminants in less harmful compounds. The main goal of bioremediation is to mineralize the contaminants to CO_2 and H_2O but, when this objective cannot be achieved with some contaminants, their transformation to less toxic components is also desirable (Margesin 2014).

Over 80% of planet earth is considered cold, having temperatures permanently below 5 °C. Microorganisms able to live in these environments must be adapted to such conditions and can be classified using the three cardinal growth temperatures (minimal, optimal, and maximal). Some organisms are classified as psychrophiles if they are able to grow between 0 and 20 °C, having an optimal growth temperature below 15 °C. Other organisms, frequently referred to as psychrotolerants or psychrotrophics, are able to grow at low temperatures, but present an optimal and maximal growth temperatures above 15 °C and 20 °C, respectively (Margesin 2009). These microorganisms present metabolic adaptations that allowed them to live under these extreme cold conditions (De Maayer et al. 2014) and therefore they play a key role in the nutrient cycle and mineralization of organic components in cold ecosystems. These activities commonly involve the production of several extracellular enzymes known as cold-adapted or cold-active enzymes. These enzymes are characterized by a higher catalytic efficiency than their mesophilic counterparts at temperatures lower than 20 °C (Collins et al. 2008; Tiquia-Arashiro and Rodrigues 2016).

Bacteria are the most studied extremophile microorganisms (Tiquia and Mormile 2010), whereas the world of fungi and yeasts has been explored in a minor proportion (Margesin and Miteva 2011). Kingdom of fungi represents a very diverse group and, considering their presence in extreme environments, it is one of the main examples of poorly explored microorganisms with biotechnological potential.

Psychrophilic yeasts are thought to be better adapted to low temperatures than bacteria. These eukaryotic microorganisms are heterogeneous in their nutritional abilities and can survive in a broad range of habitats such as deep sea, moist and uneven surfaces, polluted waters, and dry substrates and in the presence of high concentrations of salt and sugar. The survival at low temperatures (below 20 °C) is explained on the basis of the melting points of major membrane fatty acids present in yeasts; furthermore, it was proposed that the psychrophilic yeasts would be able to grow at temperatures as low as −10 °C (Shivaji and Prasad 2009). For this reason, the number of reports describing the isolation of yeasts from cold environments is increasing (Connel et al. 2008; Shivaji and Prasad 2009; Margesin and Feller 2010; Thomas-Hall et al. 2010; Carrasco et al. 2012; Rovati et al. 2013; Zalar and Gunde-Cimerman 2014; Turchetti et al. 2008). Most of these reports are focused on their biotechnological potential and putative industrial uses (Buzzini et al. 2012; Hamid et al. 2014).

26.3 Contamination in Antarctica

Antarctica is considered a remote and harsh place, harboring great and fragile cold wilderness preserved from human disturbance. Unfortunately, some Antarctic environments are no longer pristine and, like other remote regions on earth, Antarctica does not escape from the impact produced by local and global human activities (Bargagli 2008).

The development of activities on Antarctica (research, tourism, fishing, and logistic) during the last 50 years resulted in a sharp increase in human presence on this continent. Until 1980–1990, wastes produced at most Antarctic stations were dumped in landfill sites close to the stations, or alternatively disposed into the sea or ice or burnt in the open air. The Protocol on Environmental Protection to the Antarctic Treaty for protection of the Antarctic environment (ATCM 1991) provided strict guidelines for environmental management and protection and established, among several points, the compromise to clean up abandoned and polluted work sites. Several countries began the assessment of environmental pollution passives at scientific stations and the development cleanup and remediation strategies. The Protocol establishes principles for the planning and carrying out of all activities in Antarctica. However, local impacts caused by human presence are difficult to avoid. The use of fuels (for transportation and energy production), waste incineration, sewage production, and accidental oil spills are the main sources of contaminants in Antarctica (Bargagli 2008). Also, several toxic compounds, such as heavy metals, antibiotics, pesticides, and other persistent pollutants, can be transferred to the Antarctic continent through natural processes of mass transfers in the atmosphere and oceans (Bargagli 2008). Despite the strict guidelines provided by the Protocol of Environmental Protection, all these factors result in the occurrence of several contamination events produced by improper disposal and management of wastes generated at the research stations as well as those produced in the past (Corsolini 2009; Lo Giudice et al. 2013).

26.3.1 Hydrocarbon Contamination

Human activities on the Antarctic continent require, among other relevant factors, power generation. Fossil fuels are the main source to provide this energy. The transportation, storage, and use of such fuels, frequently carried out under hard climate conditions, enhance the risk of spills and contamination events (Martínez Álvarez et al. 2017).

One of the main sources of contamination in Antarctica is represented by fuel spills. Spills occur mainly by human's errors during manipulation as well as due to material obsolescence. In Antarctica, the main oil blends used diesel fuel and JP is composed primarily of C9-C14 aliphatic hydrocarbons.

Once spilled on cold soils, the persistence of hydrocarbons, including light alkanes and aromatics, is high, mainly in the subsurface, where they are not subject to evaporation and photooxidation. This fact indicates that *in situ* rates of hydrocarbon degradation are slow (de Jesús et al. 2015). Therefore, the activity of the indigenous hydrocarbon-degrading microbes is limited, likely by a combination of unfavorable conditions including low temperature and moisture, nutrient limitation, alkalinity, and presence of potentially inhibitory hydrocarbons. Based on these previous observations, cold environments can be more severely affected by contaminants than other environments, even at the same contamination level, because the required cold adaptations make these environments more sensitive (de Jesús et al. 2015).

Hydrocarbon contamination on Antarctic continent is not a large-scale problem. However, some places close to research stations are considered to be chronically contaminated (Martínez Álvarez et al. 2015, 2017). In addition, some reports established that McMurdo Station (USA Antarctic station) presents high levels of hydrocarbon contamination (Kennicutt II et al. 2010). These situations might be the result of lack of regulations or the absence of adequate treatments of the generated waste in the past, when the environmental consciousness was not a common practice. This situation, combined with the hard climate present in the continent as well as other physicochemical conditions as low evaporation, photooxidation, low humidity, and nutritional limitations, led to the persistence of those compounds for decades after the spill.

Bioremediation is a complex process and is affected by several environmental factors. As was mentioned above, low temperature represents one of the main limiting factors for biological processes in polar environments and for this reason bioremediation in Antarctica must be carried out with cold-adapted microorganisms. The low concentrations of available N and P for bacterial growth in most Antarctic soils and the imbalance in the C:N:P ratio after the introduction of large amounts of hydrocarbons are other relevant factors to take into account. Biostimulation provides adequate nutrient levels to increase degradation activity by the natural soil microflora. However, when hydrocarbon-degrading microbes (HDM) are scarce or absent, natural attenuation or biostimulation could not be enough for an efficient pollutant removal. In these cases, inoculation with previously isolated HDM (bioaugmentation) could significantly shorten the bioremediation period, providing catabolic capacity to the soil under treatment. However, bioaugmentation is not just a simple procedure which only involves inoculation of the soil with active HDB and waiting for the disappearance of hydrocarbons. Inoculum survival is not an easy task due to competition with indigenous populations and predation (Ruberto et al. 2010). It also depends on design factors such as microbial selection and inoculum size, which should be considered to avoid the frequently reported bioaugmentation failure (Coppotelli et al. 2008; Ruberto et al. 2009, 2010). In fact, in a previous report, working with chronically contaminated soils from Carlini Station, and using T-RFLP profiles for comparison of bacterial community structure from bioaugmented and non-bioaugmented land plots, we concluded that none of the two consortia used as inoculum survived at detectable levels in the soil (Vázquez et al. 2009).

26.3.2 Heavy Metals

Heavy metals (HM) naturally occur in the earth crust. Nevertheless, human activities have introduced high loads of these elements in the environment, making it difficult to differentiate between natural and anthropogenic contributions. Marine sediments allow a temporal profiling of environmental heavy metal levels, mainly because they act as a major harbor for metals, even though some sediments may also act as a natural source of contaminants (Ribeiro et al. 2011; Tiquia-Arashiro 2018). Furthermore, sediments have high physical-chemical stability and their characteristics usually represent the average condition of the system, often being representative of the average water quality. Soils and rocks are the terrigenous sources of elements to adjacent sediments and can indicate local alert signs (Lu et al. 2012; Oest et al. 2018; Patel et al. 2019).

Possible anthropogenic metal sources for Antarctic sediments are oil-derived fuel contamination, sewage disposal, and paint debris. Several metals (V, Ni, Zn, Cu, Cr, Pb, Ba, among others) are associated with petroleum contamination. Paints are another possible metal source (such as Pb, Cr, and Cu). Also, crushed batteries and scattered rubbish and buildings are sources of contamination by metals (Pb, Zn, and Cu) in soils, while burning fuel results in widespread contamination of lead. Sewage has also been considered an important source of metals for both lower latitudes and Antarctic regions (Statham et al. 2016). Near the McMurdo Station, for example, higher concentrations of metals were observed in sediments around the sewage outfall (Kennicutt II et al. 2010).

The operation of any facility on bare ground in Antarctica must be expected to leave an imprint of a variety of materials and disturbances on the soils. Most Antarctic stations lack adequate facilities for the bulking of disused machinery and building materials. Often, limited logistics did and does not allow its removal, leading to the accumulation of large amounts of metal-made objects near Antarctic stations, occasionally on bare soil. Ageing of building foundations as well as fuel storage tanks are also sources of anthropogenic metal contamination.

Nearly 40 years of human activity at Scott Base resulted in the accumulation of Ag, As, Cd, Cu, Pb, and Zn (Smykla et al. 2018). The way of introduction onto and into the soils and the chemical and physical environment within the soils have resulted in differential movement of the metals, with the mobilization agents being surface and subsurface water flow, redistribution of surface material by wind, and movement of particulates carrying adsorbed heavy metals, in a process affected by freezing and thawing cycles, as well as by permafrost presence (Curtosi et al. 2007). In this sense, Curtosi et al. (2010) referred that in some restricted areas from Potter Cove, there exist evidences of a low but detectable influence of the Carlini scientific station, which is reflected mainly by levels of Pb, Cr, and Cd. For all these reasons, and due to the possibility of biomagnification through trophic chain, heavy metals are contaminants involving a major concern in coastal Antarctica.

26.4 Antarctic Fungi with Bioremediation Abilities

Fungi and its spores can colonize a great number of different substrates and are easily dispersed through air and water, mainly for the action of winds and oceanic currents, respectively. Because of these features and its surprising metabolic flexibility, fungi are present all over the earth.

Antarctica is also inhabited by fungi. Its mycodiversity seems to be enormous and to date remains almost unknown. This is truth not only for the white continent but also for many other regions having different climate conditions. Nevertheless, most microfungi found in Antarctica are ubiquitous instead of being endemic. Some of them are transported from different areas on the planet to the Antarctic continent but are unable to grow in such hard conditions (Ruisi et al. 2007). Others, sometimes called indigenous, were able to adapt, propagate, and reproduce there, completing their life cycles and therefore becoming adapted to the harsh climatic factors of the continent. In some places, like the Antarctic Dry Valleys, fungi have shown specific physiological and morphological adaptations and hence they are considered evolved Antarctic fungi (Ruisi et al. 2007). Many Antarctic fungi are adapted to low temperatures, repeated freeze and thawing cycles, low water availability, osmotic stress, desiccation, low nutrient availability, and high UV radiation. Antarctic microorganisms must face several simultaneous stresses and they adopt different strategies at the same time to address these stresses. Sometimes, single strategies are not specific for a single stress factor and allow these microorganisms to cope with more than one unfavorable condition (Ruisi et al. 2007).

Under this context, Antarctic fungi with bioremediation capacities have been reported at a fast pace during the last 20 years or more, driven mainly by the improvement in the accessibility for researchers to work in Antarctica. Some research groups only take samples and do fungi isolation and bioprospection for remediation abilities back in their research laboratories, hundreds or thousands of kilometers away from Antarctica. Others, taking advantage of having research stations with appropriate facilities, perform several steps of the experimental work in the field, just minutes after sample is taken. In this way, bias associate with sample storing is avoided, resulting in a more representative screening. This is the case of Carlini Station, where adequate scientific facilities and marine as well as terrestrial Antarctic environment (Potter Cove and Peninsula, including the Antarctic Specially Protected Area 132) coincide in the same area.

26.5 Case of Study: Bioremediation Abilities of Yeasts Isolated from 25 de Mayo Island, Antarctica

Antarctica is one of the most suitable sites for the isolation and study of psychrophilic microorganisms. Although permanently exposed to temperatures that rarely exceed the freezing point of water, its geographic location, its difficult access, and the international diplomatic and political treatment of their lands and seas make Antarctica a very little explored region of the world in terms of microbial biodiversity (Fernández et al. 2017).

Considering this information and based on the experience of our research group, the focus of this book chapter is on yeasts isolated from several samples collected near Argentinean Scientific Research Station, Carlini, located on the Potter Cove, 25 de Mayo/King George Island, South Shetland Islands, Antarctica (62°14′18″S, 58°40′00″W) (Fig. 26.1). Thirty-one samples from areas with and without human impact (including soil near fuel storage tanks) were collected. After the isolation procedure, 60 yeasts were evaluated for their bioremediation abilities in cold environments.

26.5.1 Organic Pollutant Bioremediation Ability of Antarctic Yeasts

Yeast assimilation of two chemical compounds, considered as pollutants, was evaluated in liquid media: phenol (2.5 mM) as a model of aromatics and n-hexadecane (1 gL^{-1}) as a model of aliphatic hydrocarbons. Culture media without carbon source were included as controls. After 7 and 14 days, optical density (OD) of the cultures was measured at $\lambda = 600$ nm. Cultures with cell density exceeding by 50% or more those showed by control cultures were considered as positives. Phenol and n-hexadecane assimilation is an important feature to consider the use of these yeasts for bioremediation processes (Fernández et al. 2017). Results obtained with the isolated (60) Antarctica yeasts when evaluated for their bioremediation abilities and their classification based on its growth temperature are presented in Table 26.1.

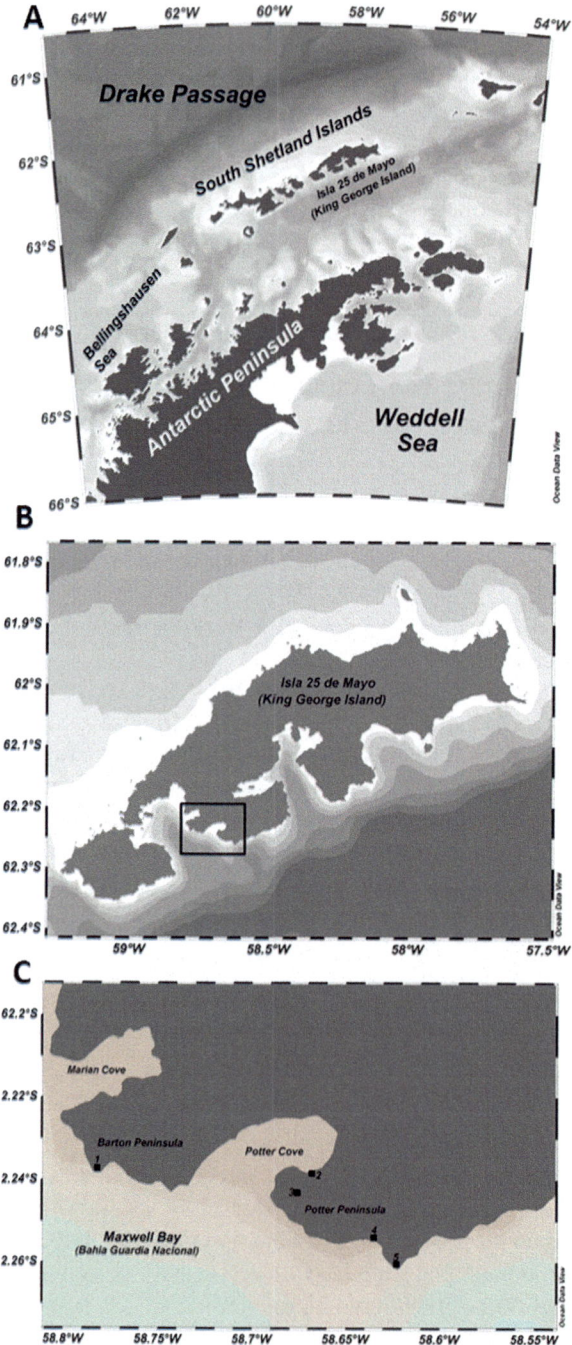

Fig. 26.1 (a, b) The studied area in King George Island/Isla 25 de Mayo, South Shetland Islands, with indication of the sampling site, Potter Peninsula (62°14′18″S, 58°40′00″W). (b) Sampling sites: *1*, nesting penguins in Barton Peninsula; *2*, Carlini Station facilities; *3*, Tres Hermanos Hill; *4*, Elephant Refuge; *5*, Stranger Point

Table 26.1 Identification of selected yeast isolates from Antarctica, assimilation of phenol and *n*-hexadecane as carbon source, tolerance of heavy metal, enzyme activity, and growth temperature

Isolate	Identification	Carbon source assimilation		Metal tolerance			Enzyme activity		Growth temperature classification
		Phenol	n-Hexadecane	$K_2Cr_2O_7$	$CdCl_2$	$CuSO_4$	Lipase	Sterase	
252	*Metschnikowia* sp.	Positive	Positive	+	+	++	–	–	Psychrotolerant
99	*Metschnikowia* sp.	Positive	Positive	–	–	++	–	–	Psychrophilic
235	*Candida smithsonii*	Positive	Positive	+++	++	+++	–	–	Psychrotolerant
243	*Cryptococcus victoriae*	Positive	Positive	–	++	++	–	–	Psychrotolerant
88	*Cryptococcus adeliensis*	Negative	Negative	+	+++	+++	3	3	Psychrotolerant
27	*Cryptococcus gastricus*	Negative	Positive	++	+++	+++	–	–	Psychrotolerant
84	*Cryptococcus gastricus*	Positive	Positive	++	++	+++	–	–	Psychrotolerant
176	*Cryptococcus gilvescens*	Negative	Positive	+	+++	+++	–	3	Psychrophilic
12	*Cryptococcus terricolus*	Positive	Negative	+	++	+++	–	–	Psychrotolerant
2	*Cryptococcus victoriae*	Negative	Positive	+	+++	+++	–	3	Psychrotolerant
6	*Cryptococcus victoriae*	Positive	Positive	–	+	++	–	6	Psychrophilic
9	*Cryptococcus victoriae*	Positive	Positive	–	+++	+++	–	5	Psychrophilic
56	*Cryptococcus victoriae*	Negative	Positive	–	–	–	–	6	Psychrophilic
103	*Cryptococcus victoriae*	Negative	Positive	–	–	–	–	2	Psychrophilic
107	*Cryptococcus victoriae*	Negative	Positive	–	–	–	–	4	Psychrophilic
131	*Cryptococcus victoriae*	Negative	Positive	–	–	–	–	4	Psychrophilic
155	*Cryptococcus victoriae*	Negative	Positive	–	++	++	–	4	Psychrophilic
163	*Cryptococcus victoriae*	Negative	Positive	–	–	–	–	5	Psychrophilic
185	*Cryptococcus victoriae*	Negative	Positive	–	+	+++	–	3	Psychrophilic
251	*Cryptococcus victoriae*	Positive	Positive	–	++	++	1	1	Psychrotolerant
278	*Cryptococcus victoriae*	Negative	Negative	+	+++	+++	–	–	Psychrophilic
291	*Cryptococcus victoriae*	Negative	Positive	–	++	+++	–	–	Psychrotolerant
293	*Cryptococcus victoriae*	Positive	Positive	++	++	+++	–	–	Psychrotolerant

(continued)

Table 26.1 (continued)

Isolate	Identification	Carbon source assimilation		Metal tolerance			Enzyme activity		Growth temperature classification
		Phenol	n-Hexadecane	$K_2Cr_2O_7$	$CdCl_2$	$CuSO_4$	Lipase	Sterase	
322	Cryptococcus victoriae	Negative	Positive	–	–	–	–	–	Psychrophilic
60	Cystobasidium laryngis	Negative	Positive	+	++	+++	–	5	Psychrotolerant
130	Cystobasidium laryngis	Negative	Negative	–	++	+++	–	4	Psychrophilic
217	Cystobasidium laryngis	Positive	Negative	++	++	+++	–	5	Psychrotolerant
309	Cystobasidium laryngis	Negative	Positive	+++	+	++	–	2	Psychrotolerant
318	Cystobasidium laryngis	Negative	Positive	+	++	+++	–	4	Psychrotolerant
341	Cystobasidium laryngis	Negative	Positive	+	+++	+++	–	–	Psychrotolerant
54	Fellomyces penicillatus	Positive	Positive	–	++	++	–	–	Psychrotolerant
28	Guehomyces pullulans	Negative	Positive	+	+++	+++	3	5	Psychrotolerant
37	Guehomyces pullulans	Positive	Negative	++	+++	+++	3	6	Psychrotolerant
53	Guehomyces pullulans	Positive	Negative	+	++	+++	3	6	Psychrotolerant
134	Holtermanniella sp.	Positive	Positive	–	+	+++	–	6	Psychrophilic
273	Leucosporidium creatinivorum	Positive	Positive	+	++	++	1	1	Psychrophilic
275	Leucosporidium creatinivorum	Negative	Positive	–	++	++	–	5	Psychrophilic
276	Leucosporidium creatinivorum	Positive	Positive	+++	+++	+++	–	–	Psychrotolerant
249	Metschnikowia sp.	Positive	Positive	++	+++	+++	–	–	Psychrotolerant
13	Meyerozyma guilliermondii	Positive	Negative	++	++	+++	–	–	Psychrotolerant
1	Mrakia frigida	Negative	Positive	–	–	–	–	–	Psychrotolerant
92	Mrakia frigida	Negative	Positive	–	++	+++	2	–	Psychrophilic
190	Mrakia frigida	Positive	Positive	+++	+++	+++	2	–	Psychrotolerant
259	Mrakia frigida	Negative	Positive	+	++	+++	2	–	Psychrophilic
260	Mrakia frigida	Negative	Positive	+	++	+++	2	–	Psychrophilic
261	Mrakia frigida	Negative	Positive	–	–	+++	2	3	Psychrophilic
263	Mrakia frigida	Negative	Positive	–	++	+++	–	–	Psychrophilic
264	Mrakia frigida	Positive	Positive	++	++	+++	2	3	Psychrophilic

153	*Mrakia frigida*	Negative	Positive	+	++	+++	2	3	Psychrophilic
162	*Mrakia frigida*	Negative	Positive	+	++	+++	1	2	Psychrophilic
296	*Nadsonia commutata*	Positive	Positive	++	++	+++	–	3	Psychrotolerant
307	*Nadsonia commutata*	Negative	Positive	–	–	+++	–	–	Psychrotolerant
314	*Nadsonia commutata*	Positive	Positive	–	–	++	–	4	Psychrotolerant
159	*Phenoliferia glacialis*	Negative	Negative	–	–	–	1	3	Psychrotolerant
166	*Phenoliferia glacialis*	Positive	Positive	+	–	++	1	–	Psychrophilic
197	*Phenoliferia glacialis*	Positive	Positive	+	++	++	–	–	Psychrotolerant
265	*Phenoliferia glacialis*	Positive	Negative	+	+++	+++	1	4	Psychrophilic
8	*Pichia caribbica*	Negative	Positive	+	++	++++	3	5	Psychrotolerant
59	*Pichia caribbica*	Positive	Positive	++	++	+++	1	2	Psychrotolerant
161	*Pichia caribbica*	Negative	Positive	++	++	+++	3	5	Psychrotolerant
168	*Pichia caribbica*	Negative	Positive	++	++	+++	–	2	Psychrotolerant
171	*Pichia caribbica*	Positive	Positive	+++	++	++	3	6	Psychrotolerant
210	*Pichia caribbica*	Positive	Positive	+	–	+++	–	–	Psychrotolerant
128	*Protomyces inouyei*	Negative	Positive	–	++	++	1	1	Psychrophilic
257	*Rhodotorula fragaria*	Positive	Positive	++	++	+++	1	–	Psychrotolerant
248	*Rhodotorula mucilaginosa*	Positive	Positive	+	++	++	–	5	Psychrotolerant
279	*Rhodotorula mucilaginosa*	Positive	Positive	+	+++	+++	2	2	Psychrotolerant
172	*Rhodotorula muscorum*	Positive	Positive	+	++	+++	1	–	Psychrotolerant
174	*Rhodotorula muscorum*	Positive	Positive	+++	++	++	2	3	Psychrotolerant
97	*Rhodotorula sp.*	Negative	Positive	+	++	+++	1	–	Psychrophilic

Special attention was paid to those yeasts able to assimilate phenol as it is a common constituent of wastewater from several industries and, for this reason, is considered as a pollutant of major concern. In our evaluation, over 48% of the yeasts were able to assimilate phenol at an initial concentration of 2.5 mM. These yeasts belonged to both asco- and basidiomycetous genera. Some of the detected genera were *Metschnikowia, Candida, Cryptococcus, Fellomyces, Guehomyces, Leucosporidum, Mrakia, Nadsonia, Pichia*, and *Rhodotorula*. Due to their toxicity to microorganisms, phenolic compounds can cause the breakdown of wastewater treatment plants by inhibition of microbial growth (Basha et al. 2010). For this reason, these phenol-degrading yeasts represent a valuable tool as potential cold-tolerant components of the microbial community for wastewater treatment plants able to deal with phenolic compounds (Viswanath et al. 2014).

In the case of n-hexadecane assimilation, 85% of the yeasts presented this ability. These yeasts were isolated not only from hydrocarbon-contaminated environments but also from pristine areas (Fernández et al. 2017), evidencing the ubiquity of these cold-adapted hydrocarbon-degrading microorganisms. Other authors reported isolation of microorganisms able to efficiently degrade crude oil-derived hydrocarbons (Das and Chandran 2011; Hassanshahian et al. 2010) and phenol (Bonfá et al. 2013) from uncontaminated environments. This ability could be related with the catabolism of natural hydrocarbons produced by different organisms present in non-contaminated sites (Schirmer et al. 2010). However, Aislabie et al. (2001) working with Scott Base and Marble Point soils detected culturable yeasts only in oil-contaminated soils but not in pristine control soils. They attributed the significant enhancement in numbers of culturable yeasts and filamentous fungi in oil-contaminated cold soils to the important role of fungi in the degradation of hydrocarbons or their metabolites. Their population would increase due to the availability of C provided by the contaminant.

The aerobic biodegradation at low temperatures of many petroleum hydrocarbon components, including n-alkanes and mono- and polycyclic aromatic hydrocarbons (PAHs), has been reported for Arctic, Alpine, and Antarctic environments (Si-Zhong et al. 2009; Yang et al. 2009). A wide variety of bacteria, fungi, and algae can metabolize aliphatic and aromatic hydrocarbons (Alexander 1999). Filamentous fungi are mainly known for their potential to degrade PAHs (Haritash and Kaushik 2009). There is, however, little information about the hydrocarbon-degradative potential of yeasts, although these microeukaryotic organisms could be a useful and efficient tool for the development of processes for the bioremediation of fuel-spilled soils from cold regions.

In our work, one of the isolates showed a wide catabolic capability. Isolate number 171 (*P. caribbica*)) was able to grow using several hydrocarbons: undecane (nC11), dodecane (nC12), tridecane (nC13), tetradecane (nC14), and Antarctic diesel fuel at 15 °C, under aerobic conditions in liquid culture. It should be noted that a negative relationship between the carbon chain length and yeast growth was observed in this case (Martorell et al. 2017). This yeast species was previously reported as a biosurfactant producer (Joshi-Navare et al. 2014), which could be related with its ability to assimilate n-alkanes and gasoil as carbon sources. Such ability could be advantageous for a better solubilization and an enhanced bioavailability of hydrocarbons in soils and waters.

26.5.2 Metal Tolerance by Antarctic Yeasts

Divalent copper and cadmium (Cu(II) and Cd(II)) and hexavalent chromium (Cr(VI)) yeast tolerance was separately evaluated in agarized YM medium containing 1 mM (final concentration) of each metal ion. Isolates were inoculated, incubated at 15 °C, and checked for growth for up to 14 days. Plates without metal ions were also inoculated as controls (Fernández et al. 2013).

The analysis of data from heavy metal tolerance assays showed that 61%, 80%, and 86% of the isolates were tolerant to Cr(VI), Cd(II), and Cu(II), respectively, but a half of the yeasts tolerated all of them. In addition, 11% could be classified as sensitive, showing no growth after 14 days in the presence of any of the metals under study.

Heavy metal-contaminated soils and water are one of the first steps for the accumulation of these harmful compounds in living organisms through the food chain, causing a negative effect on physiological activities of plants, animals, and humans (Suciu et al. 2008; Bowman et al. 2018). It is important to consider that a significant fraction of industrial plants generating phenol-rich effluents also discharge heavy metals as associate pollutant, being a complex and difficult mixture to deal with. This combination frequently results in the inhibition of growth of most phenol-degrading microorganisms used for the associated wastewater treatments (Thavamani et al. 2012; Wong et al. 2015). Thus, much attention should be paid to the phenol removal performance of these microorganisms in media with the presence of heavy metals. It was found from previous study that bacterial strains *Pseudomonas rhodesiae* and *Bacillus subtilis* could remove phenol and survive in heavy metal-polluted environments (Satchanska et al. 2015). Regarding fungi, several reports mentioned their resistance to metal ions (Fernández et al. 2013). Our results showed that almost 34% of the isolated yeasts exhibited some degree of tolerance to the three studied metals and can use phenol as carbon source. These strains, mainly those showing high levels of metal tolerance, are adequate to be used for the low-temperature treatment of effluents containing phenol and high levels of metal ions.

26.5.3 Enzymes from Antarctic Fungi Related to Bioremediation

Lipase and esterase activities were tested on the isolates growing on solid media at 15 °C. For this propose, activity was quantified as the halo diameter (of either coloration or decoloration) around the colony (Martorell et al. 2017).

Oligotrophic microorganisms are usually related to the ability to degrade a broad spectrum of substrates, while copiotrophic microorganisms are related to the efficient degradation of easily accessible substrates (Rovati et al. 2013). These concepts should be a guideline for the development of the isolation scheme and the screening process toward selecting the most promising yeasts for low-temperature biotechnological process. In a previous work, Martorell et al. (2017) reported several isolates showing multiple extracellular enzymatic activities. Those yeasts were obtained from sites with high organic matter content, either as a consequence of the human impact (soil that suffered fuel spills from the storage tanks throughout the years) or from natural origin (complex substrates as those present in soils around lichens and *Deschampsia* spp. or mud near creeks).

In this chapter only lipase and esterase producers are discussed, as these activities are related to hydrocarbon degradation (Margesin and Feller 2010) and for this reason constitute a relevant catabolic ability with potential application in bioremediation of hydrocarbon-polluted matrices in cold environments. The analysis of the locations where the lipase/esterase-positive isolates were obtained showed that several producing yeasts (*C. Adeliensis*, *G. pullulans*, *P. caribbica*, and *Ph. glacialis*) came from soils near the diesel fuel storage tanks. Although microorganisms able to efficiently degrade oil hydrocarbons have been isolated from uncontaminated environments (Si-Zhong et al. 2009), their numbers (including fungi) significantly increase in oil-contaminated soils. In this sense, Aislabie et al. (2001) attributed the significant enhancement in numbers of culturable yeasts and filamentous fungi in oil-contaminated cold soils to the important role of these microorganisms in the degradation of hydrocarbons or their metabolites, where they can take advantage of the additional C source provided by the carbonated pollutant.

Among isolates, one ascomycetous yeast was selected (*P. caribbica*), as it showed all the evaluated enzymatic activities. This strain was isolated from soil near fuel storage tanks, which contained a high amount of hydrocarbons, accumulated as a consequence of the chronic leakage of gas/oil throughout the years. This chronic presence of hydrocarbons represented a high selective pressure, determining a soil microbiota with dominance of microorganisms able to tolerate and catabolize these recalcitrant organic compounds. This observation was previously reported for the studied area (Mac Cormack et al. 2011; Martínez Álvarez et al. 2015, 2017; Ruberto et al. 2009, 2010). As was mentioned above, this isolate can assimilate several aliphatic hydrocarbons and also Antarctic diesel fuel. This yeast was selected for further studies based on their biotechnological potential, primarily for hydrocarbon bioremediation in cold environment.

26.6   Conclusions

Antarctic yeasts tested for pollutant assimilation, heavy metal tolerance, and some extracellular enzymatic activities have been reported in the past years. They belong to widely reported, cold-adapted yeast taxa, most of them included into oligotrophic, slow-growing, and metabolically diverse basidiomycetous genera. The reason for the prevalence of basidiomycetous yeast in Antarctic samples remains unclear, but could be related to soil and water oligotrophy, as well as to the isolation scheme employed. As was previously emphasized, oligotrophic microorganisms are usually related to the ability to degrade a broad spectrum of substrates, while copiotrophic microorganisms are related to the efficient degradation of easily accessible substrates. Despite the genus of yeasts isolated from cold environments, research in the field of cold-adapted yeasts from Antarctica is relatively young. It is generally accepted that information regarding cold-adapted yeasts will have a continuous increase, especially with the development of new microbiological and molecular methodologies. The tolerance to heavy metals of the phenol-degrading cold-adapted yeasts, and the production of bioremediation-related enzymes, as lipase and esterase, evidenced that the yeasts selected might be promising in treating some kinds of phenol-polluted industrial wastewater containing heavy metals, such as effluents from petroleum refineries in cold environments. Further studies on cold-tolerant yeasts isolated from Antarctica must be done in order to provide additional information for its use in bioremediation processes at low temperatures and also to infer their possible ecological role under such extreme conditions.

26.7 Future Perspectives

The bioremediation processes in the Antarctic continent are, as in other places, site and contaminant specific. They also perform better under aerobic conditions. Beyond yeast isolation and characterization, there is a need for prior studies of contaminated sites that can be done ex situ (e.g., soil analysis and preliminary studies through microcosms). Nevertheless, next step must involve field experiments performed in situ, to ensure the accuracy of the results as well as applicability of this technology. On the other hand, an ecological approach must be included to assess the possible interactions between inoculated yeast and indigenous microbiota.

The studies conducted to date reveal that despite our knowledge of the fungal microbial strains that can degrade contaminants in Antarctic soils, studies regarding its actual use are just a few. This is, to our knowledge, the next step for Antarctic soil bioremediation using psychrotolerant metal-tolerant yeast.

References

Aislabie J, Fraser R, Duncan S, Farrell RL (2001) Effects of oil spills on microbial heterotrophs in Antarctic soils. Polar Biol 24(5):308–313

ATCM (1991) Antarctic treaty consultative meeting. http://www.ats.aq/e/ep.htm

Alexander M (1999) Biodegradation and bioremediation. Gulf Professional Publishing, Houston Texas, USA, p 503

Bargagli R (2008) Environmental contamination in Antarctic ecosystems. Sci Total Environ 400(1-3):212–226

Basha KM, Rajendran A, Thangavelu V (2010) Recent advances in the biodegradation of phenol: a review. Asian J Exp Biol Sci 1(2):219–234

Bonfá MRL, Grossman MJ, Piubeli F, Mellado E, Durrant LR (2013) Phenol degradation by halophilic bacteria isolated from hypersaline environments. Biodegradation 24(5):699–709

Bowman N, Patel P, Sanchez S, Xu W, Alsaffar A, Tiquia-Arashiro SM (2018) Lead-Resistant bacteria from Saint Clair River sediments and Pb removal in aqueous solutions. Appl Microbiol Biotechnol 102:2391–2398

Buzzini P, Branda E, Goretti M, Turchetti B (2012) Psychrophilic yeasts from worldwide glacial habitats: diversity, adaptation strategies and biotechnological potential. FEMS Microbiol Ecol 82(2):217–241

Carrasco M, Rozas JM, Barahona S, Alcaíno J, Cifuentes V, Baeza M (2012) Diversity and extracellular enzymatic activities of yeasts isolated from King George Island, the sub-Antarctic region. BMC Microbiol 12:251

Collins T, Roulling F, Piette F, Marx JC, Feller G, Gerday C, D'Amico S (2008) Fundamentals of cold-adapted enzymes. In: Margesin R et al (eds) Psychrophiles: from biodiversity to biotechnology. Springer, Berlin, Heidelberg, pp 211–227

Connel L, Redman R, Craig S, Scorzetti G, Iszard M, Rodriguez R (2008) Diversity of soil yeasts isolated from South Victoria Land, Antarctica. Microb Ecol 56:448–459

Coppotelli BM, Ibarrolaza A, Del Panno MT, Morelli IS (2008) Effects of the inoculant strain *Sphingomonas paucimobilis* 20006FA on soil bacterial community and biodegradation in phenanthrene-contaminated soil. Microb Ecol 55:173–183

Corsolini S (2009) Industrial contaminants in Antarctic biota. J Chromatogr A 1216:598–612

Curtosi A, Pelletier E, Vodopivez CL, Mac Cormack WP (2007) Distribution pattern of PAHs in soil and surface marine sediments near Jubany Station (Antarctica). Possible role of permafrost as a low-permeability barrier. Sci Total Environ 383:193–204

Curtosi A, Pelletier E, Vodopivez C, St Louis R, Mac Cormack WP (2010) Presence and distribution of persistent toxic substances in sediments and marine organisms of Potter Cove, Antarctica. Arch Environ Contam Toxicol 59(4):582–592

de Jesús HE, Peixoto RS, Rosado AS (2015) Bioremediation in Antarctic soils. J Pet Environ Biotechnol 6(248):2

Das N, Chandran P (2011) Microbial degradation of petroleum hydrocarbon contaminants: an overview. Biotechnol Res Int 2011:13

De Maayer P, Anderson D, Cary C, Cowan DA (2014) Some like it cold: understanding the survival strategies of psychrophiles. EMBO Rep 15:508–517

Fernández PM, Cabral ME, Delgado OD, Fariña JI, Figueroa LIC (2013) Textile dye polluted waters as an unusual source for selecting chromate-reducing yeasts through Cr(VI)-enriched microcosms. Int Biodeter Biodegr 79:28–35

Fernández PM, Martorell MM, Blaser MG, Ruberto LAM, de Figueroa LIC, Mac Cormack WP (2017) Phenol degradation and heavy metal tolerance of Antarctic yeasts. Extremophiles 21(3):445–457

Hamid B, Rana RS, Chauhan D, Singh P, Mohiddin FA, Sahay S, Abidi I (2014) Psychrophilic yeasts and their biotechnological applications-a review. Afr J Biotechnol 13(22):2188–2197

Haritash AK, Kaushik CP (2009) Biodegradation aspects of polycyclic aromatic hydrocarbons (PAHs): a review. J Hazard Mater 169(1-3):1–15

Hassanshahian M, Emtiazi G, Kermanshahi RK, Cappello S (2010) Comparison of oil degrading microbial communities in sediments from the Persian Gulf and Caspian Sea. Soil Sediment Contam 19(3):277–291

Joshi-Navare K, Singh PK, Prabhune AA (2014) New yeast isolate *Pichia caribbica* synthesizes xylolipid biosurfactant with enhanced functionality. Eur J Lipid Sci Technol 116(8):1070–1079

Kennicutt MC II, Klein A, Montagna P, Sweet S, Wade T, Palmer T, Denoux G (2010) Temporal and spatial patterns of anthropogenic disturbance at McMurdo Station, Antarctica. Environ Res Lett 5(3):034010

Lo Giudice A, Casella P, Bruni V, Michaud L (2013) Response of bacterial isolates from Antarctic shallow sediments towards heavy metals, antibiotics and polychlorinated biphenyls. Ecotoxicology 22:240–250

Lu Z, Cai M, Wang J, Yang H, He J (2012) Baseline values for metals in soils on Fildes Peninsula, King George Island, Antarctica: the extent of anthropogenic pollution. Environ Monit Assess 184(11):7013–7021

Mac Cormack WP, Ruberto LAM, Curtosi A, Vodopivez C (2011) Human impacts in the Antarctic coastal zones: the case study of hydrocarbons contamination at Potter Cove, South Shetland Islands. In: Scott Coffen-Smout (co-ed) Ocean year book, vol 25. Brill/Martinus Nijhoff Publishers, Dalhousie University, Nova Scotia, p. 141-170.

Margesin R (2009) Effect of temperature on growth parameters of psychrophilic bacteria and yeasts. Extremophiles 13(2):257–262

Margesin R (2014) Bioremediation and biodegradation of hydrocarbons by cold-adapted yeasts. In: Cold-adapted yeasts. Springer, Berlin, Heidelberg, pp 465–480

Margesin R, Feller G (2010) Biotechnological applications of psychrophiles. Environ Technol 31:835–844

Margesin R, Miteva V (2011) Diversity and ecology of psychrophilic microorganisms. Res Microbiol 162(3):346–361

Martínez Álvarez LM, Balbo AL, Mac Cormack WP, Ruberto LAM (2015) Bioremediation of a petroleum hydrocarbon-contaminated Antarctic soil: optimization of a biostimulation strategy using response-surface methodology (RSM). Cold Reg Sci Technol 119:61–67

Martínez Álvarez LM, Ruberto LAM, Balbo AL, Mac Cormack WP (2017) Bioremediation of hydrocarbon-contaminated soils in cold regions: development of a pre-optimized biostimulation biopile-scale field assay in Antarctica. Sci Total Environ 590:194–203

Martorell MM, Ruberto LAM, Fernández PM, Figueroa LIC, Mac Cormack WP (2017) Bioprospection of cold-adapted yeasts with biotechnological potential from Antarctica. J Basic Microbiol 57(6):504–516

Oest A, Alsaffar A, Fenner M, Azzopardi D, Tiquia-Arashiro SM (2018) Patterns of change in metabolic capabilities of sediment microbial communities along river and lake ecosystems. J Int Microbiol 2018:6234931. https://doi.org/10.1155/2018/6234931

Patel D, Gismondi R, Ali A, Tiquia-Arashiro SM (2019) Applicability of API ZYM to capture seasonal and spatial variabilities in lake and river sediments. Environ Technol. https://doi.org/10.1080/09593330

Ribeiro AP, Figueira RC, Martins CC, Silva CR, França EJ, Bícego MC, Montone RC (2011) Arsenic and trace metal contents in sediment profiles from the Admiralty Bay, King George Island, Antarctica. Mar Pollut Bull 62(1):192–196

Rovati JI, Pajot HF, Ruberto LAM, Mac Cormack WP, Figueroa LIC (2013) Polyphenolic substrates and dyes degradation by yeasts from 25 de Mayo/King George Island (Antarctica). Yeast 30(11):459–470

Ruberto L, Dias R, Lo Balbo A, Vazquez SC, Hernandez EA, Mac Cormack WP (2009) Influence of nutrients addition and bioaugmentation on the hydrocarbon biodegradation of a chronically contaminated Antarctic soil. J Appl Microbiol 106(4):1101–1110

Ruberto L, Vazquez SC, Dias RL, Hernández EA, Coria SH, Levin G, Mac Cormack WP (2010) Small-scale studies towards a rational use of bioaugmentation in an Antarctic hydrocarbon-contaminated soil. Antarct Sci 22(5):463–469

Ruisi S, Barreca D, Selbmann L, Zucconi L (2007) Fungi in Antarctica. Rev Environ Sci Biotechnol 6(1–3):127–141

Satchanska G, Topalova Y, Dimkov R, Groudeva V, Petrov P, Tsvetanov C, Selenska-Pobell S, Golovinsky E (2015) Phenol degradation by environmental bacteria entrapped in cryogels. Biotechnol Biotechnol Equip 29:514–521

Schirmer A, Rude MA, Li X, Popova E, Del Cardayre SB (2010) Microbial biosynthesis of alkanes. Science 329(5991):559–562

Shivaji S, Prasad GS (2009) Antarctic yeasts: biodiversity and potential applications. In: Satyanarayana T, Kunze G (eds) Yeast biotechnology: diversity and applications. Springer, Dordrecht, pp 3–18

Si-Zhong Y, Hui-Jun J, Zhi W, Rui-Xia HE, Yan-Jun JI, Xiu-Mei LI, Shao-Peng YU (2009) Bioremediation of oil spills in cold environments: a review. Pedosphere 19(3):371–381

Smykla J, Szarek-Gwiazda E, Drewnik M, Knap W, Emslie SD (2018) Natural variability of major and trace elements in non-ornithogenic Gelisols at Edmonson Point, northern Victoria Land, Antarctica. Pol Polar Res 39(1):19–50

Statham TM, Stark SC, Snape I, Stevens GW, Mumford KA (2016) A permeable reactive barrier (PRB) media sequence for the remediation of heavy metal and hydrocarbon contaminated water: a field assessment at Casey Station, Antarctica. Chemosphere 147:368–375

Suciu I, Cosma C, Todică M, Bolboacă SD, Jäntschi L (2008) Analysis of soil heavy metal pollution and pattern in central Transylvania. Int J Mol Sci 9:434

Thavamani P, Megharaj M, Naidu R (2012) Bioremediation of high molecular weight polyaromatic hydrocarbons co-contaminated with metals in liquid and soil slurries by metal tolerant PAHs degrading bacterial consortium. Biodegradation 23:823–835

Thomas-Hall SR, Turchetti B, Buzzini P, Branda E, Boekhout T, Theelen B, Watson K (2010) Cold-adapted yeasts from Antarctica and the Italian Alps—description of three novel species: *Mrakiarobertii* sp. nov., *Mrakiablollopis* sp. nov. and *Mrakiellaniccombsii* sp. nov. Extremophiles 14:47–59

Tiquia SM, Mormile M (2010) Extremophiles–a source of innovation for industrial and environmental applications. Environ Technol 31(8–9):823

Tiquia-Arashiro SM (2018) Lead absorption mechanisms in bacteria as strategies for lead bioremediation. Appl Microbiol Biotechnol 102:5437–5444

Tiquia-Arashiro SM, Rodrigues D (2016) Thermophiles and psychrophiles in nanotechnology. In: Extremophiles: applications in nanotechnology. Springer International Publishing, Cham, Switzerland, pp 89–127

Turchetti B, Buzzini P, Goretti M, Branda E, Diolaiuti G, D'Agata C, Vaughan-Martini A (2008) Psychrophilic yeasts in glacial environments of Alpine glaciers. FEMS Microbiol Ecol 63(1):73–83

Tyagi M, da Fonseca MMR, de Carvalho CC (2011) Bioaugmentation and biostimulation strategies to improve the effectiveness of bioremediation processes. Biodegradation 22(2):231–241

Vázquez S, Nogales B, Ruberto L, Hernández E, Christie-Oleza J, Balbo AL, Mac Cormack WP (2009) Bacterial community dynamics during bioremediation of diesel oil-contaminated Antarctic soil. Microb Ecol 57(4):598

Viswanath B, Rajesh B, Janardhan A, Kumar AP, Narasimha G (2014) Fungal laccases and their applications in bioremediation. Enzyme Res 2014:21

Wong KK, Quilty B, Hamzah A, Surif S (2015) Phenol biodegradation and metal removal by a mixed bacterial consortium. Biorem J 19:104–112

Yang S, Jin H, Wei Z, He R (2009) Bioremediation of oil spills in cold environments: a review. Pedosphere 19(3):371–381

Zalar P, Gunde-Cimerman N (2014) Cold-adapted yeasts in Arctic habitats. In: Buzzini P, Margesin R (eds) Cold-adapted yeasts. Springer, Berlin, Heidelberg, pp 49–74

Chapter 27
Haloalkaliphilic Fungi and Their Roles in the Treatment of Saline-Alkali Soil

Yi Wei and Shi-Hong Zhang (ID)

27.1 Introduction

The main source of all salts in the soil is the primary minerals in the exposed layer of the earth's crust. During the soil-forming process which involves chemical, physical, and biological processes, the salt constituents are gradually released and made soluble. The released salts are transported away from their source of origin through surface or groundwater streams. The salts in the groundwater stream are gradually concentrated as the water with dissolved salts moves from the more humid to arid and semiarid areas, which is the primary cause of the soil salinization.

Soil salinization leads to serious environmental problems on a global scale (Wang et al. 2003; Yadav et al. 2011; Liang et al. 2015). The problems of soil salinity are most widespread in the arid and semiarid regions but salt-affected soils also occur extensively in subhumid and humid climates, particularly in the coastal regions where the ingress of seawater through estuaries and rivers and through groundwater causes large-scale salinization. Soil salinity is also a serious problem in areas where groundwater of high salt content is used for irrigation. The most serious salinity problems are being faced in the irrigated arid and semiarid regions of the world and it is in these very regions that irrigation is essential to increase agricultural production to satisfy food requirements.

Two main groups of the salt-affected soils have been distinguished (Szabolcs 1994): (1) Saline soils—soils containing sufficient neutral soluble salts to adversely affect the growth of most crop plants: The soluble salts are chiefly sodium chloride and sodium sulfate. (2) Saline-alkali—soils containing sodium salts capable of alkaline hydrolysis, mainly Na_2CO_3 and $NaHCO_3$: these soils have also been termed as sodic-, alkali-, or soda-affected soils once in a while. As a matter of fact, the

Y. Wei · S.-H. Zhang (✉)
College of Plant Sciences, Jilin University, Changchun, People's Republic of China
e-mail: zhang_sh@jlu.edu.cn

© Springer Nature Switzerland AG 2019
S. M. Tiquia-Arashiro, M. Grube (eds.), *Fungi in Extreme Environments: Ecological Role and Biotechnological Significance*,
https://doi.org/10.1007/978-3-030-19030-9_27

various sodium salts in nature do not occur absolutely separately, but in most cases either the neutral salts or the ones capable of alkaline hydrolysis exercise a dominant role on the soil-forming processes and therefore in determining soil properties. In most agricultural cases, the jeopardizing of soda-affected soil is more serious than that of other saline soils. It is the accumulation of solutes, primarily Na_2CO_3 and $NaHCO_3$, that induces primary soil alkalization: soda saline-alkali soil leads to many negative effects on soil organic matter decomposition and uptake of available nutrients (Rietz and Haynes 2003; Karlen et al. 2008), which subsequently affect plant survival, health, and development (Rady 2011). Therefore, accumulation of excess salts in the root zone results in a partial or complete loss of soil productivity.

Soda saline-alkali soils occur within the boundaries of at least 75 countries (Szabolcs 1994), and the severity of this issue has increased steadily in several major agricultural areas around the world (Ghassemi et al. 1995). The well-known typical saline soils are, respectively, located in Vitoria in Australia, California in the United States, Mexico City in Mexico, and Baicheng city in China (Wang et al. 2009). In Victoria, sodic soils are estimated to occupy at least 13.4 Mha, representing at least 73% of Victoria's agricultural land, and the largest sodicity class is "alkaline sodic," dominated by a diverse range of soils (Ford et al. 1993). The soil of the former Lake Texcoco in Mexico is a unique extreme environment (called a soda desert) located near one of the biggest cities in the world, Mexico City. Large parts are saline-alkaline with pH more than 10 and electrolytic conductivity (EC) more than 150 dS m^{-1} (Dendooven et al. 2010). Nowhere in China is the issue more serious than in the Songnen Plain of northeastern China (Fig. 27.1). Soil alkali is the major ecological gradient in the Songnen Plain, as well as the primary factor limiting its food security (Gao et al. 1996). Therefore, effective strategies to remediate soda saline-alkaline soil are urgently needed.

Fig. 27.1 Songyuan soda saline-alkali land in Zhenlai County, Baicheng city, Jilin province, China (the aerial photography was taken by an unmanned plane at 150 m in the air in 2016). The white snow-like crusts on the land are soda salt that has returned to the soil surface (in a soil level extending from 0 to 20 cm depth: sodium salt = 18 g/kg, pH = 10.1; measured by Dr. Shi Yang, 2016)

Physical tillage operations, chemical amendments, leaching with water, and plant-associated phytoremediation have been utilized to attempt to ameliorate soil salinity (Qadir et al. 2007). On account of the significant ecological, environmental, and economic effects of the former three techniques, phytoremediation is widely considered to be the best method for ameliorating soil salinity (Ilyas et al. 1993; Ghaly 2002; Nouri et al. 2017). Thus far, the primary factors influencing the success of phytoremediation have been the selection and application of appropriate plants, such as salt-resistant or -tolerant species, and their cropping sequence. At this point, the upper limit of plant resistance to salt restricts the application range of soil phytoremediation.

Salt-affected soils generally exhibit poor structural stability due to low organic matter content; therefore, ecosystems in severely saline soils are rather simple and fragile. Plant species are extremely scarce in severely saline soil, while microbes, including fungi, are rare. An alternative technique for saline soil remediation, which can be regarded as an auxiliary measure for phytoremediation, is the application of organic matter conditioners, which can both ameliorate salinity and increase the fertility of saline soils (Melero et al. 2007). Some studies have indicated that the structural stability of soil can be improved by the addition of organic materials (Tiquia et al. 2002; Tiquia 2003; Tejada et al. 2006; Wang et al. 2014; Oo et al. 2015). Above all, the addition of maize straw to saline soil can decrease the severity of the negative effects of salinity on mineralization and the microbial community in the soil (Wichern et al. 2006).

Soil microorganisms generally have the ability to adapt to or tolerate salinity; and examples of microbes thriving in ponds with very high salt concentrations demonstrate the evolutionary potential of microorganisms (Casamayor et al. 2002). The biodiversity of microorganisms in soda environments has indicated that abundant bacterial communities, which also act as primary producers, are usually dominated by cyanobacteria species (Antony et al. 2013). In addition, the N-fixing cyanobacterium, *Anabaena torulosa*, has been applied in remediating soil salinity during crop growth (Apte and Thomas 1997). However, fungi tend to be sensitive to salt stress, as indicated by decreasing ergosterol content in soil as its salt content is increased (Sardinha et al. 2003). In addition, it has been reported that long-term salt stress reduces fungal diversity (Bruggen et al. 2000). In general, the negative impact of elevated salinity on fungi is stronger than its effect on bacteria. The negative effects of salt stress on soil fungi reduce the microbial biomass and microbial activity of the soil and impair turnover of organic matter, which creates a vicious cycle that reduces soil fertility and eventually produces soil incapable of supporting crops. Obviously, in order to remediate the saline-alkali soil, our primary task must be to increase the beneficial fungi that can survive in the saline-alkali land.

Fortunately, within the last few decades, a series of halophilic and alkaliphilic fungi capable of living in highly saline and alkaline environments (or both) have been identified. This chapter is focused on the isolation and characterizations of extreme haloalkaliphilic fungi, and their roles in saline-alkali soil mycoremediation. In addition, we highlight the abiotic stress resistance genes and cellulase genes in extremophilic fungi, and application strategies for anti-abiostress and stable cellulose degradation genetic engineering are discussed.

27.2 Haloalkaliphilic Fungi and Their Biological Characteristics

Halophilic fungi require salt concentrations of at least 0.3 M (sodium salt, e.g., NaCl) to grow optimally, and they are capable of thriving in high-salt environments. Halotolerant fungi, however, do not necessarily require certain concentrations of salt, although they were often found in saline areas. To halotolerant fungi, salinity can directly affect sporulation and growth of fungi: at higher salinities (>5%) there tends to be increased sporulation with more chlamydospores observed, an inhibition of conidiogenesis, and fewer hyphae (Mulder et al. 1989; Mahdy et al. 1996; Mulder and El-Hendawy 1999; Mandeel 2006). On the other hand, halophilic fungi do not always have to be in saline habitats; thus there is no need to make a strict distinction between halotolerant and halophilic fungi (Arakaki et al. 2013). In this chapter, we consider halophilic fungi as a general designation.

Alkaliphilic fungi are a class of extremophilic microbes that are capable of survival in alkaline (pH roughly 8.5–11.0) environments and grow optimally even at a pH of approximately 10. Halophilic fungi growing in alkaline environments that are adapted to high pH and high concentrations of sodium ions are described as haloalkaliphilic, rather than merely halophilic or alkaliphilic. Soda-affected soils form as a result of sodium carbonate accumulation. Water evaporation reinforces the process of soda accumulation. Thus, soda soils are usually affected by both saline and alkaline as double-abiotic factors. Therefore, halophilic fungi inhabiting soda soils are most likely alkaliphilic fungi (Gunde-Cimerman et al. 2009; Grum-Grzhimaylo et al. 2016).

Most halophilic fungi live in marine aquatic bodies, seashore, and inland terrestrial soils with high salt concentrations, such as the Dead Sea, the Antarctic Ocean, and the Great Salt Plains, and a large number of studies on biodiversity and physiology have focused on the characterization of halophilic fungi present in the saline and hypersaline ecosystems, among which species of *Ascomycetes*, as well as some *Basidiomycetes*, have been described in detail (Gunde-Cimerman et al. 2000; Butinar et al. 2005a, b; Zalar et al. 2005; Evans et al. 2013; Gunde-Cimerman and Zalar 2014; Zajc et al. 2014a, b; Tiquia-Arashiro and Rodrigues 2016a; Gonçalves et al. 2017). Hypersaline fungal communities are dominated by *Aspergillus* and *Penicillium* species, with melanized dematiaceous forms commonly observed in inland lands (Moubasher et al. 1990; Grum-Grzhimaylo et al. 2016; Martinelli et al. 2017), similar to the communities observed in marine environments (Buchalo et al. 1998, 2000; Gunde-Cimerman et al. 2000; Butinar et al. 2005a, b; Kis-Papo et al. 2003, 2014; Gunde-Cimerman and Zalar 2014).

The Dead Sea, a typical high-salt habitat for microorganisms, contains 340 g/L of dissolved salt; a variety of filamentous fungi have been isolated from the Dead Sea by the Nevo group. *Gymnascella marismortui* is a remarkable salt-tolerant fungus that has been isolated from the surface water down to a depth of 300 m in the Dead Sea (Buchalo et al. 1998). *G. marismortui* grows optimally at NaCl concentrations between 0.5 and 2 M (Buchalo et al. 1998, 2000), suggesting that it is adapted to high-salt conditions and requires high salt concentrations. Among 476 fungal isolates from the Dead Sea, *Aspergillus terreus*, *Aspergillus sydowii*, *Aspergillus versicolor*, *Eurotium herbariorum*, *Penicillium westlingii*, *Cladosporium cladosporoides*, and *Cladosporium sphaerospermum* were isolated consistently and probably form the stable core of the fungal community (Kis-Papo et al. 2003, 2014), and approximately 43% of fungal isolates from the Dead Sea were found to belong to the genera *Eurotium* and *Aspergillus* (Yan et al. 2005).

The large diversity of the fungal species has been reported to inhabit high-salt environments; however, most of them can be regarded either as halotolerant or as extremely halotolerant. Halotolerant fungi can grow without NaCl added to the medium but tolerate up to saturated NaCl levels (30%) (Gunde-Cimerman et al. 2000). Up till today, only *Wallemia ichthyophaga*, *Wallemia muriae*, *Phialosimplex salinarum*, *Aspergillus baarnensis*, *Aspergillus salisburgensis* and *Aspergillus atacamensis* are obligate halophilic fungi that strictly require NaCl from 5 to 10% (Piñar et al. 2016; Martinelli et al. 2017). Actually, *Gymnascella marismortui* (Buchalo et al. 1998), *Trichosporium* spp.(Elmeleigy et al. 2010), *Aspergillus unguis* (Nazareth et al. 2012), and *Aspergillus penicillioides* (Nazareth and Gonsalves 2014) have also been reported to be obligate halophiles according to their minimum saline requirement.

Aspergillus penicillioides are commonly found in saline habitats, suggesting that the species are extensively adaptable to varied environments. Among 39 tested isolates of *A. penicillioides*, most strains had a minimum salt requirement of 5% for growth; one strain grew only on media supplemented with at least 10% solar salt (Nazareth and Gonsalves 2014). Given that *A. penicillioides* species do not reproduce sexually (Tamura et al. 1999; Gostinčar et al. 2010), which consequently inhibits their gene flow, this species has significant promise in environmental remediation applications.

As mentioned above, some halophilic fungi, such as *A. niger* and *C. cladosporoides*, have been isolated from sand and mud on the shore of salty aquatic bodies or from inflowing freshwater from floods and springs (Kis-Papo et al. 2003, 2014; Grum-Grzhimaylo et al. 2016; Martinelli et al. 2017). We also isolated the halophilic fungus *Aspergillus glaucus* CCHA from air-dried wild vegetation from the surface periphery of a solar salt field (Liu et al. 2011); this species shows extreme salt tolerance, with a salinity range of 5–32% (NaCl) required for growth (Liu et al. 2011). To our surprise, *A. glaucus* CCHA survives in solutions with a broad pH range of 2.0–11.5, indicating that it is a haloalkaliphilic fungus. Further investigation indicated that increasing the pH value (>8.0) can induce *A. glaucus* CCHA to produce a variety of organic acids, including citric acid, oxalic acid, and malic acid. In addition, *A. glaucus* CCHA shows resistance to aridity, heavy metal ions, and high temperature. The extremophilic nature of *A. glaucus* CCHA suggests that it has great promise in soil remediation applications.

Just like the proportion of the halophilic and halotolerant fungi isolated from saline environment, fewer alkaliphilic fungi have been identified in comparison with alkalitolerants. Hozzein and colleagues isolated 117 alkaliphilic and alkaline-resistant microorganisms from 30 soil samples collected from six localities around Wadi Araba, Egypt. By adjusting the pH to 10 after sterilization (using sterilized 10% Na_2CO_3 solution), they only identified 4 fungal isolates among 117 alkaliphilic and alkaline-resistant microorganisms (Hozzein et al. 2013); unfortunately, the authors did not determine the species of the isolates. Alkaliphilic fungi have also been isolated from industrial effluents. For example, *Aspergillus nidulans* KK–99 (isolated from the industrial effluents of Shreyans Paper Industry Limited, Ahmedgarh, Punjab, India) is adapted for growth in an alkalescent environment (pH 10.0) (Taneja et al. 2002). Another alkaliphilic fungus, *Myrothecium sp.* IMER1, also grows well under alkali conditions (pH 9.0) (Zhang et al. 2007).

Grum-Grzhimaylo and collaborators (2016) identified more than 100 strains of alkalitolerant and alkaliphilic fungi isolated from the alkaline soils with different degrees of salinity in Russia, Mongolia, Kazakhstan, Kenya, Tanzania, and Armenia. They found the alkaliphilic/strong alkalitolerant phenotype in about 2/3 of our recovered strains from soda soils, and uncovered that the alkaliphilic trait in filamentous fungi has evolved several times through phylogenetic analyses. Among the alkaliphilic/strong alkalitolerant fungi, the *Sodiomyces* species (*Plectosphaerellaceae*), *Acrostalagmus luteoalbus* (*Plectosphaerellaceae*), *Emericellopsis* alkaline (*Hypocreales*), *Thielavia sp.* (*Chaetomiaceae*), and *Alternaria sect. Soda* (*Pleosporaceae*) grew best at high ambient pH, but the pH tolerance of *Chordomyces antarcticum, Acrostalagmus luteoalbus,* and some other species was largely affected by the presence of extra Na ion in the growth medium, further suggesting that the frequency of alkaliphilic fungi is low, while alkalitolerants seem to be far more widespread in soil (Grum-Grzhimaylo et al. 2016).

Research aimed at isolating and characterizing halophilic fungi has progressed rapidly in China (Table 27.1). A series of promising halophilic fungi, including *A. glaucus* CCHA, have been reported. Three marine-derived isolates were collected in Wenchang, Hainan Province, China, and identified as extremely halotolerant fungi: *Wallemia sebi* PXP-89 (Peng et al. 2011a), *Penicillium chrysogenum* PXP-55 (Peng et al. 2011b), and *Cladosporium cladosporioides* PXP-49 (Xu et al. 2011). The work focusing on isolating halotolerant/alkaliphilic/haloalkaliphilic fungi is being carried out in our laboratory, and actually several halotolerant species with alkaliphilic trait, such as *Aspergillus sp.*, were recently identified based on a specimen collected from the saline-alkali soils in Songnen Plain of northeastern China. China has remarkable biodiversity and many typical hypersaline environments, including Caka Salt Lake and Qarhan Salt Lake in Qinghai, Barkol Salt Lake in Xinjiang, Yuncheng Salt Lake in Shanxi, and the Baicheng soda saline-alkali area in Jilin. All of these environments are suitable for extremophilic fungi and other microorganisms; therefore, isolating and identifying extremophilic fungi

Table 27.1 The typical halophilic, alkaliphilic, and haloalkaliphilic fungi

Species/strain	Source	[Na+] range	pH range	References
Aspergillus glaucus CCHA	Changchun, China	5–32%	2–11.5	Liu et al. (2015)
Aspergillus salisburgensis, Aspergillus atacamensis	Iquique, Chile	10–25%	NR	Martinelli et al. (2017)
Aspergillus penicillioides	Mangroves of Goa, India	10–30%	NR	Nazareth and Gonsalves (2014)
Aspergillus nidulans KK-99	Punjab, India	0–25%	4–10	Taneja et al. (2002)
Eurotium herbariorum	Dead Sea, Israel	2–31%	7–9	Butinar et al. (2005a, b)
Gymnascella marismortui	Dead Sea, Israel	5–30%	NR	Buchalo et al. (1998)
Sodiomyces sp., Acrostalagmus luteoalbus, Emericellopsis alkaline, Thielavia sp., Alternaria sect. Soda	Russia, Mongolia, Kazakhstan, Kenya, Tanzania, Armenia	NR	8.5–11	Grum-Grzhimaylo et al. (2016)
Hortaea werneckii	Ljubljana	5–31%	NR	Gunde-Cimerman et al. (2000)
Aureobasidium pullulans	Amsterdam, The Netherlands	0–17%	NR	Sterflinger et al. (1999)
Myrothecium sp. IMER1	Wuhan, China	0–5%	8–10	Zhang et al. (2007)
Myrothecium sp.GS-17	Gansu, China	NR	8–10	Liu et al. (2013)
Cladosporium cladosporioides PXP-49	Hainan, China	0–20%	5–9	Xu et al. (2011)
Wallemia sebi PXP-89	Hainan, China	0–20%	5–9	Peng et al. (2011a)
Penicillium chrysogenum PXP-55	Hainan, China	0–20%	5–9	Peng et al. (2011b)

NR No report

in China could lead to the development of promising new methods of remediating saline-alkali soil.

27.3 Saline-Alkaline Stable Enzymes Secreted from Haloalkaliphilic Fungi

Fungal and other microbial activities are central to the formation and stabilization of soil aggregates (Rietz and Haynes 2003). Soil-derived fungi produce and secrete a series of active enzymes to the soil, and these soil enzymes are closely related to soil properties, soil types, soil heath, and environmental conditions.

Salinity, sodicity, and both have extremely adverse effects not only on soil chemical and physical properties and on crop growth but also on the species and quantity of fungi, let alone on the activities of the soil enzymes and microbial biomass, and even on biochemical processes essential for maintenance of soil quality. Correspondingly, this will result in a reduction in the rate of soil organic matter decomposition and in the mineralization of carbon (C), nitrogen (N), and phosphorus (P). The resulting reduced nutrient availability will be an additional growth-limiting factor to crop production in salt-affected soils (Zhang SH et al. 2014a).

The soil enzymes are now widely used as important indicators of soil quality and soil biological activities just because most soil enzymes are not stable to the harmful salinity and sodicity. Among all the enzymes, urease, alkaline phosphatase, and catalase activity are more sensitive to soil environmental conditions. Urease specifically catalyzes the hydrolysis of nitrogen-containing organic matter. The high salinity and sodicity cause urease to be completely inactivated, and then the formation pathway of N in the soil is blocked (Liang et al. 2003, 2014). P in soil is mainly in the organic form. Alkaline phosphatase is the main enzyme involved in the cycling of P because it can transform organic P into inorganic P which is the available nutrient for plants (Dick and Burns 2011). Alkaline phosphatase reacts to external environments sensitively and is an indicator of the organic P mineralization and biological activity of soils (Krämer and Green 2000; Zhang T et al. 2014b; Zhang TB et al. 2014c). Catalase can enable the peroxide produced during metabolism to decompose, thus preventing its toxic effects on organisms. These enzyme activities play an important role in the cycling of soil C, N, and P. In addition, they participate in a great number of soil biochemical processes and they are directly involved in various biochemical reactions in the soil. The sensitivity of these enzymes to salt and alkali further illustrates the role of haloalkaliphilic fungi in soil treatment.

In the context of the C cycle, the available organic matter in soil is mainly derived from degradation of crop remains such as fallen leaves and stalks. However, as mentioned above, the microbial community of saline-alkaline land is simple and fragile, and elevated salinity reduces the abundance of fungi more effectively than that of bacteria. Reduced fungal abundance leads to decreased soil microbial biomass and activity, which further slows the turnover of organic matter. Therefore, fungi with salt and alkali resistance, as well as the ability to produce and secrete cellulose-degrading enzymes, are badly needed.

Salt and alkali resistance genes are capable of genetically improving soil fungi and enhancing their resistance to extreme environments. In order to be beneficial to the soil, fungi must possess the ability to produce and secrete a large number of hydrolytic enzymes that degrade plant organic matter (e.g., maize, wheat, or rice straw) such as cellulose, hemicellulose, lignin, and pectin (Castillo and Demoulin 1997; Santos et al. 2004; Arakaki et al. 2013; Batista-García et al. 2014; Wei and Zhang 2018).

The cellulase complex includes endoglucanases (EC 3.2.1.4), exoglucanases (EC 3.2.1.91), and β-glucosidases (EC 3.2.1.21). Endoglucanases randomly attack the internal chain of cellulose to produce cellulo-oligosaccharides. Exoglucanases catalyze the hydrolysis of crystalline cellulose from the ends of the cellulose chain to produce cellobiose, which is ultimately hydrolyzed to glucose by β-glucosidases (Béguin and Aubert 1994; Tomme et al. 1995). *Trichoderma reesei* and *Penicillium janthinellum* are known to be excellent cellulase producers, but their cellulases are not stable under alkali conditions (Mernitz et al. 1996; Wang et al. 2005; Qin et al. 2008). *Aspergillus niger*, one of the most efficient identified cellulose-degrading microorganisms, secretes large amounts of different cellulases during fermentation (Schuster et al. 2002). Endoglucanase B (EGLB), encoded by the endoglucanase gene (GenBank GQ292753) of *Aspergillus niger* BCRC31494, has been used in the fermentation industry because of its alkaline and thermal tolerance (Li et al. 2012). EGLB is a member of glycosyl hydrolase family 5 of the cellulase superfamily. When the recombinant EGLB cDNA was expressed in *Pichia pastoris*, a purified protein of 51 kDa in size was obtained. The enzyme was specific for substrates with β-1,3 and β-1,4 linkages, and it exhibited optimal activity at 70 °C and pH 4 (Li et al. 2012). Interestingly, the relative activity of recombinant EGLB at pH 9 was significantly better than that of wild-type EGLB. The advantages of endoglucanase *EGLB*, particularly its tolerance to a broad range of pH values, indicate that this enzyme has significant promise as a means of genetically improving fungi for haloalkaline soil remediation.

Based on an analysis of the genomic sequence of haloalkaliphilic fungus *A. glaucus* CCHA, we found that *A. glaucus* CCHA expresses only one gene belonging to the GH5 family, AgCel5A. The open reading frame of *Agcel5A* consists of 1509 base pairs that encode a polypeptide of 502 amino acids. AgCel5A has four potential N-glycosylation sites and three O-glycosylation sites, which indicates high similarity to the characterized GH5 β-glucosidases from *Aspergillus niger* (65%) and *Trichoderma reesei* (31%). AgCel5A was cloned and heterologously expressed in *Pichia pastoris* GS115. Recombinant AgCel5A exhibited maximal activity at pH 5.0. AgCel5A is much more stable than PdCel5C from *Penicillium decumbens* (Liu et al. 2013); it retains more than 70% of its maximum activity at pH 8.0–10.0. In addition, AgCel5A exhibited stable degradation activity under high-salt (NaCl) conditions. In the presence of 4 M NaCl, AgCel5A retained 90% activity even after 4 h of preincubation. Interestingly, the activity of AgCel5A increased as the NaCl concentration was increased. The high resistance of AgCel5A to saline and alkaline conditions suggests that the *AgCel5A* gene is an ideal candidate for genetic improvement of soil fungi and industrial applications (Zhang et al. 2016).

Few cellulase genes with tolerance to highly saline and alkaline environments from fungi have been reported, but several genes of this type have been studied in bacteria (e.g., *Paenibacillus sp.*, *Thermomonospora sp.*) (Zarafeta et al. 2016; Kanchanadumkerng et al. 2017). For example, CelDZ1, a recently identified thermotolerant and exceptionally halostable GH5 cellulase from an Icelandic *Thermoanaerobacterium* hot spring isolate, is a glycoside hydrolase with optimal activity at 70 °C and pH 5.0 (Zarafeta et al. 2016). On the other hand, CelDZ1 exhibits high halotolerance at near-saturating salt concentrations and high tolerance for metal ions and other denaturing agents (Zarafeta et al. 2016). These findings show that cellulases from extremophilic bacteria should also be considered for utilization in genetic improvement of fungal resistance to salt and alkali.

27.4 The Molecular Base of Saline-Alkali Resistance in Haloalkaliphilic Fungi

The application of soil microbes to ameliorate salinity is gaining popularity because of its effectiveness and low economic and environmental costs. The applications of haloalkaliphilic fungi like *Aspergillus glaucus* CCHA are restricted by the low number of well-characterized species. Under this realistic condition, genetic improvement of normal soil fungi is a good choice. To enable improvement of the saline and alkaline resistance of normal fungi, genes related to resistance to salt, alkali, or both stresses must first be identified in extremophilic fungi.

Debaryomyces hansenii, the multiple functional salt-loving fungus, has been extensively investigated in recent years. *D. hansenii* can accumulate high concentrations of sodium without undergoing damage; in addition, it grows well under stress factors such as high temperature and extreme pH in the presence of 0.25 M NaCl (Almagro et al. 2000). By screening *S. cerevisiae* transformants containing genes from a genomic library prepared from *D. hansenii* (Prista et al. 2002, 2005), a series of genes associated with salt tolerance were identified and characterized. The *DhGZF3* gene, which encodes GATA transcription factor homologs Dal80 and Gzf3 in *S. cerevisiae*, has been functionally analyzed in *D. hansenii*, but the gene was verified to be a negative transcription factor when it was expressed in *S. cerevisiae* (García-Salcedo et al. 2006). Using a cDNA library from stress-tolerant basidiomycetes yeast *Rhodotorula mucilaginosa*, more than 100 *S. cerevisiae* transformants with tolerance to high concentrations of various osmolytes were screened by Gostinčar and Turk (2012). Among the sequenced clones, 12 genes mediated increased stress tolerance in *R. mucilaginosa* transformants. Recently, Pereira and colleagues (Pereira et al. 2014) analyzed nine candidate polyol/H(+) symporters from the *D. hansenii* genome database via heterologous expression in *S. cerevisiae*. Five distinct polyol/H(+) symporters were confirmed, among which two symporters were specific for uncommon substrates galactitol and D-(+)-chiro-inositol.

Few stress tolerance genes have been identified in extremophilic fungi, and their functions merit additional research because they could be of significant importance in transgenic biotechnology. Most importantly, abiotic stress resistance genes isolated from extremophilic fungi generally function better than homologs from non-extremophiles in extreme environments. *EhHOG*, as mentioned above, is an *E. herbariorum* MAPK kinase gene similar to HOG1 homologs from *A. nidulans*, *S. cerevisiae*, *Schizosaccharomyces pombe*, and most other fungi (Brewster et al. 1993; Delgado-Jarana et al. 2000); however, a hog1 mutant complemented with EhHOG outperformed wild-type yeast under high-salt and freezing/thawing conditions (Yan et al. 2005).

Interestingly, several genes isolated from halophilic fungus *A. glaucus* are more resistant to osmotic stress in comparison with those of common fungi such as *S. cerevisiae* and *Magnaporthe oryzae*. A yeast expression library containing full-length cDNAs from *A. glaucus* was constructed and used to screen salt resistance transformants in our laboratory at Jilin University (Liu et al. 2011; Fang et al. 2014). Ribosomal protein L44 (RPL44), a part of the 60S large ribosomal subunit, was identified based on its association with salt resistance. In comparison with yeasts expressing MoRPL44, the RPL44 homolog of *M. oryzae*, yeasts expressing *AgRPL44* from *A. glaucus* were more resistant to salt, drought, and heavy metals. In addition, when *AgRPL44* was introduced into *M. oryzae*, the transformants displayed significantly enhanced tolerance to salt and drought, indicating that RPL44 plays a role in osmosis resistance in halophilic fungi (Liu et al. 2014; Xie 2013). Similar results were also obtained in studies of another ribosomal protein subunit, AgRPS3aE (Liang et al. 2015), as well as in studies of AgglpF (Liu et al. 2015).

ATP-dependent Lon proteases are highly conserved in diverse species and perform multiple roles. MAP1/Lon protease, the mitochondrial Lon protease homolog of *M. oryzae*, produced a positive effect on salt resistance (Li et al. 2015; Cui et al. 2015). Recently, the genetics of two ATP-dependent Lon proteases from thermophilic fungus *Thermomyces lanuginosus* were studied (Cui et al. 2017). Mitochondrial and peroxisomal Lon proteases were found to exhibit synergistic effects on resistance to multiple stressors, including salt and alkali, in *T. lanuginosus*. The common features of the genes described above are highly conserved and not specific to extremophilic fungi; however, their effects on the tolerance of transgenic cells and organisms surviving under stressful conditions are unambiguous and consistent, suggesting that additional tolerance-related genes with potential value remain to be identified and tested.

27.5 The Mycoremediation Mechanisms of Saline-Alkali Soil

The beneficial effect of microbial application on saline-alkali soil has been reported by Sahin et al. (2011). In the study, suspensions of three fungal isolates (*Aspergillus* spp. FS 9 and 11 and *Alternaria* spp. FS 8) and two bacterial strains (*Bacillus subtilis* OSU 142 and *Bacillus megaterium* M3) at 10^4 spore/mL and 10^9 CFU/mL, respectively, were mixed with leaching water and applied to the soil columns in the Igdir plain of northeastern Turkey (Sahin et al. 2011). Gypsum is an economical alternative for replacing sodium with calcium in remediating saline-alkali soils (Gharaibeh et al. 2009; Oad et al. 2002). In the experimental process, gypsum was applied for the saline-alkali soil pretreatment, and the microorganisms are not halotolerant or halophilic (Aslantas et al. 2007; Turan et al. 2006); thus the final results they obtained should not just be out of the function of microbes. Anyway, this study gives us an enlightened example for mycoremediation of saline-alkali soil by using haloalkaliphilic fungi.

Organisms at simultaneous high salt concentration and high pH value require special adaptive mechanisms, which during the course of evolution would be both facilitative and essential for life-supporting processes. Few researches focus on how haloalkaliphilic fungi cope with extremes of salt and pH value. We assume that haloalkaliphilic fungi adopt comprehensive strategies to survive in the extreme environment; in other words, under saline-alkali conditions, soil fungi must possess certain mechanisms to alleviate the influence or damage of both salt and alkali. In terms of soil effects, only reducing soil-soluble salt and regulating the pH value of soil solution can achieve the purpose of restoring saline-alkali soils.

1. Soil fungi have the ability to accumulate cation contents in cells. Saline soil will be improved with the accumulation of salt cation contents into fungal cells. *Hortaea werneckii*, the black yeast-like fungus isolated from hypersaline waters of salterns as their natural ecological niche, has been previously defined as halophilic fungus (Butinar et al. 2005a, b). *H. werneckii* cells were grown in liquid media at different salinities, ranging from 0 to 25% NaCl. The measurements of cation contents in cells grown at constant salt concentration have shown that the amounts of K^+ and Na^+ in *H. werneckii* were changing according to the NaCl concentration of the medium. When *H. werneckiio* was grown in a medium without added NaCl, it accumulated a very low amount of Na^+. But with the increasing NaCl concentration of the medium, the amounts of the Na^+ content increased and in the end reached a higher value (Kogej et al. 2005).

2. Soil fungi produce different organic acid patterns (Scervino et al. 2010). The released organic acids allow the formation of organic mineral complexes (Richardson et al. 2001); on the other hand with the release of organic acids, protons are produced that contribute to the acidification of the alkali soil solution.

   The saline-alkali soils and most cultivated soils are deficient in available forms of phosphorus. The release of these organic acids and other compounds in the rhizosphere by these microorganisms may be important in the solubilization of various inorganic phosphorus compounds (Scervino et al. 2010). In spite of this, based on the principle of acid-base neutralization, the organic acids also adjust the pH value of soil solution to a lower level.

   The reactions of the citric acid cycle are carried out by eight enzymes that completely oxidize acetate, in the form of acetyl-CoA, into two molecules each of carbon dioxide and water. Organic acids citrate, iso-citrate, succinate, fumarate, malate, and oxaloacetate are produced during each turn of the cycle. The high pH tolerance of *A. glaucus* has led to its utilization as an organic production strain (Barnes and Weitzman 1986). When *A. glaucus* CCHA was cultured in an alkaline medium, key enzymes (e.g., citrate synthase, isocitrate dehydrogenase, succinyl-CoA synthetase, malate dehydrogenase) of the citric acid cycle were significantly upregulated, suggesting that these genes contribute to the high pH tolerance of *A. glaucus* (Wei et al. 2013; Liu 2014; Zhou 2016; Wei and Zhang 2018).The case of organic acid production does not just

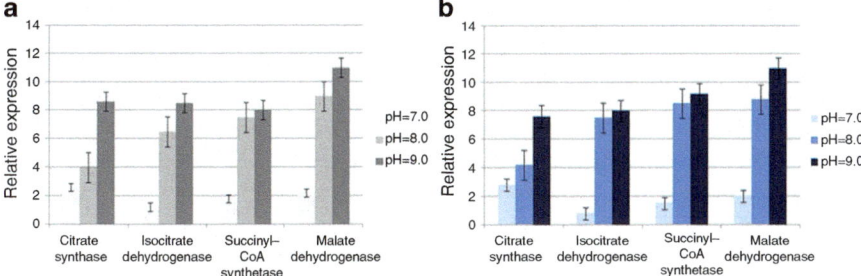

Fig. 27.2 Gene expression analysis of the citric acid cycle key enzymes at different pH value in *A. terreus* S108 (**a**) or *A. niger* S211 (**b**). *Aspergillus* strains were cultured in PD liquid medium (20% NaCl), and pH value was adjusted to 7, 8, or 9, respectively; all cultures were performed at a temperature of 35 °C for 3 days, and then mRNA was extracted for qRT-PCR; the four key enzymes of citric acid cycle (citrate synthase, isocitrate dehydrogenase, succinyl-CoA synthetase, and malate dehydrogenase) were detected through qRT-PCR. Each gene was searched from the genome sequence of *A. terreus* and *A. niger* (https://blast.ncbi.nlm.nih.gov/Blast) by using the corresponding mRNA sequence searched from the genome sequence of *A. niger* NRRL3 (http://genome. fungalgenomics.ca): citrate synthase (XM_022547762.1, homologous to NRRL3_00547), isocitrate dehydrogenase (XM_022550523.1, homologous to NRRL3_05263), succinyl-CoA synthetase (XM_022540546.1, homologous to NRRL3_00603), and malate dehydrogenase (XM_022546357.1, homologous to NRRL3_03476)

specifically occur in the CCHA strain; when other strains such as *Aspergillus terreus* S108 or *Aspergillus niger* S211 were selected as materials, we got the similar result (Fig. 27.2). Accordingly, alkali resistance might be improved in all these saline-alkali resistance fungi by overexpressing enzymes involved in the citric acid cycle.

3. Halophilic and alkaliphilic fungi are of biotechnological interest, as they produce extremozymes, which are useful in medical and environmental field because of their ability to remain active under the severe saline and alkaline conditions (Tiquia-Arashiro and Rodrigues 2016b). The enzymes secreted by haloalkaliphilic fungi possess the bioreduction effect on salt ions of soil. This bioreduction of metal particles by certain biomasses is regarded as an organism's survival mechanism against toxic metal ions and occurs via an active or passive process or a combination of both (Ibrahim et al. 2001; Durán et al. 2005). Correlation between soil properties and soil enzymes from fungi or other microorganisms has been substantiated; and these enzyme activities are now widely used as important indicators of soil quality and soil biological activities (Rietz and Haynes 2003). As described above, high salt and pH value induce the secretion of organic acids. Similar to this case, cellulases and other so-called soil enzymes are also induced with the increasing of salt concentration and pH value. When hydrolytic enzymes are secreted into soil solution, soil properties will be improved accordingly. Take cellulases for instance; on the one hand, soil cellulases can enhance the organic matters by degrading cellulose, and on the

other hand cellulases in salt soils or salt solutions have been detected to form biotical nanoparticles (Riddin et al. 2006; Tiquia-Arashiro and Rodrigues 2016b; Mohite et al. 2017). Recently, nanoparticles of varying size (10–300 nm) and shape (hexagons, pentagons, circles, squares, rectangles) were produced at extracellular levels by *Aspergillus glaucus* CCHA in our lab (unpublished data), indicating that the formation of nanoparticles by haloalkaliphilic fungus is associated with saline-alkali soil remediation.

4. The mechanisms employed by most of the soil fungi (non-mycorrhizal fungi) at the cellular level to tolerate soil salt ions are probably similar to some of the strategies employed by ectomycorrhizal fungus, namely binding to extracellular materials (Tam 1995; Aggangan et al. 2010; Gomes et al. 2018) or sequestration in the vacuolar compartment (Blaudez et al. 2000).

In brief, soil fungi buffer salinity and alkalinity by absorbing and/or constraining salt ions, secreting organic acids and/or macromolecule degradation enzymes, and providing biomass; all of these effects of fungi reduce plant stress. Therefore, haloalkaliphilic fungi are excellent biological resources for soil mycoremediation (Fig. 27.3).

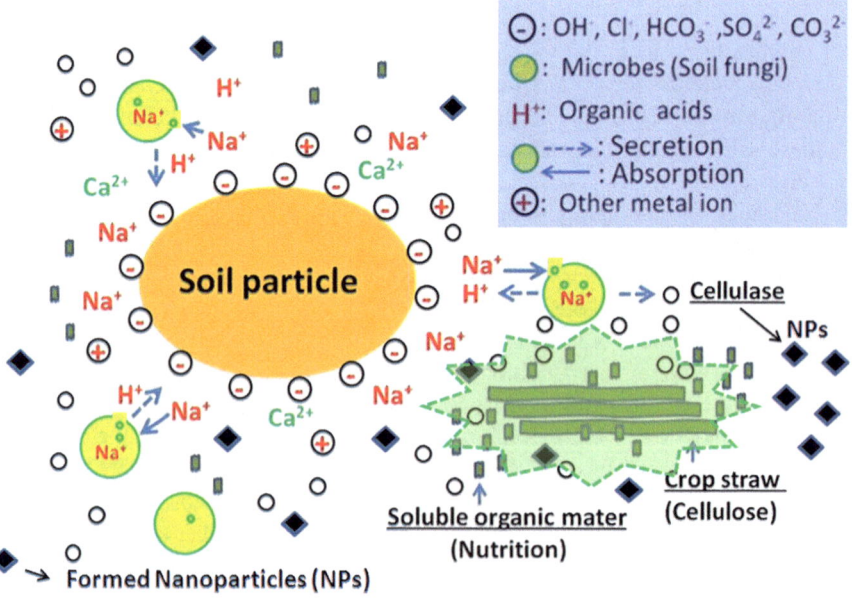

Fig. 27.3 Model of the mechanism through which mycoremediation of saline-alkali soil is achieved by adding haloalkaliphilic fungi and crop straw to soil

27.6 Conclusions and Future Directions

Bioremediation based on planting is one of the most effective methods of soil remediation because of its significant ecological, environmental, and economic effects (Ilyas et al. 1993; Ghaly 2002; Nouri et al. 2017), but the method requires persistent management to produce meaningful changes in soil characteristics. Application of organic matter conditioners, which can ameliorate and increase the fertility of saline soils, is an alternative soil remediation technique (Wang et al. 2014).

Fertile soil is a vital complex that involves numerous species and immense biomass; soil organisms have significant effects on the soil ecosystem. Soil-inhabitant fungi build a metabolic bridge between insoluble organic matter and soil nutrients by producing cellulose degradation enzymes such as cellulase, as well as performing other biological processes. However, saline-alkaline soils generally lack fungi, which ordinarily play important roles in degrading insoluble organic matter such as crop straw into soluble and easily absorbed nutrients; therefore, applying organic matter supplemented with fermentation fungi to saline-alkali soil is a feasible strategy for soil remediation.

Haloalkaliphilic fungi are excellent biological candidates for soil mycoremediation, but to date very few species with both abilities to produce effective soil enzymes and to grow in saline-alkaline environments have been reported. To get better remediation effect, natural soil fungi require to be genetically modified at their degradation ability or saline-alkali resistance. Generally, several enzymes involved in salt and/or alkali resistance, such as the alkaline-stable endoglucanases B from *Aspergillus niger* BCRC31494 (Li et al. 2012), the alkaline xylanase from *Aspergillus nidulans* KK-99 (Taneja et al. 2002), and the bilirubin oxidase from *Myrothecium sp.* IMER1 (Zhang et al. 2007), are highly abundant in fungi found in saline-alkali soil, but such fungi usually have a relatively low capacity for cellulose degradation, whereas fungi found in fertile soil show opposite characteristics. Thus, we propose two strategies to create novel haloalkaliphilic fungi with high cellulase activity: (1) Using naturally isolated haloalkaliphilic fungi as transformational receptors for cloned high-activity cellulase genes from fungi isolated from fertile soil should produce haloalkaliphilic fungi with salt/alkali resistance and high cellulase degradation activity. (2) Using fertile soil fungi as transformational receptors for salt and alkali resistance genes should produce fungi endowed with high resistance to saline-alkali environments, which would be promising candidates for saline-alkali soil remediation.

A series of salt and/or alkali resistance (or tolerance) genes have been characterized to provide a list of candidate genes to be applied in efforts to genetically improve soil fungi (Zhang 2016). In order to enhance the cellulose degradation ability of haloalkaliphilic fungi, additional cellulases with salt and alkali stability must be identified. Using cellulases with salt and alkali tolerance, two strategies can be employed to obtain saline/alkaline-resistant fungi with enhanced enzyme secretion. Indeed, natural strains remain the first choice for soil remediation;

Fig. 27.4 Mycoremediation of soda saline-alkali soil using amendments supplemented with haloalkaliphilic fungi (*A. glaucus* CCHA, *A. terreus* S108, *Eurotium herbariorum*, and *A. niger* S211 (ratio = 12:9:4:4)). The area on the right received the soil amendments mixed with haloalkaliphilic fungi, but the area on the left received salt-sensitive isolates. The experiment was conducted in soda saline-alkaline soil in Zhenlai County, which is located in the Songnen Plain of northeastern of China (see Fig. 27.1). The properties of the saline soil before organic amendments were applied indicate that it was classified as heavy soda saline-alkali soil (at a soil depth of 0–20 cm, sodium salt = 18 g/kg, pH = 10.1, measured by Dr. Shi Yang in 2016). (The aerial photography was performed by an unmanned plane at 150–200 m in the air in 2016)

therefore, isolating and screening suitable strains from extreme natural environments is still an important long-term task.

Haloalkaliphilic fungus *Aspergillus glaucus* CCHA, a fungal species with extreme tolerance to saline and alkaline conditions, has significant potential value in industrial and agricultural applications. Our group has been assessing the potential of *Aspergillus glaucus* CCHA in the mycoremediation of saline-alkaline soil in the Songnen Plain of northeastern China (one of the three most famous saline and alkaline lands in the world) for 3 years (Shi and Zhu 2016). This study primarily indicates that the applied amendments mixed with haloalkaliphilic fungi significantly encourage steady growth and yield of rice in comparison to that achieved in the control plot (Fig. 27.4).

Acknowledgements This work was partially supported by grants from the National Natural Science Foundation of China (grant nos. 31671972 and 31670141) and a project of the Ministry of Science and Technology of China (grant no. 2016YFD0300703). The authors would like to thank the Zhang Lab members, who provided the photographic pictures taken at their spare time. The authors are also grateful to former labmates Dr. Xiaodan LIU, Dr. Zheng-Qun LI, Dr. Yang SHI, and Mr. Sen-Lin ZHANG, who contributed to fungal isolation and field trials, as well as to collaborators Run-Zhi TAO, Chi ZHU, and Zhen-Dong CHEN, who provided encouragement and assistance in promoting our scientific and technological achievements regarding saline-alkali soil mycoremediation using haloalkaliphilic fungi.

References

Aggangan NS, Moon HK, Han SH (2010) Growth response of Acacia mangium Willd. seedlings to arbuscular mycorrhizal fungi and four isolates of the ectomycorrhizal fungus *Pisolithus tinctorius* (Pers.) Coker and Couch. New For 39:215–230

Almagro A, Prista C, Castro S, Quintas C, Madeira-Lopes A, Ramos J, Loureiro-Dias MC (2000) Effects of salts on *Debaryomyces hansenii* and *Saccharomyces cerevisiae* under stress conditions. Int J Food Microbiol 56(2–3):191–197

Antony CP, Kumaresan D, Hunger S, Drake HL, Murrell JC, Shouche YS (2013) Microbiology of Lonar Lake and other soda lakes. ISME J 7:468–476. https://doi.org/10.1038/ismej.2012.137

Apte SK, Thomas J (1997) Possible amelioration of coastal soil salinity using halotolerant nitrogen-fixing cyanobacteria. Plant Soil 189:205–211

Arakaki R, Monteiro D, Boscolo R, Gomes E (2013) Halotolerance, ligninase production and herbicide degradation ability of basidiomycetes strains. Braz J Microbiol 44:1207–1214

Aslantas R, Cakmakci R, Sahin F (2007) Effect of plant growth promoting rhizobacteria on young apple tree growth and fruit yield under orchard conditions. Sci Hortic 111:371–377

Barnes SJ, Weitzman PD (1986) Organization of citric acid cycle enzymes into a multienzyme cluster. FEBS Lett 201:267–270

Batista-García RA, Balcázar-López E, Miranda-Miranda E, Sánchez-Reyes A, Cuervo-Soto L, Aceves-Zamudio D, Atriztán-Hernández K, Morales-Herrera C, Rodríguez-Hernández R, Folch-Mallol J (2014) Characterization of lignocellulolytic activities from a moderate halophile strain of *Aspergillus caesiellus* isolated from a sugarcane bagasse fermentation. PLoS One 9(8):e105893. https://doi.org/10.1371/journal.pone.0105893

Béguin P, Aubert JP (1994) The biological degradation of cellulose. FEMS Microbiol Rev 13:25–58

Blaudez D, Botton B, Chalot M (2000) Cadmium uptake and subcellular compartmentation in the ectomycorrhizal fungus *Paxillus involutus*. Microbiology 146:1109–1117

Brewster JL, de Valoir T, Dwyer ND, Winter E, Gustin MC (1993) An osmosensing signal transduction pathway in yeast. Science 259:1760–1763

Bruggen AHCV, Semenov AM, Zeiss MR (2000) In search of biological indicators for soil health and disease suppression. Appl Soil Ecol 15:13–24

Buchalo AS, Nevo E, Wasser SP, Oren A, Molitoris HP (1998) Fungal life in the extremely hypersaline water of the Dead Sea: first records. Proc R Soc Lond B 265:1461–1465

Buchalo AS, Nevo E, Wasser SP, Oren A, Molitoris HP, Volz PA (2000) Fungi discovered in the Dead Sea. Mycol Res News 104:132–133

Butinar L, Sonjak S, Zalar P, Plemenitaš A, Gunde-Cimerman N (2005a) Melanized halophilic fungi are eukaryotic members of microbial communities in hypersaline waters of solar salterns. Bot Mar 48:73–79

Butinar L, Zalar P, Frisvad JC, Gunde-Cimerman N (2005b) The genus Eurotium-members of indigenous fungal community in hypersaline waters of salterns. FEMS Microbiol Ecol 51:155–166

Casamayor EO, Massana R, Benlloch S, Øvreas L, Diez B, Goddard VJ, Gasol JM, Joint I, Rodríguez-Valera F, Pedrós-Alió C (2002) Changes in archaeal, bacterial and eukaryal assemblages along a salinity gradient by comparison of genetic fingerprinting methods in a multipond solar saltern. Environ Microbiol 4:338–348

Castillo G, Demoulin V (1997) NaCl salinity and temperature effects on growth of three wood-rotting basidiomycetes from a Papua New Guinea coastal forest. Mycol Res 101:341–344

Cui X, Wei Y, Wang Y, Zheng Y, Li J, Wong F, Liu S, Yan H, Jia B, Liu J, Zhang SH (2015) Proteins interacting with mitochondrial ATP-dependent Lon protease (MAP1) in Magnaporthe oryzae are involved in rice blast disease. Mol Plant Pathol 16:847–859

Cui X, Wei Y, Xie XL, Chen LN, Zhang SH (2017) Mitochondrial and peroxisomal Lon proteases play opposing roles in reproduction and growth but co-function in the normal development, stress resistance and longevity of Thermomyces lanuginosus. Fungal Genet Biol 103:42–54

Delgado-Jarana J, Sousa S, González F, Rey M, Llobell A (2000) ThHog1 controls the hyperosmotic stress response in Trichoderma harzianum. Microbiology 152:1687–1700

Dendooven L, Alcántara-Hernández RJ, Valenzuela-Encinas C, Luna-Guido ML, Perez-Guevara F, Marsch R (2010) Dynamics of carbon and nitrogen in an extreme alkaline saline soil: a review. Soil Biol Biochem 42:865–877

Dick RP, Burns RG (2011) A brief history of soil enzymology research. Soil Science Society of America, Madison, pp 1–34

Durán N, Marcato PD, Alves OL, De Souza GI, Esposito E (2005) Mechanistic aspects of biosynthesis of silver nanoparticles by several *Fusarium oxysporum* strains. J Nanobiotechnol 3:8. https://doi.org/10.1186/1477-3155-3-8

Elmeleigy MA, Hoseiny EN, Ahmed SA, Alhoseiny AM (2010) Isolation, identification, morphogenesis and ultrastructure of obligate halophilic fungi. J Appl Sci Environ Sanit 5:201–202

Evans S, Hansen RW, Schneegurt MA (2013) Isolation and characterization of halotolerant soil fungi from the great salt plains of Oklahoma. Cryptogam Mycol 34:329–341. https://doi.org/10.7872/crym.v34.iss4.2013.329

Fang J, Han X, Xie L, Liu M, Qiao G, Jiang J, Zhuo R (2014) Isolation of salt stress-related genes from Aspergillus glaucus CCHA by random overexpression in Escherichia coli. Sci World J 39:620959

Ford GW, Martin JJ, Rengasamy P, Boucher SC, Ellington A (1993) Soil sodicity in Victoria. Aust J Soil Res 31:869–909

Gao Q, Yang XS, Yun R, Li CP (1996) MAGE, a dynamic model of alkaline grassland ecosystems with variable soil characteristics. Ecol Model 93:19–32

García-Salcedo R, Casamayor A, Ruiz A, González A, Prista C, Loureiro-Dias MC, Ramos J, Ariño J (2006) Heterologous expression implicates a GATA factor in regulation of nitrogen metabolic genes and ion homeostasis in the halotolerant yeast Debaryomyces hansenii. Eukaryot Cell 5:1388–1398

Ghaly FM (2002) Role of natural vegetation in improving salt affected soil in northern Egypt. Soil Tillage Res 64:173–178

Gharaibeh MA, Eltaif NI, Shunnar OF (2009) Leaching and reclamation of calcareous saline-sodic soil by moderately saline and moderate-SAR water using gypsum and calcium chloride. J Plant Nutr Soil Sci 172:713–719

Ghassemi F, Jakeman AJ, Nix HA (1995) Salinisation of Land and Water Resources: Human Causes, Extent, Management and Case Studies. CABI Publishing, Wallingford, UK

Gomes ECQ, Godinho VM, Silva DAS, de Paula MTR, Vitoreli GA, Zani CL, Alves TMA, Junior PAS, Murta SMF, Barbosa EC, Oliveira JG, Oliveira FS, Carvalho CR, Ferreira MC, Rosa CA, Rosa LH (2018) Cultivable fungi present in Antarctic soils: taxonomy, phylogeny, diversity, and bioprospecting of antiparasitic and herbicidal metabolites. Extremophiles. https://doi.org/10.1007/s00792-018-1003-1

Gonçalves VN, Vitoreli GA, de Menezes GCA, Mendes CRB, Secchi ER, Rosa CA, Rosa LH (2017) Taxonomy, phylogeny and ecology of cultivable fungi present in seawater gradients across the Northern Antarctica Peninsula. Extremophiles 21:1005. https://doi.org/10.1007/s00792-017-0959-6

Gostinčar C, Turk M (2012) Extremotolerant fungi as genetic resources for biotechnology. Bioengineered 3:293–297

Gostinčar C, Grube M, De Hoog S, Zalar P, Gunde-Cimerman N (2010) Extremotolerance in fungi: evolution on the edge. FEMS Microbiol Ecol 71:2–11

Grum-Grzhimaylo AA, Georgieva ML, Bondarenko SA, Debets AJM, Bilanenko EN (2016) On the diversity of fungi from soda soils. Fungal Divers 76:27–74

Gunde-Cimerman N, Zalar P (2014) Extremely halotolerant and halophilic fungi inhabit brine in solar salterns around the globe. Food Technol Biotechnol 52:170–179

Gunde-Cimerman N, Zalar P, de Hoog GS, Plemenitaš A (2000) Hypersaline waters in salterns: natural ecological niches for halophilic black yeasts. FEMS Microbiol Ecol 32:235–240

Gunde-Cimerman N, Ramos J, Plemenitas A (2009) Halotolerant and halophilic fungi. Mycol Res 113:1231–1241

Hozzein WN, Ali MIA, Ahmed MS (2013) Antimicrobial activities of some alkaliphilic and alkaline-resistant microorganisms isolated from Wadi Araba, the eastern desert of Egypt. Life Sci J 10:1823–1828

Ibrahim Z, Ahmad A, Baba B (2001) Bioaccumulation of silver and the isolation of metal-binding protein from Pseudomonas diminuta. Brazil Arch Biol Tech 44:223–225

Ilyas M, Miller RW, Qureshi RH (1993) Hydraulic conductivity of saline-sodic soil after gypsum application and cropping. Soil Sci Soc Am J 57:1580–1585

Kanchanadumkerng P, Sakka M, Sakka K, Wiwat C (2017) Characterization of endoglucanase from Paenibacillus sp. M33, a novel isolate from a freshwater swamp forest. J Basic Microbiol 57:121–131

Karlen DL, Andrews SS, Wienhold BJ, Zobeck TM (2008) Soil quality assessment: past, present and future. J Integr Biosci 6:3–14

Kis-Papo T, Grishkan I, Oren A, Wasser SP, Nevo E (2003) Survival of filamentous fungi in hypersaline Dead Sea water. Microb Ecol 45:183–190

Kis-Papo T, Weig AR, Riley R, Peršoh D, Salamov A, Sun H, Lipzen A, Wasser SP, Rambold G, Grigoriev IV, Nevo E (2014) Genomic adaptations of the halophilic Dead Sea filamentous fungus *Eurotium rubrum*. Nat Commun 5. https://doi.org/10.1038/ncomms4745

Kogej T, Ramos J, Plemenitas A, Gunde-Cimerman N (2005) The halophilic fungus *Hortaea werneckii* and the halotolerant fungus *Aureobasidium pullulans* maintain low intracellular cation concentrations in hypersaline environments. Appl Environ Microbiol 71:6600–6605

Krämer S, Green DM (2000) Acid and alkaline phosphatase dynamics and their relationship to soil microclimate in a semiarid woodland. Soil Biol Biochem 32:179–188

Li CH, Wang HR, Yan TR (2012) Cloning, purification, and characterization of a heat- and alkaline-stable endoglucanase B from *Aspergillus niger* BCRC31494. Molecules 17:9774–9789

Li J, Liang X, Wei Y, Liu J, Lin F, Zhang SH (2015) An ATP-dependent protease homolog ensures basic standards of survival and pathogenicity for Magnaporthe oryzae. Eur J Plant Pathol 141:703–716

Liang YC, Yang YF, Yang CG, Shen QR, Zhou JM, Yang LZ (2003) Soil enzymatic activity and growth of rice and barley as influenced by organic manure in an anthropogenic soil. Geoderma 115:149–160

Liang Q, Chen H, Gong Y, Yang H, Fan M, Kuzyakov Y (2014) Effects of 15 years of manure and mineral fertilizers on enzyme activities in particle-size fractions in a North China plain soil. Eur J Soil Biol 60:112–119

Liang X, Liu Y, Xie L, Liu X, Wei Y, Zhou X, Zhang SH (2015) A ribosomal protein AgRPS3aE from halophilic Aspergillus glaucus confers salt tolerance in heterologous organisms. Int J Mol Sci 16:3058–3070

Liu XD (2014) Two stress tolerance genes in halophilic Aspergillus: functional analysis and their application. Dissertation, Jilin University (in Chinese). http://kns.cnki.net/KCMS/detail/detail. aspx?dbcode=CDFD&dbname=CDFDTEMP&filename=1015507655.nh

Liu XD, Liu JL, Wei Y, Tian YP, Fan FF, Pan HY, Zhang SH (2011) Isolation, identification and biologic characteristics of an extreme halotolerant Aspergillus sp. J Jilin Univ 49:548–552. (in Chinese; abstract in English)

Liu G, Qin Y, Hu Y, Gao M, Peng S, Qu Y (2013) An endo-1,4-β-glucanase PdCel5C from cellulolytic fungus Penicillium decumbens with distinctive domain composition and hydrolysis product profile. Enzym Microb Technol 52:190–195

Liu XD, Xie L, Wei Y, Zhou XY, Jia B, Liu J, Zhang SH (2014) Abiotic stress resistance, a novel moonlighting function of ribosomal protein RPL44 in the halophilic fungus Aspergillus glaucus. Appl Environ Microbiol 80:4294–4300

Liu XD, Wei Y, Zhou XY, Pei X, Zhang SH (2015) Aspergillus glaucus aquaglyceroporin gene glpF confers high osmosis tolerance in heterologous organisms. Appl Environ Microbiol 81:6926–6937

Mahdy HM, el-Sheikh HH, Ahmed MS, Refaat BM (1996) Physiological and biochemical changes induced by osmolality in halotolerant aspergilli. Acta Microbiol Pol 45:55–65

Mandeel QA (2006) Biodiversity of the genus Fusarium in saline soil habitats. J Basic Microbiol 46:480–494

Martinelli L, Zalar P, Gunde-Cimerman N, Azua-Bustos A, Sterflinger K, Piñar G (2017) Aspergillus atacamensis and A. salisburgensis: two new halophilic species from hypersaline/arid habitats with a phialosimplex-like morphology. Extremophiles 21(4):755–773

Melero S, Madejon E, Ruiz JC, Herencia JF (2007) Chemical and biochemical properties of a clay soil under dryland agriculture system as affected by organic fertilization. Eur J Agron 26:327–334

Mernitz G, Koch A, Henrissat B, Schulz G (1996) Endoglucanase II (EGII) of Penicillium janthinellum: cDNA sequence, heterologous expression and promoter analysis. Curr Genet 29:490–495

Mohite P, Kumar AR, Zinjarde S (2017) Relationship between salt tolerance and nanoparticle synthesis by Williopsis saturnus NCIM 3298. World J Microbiol Biotechnol 33:163. https://doi.org/10.1007/s11274-017-2329-z

Moubasher A, Abdel-Hafez S, Bagy M, Abdel-Satar M (1990) Halophilic and halotolerant fungi in cultivated desert and salt marsh soils from Egypt. Acta Mycol 26:65–81

Mulder JL, El-Hendawy H (1999) Microfungi under stress in Kuwait's coastal saline depressions. Kuwait J Sci Eng 26:157–172

Mulder JL, Ghannoum MA, Khamis L, Elteen KA (1989) Growth and lipid composition of some dematiaceous hyphomycete fungi grown at different salinities. Microbiology 135:3393–3404

Nazareth S, Gonsalves V (2014) Aspergillus penicillioides—a true halophile existing in hypersaline and polyhaline econiches. Ann Microbiol 64:397–402

Nazareth S, Gonsalves V, Nayak S (2012) A first record of obligate halophilic aspergilli from the Dead Sea. Indian J Microbiol 52:22–27

Nouri H, Borujeni CS, Nirola R, Hassanli A, Beecham S, Alaghmand S, Saint C, Mulcahy D (2017) Application of green remediation on soil salinity treatment: a review on halophytoremediation. Process Saf Environ 107:94–107

Oad FC, Samo MA, Soomro A, Oad DL, Oad NL, Siyal AG (2002) Amelioration of salt affected soils. Pakistan J Appl Sci 2:1–9

Oo AN, Iwai CB, Saenjan P (2015) Soil properties and maize growth in saline and nonsaline soils using cassava-industrial waste compost and vermicompost with or without earthworms. Land Degrad Dev 26:300–310

Peng XP, Wang Y, Liu PP, Hong K, Chen H, Yin X, Zhu WM (2011a) Aromatic compounds from the halotolerant fungal strain of Wallemia sebi PXP-89 in a hypersaline medium. Arch Pharm Res 34:907–912

Peng XP, Wang Y, Sun K, Liu PP, Yin X, Zhu WM (2011b) Cerebrosides and 2-pyridone alkaloids from the halotolerant fungus Penicillium chrysogenum grown in a hypersaline medium. J Nat Prod 74:1298–1302

Pereira I, Madeira A, Prista C, Loureiro-Dias MC, Leandro MJ (2014) Characterization of new polyol/H+ symporters in Debaryomyces hansenii. PLoS One 9:e88180

Piñar G, Dalnodar D, Voitl C, Reschreiter H, Sterflinger K (2016) Biodeterioration risk threatens the 3100 year old staircase of hallstatt (Austria): possible involvement of halophilic microorganisms. PLoS One 11(2):e0148279. https://doi.org/10.1371/journal.pone.0148279

Prista C, Soeiro A, Vesely P, Almagro A, Ramos J, Loureireo-Dias MC (2002) Genes from Debaryomyces hansenii increase salt tolerance in Saccharomyces cerevisiae W303. FEMS Yeast Res 2:151–157

Prista C, Loureiro-Dias MC, Montiel V, García R, Ramos J (2005) Mechanisms underlying the halotolerant way of Debaryomyces hansenii. FEMS Yeast Res 5:693–701

Qadir M, Oster JD, Schubert S, Noble AD, Sahrawat KL (2007) Phytoremediation of sodic and saline-sodic soils. Adv Agron 96:197–247

Qin Y, Wei X, Song X, Qu Y (2008) Engineering endoglucanase II from Trichoderma reesei to improve the catalytic efficiency at a higher pH optimum. J Biotechnol 135:190–195

Rady MM (2011) Effects on growth, yield, and fruit quality in tomato (Lycopersicon esculentum Mill.) using a mixture of potassium humate and farmyard manure as an alternative to mineral-N fertilizer. J Hortic Sci Biotechnol 86:249–254

Richardson AE, Hadobas PA, Hayes JE, O'Hara CP, Simpson RJ (2001) Utilization of phosphorus by pasture plants supplied with myo-inositol hexaphosphate is enhanced by the presence of soil microorganisms. Plant Soil 229:47–56

Riddin TL, Gericke M, Whiteley CG (2006) Analysis of the inter- and extracellular formation of platinum nanoparticles by *Fusarium oxysporum* f. sp. lycopersici using response surface methodology. Nanotechnology 17:3482–3489

Rietz DN, Haynes RJ (2003) Effects of irrigation-induced salinity and sodicity on soil microbial activity. Soil Biol Biochem 35:845–854

Sahin U, Eroğlum S, Sahin F (2011) Microbial application with gypsum increases the saturated hydraulic conductivity of saline-sodic soils. Appl Soil Ecol 48:247–250

Santos SX, Carvalho CC, Bonfa MR, Silva R, Gomes E (2004) Screening for pectinolytic activity of wood-rotting Basidiomycetes and characterization of the enzymes. Folia Microbiol 49:46–52

Sardinha M, Müller T, Schmeisky H, Joergensen RG (2003) Microbial performance in soils along a salinity gradient under acidic conditions. Appl Soil Ecol 23:237–244

Scervino JM, Mesa MP, Della Mónica I, Recchi M, Moreno NS, Godeas A (2010) Soil fungal isolates produce different organic acid patterns involved in phosphate salts solubilization. Biol Fertil Soils 46:755–763

Schuster E, Dunn-Coleman N, Frisvad JC, van Dijck PW (2002) On the safety of Aspergillus niger—a review. Appl Microbiol Biotechnol 59:426–435

Shi Y, Zhu J (2016) Saline-alkali soil improvement fertilizer and preparation method and use method thereof. Chinese Patent, CN 105237293 A (in Chinese). http://patentool.wanfangdata. com.cn/Patent/Details?id=CN201510597529.0

Sterflinger K, de Hoog GS, Haase G (1999) Phylogeny and ecology of meristematic ascomycetes. Stud Mycol 43:5–22

Szabolcs I (1994) Soils and salinization. In: Pessarakli M (ed) Handbook of plant and crop stress, 1st edn. Marcel Dekker, New York, pp 3–11

Tam PCF (1995) Heavy-metal tolerance by ectomycorrhizal fungi and metal amelioration by *Pisolithus tinctorius*. Mycorrhiza 5:181–187

Tamura M, Kawasaki H, Sugiyama J (1999) Identity of the xerophilic species Aspergillus penicillioides: integrated analysis of the genotypic and phenotypic. J Gen Appl Microbiol 45:29–37

Taneja K, Gupta S, Kuhad RC (2002) Properties and application of a partially purified alkaline xylanase from an alkalophilic fungus Aspergillus nidulans KK-99. Bioresour Technol 85:39–42

Tejada M, Garcia C, Gonzalez JL, Hernandez MT (2006) Use of organic amendment as a strategy for saline soil remediation: influence on the physical, chemical and biological properties of soil. Soil Biol Biochem 38:1413–1421

Tiquia SM (2003) Evaluation of organic matter and nutrient composition of partially decomposed and composted spent pig-litter. Environ Technol 24:97–108

Tiquia SM, Lloyd J, Herms D, Hoitink HAJ, Michel FC Jr (2002) Effects of mulching and fertilization on soil nutrients, microbial activity and rhizosphere bacterial community structure determined by analysis of T-RFLPs of PCR-amplified 16S rRNA genes. Appl Soil Ecol 21:31–48

Tiquia-Arashiro SM, Rodrigues D (2016a) Halophiles in Nanotechnology. In: Extremophiles: applications in nanotechnology. Springer International Publishing, Cham, Switzerland, pp 53–58

Tiquia-Arashiro SM, Rodrigues DF (2016b) Extremophiles: applications in biotechnology. In: Springer briefs in microbiology: extremophilic microorganisms. Springer International

Publishing, Cham, Switzerland. https://doi.org/10.1007/978-3-319-45215-9. ISBN: 978-3-319-45214-2, ISBN: 978-3-319-45215-9 (eBook)

Tomme P, Warren RAJ, Gilkes NR (1995) Cellulose hydrolysis by bacteria and fungi. Adv Microb Physiol 37:1–81

Turan M, Ataoglu N, Sahin F (2006) Evaluation of the capacity of phosphate solubilizing bacteria and fungi on different forms of phosphorus in liquid culture. J Sustain Agric 28:99–108

Wang W, Vinocur B, Altman A (2003) Plant responses to drought, salinity and extreme temperatures: Towards genetic engineering for stress tolerance. Planta 218:1–14

Wang T, Liu X, Yu Q, Zhang X, Qu Y, Gao P, Wang T (2005) Directed evolution for engineering pH profile of endoglucanase III from *Trichoderma reesei*. Biomol Eng 22:89–94

Wang L, Seki K, Miyazaki T, Ishihama Y (2009) The causes of soil alkalinization in the Songnen Plain of Northeast China. Paddy Water Environ 7:259–270

Wang LL, Sun XY, Li SY, Zhang T, Zhang W, Zhai PH (2014) Application of organic amendments to a coastal saline soil in North China: effects on soil physical and chemical properties and tree growth. PLoS One 9(2):e89185. https://doi.org/10.1371/journal.pone.0089185

Wei Y, Zhang S-H (2018) Abiostress resistance and cellulose degradation abilities of haloal-kaliphilic fungi: applications for saline–alkaline remediation. Extremophiles 22(2):155–164

Wei Y, Liu XD, Jia B, Zhang SH, Liu JL, Gao W (2013) Alkaline-tolerant and halophilic *Aspergillus* strain and application thereof in environmental management. Chinese Patent, CN 103436450 A (in Chinese). http://patentool.wanfangdata.com.cn/Patent/Details?id=CN201310162347.1

Wichern J, Wichern F, Joergensen RG (2006) Impact of salinity on soil microbial communities and the decomposition of maize in acidic soils. Geoderma 137:100–108

Xie LX (2013) Cloning and stress resistance analysis of ribosomal protein genes (SpRPS3ae and SpRPL44) in extreme halotolerant *Aspergillus sp*. Dissertation, Jilin University (in Chinese). http://kns.cnki.net/KCMS/detail/detail.aspx?dbcode=CMFD&dbname=CMFDTEMP&filen ame=1013196212.nh

Xu ZH, Peng XP, Wang Y, Zhu WM (2011) (22E, 24R)-3ß,5α,9α-trihydroxyergosta-7,22-dien-6-one monohydrate. Acta Crystallogr E67:o1141–o1142

Yadav S, Irfan M, Ahmad A, Hayat S (2011) Causes of salinity and plant manifestations to salt stress: a review. J Environ Biol 32:667–685

Yan J, Song WN, Nevo E (2005) A MAPK gene from Dead Sea fungus confers stress tolerance to lithium salt and freezing–thawing: prospects for saline agriculture. Proc Natl Acad Sci U S A 102:18992–18997

Zajc J, Džeroski S, Kocev D, Oren A, Sonjak S, Tkavc R, Gunde-Cimerman N (2014a) Chaophilic or chaotolerant fungi: a new category of extremophiles? Front Microbiol 5:708-1–708-5. https://doi.org/10.3389/fmicb.2014.00708

Zajc J, Kogej T, Galinski EA, Ramos J, Gunde-Cimerman N (2014b) Osmoadaptation strategy of the most halophilic fungus, *Wallemia ichthyophaga*, growing optimally at salinities above 15% NaCl. Appl Environ Microbiol 80(1):247–256. https://doi.org/10.1128/AEM.02702-13

Zalar P, Kocuvan MA, Plemenitas A, Gunde-Cimerman N (2005) Halophilic black yeasts colonize wood immersed in hypersaline water. Bot Mar 48:323–326

Zarafeta D, Kissas D, Sayer C, Gudbergsdottir SR, Ladoukakis E, Isupov MN, Chatziioannou A, Peng X, Littlechild JA, Skretas G, Kolisis FN (2016) Discovery and characterization of a thermostable and highly halotolerant GH5 cellulase from an Icelandic hot spring isolate. PLoS One 11:e0146454. https://doi.org/10.1371/journal.pone.0146454

Zhang SH (2016) The genetic basis of abiotic stress resistance in extremophilic fungi: the genes cloning and application. In: Purchase D (ed) Fungal applications in sustainable environmental biotechnology, 1st edn. Springer Press, Switzerland, pp 29–42. isbn:978-3-319-42850-5

Zhang X, Liu Y, Yan K, Wu H (2007) Decolorization of an anthraquinone-type dye by a bilirubin oxidase-producing nonligninolytic fungus Myrothecium sp.IMER1. J Biosci Bioeng 104:104–114

Zhang HS, Zai XM, Wu XH, Qin P, Zhang WM (2014a) An ecological technology of coastal saline soil amelioration. Ecol Eng 67:80–88

Zhang T, Wan S, Kang Y, Feng H (2014b) Urease activity and its relationships to soil physiochemical properties in a highly saline-sodic soil. J Soil Sci Plant Nutr 14:302–313

Zhang TB, Kang YH, Liu SH, Liu SP (2014c) Alkaline phosphatase activity and its relationship to soil properties in a saline-sodic soil reclaimed by cropping wolfberry (*Lycium barbarum* L.) with drip irrigation. Paddy Water Environ 12:309–317

Zhang SH, Li ZQ, Wei Y, Chen LN, Liu SS, Zhou XY, Song YY, Pei X (2016) Cellulase gene from extreme saline-alkali resistant *Aspergillus* and application. Chinese Patent, CN105420259A (in Chinese). http://patenttool.wanfangdata.com.cn/Patent/Details?id=CN201610005024.5

Zhou XY (2016)Cloning and abiotic functional analysis of salt-tolerant genes in Halophilic *Aspergillus glaucus*. Dissertation, Jilin University (in Chinese). http://kns.cnki.net/KCMS/detail/detail.aspx?filename=1016091320.nh&dbname=CMFDTEMP

Chapter 28
Potential Role of Extremophilic Hydrocarbonoclastic Fungi for Extra-Heavy Crude Oil Bioconversion and the Sustainable Development of the Petroleum Industry

Leopoldo Naranjo-Briceño (ID)**, Beatriz Pernía, Trigal Perdomo, Meralys González, Ysvic Inojosa, Ángela De Sisto, Héctor Urbina, and Vladimir León**

28.1 Introduction

In spite of the laudable financial, technical, and scientific efforts focused on the development of new environmental friendly energy sources, it is unquestionable that the world is running on petroleum. In fact, the present world oil requirements to sustain the actual growth behavior of the global population are estimated in more than 95 million barrels per day (OPEC, Annual Statistical Bulletin 2017).

L. Naranjo-Briceño, PhD. (✉)
Área de Energía y Ambiente, Fundación Instituto de Estudios Avanzados (IDEA), Carretera Nacional Baruta-Hoyo de la Puerta, Valle de Sartenejas, CP, Caracas, Venezuela

Grupo de Microbiología Aplicada, Universidad Regional Amazónica Ikiam, CP, Tena, Ecuador
e-mail: leopoldo.naranjo@ikiam.edu.ec

B. Pernía
Área de Energía y Ambiente, Fundación Instituto de Estudios Avanzados (IDEA), Carretera Nacional Baruta-Hoyo de la Puerta, Valle de Sartenejas, CP, Caracas, Venezuela

Facultad de Ciencias Naturales, Universidad de Guayaquil, CP, Guayaquil, Ecuador

T. Perdomo · M. González · Y. Inojosa · Á. De Sisto · V. León
Área de Energía y Ambiente, Fundación Instituto de Estudios Avanzados (IDEA), Carretera Nacional Baruta-Hoyo de la Puerta, Valle de Sartenejas, CP, Caracas, Venezuela

H. Urbina
Área de Energía y Ambiente, Fundación Instituto de Estudios Avanzados (IDEA), Carretera Nacional Baruta-Hoyo de la Puerta, Valle de Sartenejas, CP, Caracas, Venezuela

Division of Plant Industry, Florida Department of Agriculture, Gainesville, FL, USA

© Springer Nature Switzerland AG 2019
S. M. Tiquia-Arashiro, M. Grube (eds.), *Fungi in Extreme Environments: Ecological Role and Biotechnological Significance*,
https://doi.org/10.1007/978-3-030-19030-9_28

The increasing global demand for fuel and the reduction of conventional crude reserves have generated a great interest on the exploitation of unconventional crude reserves worldwide. These unconventional sources of fossil fuel cannot be extracted, transported, and refined by conventional methods. The certified crude oil reserves worldwide are estimated in more than 1.5 billion barrels, of which 70% are constituted by unconventional crudes, such as the Orinoco Oil Belt (OOB), the biggest certified reserve of EHCO worldwide located in Venezuela with over 300 billion barrels (OPEC, Annual Statistical Bulletin 2017).

The enormous reserves of unconventional crudes contain high concentrations of toxicity pollutants and high-molecular-weight compounds, such as asphaltenes, which are heterogeneous and complex mixtures insoluble in n-heptane or n-pentane, and soluble in benzene or toluene, which contain heteroatoms (nitrogen, sulfur, and oxygen) and heavy metals such as nickel and vanadium in their structure (Waldo et al. 1991; Strauz et al. 1992; Uribe-Álvarez et al. 2011; Ayala et al. 2012; Naranjo et al. 2007; León et al. 2007).

Unconventional crude oils have been characterized as recalcitrant (very low availability and degradation by microorganisms), polar, and water insoluble, and contain sulfur and/or heavy metals in association; hence there is a high demand for the development of technologies that aim to alleviate environmental impacts during the extraction, production, and refinement of unconventional crude oils.

The sustainable development of the oil industry requires the development of novel environmental friendly technologies, which could offer both higher economic income with environmental remediation of anthropogenic intervened ecosystems and conservation of natural ecosystems. Complementary use of modern biotechnology in the oil industry provides new tools to improve their processes and products in the production chain, decreasing operational costs and increasing productive capabilities with minimum environmental impact (Naranjo et al. 2007, 2008, 2013).

Aromatic hydrocarbons ranging from the single benzene ring to the high-molecular-weight polycyclics are generally biodegraded via one or more of the three independent enzymatic systems. The intracellular P450 monooxygenases that detoxify harmful chemicals are universally present in the microsomes of eukaryotic cells, while lignin-degrading fungi specifically produce extracellular peroxidases and laccases that biodegrade aromatic hydrocarbons (Prenafeta-Boldú et al. 2018). The low functional specificity and high redox potential of peroxidases and laccases enable the oxidation of a broad range of aromatic hydrocarbons and other recalcitrant contaminants (Prenafeta-Boldú et al. 2018).

In this sense, the study and application of the powerful extracellular oxidative lignin-degrading enzyme system (LDS) secreted by fungi have a great potential as biocatalysts (as whole cells or enzymatic catalyst) in mycoremediation and EHCO bioupgrading processes, through three main pathways: (1) de-aromatization of high-molecular-complexity compounds into more soluble compounds with concomitant reduction of viscosity and enhanced bioavailability by microorganisms; (2) biodesulfurization of sulfur heteroatoms; and (3) de-metallization of heavy metals such as nickel (Ni) and vanadium (V), among others.

Here, we want to contribute and promote the sustainable development of the petroleum industry with the complementary use of the biotechnology of fungi with respect to their ability to degrade and/or transform hydrocarbons, laying special emphasis on the fungal biodiversity associated to extreme environments and the selection of promissory extremophilic and hydrocarbonoclastic fungi.

The term extremophile was first proposed by MacElroy in 1974 to describe a broad group of organisms which lived optimally under extreme conditions (MacElroy 1974; Zhang et al. 2018). Their taxonomic range has been expanded from prokaryotes to all three domains Archaea, Bacteria, and Eukarya (Zhang et al. 2018). An extreme environment is a place that contains conditions that are hard to survive in for most known life forms. These conditions may be extremely high or low temperature (extremely hot or cold), high concentration of a salt (hypersaline), high acidity or alkalinity (acidic or alkaline), desiccation (without water), extremely high pressure (under pressure), high or low content of oxygen or carbon dioxide in the atmosphere (with or without O_2 or CO_2), high levels of radiation (UV emission, radioactivity), and places anthropogenically impacted, such as soil, sediment, or water contaminated by petroleum or other toxic and contaminant substances (altered by humans).

According to different extreme habitats, extremophiles are classified into seven categories (Arulazhagan et al. 2017; Zhang et al. 2018). Organisms whose optimal growth temperature ranges from 50 to 80 °C or exceeds 80 °C are called thermophiles or hyperthermophiles, respectively. Psychrophiles are organisms that grow at low temperatures ranging from 0 to 15 °C. Halophiles require >3% of NaCl to grow and are classified as halotolerant or slight halophiles (2–5% NaCl), moderate halophiles (5–20% NaCl), and extreme halophiles (20–35% NaCl). Acidophiles or alkaliphiles show optimal growth at pH values 1–5 and pH >9, respectively. Piezophiles or barophiles reside under high hydrostatic pressure which have been isolated from the deep-sea sediments (>3000 m depth and pressures of up to 110 MPa). Finally, xerophiles are organisms that grow under low water content (aw 0.60–0.90). In addition, these organisms are normally polyextremophiles and are adapted to live in habitats where various physicochemical parameters reach extreme values (Rampelotto 2013).

In extreme habitats, microorganisms require a large adaptation process until reaching optimal growth and reproduction. This evolutional redesign involves novel morphophysiological characteristics and modifications of genes and proteins, with subsequent changes in regulatory and metabolic pathways until epigenetic modifications, which have a great interest for biotechnological purposes. In fact, Zhang et al. (2018) reported a total of 314 new bioactive fungal natural products from 56 Ascomycota extremophilic fungi (asexual stage), including terpenoids/steroids, alkaloids/peptides/amides, quinones/phenols, esters/lactones, xanthones, and polyketides. Likewise, this is particularly true for their enzymes, which remain catalytically active under extremes of temperature, salinity, pH, and solvent conditions. Interestingly, some of these enzymes display polyextremophilicity (i.e., stability and activity in more than one extreme condition) making them widely functional in industrial biotechnology (Rampelotto 2013).

The exploitation, production, refining, and transportation of oil and its derivatives occasionally lead to technical and operational accidents with serious harm to the environment, some with irreversible destruction (León et al. 2009; Pernía et al. 2012, 2018). Polycyclic aromatic hydrocarbons are generated from both natural and anthropogenic processes, and are ubiquitous environmental pollutants with cytotoxicity, mutagenicity, and carcinogenicity capabilities. Due to their hydrophobic nature, they persist in the environments. More than two decades ago, the United States Environmental Protection Agency (USEPA) considered that some PAHs are toxic and possibly human carcinogens (Nadon et al. 1995).

There is wide fungal biodiversity with diverse enzymatic mechanisms that transform different hydrocarbon chemical structures, from short-chain aliphatics to heavy-weight polycyclic aromatics (Prenafeta-Boldú et al. 2018). The hydrocarbonoclastic fungi are a fascinating group of microorganisms with the unique ability to metabolize hydrocarbons as a sole source of carbon and energy, despite their low biodegradability due to their littler solubility and high hydrophobicity that limit their transport into microbial cells (Arulazhagan et al. 2017).

The term extremophilic hydrocarbonoclastic fungi is proposed here to describe a large and heterogeneous group of cultivable fungi which live optimally under extreme conditions, as well as are characterized by having a high ability to grow using hydrocarbons as the sole carbon source and energy. Usually, these fungi are isolated from soils, sediments, fluids, vapor, or water impregnated by petroleum or its derivatives. These extreme environments are mainly the consequence of anthropogenic activities, and usually have hard conditions to survive in for most known forms of life, such as high concentration of salt (hypersaline); high acidity or alkalinity (acidic or alkaline); high concentrations of high-molecular-weight compounds and toxicity pollutants (i.e., asphaltenes); heavy metals and heteroatoms such as sulfur; and high levels of PAHs with high toxicity. Thus, extremophilic hydrocarbonoclastic fungi are normally polyextremophiles, adapted to live in habitats under pressure of various physicochemical conditions considered "extreme."

Besides their polyextremophiles characteristics, these fungi have the unique ability to use hydrocarbons as a sole source of carbon and energy, despite their high toxicity and hydrophobicity properties; hence extremophilic hydrocarbonoclastic fungi play a pivotal role in the degradation/transformation of petroleum and its derivatives. This is the reason why the use of extremophilic hydrocarbonoclastic fungi as biocatalysts at great scale requires the maintenance and vegetative reproduction in the laboratory and industrial levels in the absence of a sexual stage. The asexual reproduction by mitotic division in fungi is commonly used to produce fungal mycelium to colonize the environments, which generates new identical individuals by remaining haploid, resulting in a progeny with the same genetic information as its own parental inoculum (Moore et al. 2011). Hence, the mitotic division guarantees researchers a simple, fast, and profuse vegetative reproduction of the fungal biocatalysts at the same time keeping their original hydrocarbonoclastic and polyextremophile characteristics.

The scope of this chapter is to demonstrate the application of the fungal biotechnology in the degradation and bioconversion of unconventional crude and its possible applications for the sustainable development of the petroleum industry. Contrary to other publications that are focused on bioremediation of light crude oils, here we present promising fungi with degradation potential of recalcitrant crudes like EHCO, a type of crude oil that is considered nonbiodegradable by microorganisms. At the same time, we demonstrate for the first time the potential of hydrocarbonoclastic fungi isolated from extreme environments (EHCO, hydrocarbon pits, crude distribution pipes, and the natural asphaltene Lake of Guanoco) to tolerate high concentration of EHCO, dibenzothiophene (DBT), phenanthrene, naphthalene, and pyrene in vitro conditions. We also described the relationship between lignin-degrading enzyme system (LDS) and the EHCO bioconversion by several extremophilic hydrocarbonoclastic fungi studied.

28.2 Biotechnology as the Measurement of Human Being and Biodiversity: Potential Applications for Sustainable Development of the Petroleum Industry

Since ancestral times, and based on its own essence, biotechnology started from the logic of human being, the observation, learning, and exploiting of nature to a benefit, pursuing a better quality life. This initiative was originally contemplative and through the centuries, based on success and error testing, formidably was ahead of its time and was evolving in the same way that knowledge evolved through the course of time. Thus, it could be considered that biotechnology occurs empirically from the activities of daily living of our ancestors, becoming implicit to the human being as a thinking and dynamic element of nature, which blooms before his eyes, building a magical altar of incommensurable bio-possibilities, helping to solve their most basic needs. Therefore, biotechnology dates from the same origin of human evolution as a thinking creature that expands human possibilities. Likewise, biotechnology could be considered as the oldest technology ever practiced by humans. In this process of singular complexity, the human being as subjects and from its knowledge obtains benefits from nature and its biodiversity, while, in the same act, nature *as subject and not as an object* is observed, acquired, and learned by humans. In this unique process, the bio-possibilities were greater and greater as the knowledge of the human kind was approaching to nature, and its exuberance biodiversity.

The harmonizing use of conventional oil production technologies with petroleum biotechnology emerges as an innovative strategy to help to ensure a sustainable development and the conservation of the environment. However, very little has been described concerning the medullary areas inside the petroleum industry where the biotechnology could have potential applications (Foght 2004). To identify medullary areas not only the products but also the operational processes associated to the petroleum industry in its whole-value chain should be known and the clear increase of the exploitation, production, and processing of unconventional crude oils should be considered. However, it is also important to understand the essence that fundamentally supports any biotechnological development: a specific problem or requirement to be solved, knowledge and technology, and biodiversity.

In principle, the petroleum biotechnology could be defined as *the study and rational use of the biodiversity, its genetic, enzymatic, and metabolic resources, to give high-added value to the products and processes of the petroleum industry in its whole-value chain, and to contribute to mitigate the environmental impact of its operational activities.* Major aims of the possible contribution of the complementary use of biotechnology in the petroleum industry would be focused on (1) the improvement of its productive capabilities (efficiency and productivity); (2) a comprehensive security of its operational processes (sustainability, safeguarded processes, and productivity); and (3) the remediation of its associated environmental liabilities (environmental sanitation, conservation, and sustainability).

Prenafeta-Boldú et al. (2018) commented that applied research on hydrocarbonoclastic fungi includes dedication to preventing biodeterioration as well as the potential application of fungal enzymatic capabilities for bioremediation purposes. Although there are numerous benefits and applications of hydrocarbonoclastic fungi, in this chapter only four main research and development areas are discussed (Fig. 28.1):

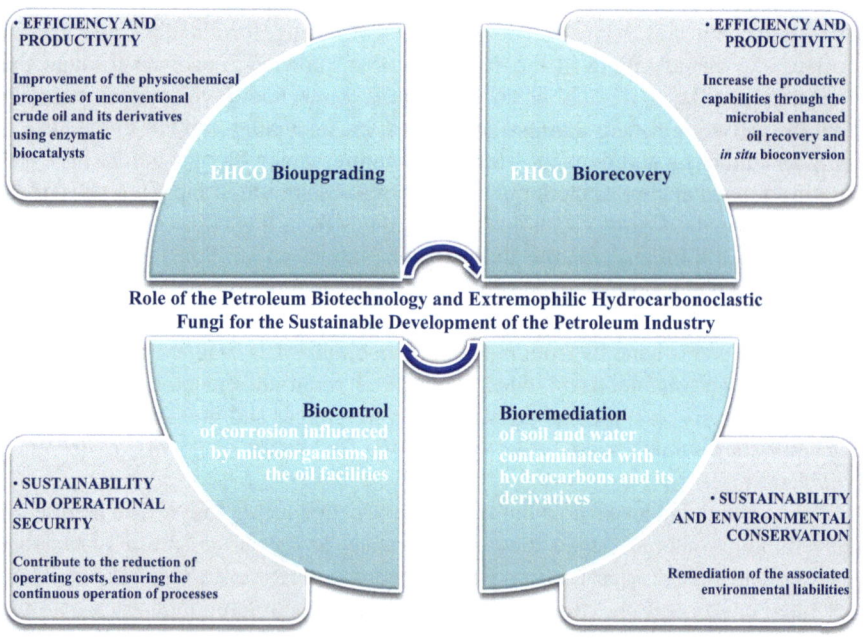

Fig. 28.1 Some contributions of the complementary use of biotechnology in the petroleum industry

1. Improvement of the physicochemical properties of unconventional crude oil and its derivatives using enzymatic biocatalysts, nanostructured support, and other coupled processes of bioconversion (**bioupgrading**)
2. Improvement of the recovery factor of oil in reservoirs by using promissory extremophile microorganisms and biodegradable polymers (MEOR, microbial enhanced oil recovery) (**biorecovery**)
3. Bioremediation of soil and water contaminated with hydrocarbons, oil pits, drilling waste, and industrial pollutants (**bioremediation**)
4. Contribution to the reduction of operating costs, ensuring the continuous operation of processes by the detection, monitoring, and control of microbial influenced corrosion, biofouling, and biodeterioration in oil facilities (**biocontrol**)

28.3 Fungal Biodiversity with Hydrocarbonoclastic Potential: From the Meta-analysis to the Tangible Reality Show

The work of Pernía et al. (2012) represents the first study where a meta-analysis was applied to the study of fungal diversity isolated from crude oil, its derivatives, and hydrocarbon-contaminated environments, including their hydrocarbonoclastic capabilities, through a detailed review of scientific literature published during the last century (from 1900 to 2012) in this passionate field of science. The results of this meta-analysis showed that the substrates with the highest fungal diversity were soils impacted with crude and natural asphalt, obtaining the lowest diversity in soil contaminated with diesel and gasoline in agreement with their high toxicity. The total fungal diversity isolated was constituted mainly by phylum Ascomycota (83%) followed by Zygomycota (10%), Basidiomycota (1%), and others (6%). In the phylum Ascomycota, the most common orders obtained were the Eurotiales (54%), followed by the Hypocreales (18%), Microascales (6%), and Saccharomycetales (5%), and the most predominant genera were *Penicillium* (18%), *Aspergillus* (17%), and *Fusarium* (6%).

Based on the percentage of degradation of total petroleum hydrocarbons (TPH), the group that presented greater degradation, TPH: $52 \pm 3.53\%$, was formed by the genera *Aspergillus, Bjerkandera, Coriolus, Emericella, Phanerochaete, Pleurotus, Rhizopus,* and *Trametes*; with respect to degradation rate of the saturated and aromatic fractions, two groups were obtained by the genera: (1) *Beauveria, Coriolopsis, Emericella, Fusarium, Phanerochaete, Pleurotus,* and *Trametes*, ($74.43 \pm 3.40\%$), and (2) *Coriolopsis, Fusarium, Pleurotus,* and *Trametes* ($97.75 \pm 2.25\ \%$). Likewise, evidence of degradation of resin and asphaltene fractions (10–28% and 10–40%, respectively) was found by the genera *Aspergillus, Candida, Emericella, Eupenicillium, Fusarium, Graphium, Neosartorya, Paecilomyces,* and *Meyerozyma* (Pernía et al. 2012).

28.3.1   Exploring Culturable Extremophilic Fungal Biodiversity with Potential Applications in the Sustainable Development of the Petroleum Industry: Knowing the Bio-possibilities

Fungi have essential roles in natural ecosystems through diverse forms of lifestyles, such as parasitism, mutualistic symbiotic associations with plant roots (e.g., arbuscular mycorrhizal fungi, ectomycorrhizal fungi, Orchid- and Ericoid-mycorrhizal fungi) and other organisms, as well as decomposers of the organic matter; hence fungi directly contribute to biogeochemical cycles (e.g., carbon, nitrogen, and phosphorous nutrients) and influence the greenhouse gas balance in the atmosphere on the global scale. In fact, the incorporation of mycorrhizae in global carbon cycle models is crucial in order to accurately predict ecosystem responses and feedbacks to climate change (Terrer et al. 2016).

The fungal decomposers can downgrade complex organic molecules such as cellulose, hemicellulose, lignin, pectin, starch, and non-synthesized xenobiotic compounds by natural metabolic processes, whose toxicity lies in their chemical nature and is persistent in the environment due to their low bioavailability (Dávila and Vázquez-Duhalt 2006, Naranjo et al. 2007, Pernía et al. 2018). The metabolism of organic compounds to less structurally complex products can directly or indirectly affect the growth of other surrounding microorganisms (Amund et al. 1987, Bartha and Atlas 1977, Brock 2015, Obire 1993, Odokuma and Opokwasili 1993, Pernía et al. 2018).

In the oil production value chain (processes of exploitation, production, refining, transportation of petroleum and its derivatives) there are occasional technical and operational accidents that release xenobiotic compounds into the environment promoting a selective pressure on the microbiota (Atlas and Bartha 1972; Calomiris et al. 1986; Pernía et al. 2018). The fungal catabolic activity (intracellular or extracellular) modifies bio-availability features of the xenobiotic compounds and derivatives, affecting the dynamics of autochthonous microbial communities (Coyne 2000). Subsequently, a selective enrichment process occurs for certain degraders and/or tolerant species, and the disappearance of those that do not have these capabilities (Benka-Coker and Ekundayo 1997).

In the study of the autochthonous fungal communities associated with oil-polluted environments it's essential to describe their microbial and functional biodiversity. However, for the selection of powerful extremophilic fungi as biocatalysts, the studies of both hydrocarbonoclastic potential and their tolerance to xenobiotic and toxic compounds abilities are crucial. Despite the unquestionable importance of the microbial and functional diversity studies that include the cultivable and non-cultivable microorganisms by means of metagenomics tools, in this work our effort was focused on the analyses of the cultivable extremophilic fungi from the extreme environments due to their potential use as promising biocatalysts for the sustainable development of the petroleum industry.

In this way, our research team started an exhaustive study on the fungal diversity associated with different types of samples: (1) Carabobo EHCO (samples taken directly from an oil well), (2) oil-polluted soil adjacent to an oil pit, (3) the natural asphalt Lake of Guanoco, and (4) petroleum naphtha distribution system (Fig. 28.2). The tolerance to EHCO and polyaromatic hydrocarbon (PAH) compound models, such as naphthalene, phenanthrene, DBT, and pyrene, was also investigated at inter- and intraspecific species levels. The isolation, maintenance, growth, and molecular identification of the fungal strains were performed according to Naranjo et al. (2007, 2013). The descriptive analysis into the functional group of fungi was carried out according to Pernía et al. (2012).

28.3.2 Isolation of Cultivable Fungal Biodiversity from Extreme Environments: Identifying the Cultivable Fungal Communities

At first, the relative frequencies of the different phylum and orders of the fungal strains isolated from the extreme environments were determined as mentioned above. All sites or samples studied were considerate extreme environments because they showed various physicochemical conditions unfavorable for survival of most known life forms.

In the case of the natural samples of asphalt and EHCO, these contain a heterogeneous mixture of organic compounds with a high concentration of asphaltenes and toxicity pollutants such as heavy metals and sulfur. Asphaltene's recalcitrancy is explained by the high degree of aromaticity combined with the presence of short alkyl chains (Naranjo et al. 2013). Carabobo EHCO assays have heavy American Petroleum Institute (API) gravity of 8.5°, sulfur concentration of 3.9%, and heavy metal Ni and V concentration of 480 mg/L (Pernía et al. 2018).

Oil-polluted soil from oil pits possesses a significant risk to wildlife and humans, due to its enormous mixture of organic compounds and chemical pollutants with high toxicity levels. During extraction and production processes in the Orinoco Oil Belt in Venezuela, oil pits generally store high volumes of sludge, oil spill, chemical treaters, formation water (brine), and drilling waste impregnated with EHCO and highly salinized water-based drilling fluids, which are mainly constituted by a wide range of corrosive compounds such as sodium bicarbonate, sodium carbonate, potassium chloride, potassium hydroxide, glycol, sodium hydroxide, thickeners, and lubricants, among others, xenobiotics that radically increase the alkalinity and salinity levels of the soil destroying its structure (Arellano et al. 2008, Naranjo et al. 2013).

Petroleum naphtha is a highly volatile and flammable intermediate hydrocarbon liquid stream derived from the refining of crude oil that contains paraffins, naphthenes (cyclic paraffins), and aromatic hydrocarbons. In this case, we refer to olefin-containing naphtha derived from the fluid catalytic cracking, visbreakers, and

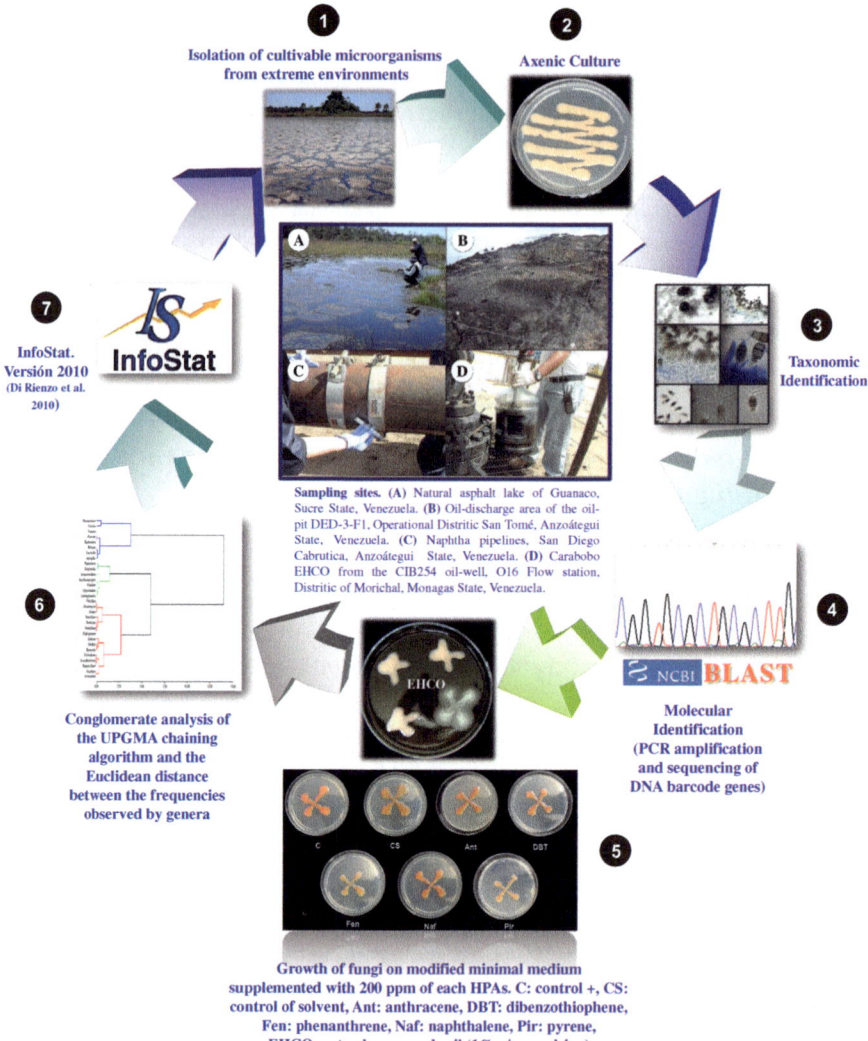

Fig. 28.2 Sampling sites and biotechnological strategy proposed. The samples were taken from (a) the natural asphalt lake of Guanaco, Sucre State, Ven; (b) the oil-discharge area of the oil pit DED-3-F1, Operational District San Tomé, Anzoátegui State, Ven; (c) the naphtha pipeline distribution system, San Diego Cabrutica, Anzoátegui State, Ven; (d) Carabobo EHCO from the CIB254 oil well, O16 Flow station, District of Morichal, Monagas State, Venezuela

coking processes used in many refineries (named cracked naphtha) that is broadly used by the Venezuelan petroleum industry as a solvent that decreases the high viscosity levels of the EHCO from OOB. According to the National Institute for Occupational Safety and Health (NIOSH), petroleum naphtha is immediately dangerous to life and health (CDC-NIOSH. 2015).

The biotechnological strategy proposed for identification of fungal communities from extreme environments is indicated in Fig. 28.2. Results showed, in terms of biodiversity, the most diverse substrate was the natural asphalt Lake of Guanoco, where 11 different fungal genera were isolated (including one basidiomycete fungus), followed by the oil-polluted soil (7 genera), and EHCO and naphtha, where only 3 genera and 1 genus were isolated, respectively (Table 28.1).

As reported by Pernía et al. (2012), the most isolated fungal group was phylum Ascomycota (91.66%) and only 8.33% belong to phylum Basidiomycota (Fig. 28.3a). In the phylum Ascomycota, the most common orders obtained were the Eurotiales (61.36%), followed by the Hypocreales (20.45%), Saccharomycetales (6.81%), Dothidiales (2.27%), Xylariaceae (2.27%), Pleosporales (2.27%), Microascales (2.27%), and incertae sedis (2.27%), which in this case correspond to the genus *Staphylotrichum* (Fig. 28.3b). Likewise, the Ascomycota genera with a higher number of species (in parenthesis) were *Aspergillus* (11), *Penicillium* (7), *Fusarium* (6), *Neosartorya* (4), *Trichoderma* (3), *Candida* (2), and *Byssochlamys* (2). The rest of the fungal strains were represented by only one species. The phylum Basidiomycota was represented by two orders Sporidiobolales (75%) and Polyporales (25%), which were comprised by the yeast-like fungi *Rhodotorula* (3) and *Pycnoporus sanguineus* (1), respectively (Fig. 28.3c).

From the list of isolated fungi (Table 28.1), some species have been reported as cosmopolitan fungi with worldwide distribution, such as *Aspergillus fumigatus* and *Aspergillus terreus*, *Penicillium glabrum* (formerly *Penicillium frequentans*) and *Penicillium oxalicum*, *Neocosmospora* (=*Fusarium*) *solani*, *Cladosporium sphaerospermum*, *Trichoderma viride*, and *Trichoderma inhamatum* (Domsch et al. 1980). *Penicillium oxalicum* was isolated from crude oil-impacted soil in the Borneo Islands (Chaillan et al. 2004). *Aspergillus fumigatus* has been widely reported as an inhabitant of soils polluted with hydrocarbons such as crude oil, kerosene, and diesel (April et al. 1998; Bento and Gaylarde 2001; Chaillan et al. 2004; Gesinde et al. 2008; Hemida et al. 1993; Oudot et al. 1993). In addition, *A. fumigatus* was reported as capable of degrading 20% of crude oil (Oudot et al. 1993). *Aspergillus terreus* was isolated from oil-polluted soils, gasoline, and kerosene (Chaillan et al. 2004; Colombo et al. 1996; Hemida et al. 1993; Uzoamaka et al. 2009), and possesses the capability of degrading up to 30% of crude oil (Algounaim et al. 1995). *Neocosmospora solani* was isolated from oil-polluted soils, asphalt, and kerosene (Colombo et al. 1996; Hemida et al. 1993; Naranjo et al. 2007, 2008) and could degrade up to 19% of the crude oil in 30 days (Chaineau et al. 1999).

Interestingly, only yeast-like fungi were isolated from EHCO (sample taken directly from an EHCO well), such as *Candida tropicalis, Candida viswanathii, Rhodotorula mucilaginosa*, and *Cyberlindnera* (=*Williopsis*) *saturnus*. *Candida tropicalis* was isolated previously from oil-polluted soils (April et al. 1998; Chaillan et al. 2004), in agreement with a prior report in which the yeasts *Candida palmioleophila* and *Meyerozyma* (=*Pichia*) *guilliermondii* were capable of the degradation of resins and asphaltenes (Chaillan et al. 2004).

Table 28.1 List of genus and species isolated from extreme environments

Diversity in Ascomycota Phylum

Genus	Species	Percentage (%)	Order	Species isolated	Sample's sites	Reference
Aspergillus	aureolus	25	Eurotiales	1	Natural asphalt lake	Naranjo et al. (2007)
	fumigatus			1	Natural asphalt lake	Naranjo et al. (2007)
	terreus			3	Natural asphalt lake	Naranjo et al. (2007)
	sp. 1			1	Oil-polluted soil	This manuscript
	sp. 2			1	Oil-polluted soil	This manuscript
	sp. 3			1	Oil-polluted soil	This manuscript
	sp. 4			1	Oil-polluted soil	This manuscript
	sp. 5			1	Oil-polluted soil	This manuscript
	sp. 6			1	Oil-polluted soil	This manuscript

Penicillium	aculeatum (=Talaromyces aculeatus)	15.91	Eurotiales	1	Natural asphalt lake	Naranjo et al. (2007)
	glabrum (formerly Penicillium frequentans)			1	Oil-polluted soil	This manuscript
	indonesiae			1	Natural asphalt lake	Naranjo et al. (2007)
	marneffei			1	Natural asphalt lake	Naranjo et al. (2007)
	oxalicum			2	Natural asphalt lake	Naranjo et al. (2007)
	sp. 37			1	Natural asphalt lake	Naranjo et al. (2007)
Fusarium	proliferatum	13.63	Hypocreales	2	Natural asphalt lake	Naranjo et al. (2007)
	solani (=Neocosmospora solani)			2	Natural asphalt lake	Naranjo et al. (2007)
	venenatum			2	Natural asphalt lake	Naranjo et al. (2007)
Neosartorya	spinosa	9.1	Eurotiales	2	Natural asphalt lake	Naranjo et al. (2007)
	pseudofischeri			2	Natural asphalt lake	Naranjo et al. (2007)
Cladosporium	sphaerospermum	2.27	Dothidiales	1	Natural asphalt lake	Naranjo et al. (2007)
Trichoderma	viride	6.81	Hypocreales	2	Natural asphalt lake	Naranjo et al. (2007)
	inhamatum			1	Natural asphalt lake	Naranjo et al. (2007)

(continued)

Table 28.1 (continued)

Diversity in Ascomycota Phylum

Genus	Species	Percentage (%)	Order	Species isolated	Sample's sites	Reference
Candida	viswanathii	4.54	Saccharomycetales	1	EHCO	Naranjo et al. (2008)
	tropicalis			1	EHCO	Naranjo et al. (2008)
Fennelia	nivea var. indica	2.27	Eurotiales	1	Oil-polluted soil	This manuscript
Byssochlamys	nivea	4.54	Eurotiales	2	Oil-polluted soil	This manuscript
Eupenicillium	javanicum	2.27	Eurotiales	1	Oil-polluted soil	This manuscript
Paecilomyces	sp.	2.27	Eurotiales	1	Natural asphalt lake	Naranjo et al. (2007)
Pestalotiopsis	palmarum	2.27	Xylariaceae	1	Natural asphalt lake	Naranjo et al. (2007, 2011)
Phoma	glomerata (=Didymella glomerata)	2.27	Pleosporales	1	Natural asphalt lake	Naranjo et al., (2007)
Pseudallescheria	angusta	2.27	Microascales	1	Natural asphalt lake	Naranjo et al., (2007)
Staphylotrichum	sp.	2.27	Incertae sedis	1	Oil-polluted soil	This manuscript
Williopsis	saturnus (=Cyberlindnera saturnus)	2.27	Saccharomycetales	1	EHCO	Naranjo et al. (2008)

Diversity in Basidiomycota Phylum

Rhodotorula	mucilaginosa	75	Sporidiobolales	1	EHCO	This manuscript
	mucilaginosa			1	Oil-polluted soil	This manuscript
	mucilaginosa			1	Naphtha pipelines	Naranjo et al. (2015)
Pycnoporus	sanguineus	25	Polyporales	1	Natural asphalt lake	Urbina et al. (2007); Pernía et al. (Non-published)

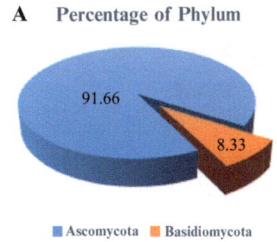

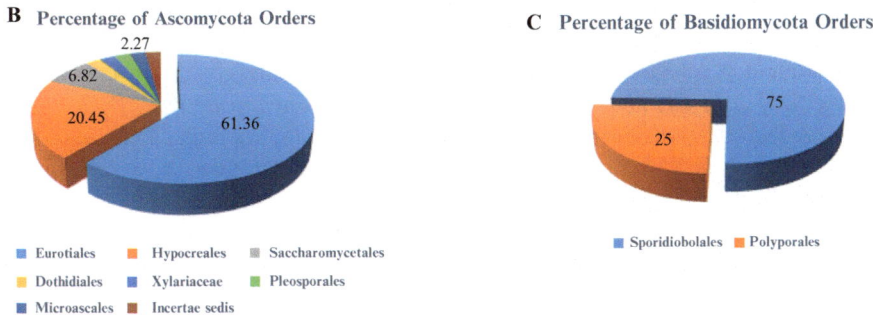

Fig. 28.3 Percentage of fungi isolated from extreme environments according to phylum and order

Surprisingly, the common species isolated from all petroleum substrates was *R. mucilaginosa* which, as further discussed, is one of the most tolerant species to EHCO and PAHs. *Rhodotorula mucilaginosa* was also isolated by Gallego et al. (2007), reporting that it could degrade linear alkanes, C11-C33, branched alkanes, isoprenoids, and cycloalkanes. Other investigations have described *R. mucilaginosa* as halotolerant (it's able to grow in concentrations higher than 2.5 M of NaCl); it tolerates a wide pH range of 2–10, and is capable of degrading nitrobenzene (Gross and Robbins 2000; Lahav et al. 2002; Turk et al. 2011; Zheng et al. 2009; Urbina and Aime 2018). More recently, Naranjo et al. (2015) described the isolation and molecular identification of fungal strains from naphtha systems in the oil industry, where the naphtha-tolerant *R. mucilaginosa* was the most predominant yeast-like fungal species.

The rest of the fungal species, such as *Byssochlamys lagunculariae* (formerly *Byssochlamys nivea*), *Fusarium venenatum*, *Trametes coccinea*, and *Staphylotrichum* sp., are new reports for science as species capable of inhabiting EHCO or oil-polluted soils that also have a great potential to be used for mycoremediation or EHCO bioconversion purposes.

28.3.3 Fungal Screening to Determine Hydrocarbonoclastic Potential and Tolerance to EHCO and HPAs: Obtaining the Powerful Biocatalysts

Our in vitro results show that all isolated fungi were tolerant to 1% EHCO. The most toxic compounds for fungi were DBT and phenanthrene, which showed a growth inhibition of more than 50%. These findings are presented in a cladogram where fungal species are discriminated by functional groups according to their tolerance to EHCO and PAHs (Fig. 28.4) as follows:

Group No. 4 is the most tolerant capable of growth at 100% rate in all hydrocarbons studied in comparison to the control medium. Interestingly, this group is constituted by the filamentous fungi species *B. lagunculariae* 87, *Penicillium* (=*Eupenicillium*) *javanicum*, and *Penicillium* sp. 37, and the unicellular fungi *C. tropicalis*, *C. viswanathii*, *R. mucilaginosa* (three strains), and *Ci. saturnus*. All these unicellular fungi were isolated from EHCO and, in the case of *R. mucilaginosa*, were also isolated from oil-polluted soils and naphtha distribution system. These results strongly suggested that the fungal strains with greater hydrocarbonoclastic and tolerance capabilities are associated with EHCO wells and belong to unicellular fungi.

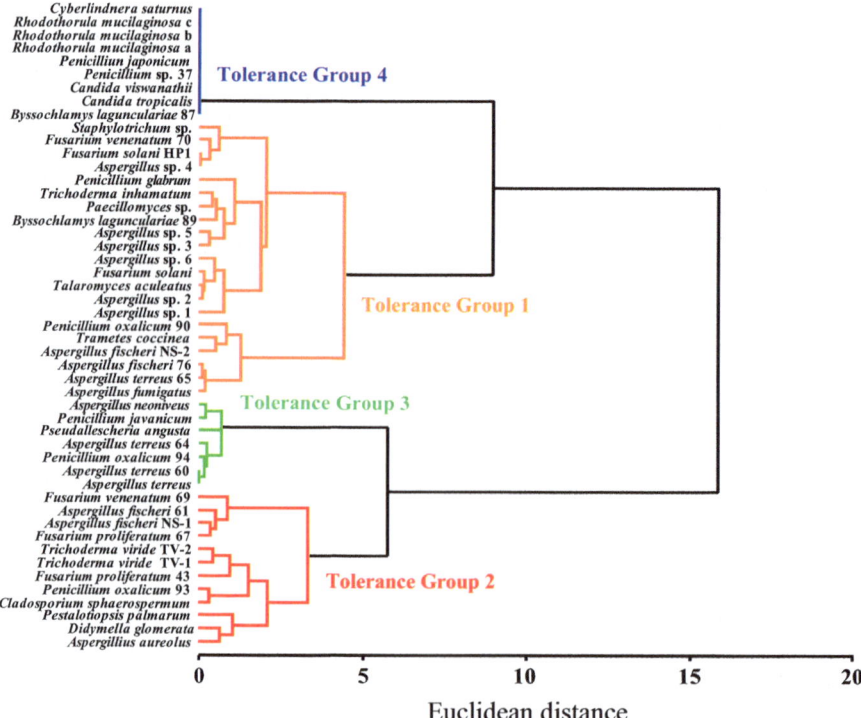

Fig. 28.4 Dendrogram of tolerance groups of fungal species according to their tolerance to PAHs

Group No. 1, the second most tolerant group, comprised of fungi that could grow at 95.81 ± 2.35% rate in EHCO, 92.62 ± 2.32% rate in naphthalene, and 80.14 ± 3.52% rate in pyrene. However, their growth was inhibited at approximately 50% rate in DBT and phenanthrene. This group clustered the major amount of fungal strains studied and was constituted by *Aspergillus* sp. (1–6), *A. fumigatus*, *A. terreus*, *B. lagunculariae* 89, *N. solani*, *Fusarium venenatum* 70, *A. fischeri* (formerly *Neosartorya pseudofischeri* and *Neosartorya spinosa*), *Paecilomyces* sp., *Talaromyces* (=*Penicillium*) *aculeatum*, *Penicillium glabrum* (formerly *Penicillium frequentans*), *Penicillium oxalicum*, *Trametes coccinea*, *Staphylotrichum* sp., and *Trichoderma inhamatum*. The rest of the tolerance groups (Nos. 2 and 3) were the less tolerant, and their growth was inhibited to more than 50% rate in the same conditions.

According to ranges of HPA tolerance, Group No. 4 has a 100% tolerance for all PAHs studied. Group No. 3 tolerates CEP and DBT 100% but it is less tolerant to phenanthrene (12–45%), naphthalene (38–57%), and pyrene (38–54%). Group No. 2 tolerates CEP in 19–61%, DBT (0–57%), phenanthrene (0–60%), and pyrene (0–100%). Finally, Group No. 1 showed a low tolerance to CEP (11.25–62%) but with a wide range of tolerance to the rest of the PAHs studied: DBT (0–100%), phenanthrene (9–100%), naphthalene (75–100%), and pyrene (50–100%).

The discrimination of the different tolerance groups of fungi according to their tolerance rate to EHCO and PAH compounds was also evaluated (Fig. 28.5). Group No. 4 clustered fungi with the same proportion tolerance for all the hydrocarbons, while Group No. 3 showed a greater tolerance for EHCO, and finally fungal species in Groups No. 1 and 2 had greater tolerance to naphthalene alone. Likewise, this analysis corroborates the fact that DBT and phenanthrene were the most toxic PAHs for the fungi isolated in this study.

Another important result observed at the present work was the difference observed in tolerance to hydrocarbons between species of the same genus, including the difference between strains of the same species (Fig. 28.6). It is known that a functional group is composed of microorganisms that, regardless of their taxonomic classification, present an identical pattern of biochemical responses in the use and transformation of organic substrates.

In the case of the genus Aspergillus, any of its species could tolerate DBT and phenanthrene, and their growth was inhibited to more than 50% rate (Fig. 28.6a). Among Aspergillus species, the most tolerant were *Aspergillus* sp. 4–6, which were isolated from an oil-polluted soil adjacent to an oil pit. As can be appreciated, the species isolated from this site also showed a higher tolerance for DBT, phenanthrene, and naphthalene, in comparison to the species isolated from the natural asphalt Lake of Guanoco.

In the case of Fusarium species, the most tolerant species was *N. solani*, widely reported as a hydrocarbon degrader (León et al. 2007; Naranjo et al. 2007, 2008), and *F. venenatum*, which has not been previously reported as a hydrocarbon degrader (Fig. 28.6b). Interestingly, different strains of *F. venenatum* isolated from natural asphalt Lake of Guanoco showed different degrees of tolerance to EHCO as strain 70 grew at 100% rate while strain 69 only grew at 19% rate, compared to the control medium.

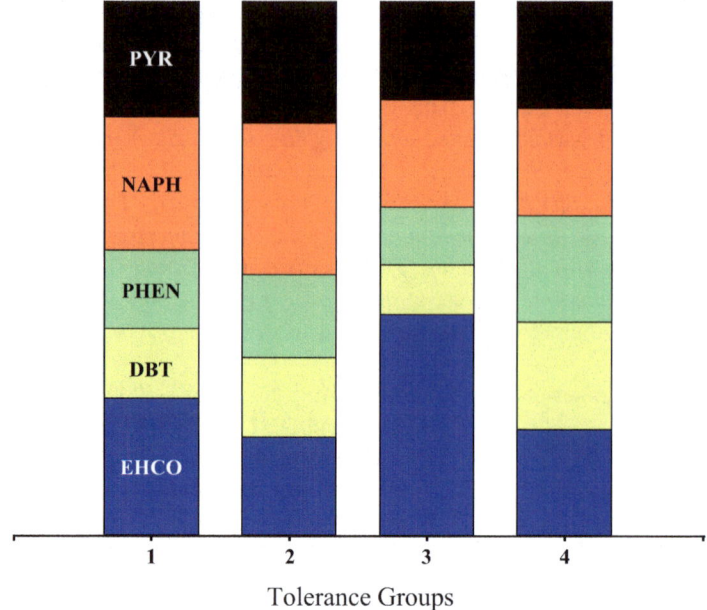

Fig. 28.5 Proportion of tolerance to the different PAHs by fungal group corresponding to the dendrogram. *PYR* Pyrene, *NAPH* Naphthalene, *PHEN* Phenanthrene, *DBT* Dibenzothiophene, *EHCO* Extra-heavy crude oil

In the case of Penicillium species, the most tolerant species were *Penicillium* sp. 37 and *Penicillium javanicum*, both able to grow at 100% rate in all hydrocarbons tested (Fig. 28.6c). However other species of this same genus (such as *P. oxalicum*) were not able to tolerate hydrocarbons with growth inhibitions greater than 50% rate.

The results showed here strongly suggested that the fungal strains belonging to the same species do not necessarily have the same hydrocarbonoclastic abilities and characteristics to tolerate xenobiotic compounds derived from petroleum. Likewise, fungal strains belonging to the same species but isolated from different places do not necessarily have the same characteristic and behavior.

28.4 Relationship Between the LDS and EHCO Bioconversion in Mitosporic Extremophilic Hydrocarbonoclastic Fungi: Inducing a Powerful Exoenzymatic System

The LDS includes a large range of oxidoreductases and hydroxylases, such as laccases (LACs) and high-redox-potential ligninolytic peroxidases like lignin peroxidase (LIPs), manganese peroxidases (MNPs), versatile peroxidases (VEPs), and others. Lignin peroxidases are able to directly oxidize non-phenolic units, while

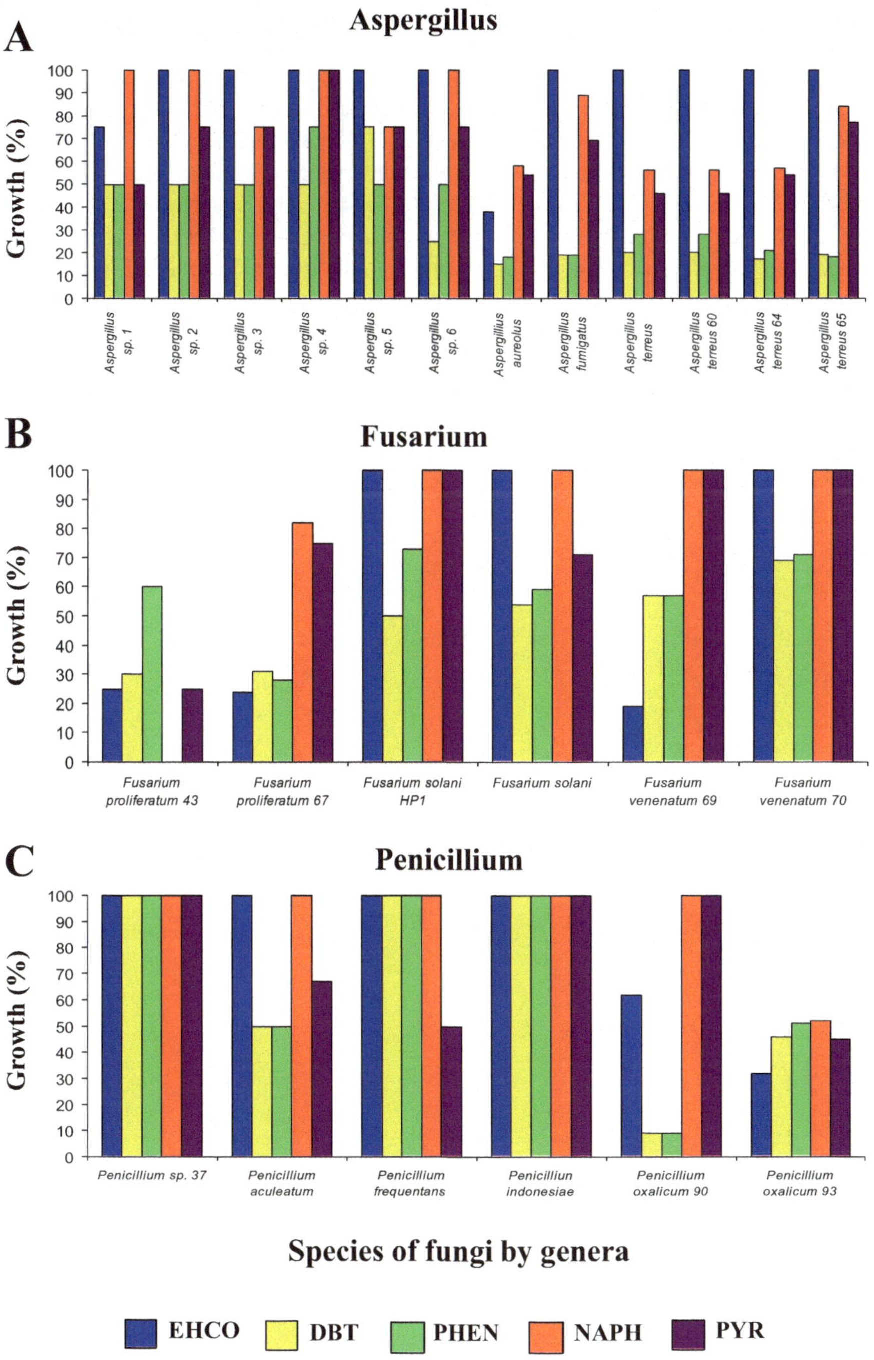

Fig. 28.6 Percentages of growth in the presence of PAHs compared with the control of fungal species of the genera: (**a**) *Aspergillus*, (**b**) *Fusarium*, and (**c**) *Penicillium*. *PYR* Pyrene, *NAPH* Naphthalene, *PHEN* Phenanthrene, *DBT* Dibenzothiophene, *EHCO* Extra-heavy crude oil

MNPs and LACs oxidize preferentially phenolic units, but also act on non-phenolic units when mediators (e.g., 2,2′-azino-bis(3-ethylbenzothiazoline-6-sulfonic acid), commonly known as ABTS) are present in the reaction mixture, whereas VEPs are able to combine the catalytic properties of LIPs and MNPs (Martínez et al. 1996, 2005; Ruiz-Dueñas et al. 1999; Saparrat et al. 2002; Gianfreda and Rao 2004).

Concurrently, research is focused on the conversion of heavy oils to lighter oils mediated by biocatalyst. A few of the biotechnological strategies to improve the EHCO physical-chemical properties have promoted an increase in their commercial value. Our research team proposed 13 years ago the use of the extracellular oxidative enzymes (EOE) of the LDS present in extremophilic fungi as biocatalysts (Naranjo et al. 2007, 2008; León et al. 2007) based on the following facts: (1) lignin is a complex biopolymer structurally similar to asphaltenes, molecules responsible for the high viscosity of EHCO (Fig. 28.7), and (2) ligninolytic fungi have a unique and powerful LDS that includes a wide range of unspecific oxidoreductases and hydrolases which are involved in the transformation and degradation processes of

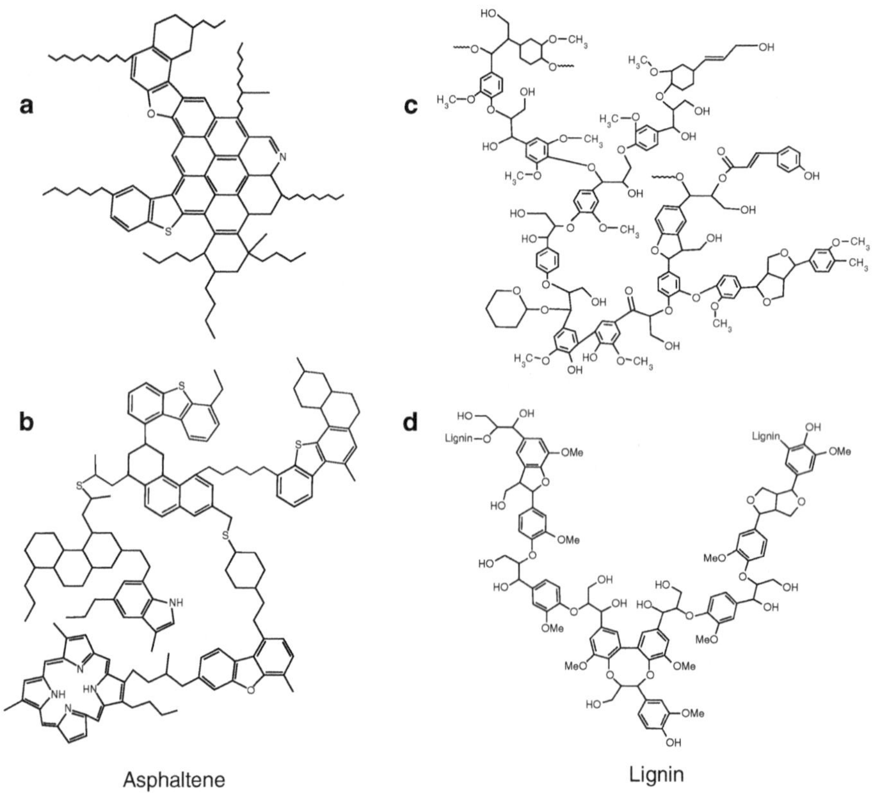

Asphaltene

Lignin

Fig. 28.7 Representative molecular structures of asphaltenes and lignin. (**a**) A single condensed polycyclic aromatic core; (**b**) multiple smaller polycyclic aromatic cores with aliphatic bridges; (**c**, **d**) lignin structures

polymeric substances in partially degraded and oxidized soluble products that can be easily assimilated by microorganisms (Gianfreda and Rao 2004, Naranjo et al. 2007, 2013). Studies dealing with the interactions between extremophilic microorganisms and crude oil have led to the identification of biocatalysts which, through multiple biochemical pathways, carry out the desulfurization, denitrogenation, and demetallation reactions in oils.

The efficiency of the degradation process by biological systems depends on the susceptibility of these compounds to be degraded and converted into less toxic compounds; therefore, supposed susceptibility is directly related to the chemical structure, concentration, and physical properties of the hydrocarbon (Foght 2004, Gianfreda and Rao 2004; Naranjo et al. 2007, 2013). If the hydrocarbons are soluble, they can be assimilated by the microorganisms more easily and, if they are insoluble they must be transformed into soluble compounds so that they can be used later. However, the first step in the transformation of insoluble compounds is usually catalyzed by extracellular oxidative enzymes of microorganisms, which are released by the cells to the surrounding environment (Foght 2004).

The fungal mineralization and transformations of EHCO and/or its asphaltene fraction by fungi have been documented previously; however they are still controversial. The ability of the fungi *Aspergillus flavipes*, *Aspergillus* (=*Emericella*) *nidulans*, *P. javanicum*, and *Parascedosporium putredinis* (=*Graphium*) *putredinis* to degrade resins (15–28%) and asphaltenes (15–40%) was described by Oudot et al. (1993). Chaillan et al. (2004) reported the ability of filamentous fungi *Albonectria rigidiuscula* (formerly *Fusarium decemcellulare*) and *Paecilomyces variotii*, and the yeasts *Candida palmioleophila* and *M. guilliermondii* to degrade a range of 10–15% of resins and asphaltenes. Transformations of petroporphyrins and asphaltenes by chloroperoxidase (CPO) of *Caldariomyces fumago*, a protein with high peroxidase activity and versatility that can halogenate aromatic molecules like polycyclic aromatic hydrocarbons (PAHs), were described by Fedorak et al. (1993).

Later, García-Arellano et al. (2004) showed through Fourier-transform infrared spectroscopy (FT-IR) that a chemically modified cytochrome C catalyzed the oxidation of carbon and sulfur atoms in the rich fraction of petroporphyrins of the asphaltenes, with 74% and 95% Ni and V removal, respectively. The chemical modification of the surface of the protein with polyethylene glycol resulted in the formation of a conjugate polymer-protein soluble in organic solvents, and the methyl esterification of the heme group increasing the hydrophobicity of its active site. In fact, the highest activity was detected in a tertiary mixture of solvents with 5% water. In this case, the enzymatic biotreatment of asphaltenes represents an interesting alternative for the elimination of heavy metals and the reduction of poisoning catalysts after cracking and hydrocracking of crude oils.

Recently, Uribe-Álvarez et al. (2011) described for the first time the ability of *A. fischeri*, isolated from natural asphalt Lake of Guanoco in Venezuela, of growing on asphaltenes as the sole carbon and energy source and to mineralize 13.2% of asphaltenes. Ayala et al. (2012) described the biotransformation of porphyrin-free asphaltene fraction catalyzed by CPO-based biocatalyst to reduce coke formation during thermal decomposition in the oil industry. On the other hand, Hernández-

López et al. (2016) evaluated the capacity of *A. fischeri* to transform high-molecular-weight polycyclic aromatic hydrocarbons (HMW-PAHs) and asphaltenes as the sole carbon source by reverse-phase high-performance liquid chromatography (HPLC), nano-LC mass spectrometry, and IR spectrometry, together with a comparative microarray study and the complete genome of *A. fischeri*. The formation of hydroxy and ketone groups on the PAH molecules and the internalization of aromatic substrates into fungal cells suggested oxidation of these recalcitrant compounds mediated by the cytochrome P450 system.

Finally, Pourfakhraei et al. (2018) described using saturate, asphaltene, resin, and aromatic (SARA) analysis chromatograms that the wood-decaying fungus *Daedaleopsis* sp. can degrade asphaltene and aromatic fraction with a reduction of 88.7% and 38%, respectively, with an increase of 44.8% in the saturate fraction. They also reported that LAC-, LIP-, and MNP-specific activities from LDS were induced in the presence of heavy crude oil (HCO) as the sole carbon source and energy.

In order to confirm the relationship between LDS and bioconversion of EHCO claimed by Naranjo et al. (2013), the study of the enzymatic activities of specific LACs, LIPs, and MNPs induced in minimal medium Czapek supplemented with EHCO as the sole source of carbon and energy was performed. The fungi *Fusarium proliferatum* (BM-02), *Pestalotiopsis palmarum* (BM-04), *Aspergillus terreus* (BM-36), and *Pseudallescheria angusta* (BM-39) were randomly selected from the list of fungal strains isolated from extreme environments studied (Table 28.1). At first, a screening was carried out to confirm their hydrocarbonoclastic capabilities against several PAH compounds (naphthalene, phenanthrene, dibenzothiophene, and pyrene) according to Naranjo et al. (2013). The phenotypic results showed that all fungi were able to grow using EHCO and all PAHs with the exception of *F. proliferatum* (BM-02) and *Ps. angusta* (BM-39) that were unable to grow using phenanthrene (Table 28.2, Fig. 28.8). Regarding the study of the enzymatic activities, the results showed that all fungi induced the enzymatic activities studied in the presence of EHCO compared with the control (Fig. 28.9a, b and c). However, interest-

Table 28.2 Filamentous fungi isolated from extreme environments and their abilities to grow using PAHs or EHCO as the sole source of carbon and energy

Specie	Strain No.	Growth in modified minimal culture medium Czapek compared with the control					
		Naphthalene	Phenanthrene	Pyrene	DBT	EHCO	Sucrose (+)
Fusarium proliferatum	BM-02	++	−	++	++	++	++
Pestalotiopsis palmarum	BM-04	+	+	+	+	+	+
Aspergillus terreus	BM-36	++	++	++	++	++	+++
Pseudallescheria angusta	BM-39	++	−	+++	+++	+++	+++

No growth: −; weak growth: +; moderate growth: ++; good growth: +++

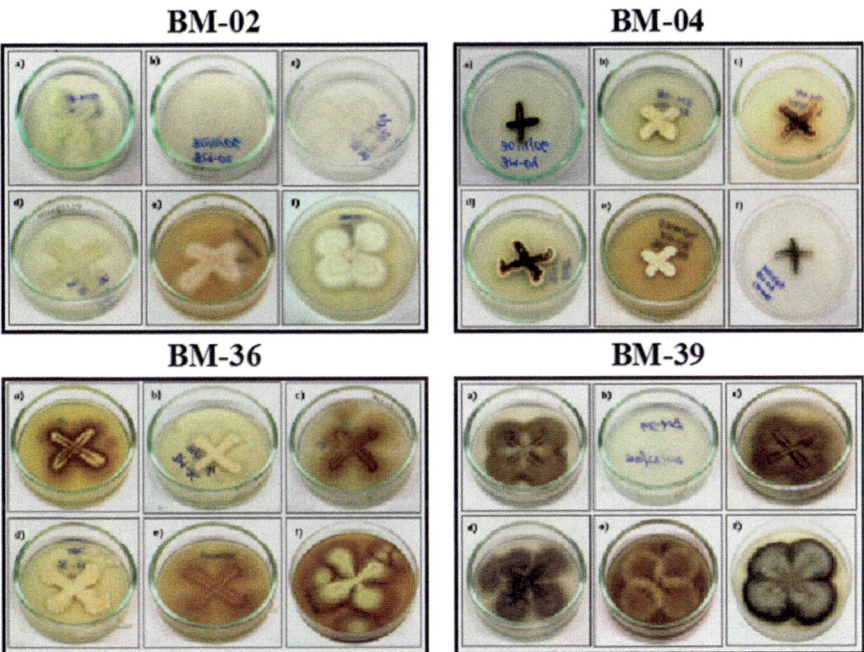

Fig. 28.8 Growth in minimal modified culture medium Czapek supplement with several PAHs as the sole carbon source, such as naphthalene, phenanthrene, pyrene, dibenzothiophene (DBT), and extra-heavy crude oil (EHCO). The fungi *Fusarium proliferatum* (BM-02), *Pestalotiopsis palmarum* (BM-04), *Aspergillus terreus* (BM-36), and *Pseudallescheria angusta* (BM-39) were randomly selected from the list of extremophilic fungi isolated from extreme environments

ingly, the lignin peroxidase was strongly induced with EHCO in the fungi *Pestalotiopsis palmarum* (BM-04), *Aspergillus terreus* (BM-36), and *Pseudallescheria angusta* (BM-39), with the exception of *Fusarium proliferatum* (BM-02), which showed a few inductions in all activities studied (Fig. 28.9c).

28.5 Conclusions and Future Perspectives

It is unquestionable that the modern world was created to the image of petroleum with a concomitant reduction of conventional crude oil reserves and a rising of global demand for fuels and oil derivatives. Consequently, with the increasing of exploitation of unconventional crude reserves, the development and improvement of clean-alternative fuel technologies are required with the mandatory establishment of innovative protocols for unconventional hydrocarbon exploitation. The studying and application of the petroleum biotechnology with promissory microorganisms, specially extremophilic hydrocarbonoclastic fungi and their powerful LDS, in the

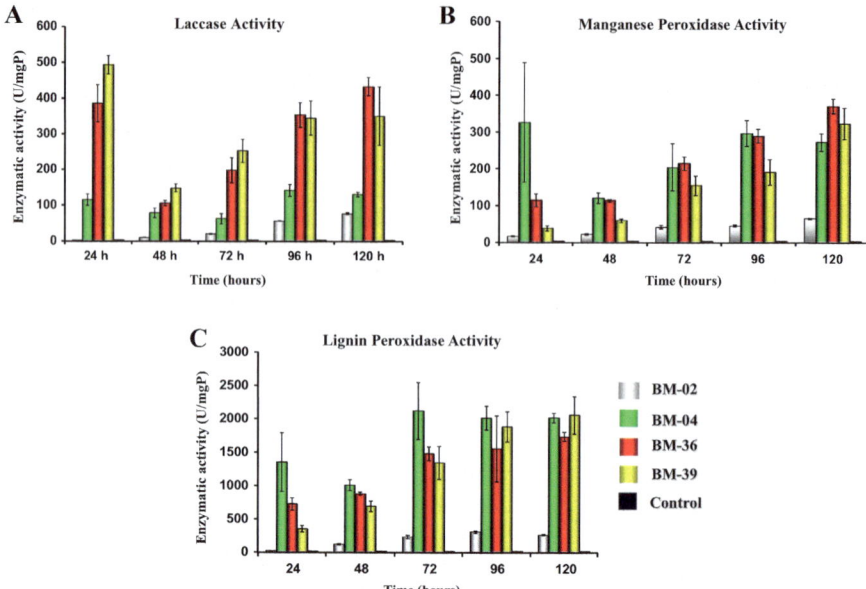

Fig. 28.9 Relationship between the LDS and the bioconversion of EHCO in the fungi *Fusarium proliferatum* (BM-02), *Pestalotiopsis palmarum* (BM-04), *Aspergillus terreus* (BM-36), and *Pseudallescheria angusta* (BM-39)

whole oil industry production chain, is a pathway for improving products and processes, guaranteeing the decreasing of operational costs with increased productivity capabilities with minimum environmental impact. Here, we introduced the term extremophilic hydrocarbonoclastic fungi as a large and heterogeneous group of cultivable fungi which can live optimally under extreme conditions and are characterized by having a high ability to grow using hydrocarbons as the sole carbon source and energy.

Treatment of the unconventional hydrocarbons under extreme conditions plays a vital role to save the ecosystems for anthropogenic intervention which leads to sustainable development. This book chapter shows, through a sequential and comprehensible explanation, a biotechnological strategies to study cultivable fungal biodiversity inhabiting in extreme environments for the isolation of powerful biocatalysts, following a simple and fast screening to determine both their hydrocarbonoclastic potential and tolerance to EHCO and HPAs.

The exploration of cultivable fungal communities from the extreme environments studied allows us to identify phylum Ascomycota as the most isolated fungal group, being Eurotiales the most common order obtained, followed by Hypocreales and Saccharomycetales, and the genera with the most number of species *Aspergillus* (11), *Penicillium* (7), *Fusarium* (6), *Neosartorya* (4), *Trichoderma* (3), *Byssochlamys* (2), and *Candida* (2), respectively.

Only a few strains of the phylum Basidiomycota were isolated, composed mainly by the orders Sporidiobolales followed by the Polyporales, which are constituted by the yeast-like fungi *Rhodotorula* (3) and *Pycnoporus sanguineus* (1), respectively. However, surprisingly the more common species isolated from all of the extreme environments studied was *R. mucilaginosa*.

It is well understood that the study of microorganisms with potential in oil biotechnology requires research of both their hydrocarbonoclastic potential and tolerance abilities to EHCO and HPAs. These abilities have been pointed out here as crucial for the appropriated selection of powerful hydrocarbonoclastic extremophilic fungi as biocatalysts. In our research, the fungal strains with greater hydrocarbonoclastic and tolerance capabilities were isolated from Carabobo-EHCO wells and belonging mainly to unicellular fungi, such as *C. tropicalis*, *C. viswanathii*, *Ci. saturnus*, and *R. mucilaginosa*. These yeast-like fungi are the most tolerant species to EHCO and PAHs.

Another important result observed during our investigations was the difference obtained in tolerance to hydrocarbons between species of the same genus, including the difference between strains of the same species, such as *Aspergillus*, *Fusarium*, and *Penicillium*. Likewise, fungal strains belonging to the same species but isolated from different extreme environments do not necessarily have the same hydrocarbonoclastic characteristic and behavior.

On the other hand, the relationship between the powerful LDS and EHCO bioconversion in extremophilic hydrocarbonoclastic fungi was confirmed, where a strong induction of the lignin peroxidase activity and a low induction of the laccase activity were obtained repetitively, in all the fungal strains studied.

Further research is required to develop new visions and perspectives, novel clean technologies, confident alternatives, and strategies to ensure an economically profitable development with social justice, and ecological sustainability, safeguarding the future and quality of life for the next generations. The perspective of this chapter is the potential application of the promissory extremophilic hydrocarbonoclastic fungi to be used as biocatalysts for mycoremediation or EHCO bioupgrading process under stressed conditions to increase the revenues for the industry dedicated to the exploitation of unconventional crudes.

Acknowledgments This work was supported by the Projects FONACIT No. 2005000440, Sub-Project 3: MISIÓN CIENCIA No. 2007001401 and FONACIT No. G- 2011000330. The authors recognize Dr. V. León for being a pioneer of the Petroleum Biotechnology in Venezuela, and dedicate this work to the memory of Dr. J. Demey, rest in peace. The authors thank Judith Nyisztor and Aitana Naranjo for grammatical support.

References

AlGounaim MY, Diab A, AlAbdulla R, AlZamin N (1995) Effect of petroleum oil pollution on the microbiological populations of the desert soil of Kuwait. Arab Gulf J Sci Res 13(3):653–672

Amund O, Adebowale AA, Ugoji EO (1987) Occurrence and characteristics of hydrocarbon-utilizing bacteria in Nigerian soils contaminated with spent motor oil. Indian J Microbiol 27:63–67

April TM, Abbott SP, Foght JM, Currah RS (1998) Degradation of hydrocarbons in crude oil by the ascomycete *Pseudallescheria boydii* (Microascaceae). Can J Microbiol 44:270–278

Arellano T, Infante C, Naranjo L (2008) Manejo integral de fosas de hidrocarburos generadas por la actividad petrolera venezolana. Thesis for Magister in Environmental Management, UNEFA 1–250

Arulazhagan P, Mnif S, Rajesh Banu J, Huda Q, Jalal MAB (2017) HC-0B-01: biodegradation of hydrocarbons by extremophiles. In: Heimann K, Karthikeyan O, Muthu S (eds) Biodegradation and bioconversion of hydrocarbons. Environmental footprints and eco-design of products and processes. Springer, Singapore, pp 137–162

Atlas RM, Bartha R (1972) Degradation and mineralization of petroleum by two bacteria isolated from coastal waters. Biotechnol Bioeng 14:297–308

Ayala M, Hernández-López EL, Perezgasga L, Vázquez-Duhalt R (2012) Reduced coke formation and aromaticity due to chloroperoxidase-catalyzed transformation of asphaltenes from Maya crude oil. Fuel 92:245–249

Bartha R, Atlas RM (1977) The microbiology of aquatic oil spills. Adv Appl Microbiol 22:225–226

Benka-Coker MO, Ekundayo JA (1997) Applicability of evaluating the ability of microbes isolated from an oil spill site to degrade oil. Environ Monit Assess 45:259–272

Bento FM, Gaylarde CC (2001) Biodeterioration of stored diesel oil: studies in Brazil. Int Biodeter Biodegr 47:107–112

Calomiris JJ, Austin B, Walker JD, Colwell RR (1986) Enrichment for estuarine petroleum-degrading bacteria using liquid and solid media. J Appl Bacteriol 42:135–144

CDC-NIOSH (2015) Pocket guide to chemical hazards-petroleum distillates (naphtha). www.cdc.gov

Chaillan F, Flèche AL, Bury E, Phantavong Y, Grimont P, Saliot A, Oudot J (2004) Identification and biodegradation potential of tropical aerobic hydrocarbon-degrading microorganisms. Res Microbiol 155:587–595

Chaineau CH, Morel J, Dupont J, Bury E, Oudot J (1999) Comparison of the fuel oil biodegradation potential of hydrocarbon-assimilating microorganisms isolated from a temperate agricultural soil. Sci Total Environ 227:237–247

Colombo JC, Cabello M, Arambarri AM (1996) Biodegradation of Aliphatic and aromatics hydrocarbons by natural soil microflora and pure cultures of imperfect and lignolytic fungi. Environ Pollut 94:355–362

Coyne M (2000) Microbiología del Suelo: un enfoque exploratorio, 1ª Edición edn. Editorial Paraninfo SA, Madrid, Spain, pp 1–440

Dávila A, Vázquez-Duhalt R (2006) Enzimas ligninolíticas fúngicas para fines ambientales. Mensaje Bioquímico 30:29–55

Domsch KH, Gams W, Anderson TH (1980) Compendium of soil fungi, vol 1, pp 1–860

Fedorak PM, Semple KM, Vazquez-Duhalt R, Westlake DWS (1993) Chloroperoxidase mediated modifications of petroporphyrins and asphaltenes. Enzyme Microb Technol 15:429–437

Foght JM (2004) Whole-cell bio-processing of aromatic compounds in crude oil and fuels. In: Petroleum biotechnology: developments and perspectives. Elsevier Science, Amsterdam, pp 145–175

Gallego JLR, García-Martınez MJ, Llamas JF, Belloch C, Pelaez AI, Sanchez J (2007) Biodegradation of oil tank bottom sludge using microbiol Consortia. Biodegradation 18:269–281

García-Arellano H, Buenrostro-Gonzalez E, Vazquez-Duhalt R (2004) Biocatalytic transformation of petroporphyrins by chemical modified cytochrome c. Biotechnol Bioeng 85:790–798

Gesinde AF, Agbo EB, Agho MO, Dike EFC (2008) Bioremediation of some Nigerian and Arabian crude oils by fungal isolates. Int J Pure Appl Sci 2:37–44

Gianfreda L, Rao MA (2004) Potential of extra-cellular enzymes in remediation of polluted soils: a review. Enzyme Microb Technol 35:339–354

Gross S, Robbins EI (2000) Acidophilic and acid-tolerant fungi and yeasts. Hydrobiologia 433:91–109

Hemida SK, Bagy MMK, Khallil AM (1993) Utilization of hydrocarbons by fungi. Cryptogamie Mycologie 14:207–213

Hernández-López EL, Perezgasga L, Huerta-Saquero A, Vazquez-Duhalt R (2016) Biotransformation of petroleum asphaltenes and high molecular weight polycyclic aromatic hydrocarbons by *Neosartorya fischeri*. Environ Sci Pollut Res Int 23:10773–10784

Lahav R, Nejidat A, Abeliovich A (2002) The identification and characterization of osmotolerant yeast isolates from chemical wastewater evaporation ponds. Microb Ecol 43:388–396

León V, Córdova J, Muñoz S, De Sisto A, Naranjo L (2007) Process for the upgrading of heavy crude oil, extra-heavy crude oil or bitumens through the addition of a biocatalyst. United States Patent Application 20070231870

León Y, De Sisto A, Inojosa Y, Malaver N, Naranjo-Briceño L (2009) Identificación de biocatalizadores potenciales para la remediación de desechos petrolizados de la Faja Petrolif era del Orinoco. RET 1:12–25

Macelroy RD (1974) Some comments on the evolution of extremophiles. BioSystem 6:74–75

Madigan MT, Martinko JM, Bender KS, Buckley DH, Stahl DA (2015) Brock biology of microorganisms, 14th edn. Pearson, Boston

Martínez MJ, Ruiz-Dueñas FJ, Guillén F, Martínez AT (1996) Purification and catalytic properties of two manganese-peroxidase isoenzymes from *Pleurotus eryngii*. Eur J Biochem 237:424–432

Martínez AT, Speranza M, Ruiz-Duenas FJ, Ferreira P, Camarero S, Guillen F, Martınez MJ, Gutierrez A, del Río JC (2005) Biodegradation of lignocellulosics: microbial, chemical, and enzymatic aspects of the fungal attack of lignin. Int Microbiol 8:195–204

Moore D, Robson GD, Trinci AP (2011) 21st century guidebook to fungi with CD. Cambridge University Press, New York, pp 1–640

Nadon L, Siemiatycki J, Dewar R, Krewski D, Gérin M (1995) Cancer risk due to occupational exposure to polycyclic aromatic hydrocarbons. Am J Ind Med 28(3):303–324

Naranjo L, Urbina H, De Sisto A, Leon V (2007) Isolation of autochthonous non-white rot fungi with potential for enzymatic upgrading of Venezuelan extra-heavy crude oil. Biocatal Biotransformation 25:341–349

Naranjo L, Urbina H, González M, Córdova J, Muñoz S, León V. (2008) Potential of autochthonous non-white rot fungi for partial enzymatic conversion (PEC-IDEA Technology) of Venezuelan extra-heavy crude oil. In: Proceeding of the 6th international symposium on fuels and lubricants (ISFL). New Delhi, India. Paper No. 128

Naranjo-Briceño L, Perniá B, Guerra M, Demey JR, González M, De Sisto A, Inojosa Y, Fusella E, Freites M, Yegres JF (2013) Potential role of oxidative exoenzymes of the extremophilic fungus Pestalotiopsis palmarum BM-04 in biotransformation of extra-heavy crude oil. Microb Biotechnol 6(6):720–730

Naranjo L, Pernía B, Inojosa Y, Rojas D, Sena D'Anna L, González M, De Sisto A (2015) First evidence of fungal strains isolated and identified from naphtha storage tanks and transporting pipelines in Venezuelan oil facilities. Adv Microbiol 5:143–154

Obire O (1993) The suitability of various Nigerian petroleum fractions as substrate for bacterial growth. Discov Innov 5:45–49

Odokuma LO, Okpokwasili GC (1993) Seasonal ecology of hydrocarbon-utilizing microbes in the surface waters of a river. Environ Monit Assess 27(3):175–191

Oudot JP, Dupont J, Haloui S, Roquebert MF (1993) Biodegradation potential of hydrocarbon-degrading fungi in tropical soil. Soil Biol Biochem 25:1167–1173

OPEC (2017) OPEC annual statistical bulletin 2017. http://www.opec.org/opec_web/en/

Pernía B, Demey JR, Inojosa Y, Naranjo L (2012) Biodiversidad y potencial hidrocarbonoclástico de hongos aislados de crudo y sus derivados: un meta-análisis. Latinoam Biotecnol Amb Algal 3:1–40

Pernía B, Rojas-Tortolero D, Sena L, De Sisto A, Inojosa Y, Naranjo L (2018) Fitotoxicidad de HAP, crudos extra pesados y sus fracciones en *Lactuca sativa*: una interpretación integral utilizando un índice de toxicidad modificado. Rev Int Contam Ambient 34:79–91

Pourfakhraei E, Badraghi J, Mamashli F, Nazari M, Saboury AA (2018) Biodegradation of asphaltene and petroleum compounds by a highly potent *Daedaleopsis* sp. J Basic Microbiol:1–14

Prenafeta-Boldú FX, de Hoog GS, Summerbell RC (2018) Fungal communities in hydrocarbon degradation. In: McGenity T (ed) Microbial communities utilizing hydrocarbons and lipids: members, metagenomics and ecophysiology. Handbook of hydrocarbon and lipid microbiology. Springer, Cham, pp 1–36

Rampelotto PH (2013) Extremophiles and extreme environments. Life 3:482–485

Ruiz-Dueñas FJ, Martínez MJ, Martínez AT (1999) Molecular characterization of a novel peroxidase isolated from the lignolytic fungus *Pleurotus eryngii*. Mol Microbiol 31:223–235

Saparrat MCN, Guillén F, Arambarri AM, Martínez AT, Martínez MJ (2002) Induction, isolation, and characterization of two laccases from the with rot basidiomycete *Coriolopsis rigida*. Appl Environ Microbiol 68:1534–1540

Strausz OP, Mojelsky TW, Lown EM (1992) The molecular structure of asphaltenes: an unfolding story. Fuel 71:1355–1363

Terrer C, Vicca S, Hungate BA, Phillips RP, Colin Prentice I (2016) Mycorrhizal association as a primary control of the CO_2 fertilization effect. Science 353:72–74

Turk M, Plemenitaš A, Gunde-Cimerman N (2011) Extremophilic yeasts: plasma-membrane fluidity as determinant of stress tolerance. Fungal Biol 115:950–958

Urbina H, Reyes A, Fusella E, González M, León V, Naranjo L (2007) Pycnoporus sanguineus IDEA, a laccase-overproducing fungi with high potential in partial enzymatic conversion (PEC-Technology) of Venezuelan extra-heavy crude oil. J Biotechnol 131(2 Supplement 1):S94–S95

Urbina H, Aime MC (2018) A closer look at Sporidiobolales: ubiquitous microbial community members of plant and food biospheres. Mycologia 110:79–92

Uribe-Álvarez C, Ayala M, Perezgasga L, Naranjo L, Urbina H, Vazquez-Duhalt R (2011) First evidence of mineralization of petroleum asphaltenes by a strain of *Neosartorya fischeri*. J Microbial Biotechnol 4:663–672

Uzoamaka GO, Floretta T, Florence MO (2009) Hydrocarbon degradation potentials of indigenous fungal isolates from petroleum contaminated soils. J Phys Nat Sci 3:1–6

Waldo GS, Carlson RM, Moldowan JM, Peters KE, Penner-Hahn JE (1991) Sulfur speciation in heavy petroleums: information from X-ray absorption near-edge structure. Geochim Cosmochim Acta 55:801–814

Zhang X, Li SJ, Li JJ, Liang ZZ, Zhao CQ (2018) Novel natural products from extremophilic fungi. Mar Drugs 16(6):4

Zheng C, Zhou J, Wang J, Qu B, Wang J, Lu H, Zhao H (2009) Aerobic degradation of nitrobenzene by immobilization of *Rhodotorula mucilaginosa* in polyurethane foam. J Hazard Mater 168:298–303

Chapter 29
Thermophilic Fungi in Composts: Their Role in Composting and Industrial Processes

Sonia M. Tiquia-Arashiro 🆔

29.1 Introduction

Composting is a natural biological process, carried out under controlled aerobic conditions (requires oxygen) (Tiquia and Tam 1998a, 2002; Tiquia et al. 2000; Zhang et al. 2016; Wei et al. 2018). In this process, the organic matter is transformed into a more stable organic matter, with a final product sufficiently stable for storage and use in agriculture as fertilizer, in gardening, or in landscaping (Richard and Tiquia 1999; Tiquia et al. 2002a; Krause et al. 2003; Pampuro et al. 2017). A typical aerobic composting is a self-heating process in which microbial metabolism drives the temperature above 50 °C, followed by sustained high temperatures between 60 and 80 °C, and then followed by gradual cooling of the compost pile (Tiquia et al. 1996, 1997a; Tiquia 2005a, b; Kumar 2011). The high temperature (50–80 °C) oxidizes phytotoxins and destroys animal and plant pathogens (Senesi 1989; Tam and Tiquia 1994; Tiquia 2000, 2010a; Tiquia and Tam 1998b; Tiquia et al. 1998a). The composting process represented a combined activity of a wide succession of environments, as one enzyme/microbial group overlapped the other and each emerged gradually due to the continual change in temperature and progressive breakdown of complex compounds to simpler ones (Tiquia et al. 1997b, 2001; Tiquia 2002a; Yu et al. 2018). Composting has been suggested as a potential strategy to eliminate antibiotic residues (Gou et al. 2018; Liu et al. 2018a).

Microbes play a key role as degraders during the composting process; the mesophilic microorganisms constitute the pioneer microflora, while thermophilic microorganisms are the dominant microflora that contribute significantly to the quality of compost (Tiquia et al. 1998c; Tiquia 2003, 2005b; Liu et al. 2018b). These mesophilic and thermophilic microbial consortia have distinct physiological requirements

S. M. Tiquia-Arashiro (✉)
Department of Natural Sciences, University of Michigan-Dearborn, Dearborn, MI, USA
e-mail: smtiquia@umich.edu

© Springer Nature Switzerland AG 2019
S. M. Tiquia-Arashiro, M. Grube (eds.), *Fungi in Extreme Environments:*
Ecological Role and Biotechnological Significance,
https://doi.org/10.1007/978-3-030-19030-9_29

587

and tolerances, consistent with the continuously changing environment throughout composting (Tiquia et al. 2002b; Tiquia 2005a; Federici et al. 2011; Jurado et al. 2014; Waqas et al. 2018). Bacteria including those that belong to the groups *Proteobacteria, Firmicutes, Bacteroidetes,* and *Actinobacteria* are by far the most important decomposers during the most active stages of composting (Partanen et al. 2010; Neher et al. 2013; Zhang et al. 2016), partly because of their availability to grow rapidly on soluble proteins, and other readily available substrates, and partly because they are the most tolerant of high temperatures (Kuok et al. 2012). Most fungi are eliminated above 50 °C; only a few have been recovered that can grow at all up to 62 °C (Tiquia 2005a; Langarica-Fuentes et al. 2014), which suggests that their degradative activities during the thermophilic stages of composting are minor compared to those of bacteria (Martins et al. 2013; Langarica-Fuentes et al. 2014). As peak heats are attained in composts, fungi tend to disappear from the central zone of the compost. In grass and straw composts where a peak heat of 70 °C was recorded, the thermophilic fungi disappear from the compost core for a period of 3 days (Chang and Hudson 1967). As compost temperatures fall below 60 °C, thermophilic fungi reappear in the middle of the compost (Chang and Hudson 1967). While current understanding tells us that bacteria are the dominant degraders in thermophilic composting processes, there is much to be said about the minority of thermophilic fungi during the composting process.

Composting is a promising source of new organisms and thermostable enzymes (Dougherty et al. 2012; D'Haeseleer et al. 2013; Nguyen et al. 2013; Tiquia-Arashiro 2014; Habbeche et al. 2014; Pomaranski and Tiquia-Arashiro 2016) that may be helpful in environmental management and industrial processes (Tiquia and Mormile 2010; Tiquia-Arashiro and Mormile 2013; Salar 2018). Fungi are known to have an important role in the composting process as degraders of recalcitrant materials such as cellulose and lignin and thermophilic fungi have been suggested as the main contributors to lignocellulose degradation. Despite the relevance of fungi in composting, especially the thermophilic fraction, most of the research on the diversity, composition, and succession of these microorganisms had been conducted several decades ago using classical culture-based methods (Chang and Hudson 1967; Kane and Mullins 1973; Klamer and Sochting 1998).

This chapter covers the diversity of thermophilic fungi during composting, their role, and potential applications in biotechnology. Readers may find that the available information on several aspects of compost ecosystem is scanty which is due to horizontal advancements in some areas and because compost represents a complex ecological system from the viewpoint of microbial distribution and activity.

29.2 Fungal Communities in Composts

Culture-independent methods, including denaturing gradient gel electrophoresis (DGGE) of PCR-amplified DNA fragments, terminal restriction fragment length polymorphism analysis (T-RFLP), clone library analysis, and more recently high-throughput sequencing, have been used extensively to investigate microbial successions in composts (Ishii and Takii 2003; Tiquia 2005a, 2010b; Tiquia and Michel Jr. 2002; Tiquia et al. 2005; Szekely et al. 2009; De Gannes et al. 2013). However, few investigations have focused on fungal populations of large-scale composting processes using molecular techniques. Bonito et al. (2010) used DGGE to study windrow-type systems; Hultman et al. (2009) and Hansgate et al. (2005) utilized a clone library approach to study fungi in a rotating drum system and a reactor-type system, respectively; and Langarica-Fuentes et al. (2014) took advantage of high-throughput sequencing to monitor fungal succession in an in-vessel composting system. Gu et al. (2017) used the Dirichlet multinomial mixtures mode to analyze Illumina sequencing data to reveal both temporal and spatial variations of the fungi community present in the aerobic composting.

Bonito et al. (2010) identified fungi microflora associated with composting organic municipal wastes to gain a better understanding of the diversity of fungi at different stages of composting. A disproportionate number of yeast sequences have been detected from day-0 clone libraries, including the human pathogens *Candida tropicalis* and *Candida krusei* (*Saccharomycetales*). *Basidiomycetes* account for over half of the clones from the day-210 compost sample while *Cercophora* and *Neurospora* species account for most of the fungal clones from day-410 sample. Surprisingly, no *Zygomycetes* or *Aspergillus* species were detected.

Hansgate et al. (2005) employed F-ARISA (fungal-automated rRNA intergenic spacer analysis) and 18S rRNA gene cloning and sequencing to examine changes in fungal community structure during composting. Sequencing of the 18S rRNA portion of cloned F-ARISA products revealed the presence of four distinct fungal genera including *Backusella* sp., *Mucoraceae*, *Geotrichum* sp., and the yeast *Pichia* sp. Clone libraries constructed using fungus-specific 18S rRNA primers contained sequences similar to several other fungal genera including *Penicillium* sp., *Aspergillus* sp., *Hamigera* sp., *Neurospora* sp., and the yeast *Candida* sp.

Langarica-Fuentes et al. (2014) characterized the fungal community composition at different stages of in-vessel composting process. A complex succession of fungi is revealed, with 251 fungal OTUs identified throughout the monitoring period. The *Ascomycota* are the dominant phylum (82.5% of all sequences recovered), followed by the *Basidiomycota* (10.4%) and the subphylum *Mucoromycotina* (4.9%). In the early stages of the composting process, yeast species from the order *Saccharomycetales* are abundant, while in later stages and in the high-temperature regions of the pile, fungi from the orders *Eurotiales*, *Sordariales*, *Mucorales*, *Agaricales*, and *Microascales* are the most prominent. This study presents an in-depth view on the succession of fungi during the composting of municipal solid waste and provides a guide to those species that drive an in-vessel composting process towards a satisfactory product. Similar communities are likely to be observed in other composting plants where municipal solid waste is processed; however, differences in the process nature, length of composting, and conditions achieved (temperature, pH, water content, etc.) are likely to determine the exact succession and communities present.

Gu et al. (2017) characterized fungal diversity in the aerobic composting with Illumina sequencing. A total of 670 operational taxonomic units (OTUs) were detected, and the dominant phylum was *Ascomycota*. There were four types of samples of fungi communities during the composting process. Samples from the early composting stage (type I) were dominated by *Saccharomycetales* sp. Fungi in the medium composting stage (types II and III) were dominated by *Sordariales* spp. and *Acremonium alcalophilum*, *Saccharomycetales* sp., and *Scedosporium minutisporum*. Samples from the late composting stage (IV) were dominated by *Scedosporium minutisporum*. The results of their study indicate that time and depth influence fungal distribution and variation in the waste during static aerobic composting.

29.3   Thermophilic Fungi in Composts

Cooney and Emerson (1964) define thermophilic fungi as fungi with a maximum growth temperature of 50 °C or higher and a minimum growth temperature of 20 °C or higher. Thermotolerant species have a maximum growth temperature of about 50 °C and a minimum well below 20 °C (Cooney and Emerson 1964; Awasthi et al. 2014). Crisan (1973), however, defines thermophilic fungi as fungi with a temperature optimum of 40 °C or higher. Most thermophiles are isolated from composts (Tansey and Brock 1978; Awasthi et al. 2014; López-González et al. 2015; Sebők et al. 2016; Ahirwar et al. 2017; Wang et al. 2018); their prevalence in composts can be explained by the high temperatures, humidity, and aerobic conditions within the composts. Moreover, the supply of carbohydrates and nitrogen in composts favors the development of thermophilic microflora (Cooney and Emerson 1964). During the composting process, various organic materials are converted into simpler units of organic carbon and nitrogen (Tiquia 2002a, b, 2003; Tam and Tiquia 1999; Tiquia and Tam 2000; Tiquia et al. 1998b, 2002c). The overall efficiency of organic material degradation depends on the microbes and their activities (Tiquia et al. 2002b, c). Thermophilic fungi promote the degradation of organic materials by secreting various types of cellulolytic and xylanolytic enzymes. These fungi might have enzymes that maintain their activities at high temperatures. *Aspergillus*, *Chaetomium*, *Humicola*, *Mucor*, *Penicillium*, and *Thermomyces* spp. are the dominant fungi of compost ecosystems. Species of *Aspergillus* and *Mucor* are predominant in composting of biowaste (Ryckeboer et al. 2003). *Aspergillus fumigatus* and *Humicola grisea* var. *thermoidea* have been reported to be the dominant members of the spent mushroom compost. Other fungi reported from spent mushroom compost are *Aspergillus flavus*, *Aspergillus nidulans*, *Aspergillus terreus*, *Aspergillus versicolor*, *Chrysosporium luteum*, *Malbranchea cinnamomea* NFCCI 3724, *Melanocarpus albomyces*, *Mucor* spp., *Myceliophthora thermophila*, *Nigrospora* spp., *Oidiodendron* spp., *Paecilomyces* spp., *Penicillium chrysogenum*, *Penicillium expansum*, *Trichoderma viride*, and *Trichuris* spp. (Kleyn and Wetzler 1981; Ahirwar et al. 2017; Kertesz and Thai 2018).

Several known thermophilic fungi have been found in mushroom composting. Mushroom composting represents an interesting example of thermogenic solid-state fermentation process that results from succession of microbial communities. The composting process consists of two phases. Phase I is an outdoor fermentation process during which the raw materials are mixed, wetted, and stacked with considerable dry mass losses. Phase II is an indoor process of pasteurization to produce a selective and pathogen-free substrate (Noble and Gaze 1994). During phase I, fungal and bacterial activities produce large quantities of heat. Temperature ranges between ambient and 80 °C in distinct zones within the cross sections of the compost stack and ammonia disappear most rapidly in the range of 40–45 °C. Mushroom compost is an interesting example of a complete spectrum of microbial diversity. It is a rich reservoir of microbial types, comprising of mesophilic and thermophilic bacteria, fungi, and actinomycetes. In phase I, the pioneer thermophilic mycoflora of mushroom compost comprises fast-growing and rapidly sporulating fungi such as *Aspergillus fumigatus* and *Rhizomucor* spp. with a pH optimum below 7.0 and temperature optima of about 40 °C. When self-heating and ammonification start and pH reaches 9.0, the pioneer flora disappears and paves way for *Talaromyces thermophilus* and *Thermomyces lanuginosus*; during massive heat production these fungi possess moderate growth rate, as they exhibit high thermal death point and pH tolerance, but do not degrade cellulose. At the end of the composting process, about 50–70% of the compost biomass is constituted by thermophilic fungi (Sparling et al. 1982; Weigant 1991). While most of the species are eliminated, *Sporotrichum thermophilum* appears as near-exclusive species after phase II composting and constitutes a climax species in the mushroom compost along with thermophilic actinomycetes (Straatsma et al. 1994). The number of CFU of *S. thermophilum* in fresh matter of phase II is about 10^6 g^{-1} compost (Bilai 1984); however, actinomycetes and bacteria appear to play a decisive role in successful colonization by this thermophile. In the beginning of phase II of mushroom composting, thermophilic fungi and actinomycetes extensively colonize the plant matter until temperature reaches 60 °C, as an outcome of slow peak heating for about 2 days (Straatsma et al. 1994). The high temperature of the first indoor period of phase II kills most of the pathogenic and nonpathogenic microorganisms, except the spores of actinomycetes and thermophilic fungi such as *Scytalidium thermophilum* (Straatsma et al. 1991). Klamer et al. (1998) reported *A. fumigatus* and *Rhizomucor pusillus* as predominant species before peak heating and *P. variotii*, *S. thermophilum*, and *T. lanuginosus* as dominant forms after peak heating. Tewari (2000) reported the presence of *H. lanuginosa* and *S. thermophilum* during peak-heat stage of phase II composting. *S. thermophilum* is a natural inhabitant of compost ingredients, including drainage from compost, and has been documented to be present throughout composting. Dominance of *S. thermophilum* has been reported by several workers (Straatsma et al. 1991; Vijay 1996; Klamer et al. 1998; Rajni 1999), while *H. grisea* var. *thermoidea* and *H. insolens* have been described by others (Fergus 1964). They are inherently close partners in the degradation processes in compost and provide selectivity to compost (Straatsma et al. 1989; Opden Camp et al. 1990). Rajni (1999) and Rawat (2004) observed nearly similar microbial distribution patterns in compost as reported by Straatsma et al. (1991), with predominance of *S. thermophilum* (Kertesz and Thai 2018). In mushroom compost, thermophilic fungi are responsible for the degradation of lignocellulose, which is a prerequisite for the growth of the edible fungus (Sharma 1989; Kertesz and Thai 2018). Thermophilic fungi grow extensively during the last phase of composting in mushroom compost from the spores that survive the pasteurization temperature (Straatsma et al. 1989). Thus, they contribute significantly towards the quality of compost.

29.3.1 Thermophilic Fungi in Straw Compost

Thermophilic fungi of wheat straw compost were studied in detail by Chang and Hudson (1967). Initial high population of mesophilic fungi results in peak heating in the central region of the pile wherein temperature rises rapidly and reaches a plateau around 50 °C. Thermophilic fungi rapidly develop replacing the mesophilic population and persist until the compost cools down. In wheat and broad bean straw composts, thermophilic fungi are not present at peak high temperature. However, when the composts cool down to 51.5 °C, *Penicillium dupontii*, *Myriococcum albomyces*, *Thermomyces lanugionsus*, and *Sporotrichum thermophile* are found in abundance (Moubasher et al. 1982; Zhang et al. 2015). Several critical factors reported to influence the colonization by thermophilic fungi include (1) existence of suitably high temperature to promote germination and growth arid multiplication of propagules; (2) ability of thermophilic fungi to break down complex carbon substances; and (3) absence of repressive activity among the compost-inhabiting organisms. In a complex of microbial interactions such as above, succession of individual species is governed by their traditional requirements and availability of suitable temperature and pH conditions. For example, due to simple nutritional requirements thermophilic mucoraceous members appear early in the composting process. *Humicola lanuginosa* develops early but exists throughout the composting process as it lives as a commensal with other thermophilic organisms (Hedger and Hudson 1974; Salar 2018). Besides, this organism can tolerate a wide range of temperatures on either side of optima and elaborates a variety of hydrolytic enzymes that help in continuous presence.

29.3.2 Thermophilic Fungi in Municipal Waste Composts

Municipal wastes generally contain, among other things, substrates rich in lignohemicellulose. Thermophilic fungi play a significant role in the conversion of these materials. Some species are unique in their ability to degrade plastic substances and hence special interest has been envisioned in their study from municipal waste compost. Thermophilic fungi isolated from municipal waste composts include

Thermomucor (Subrahmanyam et al. 1977; Singh et al. 2016), *Thermoascus auran-tiacus* (Cooney and Emerson 1964; Sebők et al. 2016), and *Myceliophthora thermophila* (Sen et al. 1980; Sebők et al. 2016).

29.3.3 Thermophilic Fungi in Paddy Straw Composts

Paddy straw is an excellent substrate for the colonization of thermophilic fungi. In an extensive controlled study of this substrate, Satyanarayana and Johri (1984) observed that colonizing ability of thermophilic fungi on paddy straw was directly proportional to the inoculum concentration. For example, colonization by *Humicola lanuginosa*, *Sporotrichum thermophile*, and *Torula thermophila* (*Scytalidium thermophilum*) increased with higher inoculum concentration. *Aspergillus fumigatus* showed a strong competitive ability both in pure and mixed cultures. Decomposing ability of these organisms varied with C:N ratio and the length of paddy straw pieces. During peak heating period, only a few thermophilic fungal propagules were present but these exhibited high rate of respiration as suggested by the evolution of carbon dioxide.

29.4 Industrial Applications

The biotechnological applications of thermophilic fungi are numerous. Pure culture studies of thermophilic fungi have provided clear evidence that they possess a variety of extracellular enzymes capable of hydrolyzing polymers such as starch, protein, pectin, hemicellulose, cellulose, and lignin. They have also been reported to produce, among others, many antibacterial and antifungal substances, extracellular phenolic compounds, and organic acids. Some thermophilic fungi have already been used in industries involving food processing, bioconversion of organic materials, biodegradation of plastics, biosorption of metals/radionuclides, cancer treatment, and synthesis of nanoparticles (Bengtsson et al. 1995; Zafar et al. 2013; Aydi Ben Abdallah et al. 2015; Tiquia-Arashiro and Rodrigues 2016a; Salar 2018).

29.4.1 Production of Thermostable Enzymes

Thermostable enzymes have become the focus of biotechnological interest because they are more tolerant to the conditions in industrial processes and storage. The production of thermostable enzymes has grown through advances in isolating many thermophilic microorganisms. The advantage of the use of thermostable enzymes is the possibility of conducting biotechnological processes at elevated temperatures and thus reducing the risk of contamination by mesophilic microorganisms, decreasing the viscosity of the reaction medium, increasing the bioavailability and solubility of organic compounds, and increasing the diffusion coefficient of substrates and products resulting in higher reaction rates (Kumar and Nussinov 2001).

Cellulose is one of the main components of plant cell wall material and is the most abundant and renewable nonfossil carbon source on earth. Degradation of cellulose to its constituent monosaccharides has attracted considerable attention to produce food and biofuels. Cellulose can be hydrolyzed to glucose and other soluble sugars by using cellulase enzymes of bacteria and fungi (Plecha et al. 2013). Thermophilic cellulases are key enzymes for efficient biomass degradation. Their importance stems from the fact that cellulose swells at higher temperatures, thereby becoming easier to break down. In industrial processes, cellulolytic enzymes have been employed in the extraction of pigments and flavor compounds in fruit juice and wine production; as additive of detergents for washing jeans; in the pretreatment of biomass to improve the nutritional quality of forage for animal feed; in the textile industry in the polishing process of cotton fibers; and for saccharification of lignocellulosic residues to obtain reducing sugar (Ando et al. 2002; Baffi et al. 2013). The interest in the use of cellulases to produce fermentable sugars from cellulosic wastes at present is focusing on biofuel production such as biogas, bioethanol, biodiesel, and fuel cells. The use of whole biomass to obtain alcohol-based fuels requires an efficient conversion of lignocellulosic material into fermentable pentose and hexose sugars. Thermal stability of several commercial cellulase preparations is an important parameter for the success of the process. Thus, the industries have been developing cellulases with higher thermal stability and especially stable at industrially relevant conditions. Many thermophilic fungi from composts (*Myriococcum thermophilum, Sporotrichum thermophile, Thermoascus aurantiacus*, and *Thermomyces lanuginosus*) have been isolated in recent years and the cellulases produced by these eukaryotic microorganisms have been purified and characterized at both structural and functional level (Lee et al. 2014; de Cassia Pereira et al. 2015; Mehta et al. 2016; Jain et al. 2017).

Several studies have reported the production of thermostable xylanase from thermophilic and hyperthermophilic organisms, prokaryotes, and eukaryotes. Among thermophilic compost fungi, *Mycothermus thermophilus* (Lee et al. 2014; Ma et al. 2017), *Talaromyces thermophilus* (Maalej et al. 2009)*, Thermomyces lanuginosus* (Jiang et al. 2005; Lee at al. 2014), *Thermoascus aurantiacus* (Lee at al. 2014), and *Rhizomucor miehei* (Zhou et al. 2014) produce thermostable xylanases with action from 50 °C up to 80 °C. A large variety of xylanases produced by these thermophilic fungi have become a major group of industrial enzymes that are capable of degrading xylan to renewable fuels and chemicals, in addition to their use in food, paper, and pulp industries.

Pectinases are a group of enzymes that catalyze the degradation of pectic substances by depolymerization reaction and by de-esterification reactions. One of the most common applications of pectinases is in fruit processing for various purposes like musts, juices, pastes, and purées. These extraction processes are carried out at temperatures greater than 65 °C and subsequently cooled to 50 °C (Lea 1995); thus, the use of thermostable pectinases avoids the cooling step and so it could reduce the time and cost of processes (Zhang et al. 2011). Thermostable pectinases are also very useful in the degradation of pectin waste from processing plant material industry, reducing BOD and COD (Kapoor et al. 2000). Pectinase from *Penicillium echinulatum* is associated with a cellulolytic enzyme complex and has improved sugarcane bagasse saccharification, suggesting a new application for these enzymes (Delabona et al. 2013). Several pectinolytic thermophilic fungi have been isolated so far including those belonging to the genera *Thermomyces*, *Aspergillus*, *Monascus*, *Chaetomium*, *Neosartoria*, *Scopulariopsis*, and *Thermomucor* (Martin et al. 2010). The thermophilic *Thermoascus aurantiacus* produces considerable amounts of pectinase in media based on citrus peel (Martins et al. 2002), which showed optimal activity at 70 °C and stability at 60 °C for 2 h.

In nature, lignocellulose accounts for the major part of biomass and, consequently, its degradation is essential for the operation of the global carbon cycle (Sánchez 2009). Lignocellulose, such as wood, is mainly composed of a mixture of cellulose (ca. 40%), hemicellulose (ca. $20 \pm 30\%$), and lignin (ca. $20 \pm 30\%$) (Bajpai 2016). Lignin is an integral cell wall constituent, which provides plant strength and resistance to microbial degradation (Ochoa-Villarreal et al. 2012). The ligninolytic capacity of most thermophilic fungi is largely known. However most of them are known to be able to degrade wood or other lignocelluloses, celluloses, or hemicelluloses (Sharma 1989; Kuhad et al. 1997; Dashtban et al. 2009). The thermophilic fungus *Thermoascus aurantiacus* has a high ligninolytic capacity (McClendon et al. 2012), and it has been isolated from composts.

29.4.2 Plastic Biodegradation

Polyurethanes (PUs) are synthetic plastics with a wide range of applications in the medical, automotive, construction, furnishing, and industrial sectors (Krasowska et al. 2012). They are known to be vulnerable to microbial attack as they contain ester linkages within the backbone of the polymer that are naturally vulnerable to esterases (Zafar et al. 2013). In contrast, polyether PUs, which contain ether linkages within the polymer backbone, are reported to be far more recalcitrant (Darby and Kaplan 1968). It has been reported that a number of fungal isolates are able to degrade impranil (liquid dispersion of PU) including thermotolerant and thermophilic fungi (Zafar et al. 2013), and a number of fungal species that are capable of degrading PU have been isolated and identified (Darby and Kaplan 1968; Pathirana and Seal 1984; Cosgrove et al. 2007; Mathur and Prasad 2012). Zafar et al. (2013) demonstrated that polyester PU is susceptible to fungal biodegradation in compost under both thermophilic (thermophilic stage) and mesophilic (maturation phase) conditions and that positive selection for rare taxa from the existing compost community on the PU surface occurs. The most dominant fungi identified from the surfaces of PU coupons by pyrosequencing was *Fusarium solani* at 25 °C (mesophilic phase), while at both 45 °C and 50 °C (thermophilic phase) *Candida ethanolica* was the dominant species. The diversity in the fungal community recovered from polyester PU coupons buried at the surface of compost pile was dependent on the incubation temperature (Zafar et al. 2014). At 37 °C, *Acremonium flavum* and

Candida rugosa are consistent mesophilic species with dominant *Arthrographis kalrae* on day 28. At 45 °C on day 2, the biomass obtained from the surface of buried polyester PU coupons are dominated by *Aspergillus* spp. and on day 28 a mixed community of *Lichtheimia* sp. and *Aspergillus fumigatus* with occasional isolates of *Malbranchea cinnamomea* and *Emericella nidulans* are found. *A. fumigatus* and *E. nidulans* have previously been isolated as potential polyester PU degraders (Barratt et al. 2003). *M. cinnamomea* and *A. fumigatus* have also been recovered in the compost at 50 °C. The major population at 50 and 55 °C is *Thermomyces lanuginosus*, a PU degrader (Zafar et al. 2014).

29.4.3 Remediation of Metals and Radionuclides

The use of biological materials for metal removal and recovery technologies has gained important credibility during the past decade, because of the good performance and low cost of this complexing material (Wu et al. 2005; Cho et al. 2012; Lakherwal 2014; Bowman et al. 2018). The natural affinity of biological compounds for metallic elements could contribute to economically purifying heavily metal-loaded wastewater. Among the various resources in biological wastes, dead biomass of microorganisms (bacteria, yeasts, fungi, algae) exhibits particularly interesting metal-binding capacities (Cho et al. 2010). For instance, *Rhizopus arrhizus*, a Mucorale filamentous fungus, can accumulate lead or uranium, up to I (1% and 16% of its own dry mass, respectively) (Tobin et al. 1984). These properties are attributed to the high content of complexing functional groups in their cellular wall (e.g., amino, amide, hydroxyl, carboxyl, sulfhydryl, phosphate radicals) (Tiquia-Arashiro 2018). Residual biomass, produced by the thermophilic fungus, *Talaromyces emersonii* CBS 814.70, following growth on glucose-containing media, was examined for its ability to take up uranium from aqueous solution (Bengtsson et al. 1995). It was found that the biomass had a relatively high observed biosorption capacity for the uranium (280 mg/g dry weight biomass). The calculated maximum biosorption capacity obtained by fitting the data to a Langmuir model was calculated to be 323 mg uranium/g dry weight biomass. Some of the critical biosorption parameters have already been identified, and pH was shown to influence to a large extent the formation of metal-biosorbent complexes. pH variation can modify the speciation and the availability of the metallic elements in solution and also the chemical state of the chemical functional groups responsible for metal binding in the biomass.

29.4.4 Cancer Treatment

Aspergillus terreus, a thermophilic fungus abundant in composts (Aydi Ben Abdallah et al. 2015), produces asperjinone, a nor-neolignan, and terrein, a suppressor of ABCG2-expressing breast cancer cells, which can restore drug sensitivity and could be the key to improve breast cancer therapeutics. Terrein displayed strong cytotoxicity against breast cancer MCF-7 cells. Treatment with terrein significantly suppressed the growth of ABCG2-expressing breast cancer cells. This suppressive effect was achieved by inducing apoptosis via activating the caspase-7 pathway and inhibiting the Akt signaling pathway, which led to a decrease in ABCG2-expressing cells and a reduction in the side-population phenotype (Liao et al. 2012).

29.4.5 Nanoparticle Synthesis

Some microorganisms have developed the ability to resort to specific defense mechanisms to quell stresses like toxicity of heavy metal ions or metals (Tiquia-Arashiro 2018; Bowman et al. 2018; Tiquia-Arashiro and Rodrigues 2016a). The microorganisms can survive and grow even at high metal ion concentrations and are capable of binding large quantities of metallic cations (Tiquia-Arashiro and Rodrigues 2016a, b). The remarkable ability of these group of microbes to reduce heavy metal ions makes them one of the best candidates for nanoparticle synthesis (Tiquia-Arashiro and Rodrigues 2016b, c, d, e, f). Syed et al. (2013) elucidated the biosynthesis of silver nanoparticles (AgNPs) by the thermophilic fungus *Humicola* sp., a dominant fungus in compost ecosystems. The fungus when reacted with Ag^+ ions reduces the precursor solution and leads to the formation of extracellular nanoparticles. The uniqueness of this study is that the investigators achieved superior control over the size of these nanoparticles, focusing upon them to be in the size range of 5–25 nm, so that these AgNPs when employed in biomedical applications will not block the glomerulus of the kidneys and will easily pass through urine within a short period of time. The AgNPs synthesized are nontoxic to cancer and normal cells up to concentrations of 50 μg/mL and thus will find various applications in drug and targeted drug delivery systems (Syed et al. 2013).

Gadolinium oxide nanoparticles are very important as nuclear, electronic, laser, optical, catalyst, and phosphor materials (Tiquia-Arashiro and Rodrigues 2016a, b). Many organic compounds use Gd_2O_3 for their dimerization (Gündüz and Uslu 1996). It is also used in imaging plate neutron detectors, as neutron reactors (Gündüz and Uslu 1996), and as an additive in ZnO_2 to enhance its toughness. Gd_2O_3 has several potential applications in biomedicine, too. For example, it is used in magnetic resonance imaging, since it exhibits superparamagnetism and involves T1 relaxation, and can be useful as a multimodal contrast agent for in vivo imaging (Bridot et al. 2007). It can also be easily doped with other lanthanides and exploited as a fluorescent tag, thus replacing other fluorescent organic mrolecules. Khan et al. (2014) showed that the thermophilic fungus *Humicola* sp. can be used for the synthesis of Gd_2O_3 nanoparticles at 50 °C. AsGdCl$_3$ is dissolved in water along with fungal biomass, and GdCl$_3$ ionizes to Gd^{3+} and $3Cl^-$. The Gd^{3+} ions are then attracted towards anionic proteins, which are secreted by *Humicola* sp. in solution. Reductase enzymes present in the anionic protein fraction act on Gd^{3+} and convert it to Gd^{2+}. Oxidase enzymes, which are also secreted by the fungus in the solution mixture, act on these Gd^{2+} ions resulting in the formation of Gd_2O_3 nanoparticles. The $GdCl_3$ NPs are irregular in shape, presenting an overall quasi-spherical morphology. Particle size distribution analysis of Gd_2O_3 nanoparticles confirmed that the nanoparticles are in the range of 3–8 nm with an average size of 6 nm. Since Gd_2O_3 nanoparticles have proved their value in site-specific drug delivery systems for cancer therapy, Khan et al. (2014) extended the work of biosynthesis of Gd_2O_3 nanoparticles to bioconjugation with taxol. Bioconjugation of taxol with gold and iron oxide nanoparticles has also been reported (Gibson et al. 2007; Hwu et al. 2009). Taxol is one of the most important anticancer drugs used for breast, ovarian, and lung cancers. The potent anticancer effect of taxol is mainly attributed to its mechanism of action. It stabilizes microtubules by preventing their depolymerization Khan et al. (2014). However, taxol is a hydrophobic drug and less specific to certain tumors due to its low solubility in water. To counter these problems, we carried out the bioconjugation of chemically modified taxol with biocompatible Gd_2O_3 nanoparticles, which may result in an enhancement of the hydrophilicity of taxol and may render it more potent in killing tumor/cancer cells (Khan et al. (2014).

29.5 Conclusions

Thermophilic fungi occur widely in composts, manures, and decomposing plant materials. They play an important role in the decomposition of organic matter due to their avidity for degrading various components of organic matter such as starch, pectin, hemicellulose, cellulose, and, to a lesser extent, lignin. While thermophilic fungi have long been known to be involved in composting and humification, the mechanisms involved in the accelerated decomposition of biomass are not well understood. This literature survey shows that although several thermophilic fungi have been isolated and identified, little knowledge about the physiology of this group is available. The role of thermophilic fungi in decomposition during composting suggests that thermophilic fungi may be good sources of thermostability of enzymes that can be applied in many industrial processes.

References

Ahirwar S, Soni H, Prajapati BP, Kango N (2017) Isolation and screening of thermophilic and thermotolerant fungi for production of hemicellulases from heated environments. Mycology 8:125–134

Ando S, Ishida H, Kosugi Y, Ishikawa K (2002) Hyperthermostable endoglucanase from *Pyrococcus horikoshii*. Appl Environ Microbiol 68:430–433

Awasthi MK, Pandey AK, Khane J, Singh P, Wong JCW, Selvam A (2014) Evaluation of thermophilic fungal consortium for organic municipal solid waste composting. Bioresour Technol 168:214–221

Aydi Ben Abdallah R, Jabnoun-Khiareddine H, Mejdoub-Trabelsi B, Daami-Remadi M (2015) Soil-borne and compost-borne *Aspergillus* species for biologically controlling post-harvest diseases of potatoes incited by *Fusarium sambucinum* and *Phytophthora erythroseptica*. J Plant Pathol Microbiol 6:313. https://doi.org/10.4172/2157-7471.1000313

Baffi M, Tobal T, Lago J, Boscolo M, Gomes E, Da-Silva R (2013) Wine aroma improvement using a β-glucosidase preparation from *Aureobasidium pullulans*. Appl Biochem Biotechnol 169:493–501

Bajpai P (2016) Structure of lignocellulosic biomass. In: Bajpai P (ed) Pretreatment of lignocellulosic biomass for biofuel production, SpringerBriefs in green chemistry for sustainability. Springer Publishers, New York, pp 7–12

Barratt SR, Ennos AR, Greenhalgh M, Robson GD, Handley PS (2003) Fungi are the predominant micro-organisms responsible for degradation of soil-buried polyester polyurethane over a range of soil water holding capacities. J Appl Microbiol 95:78–85

Bengtsson L, Johansson B, Hackett TJ, McHale L, McHale AP (1995) Studies on the biosorption of uranium by *Talaromyces emersonii* CBS 814.70 biomass. Appl Microbiol Biotechnol 42:807–811

Bilai VT (1984) Thermophilic micromycete species from mushroom composts. Mikrobiol Zh (Kiev) 46:35–38

Bonito G, Isikhuemhen OS, Vilgalys R (2010) Identification of fungi associated with municipal compost using DNA-based techniques. Bioresour Technol 101:1021–1027

Bowman N, Patel P, Sanchez S, Xu W, Alsaffar A, Tiquia-Arashiro SM (2018) Lead-resistant bacteria from Saint Clair River sediments and Pb removal in aqueous solutions. Appl Microbiol Biotechnol 102:2391–2398

Bridot JL, Faure AC, Laurent S, Rivière C, Billotey C, Hiba B, Janier M, Josserand V, Coll JL, Elst LV, Muller R, Roux S, Perriat P, Tillement OJ (2007) Hybrid gadolinium oxide nanoparticles: multimodal contrast agents for in vivo imaging. Am Chem Soc 129:5076–5084

de Cassia Pereira J, Paganini Marques N, Rodrigues A, Brito de Oliveira T, Boscolo M, da Silva R, Gomes E, Bocchini Martins DA (2015) Thermophilic fungi as new sources for production of cellulases and xylanases with potential use in sugarcane bagasse saccharification. J Appl Microbiol 118:928–939

Chang Y, Hudson HJ (1967) The fungi of wheat straw compost: I. Ecological studies. Trans Br Mycol Soc 50:649–666

Cho DH, Kim EY, Hung YT (2010) Heavy metal removal by microbial biosorbents. In: Wang L, Tay JH, Tay S, Hung YT (eds) Environmental bioengineering. Handbook of environmental engineering, volume 11. Humana Press, Totowa, NJ, pp 375–402

Cho K, Zholi A, Frabutt D, Flood M, Floyd D, Tiquia SM (2012) Linking bacterial diversity and geochemistry of uranium-contaminated groundwater. Environ Technol 33:1629–1640

Cooney DG, Emerson R (1964) Thermophilic fungi: an account of their biology, activities, and classification. W. H. Freeman and Company. USA, San Francisco, p 188

Cosgrove L, McGeechan PL, Robson GD, Handley PS (2007) Fungal communities associated with degradation of polyester polyurethane in soil. Appl Environ Microbiol 73:5817–5824

Crisan EV (1973) Current concepts of thermophilism and the thermophilic fungi. Mycologia 65:1171–1198

D'haeseleer P, Gladden JM, Allgaier M, Chain PS, Tringe SG, Malfatti SA, Aldrich JT, Nicora CD, Robinson EW, Paša-Tolić L, Hugenholtz P, Simmons BA, Singer SW (2013) Proteogenomic analysis of a thermophilic bacterial consortium adapted to deconstruct switchgrass. PLoS One 8(7):e68465. https://doi.org/10.1371/journal.pone.0068465

Darby RT, Kaplan AM (1968) Fungal susceptibility of polyurethanes. Appl Microbiol 16:900–905

Dashtban M, Schraft H, Qin W (2009) Fungal bioconversion of lignocellulosic residues: opportunities and perspectives. Int J Biol Sci 5:578–595

De Gannes V, Eudoxie G, Hickey WJ (2013) Prokaryotic successions and diversity in composts as revealed by 454-pyrosequencing. Bioresour Technol 133:573–580

Delabona Pda S, Cota J, Hoffmam ZB, Paixão DA, Farinas CS, Cairo JP, Lima DJ, Squina FM, Ruller R, Pradella JG (2013) Understanding the cellulolytic system of Trichoderma harzianum P49P11 and enhancing saccharification of pretreated sugarcane bagasse by supplementation with pectinase and α-Larabinofuranosidase. Bioresour Technol 2013 131:500–507

Dougherty MJ, D'haeseleer P, Hazen TC, Simmons BA, Adams PD, Hadi MZ (2012) Glycoside hydrolases from a targeted compost metagenome, activity-screening and functional characterization. BMC Biotechnol 12:38. https://doi.org/10.1186/1472-6750-12-38

Federici E, Pepi M, Esposito A, Scargetta S, Fidati L, Gasperini S, Cenci G, Altieri R (2011) Two-phase olive mill waste composting: community dynamics and functional role of the resident microbiota. Bioresour Technol 102:10965–10972

Fergus CL (1964) Thermophilic and thermotolerant molds and actinomycetes of mushroom compost during peak heating. Mycologia 56:267–284

Gibson J, Khanal BP, Zubarev ER (2007) Paclitaxel-functionalized gold nanoparticles. J Am Chem Soc 129:11653–11661

Gou M, Hua HW, Zhang YJ, Wang JT, Hayden H, Tang YQ, He JH (2018) Aerobic composting reduces antibiotic resistance genes in cattle manure and the resistome dissemination in agricultural soils. Sci Total Environ 612:1300–1310

Gu W, Lu Y, Tan Z, Xu P, Xie K, Li X, Sun L (2017) Fungi diversity from different depths and times in chicken manure waste static aerobic composting. Bioresour Technol 239:447–453

Gündüz G, Uslu IJ (1996) Powder characteristics and microstructure of uranium dioxide and gadolinium oxide fuel. Nucl Mater 231:113–120

Habbeche A, Saoudi B, Jaouadi B, Haberra S, Kerouaz B, Boudelaa M, Badis A, Ladjama A (2014) Purification and biochemical characterization of a detergent-stable keratinase from a newly thermophilic actinomycete *Actinomadura keratinilytica* strain Cpt29 isolated from poultry compost. J Biosci Bioeng 117:413–421

Hansgate AM, Schloss PD, Hay AG, Walker LP (2005) Molecular characterization of fungal, community dynamics in the initial stages of composting. FEMS Microbiol Ecol 51:209–214

Hedger IN, Hudson HJ (1974) Nutritional studies of *Thermomyces lanuginosus* from wheat straw compost. Trans Br Mycol Soc 62:129–141

Hultman J, Vasara T, Partanen P, Kurola J, Kontro MH, Paulin L, Auvinen P, Romantschuk M (2009) Determination of fungal succession during municipal solid waste composting using a cloning-based analysis. J Appl Microbiol 108:472–487

Hwu JR, Lin YS, Josephrajan T, Hsu MH, Cheng FY, Yeh CS, Su WC, Shieh DB (2009) Targeted paclitaxel by conjugation to iron oxide and gold nanoparticles. J Am Chem Soc 131:66–68

Ishii K, Takii S (2003) Comparison of microbial communities in four different composting processes as evaluated by denaturing gradient gel electrophoresis analysis. J Appl Microbiol 95:109–119

Jain KK, Kumar S, Deswal D, Kuhad RC (2017) Improved production of thermostable cellulase from *Thermoascus aurantiacus* RCKK by fermentation bioprocessing and its application in the hydrolysis of office waste Paper, algal pulp, and biologically treated wheat straw. Appl Biochem Biotechnol 181:784–800

Jiang ZQ, Yang SQ, Tan SS, Li LT, Li XT (2005) Characterization of a xylanase from the newly isolated thermophilic *Thermomyces lanuginosus* CAU44 and its application in bread making. Lett Appl Microbiol 41:69–76

Jurado M, López MJ, Suárez-Estrella F, Vargas-García MC, López-González JA, Moreno J (2014) Exploiting composting biodiversity: study of the persistent and biotechnologically relevant microorganisms from lignocellulose-based composting. Bioresour Technol 162:283–293

Kane BE, Mullins JT (1973) Thermophilic fungi in a municipal waste compost system. Mycologia 65:1087–1100

Kapoor M, Khalil Beg Q, Bhushan B, Dadhich KS, Hoondal GS (2000) Production and partial purification and characterization of a thermo-alkali stable polygalacturonase from Bacillus sp. MG-cp-2. Process Biochem 36:467–73

Kertesz MA, Thai M (2018) Compost bacteria and fungi that influence growth and development of *Agaricus bisporus* and other commercial mushrooms. Appl Microbiol Biotechnol 102:1639–1650

Khan SA, Gambhir S, Ahmad A (2014) Extracellular biosynthesis of gadolinium oxide (Gd_2O_3) nanoparticles, their biodistribution and bioconjugation with the chemically modified anticancer drug taxol. Beilstein J Nanotechnol 5:249–257

Klamer M, Sochting U (1998) Fungi in a controlled compost system with special emphasis on the thermophilic fungi. In: Proceedings of the international symposium on composting & use of composted material in horticulture. Ayr, UK, 5–11 Apr 1997, pp 405–413

Klamer M, Sochting U, Szmidt RAK (1998) Fungi in a controlled compost system with special emphasis on the thermophilic fungi. In: Proceedings of the international symposium on composting and use of composted materials for horticulture, Ayr, UK, 5–11 Apr 1997, pp 405–412

Kleyn JG, Wetzler TF (1981) The microbiology of spent mushroom compost and its dust. Can J Microbiol 27:748–753

Krasowska K, Janik H, Gradys A, Rutkowska M (2012) Degradation of polyurethanes in compost under natural conditions. J Appl Microbiol 125:4252–4260

Krause MS, De Ceuster TJJ, Tiquia SM, Michel FC Jr, Madden LV, Hoitink HAJ (2003) Isolation and characterization of *Rhizobacteria* from composts that suppress the severity of bacterial leaf spot of radish. Phytopathology 93:1292–1300

Kuhad RC, Singh A, Eriksson KEL (1997) Microorganisms and enzymes involved in the degradation of plant fiber cell walls. In: Eriksson KEL (ed) Advances in biochemical engineering/biotechnology, volume 57. Springer, Germany, pp 46–125

Kumar S (2011) Composting of municipal solid waste. Crit Rev Biotechnol 31:112–136

Kumar S, Nussinov R (2001) How do thermophilic proteins deal with heat? Cell Mol Life Sci 58:1216–1233

Kuok F, Mimoto H, Nakasaki K (2012) Effects of turning on the microbial consortia and the in situ temperature preferences of microorganisms in a laboratory-scale swine manure composting. Bioresour Technol 116:421–427

Lakherwal D (2014) Adsorption of heavy metals: a review. Int J Environ Res Dev 4:41–48

Langarica-Fuentes A, Zafar U, Heyworth A, Brown T, Fox G, Robson GD (2014) Fungal succession in an in-vessel composting system characterized using 454 pyrosequencing. FEMS Microbiol Ecol 88:296–308

Lea AGH (1995) Enzymes in the production of beverages and fruit juices. In: Tucker GA, Woods LFJ (eds) Enzymes in food processing. Springer, New York, USA, pp 223–249

Lee H, Lee YM, Jang Y, Lee S, Ahn BJ, Kim GH, Kim JJ (2014) Isolation and analysis of the enzymatic properties of thermo philic fungi from compost. Mycobiology 42:181e184

Liao WY, Shen CN, Lin LH, Yang YL, Han HY, Chen JW, Kuo SC, Wu SH, Liaw CC (2012) Asperjinone, a nor-neolignan, and terrein, a suppressor of ABCG2-expressing breast cancer cells, from thermophilic Aspergillus terreus. J Nat Prod 75:630–635

Liu Y, Feng Y, Cheng D, Xue J, Wakelin S, Li Z (2018a) Dynamics of bacterial composition and the fate of antibiotic resistance genes and mobile genetic elements during the co-composting with gentamicin fermentation residue and lovastatin fermentation residue. Bioresour Technol 261:249–256

Liu L, Wang S, Guo X, Zhao T, Zhang B (2018b) Succession and diversity of microorganisms and their association with physicochemical properties during green waste thermophilic composting. Waste Manag 73:101–112

López-González JA, del Carmen Vargas-García M, López MJ, Suárez-Estrella F, del Mar Jurado M, Moreno J (2015) Biodiversity and succession of mycobiota associated to agricultural lignocellulosic waste-based composting. Bioresour Technol 187:305–313

Ma R, Bai Y, Huang H, Luo H, Chen S, Fan Y, Cai L, Yao B (2017) Utility of thermostable xylanases of *Mycothermus thermophilus* in generating prebiotic xylooligosaccharides. J Agric Food Chem 65:1139–1145

Maalej I, Belhaj I, Masmoudi NF, Belghith H (2009) Highly thermostable xylanase of the thermophilic fungus *Talaromyces thermophilus*: purification and characterization. Appl Biochem Biotechnol 158:200–212

Martin N, Guez MAU, Sette LD, Da Silva R, Gomes E (2010) Pectinase production by a Brazilian thermophilic fungus *Thermomucor indicae-seudaticae* N31 in solid-state and submerged fermentation. Microbiology 79:306–313

Martins EDS, Silva D, Da Silva R, Gomes E (2002) Solid state production of thermostable pectinases from thermophilic *Thermoascus aurantiacus*. Process Biochem 37:949–954

Martins LF, Antunes LP, Pascon RC, de Oliveira JCF, Digiampietri LA, Barbosa D, Peixoto BM, Vallim MA, Viana-Niero C, Ostroski EH, Telles GP, Dias Z, da Cruz JB, Juliano L, Verjovski-Almeida S, da Silva AM, Setubal JC (2013) Metagenomic analysis of a tropical composting operation at the Sao Paulo Zoo park reveals diversity of biomass degradation functions and organisms. PLoS One 8(4):e61928. https://doi.org/10.1371/journal.pone.0061928

Mathur G, Prasad R (2012) Degradation of polyurethane by *Aspergillus flavus* (ITCC 6051) isolated from soil. Appl Biochem Biotechnol 167:1595–1602

McClendon SD, Batth T, Petzold CJ, Adams PD, Simmons BA, Singer SW (2012) *Thermoascus aurantiacus* is a promising source of enzymes for biomass deconstruction under thermophilic conditions. Biotechnol Biofuels 5:54. https://doi.org/10.1186/1754-6834-5-54

Mehta R, Singhal P, Singh H, Damle D, Sharma AK (2016) Insight into thermophiles and their wide-spectrum applications. Biotech 6:81. https://doi.org/10.1007/s13205-016-0368-z

Moubasher AH, Aboel-Hafez SII, Aboelfattah HM, Moharrarh AM (1982) Fungi of wheat and broad bean straw composts 2. Thermophilic fungi. Mycopathologia 84:61–72

Neher DA, Weicht TR, Bates ST, Leff JW, Fierer N (2013) Changes in bacterial and fungal communities across compost recipes, preparation methods, and composting times. PLoS One 8(11):e79512. https://doi.org/10.1371/journal.pone.0079512

Nguyen S, Ala F, Cardwell C, Cai D, McKindles KM, Lotvola A, Hodges S, Deng Y, Tiquia-Arashiro SM (2013) Isolation and screening of carboxydotrophs isolated from composts and their potential for butanol synthesis. Environ Technol 34:1995–2007

Noble R, Gaze RH (1994) Controlled environment composting for mushroom cultivation: substrates based on wheat and barley straw and deep litter poultry litter. J Aric Sci 123:71–79

Ochoa-Villarreal M, Aispuro-Hernández E, Vargas-Arispuro I, Martínez-Téllez MA (2012) Plant cell wall polymers: Function, structure and biological activity of their derivatives, polymerization. In: De Souza Gomes A (ed) InTech. https://doi.org/10.5772/46094. Available from: https://www.intechopen.com/books/polymerization/plant-cell-wall-polymers-function-structure-and-biological-activity-of-their-derivatives

Opden Camp HJM, Stumm CK, Straatsma G, Derikx PJL, Van Griensve LJLD (1990) Microb Ecol 19:303–309

Pampuro N, Bertora C, Sacco D, Dinuccio E (2017) Fertilizer value and greenhouse gas emissions from solid fraction pig slurry compost pellets. J Agric Sci 155:1646–1658

Partanen P, Hultman J, Paulin L, Auvinen P, Romantschuk M (2010) Bacterial diversity at different stages of the composting process. BMC Microbiol 10:94. https://doi.org/10.1186/1471-2180-10-94

Pathirana RA, Seal KJ (1984) Studies on polyurethane deteriorating fungi. Part 1. Isolation and characterization of the test fungi employed. Int Biodeterior 20:163–168

Plecha S, Hall D, Tiquia-Arashiro SM (2013) Screening and characterization of soil microbes capable of degrading cellulose from switchgrass (*Panicum virgatum* L.). Environ Technol 34:1895–1904

Pomaranski E, Tiquia-Arashiro SM (2016) Butanol tolerance of carboxydotrophic bacteria isolated from manure composts. Environ Technol 37:1970–1982

Rajni J (1999) Molecular ecology of *Scytalidium thermophilum*. Ph.D. thesis, G.B. Pant University of Agriculture and Technology, Pantnagar, India, p 88

Rawat S (2004) Microbial diversity of mushroom compost and xylanase of *Scytalidium thermophilum*. Ph.D. thesis, G.B. Pant University of Agriculture and Technology, Pantnagar, India, p 199

Richard TL, Tiquia SM (1999) In-barn composting system plus a warm bed for pigs. BioCycle 40:44

Ryckeboer J, Mergaert J, Vaes K, Klammer S, De Clercq D, Coosemans J, Insam H, Swings J, De Clercq D, Coosemans J, Insam H, Swings J (2003) A survey of bacteria and fungi occurring during composting and self-heating processes. Ann Microbiol 53:349–410

Salar RK (2018) Thermophilic fungi. Basic Concepts and Biotechnological Applications. CRC Press, Boca Raton, FL, p 352

Sánchez C (2009) Lignocellulosic residues: biodegradation and bioconversion by fungi. Biotechnol Adv 27:185–194

Satyanarayana T, Johri BN (1984) Ecology of thermophilic fungi. In: Mukerji et al (eds) Progress in microbial ecology. Print House, Lucknow, India, pp 349–361

Sebők F, Dobolyi C, Bobvos J et al (2016) Thermophilic fungi in air samples in surroundings of compost piles of municipal, agricultural and horticultural origin. Aerobiologia 32:255–263

Sen TL, Abraham TK, Chakrabarthy SL (1980) Utilization of cellulolytic waste by thermophilic fungi. In: Young MM (ed) Advances in biotechnology. Volume II. Pergamon Press, Oxford, UK, pp 633–638

Senesi N (1989) Composted materials as organic fertilizers. Sci Total Environ 81(81):521–542

Sharma HSS (1989) Economic importance of thermophilic fungi. Appl Microbiol Biotechnol 31:1–10

Singh B, Poças-Fonseca MJ, Johri BN, Satyanarayana T (2016) Thermophilic molds: biology and applications. Crit Rev Microbiol 42:985–1006

Sparling GP, Fermor TR, Wood DA (1982) Measurement of the microbial biomass in composted wheat straw, and the possible contribution of the biomass to the nutrition of *Agaricus bisporus*. Soil Biol Biochem 14:609–611

Straatsma G, Gerrits JPG, Augustijn MPAM, OpDenCamp HJM, Vogels GD, Van Griensven LJLD (1989) Population dynamics of *Scytalidium thermophilum* in mushroom compost and stimulatory effects on growth rate and yield of *Agaricus bisporus*. J Gen Microbiol 135:751–789

Straatsma G, Gerrits JPG, Gerrits TM, Op den Camp HJM, Van Griensven LJLS (1991) Growth kinetics of *Agaricus bisporus* mycelium on solid substrate (mushroom compost). J Gen Microbiol 137:1471–1477

Straatsma G, Samson RA, Olijnsma TW, Op-Den-Camp HJM, Gerrits JPG, Griensven LJLD, Van-Griensven LJLD (1994) Ecology of thermophilic fungi in mushroom compost, with emphasis on *Scytalidium thermophilum* and growth stimulation of *Agaricus bisporus* mycelium. Appl Environ Microbiol 60:454–458

Subrahmanyam A, Mehrotra BS, Thirumalacher M (1977) *Thermomucor*, a new genus of *Mucorales*. Geor J Sci 35:1–6

Syed A, Saraswati S, Kundu GC, Ahmad A (2013) Biological synthesis of silver nanoparticles using the fungus *Humicola* sp. and evaluation of their cytotoxicity using normal and cancer cell lines. Spectrochim Acta Part A: Mol Biomol Spectr 114:144–147

Szekely A, Sipos R, Berta B, Vajna B, Hajdu C, Marialigeti K (2009) DGGE and T-RFLP analysis of bacterial succession during mushroom compost production and sequence-aided T-RFLP profile of mature compost. Microb Ecol 57:522–533

Tam NFY, Tiquia SM (1994) Assessing toxicity of spent pig litter using a seed germination technique. Res Conser Recycl 11:261–274

Tam NFY, Tiquia SM (1999) Nitrogen transformation during co-composting of spent pig manure, sawdust litter and sludge in forced-aerated system. Environ Technol 20:259–267

Tansey ÌR, Brock ÔD (1978) Microbial life at high temperatures: ecological aspects. In: Kushner DJ (ed) Microbial life in extreme environments. Academic Press, London, UK, pp 159–216

Tewari P (2000) Thermophilic microorganisms from mushroom compost and their polysaccharolytic activities. M.Sc. Thesis, G.B. Pant University of Agriculture and Technology, Pantnagar. India, p 99

Tiquia SM (2000) Evaluating phytotoxicity of pig manure from the pig-on-litter system. In: Warman PR, Taylor BR (eds) International composting symposium (ICS 1999). Volume II. CBA Press, Inc., Nova Scotia, Canada, pp 625–647

Tiquia SM (2002a) Evolution of enzyme activities during manure composting. J Appl Microbiol 92:764–775

Tiquia SM (2002b) Microbial transformation of nitrogen during composting. In: Insam H, Riddech N, Klammer S (eds) Microbiology of composting. Springer Verlag, Heidelberg, Germany, pp 237–246

Tiquia SM (2003) Evaluation of organic matter and nutrient composition of partially decomposed and composted spent pig-litter. Environ Technol 24:97–108

Tiquia SM (2005a) Microbial community dynamics in manure composts based on 16S and 18S rDNA T-RFLP profiles. Environ Technol 26:1104–1114

Tiquia SM (2005b) Microbiological parameters as indicators of compost maturity. J Appl Microbiol 99:816–828

Tiquia SM (2010a) Reduction of compost phytotoxicity during the process of decomposition. Chemosphere 79:506–512

Tiquia SM (2010b) Using terminal restriction fragment length polymorphism (T-RFLP) analysis to assess microbial community structure in compost systems. Methods Mol Biol 599:89–101. In: Cummings SP (Ed)

Tiquia SM, Michel FC Jr (2002) Bacterial diversity in livestock manure composts as characterized by terminal restriction fragment length polymorphisms (T-RFLP) of PCR-amplified 16S rRNA gene sequences. In: Insam H, Riddech N, Klammer S (eds) Microbiology of composting. Springer Verlag, Heidelberg, Germany, pp 64–82

Tiquia SM, Mormile M (2010) Extremophiles—a source of innovation for industrial and environmental applications. Environ Technol 31(8-9):823

Tiquia SM, Tam NFY (1998a) Composting of spent pig litter in turned and forced aerated piles. Environ Pollut 99:329–337

Tiquia SM, Tam NFY (1998b) Composting of pig manure in Hong Kong. BioCycle 39(2):78–79

Tiquia SM, Tam NFY (2000) Fate of nitrogen during composting of chicken litter. Environ Pollut 110:535–541

Tiquia SM, Tam NFY (2002) Characterization and composting of poultry litter in forced-aeration piles. Process Biochem 37:869–880

Tiquia SM, Tam NFY, Hodgkiss IJ (1996) Microbial activities during composting of spent pig-manure sawdust litter at different moisture contents. Bioresour Technol 55:201–206

Tiquia SM, Tam NFY, Hodgkiss IJ (1997a) Effects of turning frequency on composting of spent pig-manure sawdust litter. Bioresour Technol 62:37–42

Tiquia SM, Tam NFY, Hodgkiss IJ (1997b) Effects of bacterial inoculum and moisture adjustment on further composting of pig manure. Environ Pollut 96:161–171

Tiquia SM, Tam NFY, Hodgkiss IJ (1998a) *Salmonella* elimination during composting of spent pig litter. Bioresour Technol 63:193–196

Tiquia SM, Tam NFY, Hodgkiss IJ (1998b) Changes in chemical properties during composting of spent litter at different moisture contents. Agric Ecosyst Environ 67:79–89

Tiquia SM, Tam NFY, Hodgkiss IJ (1998c) Composting of spent pig litter at different seasonal temperatures in subtropical climate. Environ Pollut 98:97–104

Tiquia SM, Richard TL, Honeyman MS (2000) Effects of windrow turning and seasonal temperatures on composting of hog manure from hoop structures. Environ Technol 21:1037–1046

Tiquia SM, Wan JHC, Tam NFY (2001) Extracellular enzyme profiles during co-composting of poultry manure and yard trimmings. Process Biochem 36:813–820

Tiquia SM, Lloyd J, Herms D, Hoitink HAJ, Michel FC Jr (2002a) Effects of mulching and fertilization on soil nutrients, microbial activity and rhizosphere bacterial community structure determined by analysis of T-RFLPs of PCR-amplified 16S rRNA genes. Appl Soil Ecol 21:31–48

Tiquia SM, Wan JHC, Tam NFY (2002b) Microbial population dynamics and enzyme activities during composting. Compost Sci Util 10:150–161

Tiquia SM, Richard TL, Honeyman MS (2002c) Carbon, nitrogen, and mass loss during composting. Nutr Cycl Agroecosyst 62:15–24

Tiquia SM, Ichida JM, Keener HM, Elwell D, Burt E Jr, Michel FC Jr (2005) Bacterial community structure on feathers during composting as determined by terminal restriction fragment length polymorphism analysis of 16S rDNA genes. Appl Microbiol Biotechnol 67:412–419

Tiquia-Arashiro SM (2014) Biotechnological applications of thermophilic carboxydotrophs. In: Tiquia-Arashiro SM (ed) Thermophilic carboxydotrophs and their applications in biotechnology. Chapter 4. Springer International Publishing, pp 29–101

Tiquia-Arashiro SM (2018) Lead resistance mechanisms in bacteria as strategies for lead bioremediation. Appl Microbiol Technol 102:5437–5444

Tiquia-Arashiro SM, Mormile M (2013) Sustainable technologies: bioenergy and biofuel from biowaste and biomass. Environ Technol 34(13-14):1637–1638

Tiquia-Arashiro SM, Rodrigues D (2016a) Extremophiles: applications in nanotechnology. Springer International Publishing, Cham, Switzerland, p 193

Tiquia-Arashiro SM, Rodrigues D (2016b) Nanoparticles synthesized by microorganisms. In: Tiquia SM, Rodrigues D (eds) Extremophiles: applications in nanotechnology. Springer International Publishing, Cham, Switzerland, pp 1–51

Tiquia-Arashiro SM, Rodrigues D (2016c) Thermophiles and psychrophiles in nanotechnology. In: Tiquia SM, Rodrigues D (eds) Extremophiles: applications in nanotechnology. Springer International Publishing, Cham, Switzerland, pp 89–127

Tiquia-Arashiro SM, Rodrigues D (2016d) Applications of nanoparticles. In: Tiquia-Arashiro SM, Rodrigues D (eds) Extremophiles: applications in nanotechnology. Springer International Publishing, Cham, Switzerland, pp 163–193

Tiquia-Arashiro SM, Rodrigues D (2016e) Halophiles in nanotechnology. In: Extremophiles: applications in nanotechnology. Springer International Publishing, Cham, Switzerland, pp 53–58

Tiquia-Arashiro SM, Rodrigues D (2016f) Thermophiles and psychrophiles in nanotechnology. In: Extremophiles: applications in nanotechnology. Springer International Publishing, Cham, Switzerland, pp 89–127

Tobin JM, Cooper DG, Neufeld RJ (1984) Uptake of metal ions by *Rhizopus arrhizus* biomass. Appl Environ Microbiol 47:821–824

Vijay B (1996) Investigations on compost mycoflora and crop improvement in *Agaricus bisporus* (Lange). Ph.D. thesis, H.P. University, Shimla, India

Wang K, Yin X, Mao H, Chu C, Tian Y (2018) Changes in structure and function of fungal community in cow manure composting. Bioresour Technol 255:123–130

Waqas M, Almeelbi T, Nizami AS (2018) Resource recovery of food waste through continuous thermophilic in-vessel composting. Environ Sci Pollut Res 25:5212–5222

Wei H, Wang L, Hassan M, Xie B (2018) Succession of the functional microbial communities and the metabolic functions in maize straw composting process. Bioresour Technol 256:333–341

Weigant WM (1991) A simple method to estimate the biomass of thermophilic fungi in composts. Biotech Tech 5:421–426

Wu W, Gu B, Fields MW, Gentile M, Ku YK, Yan H, Tiquia SM, Yan T, Nyman J, Zhou J, Jardine PM, Criddle CS (2005) Uranium (VI) reduction by denitrifying biomass. Biorem J 9:1–13

Yu M, He X, Liu J, Wang Y, Xia B, Li D, Zhang H, Yang C (2018) Characterization of isolated fractions of dissolved organic matter derived from municipal solid waste compost. Sci Total Environ 635:275–283

Zafar U, Houlden A, Robson GD (2013) Fungal communities associated with the biodegradation of polyester polyurethane buried under compost at different temperatures. Appl Environ Microbiol 79:7313–7324

Zafar U, Nzeram P, Langarica-Fuentesa A, Houlden A, Heyworth A, Saiani A, Robso GD (2014) Biodegradation of polyester polyurethane during commercial composting and analysis of associated fungal communities. Bioresour Technol 158:374–377

Zhang L, Ma H, Zhang H, Xun L, Chen G, Wang L (2015) *Thermomyces lanuginosus* is the dominant fungus in maize straw composts. Bioresour Technol 197:266–275

Zhang LL, Zhang HQ, Wang ZH, Chen GJ, Wang LS (2016) Dynamic changes of the dominant functioning microbial community in the compost of a 90-m³ aerobic solid state fermentor revealed by integrated meta-omics. Bioresour Technol 203:1–10

Zhang H, Woodams EE, Hang YD (2011) Influence of pectinase treatment on fruit spirits from apple mash, juice and pomace. Process Biochem 186:1909–1913

Zhou P, Zhang G, Chen S, Jiang Z, Tang Y, Henrissat B, Yan Q, Yang S, Chen CF, Zhang B, Du Z (2014) Genome sequence and transcriptome analyses of the thermophilic zygomycete fungus *Rhizomucor miehei*. BMC Genomics 15:294

Index

© Springer Nature Switzerland AG 2019
S. M. Tiquia-Arashiro, M. Grube (eds.), *Fungi in Extreme Environments:*
Ecological Role and Biotechnological Significance,
https://doi.org/10.1007/978-3-030-19030-9

Printed by Printforce, the Netherlands